Lecture Notes in Artificial Intelligence   11856

Subseries of Lecture Notes in Computer Science

More information about this series at http://www.springer.com/series/1244

Maosong Sun · Xuanjing Huang ·
Heng Ji · Zhiyuan Liu · Yang Liu (Eds.)

Chinese Computational Linguistics

18th China National Conference, CCL 2019
Kunming, China, October 18–20, 2019
Proceedings

Editors
Maosong Sun
Tsinghua University
Beijing, China

Xuanjing Huang
Fudan University
Shanghai, China

Heng Ji
University of Illinois
at Urbana Champaign
Illinois, USA

Zhiyuan Liu
Tsinghua University
Beijing, China

Yang Liu
Tsinghua University
Beijing, China

ISSN 0302-9743   ISSN 1611-3349   (electronic)
Lecture Notes in Artificial Intelligence
ISBN 978-3-030-32380-6   ISBN 978-3-030-32381-3   (eBook)
https://doi.org/10.1007/978-3-030-32381-3

LNCS Sublibrary: SL7 – Artificial Intelligence

This Springer imprint is published by the registered company Springer Nature Switzerland AG
The registered company address is: Gewerbestrasse 11, 6330 Cham, Switzerland

Preface

Welcome to the proceedings of the 18th China National Conference on Computational Linguistics (18th CCL). The conference and symposium were hosted by Kunming University of Science and Technology located in Kunming, Yunnan Province, China.

CCL is an annual conference (bi-annual before 2013) that started in 1991. It is the flagship conference of the Chinese Information Processing Society of China (CIPS), which is the largest NLP scholar and expert community in China. CCL is a premier nation-wide forum for disseminating new scholarly and technological work in computational linguistics, with a major emphasis on computer processing of the languages in China such as Mandarin, Tibetan, Mongolian, and Uyghur.

The Program Committee selected 147 papers (91 Chinese papers and 56 English papers) out of 371 submissions from China, Macau (region), Singapore, and the USA for publication. The accepted rate is 39.62%. The 56 English papers cover the following topics:

- Linguistics and Cognitive Science (4)
- Fundamental Theory and Methods of Computational Linguistics (1)
- Information Retrieval and Question Answering (4)
- Text Classification and Summarization (8)
- Knowledge Graph and Information Extraction (8)
- Machine Translation and Multilingual Information Processing (4)
- Minority Language Processing (6)
- Language Resources and Evaluation (2)
- Social Computing and Sentiment Analysis (2)
- NLP Applications (17)

The final program for the 18th CCL was the result of a great deal of work by many dedicated colleagues. We want to thank, first of all, the authors who submitted their papers, and thus contributed to the creation of the high-quality program. We are deeply indebted to all the Program Committee members for providing high-quality and insightful reviews under a tight schedule. We are extremely grateful to the sponsors of the conference. Finally, we extend a special word of thanks to all the colleagues of the Organizing Committee and secretariat for their hard work in organizing the conference, and to Springer for their assistance in publishing the proceedings in due time.

We thank the Program and Organizing Committees for helping to make the conference successful, and we hope all the participants enjoyed a remarkable visit to Kunming, a beautiful city in South China.

August 2019

Maosong Sun
Xuanjing Huang
Heng Ji

Organization

Program Committee

18th CCL Program Chairs

Maosong Sun	Tsinghua University, China
Xuanjing Huang	Fudan University, China
Heng Ji	University of Illinois at Urbana Champaign, USA

18th CCL Area Co-chairs

Linguistics and Cognitive Science

Weiwei Sun	Peking University, China
John Sie Yuen Lee	City University of Hong Kong, SAR China

Basic Theories and Methods of Computational Linguistics

Yue Zhang	Westlake University, China
Derek Wong	University of Macau, SAR China

Information Retrieval and Question Answering

Zhicheng Dou	Renmin University of China, China
Chenyan Xiong	Microsoft Research, USA

Text Classification and Abstracts

Yansong Feng	Peking University, China
Fei Liu	University of Central Florida, USA

Knowledge Graph and Information Extraction

Yubo Chen	Institute of Automation, CAS, China
Feifan Liu	University of Massachusetts Medical School, USA

Machine Translation

Zhaopeng Tu	Tencent AI Lab, China
Rui Wang	National Institute of Information and Communications Technology, NICT, Japan

Minority Language Information Processing

Cunli Mao   Kunming University of Science and Technology, China
Yi Fang   Santa Clara University, USA

Language Resources and Evaluation

Sujian Li   Peking University, China
Wenjie Li   The Hong Kong Polytechnic University, SAR China

Social Computing and Sentiment Analysis

Xiao Ding   Harbin Institute of Technology, China
Jiang Guo   Massachusetts Institute of Technology, USA

NLP Applications

Xipeng Qiu   Fudan University, China
Weilian Wang   University of California, Santa Barbara, USA

18th CCL Local Arrangement Chairs

Zhengtao Yu   Kunming University of Science and Technology, China
Yang Liu   Tsinghua University, China

18th CCL Evaluation Chairs

Ting Liu   Harbin Institute of Technology, China
Ruifeng Xu   Harbin Institute of Technology, China

18th CCL Publications Chairs

Zhiyuan Liu   Tsinghua University, China
Shizhu He   Institute of Automation, CAS, China

18th CCL Workshop Chairs

Minlie Huang   Tsinghua University, China
Kang Liu   Institute of Automation, CAS, China

18th CCL Sponsorship Chairs

Jiajun Zhang   Institute of Automation, CAS, China
Binyang Li   University of International Relations, China

18th CCL Publicity Chairs

Chenliang Li   Wuhan University, China
Xia Hu   Texas A&M University, USA

18th CCL System Demonstration Chairs

Xianpei Han   Institute of Automation, CAS, China
Changliang Li   Jinshan AI Lab, China

18th CCL Student Seminar Chairs

Yang Feng   Institute of Computing Technology, CAS, China
Huimin Chen   Tsinghua University, China

18th CCL Finance Chair

Yuxing Wang   Tsinghua University, China

18th CCL Organizers

Chinese Information Processing Society of China

Tsinghua University

Kunming University of Science and Technology

Publishers

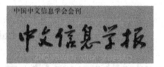

Lecture Notes in Artificial
Intelligence, Springer

Journal of Chinese Information
Processing

Science China

Journal of Tsinghua University
(Science and Technology)

Sponsoring Institutions

Platinum

Gold

Contents

Text Classification and Summarization

Knowledge Graph and Information Extraction

Machine Translation and Multilingual Information Processing

Minority Language Processing

Language Resources and Evaluation

Social Computing and Sentiment Analysis

NLP Applications

Linguistics and Cognitive Science

Colligational Patterns in China English:
The Case of the Verbs of Communication

Lixin Xia[1], Yun Xia[2(✉)], and Qian Li[1]

[1] Laboratory of Language Engineering and Computing/Center
for Lexicographical Studies of Guangdong, University of Foreign Studies,
Guangzhou 510420, China
[2] Nanfang College of Sun Yat-Sen University, Guangzhou 510970, China
344192304@qq.com

Abstract. This present study aims to investigate the colligational structures in China English. A corpus-based and comparative methodology was adopted in which three verbs of communication (*discuss*, *communicate* and *negotiate*) were chosen as the research objects. The Corpus of China English (CCE) was built with 13.9 million tokens. All the colligational patterns of the three verbs were extracted from the CCE by the tool Colligator, and analyzed in comparison with those from the British National Corpus (BNC). The major findings of the study are: (1) China English and British English share some common core colligational patterns, for example, the *V deter.* pattern for the verb *discuss*, the *V prep.* and *V adv.* patterns for the verb *communicate,* and the *V prep.* and *V deter.* patterns for the verb *negotiate* are the most frequently used patterns of all the colligations in both corpora. (2) The number of occurrences of some colligational patterns in CCE is significantly greater than that in BNC. They are *V pl. n., V conj.* and *V prop. n.* for the verb *discuss, V prep.* for the verb *communicate,* and *V infin. to* for the verb *negotiate.* (3) Some colligational patterns occur less frequently in CCE than those in BNC, such as the patterns *V deter., V pers. pron.* and *V poss. pron* for the verb *discuss, V sing. n.* and *V deter.* for the verb *communicate,* and *V adj.* for the verb *negotiate.* (4) No new colligational patterns have been found in China English. The study extends research on World Englishes, and its findings can attest to the process of structural nativisation of the English language used in the contexts of China.

Keywords: Colligation · China English · Language contact · Corpus of China English · Computational linguistics · Corpus linguistics

1 Introduction

Since Ge's [1] introduction of the term 'China English', it has aroused growing interest of scholars in China, and more recently, abroad. At the earlier stage, scholars debated whether there was such a language variety called China English, and some of them tried to offer a definition of it [2–6]. After that, scholars began to look at the linguistic features of China English [6–13]. Most of these studies focused on the phonological and lexical features of China English. It was not until recently that structural nativization has become a topic for researchers on the variety of China English. These

© Springer Nature Switzerland AG 2019
M. Sun et al. (Eds.): CCL 2019, LNAI 11856, pp. 3–14, 2019.
https://doi.org/10.1007/978-3-030-32381-3_1

recent studies [14–17] have identified several new linguistic structures in the English language when it is used in China, including new collocational structures, ditransitive verbs, and verb-complementation. However, no previous studies have approached structural nativization through colligations in the field of China English. Therefore, it is of significance to explore colligational patterns in China English. This current study will extend the research on China English, and its findings can attest to the process of structural nativisation.

Colligation refers to the grammatical company a word or a word category keeps. Put simply, colligation is the co-occurrence of syntactic categories, usually within a sentence. For example, the verbs of *agree* (including *choose, decline, manage*, etc.) co-occur with an infinitive with "to", but not a gerund. The difference between a collocation and colligation is that "while collocation relates to the lexical aspect of a word's selection of neighbouring words, colligation can be described as relating to the grammatical aspect of this selection" [18]. The reasons to choose colligational patterns for this study is that as a basic grammatical category in English, colligational patterns are not given due attention in the field of world Englishes, including China English. Second, thanks to existing tools and corpora, colligational patterns can be examined by analyzing large amounts of corpus data. Thirdly, the difference between China English and "standard" English in colligational construction is indicative of structural nativization of the English language used in the contexts of Chinese culture. Finally, colligational patterns carry constructional meanings, which is significant for the task of natural language processing.

The following research questions will be addressed in the present study:

(1) Are there any common colligational patterns shared by both China English and British English?
(2) Are there any preferences for certain colligational patterns in China English? If so, what are they?
(3) Are there any innovative colligational patterns in China English? If so, what are they?

In this present study, a corpus-based methodology will be adopted in order to look at the differences between China English and British English in the use of colligational constructions. The subsequent sections of this paper will be devoted to discussing the data obtained from the corpora in order to provide answers to the above research questions.

2 Literature Review

Structural nativization is defined as "the emergence of locally characteristic linguistic patterns" [19]. Studies of structural nativization in the field of world Englishes focus mainly on new ditransitive verbs and verb-complementation in new Englishes, such as Indian English, Singapore English, Fijian English, Jamaican English, etc. For example, Mukherjee and Hoffmann [20] did a case study of the verb complementation of "give", and found that while in British English the verb is most frequently used with the ditransitive construction, it is most frequently used with monotransitive one in Indian

English. Schilk [21] compared the collocation and verb-complementational patterns of three verbs between Indian English and British English. Schilk et al. [22] also compared verb-complementation patterns between two varieties located in the Outer Circle: Indian English and Sri Lankan English.

Although much has been done to investigate structural nativizatoin in New Englishes, such as Indian English, and Singapore English, fewer such studies have been done in China English. Yu [23] conducted a comparative study of the adjective "foreign" in China's English newspapers and British newspapers. He argues that the linguistic use of the word in China's news reports show clear evidence of nativisation which is a result of Chinese people's way of thinking. Using the same data as Yu's study [23], Yu and Wen [17] examined the collocations of 20 adjectives, and reached similar conclusion that English used in China's news reports shows a distinct tendency towards systematic nativisation at the collocational level. They argue that this different and innovative use of certain collocational patterns is a result of language contact.

Other scholars investigate into structural nativization in the grammatical structures in China English. Using three sets of data (the interview data, the newspaper data, and the short-story data), Xu [6] identified eight categories of syntactic features in the interview data: adjacent default tense, null-subject/object utterances, co-occurrence of connective pairs, subject pronoun copying, yes–no response, topic–comment, unmarked OSV, and inversion in subordinate finite wh-clauses; three categories in the newspaper data: nominalization, multiple-coordinate construction and modifying-modified sequencing; and the use of imperatives and tag variation strategy in the short-story data. Xu's study is the most comprehensive one of such studies, and he identified distinctive grammatical patterns with Chinese characteristics.

Hu and Zhang [15] collected data from the *China Daily* and built a corpus with about 1 million tokens. They compared the collocation and colligation patterns of some high-frequency verbs of *transformation* in the corpus with those in the Corpus of Contemporary American English (COCA). It is found that the intransitive use of the verb "grow" has a greater frequency in China's English newspapers than in American news reports, and the transitive use of the verb "develop" and "increase" has a smaller frequency.

Using corpus data collected from an online discussion forum, Ai and You [14] examined several locally emergent linguistic patterns in China English, and found some new ditransitive verbs, verb-complementation and collocation patterns in China English.

Their studies have contributed new knowledge to the grammatical features of China English. However, they have some common limitations. Xu's [6] data is small in size. His written data only consists of 20 newspaper articles and 12 short stories. Considering the availability of large scale corpora nowadays, his collection of text is too small for linguistic studies, especially for identifying grammatical patterns. The other studies [15, 17, 23] have the same problems with the data size. Besides, Xu's [6] spoken data come from interviews with Chinese English learners which, in his own words, "is difficult to distinguish between the syntactic features of Chinese learners' English and those of China English" [6]. That might be the reason why more distinct grammatical features were detected in the interview data than those in the written data, the latter of which were written by more competent English users. Ai and You's [14] corpus has about 7 million tokens. However, it only consists of data from Chinese English learners in an online discussion forum. From the point of view of language acquisition, these "creative uses are in fact instances of erroneous use" [14]. Therefore, their findings of these new grammatical patterns in China English could be ungrammatical use of English by English learners rather than accepted new syntactic structures.

In a word, these previous studies leave some room for improvement. First, no previous studies examined systematically the processes of acculturation of English in China through colligation. Next, the data for previous studies was either too limited in size or not representative of China English. The data for most studies came from a single source, either from newspapers [15, 17, 23], or from learners' production [6, 14]. Therefore, this present study intends to fill up these gaps, and aims to explore structural nativization through colligation in a larger corpus of China English.

3   Methods

Owing to the availability of large corpora and corpus-enquiry softwares, it is now possible to identify linguistic features of a language based on large volumes of data. In this present study, a corpus-based and comparative methodology will be adopted to investigate the uses of colligational patterns in different English varieties. In this section, the source of the data and research procedures will be introduced.

3.1   Source of the Data

In order to facilitate the studies on China English, the Corpus of China English (CCE) was built with 13,962,102 tokens. It collects written texts from the following four genres: magazine, newspaper, fiction and academic, each of which contains about 3.5 million tokens. So it has the same genres as COCA except the spoken parts. To guarantee its representativeness, only the texts written by Chinese speakers of English were collected in the corpus. They include Chinese journalists, magazine writers, novelists, scholars, and others who write and communicate in English. As proficient English speakers, their production can be regarded as China English. BNC was chosen as the reference corpus in this study which has about 96,134,547 tokens.

In the present study, the communicative verbs are to be selected as the objects of study. Verbs, as the core member of a sentence, have been a focus of linguistic research in the field of semantics and syntax. Verbs of communication are considered as one of the basic categories of verbs as they represent the most essential objectives and motivations in human being's communication [24]. In the FrameNet, 435 verbs of communication are collected and classified into 37 groups, such as discussion, encoding, reasoning, verdict, commitment, request, questioning, reporting, encoding, statement, response, summarizing, etc. In this study, the sub-category of discussion was selected, which includes the following verbs: discuss, confer, communicate, debate, negotiate, parley, consult, bargain, etc. The numbers of occurrence of the verbs are 792, 17, 251, 36, 174, 0, 93 and 19 respectively in CCE. As frequency counts in the analysis of linguistic data, the verbs *discuss* (792), *communicate* (251), and *negotiate* (174) were chosen for this study.

3.2   Procedures

After the CCE has been built, data extraction and analysis were carried out in the following steps:

First, the CCE was tagged by CLAWS7. It is a prerequisite for extracting colligational patterns with the tool Colligator 2.0 developed by members of the Foreign Language Education Research Informed by Corpora (FLERIC) team at Beijing Foreign Studies University [25].

Second, the colligation patterns for the three selected verbs *discuss*, *communicate* and *negotiate* were extracted from the CCE by the tool Colligator 2.0.

Third, from the BNC, the concordance lines of the three verbs were retrieved, out of which 1000 concordance lines were sampled for each verb. The colligational patterns of these verbs were also extracted by Colligator 2.0.

Fourth, an analysis was done to investigate the differences between China English and British English with the colligational patterns. The significance of the differences was calculated by the log-likelihood ratio calculator developed by members of the FLERIC team [25]. If the significance value is less than 0.05, then the result is statistically significant.

4   Results

In this section, the data of the three verbs *discuss*, *communicate*, and *negotiate* in the CCE and BNC will be given for further discussion in the next section. The number of occurrences of the three verbs in the CCE and BNC will be given first. Then the top 10 colligational patterns in CCE will be listed.

4.1   Number of Occurrences of the Three Verbs

Both the original frequency and normative frequency of the three verbs in CCE and BNC are given in Table 1. The normative frequency is the original frequency divided by the total number of the tokens in a corpus, expressed in terms of percents. Their log-likelihood and significance values are calculated by the log-likelihood ratio calculator.

Table 1. Number of occurrences of the three verbs in the two corpora

Verbs	f./n.f. in CCE	f./n.f. in BNC	Log-likelihood	Significance
Discuss	792/0.0057%	5505/0.0057%	−0.061959095939	0.803430000000
Communicate	251/0.0018%	1507/0.0016%	3.907184554432	0.048080000000
Negotiate	174/0.0012%	1295/0.0013%	−0.947203963098	0.330430000000

In Table 1, the log-likelihood and significance values show that the verbs *communicate* occur more frequently in CCE than in BNC. The difference is significant as the p-values are less than 0.05. The other two verbs *discuss* and *negotiate* are used less frequently in CCE, but the difference is not significant.

4.2   Colligational Patterns of *Discuss*

The frequencies of the colligational patterns of the word *discuss* in CCE are shown in Table 2. The lemma *discuss* has four inflections in the corpus, i.e. *discuss, discusses, discussing* and *discussed*. For the sake of easy comparison and space limit, only the word form of *discuss* was retrieved from the two corpora by the Colligator 2.0. In column 1 and 2, the ten patterns used most frequently in CCE are listed with frequencies and normative frequencies. In column 3, the corresponding number of occurrences of the patterns in BNC is given, with the first number being the number of occurrences of the word form out of its 1000 samples, and the second one being the total number of occurrences of the patterns in BNC. For example, the colligation pattern *V deter.* occurs 123 times out of the 310 concordance lines of the word form "discuss" in the 1,000 samples. Then in all of the 5,505 concordance lines of the lemma "discuss" in BNC, the colligational pattern *V deter* should be a × b/c, in which a is 5505, b is 123, c is 310, and the result is approximately 2184. In column 4 and 5 the log-likelihood and significance values are calculated by the log-likelihood ratio calculator.

Table 2. Colligational patterns of "discuss" in CCE and BNC

Colligations	f./n.f. in CCE	f. in the samples/f./ n.f. in BNC	Log-likelihood	Significance
V deter.	204/0.0015%	123/2184/0.0023%	−41.097179372915	0.000000000145
V pl. n.	77/0.0006%	23/408/0.0004%	4.185115597545	0.040780000000
V sing. n.	60/0.0005%	18/319/0.0003%	3.181355764705	0.074483000000
V adj.	53/0.0004%	18/319/0.0003%	0.796199732573	0.372230000000
V prep.	39/0.0003%	20/355/0.0004%	−2.950859085114	0.085832000000
V pers. pron.	34/0.0002%	21/373/0.0004%	−7.701856150918	0.005516400000
V poss. pron.	23/0.0002%	26/462/0.0005%	−35.164334708706	0.000000003030
V conj.	23/0.0002%	3/53/0.0001%	16.177404440166	0.000057678000
V inter. pron.	22/0.0002%	6/107/0.0001%	2.040675202579	0.153140000000
V prop. n.	17/0.0001%	2/35/0.0001%	13.977368116221	0.000185020000

In Table 2, the colligational pattern *V deter.* means the verb is followed by a determiner, such as "the" or "a", *V pl. n.* followed by a plural noun, *V sing. n.* by a singular noun, *V adj.* by an adjective, *V prep.* by a preposition, *V pers. pron.* by a personal pronoun, *V poss. pron.* by a possessive pronoun, *V conj.* by a conjunction, *V inter. pron.* by an interrogative pronoun, and *V prop. n.* by a proper noun.

From Table 2, one can see that the pattern *V deter.* is used most frequently of all the patterns in the two corpora. It is the common colligational structure for the verb *discuss* that is shared by China English and British English. However, there are marked differences between China English and British English with colligational patterns. On the one hand, three colligations (*V pl. n.*, *V conj.* and *V prop. n.*) are used more frequently in China English than in British, but on the other hand, 3 colligations (*V deter.*, *V pers.*

pron. and *V poss. pron.*) occur less frequently. These differences are statistically significant as the p-value is less than 0.05. The other patterns are used differently, but are not statistically significant.

4.3   Colligational Patterns of *Communicate*

The verb *communicate* ranks the second in frequency in the sub-category of the communicative verbs in CCE. Its frequencies of the colligational patterns in the two corpora are given in Table 3. The frequencies of the colligational patterns in BNC are calculated in the same way as discussed in Sect. 4.2. There are 1507 hits of the lemma "communicate" in BNC, and 555 hits of the word form "communicate" in the 1,000 samples from BNC. So the colligation patterns of the verb should be a × b/c, in which *a* is the total number of the lemma in BNC, *b* is the number of the colligational patterns out of the 1000 samples, and c is the word form of "communicate" in the 1,000 samples.

Table 3. Colligational patterns of "communicate" in CCE and BNC

Colligations	f./n.f. in CCE	f. in the samples/f./ n.f. in BNC	Log-likelihood	Significance
V prep.	135/0.0010%	225/611/0.0006%	17.770560822985	0.000024921000
V adv.	26/0.0002%	48/130/0.0001%	2.064089252046	0.150800000000
V conj.	16/0.0001%	24/65/0.0001%	3.202086039735	0.073545000000
V,	11/0.0001%	21/57/0.0001%	0.698280194051	0.403360000000
V pers. pron.	8/0.0001%	27/73/0.0001%	−0.616611597380	0.432310000000
V prep. to	5/0.0000%	25/68/0.0001%	−2.633409852916	0.104640000000
V adj.	3/0.0000%	11/30/0.0000%	−0.420681212860	0.516600000000
V pl. n.	3/0.0000%	7/19/0.0000%	0.017733582690	0.894060000000
V sing. n.	2/0.0000%	22/60/0.0001%	−6.862504133256	0.008802300000
V deter.	2/0.0000%	52/141/0.0001%	−25.451343505486	0.000000453680

In Table 3, the colligational patterns are named the same as in Table 2. The pattern *V,* means the verb *communicate* is used as an intransitive verb and followed by a comma, e.g. "By contrast, all of us need to learn how to communicate, and to understand the language we use.". The pattern *V prep. to* means the verb is followed by a preposition "to", e.g. "I try to communicate **to** the audience why I think they should bother to listen to what I've got to say.".

From Table 3, it is found that the pattern "*V prep.*" occurs most frequently of all the patterns in both CCE and BNC. The greatest difference between China English and British English in the use of the colligational patterns might be the pattern "*V deter.*". It is the second most frequently used pattern in BNC, but it occurs only twice in CCE. Moreover, the pattern "*V sing. n.*" is also used much less frequently in CCE. Both patterns occur less in China English. The number of occurrence of other patterns is different at various degrees, but not statistically significant.

4.4   Colligational Patterns of *Negotiate*

The verb *negotiate* is the third most frequently used verbs in the sub-category of the communicative verbs in CCE. Its frequencies of the colligational patterns in the two corpora are given in Table 4. The frequencies of the colligational patterns in BNC are calculated in the same way as discussed in Sects. 4.2 and 4.3. There are 1259 hits of the lemma "negotiate" in BNC, and 328 hits of the word form "negotiate" in the 1,000 samples from BNC. So the colligation patterns of the verb should be a × b/c, in which *a* is the total number of the lemma in BNC, *b* is the number of the colligational patterns out of the 1000 samples, and c is the word form of "negotiate" in the 1,000 samples.

Table 4. Colligational patterns of "negotiate" in CCE and BNC

Colligations	f./n.f. in CCE	f. in the samples/f./ n.f. in BNC	Log-likelihood	Significance
V prep.	64/0.0005%	89/342/0.0004%	3.267490811110	0.070666000000
V deter.	49/0.0004%	98/387/0.0004%	−0.848272482169	0.357040000000
V conj.	10/0.0001%	14/55/0.0001%	0.405331352618	0.524350000000
V,	7/0.0001%	13/51/0.0001%	−0.019940989003	0.887700000000
V sing. n	6/0.0000%	15/59/0.0001%	−0.762294469381	0.382610000000
V pl. n.	5/0.0000%	11/43/0.0000%	−0.234859968693	0.627940000000
V adj.	5/0.0000%	21/83/0.0001%	−4.771997876957	0.028926000000
V poss. pron.	2/0.0000%	8/32/0.0000%	−1.726250032201	0.188890000000
V infin. to	2/0.0000%	0/0/0.0000%	8.260047792932	0.004052700000
V pers. pron.	1/0.0000%	3/12/0.0000%	−0.333736976783	0.563470000000

In Table 4, the colligational patterns are named the same as in Tables 2 and 3. The data in Table 4 indicates that the colligational patterns of "negotiate" in China English tend to be more identical with those in British English when compared with the colligational patterns of the verbs "discuss" and "communicate". The patterns "*V prep.*" and "*V deter.*" are top two patterns in both corpora although they are in alternating orders, i.e. China English has more "*V prep.*", and British English more "*V deter.*". However, the most striking difference between China English and British English is the use of the colligational pattern of "*V infin. to*", which occurs twice in CCE, but nil in BNC. Another significant difference is the use of "*V adj.*" pattern, which is used less frequently in China English. The log-likelihood shows that 7 out of the 10 patterns have a lower frequency in China English than in British English. This may indicate that China English tends to heavily rely on some patterns, and use less varying patterns.

5 Discussion

With the results shown in Sect. 4, it is now possible to answer the research questions listed in Sect. 1. The answer to the first research question is affirmative. Although the three verbs are used with different grammatical structures, each of them has some common colligational patterns shared by both China English and British English. For the verb "discuss", it is the pattern *V deter.* A verb's colligational patterns are closely related to its transitivity. As the verb "discuss" is mainly used as a transitive verb, the most frequent structure might be "discuss sth.", in which a determiner is needed when required by the noun. However, the colligational pattern is used much less in China English than in British English with a significant value less than 0.05. This may suggest speakers of China English tend to omit the determiner or use other forms instead. For example, the pattern "*V pl. n.*" is used significantly more often in China English than in British English (See Table 2).

For the verb "communicate", the common colligational patterns are *V prep.* and *V adv.* as shown by Ex.1 and Ex. 2, which are top two patterns used both in China English and British English. So the verb is mainly used as an intransitive one.

Ex. 1. Zhou required that that the courts expand their use of technology to better communicate **with** the public (from CCE).

Ex. 2. The addresser and the addressee must communicate **simultaneously** at two levels (from CCE).

For the verb "negotiate", the top two colligational patterns shared by China English and British English are *V prep.* and *V deter.* The verb is used both as a transitive verb and an intransitive one. However, in China English, it is more frequently used as an intransitive one because the colligational pattern of "*V prep.*" occur more often in CCE than in BNC. On the contrary, native English speakers use more transitive colligational patterns of "*V deter.*".

The answer to the second research question is positive. Although China English and British English share some common colligational patterns, there are marked differences between China English and British English in the use of colligational patterns with regard to frequency. Speakers of China English tend to use some colligational patterns more frequently but other patterns less frequently than their British counterparts. For the verb "discuss", the following colligational patterns have a greater frequency value in China English than in British English: *V pl. n.*, *V conj.*, and *V prop. n.*, and the following ones a smaller frequency value: *V deter.*, *V pers. pron.*, and *V poss. pron.*. For the verb "communicate", the pattern "*V prep.*" is used more frequently in China English, but the patterns "*V sing. n.*" and "*V deter.*" are used less frequently. For the verb "negotiate", the pattern "*V adj.*" has a significantly smaller frequency in CCE than in BNC. This may suggest that Chinese speakers of English tend to use less often the structure of "negotiate +adj. +n.", such as "The case of minjian online writers has shown that both the Chinese state and Internet users constantly negotiate **new** boundaries in this new domain" (from CCE).

The answer to the third research question is negative. Previous studies [14] claimed that new grammatical structures such as "discuss about sth." were founded in China English. But in the present study, no such colligational patterns have been found. This might be caused by the source of corpus data. Ai and You's [14] data came from Chinese English learners in an online forum. So that grammatical structure is more an erroneous use than a creative one.

In the present study, only one colligational pattern has been found in CCE but not in the sampled data of BNC, i.e. "*V infin. to*", in which the verb "negotiate" is followed by an infinitive with "to" as in Ex. 3 and Ex. 4.

Ex. 3. Sometimes he can negotiate **to** have a larger loan because the future interest provides him growing collateral beyond the 10 million yuan (from CCE).

Ex. 4. If one party of the married couple signs the purchase contract of a real estate property and makes the down payment with his or her savings prior to marriage, and the couple jointly repay the mortgage within the marriage, in case of divorce, the couple should negotiate **to** divide the property (from CCE).

However, similar use of the pattern was found by a thorough examination in BNC although there were only a few examples. The sampled data in the present study is not large enough to include this use. Therefore, the colligational pattern "*V infin. to*" is used both in China English and British English. The difference is only quantitative.

6 Conclusion

The present paper uses corpus data to examine colligational patterns in China English and British English in order to identify the differences between these two English varieties in the use of grammatical constructions. Research results show that there are distinct differences between them with regard to the frequency. These results are a further proof of structural nativization in China English. Mukherjee and Gries [26] argue that "structural nativisation not only refers to entirely new and innovative forms and structures in individual varieties, but also covers quantitative differences between varieties of English in the use of forms and structures that belong to the common core that is shared by all Englishes". Below is a brief summary of the study's major contributions:

(1) It sheds light on the differences between China English and British English in the use of colligational patterns. No literature has been found to study structural nativization through colligations in the field of world Englishes. (2) The Corpus of China English was built for the studies of China English, which is a valuable language resource for natural language processing, computational linguistics, and the study of world Englishes. (3) It has explored the quantitative differences in the use of colligational patterns in China English by analyzing very large amounts of natural data, which is a methodological contribution to the study of China English.

The findings of the study show that speakers of China English have a clear preference for certain colligational patterns. However, the present study has some limitation. For each investigated verb in the BNC, only 1,000 concordance lines were selected by random sampling. It might not be a big problem for the verb "communicate" or "negotiate" because there are 1507 and 1295 hits for the two verbs respectively. But there are 5505 hits for the verb "discuss". Therefore, some use of colligational patterns might be missing in the sampled data. Besides, only three verbs of communication were chosen as the objects of study due to time and space limit. It is better if more communicative verbs or other types of verb are investigated.

Acknowledgements. This work was partly supported by a grant from the Laboratory of Language Engineering and Computing of Guangdong University of Foreign Studies under the project entitled "A Study on Grammatical Nativization of China English" (Grant No. LEC2017ZBKT002), a grant from the Humanities and Social Science Planning and Funding Office of Ministry of Education of the People's Republic of China under Grant No. 15YJA740048 (China English or Chinglish: A Study Based on China English Corpus), and a grant from Guangdong Planning Office of Philosophy and Social Science (Grant No. GD14XWW20). It was also part of the research achievements of the special innovation project entitled "A Study on the Definitions of Chinese-English Dictionaries for CFL Learners from the User's Perspective" (Grant No. 2017WTSCX32) granted by the Department of Education of Guangdong Province.

References

1. Ge, C.: Random thoughts on some problems in Chinese-English translation. Chin. Transl. J. **2**, 1–8 (1980)
2. Huang, J.: The positive role of 'Sinicism' in the English-translated version. Chin. Transl. J. **1**, 39–45 (1988)
3. Wang, R.: Chinese English is an objective reality. J. PLA Foreign Lang. Inst. **1**, 2–7 (1991)
4. Jia, G., Xiang, Y.: In defence of Chinese English. Foreign Lang. Teach. **5**, 34–38 (1997)
5. Li, W.: China English and Chinglish. Foreign Lang. Teach. Res. **4**, 18–24 (1993)
6. Xu, Z.: Chinese English: Features and Implications. Open University of Hong Kong Press, Hong Kong (2010)
7. Cannon, G.: Chinese Borrowings in English. Am. Speech **63**, 3–33 (1988)
8. Jiang, Y.: China English: issues. Stud. Featur. Asian Engl. **5**(2), 4–23 (2002)
9. Du, R., Jiang, Y.: "China English" in the past 20 years. Foreign Lang. Teach. Res. **33**(1), 37–41 (2001)
10. Bolton, K.: China Englishes: A Sociolinguistic History. Cambridge University Press, Cambridge (2003)
11. Kirkpatrick, A.: Research China English as a lingual franca in Asia: the Asian Corpus of English (ACE) project. Asian Engl. **13**(1), 4–8 (2010)
12. Liu, Y., Zhao, Y.: English spelling variation and change in newspapers in mainland China, Hong Kong and Taiwan: which English spelling system is preferred in the Chinese news media: British or American? Engl. Today **31**(4), 5–14 (2015)
13. Stadler, S.: Conventionalized politeness in Singapore Colloquial English. World Engl. **37**, 307–322 (2018)
14. Ai, H., You, X.: The grammatical features of English in a Chinese Internet discussion forum. World Engl. **34**(2), 211–230 (2015)
15. Hu, J., Zhang, P.: A corpus-based study on nativization of high-frequency verbs of transformation in China's English newspapers. Corpus Linguist. **2**(1), 59–70 (2015)
16. Ma, Q., Xu, Z.: The nativization of English in China. In: Xu, Z., He, D., Deterding, D. (eds.) Researching Chinese English: The State of the Art. ME, vol. 22, pp. 189–201. Springer, Cham (2017). https://doi.org/10.1007/978-3-319-53110-6_13
17. Yu, X., Wen, Q.: Collocation patterns of evaluative adjectives of English in Chinese newspapers. Foreign Lang. Teach. Res. **254**(5), 23–28 (2010)

18. Svensén, B.: A Handbook of Lexicography: The Theory and Practice of Dictionary-Making. Cambridge University Press, Cambridge (2009)
19. Schneider, E.W.: Postcolonial English: Varieties Around the World. Cambridge University Press, Cambridge (2007)
20. Mukherjee, J., Hoffmann, S.: Describing verb-complementational profiles of new Englishes: a pilot study of Indian English. Engl. World-Wide **27**, 147–173 (2006)
21. Schilk, M.: Structural Nativization in Indian English Lexicogrammar. John Benjamins, Amsterdam (2011)
22. Schilk, M., Bernaisch, T., Mukherjee, J.: Mapping unity and diversity in South Asian English lexicogrammar: verb-complementational preferences across varieties. In: Hundt, M., Gut, U. (eds.) Mapping Unity and Diversity World-wide, pp. 137–165. John Benjamins, Amsterdam (2012)
23. Yu, X.: The use of the word "foreign" in China's English newspapers: a corpus-based study. Foreign Lang. Educ. **27**(6), 23–26 (2006)
24. Liu, M., Wu, Y.: Beyond frame semantics: insight from Mandarin verbs of communication. In: 4the Chinese Lexical Semantic Workshop. Hongkong: Department of Chinese, Translation and Linguistics, City University of Hongkong (2003)
25. Liang, M., Li, W., Xu, J.: Introduction to Corpora. Foreign Languages Teaching and Researching Press, Beijing (2010)
26. Mukherjee, J., Gries, S.: Collostructional notification in New Englishes: verb-construction associations in the International Corpus of English. Engl. World-Wide **30**(1), 27–51 (2009)

Testing the Reasoning Power
for NLI Models with Annotated
Multi-perspective Entailment Dataset

Dong Yu[1(✉)], Lu Liu[1], Chen Yu[1], and Changliang Li[2]

[1] Beijing Language and Culture University, Beijing, China
yudong@blcu.edu.cn, luliu.nlp@gmail.com, yuchen7312@gmail.com
[2] Kingsoft AI Lab, Beijing, China
lichangliang@kingsoft.com

Abstract. Natural language inference (NLI) is a challenging task to determine the relationship between a pair of sentences. Existing Neural Network-based (NN-based) models have achieved prominent success. However, rare models are interpretable. In this paper, we propose a Multi-perspective Entailment Category Labeling System (METALs). It consists of three categories, ten sub-categories. We manually annotate 3,368 entailment items. The annotated data is used to explain the recognition ability of four NN-based models at a fine-grained level. The experimental results show that all the models have poor performance in the commonsense reasoning than in other entailment categories. The highest accuracy difference is 13.22%.

Keywords: Natural Language Inference · Multi-perspective Entailment Category Labeling System · Entailment categories

1 Introduction

NLI is an important subtopic in Natural Language Understanding. NLI is to identify the entailment relation between two sentences, commonly formalized as 3-way classification task. Previous works devote to the development of NLI datasets and models [2,8,9].

In 2015, the SNLI [4] corpus is released by Bowman. It provides more than 570K P-H sentence pairs. After that, other datasets have also emerged, such as MultiNLI [33], Dialogue NLI [32], and QA-NLI [10]. Those datasets have promoted the development of many NN-based NLI models, especially SNLI. Since the release of this dataset, the highest accuracy of it is constantly improved [8,11,21,24,25,31], and the recognition ability of some models now have exceed human beings. However, those models are hard to interpret. It's often not clear why they work or how they exactly work. Consequently, some researches set out to explore the nature of NLI tasks in novel ways [5,15,22].

© Springer Nature Switzerland AG 2019
M. Sun et al. (Eds.): CCL 2019, LNAI 11856, pp. 15–26, 2019.
https://doi.org/10.1007/978-3-030-32381-3_2

Some studies begin to do more fine-grained classification [2,14,27]. But the sentence pairs that contain multiple entailment categories are still unexplained. The categories of entailment are still not fine-grained enough.

In this paper, we propose METALs aiming at the entailment data (in Sect. 4.2). The METALs is set into three categories, ten sub-categories. We chose entailment because it is more regular than neutral, less studied than contradiction. We manually annotate 3368 entailment items in SNLI test set. The current version of our labeled corpus is freely available at https://github.com/blcunlp/Multi-perspective-Entailment-Dataset.

To interpret the recognition ability of NN-based models at a fine-grained and multi-perspective level, we conduct experiments based on four NLI models (in Sect. 5.1). The experimental results demonstrate that all the models are excellent in the entailment data due to the reduction of information or the change of syntactic structure. But they are inadequate in the entailment examples which contain inference (in Sect. 5.2).

2   Related Works

Since the NLI task is presented, many NLI datasets have released. An early attempt is Recognizing Textual Entailment (RTE) Challenge [9]. There are several larger benchmarks inspired by this work. The Sentences Involving Compositional Knowledge (SICK) [18] benchmark is proposed in 2014 and collect about 10k sentence pairs. In 2015, the SNLI [4] corpus is released by Bowman. It provides more than 570K P-H sentence pairs. After that, other datasets have appeared, such as MultiNLI [33], Dialogue NLI [32], and QA-NLI [10]. Those datasets promote the development of many NN-based NLI models, especially SNLI. And our work focus on SNLI models.

There are two main categories of the NN-based models on SNLI: one is encoding sentence embeddings and integrating the two sentence representations to predict [6,8,17,19,21,28], and the other is cross sentence attention-based models which concern more about the interaction between each sentence pair [13,23,24,31]. Recently, some pre-training models [11,16,25,34] achieve good results in NLI tasks.

Although some models have got high accuracy, how good the model is, where it is, and why it is good are still inconclusive. Recently, some works devote to explain the performance of models. [27] proposes a detailed evaluation and sketches a process to generate its annotation. They label a subset of 210 examples and utilize a series of experiments to prove that these annotations are useful. [14] creates a new NLI test set that shows the deficiency of models in inferences which require lexical and world knowledge. [2] proposes a methodology for the creation of specialized data sets for NLI and experiments it over a sample of pairs taken from the RTE-5 data set [3].

Early studies have shown that the relationship between P and H is identified by analyzing and utilizing lexical, syntax, and world knowledge. [29] predicts the datasets completely based on the grammatical clues, and puts forward more than

50 grammatical tags. [7] extracts 100 entailment samples and classifies entail phenomena. The conclusion is that lexical and world knowledge is needed for RTE. [12] proposes an annotation method, ARTE, which extract 23 entailment relations. Subsequently, researches begin to focus on the specific phenomenon of NLI. [26] studies motion-space reasoning of entailment and constructs reasoning corpus. [1] shows how to improve classification accuracy by using the hypostatic relation.

Inspired by the previous work, we propose a Multi-perspective Entailment Category Labeling System (METALs) and experiment it over a sample of pairs from SNLI. We select several models with high classification accuracy and analyze them below.

3 Categories

In order to detect the recognition ability of the model more finely, we consider it from three aspects. It refers to widely accepted linguistic categories in the literature [12]: lexical, syntactic, and reasoning.

At the lexical level, We regard the ontological knowledge relation between words as the inference basis. The entailment is mainly based on language conversion, addition, deletion, and replacement at the level of syntactic structure. After lexical and syntactic level mining, there are still some sentence pairs that cannot be categorized. These pairs need additional world knowledge, so we sum up them as commonsense reasoning. In addition, some of the data is classified as discard due to they contain many errors, such as spelling errors, adding irrelevant ingredients, capital letter, and punctuation change, etc.

From the three levels, we divide the entailment types into three categories, ten sub-categories. The categories are not independent. They can be applied in combinations so that they can achieve a result as informative as possible. We briefly introduce the definitions and main ideas associated with each component of the architecture in this section.

3.1 Lexical and Phrasal Level (LP)

The lexical level is intended to capture basic lexical's inherent properties of the entailment phenomenon. It also called as ontological relations. Ontological involves three kinds of lexical ontological relations, which are drawn from Word-Net [20]. We chose three relations that are typically associated with the notion of semantic similarity.

Hypernymy (Hyp) refers to two entail words linked by the is-a-kind-of relation. A hyponym is a word or phrase whose semantic field is more specific than its hypernym. "poodle-dog-animal" is a typical example of transitive relation in this category.

Synonymy (Syn) refers to words or phrase that have the same meaning or similar meaning in the sentence pair. These two words may have slight differences in using range, emotional color, or colloquial style. But their basic meanings are mostly same. "screaming" and "yelling" have a similar meaning. They are interchangeable within the context in which they appear, such as "A man is X at a camera".

Meronymy (Mer) is often expressed as "part-of". A meronym denotes a constituent part of, or a member of something, such as "tree" and "tree branch".

3.2   Syntactic Structure (SS)

The syntactic structure in entailment is reflected in three aspects. The first one is changing the surface structure of the sentence and keeping the meaning in the same. The second is the extraction and reorganization of sentence components. The third is the ellipsis of syntactic components.

Syntactic Transformations (ST) is using different language forms to express synonymous meaning. There are four typical transformations: changes in active-passive voice, changes of word order, uses double negatives, changes between simple-compound sentences.

Extraction (Ext) refers to extract a certain component of a sentence and reassembling it into a new sentence. The new sentence is H, which contains part of semantic information of the original sentence. The extracted part can be a syntactic component or a complete clause. The existential sentence which often appears in the H sentence is one of the reorganizing sentence forms.

Ellipsis (Elli) refers to delete or omit some structures in a sentence. Language recursion makes the language structure layer upon layer without causing structural confusion. Therefore, the overall structure of the sentence remains unchanged and the semantic information is extracted, when some structures in a sentence are deleted or omitted. Reduction information is an universal entailment.

3.3   Commonsense Reasoning (CR)

Understand natural language requires not only linguistic knowledge but also human commonsense and social experience. Commonsense reasoning includes judgments about the physical properties, purpose, intentions, and behavior of people and objects as well as possible outcomes of their actions and interactions.

Spatial Reasoning (SR) refers to the inference obtained by judging the absolute location like "on the beach" and "in the sand" or relative location like "in front of" and "behind". The judgment of the spatial relationship is complicated.

Quantities Reasoning (QR) includes not only the addition, subtraction, and numbers comparison but also the judgment of cardinal numbers with approximate numbers.

Emotion Reasoning (EmoR) is to infer emotions, emotional states, and psychological changes through the character's actions, facial expressions, and other external information. Emotion is one of the inherent affective phenomena which is intrinsic in human experience.

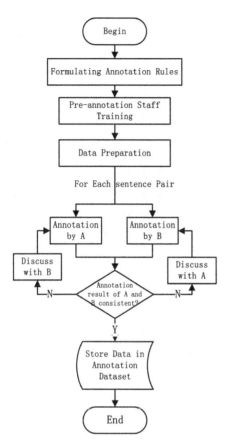

Fig. 1. The annotation flow diagram.

Table 1. Percentages for all categories on pre-labeled one hundred data.

Category	Percentage
LP	2
SS	38
CR	17
AllSingle	57
Complex	**43**
Discard	1

Table 2. Percentages for all categories for all categories and sub-categories on annotated 3368 sentence pairs.

Category	Sub-category	Num	Percent
LP	Hyp	296	8.79
	Syn	608	18.05
	Mer	5	0.15
	LP-com	75	2.23
SS	Elli	1646	48.87
	Ext	361	10.72
	ST	78	2.32
	SS-com	144	4.28
CR	EmoR	22	0.65
	QR	115	3.41
	SR	441	13.09
	EveR	803	23.84
	CR-com	73	2.17
Diacard		86	2.55

Event Reasoning (EveR) is the judgment of the attributes, purposes, intentions, and behaviors of people as well as the possible results of their interactions. We describe the event from the general event relationship, event framework and event environment, and background.

4   Data Annotations

4.1   Data Selection

We select the open source dataset SNLI [4] as a data source. SNLI is a current mainstream NLI dataset. It consists of 570k pairs of sentences. Each sentence pair includes a P, an H, and a label describing the relationship between them (entailment, neutral, or contradiction). Its data are manually labeled through crowdsourcing.

The crowdsourcing workers obtain a description of the photo scene as a P. And they are asked to write different H according to the three requirements, corresponding to three inference labels.

SNLI test set consists of 10k sentence pairs. After removing the invalid data (The workers are inconsistent in the labeling of certain sentence pairs, and these sentences are ultimately unlabeled.) in the test set, we get 9824 samples. 3368 sentence pairs are labeled as entailment among them. Those are the data source for the annotation data in this paper.

4.2 Multi-perspective Entailment Category Labeling System

After analyzing 100 pre-labeled sentence pairs, we find that nearly half of P-H pairs have more than one cause of entailment relationship. The statistical results are shown in Table 1. "AllSingle" are sentence pairs with one and only one category. "Complex" are sentence pairs with more than two categories. Differing from the previous method of labeling only one category for each P-H pairs [22], we adopt a **METALs** to maintain the integrity of the entailment categories of each P-H pair. If a sentence pair belongs to both category A and category B, we label it on both A and B. METALs has three categories and ten sub-categories (in Sects. 3.1 to 3.3). The following is a specific example of an annotation.

Example 1. P: African woman walking through field. H: There is women navigating through the fields.

The example contains three perspectives of judgments, so we assign it three labels. Firstly, omitting the attribute modifier "Africa" is labeled as *Elli* in *SS*. Secondly, "Walking-navigating" with similar meanings is labeled as *Syn* in *LP*. Thirdly, "There be" is a sentence pattern transformation, we label it as *ST* in *SS*.

METALs avoids the influence of subjective judgment on objective results in single labeling. Our comprehensive, accurate and highly operable system can solve the ambiguous problems in classification, and make a more fine-grained evaluation of the causes of entailment.

4.3 Data Collection

We manually annotate all 3368 items selected from SNLI test set with MET-ALs. We formulate annotation rules before labeling. In the labeling process, the annotator should first judge the category and find the entailment fragment in the sentence. Then they should match the entailment fragment to sub-categories. The annotation flow diagram is shown in Fig. 1.

In order to ensure the quality of the annotations, we recruit two graduate students with linguistic learning background to annotate. They begin to label the data after the training. In the labeling process, two annotators judge the entailment type under the guidance of the annotation specification. A consistency assessment is then performed. Sentence pairs that are consistent with the

Table 3. Experimental results on four models. "Author-acc" is the accuracy given in the original paper. "Our-acc" is the accuracy of our retraining model.

Model	Author-acc	Our-acc
ESIM	88.0	88.0
GPT	89.9	89.8
BERT_base	-	89.3
MT-DNN_base	91.1	91.0

Table 4. The classification accuracy (%) of ESIM, GPT, BERT and MT-DNN in entailment categories.

Category	ESIM	GPT	BERT	MT-DNN
LP	82.76	85.52	85.52	**89.66**
SS	96.73	**97.16**	96.90	96.81
CR	84.09	**84.79**	83.68	83.54
LP&SS	94.06	94.44	95.40	**95.59**
LP&CR	**87.03**	**87.03**	85.95	**87.03**
SS&CR	87.44	88.65	**89.13**	87.92
LP&SS&CR	91.67	92.42	**93.18**	**93.18**
Discard	76.74	74.42	**80.23**	75.58

evaluation are identified as the final labeled data and stored in the final labeled dataset. Sentence pairs that assess inconsistencies would be discussed by two annotators and relabeled until consistent results are achieved.

The annotation process is not linear. Multiple iterations improve correctness and credibility of the annotation results.

4.4 Data Annotation Results

Table 2 exhibits the percentage of categories and sub-categories on total entailment samples. "*-com" is the complex sub-categories of category *.

In the annotation process, the entailment relationship of some sentence pairs contains a complex of two sub-categories under the same category. For example, sentence pair "P: A *younger* man *dressed* in *tribal attire*. H: A *young* man *wears clothing*." contains two kinds of entailment relations, including sub-category *HH* (tribal attire-clothing) and sub-category *Syn* (younger-young and dressed-wears) below category *LP*. Sub-categories have many combinations. But the number of each combination is rare. We name the combination of the different sub-categories under the same category as "*-com" and analyze them as a whole.

5   Experiments

5.1   Experimental Setup

We concentrate on several classic and popular models on SNLI, which attain strong performance.

Enhanced Sequential Inference Model (ESIM) [24] is a strong benchmark on SNLI. It achieves 88.0% in accuracy on SNLI and exceeds the human performance (87.7%) for the first time.

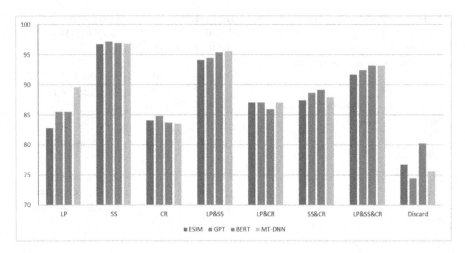

Fig. 2. The classification accuracy (%) of ESIM, GPT, BERT and MT-DNN in entailment categories.

Generative Pre-trained Transformer (GPT) [25] explores a semi-supervised approach for language understanding tasks using a combination of unsupervised pre-training and supervised fine-tuning. It trains a left-to-right Transformer Language Model [30].

Encoder Representations from Transformers (BERT) [11] addresses unidirectional constraints by proposing a new pre-training objective: the "masked language model". It bases on a multi-layer bidirectional Transformer encoder [30]. In this paper, we retraining the BERT_base model.

Multi-task Deep Neural Network (MT-DNN) [16] is a multitask deep neural network model for learning universal language embedding. MT-DNN integrates the advantages of multitask learning and BERT, and outperforms BERT in all ten natural language understanding tasks. MT-DNN_base are fine-tuned based on the pre-trained BERT_base.

We train above four models on SNLI. Our experimental results are shown in Table 3. The results obtained from our retraining almost all reach the results given by the author, which fully demonstrate that our analysis in the next chapter is credible.

5.2 Results and Analysis

In order to explore the performance of these models on different entailment categories, we test them on our annotated dataset using the reproductive models.

Performance on Coarse-Grain. The results of the four models in three categories and *Discard* are presented in Table 4 and Fig. 2. Since we use METALs, some sentence pairs may be labeled as two categories at the same time. The

Table 5. The classification accuracy (%) of ESIM, GPT, BERT and MT-DNN in entailment sub-categories. "*-com" is the complex sub-categories of category *.

Category	Sub-category	ESIM	GPT	BERT	mt-dnn
LP	Hyp	93.58	94.26	93.92	**94.59**
	Syn	89.64	90.13	90.79	**91.45**
	Mer	**100.00**	80.00	**100.00**	**100.00**
	LP-com	88.00	92.00	92.00	**96.00**
SS	Elli	93.74	94.71	**94.96**	94.65
	Ext	95.84	**96.40**	95.57	95.57
	ST	91.03	85.90	**92.31**	91.03
	SS-com	**95.14**	94.44	93.75	94.44
CR	EmoR	77.27	86.36	**95.45**	90.91
	QR	91.30	88.70	**93.04**	92.17
	SR	93.88	**95.01**	94.56	93.88
	EveR	81.07	**81.82**	80.57	80.57
	CR-com	89.04	**90.41**	87.67	89.04

"*LP&SS*" means that the sentence pairs are entailment caused by *LP* and *SS* at the same time.

For sentences that contain only one category, we can easily observe from Fig. 2 that all models are excellent in *SS*. But models have poor performance in both *CR* and *Discard*. In Table 4, the classification accuracy of the four models is 19.99%, 22.74%, 16.67%, and 21.23% higher in *SS* than in *Discard*, and 12.63%, 12.37%, 13.22%, and 13.27% higher in *SS* than in *CR*. This stems from the model's stronger capability in recognizing information reduction and semantic structural change in *SS*. On the contrary, the cases in *CR* are more complex, because they require additional commonsense knowledge for reasoning and imagination. The results of all models are not satisfactory in *LP*. And the classification accuracy of all pre-training models outperforms ESIM in this category. In particular, MT-DNN is 6.90% higher than ESIM. It shows that pre-training models bring a lot of knowledge about words and phrases.

The classification accuracy of these four models in *LP&SS* is significantly higher than that in *CR&SS*, *CR&LP*, and *CR&LP&SS*. This further indicates that the performance of the models on *CR* is poor, so the classification accuracy is still relatively low when combined with other categories.

The results for *Discard* are significantly lower than the other categories, and even the highest is only 80.23%. We illustrate this problem with a concrete example. Given a P: *A woman holds a newspaper that says "Real change"* and an H: *A woman **on a street** holding a newspaper that says "real change"*, the extra ingredients, like on a street, add in H are often hard to be inferred by models, even human. In essence, this kind of example should not be considered as entailment, but the noise mix into the dataset when it is set up. The noise results from the flexible annotation rules of SNLI.

Performance on Fine-Grained. The classification performance of all four models on all sub-categories is displayed in Table 5. We label sentence pairs that have multiple subcategories under the same category as complex entailment in that category.

In category *LP*, because the *Mer* only have five samples in the labeled test dataset, the results of four models in this sub-category valueless to analyze. MT-DNN outperforms other models in all sub-categories. This model is good at recognizing sentence pairs with word relations and phrase relations. And all three pre-training models perform better than ESIM. It indicates that the pre-training model has mastered more lexical and phrasal information in the process of pre-training.

In category *SS*, the classification accuracy of each model on *Ext* is more than 95%. In this sub-category, the H is part of the information proposed from the P, so it is easy for models to recognize the entailment relationship. All models have higher classification accuracy on almost all sub-categories of category *SS* than they do on all test sets.

In category *CR*, it is obvious that all pre-training models are excellent in *EmoR*, the classification accuracy of the three pre-training models is 9.09%, 18.18%, and 13.64% higher than ESIM. Pre-trained language models enable texts with similar emotions to have an associated representation, such as reasoning from "laugh/grin/smile" to "happy". Each model performs well on both *QR* and *SR*. This proves that the entailment of *QR* and *SR* are relatively easy to identify. The *EveR* makes up a big proportion of the total entailment samples, reaching 23.84%. However, the classification accuracy of all models in this category is relatively low. This leads to a decrease in the classification accuracy of the model for the entire datasets. *EveR* is difficult for all models because the recognition of such entailment requires a strong knowledge, such as common sense knowledge. It is urgent to study how to improve the ability of models for *EveR* sentence pairs recognition.

From the above analysis, we can draw the conclusion that current models are better at recognizing the entailment relationship of *SS*. *CR* and *LP* are a bit difficult for the model, especially *CR*. Future work should focus on the representation and use of common sense knowledge in the training model. We have already seen the promotion of the pre-training model *LP*. In the future, we should also explore how to increase it further.

6   Conclusion

In this paper, we propose a complete entailment category annotation system (METALs). It has three categories, ten sub-categories. We manually annotate 3368 items of SNLI test set. To granularly examine the recognition ability of NN-based NLI models, We conduct experiments with four models on the labeled dataset for more fine-grained comparison of the strengths and weakness of each model. The experimental results demonstrate that all the models are excellent in *SS*. But results are not satisfactory in both *CR* and *LP*.

Acknowledgments. This work is funded by National Key R&D Program of China, "Cloud computing and big data" key projects (2018YFB1005105).

References

1. Akhmatova, E., Dras, M.: Using hypernymy acquisition to tackle (part of) textual entailment. In: Proceedings of the 2009 Workshop on Applied Textual Inference, pp. 52–60. Association for Computational Linguistics (2009)
2. Bentivogli, L., Cabrio, E., Dagan, I., Giampiccolo, D., Leggio, M.L., Magnini, B.: Building textual entailment specialized data sets: a methodology for isolating linguistic phenomena relevant to inference. In: LREC. Citeseer (2010)
3. Bentivogli, L., Clark, P., Dagan, I., Giampiccolo, D.: The fifth PASCAL recognizing textual entailment challenge. In: TAC (2009)
4. Bowman, S.R., Angeli, G., Potts, C., Manning, C.D.: A large annotated corpus for learning natural language inference. In: Proceedings of the 2015 Conference on Empirical Methods in Natural Language Processing, pp. 632–642. Association for Computational Linguistics, Lisbon (2015). https://doi.org/10.18653/v1/D15-1075
5. Carmona, V.I.S., Mitchell, J., Riedel, S.: Behavior analysis of NLI models: uncovering the influence of three factors on robustness (2018)
6. Chen, Q., Zhu, X., Ling, Z.H., Wei, S., Jiang, H., Inkpen, D.: Recurrent neural network-based sentence encoder with gated attention for natural language inference. arXiv preprint arXiv:1708.01353 (2017)
7. Clark, P., Murray, W.R., Thompson, J., Harrison, P., Hobbs, J., Fellbaum, C.: On the role of lexical and world knowledge in RTE3. In: Proceedings of the ACL-PASCAL Workshop on Textual Entailment and Paraphrasing, pp. 54–59. Association for Computational Linguistics (2007)
8. Conneau, A., Kiela, D., Schwenk, H., Barrault, L., Bordes, A.: Supervised learning of universal sentence representations from natural language inference data (2017)
9. Dagan, I., Glickman, O., Magnini, B.: The PASCAL recognising textual entailment challenge. In: Quiñonero-Candela, J., Dagan, I., Magnini, B., d'Alché-Buc, F. (eds.) MLCW 2005. LNCS (LNAI), vol. 3944, pp. 177–190. Springer, Heidelberg (2006). https://doi.org/10.1007/11736790_9
10. Demszky, D., Guu, K., Liang, P.: Transforming question answering datasets into natural language inference datasets (2018)
11. Devlin, J., Chang, M.W., Lee, K., Toutanova, K.: BERT: pre-training of deep bidirectional transformers for language understanding. arXiv preprint arXiv:1810.04805 (2018)
12. Garoufi, K.: Towards a better understanding of applied textual entailment. Ph.D. thesis, Citeseer (2007)
13. Ghaeini, R., et al.: DR-BiLSTM: dependent reading bidirectional LSTM for natural language inference. arXiv preprint arXiv:1802.05577 (2018)
14. Glockner, M., Shwartz, V., Goldberg, Y.: Breaking NLI systems with sentences that require simple lexical inferences. arXiv preprint arXiv:1805.02266 (2018)
15. Gururangan, S., Swayamdipta, S., Levy, O., Schwartz, R., Bowman, S.R., Smith, N.A.: Annotation artifacts in natural language inference data (2018)

16. Liu, X., He, P., Chen, W., Gao, J.: Multi-task deep neural networks for natural language understanding. arXiv preprint arXiv:1901.11504 (2019)
17. Liu, Y., Sun, C., Lin, L., Wang, X.: Learning natural language inference using bidirectional LSTM model and inner-attention. arXiv preprint arXiv:1605.09090 (2016)
18. Marelli, M., Bentivogli, L., Baroni, M., Bernardi, R., Menini, S., Zamparelli, R.: Semeval-2014 task 1: evaluation of compositional distributional semantic models on full sentences through semantic relatedness and textual entailment. In: Proceedings of the 8th International Workshop on Semantic Evaluation (SemEval 2014), pp. 1–8 (2014)
19. McCann, B., Keskar, N.S., Xiong, C., Socher, R.: The natural language decathlon: multitask learning as question answering. arXiv preprint arXiv:1806.08730 (2018)
20. Miller, G.A.: WordNet: a lexical database for English. Commun. ACM **38**(11), 39–41 (1995)
21. Mou, L., et al.: Natural language inference by tree-based convolution and heuristic matching. In: Meeting of the Association for Computational Linguistics (2016)
22. Naik, A., Ravichander, A., Sadeh, N., Rose, C., Neubig, G.: Stress test evaluation for natural language inference (2018)
23. Parikh, A.P., Täckström, O., Das, D., Uszkoreit, J.: A decomposable attention model for natural language inference. arXiv preprint arXiv:1606.01933 (2016)
24. Qian, C., Zhu, X., Ling, Z.H., Si, W., Inkpen, D.: Enhanced LSTM for natural language inference (2017)
25. Radford, A., Narasimhan, K., Salimans, T., Sutskever, I.: Improving language understanding with unsupervised learning, Technical report, OpenAI (2018)
26. Roberts, K.: Building an annotated textual inference corpus for motion and space. In: Proceedings of the 2009 Workshop on Applied Textual Inference, pp. 48–51. Association for Computational Linguistics (2009)
27. Sammons, M., Vydiswaran, V., Roth, D.: Ask not what textual entailment can do for you... In: Proceedings of the 48th Annual Meeting of the Association for Computational Linguistics. pp. 1199–1208. Association for Computational Linguistics (2010)
28. Tay, Y., Tuan, L.A., Hui, S.C.: Compare, compress and propagate: enhancing neural architectures with alignment factorization for natural language inference. arXiv preprint arXiv:1801.00102 (2017)
29. Vanderwende, L., Dolan, W.B.: What syntax can contribute in the entailment task. In: Quiñonero-Candela, J., Dagan, I., Magnini, B., d'Alché-Buc, F. (eds.) MLCW 2005. LNCS (LNAI), vol. 3944, pp. 205–216. Springer, Heidelberg (2006). https://doi.org/10.1007/11736790_11
30. Vaswani, A., et al.: Attention is all you need. In: Advances in Neural Information Processing Systems, pp. 5998–6008 (2017)
31. Wang, Z., Hamza, W., Florian, R.: Bilateral multi-perspective matching for natural language sentences (2017)
32. Welleck, S., Weston, J., Szlam, A., Cho, K.: Dialogue natural language inference (2018)
33. Williams, A., Nangia, N., Bowman, S.R.: A broad-coverage challenge corpus for sentence understanding through inference (2017)
34. Zhang, Z., et al.: I know what you want: semantic learning for text comprehension. arXiv preprint arXiv:1809.02794 (2018)

Enhancing Chinese Word Embeddings from Relevant Derivative Meanings of Main-Components in Characters

Xinyu Su, Wei Yang$^{(\boxtimes)}$, and Junyi Wang

School of Computer Science and Technology,
University of Science and Technology of China, Hefei 230027, China
{sa517303,sa517352}@mail.ustc.edu.cn, qubit@ustc.edu.cn

Abstract. Word embeddings have a significant impact on natural language processing. In morpheme writing systems, most Chinese word embeddings take a word as the basic unit, or directly use the internal structure of words. However, these models still neglect the rich relevant derivative meanings in the internal structure of Chinese characters. Based on our observations, the relevant derivative meanings of the main-components in Chinese characters are very helpful for improving Chinese word embeddings learning. In this paper, we focus on employing the relevant derivative meanings of the main-components in the Chinese characters to train and enhance the Chinese word embeddings. To this end, we propose two main-component enhanced word embedding models named MCWE-SA and MCWE-HA respectively, which incorporate the relevant derivative meanings of the main-components during the training process based on the attention mechanism. Our models can fine-grained enhance the precision of word embeddings without generating additional vectors. Experiments on word similarity and syntactic analogy tasks are conducted to validate the feasibility of our models. Furthermore, the results show that our models have a certain improvement in the similarity task over most baselines, and have nearly 3% improvement in Chinese analogical reasoning dataset compared with the state-of-the-art model.

Keywords: Relevant derivative meaning · Component level ·
Enhanced word embedding

1 Introduction

Data representation is the basic work in natural language processing (NLP), and the quality of data representation directly affects the performance of the entire system. Word embedding, which is also called distributed word representation, has been an important foundation in the field of NLP. It encodes the semantic meaning of a word into a real-valued low-dimensional vector, which performs better in many tasks such as text classification [1,2], machine translation [3,4], sentiment analysis [5,6] and question answering [7,8] over traditional one-hot

© Springer Nature Switzerland AG 2019
M. Sun et al. (Eds.): CCL 2019, LNAI 11856, pp. 27–39, 2019.
https://doi.org/10.1007/978-3-030-32381-3_3

representations. Among many word embedding models, Continuous Bag-of-Word (CBOW) [9], Skip-gram [9] and Global Vectors(GloVe) [12] are popular because of their simplicity and efficiency. The idea of those algorithms is mainly based on the distributed hypothesis which means words that are used and occur in the same contexts tend to purport similar meanings [13]. However, these models only focus on word level information, and do not pay attention to the fine-grained morphological information inside the words or the characters, such as components of Chinese characters or English morphemes.

Different from English words, Chinese characters are glyphs whose components may depict objects or represent abstract notions. Usually a character consists of two or more components which may have meanings related to the character, using a variety of different principles[1]. That means Chinese words themselves are often composed of Chinese characters and subcharacter components, including abundant semantic information.

Previous researchers have done some work by using the abundant information inside Chinese for word embeddings enhancement with internal morphological semantics. Li et al. [14] used the radicals to enhance the Chinese character embeddings. Chen et al. [15] proposed the CWE model to improve the quality of Chinese word embeddings by exploiting character level information. For a more fine-grained combination of Chinese character and radical, Yin et al. [16] proposed methods to enhance Chinese character embeddings based on CWE model. Jian et al. [17] used external language to calculate the similarity between Chinese words and characters to enhance quality of word vectors based on the rich internal structure information of Chinese words. Huang et al. [19] proposed the GWE model, a pixel-based model that learns character features from font images to enhance the representation of words. Yu et al. [18] used Chinese characters and subcharacter components to improve Chinese word embeddings and proposed the JWE model to jointly learn Chinese word and character embeddings. Cao et al. [20] proposed cw2vec model, which exploits stroke-level information to improve the learning of Chinese word embeddings.

However, the subcharacter components of a character contain a lot of extra noise information, so we explore a new direction for better quality of word embeddings through integrating several word components into main-component. Chinese characters are composed of components and radicals, and a component of the complex subword item may be a simple Chinese character. In our models, the subcharacter components of characters can be roughly divided into two types: main-component and radical. The main-component which consists of several components indicates the basic meaning of a character while the radical indicates some attribute meanings of a character. For example, as Fig. 1 shows, "慧" (intelligent) is divided into the subcharacter components "丰" (abundant), "彐" (snow) and "心" (heart) where "丰" (abundant) and "彐" (snow) are the components and "心" (heart) is the radical. All these subcharacter components mentioned above may be not relevant to the semantics of the character "慧"

---

[1] https://en.wikipedia.org/wiki/Written_Chinese.

(intelligent). However, the main-component "彗" (clever) consisting of several small components is closely related to the meaning of the character.

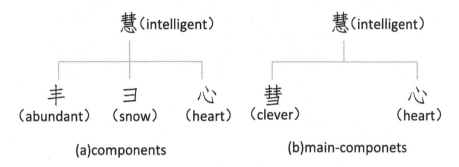

Fig. 1. An illustrative example to show the derivative meanings of the main-components are more relevant to the target word than the components.

The attention mechanism [21] originates from the way of imitating human thinking. It allocates limited information processing resources to the important part, that is, assigns different weights to different parts of the input. Soft attention concentrates on each input value and assigns a weight between 0 and 1. On the other hand, hard attention only focuses on the components that need the most attention, selectively deletes some low-associative values, and assigns weights to 0 or 1.

In this paper, we modify the CBOW model in the word2vec source code[2] and introduce the information of the relevant derivative meanings of the Chinese main-components, and propose two efficient models called MCWE-SA and MCWE-HA. The learned Chinese word embeddings not only contain more morphological information, but also have a higher similarity to the synonym. Our models directly modify embeddings of the target words, without generating and training extra embeddings for relevant derivative meanings. In addition, we create a derivative meaning table to describe the relationship between Chinese words and the relevant derivative meanings of their main-components. Through our models, the rich implicit information in Chinese is fully utilized and the similarity of related words is improved. Our contributions of this paper can be summarized as follows:

- Rather than directly leverage the components of the word itself, we provide a method to use the relevant derivative meanings of the main-components to train the word embeddings. In order to verify the feasibility of our method, two models named MCEW-SA and MCWE-HA, are proposed to incorporate the extra meanings.

---

[2] https://code.google.com/p/word2vec.

- We put forward a method to assign the weights of relevant derivative meanings at input layer based on the attention scheme. Through the attention mechanism, the derivative meanings negatively related to the target word will be filtered in order to improve the accuracy of word embedding.
- We evaluate the quality of word embeddings learned by our models and the state-of-the-art models, using a medium-sized corpus through word similarity tasks and word analog tasks. The results show that all of our models have performance improvements and outperform all baselines on the word analog task.

2   Main-Component Word Embedding Models

In this section, we introduce Main-Component Word Embedding models, named MCWE-Soft Attention (MCWE-SA) and MCWE-Hard Attention (MCWE-HA), which is based on CBOW model [9]. It should be noted that our models use the internal meanings of the Chinese characters, rather than directly using the characters or components of the word themselves. In particular, some radicals are also main-components, so we treat these radicals as the main-components and have the same contribution in our models. Most of the main-components are frequently-used Chinese words which may contain some ambiguous information. To address this concern, we propose the MCWE-SA which bases on the soft attention scheme. The MCWE-SA assumes that the relevant derivative meanings of main-components have their own weights, and assigns higher weights to the meanings closely related to the target words, so that they have greater impact on the word embeddings. We treat soft attention as a filter to remove the negative correlation and add positive contributions to the target word vector. What's more, MCWE-HA which bases on the hard attention scheme only focuses on the relevant derivative meaning of the main-component with the highest similarity to the target word. When backpropagating the target word vector parameters, we introduce the vector update method for related derivative meanings. In what follows, we will introduce each of our models in detail.

2.1   MCWE-SA

Through observation, we discover that most Chinese words have more than one relevant derivative meaning with their main-components, but some relevant derivative meanings have low correlation with the corresponding word. For example, main-component "知(know)" means "学识(knowledge)" and "了解(understand)". As Fig. 2 shows, for the item "智慧(intelligence)" ↦ {学识(knowledge), 了解(understand), 太阳(sun), 时间(time), 白天(day), 聪慧(clever), 扫把(broom), 思想(thought), 心脏(heart), 感情(feeling)}, each relevant derivative meaning has a bias on the word "智慧(intelligence)". Therefore, we assign different weights to each relevant derivative meaning based on the idea of soft attention model. We measure the weights of relevant derivative meanings by calculating the cosine similarity between the corresponding relevant derivative

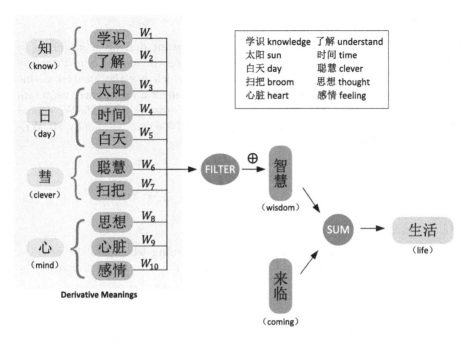

Fig. 2. An example of MCWE-SA. We take the sentence *"智慧生活来临 (wisdom life coming)"* as an example. When we select *"智慧(wisdom)"* as the input word and calculate its word vector, we pick out the relevant derivative meanings of the main-components of *"智慧"* from the derivative meaning table, and add all the word vectors of the relevant derivative meanings which have a positive correlation to the vector of *"智慧"* according to the correlation weights.

meanings and the target word where cosine similarity is usually used to measure the similarity between word embeddings.

We denote $W = (w_1; w_2; ....; w_i)$ as the vocabulary of words, $M_i = (m_1; m_2; ...; m_k)$ as the relevant derivative meanings in the derivative meaning table for each word w_i. The item for w_i in the derivative meaning table is $w_i \mapsto M_i$ where M_i is a collection of the relevant derivative meanings of w_i's main-components. We denote $sim(\cdot)$ as a method to measure the similarity between Chinese words and their relevant derivative meanings. Furthermore, we remove the negatively correlated derivative meanings. Hence, at the input layer, the modified embedding of w_i can be expressed as

$$\hat{v}_{w_i} = \frac{1}{2}\{v_{w_i} + \frac{1}{N_i}\sum_{k=1}^{N_i} sim(w_i, m_k) \cdot m_k\},$$

$$sim(w_i, m_k) = cos(v_{w_i}, v_{m_k}), \quad s.t. \ \cos(v_{w_i}, v_{m_k}) > 0$$

(1)

where v_{w_i} is the original word embedding of w_i, $\hat{v}_{w_i}$ is the modified word embedding of w_i and v_{m_k} indicates the vector of relevant derivative meaning m_k that positively contributes to w_i. N_i denotes the number of v_{m_k} whose cosine similarity between v_{w_i} is greater than 0.

2.2 MCWE-HA

In order to further reduce the impact of low-relevant derivative meanings to the word, we propose MCWE-HA which is based on the idea of hard attention model. We only choose the derivative meaning which has the greatest cosine similarity to the target word. According to experimental experience, we retain the derivative meanings of similarity with token w_i greater than 0.9. We denote M_i as a collection of relevant derivative meanings of the word w_i, finally MCWE-HA is mathematically defined as

$$\hat{v}_{w_i} = \frac{1}{2}(v_{w_i} + v_{m_{max}}),$$
$$m_{max} = \underset{m_k}{argmax}\, cos(v_{w_i}, v_{m_k}), m_k \in M_i \qquad (2)$$
$$s.t.\ cos\,(v_{w_i}, v_{m_k}) > 0.9$$

where m_{max} denotes the vector of derivative meaning which has the greatest cosine similarity to w_i. We denote a paradigm of Fig. 3 to better illustrate the MCWE-HA model.

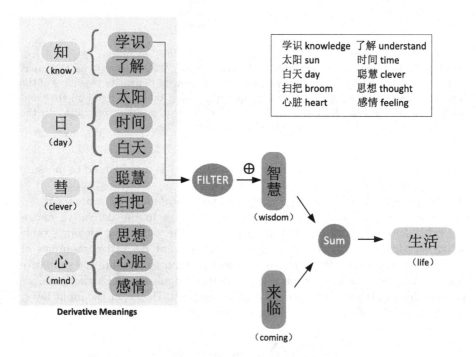

Fig. 3. An example of MCWE-HA. When we choose "智慧 *(wisdom)*" as the input word and calculate its word vector, we select the relevant derivative meaning of the main-component with the highest similarity to "智慧 *(wisdom)*" from the derivative meaning table, here we choose the vector of "学识 *(knowledge)*". If the angle between the vector of "学识" and "智慧" is not greater than approximately 25°, we add the two vectors and get the average of their sum.

2.3 Update Rules for Relevant Derivative Meanings

Considering that the relevant derivative meanings are highly correlated with the target word embeddings and this relationship does not change in the current training round, when we update the embedding of the target word, we also choose to update the implied words which have great similarity with the corresponding word. According to the experimental experience, the cosine similarity between the derivative meanings and the target word is set to be greater than 0.9, which means the derivative meanings' word vector whose the angle between the two vectors not exceeding approximately $25°$ is updated. Our models aim to maximize the log-likelihoods function as follows:

$$L(v_{m_k}) = \sum_{m_k \in M} \log P(v_{m_k} | context(v_{w_i}))$$
$$s.t. \ \cos{(v_{w_i}, v_{m_k})} > 0.9$$

(3)

where $context(v_{w_i})$ represents the context window of the word v_{w_i}, which is the composition of context words, and M represents the set of all relevant derivative meanings in derivative meaning table. $P(v_{m_k}|context(v_{w_i}))$ is conditional probability which is defined by the softmax function. When updating v_{w_i} in backpropagation, we also update v_{m_k} with the same weight.

3 Experiments and Analysis

In this section, we evaluate the performance of our models in generating high-quality word embeddings on word similarity evaluation and word analogy task.

3.1 Experimental Settings

Training Corpus: We adopt a medium-sized Chinese corpus which is downloaded from Chinese Wikipedia Dump[3] to train all word embedding models. We utilize a script named WikiExtractor[4] to convert data from XML into text format. Moreover, we use THULAC[5] for Chinese word segmentation and normalize all characters as simplified Chinese. In pre-processing, we filter all digits, punctuation masks and non-Chinese characters in order to improve the efficiency of model training. To get better quality of the word embeddings, we remove the Chinese common stop words[6] in the corpus. Finally, we obtained a training corpus of approximately 2GB in size, containing 354,707,204 tokens and 1,090,983 unique words.

---

[3] https://download.wikipedia.com/zhwiki.
[4] https://github.com/attardi/wikiextractor.
[5] http://thulac.thunlp.org/.
[6] https://github.com/goto456/stopwords.

Derivative Meaning Table: The Modern Chinese Word List[7] is divided into two parts: 2500 common characters and 1000 secondary common characters, and the coverage of those words in most corpora reached 99.48% which means mastering the Chinese common and secondary words has reached the basic requirements for using Chinese[8]. Hence, we use crawler script to obtain the main-components and radicals' information of those Chinese characters which is a total of 3500 from HTTPCN[9]. In this step, we finally obtained 3491 main-components.

To create the derivative meaning table, we need to obtain the relevant derivative meanings of the main-components of each character. Although the main-components are part of the characters, they are also simple Chinese characters with their own meanings. Therefore, we crawl the Chinese interpretations of 3491 main-components from HTTPCN and obtain the relevant derivative meanings. By manually labeling, we simplify the interpretation of each main-component into a core phrase which may appear in the corpus. Although this process costs manpower and time, it could be done once and for all for each language because it has the same knowledge base. We traverse the entire corpus, pick out all the different words and put their main-components' derivative meanings in the table, so as to improve efficiency by looking up the table during model training. When we choose a Chinese word during training, its main-components will be determined and can be further replaced by derivative meanings as an intermediate variable by looking up the table.

Baselines: For comparison, we choose two classic models including CBOW [9] and GloVe [12] and two component-level state-of-the-art character embedding models including CWE [15] and GWE [19]. In addition, we introduce a model named Latent Meaning Model-Average(LMM-A) [22], which uses the latent meanings of English morphemes to enhance the word embeddings. LMM-A employs the morpheme embeddings to adjust the word embeddings of the target word during training, and assumes that all latent meanings have the same contribution to the target word. We modified the source code of the LMM-A model to match our experiments. In order to better verify the word analog performance of our models, we also selected two additional models named cw2vec [20] and JWE [18] respectively for word analogy experiment.

Parameter Settings: For the sake of fairness, we use the same hyperparameter settings for all models. In order to speed up the training process, we adopt negative sampling techniques for CBOW, CWE, GWE, LMM-A and our models. What's more, we set the word vector dimension as 200, the window size as 5, the negative samples as 10, the training iteration as 15, the learning rate as 0.025 and the subsampling parameter as 1e-4.

---

[7] https://en.wikipedia.org/wiki/Simplified_Chinese_characters.

[8] https://en.wikipedia.org/wiki/List_of_Commonly_Used_Characters_in_Modern_Chinese.

[9] http://tool.httpcn.com/zi.

3.2 Word Similarity

This experiment is used to evaluate the semantic relevance of generated word embeddings in word pairs. For Chinese word similarity task, we employ two different manually-annotated datasets named wordsim-240 and wordsim-296 provided by Chen et al. [15]. These datasets are composed of word pairs and manually labeled with the scores to measure the similarity of word pairs. We utilize the cosine similarity to measure the similarity of each word pair, and the Spearman' s rank correlation coefficient (ρ) is employed to evaluate our calculation results and human scores. More details of results are shown in Table 1.

The performance of our models on the Wordsim-240 dataset exceeds the original CBOW model, which indicates that adding implicit information to the input layer during the model training can indeed improve the quality of the word vector. The performance of the MCWE-SA model exceeds all baselines on the Wordsim-296, indicating that adding the derivative meanings of positive elements can effectively enhance the similarity between words. On the other hand, the LMM-A model does not perform well on both datasets, indicating that the method of averaging all the latent meanings' weights is not desirable because it contains a large amount of negative correlation information. Our strategy uses the attention mechanism to select information with positive meanings and filter out the negative correlation vector, which finally improves the quality of Chinese word embeddings. Although our models have little improvement in the word similarity task compared to the CBOW model, we verify the method that adding derivative meanings is a feasible direction, and we can continue to improve performance by optimizing the derivative meaning table.

3.3 Syntactic Analogy

This task examines the quality of word embeddings by discovering the semantic inferential capability between pairs of words. The core task of syntactic analogy is to answer the questions like "雅典 (Athens) is to 希腊 (Greece) as 东京 (Tokyo) is to 日本 (Japan)" where 日本 (Japan) is the answer we hope to get. This means that the model gets the answer correctly if the similarity of "vector (希腊) - vector (雅典) + vector (东京)" and "vector (日本)" is the largest among all words. We utilize the Chinese analogical reasoning dataset created by Chen et al. [15], which contains 3 analogy types: some capitals and their countries (677 groups), some cities and their states (175 groups) and family relationships (272 groups). Each strategy contains four phrases, and we use the information from the first three phrases to predict the fourth and calculate the accuracy of the final result. The results in Table 2 show that our models outperform the comparative baselines in all classifications. The JWE model uses subcharacter components for word embedding enhancements, which has a good effect compared to most baselines but ignores the implicit information inside the characters, so the final performance is still not as good as our models. The CBOW model is stable and exhibits high performance, but still weaker than our models. Since our models make the spatial distance of words with the same main-components closer and

Table 1. Spearman' s Coefficients of word similarity on wordsim-240 and wordsim-296 ($\rho \times 100$). The higher the values, the better the performance.

Model	Wordsim-240	Wordsim-296
CBOW	51.17	57.46
GloVe	50.15	42.07
CWE	**53.40**	57.06
GWE	52.07	56.98
LMM-A	39.45	43.01
MCWE-SA	52.23	**57.95**
MCWE-HA	51.21	57.23

Table 2. Evaluation accuracies (%) on word analogy reasoning. The higher the values, the better the performance.

Model	Total	Capital	State	Family
CBOW	79.92	86.91	85.14	60.66
GloVe	50.84	55.08	68.00	30.14
CWE	59.00	66.72	65.71	37.13
GWE	64.54	71.24	77.14	41.17
LMM-A	33.87	31.01	28.00	44.12
JWE	73.07	78.83	82.28	54.04
cw2vec-subword	39.11	37.64	71.42	21.69
MCWE-SA	**81.43**	**89.18**	88.00	59.56
MCWE-HA	81.33	87.40	**89.72**	**62.13**

the words with the same main-components have similar derivative meanings, our word embeddings achieve high performance in semantic analogy.

3.4   The Impacts of Corpus Size

The parameter settings in word embedding training have a great impact on final result. The larger corpus size, the more semantic information it contains, which can improve the quality of the word vector. We take a task to investigate the impact of corpus size for word embeddings. In the analysis of corpus size, we set the same hyperparameters as before. We select the Chinese analogical reasoning dataset as the evaluation standard of syntactic analogy task. The entire corpus previously mentioned is divided into $\frac{1}{4}$, $\frac{2}{4}$, $\frac{3}{4}$, and $\frac{4}{4}$ respectively as our new corpus for the task. As shown in Fig. 4, the performance of the MCWE-SA model is better than the CBOW model in each corpus. Although the MCWE-HA model has weaker performance than CBOW at the beginning, with the increment of the corpus, the performance exceeds that of CBOW from the point that the corpus's size is about 800 MB.

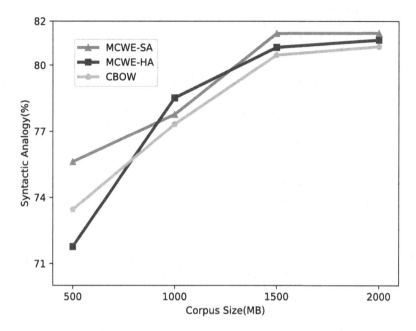

Fig. 4. Results on Chinese analogical reasoning dataset with different corpus size.

4   Conclusion

In this paper, we explored a new direction of using the relevant derivative meanings of Chinese internal components instead of themselves to enhance the Chinese word embeddings. We proposed two models named MCWE-SA and MCWE-HA, which make full use of subword information. The attention model was used to dynamically adjust the weights of derivative meanings of the main-components in Chinese characters. Experimental results showed that our models have obvious advantages in syntactic analogy compared to all baselines.

Acknowledgments. The authors are grateful to the reviewers for constructive feedback. We would like to thank the anonymous reviewers for their insightful comments and suggestions. This work was supported by the National Natural Science Foundation of China (No. 61572456) and the Anhui Initiative in Quantum Information Technologies (No. AHY150300).

References

1. Tang, D., Wei, F., Yang, N., Zhou, M., Liu, T., Qin, B.: Learning sentiment-specific word embedding for twitter sentiment classification. In: ACL, pp. 1555–1565 (2014)
2. Conneau, A., Schwenk, H., Barrault, L., Lecun, Y.: Very deep convolutional networks for natural language processing. arXiv preprint arXiv:1606.01781 (2016)
3. Jean, S., Cho, K., Memisevic, R., Bengio, Y.: On using very large target vocabulary for neural machine translation. arXiv preprint arXiv:1412.2007 (2015)
4. Kyunghyun, C., et al.: Learning phrase representations using RNN encoder-decoder for statistical machine translation. arXiv: 1406.1078 (2014)
5. Bonggun, S., Timothy, L., Jinho, D.: Lexicon integrated cnn models with attention for sentiment analysis. arXiv preprint arXiv:1610.06272 (2016)
6. Tang, D., Wei, F., Qin, B., Yang, N., Liu, T., Zhou, M.: Sentiment embeddings with applications to sentiment analysis. In: IEEE Transactions on Knowledge and Data Engineering, pp. 496–509 (2016)
7. Guangyou, Z., Tingting, H., Jun, Z, Po, H.: Learning continuous word embedding with metadata for question retrieval in community question answering. In: ACL, pp. 250–259 (2015)
8. Antoine, B., Sumit, C., Jason, W.: Question answering with subgraph embeddings. arXiv preprint arXiv:1406.3676 (2014)
9. Tomas, M., Kai, C., Greg, C., Jeffrey, D.: Efficient estimation of word representations in vector space. arXiv preprint arXiv:1301.3781 (2013a)
10. Tomas, M., Ilya, S., Kai, C., Greg, S., Jeff, D.: Distributed representations of words and phrases and their compositionality. In: Advances in Neural Information Processing Systems, pp. 3111–3119 (2013b)
11. Tomas, M., Wen-tau, Y., Geoffrey, Z.: Linguistic regularities in continuous space word representations. In: Proceedings of the 2013 Conference of the North American Chapter of the Association for Computational Linguistics, pp. 746–751 (2013c)
12. Jeffrey, P., Richard, S., Christopher, M.: Glove: global vectors for word representation. In: Proceedings of the 2014 conference on empirical methods in natural language processing (EMNLP), pp. 1532–1543 (2014)
13. Harris, Z.: Distributional structure. Word **10**(23), 146–162 (1954). https://doi.org/10.1080/00437956.1954.11659520
14. Li, Y., Li, W., Sun, F., Li, S.: Component-enhanced Chinese character embeddings. In: Proceedings of EMNLP, pp. 829–834 (2015)
15. Chen, X., Xu, L., Liu, Z., Sun, M., Luan, H.: Joint learning of character and word embeddings. In: IJCAI, pp. 1236–1242 (2015)
16. Rongchao, Y., Quan, W., Peng, L., Rui, L., Bin, W.: Multi-granularity Chinese word embedding. In: Proceedings of EMNLP, pp. 981–986 (2016)
17. Xu, J., Liu, J., Zhang, L., Li, Z., Chen, H.: Improve Chinese word embeddings by exploiting internal structure. In: NAACL, pp. 1041–1050 (2016)
18. Yu, J., Jian. X., Xin, H., Song, Y.: Joint embeddings of Chinese words, characters, and fine-grained subcharacter components. In: EMNLP, pp. 286–291 (2017)
19. Tzu-Ray, S., Hung-Yi, L.: Learning Chinese word representations from glyphs of characters. In: EMNLP (2017)
20. Cao, S., Lu, W., Zhou, J., Li, X.: cw2vec: Learning Chinese word embeddings with stroke n-gram information. In: AAAI (2018)
21. Kelvin, X., et al.: Show, attend and tell: neural image caption generation with visual attention. In: ICML (2015)

22. Xu, Y., Liu, J., Yang, W., Huang, L.: Incorporating latent meanings of morphological compositions to enhanceword embeddings. In: ACL, pp. 1232–1242 (2018)
23. Lai, S., Liu, K., Xu, L., Zhao, J.: How to generate a good word embedding?. arXiv preprint arXiv:1507.05523 (2015)
24. Piotr, B., Edouard, G., Armand, J., Tomas, M.: Enriching word vectors with subword information. In: ACL, pp. 135–146 (2017)
25. Chen, Z., Hu, K.: Radical enhanced Chinese word embedding. In: CCL(The Seventeenth China National Conference on Computational Linguistics) (2018)

Association Relationship Analyses of Stylistic Syntactic Structures

Haiyan Wu and Ying Liu(✉)

Department of Chinese Language and Literature, Tsinghua University, Beijing, China
{wuhy17,yingliu}@mail.tsinghua.edu.cn

Abstract. Exploring linguistic features and characteristics helps better understand natural language. Recently, there have been many studies on the internal relationships of linguistic features, such as collocation of morphemes, words, or phrases. Although they have drawn many useful conclusions, some summarized linguistic rules lack physical verification of large-scale data. Due to the development of machine learning theories, we are now able to use computer technologies to process massive corpus automatically. In this paper, we reveal a new methodology to conduct linguistic research, in which machine learning algorithms help extract the syntactic structures and mine their intrinsic relationships. Not only the association of parts of speech (POS), but also the positive and negative correlations of syntactic structures, linear and nonlinear correlation are considered, which have not been well studied before. Combined with the linguistic theory, detailed analyses show that the association between parts of speech and syntactic structures mined by machine learning method has an excellent stylistic explanatory effect.

Keywords: Syntactic structure · Part of speech collocation · Positive and negative correlation · Linear and nonlinear relationship · Machine learning

1 Introduction

In recent years, many scholars have summarized linguistic rules or features from different levels, as well as their internal relationships [3,4,7,12,13,26]. For example, Dexi Zhu pointed out that the constituent units of the style are morphemes, words, phrases and sentences, and how morphemes form words, how words form phrases and how they form sentences [6,21]. The process is concluded as a collocation relationship between features. As early as 1957, Firth [8] proposed the concept of collocation. He believed that the words near a word can reflect the

This work is supported by 2018 National Major Program of Philosophy and Social Science Fund "Analyses and Researches of Classic Texts of Classical Literature Based on Big Data Technology" (18ZDA238) and Project of Humanities and Social Sciences of Ministry of Education in China (17YJAZH056).

M. Sun et al. (Eds.): CCL 2019, LNAI 11856, pp. 40–52, 2019.
https://doi.org/10.1007/978-3-030-32381-3_4

meaning of it. Some researchers regarded the word collocation as an abstraction in syntactic level, and others thought it was vocabulary level [8]. Most of the current methods studying the linguistic features [1,9,19] are based on the experience of linguistic knowledge and the analyses of small-scale data. With the development of computer technology, we now can perform automatic large-scale analysis on syntax. Although its performance is not as good as linguistics experts, it is believed that in the massive corpus, some statistical conclusions can be drawn. Generally, associated relationship mining can be divided into the following two aspects.

- **Linguistic Methods.** It is regarded that linguists are the pioneers of the theory of language association. For example, as the study of the common collocation of words in Red Sorghum, WenZhu found out the distribution, semantic changes, and rhetoric effect of collocation variation in five basic syntactic structures [20]. Xianghui Cheng [26] pointed out that a register was a collection of language characteristics people used in a specific scene, including vocabulary, syntax, and rhetoric.

  Linguistic-based researches usually have manually selected examples and word-for-word analyses. The advantage is that there is an excellent theoretical basis. The shortcoming is that these studies lack the verification on the large-scale corpus.
- **Statistical Methods.** With the development of computer technology, many scholars have applied computational methods to the field of linguistics research. Such as, Seretan designed the hybrid system that combined statistical methods to do multilingual parsing for detecting accurate collocational information. Experiments showed that for different languages, this system had good results [14,17]. Besides, Seretan thought that automatic acquisition of collocations from text corpora by machine learning was essential. Then he proposed a framework for collocation identification based on syntactic criteria. It was shown that the results obtained were more reliable than those of traditional methods based on linear proximity constraints [15,16]. Xiaoli Huo considered the stylistic features were related to the specific linguistic material information. It is necessary to express the characteristics of different registers through lexical selection, phrase construction, sentence pattern transformation, and tone adjustment, to meet the communication needs [24].

However, these methods usually are specific for some features or materials, which are not flexible enough to generalize different features of the different registers.

In this work, we focus on the collocations of POS and syntactic structures. Instead of manually annotating, we utilize machine learning methods on the corpus to help analyze the features. The results verify some interpretation of linguistic theories. Furthermore, some potential laws or internal collocations are explored. The process of our study and experiments can be concluded as a new methodology of verifying linguistic theories with the help of computer technology, which is conducive to the development of linguistic theories.

2   Algorithms

2.1   Pearson Correlation Coefficient

In this paper, we employ *Pearson Correlation Coefficient* [2] to calculate the positive and negative correlation between syntactic structures. This algorithm is used to measure whether two variables have a linear correlation or not.

$$\gamma = \frac{\sum_i (x_i - \bar{x}_\iota)(y_i - \bar{y}_\iota)}{\sqrt{\sum_i (x_i - \bar{x}_\iota)^2}\sqrt{\sum_i (y_i - \bar{y}_\iota)^2}} \tag{1}$$

Where $\bar{x}_\iota$ and $\bar{y}_\iota$ represent the mean of sample X and sample Y respectively. Here, x and y can denote occurrence of different POS or syntactic structures. In following sections, we calculate the relationship between POS and the syntactic structures, and use *Heatmap*[1] to visualize their correlation coefficients (Figs. 2 and 3). In *Heatmap*, the deeper the color is, the higher their correlation is.

2.2   Hierarchical Cluster

To find the linear correlation of POS and syntactic structures, we use a clustering method *Hierarchical Cluster* [11], which brings together attributes with higher correlation. Euclidean distance is used as the distance metric, shown as Formula 2.

$$\|a - b\|_2 = \sqrt{\sum_i (a_i - b_i)^2} \tag{2}$$

where a and b represent the vector representations of two different features (POS or the syntactic structures)respectively. The clustering results are visualized together with the *Heatmap*.

2.3   LassoCV Algorithm

Most of the previous methods focus on the linear correlation of syntactic structures. But nonlinear relations are also crucial in the study of stylistic features. Therefore, we use *LassoCV* algorithm to consider the nonlinear relationship of syntactic structures. *LassoCV* algorithm judges and eliminates the collinearity between features by adding the penalty function [18], and the loss function is as follows in Formula 3.

$$\frac{\|y - Xw\|_2^2}{2n} + \gamma\|w\|_1 \tag{3}$$

where w are the parameters, n is the total number of samples, and γ is the hyperparameter of L2-regularization. In our paper, due to differentiate three registers, we choose *MultiTaskLassoCV* which is the extension algorithm of *LassoCV*.

For all the above algorithms, we use *Python* and *Scikit-Learn Toolkit*[2].

---

[1] https://seaborn.pydata.org/generated/seaborn.heatmap.html.
[2] https://scikit-learn.org/.

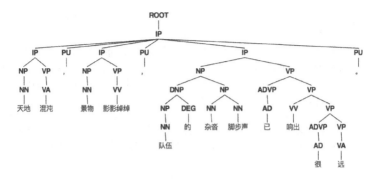

Fig. 1. Syntax tree

2.4 Relative Definition

To better describe the syntactic structure, we give such definitions. Take VP →
[VV, AS] as an example, VP on the left of the arrow is the **parent (father)
node**, VV and AS on the right of the arrow are the **child nodes**, and the
number of child nodes is called the **degree** of the syntactic structure VP →[VV,
AS] or the **out-degree** of the phrase VP. Here, the out-degree of the phrase VP
is 2.

3 Dataset and Statistic Information

3.1 Datasets

Our experiments are based on *Novel*, *News*, and *TextBook*. The detailed information is shown as follows.

- **Novel** comes from Mo Yan and Yu Hua, a total of 20 articles, of which 12
  are Mo Yan and 8 are Yu Hua.
- **News** is a public dataset[3], which covers ten main topics including domestic
  and foreign news, stock news, financial news, breaking news, entertainment
  news and so on.
- **TextBook** consists of some Chinese textbooks in elementary, middle, and
  high schools. The total of 480 articles is mainly proses and novels.

3.2 Syntactic Structures

Avram Noam Chomsky pointed out that each sentence should conform to its
syntactic rules [5]. Since we study the association of syntactic structures, we use
Syntax Tree to convert each sentence in the corpus into a corresponding syntactic
structure representation, as shown in Fig. 1.

---

[3] https://www.sogou.com/labs/resource/cs.php.

For each sentence, we use *Stanford CoreNLP*[4] to construct its corresponding syntax tree. Take a sentence from *Novel* as an example, e.g. "天地混沌，景物影影绰绰，队伍的杂沓脚步声已响出很远。(It is too dark to see the scenery, and the footsteps of the team have been far away.)", whose corresponding syntactic tree is shown in Fig. 1. In Fig. 1, the syntax structures are extracted IP → [IP, PU, IP, PU, IP, PU], IP →[NP, VP], IP → [NP, VP], IP → [NP, VP], NP → NN, VP → VA, VP → VV, NP →[DNP, NP] and so on. The abbreviations of these syntactic structures are marked by the Pennsylvania tree library [25]. The meaning of the abbreviation of the parent nodes are shown in Table 1.

Table 1. Penn Treebank tags for phrase structures

Tag	Description	Tag	Description
ADJP	Adjective phrase	LCP	Phrase formed by "XP+LC"
ADVP	Adverbial phrase headed by AD	LST	List marker
CLP	Classifier phrase	NP	Noun phrase
CP	Clause headed by C (complementizer)	PP	Preposition phrase
DNP	Phrase formed by "XP+DEG"	PRN	Parenthetical
DP	Determiner phrase	QP	Quantifier phrase
DVP	Phrase formed by "XP+DEV"	UCP	Unidentical coordination phrase
FRAG	Fragment	VP	Verb phrase
IP	Simple clause headed by INFL	...	...

By analogy, the same operation is performed for each sentence of each register, and the statistical information is shown in Table 2.

Table 2. Dataset statistic information

Dataset	Novel	News	TextBook
Total # of sentences	117,353	29,754	24,119
Average length of sentences	19.36	27.85	17.41
Total # of syntactic structures	2,371,399	800,338	431,792
# of different syntactic structures	5,404	2,502	3,188
# of different NP-phrase	1,570	707	1,261
# of different VP-phrase	1,612	769	821
# of average species	15.5081	20.0250	14.3592
Average height	9.6474	11.3023	9.3987
Average width	19.1002	27.2606	17.2163
Syntactic richness	0.5934	0.5855	0.6031

---

[4] https://github.com/stanfordnlp/CoreNLP.

From Table 2, we find that the number of sentences in *News* is close to *TextBook*, but the number of syntactic structure in *News* is larger than that in *TextBook*. It is because the sentence length of *News* is the longest, which leads to more syntactic structures in the sentence. Besides, since NP and VP phrases appear in large numbers in our corpus, we count their number for further analyses.

4   Association Relationship Mining

In a syntax structure, children POS co-occurrence can reflect the meaning of their father nodes. The positive and negative collinearity of POS and syntax structures reflect the characteristics of the language.

4.1   Parts of Speech (POS) Collocation Mining

To study the syntactic structure, we utilize the hierarchical clustering in Sect. 2.2 to cluster the correlation values of POSs. The result is shown in Fig. 2, in which the meaning of these acronyms corresponds to the mark of the Pennsylvania Tree Library [22]. In Fig. 2, we conclude that some POSs have higher associated values, e.g., *CD+M, AD+VV, VA+DEC, VC+SP, P+NN, MSP+VV+DEC/DEG, NP/VV+AS, VV/NN+SP, NT+NR, VV + P + NN/NR + LC*, etc.

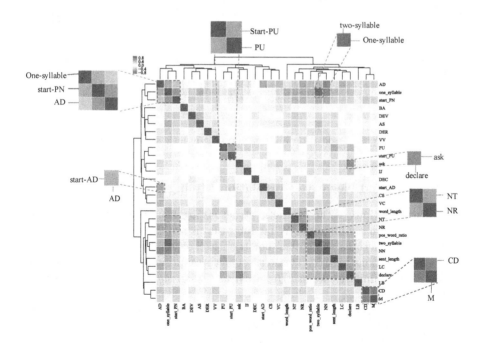

Fig. 2. POS collocation relationship

We can find that the POSs having positive collocations usually appear in the same syntactic structure. Using the collocation of *VV + AS* as an example, in Chinese, we often encounter the following sentence represents a certain state [23], e.g. "VP + 了" ("下雨了" (it rained)). In this case, only if *VV* and *AS* appear at the same time can they indicate a certain state. They form a syntactic structure VP → VV AS. The linear relationships between the POSs recover some specific syntax structures. It can help us understand the internal association of the syntactic structures.

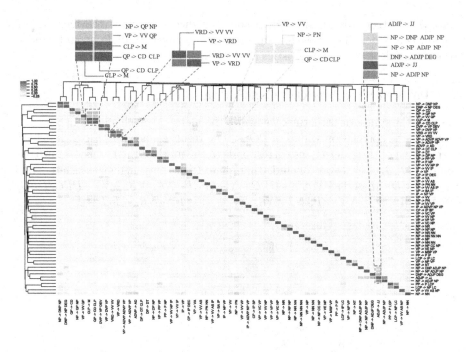

Fig. 3. Association relationship of syntactic structures

4.2   Positive and Negative Correlation Mining

The internal relations of the syntactic structure are from the POS, which have been studied by many linguists. What interests us are the relationship between the syntactic structures, which has not been well studied. We consider the two groups of correlations between syntactic structures: positive and negative, linear and non-linear. Similarly, we use *Hierarchical Clustering*. The result is shown in Fig. 3.

In Fig. 3, generally, the correlation between syntactic structures is not as strong as POSs. But there are still some associations between syntactic structures. Taking ADJP → JJ as an example, it has a positive relationship with NP, DNP (upper right in Fig. 3). ADJP → JJ is a sub-tree of the associated syntactic structures. For example, ADJP → JJ is an extention of the ADJP in the NP →

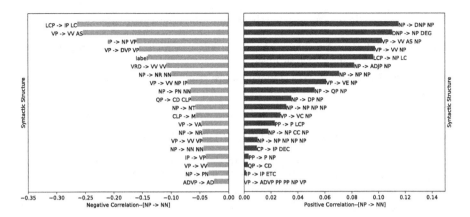

Fig. 4. NP → NN related syntactic structures

Table 3. Positive and negative correlation of syntactic structures

VP→VV	IP → NPVP	QP → CD
VP→DVP VP	NP→PN	NP→QP NP
VP→VV VP	VP→VV AS IP	NP→NP PRN
VP→VP VP	CP→IP DEC	NP→NR PRN
VP→MSP VP	VRD→VV VV	NP→NP NP NP
VP→VV NP IP	LCP→IP LC	NP→NP CC NP
VP→VA	*IP→VP*	*NP→PN*
NP→QP NP	*CLP→M*	*ADVP→AD*
VP→VV AS NP	*PP →P NP*	*VP→VV*
QP→ CD CLP	*ADVP→AD*	*IP→VP*
DNP→ NP DEG	*DNP → NP DEG*	*NP → DNP NP*
NP → DNP NP	*QP→CD CLP*	*CP→IP DEC*
NP→NN	*LCP→NP LC*	*IP → NP VP*
P → VV NP	*NP → DNP NP*	*VP→VA*

[DNP, ADJP]. Other correlations between syntactic structures are similar, and their co-occurrences usually reflect specific sentence patterns.

Since VP and NP phrases are the most, it is essential to analyze them better. We calculate the correlation of different syntactic structures with NP → NN, shown in Fig. 4. The red part indicates negative correlations, and the blue part indicates positive correlations. We only list a part based on the value of the correlation value. The positively related structures are determiner or adjectival phrases and verb-object constructions, which usually co-occur with NP phrase.

Similarly, we list the positive (regular font) and negative (bold italics) related phrases of the other syntactic structures with the largest number of categories, shown in Table 3.

Through statistical analysis, we draw the following two conclusions:

1. Positively related syntactic structures usually can form a larger syntactic tree, and usually are father and children nodes;
2. Negatively related syntactic structures are conflicting and usually can not be in the same syntax tree.

4.3   Linear and Nonlinear Correlation Mining

Since the positive and negative correlation of the syntactic structures is based on *Pearson Correlation Coefficient*, it is a linear correlation. However, for complex registers, it is necessary to study the nonlinear relationship between syntactic structures. Next, we use *MultiTaskLassoCV* in Sect. 2.3 to get the nonlinear correlation of the syntactic structure as shown in Table 4.

From Table 4, we use black bold italics to represent nonlinear relationships, and the rest are linear correlations. We find that most of the syntactic structures are linearly related, and the nonlinear correlations are the phrases of IP and CP, as shown in Fig. 5. We take IP phrases as an example to analyze.

In Fig. 5, both sentences belong to the IP phrase, and in which IP phrase can be expressed in these forms: IP→[IP, PU], IP→[NP, VP], IP→VP, IP→[NP, VP, PU, PU], and IP→[PU, VP], etc. In Fig. 5, since some IP phrases are located in the same layer and some are nested in the inner layer. Recursive loops can occur in the tree to form complex IP phrases. Therefore, the relationship between such IP phrases is non-linear.

By statistical analysis, *Novel* contains *IP*, *CP*, *NP*, etc. and their corresponding percents are 77.59%, 7.7%, 6%. *News* includes 84% *IP*, 5% *CP* and 7.6% *NP*, etc. Similarly, *Textbook* mainly includes *IP*, *CP* and *NP*, and the corresponding percents are 74.55%, 8.93%, 6.43%. For the phrases of *IP*, *CP* and *NP* in these three registers, *IP* in *Novel* is mainly IP → VV, of which most are momentary verbs e.g. "笑" (Smile) "说" (Say); *IP* of *News* is mainly IP → [NP, VP] and IP →[NP, NP], e.g. "政府指出 (Government Points)......", "市场经济 (Market economy)......"; and *Textbook* is IP → VP, e.g. "踢这个蓝色的球" (Kick this blue ball). For the three registers, whether from the height of *IP* in the syntax tree or the length of the sentence involved, *IP* phrase of *News* is the highest, with an average height of 2.8 (*TextBook*: 1.84; *Novel*: 1.81). Usually, the higher the syntax tree is, the more intermediate nodes restrict the central words. Besides, The average length of the sentence involved in *IP* phrase of *Novel* is 9.19, *News* is 15.92, *TextBook* is 12.83. Combined with the lengths, we can conclude that *News* contains the most restrictive words followed by *TextBook*, and *Novel*.

Except for the phrases *IP* and *CP*, the other structures usually have linear correlations, and the results are consistent with the clustering results.

Table 4. Linear and non-linear relationship of syntactic structures

NP→NN NN	NP→NP NP	NP→NR	VP→VV AS
NP→NT	NP→NP CC NP	NP →NR NR NR	NP→QP NP
NP→NR PRN	VP→VV VP	VP→NP VP	NP→NP PRN
VP→VV AS NP	VP→VP VP	VP→VV	NP→NR PRN
NP→NN	VP→MSP VP	VRD→VV VV	NP→NP NP NP
LCP→NN LC	NP→ADJP NP	LCP→NP LC	NP→NP CC NP
ADVP→AD	VP→VA	NP→PN	NP→NN PN
NP→PN	NP→QP NP	CLP→M	ADVP→AD
VP→VA	VP→VV AS NP	PP→P NP	VP→VV
NP→DT DEG	DNP→NP DEG	DNP→NP DEG	NP→DNP NP
VP→VV VP	NP→DNP NP	QP→CD CLP	NP→NR
LCP → IP LC	*IP→NP VP*	*IP→VP*	*IP→IP ETC*
CP→IP DEC	*PP → P IP*		

5 Linguistic Relevance Analyses

In the previous sections, we analyze the relationship of syntactic structures from the perspective of machine learning. In this part, we will explain the conclusions we have obtained from the registers itself. Combined with Fig. 3, Tables 3, and 4, we choose the following sets of syntactic structures for analysis and explanation. These groups of structures come from *Novel*, *News*, and *TextBook*.

1. LCP →[NP, LCP] + PP →[PP, LCP], from the linguistic point of view, LCP →[NP, LCP] literally means "the inside of *NP*", and PP →[PP, LCP] means "*PP* (in)...", so the phrase LCP →[NP, LCP] combines with PP →[PP, LCP] together to form PP →[PP, NP, LCP]. From this perspective, the formation process of PP →[PP, NP, LCP] is a linear association. In the register, they often form a larger prepositional phrase to form a sentence adverbial by co-occurrence. Combined with Fig. 3, we find that the phrases LCP →[NP, LCP], NP →[NP, NP] and PP →[PP, LCP]are clustered together, and are red, which indicates that they have a positive correlation. In other words, the probability of co-occurrence is relatively high. This is consistent with the interpretation of linguistics, the sentences with such syntactic structures often appear in the register, e.g.

   a. "他的两头羊在羊堆里拱出来。" (His two sheep were arched out in the sheep.) from Hua Yu's *To Live*, the phrase *CLP*: 在 (*PP*) 羊堆 (*NP*) 里 (*LCP*)".

   b. "她在父亲身边跪下，轻轻地把父亲的裤子褪下来。" (She kneels beside her father and gently fades her father's pants down.) from Mo Yan's *Red Sorghum*, the phrase *CLP*: "在 (*PP*) 父亲 (*NP*) 身边 (*LCP*)".

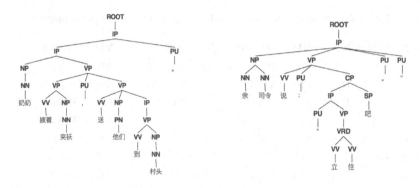

Fig. 5. Syntax trees of IP and CP

2. VP →[ADVP, VP]+ VP →[VV, VV], from the perspective of mathematical formulas, the phrases VP →[ADVP, VP] and VP →[VV, VV] can be combined into VP →[ADVP, VV, VV]. In linguistics, such a phrase structure is reasonable. Besides, KeshengLi held that the two VV in this phrase have the same syntactic status [10], e.g.

   a. 打算回家 (Going home) [Predicate and Object Structure]
   b. 研究结束 (End of study) [Subject-Predicate Structure]
   c. 挖掘出来 (Dig out) [Predicate-Complement Structure]

From Fig. 3, we find that these two phrases are clustered together and are red, which is consistent with their semantic interpretation.

In our chosen corpus, there are indeed a large number of sentences with these structures, e.g.

   a. " 中国证监会今日正式发布实施 ……" (The China Securities Regulatory Commission officially released the implementation today...), the phrase VP is 正式 (ADVP) 发布 (VV) 实施 (VV).
   b. " 大力地促进发展农产业 ..." (Vigorously promote the development of the agricultural industry). the phrase VP is 大力地 (ADVP) VV (促进) VV (发展).

It can be seen from these two examples, and we think that in language expressions, such a combination is reasonable. From Fig. 3, VP → [ADVP, VP] and VP →[VV, VV] are clustered together, which shows these two phrases have a positive correlation,ie, co-occurrence.

3. DP → [DT, QP] + DP → [DT, CLP] + NP →[DP, CP], the phrases DP →[DT, QP] and DP → [DT, CLP] act as modifiers, while *NP* is the central language. They can combine two groups of phrases NP → [DT, QP, CP] and NP → [DT, CLP, CP], where NP → [DT, QP, CP] emphasizes quantity and NP → [DT, CLP, CP] emphasizes location, e.g.

   a. " 把那十件红色的裙子拿个过来。" (Take the ten red skirts over.), where NP → [DT, QP, CP] is " 那 (DT) 十件 (QP) 红色的裙子 (CP)"
   b. " 那个放在椅子上的书叫什么名字? " (What is the name of the book on the chair?), where NP → [DT,CLP,CP] is " 那个 (DT) 放在椅子上的书 (CLP) 叫什么名字 (CP)?"

From Fig. 3, we discover that the phrases DP → [DT, QP], DP →[DT, CLP] and NP →[DP, CP] are clustered together, which proves that there is a correlation between them. This fully explains the rationality of linguistic interpretation.

Through mining the associations of syntactic structures, we can conclude that studying the relevance of syntactic structures helps us to analyze registers thoroughly and their differences and connections at the syntactic structure level.

6   Conclusion

In this paper, we study the syntax structures of the different registers. Instead of manually efforts, we use machine learning methods to explore the associations inside and between syntactic structures, including positive and negative correlations, linear, and nonlinear correlations. Combined with the theory of linguistics, we carry out a detailed analysis of the syntactic structures and part of speech (POS). Through analyses, we find that the associations between POS and syntactic structure explored by machine learning methods have a good interpretation effect, which provide a insight to study the theory of stylistic features. However, current discoveries by the machine learning methods are more preliminary and plain than those linguistic theories proposed by linguistics. Our further work includes a combination of more powerful machine learning methods with more profound linguistic theories.

References

1. Allen, J.F.: Towards a general theory of action and time. Artif. Intell. **23**(2), 123–154 (1984)
2. Benesty, J., Chen, J., Huang, Y., Cohen, I.: Pearson correlation coefficient. In: Cohen, I., Huang, Y., Chen, J., Benesty, J. (eds.) Noise Reduction in Speech Processing, pp. 1–4. Springer, Heidelberg (2009). https://doi.org/10.1007/978-3-642-00296-0
3. Bernstein, J.B.: Topics in the syntax of nominal structure across romance (1994)
4. Chang, H.W.: The acquisition of Chinese syntax. In: Advances in Psychology, vol. 90, pp. 277–311. Elsevier (1992)
5. Chomsky, N.: Aspects of the Theory of Syntax, vol. 11. MIT Press, Cambridge (2014)
6. Zhu, D.: Grammar Printed Lecture. Commercial Press, Beijing (1982)
7. Ferguson, C.A.: Dialect, register, and genre: working assumptions about conventionalization. In: Sociolinguistic Perspectives on Register, pp. 15–30 (1994)
8. Firth, J.R.: Papers in Linguistics 1934-1951: Repr. Oxford University Press (1961)
9. Foley, W.A., et al.: Functional Syntax and Universal Grammar. Cambridge University Press, Cambridge (2009)
10. Li, K., Man, H., et al.: Boundedness of VP and linked event structure. Ph.D. thesis (2013)
11. Langfelder, P., Zhang, B., Horvath, S.: Defining clusters from a hierarchical cluster tree: the dynamic tree cut package for R. Bioinformatics **24**(5), 719–720 (2007)

12. Pollock, J.Y.: Verb movement, universal grammar, and the structure of IP. Linguist. Inq. **20**(3), 365–424 (1989)
13. Rorat, T.: Plant dehydrins–tissue location, structure and function. Cell. Mol. Biol. Lett. **11**(4), 536 (2006)
14. Seretan, V.: Induction of syntactic collocation patterns from generic syntactic relations (2005)
15. Seretan, V.: Collocation extraction based on syntactic parsing. Ph.D. thesis, University of Geneva (2008)
16. Seretan, V.: Syntax-Based Collocation Extraction, vol. 44. Springer, Heidelberg (2011). https://doi.org/10.1007/978-94-007-0134-2
17. Seretan, V., Wehrli, E.: Accurate collocation extraction using a multilingual parser. In: Proceedings of the 21st International Conference on Computational Linguistics and the 44th Annual Meeting of the Association for Computational Linguistics, pp. 953–960. Association for Computational Linguistics (2006)
18. Tibshirani, R.: Regression shrinkage and selection via the lasso. J. R. Stat. Soc.: Ser. B (Methodol.) **58**(1), 267–288 (1996)
19. Watson, D., Tellegen, A.: Toward a consensual structure of mood. Psychol. Bull. **98**(2), 219 (1985)
20. Zhu, W.: A study on the syntactic structure of the study of word collocation variation in "red sorghum". Chin. Teach. **5**, 153–155 (2014)
21. Wright, W.: Lectures on the Comparative Grammar of the Semitic Languages, vol. 43. University Press (1890)
22. Xia, F.: The part-of-speech tagging guidelines for the Penn Chinese Treebank (3.0). IRCS Technical reports Series, p. 38 (2000)
23. Fan, X.: Sentence meaning. Chin. Linguist. **3**, 2–12 (2010)
24. Huo, X.L.: Research on stylistic features and its influence variables. Master's thesis, Nanjing University (2014)
25. Xue, N., Xia, F., Chiou, F.D., Palmer, M.: The penn chinese TreeBank: phrase structure annotation of a large corpus. Natural language engineering **11**(2), 207–238 (2005)
26. Yip, M.J.: The tonal phonology of Chinese. Ph.D. thesis, Massachusetts Institute of Technology (1980)

Fundamental Theory and Methods of Computational Linguistics

Adversarial Domain Adaptation for Chinese Semantic Dependency Graph Parsing

Huayong Li, Zizhuo Shen, DianQing Liu, and Yanqiu Shao$^{(\boxtimes)}$

Information Science School, Beijing Language and Culture University,
Beijing 100083, China
yqshao163@163.com

Abstract. The Chinese Semantic Dependency Graph (CSDG) Parsing reveals the deep and fine-grained semantic relationship of Chinese sentences, and the parsing results have a great help to the downstream NLP tasks. However, most of the existing work focuses on parsing in a single domain. When transferring to other domains, the performance of the parser tends to drop dramatically. And the target domain often lacks the annotated data, so it is difficult to train the parser directly in the target domain. To solve this problem, we propose a lightweight yet effective domain adaptation component for CSDG parsing that can be easily added to the architecture of existing single domain parser. It contains a data sampling module and an adversarial training module. Furthermore, we present CC SD, the first Chinese Cross-domain Semantic graph Dependency dataset. Experiments show that with the domain adaptation component we proposed, the model can effectively improve the performance in the target domain. On the CCSD dataset, our model achieved state-of-the-art performance with significant improvement compared to the strong baseline model.

Keywords: Chinese Semantic Dependency Graph Parsing ·
Adversarial domain adaptation · Cross-domain parsing

1 Introduction

Chinese Semantic Dependency Graph (CSDG) Parsing is one of the key technologies in Chinese natural language processing. CSDG Parsing focuses on the analysis of the deep semantic relationship between words in Chinese sentences. Unlike the restricted representation of tree structure, CSDG Parsing allow a word to have multiple parent nodes (***non-local***), and dependent arcs to cross each other (***non-projection***). Therefore, CSDG Parsing can fully and naturally represent the linguistic phenomenon of natural language [5]. Figure 1 shows an example of CSDG Parsing.

In recent years, semantic dependency parsing has made great progress [11,18]. The neural network approach has become the mainstream technology for this

© Springer Nature Switzerland AG 2019
M. Sun et al. (Eds.): CCL 2019, LNAI 11856, pp. 55–66, 2019.
https://doi.org/10.1007/978-3-030-32381-3_5

task. Transition-based methods [4,6,19,20] and graph-based methods [10,11, 17] have been successfully applied to dependency parsing tasks. Especially, the Biaffine network [11], has gained more and more attention in the field of semantic dependency parsing.

Most studies on semantic dependency parsing focused on single domain parsing, which means the data used in the research comes from only one domain. Yet, the phenomena of natural language in different domains vary greatly. Therefore, even if the model achieves a high *in-domain* score, it tends to be much worse on the *out-domain* dataset, which greatly limits the application value of the results of semantic dependence parsing [8]. In order to make semantic dependence parsing practical, it is necessary to solve the problem of domain adaptation [12]. Recently, the semi-supervised and few-shot domain adaptation has received more and more attention. However, building a robust cross-domain semantic dependency parser is still a very challenging job.

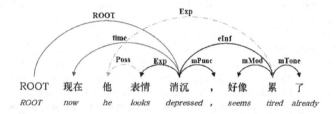

Fig. 1. An example of CSDG Parsing. Here, "他" is the argument of "表情" and it is also an argument of "累" (*non-local*). Arc "累" → "他" and "消沉" → "表情" cross each other (*non-projection*).

In this paper, we propose a domain adaptation component for CSDG parsing by integrating shared information in different domains. We address the lack of out-domain annotation using adversarial cross-domain learning to effectively utilize the annotated data in the source domain. Since there is no cross-domain semantic dependency parsing dataset in Chinese, we release the CCSD, the first Chinese Cross-domain Semantic graph Dependency dataset. Experiments show that our model can effectively improve the performance in the target domain. Our contributions can be summarized as follows:

- We are the first to apply adversarial domain adaptation for Chinese Semantic Dependency Graph Parsing;
- We release the first Chinese Cross-domain Semantic Graph Dependency Parsing dataset, the dataset and code are available on Github[1];
- We propose two data sampling strategies and three adversarial learning methods and analyze their performance;

---

[1] https://github.com/LiangsLi/Domain-Adaptation-for-Chinese-Semantic-Dependency-Graph-Parsing.

2 Related Work

2.1 Semantic Dependency Graph Parsing

In order to be able to express more linguistic phenomena, [5] extended semantic dependency tree to graph structure. To parse semantic dependency graphs, [9,18] proposed a neural transition-based approach, using a variant of list-based arc-eager transition algorithm, and [11] proposed a graph-based model with Biaffine attention mechanism.

2.2 Domain Adaptation

Most of studies on domain adaptation in the field of dependency parsing are about syntactic parsing. In the domain adaptation track of CoNLL 2007 shared task [13], [3] applied a tree revision method which learns how to correct the mistakes made by the base parser on the adaptation domain, and [14] used two models to parse unlabeled data in the target domain to supplement the labeled out-of-domain training set. [21] proposed producing dependency structures using a large-scale HPSG grammar to provide general linguistic insights for statistical models, which achieved performance improvements on out-domain tests. [8] proposed a data-oriented method to leverage ERG, a linguistically-motivated, hand-crafted grammar, to improve cross-domain performance of semantic dependency parsing.

In other tasks, [12] proposed an adversarial approach to domain adaptation. [7] proposed adversarial multi-criteria learning for Chinese word segmentation. [16] applied adversarial domain adaptation to the problem of duplicate question detection across different domains.

3 Method

3.1 Deep Biaffine Network for Single Domain CSGD Parsing

Before introducing cross-domain parsing method, we need to review the current best single domain parsing network. Following the work of [11], we use the Biaffine network as our general architecture for single domain CSGD parsing.

Representation Layer. Each word x_i is represented as the concatenation of word embedding e_i^{word}, POS tag embedding e_i^{pos} and chars' representation h_i^{char}:

$$x_i = e_i^{word} \oplus e_i^{pos} \oplus h_i^{char} \tag{1}$$

Feature Extraction Layer. We use Bi-LSTM as feature extraction layer. In order to improve the efficiency of training, Highway mechanism [22] is used:

$$h_t^{lstm} = HighwayLSTM\left(x_i; W_H; b_H\right) \tag{2}$$

where, h_t^{lstm} represents the output of the Highway LSTM, and W_H, b_H represents the parameters of the Highway LSTM, respectively.

Biaffine Scorer Layer. We use Biaffine network [10] to predict dependency edges and dependency labels respectively. For predicting dependency edge, we first feed the Bi-LSTM encoded representation h_i^{lstm} into two single-layer feed-forward networks (FNN) to get the head representation and the dependent representation.

$$h_i^{edge-head} = FNN^{edge-head}\left(h_i^{lstm}\right) \tag{3}$$

$$h_i^{edge-dep} = FNN^{edge-dep}\left(h_i^{lstm}\right) \tag{4}$$

Then, we use Biaffine transformation (Eq. 5) to obtain the scoring matrix of all possible edges in a sentence. Finally, we calculate the probability of each edge.

$$Biaffine\left(x_1, x_2\right) = x_1^T U x_2 + W\left(x_1 \oplus x_2\right) + b \tag{5}$$

$$s_{i,j}^{edge} = Biaffine^{edge}\left(h_i^{edge-dep}, h_j^{edge-head}\right) \tag{6}$$

$$p_{i,j}^{*edge} = sigmoid\left(s_{i,j}^{edge}\right) \tag{7}$$

The method of dependency label prediction is nearly similar to that of dependency edge prediction. It's worth noting that we use softmax function to calculate the label probability (Eq. 8).

$$p_{i,j}^{*label} = softmax\left(s_{i,j}^{label}\right) \tag{8}$$

3.2   Cross-Domain CSDG Parsing Method

The task of Cross-Domain CSDG Parsing is to transfer the CSDG parser from the source domain to the target domain. Since the amount of annotated training data in the target domain is very limited, some of which may not even have annotated training data. Therefore, the *Shared-Private* (or *Global-Local*) network architecture [7] is difficult to apply to our task because there is not enough data to train the private (or local) module for each target domain. In this paper, we propose an effective domain adaptation component that can be easily added to the single domain parser described in Sect. 3.1. As shown in Fig. 2, the component consists of two parts: *data sampling module* and *adversarial training module*.

Data Sampling Module. In cross-domain parsing, we first need to decide how to sample data in two domains. Here we propose two approaches of data sampling, uniform and proportional sampling [15]. We choose which domain the data comes from based on the probability of source P_{source} and the probability of target P_{target}.

In uniform sampling, P_{source} and P_{target} are equal, both equal to 0.5. In proportional sampling, the probability of sampling a domain is defined as follow:

$$P_{source} = \frac{N_{source}}{N_{source} + N_{target}} \tag{9}$$

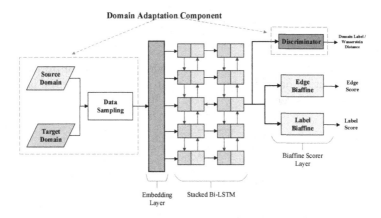

Fig. 2. Architecture of cross-domain CSDG paring with adversarial training

$$P_{target} = \frac{N_{target}}{N_{source} + N_{target}} \tag{10}$$

Where N_{source} indicates the size of source domain training set, N_{target} indicates the size of target domain training set.

Adversarial Training Module. The adversarial training [1,7,12,16] module contains a discriminator and three different adversarial training strategies. We adopt two main approaches to achieve adversarial learning. One is based on domain classification and the other is based on *Wasserstein distance*. In particular, the classification-based approach contains two different adversarial training strategies.

In the classification-based approach, we adopt a domain classifier as the discriminator which is optimized to correctly distinguish which domain the data comes from. The loss of the discriminator needs to be minimized when training the discriminator as follows:

$$\min_{\theta^{dis}} J_{dis}\left(\theta^{dis}\right) = \sum_{i=1}^{n}\sum_{j=1}^{m} \log p\left(L_j|X_i, \Theta^f, \theta^{dis}\right) \tag{11}$$

where Θ^f indicates the parameters of feature extraction layer, Θ^{dis} indicates the parameters of discriminator.

In contrast, the feature extraction layer is optimized to confuse the discriminator in order to make the Bi-LSTM encoded feature representation more general. Therefore, we need to maximize the adversarial loss to optimize the parameters of feature extraction layer. Inspired by [7,12,16], we use two different adversarial training strategies that contain two different adversarial losses.

The first is the cross-entropy loss training strategy[2], which is to maximize the cross-entropy of the discriminator when training the parameters of feature extraction layer:

---

[2] This method is also known as *Gradient Reversal*.

$$\max_{\Theta^f} J_{adv}\left(\Theta^f\right) = \sum_{i=1}^{n}\sum_{j=1}^{m}\log p\left(L_j|X_i,\theta^f,\theta^{dis}\right) \tag{12}$$

The second is the entropy training strategy, which is to maximize the entropy of domain distribution predicted by discriminator when training the parameters of feature extraction layer:

$$\max_{o^f} J_{adv}\left(\Theta^f\right) = \sum_{i=1}^{n}\sum_{j=1}^{m} H\left(p\left(L_j|X_i,\Theta^f,\Theta^{dis}\right)\right) \tag{13}$$

$$H(p) = -\sum_{i} p_i \log p_i \tag{14}$$

where $H(p)$ is the entropy of domain distribution p.

In the Wasserstein approach [1,2], we minimize the Wasserstein distance between source domain distribution P_s and target domain distribution P_t in order to train discriminator.

$$W\left(P_s, P_t\right) = \sup_{\|f\|_L \le 1} E_{x \sim P_s}[f(x)] - E_{x \sim P_t}[f(x)] \tag{15}$$

$$\min_{\Theta^{dis}} J_{dis}\left(\Theta^{dis}\right) = \min W\left(P_S, P_t\right) \tag{16}$$

where Θ^{dis} indicates the parameters of discriminator, and f is a *Lipschitz-1 continuous function*.

In contrast, the feature extraction layer needs to be trained by maximizing Wasserstein distance:

$$\max_{\Theta^f} J_{adv}\left(\Theta^f\right) = \max W\left(P_s, P_t\right) \tag{17}$$

According to [1], the adversarial training based on Wasserstein distance can obtain more stable training process.

3.3   Jointly Training

First, we calculate the loss of the parser J_{parser}, which consists of two parts, the loss of edge Biaffine J_{edge} and the loss of label Biaffine J_{label}:

$$J_{parser}\left(\Theta^p\right) = \beta J_{label}\left(\Theta^p\right) + (1 - \beta)J_{edge}\left(\Theta^p\right) \tag{18}$$

where Θ^p indicates the parameters of CSDG parser (note that the parameters of the discriminator are not included in Θ^p), β is the combined ratio of two losses. The loss of edge Biaffine and the loss of label Biaffine is calculated as follows:

$$J_{edge}\left(\Theta^p\right) = -p_{i,j}^{edge}\log p_{i,j}^{*edge} - \left(1 - p_{i,j}^{edge}\right)\log\left(1 - p_{i,j}^{*edge}\right) \tag{19}$$

$$J_{label}\left(\Theta^p\right) = -\sum_{label}\log p_{i,j}^{*label} \tag{20}$$

Algorithm 1. Adversarial domain adaptation learning for CSDG parsing

Input: source domain data and a target domain data $D \in (X_s \cup X_t)$

Hyper-parameters: the learning rates of parser and adversarial competent α_1, α_2; adversarial interpolation λ

Parameters to be trained: $\Theta^p, \Theta^f, \Theta^{dis}$

Sampling probability: P_{sampling}

Loss function: $J_{parser}, J_{dis}, J_{adv}$

1: **while** Θ^p and Θ^f not converge **do**
2:     Pick a batch of data from D according to P_{sampling}
3:     Calculate the parser loss J_{parse} using θ^p for this batch
4:     Calculate the adversarial loss J_{adv} using Θ^{dis}, Θ^f for this batch
5:     $\Theta^p = \Theta^p - \alpha_1 \nabla_\Theta^p J_{\text{parser}}$
6:     $\Theta^{dis} = \Theta^a - \alpha_2 \nabla_\Theta^a J_{\text{dis}}$
7:     $\Theta^p = \Theta^f + \alpha_1 \nabla_\Theta^f J_{\text{adv}}$

Finally, the entire model is optimized by the a jointly losses as follow:

$$J(\Theta) = J_{parser}\left(\theta^p\right) + J_{dis}\left(\theta^{dis}\right) + \lambda J_{adv}\left(\Theta^f\right) \tag{21}$$

Where λ is the weight that controls the interaction of the loss terms. The details of training procedure are described as Algorithm 1.

4   CCSD Dataset

Since there is no cross-domain semantic dependency parsing dataset in Chinese, we present a Chinese Cross-domain Semantic Dependency dataset, named *CCSD*. It contains one source domain and four target domains.

4.1   Corpus Collection

The data sources for each domain are as follows:

Source Domain. Source domain contains a lot of balanced corpus. The sentences in source domain are selected from the SemEval-2016 Task 9 dataset [5] and the textbook *Boya Chinese* (Chinese: 博雅汉语).

Target Domain. Target area consists of four different domains. The data sources for different target domains are as follows:

- **Novel.** Novel domain contains 1562 sentences selected from the short story *the Little Prince* (Chinese: 小王子) and 3438 sentences selected from novel *Siao Yu* (Chinese: 少女小渔).
- **Drama.** Drama domain contains 5000 sentences selected from drama *My Own Swordsman* (Chinese: 武林外传).

- **Prose**. Prose domain contains 5000 sentences selected from the prose collection *Cultural Perplexity in Agonized Travel* (Chinese: 文化苦旅). This domain contains a lot of rhetorical figures.
- **Inference**. Inference domain contains 22,308 sentences, selected from the *CNLI* (Chinese Natural Language Inference) dataset. Compared with other domains, the sentences in this domain are much simpler.

4.2   Dataset Construction and Statistics

For each domain, we divide the data into training set, validation set, and test set. The detailed statics are described in Table 1.

Table 1. Statistics about the dataset. The table shows the number of sentences. It is worth noting that there is no annotated training data in the *Inference* domain.

	Domain	Train set	Dev set	Test set	Unannotated
Source	*Source*	24,003	2,000	2,000	—
Target	*Novel*	3,000	1,000	1,000	—
	Prose	3,000	1,000	1,000	—
	Drama	3,000	1,000	1,000	—
	Inference	—	2,000	2,000	18,308

It is worth noting that we have manually annotated all the data in the Source domain, Novel domain, Prose domain and the Drama domain. But for the data in the Inference domain, we only annotated the data of the validation set and the test set. However, Inference domain contains a lot of unannotated data.

Table 2. Statistics of the four target domains. **Ave-len** represents the average sentence length of the corpus of each domain. **Non-local** represents the proportion of non-local phenomena, and **Non-projective** represents the proportion of non-projective phenomena. **Unigrams** and **Bigrams** show the proportion of n-grams that shared between the source and each target domains.

Domain	Ave-len	Non-local	Non-projective	Unigrams	Bigrams
Novel	12.49	12.02%	5.23%	97.42%	35.48%
Drama	10.16	6.47%	2.40%	97.06%	30.06%
Prose	17.22	13.98%	4.14%	94.61%	28.88%
Inference	11.72	5.29%	2.22%	96.98%	25.38%

As shown in Table 2, We compared the distribution of data in different domains. According to the statistical results, the data in the Inference domain is the simplest, and the data distribution of Novel domain and Drama domain is closest to the source domain.

5   Experiments

5.1   Setup

Baselines and Evaluation Metrics. we use two baseline models: Non-transfer model and Pre-trained model. Non-transfer model is trained directly in the target domain. Pre-trained model is to train a fundamental model using source domain data, and then fine-tune model using target domain data. We use the labeled attachment score (LAS) in the target domain as evaluation metric.

Hyperparameters. The dimension of LSTM is 400. We dropout word embedding with 20% probability and the inputs of LSTM and Biaffine network with 33% probability. The joint optimization ratio of biaffine classifiers is 0.5. We set the interpolation coefficients λ of adversarial loss to 0.1. Other configurations are same as [11].

5.2   Overall Results

Table 3 shows the experiment results of our proposed models on test sets of target domains. Note that we do not show the metrics in source domain, because in this task we use the metrics of the target domain to measure the pros and cons of the model.

Since there is no training data in Inference domain, we only use the unsupervised data of the target domain to train the discriminator, and train the parser with the data of the source domain. In Novel Domain, our best model gains 6.41% and 1.67% improvement compared with two strong baseline models. In Prose Domain, our best model gains the 4.77% and 2.79% improvement

Table 3. Results of proposed models on target domains. In the Sampling column, U, P indicate uniform sampling, and proportional sampling. In the Adv Learning column, E, CE, W indicate entropy method, cross-entropy method, Wasserstein distance method.

Model	Sampling	Adversarial	Result (LAS)			
			Novel	*Prose*	*Drama*	*Inference*
Non-transfer	–	–	68.75	68.22	68.01	–
Pre-trained	–	–	73.49	70.2	69.42	87.1
Our model	U	–	74.51	71.97	71.22	–
	U	E	73.85	72.26	71.72	**87.47**
	U	CE	73.52	72.22	71.14	84.1
	U	W	74.13	71.95	70.96	86.13
	P	–	74.62	72.41	71.4	–
	P	E	74	72.1	**72.19**	77.71
	P	CE	**75.16**	**72.99**	72.01	87.05
	P	W	74.36	72.31	71.79	83.38

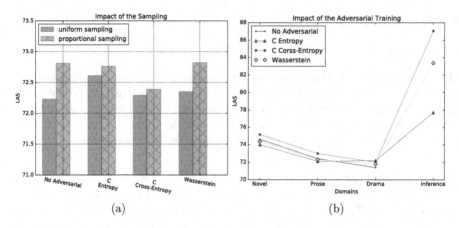

Fig. 3. Results of the comparative experiment. (a) is comparison between different sampling strategies. (b) is comparison between different adversarial training strategy

respectively. In Drama Domain, our best model gains the 4.18% and 2.77% improvement respectively. In Inference domain, our best model gains the 0.37% improvement on LAS.

The experimental results can strongly prove the ability of our proposed models. Due to the differences of language characteristics, the optimal domain adaption methods of each domain are not same. According to the experimental results, the proportional sampling and adversarial learning based on cross-entropy method achieves the highest LAS in most cases.

Analysis on the Data Sampling Strategy. Figure 3(a) studies the impact of sampling strategies on parser performance. In order to better compare the impact of sampling strategies, we performed experiments of uniform sampling and proportional sampling in four target areas based on different adversarial strategies. Then we average the results of the four target domains.

As shown in Fig. 3(a), no matter which adversarial strategy is used, the probability sampling method is obviously better than the uniform sampling method. We believe that this is due to the limited amount of training data in the target domain. When using uniform sampling, the model is easy to overfit the training data, thus affecting the generalization ability of the model.

Analysis on the Adversarial Training Strategy. Figure 3(b) studies the impact of the choice of adversarial strategy. All of the above comparison experiments were done under proportional sampling.

As shown in Fig. 3(b), the performance of the model using the classification-based cross-entropy method is the best in all four areas. And its performance is significantly better than the model without the adversarial strategy. It is worth noting that on Novel and Prose domain, the performance of the model using the other two adversarial is not much different or slightly worse than that of the model without the adversarial strategy. This is because the data distribution of Novel and Prose is very close to the source domain, so the effect of adversarial is less obvious. Because the sentences in Inference domain are relatively simple, even in the unsupervised case, the model still achieves good performance after using the adversarial strategy.

Conclusions as a result, the adversarial strategy, especially classification-based cross-entropy method, can significantly improve the performance of the model in the target domain.

6 Conclusions

Performance of Chinese Semantic Dependency Graph Parser tends to drop significantly when transferring to new domains, especially when the amount of data in the target domain is small. In order to solve this problem, we propose an effective domain adaptation component, which has two parts: data sampling module and adversarial training module. Since there is no open Chinese cross-domain semantic dependency graph parsing dataset, we release a new dataset CCSD. Experiments show that our model performance is significantly improved compared to the strong baseline model. On the CCSD dataset, the model using proportional sampling and cross-entropy adversarial learning achieves the highest LAS score in most cases.

Acknowledgement. This research project is supported by the National Natural Science Foundation of China (61872402), the Humanities and Social Science Project of the Ministry of Education (17YJAZH068), Science Foundation of Beijing Language and Culture University (supported by the Fundamental Research Funds for the Central Universities) (18ZDJ03).

References

1. Arjovsky, M., Chintala, S., Bottou, L.: Wasserstein GAN. arXiv preprint arXiv:1701.07875 (2017)
2. Arjovsky, M., Chintala, S., Bottou, L.: Wasserstein generative adversarial networks. In: International Conference on Machine Learning, pp. 214–223 (2017)
3. Attardi, G., Dell'Orletta, F., Simi, M., Chanev, A., Ciaramita, M.: Multilingual dependency parsing and domain adaptation using DeSR. In: Proceedings of the 2007 Joint Conference on Empirical Methods in Natural Language Processing and Computational Natural Language Learning (EMNLP-CoNLL), pp. 1112–1118 (2007)
4. Che, W., Liu, Y., Wang, Y., Zheng, B., Liu, T.: Towards better UD parsing: deep contextualized word embeddings, ensemble, and Treebank concatenation. arXiv preprint arXiv:1807.03121 (2018)
5. Che, W., Zhang, M., Shao, Y., Liu, T.: Semeval-2016 task 9: Chinese semantic dependency parsing, pp. 378–384, June 2012. https://doi.org/10.18653/v1/S16-1167

6. Chen, D., Manning, C.: A fast and accurate dependency parser using neural networks. In: Proceedings of the 2014 Conference on Empirical Methods in Natural Language Processing (EMNLP), pp. 740–750 (2014)
7. Chen, X., Shi, Z., Qiu, X., Huang, X.: Adversarial multi-criteria learning for Chinese word segmentation. arXiv preprint arXiv:1704.07556 (2017)
8. Chen, Y., Huang, S., Wang, F., Cao, J., Sun, W., Wan, X.: Neural maximum subgraph parsing for cross-domain semantic dependency analysis. In: Proceedings of the 22nd Conference on Computational Natural Language Learning, pp. 562–572 (2018)
9. Ding, Y., Shao, Y., Che, W., Liu, T.: Dependency graph based Chinese semantic parsing. In: Sun, M., Liu, Y., Zhao, J. (eds.) CCL/NLP-NABD-2014. LNCS (LNAI), vol. 8801, pp. 58–69. Springer, Cham (2014). https://doi.org/10.1007/978-3-319-12277-9_6
10. Dozat, T., Manning, C.D.: Deep biaffine attention for neural dependency parsing. arXiv preprint arXiv:1611.01734 (2016)
11. Dozat, T., Manning, C.D.: Simpler but more accurate semantic dependency parsing. arXiv preprint arXiv:1807.01396 (2018)
12. Ganin, Y., et al.: Domain-adversarial training of neural networks. J. Mach. Learn. Res. 17(1), 2030–2096 (2016)
13. Nivre, J., et al.: The CoNLL 2007 shared task on dependency parsing. In: Proceedings of the 2007 Joint Conference on Empirical Methods in Natural Language Processing and Computational Natural Language Learning (EMNLP-CoNLL) (2007)
14. Sagae, K., Tsujii, J.I.: Dependency parsing and domain adaptation with data-driven LR models and parser ensembles. In: Bunt, H., Merlo, P., Nivre, J. (eds.) Trends Parsing Technol., pp. 57–68. Springer, Heidelberg (2010)
15. Sanh, V., Wolf, T., Ruder, S.: A hierarchical multi-task approach for learning embeddings from semantic tasks. Proc. AAAI Conf. Artif. Intell. 33, 6949–6956 (2019)
16. Shah, D.J., Lei, T., Moschitti, A., Romeo, S., Nakov, P.: Adversarial domain adaptation for duplicate question detection. arXiv preprint arXiv:1809.02255 (2018)
17. Wang, W., Chang, B.: Graph-based dependency parsing with bidirectional LSTM. In: Proceedings of the 54th Annual Meeting of the Association for Computational Linguistics (Volume 1: Long Papers), vol. 1, pp. 2306–2315 (2016)
18. Wang, Y., Che, W., Guo, J., Liu, T.: A neural transition-based approach for semantic dependency graph parsing. In: Thirty-Second AAAI Conference on Artificial Intelligence (2018)
19. Yu, J., Elkaref, M., Bohnet, B.: Domain adaptation for dependency parsing via self-training. In: Proceedings of the 14th International Conference on Parsing Technologies, pp. 1–10 (2015)
20. Zhang, X., Du, Y., Sun, W., Wan, X.: Transition-based parsing for deep dependency structures. Comput. Linguist. 42(3), 353–389 (2016)
21. Zhang, Y., Wang, R.: Cross-domain dependency parsing using a deep linguistic grammar. In: Proceedings of the Joint Conference of the 47th Annual Meeting of the ACL and the 4th International Joint Conference on Natural Language Processing of the AFNLP, pp. 378–386 (2009)
22. Zilly, J.G., Srivastava, R.K., Koutník, J., Schmidhuber, J.: Recurrent highway networks. In: Proceedings of the 34th International Conference on Machine Learning-Volume 70, pp. 4189–4198. JMLR. org (2017)

Information Retrieval and Question Answering

Encoder-Decoder Network with Cross-Match Mechanism for Answer Selection

Zhengwen Xie, Xiao Yuan, Jiawei Wang, and Shenggen Ju[(✉)]

College of Computer Science, Sichuan University, Chengdu 610065, China
jsg@scu.edu.cn

Abstract. Answer selection (AS) is an important subtask of question answering (QA) that aims to choose the most suitable answer from a list of candidate answers. Existing AS models usually explored the single-scale sentence matching, whereas a sentence might contain semantic information at different scales, e.g. Word-level, Phrase-level, or the whole sentence. In addition, these models typically use fixed-size feature vectors to represent questions and answers, which may cause information loss when questions or answers are too long. To address these issues, we propose an Encoder-Decoder Network with Cross-Match Mechanism (EDCMN) where questions and answers that represented by feature vectors with fixed-size and dynamic-size are applied for multiple-perspective matching. In this model, Encoder layer is based on the "Siamese" network and Decoder layer is based on the "matching-aggregation" network. We evaluate our model on two tasks: Answer Selection and Textual Entailment. Experimental results show the effectiveness of our model, which achieves the state-of-the-art performance on WikiQA dataset.

Keywords: Answer selection · Multi-Perspective · Cross-Match Mechanism

1 Introduction

Answer selection is an important subtask of question answering (QA) that enables choosing the most suitable answer from a list of candidate answers in regards to the input question. In general, a good answer has two characteristics: First, the question affects the answer, therefore the good answer must be related to the question; Second, the good answer does not require strict word matching, but show better semantic relevance. These characteristics consequently make traditional feature engineering techniques less effective compared to neural models [5, 13, 16, 26].

Previous neural models can be divided into two kinds of frameworks. The first one is the "Siamese" network [17, 27]. It usually utilizes either Recurrent Neural Network (RNN) or Convolutional Neural Networks (CNN) to generates sentence representations, and then calculate similarity score solely based on the two sentence vectors for the final prediction. The second kind of framework is called "matching-aggregation" network [14, 23], it matches two sentences at Word-level by the fixed-size feature vector, and then aggregates the matching results to generate a final vector for prediction. Studies on benchmark QA datasets show that the second one performs better [23].

© Springer Nature Switzerland AG 2019
M. Sun et al. (Eds.): CCL 2019, LNAI 11856, pp. 69–80, 2019.
https://doi.org/10.1007/978-3-030-32381-3_6

Despite the second framework has made considerable success on the answer selection task, there are still some limitations. First, [14, 24] only explored the single-scale matching, whereas a sentence might contain semantic information at different scales, e.g. Word-level, Phrase-level, or the whole sentence. To deal with this issue, [23] applied the multi-perspective matching technique to match two sentences representations at Word-level and Sentence-level, but ignored the Phrase-level information. Second, [18, 19, 23] just aggregate questions and answers which have different lengths into fixed-size feature vectors for matching, which may lose a large amount of rich information contained in sentence compared to matching with feature vectors with dynamic fixed-size.

To address these issues, we propose an Encoder-Decoder Network with Cross-Match Mechanism (EDCMN) for answer selection. Our model is a new framework where the Encoder layer is based on the "Siamese" network and the Decoder layer is based on the "matching-aggregation" network. We first obtain three semantic representations which capture coarse-to-fine information including Sentence-level, Phrase-level and Word-level in the Encoder layer. After that, to get the interaction between QA pairs, we utilize three feature augmentation methods to match both on feature vectors with dynamic-size and fixed-size. And then we obtain six matching vectors and have a concatenation on them. Finally, we compress the concatenation vector to generate a final vector for prediction. The Cross-Match Mechanism in our Encoder-Decoder framework enables capturing multiple perspective information which is suitable to identify complex relations between questions and the answers.

The main contributions of our paper can be summarized as follows:

- We propose an Encoder-Decoder Network with Cross-Match Mechanism which based on two classical frameworks. Our model is the first to apply Encoder-Decoder architecture on the answer selection task, and it requires no additional information and relies solely on the original text.
- The Cross-Match Mechanism in EDCMN captures information of sentence at different scales and matching on multiple perspectives.
- In comparison to other state-of-the-art representation learning approaches with attention, our approach achieves the best results and significantly outperforms various strong baselines.

2   Related Work

Answer selection (AS) has been studied for many years. The previous work focused on designing hand-craft features to capture n-gram overlapping, word reordering and syntactic alignments phenomena [6, 25]; This kind of method can work well on a specific task or dataset, but it's hard to generalize well to other tasks.

Recently, researchers started using deep neural networks for answer selection, the first kind of framework is based the "Siamese" architecture [4, 11, 17, 20, 27], In this framework, the same neural network encoder (e.g., a CNN or an RNN) is applied to two input sentences individually. However, there is no explicit interaction between the two sentences during the encoding procedure. The second kind of framework is called "matching-aggregation" model [14, 22, 23]. Under this framework, smaller units of the two sentences are firstly matched, and then the matching results are aggregated into a vector to make the final decision. It captures interactive features between two sentences at Word-level, but ignored other granular matchings.

[11] uses self-attention to focus on the important parts of the sentence, but ignore the interaction between sentence pairs. [1, 17] apply additive attention to solve the problem of lack of interaction between question and answer and this approach only gets the coarse-grained information. [12, 24, 27] employ models which form an interactive Word-level matrix of questions and answers to discover fine-grained alignment of two sentences, whereas it only used one matching method. [14, 19] utilize multiple attentions to focus on different parts of the semantic representation, then compress them into fixed-size feature vectors for matching, which may lose rich information in dynamic-size feature vectors.

Our work is also inspired by the idea of Multi-Perspective Matching models [19, 23] in NLP. We propose the EDCMN model which achieve step by step learning. We are the first to apply Encoder-Decoder architecture to the answer selection task. Our model not only employs the Multi-Perspective Matching model to identify complex relations between questions and answers but also captures information of sentence at different scales in Encoder which results in a more close match in the Decoder layer.

3 Model

We introduce EDCMN and its detailed implementation in this section. We first cover the model architecture and the Cross-Match Mechanism which is the core innovation in this paper, in Sect. 3.1. We then introduce the Encoder layer and Decoder layer in Sects. 3.2 and 3.3, respectively.

3.1 Cross-Match Mechanism

The Cross-Match Mechanism is inspired by ResNet [28] which reveals that the information at different scales can be extracted as the network get deeper, in addition, hierarchical information can be combined when it crosses different levels. As shown in Fig. 1, the cross-matching mechanism is mainly composed of an encoder layer and a decoder layer. The encoder layer obtains sentence representations at Word-level, Phrase-level and Sentence-level by LSTM, CNN and Self-attention components respectively. The decoder layer first applies Match functions to get six interaction representations about QA pairs, and then merge them to the final vector with the Compress function.

The biggest benefit of the Cross-Match Mechanism is that it realizes short-circuit connection through concatenate feature, which makes some of the features extracted from earlier layers may still be used directly by deeper layers, meanwhile, it can match vectors in multiple ways.

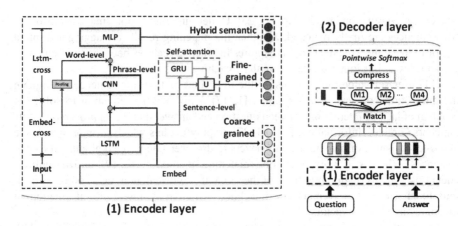

Fig. 1. The Architecture for Encoder-Decoder Network with Cross-Match Mechanism (EDCMN), the left-hand side shows the details about the Encoder layer.

3.2   Encoder Layer

Embed Cross. With pre-trained d dimension word embedding, we can obtain sentence representations $H_q = [s_q^1, \ldots, s_q^m]$ and $H_a = [s_a^1, \ldots, s_a^n]$ where $s_q^i \in \mathbb{R}^d$ is the embedding of the i-th word in the sentence H_q. m and n are the lengths of H_q and H_a respectively. The model obtains the context information of the two vectors entered through the bidirectional Long Short-Term Memory (LSTM) [2] Network.

After the Bi-LSTM, the model obtains the sentence representation of the question $T_q = \{t_q^1, \ldots, t_q^m\}$ and $T_a = \{t_a^1, \ldots, t_a^n\}$ respectively. The coarse-grained semantic representations of QA pairs are obtained.

$$R_Q^1 = \{t_q^1, \ldots, t_q^m\} \tag{1}$$

$$R_A^1 = \{t_a^1, \ldots, t_a^n\} \tag{2}$$

Where $R_Q^1 \in \mathbb{R}^{m \times 2d}$, $R_A^1 \in \mathbb{R}^{n \times 2d}$, d is the dimension of pre-trained word embedding.

In order to make full use of the information of word embedding, the model has a concatenation on Bi-LSTM's output vectors and Embed Cross vectors. We apply this method to the question vector and then obtain a splicing vector $Cross_Q^1 \in \mathbb{R}^{m \times 3d}$. In the same way, the answer vector $Cross_A^1 \in \mathbb{R}^{n \times 3d}$ is got.

LSTM Cross. This component consists of two parts: Fine-grained representations and Hybrid semantic representations.

Fine-Grained Representations. We first create an encoding of the importance for each segment in the unpooled representation T which is the output of the Bi-LSTM by applying an additional, separate Bi-GRU and obtain the concatenated output states $P \in \mathbb{R}^{l \times 2d}$ of this Bi-GRU where the i-th row in P. We then reduce each row P_i to a scalar v_i and apply *softmax* on the vector v to obtain scaled importance values:

$$P = BiGRU(T) \tag{3}$$

$$v_i = w^T P_i \tag{4}$$

$$\alpha = soft \max(v) \tag{5}$$

Where $w \in \mathbb{R}^{2d}$, are learned network parameters for the reduction operation, $v_i \in \mathbb{R}$ is the (unscaled) importance value of the i-th segment in P, and $\alpha \in \mathbb{R}^l$ is the attention vector. as mentioned before, self-attention [9, 15] can solve the long-range dependency problem and choose the relevant information for sentence semantic. by utilizing this operation, the model can grasp the most relevant parts for inference relation precisely and make the correct decision. Unlike before, we retain the length of the sentence to get more fine-grained information. Finally, the sentence-level representation U according to our importance vector α is obtained:

$$u_{i,j} = \alpha P_{i,j} \tag{6}$$

$$R_Q^2 = SelfAttention_Q = U^Q \tag{7}$$

$$R_A^2 = SelfAttention_A = U^A \tag{8}$$

Hybrid Semantic Representations. From the perspective of a word, each word itself may have many meanings, but a phrase made up of many words makes the word less ambiguous. For example, it's difficult to know whether the word "apple" refers to a fruit or a company. If it is the phrase "apple company", we can clearly know the true meaning of apple. Inspired by this observation, the model utilizes 1-D convolutions (Conv1D) with multiple-windows to capture different Phrase-level information in sentences, then apply the max-pooling operation to select the most significant properties in the sentence.

$$CNN_{\max}^i = Conv1D_{Max-pooling}(Cross^1) \tag{9}$$

$$CNN_{feature} = [CNN_{\max}^1, \ldots, CNN_{\max}^k] \tag{10}$$

Therefore, we can obtain $CNN_Q \in \mathbb{R}^{ku}$ and $CNN_A \in \mathbb{R}^{ku}$, which extract the phrase structure information from questions and answers. k is the number of windows and u is the number of filters.

Furthermore, to take Word-level semantic into consideration, we first utilize the concatenation on the Conv1D feature $CNN_{feature} \in \mathbb{R}^{ku}$ and LSTM cross feature $Cross^2 \in \mathbb{R}^{2d}$. Then, we apply a multi-layer perceptron (MLP) with *relu* activation function on concatenation vectors to get the hybrid semantic representations $R_Q^3 \in \mathbb{R}^d$ and $R_A^3 \in \mathbb{R}^d$ which denotes the hierarchical semantic fusion in questions and answers separately.

$$Cross^2 = Max(t^1, \ldots, t^l) \tag{11}$$

$$Hybird_{feature} = [Cross^2; CNN_{feature}] \tag{12}$$

$$R_Q^3 = R_A^3 = MLP_1(Hybird_{feature}) \tag{13}$$

Where $[.; .]$ represents the concatenation operation, l is the length of the sentence.

3.3 Decoder Layer

Match and Interaction. After getting the multi-granularity feature vectors, we employ three different ways to explore the multi-scale matching.

Vectors with Dynamic-Size. For coarse-grained semantic representations and fine-grained semantic representations, their vectors both with dynamic-size. We construct a similarity matrix to further match the similarity word by word. Then, the similarity matrix representation is transformed into high-dimensional vector which hidden dimensions is $m \times n$. At last, we convert it into the compressed vector. This method can make the model more robust, because dimensions of the vector get higher as sentence gets longer. Let f denote a match function that matching two semantic representations in different ways, $M_1 \in \mathbb{R}^d$ and $M_2 \in \mathbb{R}^d$ are as follows:

$$M_1 = f(R_Q^1, R_A^1) = relu(W_1(R_Q^1 \otimes (R_A^1)^T) + b_1) \tag{14}$$

$$M_2 = f(R_Q^2, R_A^2) = relu(W_2(R_Q^2 \otimes (R_A^2)^T) + b_2) \tag{15}$$

Vectors with Fixed-Size. For the hybrid semantic vector, in order to measure the gap between the question representation and answer representation, a direct strategy is to compute the absolute value of their difference. $M_3 \in \mathbb{R}^d$ is approximating the Euclidean distance between the two vectors. Then, their cosine distance $M_4 \in \mathbb{R}^d$ is calculated by the element-wise product, $\odot$ means element-wise product:

$$M_3 = f(R_Q^3, R_A^3) = R_Q^3 - R_A^3 \tag{16}$$

$$M_4 = f(R_Q^3, R_A^3) = R_Q^3 \odot R_A^3 \tag{17}$$

The hybrid hierarchical semantic vectors R_Q^3 and R_A^3 without matching directly are also as outputs, which can preserve the semantic integrity to the greatest and make the network more robust.

Compression and Label Prediction. To be specific, we concatenate those six matching vectors by rows, then MLP with activation function are applied to them to calculate the probability distribution of the matching relation between the QA pair. The final compression vector $Score \in \mathbb{R}^{6d}$ and the output of this layers are as follows:

$$Score = MLP_2([R_Q^3; R_A^3; M_1; M_2; M_3; M_4]) \qquad (18)$$

$$P(y|H_q, H_a) = Soft \max(Score) \qquad (19)$$

We regard the answer selection task as the binary classification problem and the training objective is to minimize the negative loglikelihood in training stage:

$$Loss = -\frac{1}{N}\sum_{i=1}^{N} y_i \log P(y|H_q, H_a) + R \qquad (20)$$

Where y_i is the one-hot representation for the true class of the i-th instance, N represents the number of training instances, and $R = \lambda \parallel \theta \parallel_2^2$ is the L2 regularization.

4 Experimental Setup

4.1 Datasets and Evaluation Metric

Answer Selection Task. For answer selection task, we experiment on two benchmark datasets. We evaluate models by mean average precision (MAP) and mean reciprocal rank (MRR). Statistical information of QA datasets is shown in Table 1.

Table 1. The statistics of Answer Selection datasets and Textual Entailment datasets.

Dataset	WikiQA	TREC-QA	SCITAIL
# of questions/premise (train/dev/test)	873/126/243	1162/65/68	1542/121/171
Avg length of questions/premise	6	8	11
Avg length of answers/hypothesis	25	28	7
Avg # of candidate answers	9	38	–

WikiQA [10] is a recent popular benchmark dataset for open-domain question answering, based on factual questions from Wikipedia and Bing search logs. each question is selected from Wikipedia and used sentences in the summary paragraph as candidates.

TREC-QA [21] is a well-known benchmark dataset collected from the TREC Question Answering tracks. The dataset contains a set of factoid questions, where candidate answers are limited to a single sentence.

Textual Entailment Task. For textual entailment task, we experiment on one new dataset that is created from a QA task rather than sentences authored specifically for the entailment task. which is more challenging. Following previous works [7], we use accuracy (ACC) as evaluation metrics. Statistical information of the dataset is shown in Table 1.

SCITAIL [7] is the first entailment set that hypotheses from science questions and the corresponding answer candidates, and premises from relevant web sentences retrieved from a large corpus. combined with the high lexical similarity of premise and hypothesis for both entailed and non-entailed pairs, makes this new entailment task particularly difficult.

4.2   Implementation Details

We initialized word embedding with 300d-GloVe vectors pre-trained from the 840B Common Crawl corpus [8], while the word embeddings for the out-of-vocabulary words were initialized randomly.

WikiQA: To train our model in mini-batch, we truncate the question to 25 words, the answer to 90 words and batch size to 32. We add 0 at the end of the sentence if it is shorter than the specified length. We resort to Adam algorithm as the optimization method and update the parameters with the learning rate as 0.001. The CNN windows are [1, 2, 3, 4, 5, 6] and the number of CNN filters is 300. We set a dropout rate as 0.5 at the encoder layer and 0.7 at the decoder layer. We add the L2 penalty with the parameter λ as 10^{-5}.

TREC-QA: The experiment settings are the same as WikiQA.

SCITAIL: We truncate both questions and answers to 20 words and change the number of CNN filters to 100, the other experiment settings are the same as WikiQA.

4.3   Experimental Results

WikiQA and TREC-QA. Table 2 reports our experimental results and compared models on these two datasets. We selected 9 models for comparison. On WikiQA, our model achieves state-of-the-art performance. More specifically, Compared with MAN [19] which matching only on fixed-size vectors, our EDCMN model obtains a fine improvement (1.8%) by achieving 74% in MAP, and we outperform it by 1.4% on MRR. Both BiMPM [23] and our model benefit from Multi-Perspective Matching, however, EDCMN is better than BiMPM by 2.2% in terms of MAP and 2.1% in terms of MRR. The reason is that we take Phrase-level information into consideration, but BiMPM ignored it. On TREC-QA (clean), Our EDCMN model is better than most strong baselines such as ABCNN [26], LDC [23], MAN [18], which further indicates that our Cross-Match Mechanism is very effective for matching vectors.

SCITAIL. Table 3 presents the results of our models and the previous models on this dataset. We compare the results reported in the original paper [7]: Majority class, n-gram, decomposable attention, ESIM, and DGEM. Compared with ESIM [3] which had got a huge success on the NLI task, our EDCMN model outperforms 7.6% in test accuracy. We also obtain a fine improvement by achieving 78.2% contrast to the DGEM in test accuracy. This ascertains the effectiveness of the EDCMN model.

Table 2. Performance for answer sentence selection on WikiQA and TREC-QA test set.

Model	WikiQA		TREC-QA (clean)	
	MAP	MRR	MAP	MRR
AP-BiLSTM [12]	0.671	0.684	0.713	0.803
ABCNN [27]	0.692	0.711	0.777	0.836
PWIM [5]	0.709	0.723	0.738	0.827
LDC [24]	0.706	0.723	0.771	0.845
IARNN [20]	0.734	0.742	–	–
BiMPM [23]	0.718	0.731	0.802	0.875
IWAN [14]	0.733	0.750	0.822	0.889
MCAN-SM [18]	–	–	0.827	0.880
MAN [19]	0.722	0.738	0.813	0.893
EDCMN (proposed)	**0.740**	**0.752**	**0.811**	**0.896**

4.4 Ablation Analysis

This section shows the impacts and contribution of the different components of our EDCMN model. Table 4 presents the results on the WikiQA test set. 'w/o' stands for without; 'fea' stands for feature; 'dyna' stands for Dynamic.

In the Encoder layer, we take four steps. First, we take away Embed-Cross feature and just let the Bi-LSTM's output as the input of Conv1D. The influence is small, causing a MAP to drop by 1% and MRR to drop by 0.9%, but it indicates that the effectiveness of the features reuse. Second, we remove Word-level feature, which is also called LSTM cross feature, The MAP drops by 1.5% and MRR drop by 1.9%. It shows that Word-level information is a useful supplement to the model. Third, we get rid of the Conv1D segment and find that the influence is huge, the MAP and MRR drop by 3.1% and 4.1% respectively which confirms the effectiveness of Phrase-level feature. At last, we abandon Sentence-level feature which is generated by self-attention, the influence of Sentence-level feature is close to Word-level feature, the MAP and MRR only drop by 1.6% and 1.7% respectively.

In the Decoder layer, we also take four ablation experiments. if we abandon the Match component and compress Encoder layer's output into the final vector directly for prediction, we find that our model has the worst performance on the dataset. Both MAP and MRR drop nearly by 8%. We can observe that Mat feature reduction had caused the most enormous influence (6%) in these match methods, indicating matching feature vectors with dynamic-size successfully acquires many rich characteristics of sentences.

Ablation analysis shows that different-scale information and dynamic-size feature vectors are important for matching, which confirms the effectiveness of our model (Fig. 2).

Table 3. Performance for Textual Entailment on SCITAIL train and test set.

Models	Dev	Test
Majority class	63.3	60.3
DecompAtt	75.4	72.3
ESIM [3]	70.5	70.6
Ngram	65.0	70.6
DGEM [7]	79.6	77.3
EDCMN	**77.9**	**78.2**

Table 4. Ablation analysis for Answer Selection On WikiQA test set.

Model structure	MAP	MRR
Full Model	0.740	0.752
(1) w/o Embed-cross fea	0.731	0.739
(2) w/o Word-level fea	0.725	0.733
(3) w/o Phrase-level fea	0.709	0.711
(4) w/o Sentence-level fea	0.724	0.735
(5) w/o Sub fea (fixed-size)	0.720	0.733
(6) w/o Mul fea (fixed-size)	0.722	0.732
(7) w/o Mat-fea (dyna-size)	0.681	0.695
(8) w/o Match	0.672	0.675

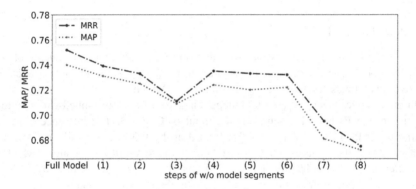

Fig. 2. Ablation Analysis about different components of model on WikiQA test set

5 Conclusion

In this paper, we propose an Encoder-Decoder Network with Cross-Match Mechanism, where the encoder layer is based on the "Siamese" network and decoder layer is based on the "matching-aggregation" network. The Cross-Match Mechanism which is the core innovation captures information of sentences at different scales including Sentence-level, Phrase-level, and Word-level. In addition, it explores sentence matching both on vectors with dynamic-size and fixed-size and it is more suitable for identifying complex relations between questions and the answers. In the experiments, we show that proposed model achieves state-of-the-art performance on the WikiQA dataset. In future work, we will incorporate external knowledge bases into our model to improve its performance. Furthermore, unlabeled data is much easier to obtain than labeled data. We will explore unsupervised methods for answer selection.

Acknowledgements. This research was partially supported by the Sichuan Science and Technology Program under Grant Nos. 2018GZ0182, 2018GZ0093 and 2018GZDZX0039.

References

1. Bachrach, Y., et al.: An attention mechanism for neural answer selection using a combined global and local view. In: Proceedings of International Conference on Tools with Artificial Intelligence, ICTAI, November 2017, pp. 425–432 (2018). https://doi.org/10.1109/ICTAI.2017.00072
2. Hochreiter, S.: Recurrent cascade-correlation and neural sequence chunking. LSTM, vol. 9, no. 8, pp. 1–32 (1997)
3. Chen, Q., et al.: Enhanced LSTM for Natural Language Inference, pp. 1657–1668 (2016)
4. Feng, M., et al.: Applying deep learning to answer selection: a study and an open task. In: 2015 Proceedings of IEEE Workshop on Automatic Speech Recognition and Understanding, ASRU 2015, pp. 813–820 (2016). https://doi.org/10.1109/ASRU.2015.7404872
5. He, H., Lin, J.: Pairwise word interaction modeling with deep neural networks for semantic similarity measurement, pp. 937–948 (2016). https://doi.org/10.18653/v1/n16-1108
6. Heilman, M., Smith, N.: Tree edit models for recognizing textual entailments, paraphrases, and answers to questions. In: Human Language Technologies: The 2010 Annual Conference of the North American Chapter of the Association for Computational Linguistics, pp. 1011–1019, June 2010
7. Khot, T., et al.: SCITAIL: a textual entailment dataset from science question answering. In: AAAI Conference on Artificial Intelligence (2018)
8. Lin, W.S., et al.: Utilizing different word representation methods for twitter data in adverse drug reactions extraction. In: 2015 Conference on Technologies and Applications of Artificial Intelligence, TAAI 2015, pp. 260–265 (2016). https://doi.org/10.1109/TAAI.2015.7407070
9. Lin, Z., et al.: A structured self-attentive sentence embedding, pp. 1–15 (2017)
10. Yang, Y., Yih, W.T., Meek, C.: WIKI QA : a challenge dataset for open-domain question answering, September 2015, pp. 2013–2018 (2018)
11. Rücklé, A., Gurevych, I.: Representation learning for answer selection with LSTM-based importance weighting. In: Proceedings of 12th International Conference on Computational Semantics (IWCS 2017) (2017, to appear)
12. dos Santos, C., et al.: Attentive pooling networks. Cv (2016)
13. Severyn, A.: Rank with CNN (2014). https://doi.org/10.1145/2766462.2767738
14. Shen, G., et al.: Inter-weighted alignment network for sentence pair modeling, pp. 1179–1189 (2018). https://doi.org/10.18653/v1/d17-1122
15. Shen, T., et al.: DiSAN: directional self-attention network for RNN/CNN-free language understanding (2017)
16. Shen, Y., et al.: A latent semantic model with convolutional-pooling structure for information retrieval. In: Proceedings of 23rd ACM International Conference on Conference on Information and Knowledge Management - CIKM 2014, pp. 101–110 (2014). https://doi.org/10.1145/2661829.2661935
17. Tan, M., et al.: LSTM-based deep learning models for non-factoid answer selection, vol. 1, pp. 1–11 (2015)
18. Tay, Y., et al.: Multi-cast attention networks for retrieval-based question answering and response prediction (2018)

19. Tran, N.K., Niedereée, C.: Multihop attention networks for question answer matching, pp. 325–334 (2018). https://doi.org/10.1145/3209978.3210009
20. Wang, B., et al.: Inner attention based recurrent neural networks for answer selection, pp. 1288–1297 (2016). https://doi.org/10.18653/v1/p16-1122
21. Wang, M., et al.: What is the jeopardy model? A quasi-synchronous grammar for QA. In: Proceedings of the 2007 Joint Conference on Empirical Methods in Natural Language Processing and Computational Natural Language Learning, pp. 22–32, June 2007
22. Wang, S., Jiang, J.: A compare-aggregate model for matching text sequences, pp. 1–11 (2016)
23. Wang, Z., et al.: Bilateral multi-perspective matching for natural language sentences. In: International Joint Conference on Artificial Intelligence, pp. 4144–4150 (2017)
24. Wang, Z., et al.: Sentence similarity learning by lexical decomposition and composition. Challenge 2 (2016)
25. Wang, Z., Ittycheriah, A.: FAQ-based question answering via word alignment, p. 1 (2015)
26. Yang, L., et al.: aNMM: ranking short answer texts with attention-based neural matching model (2018)
27. Yin, W., et al.: ABCNN: attention-based convolutional neural network for modeling sentence pairs (2015)
28. He, K., et al.: Deep residual learning for image recognition. In: Proceedings of the IEEE Conference on Computer Vision and Pattern Recognition, pp. 770–778, December 2016. https://doi.org/10.1109/CVPR.2016.90

BB-KBQA: BERT-Based Knowledge Base Question Answering

Aiting Liu$^{(\boxtimes)}$ (ID), Ziqi Huang (ID), Hengtong Lu (ID), Xiaojie Wang,
and Caixia Yuan

Beijing University of Posts and Telecommunications, Beijing, China
{aitingliu,huangziqi,luhengtong,xjwang,yuancx}@bupt.edu.cn

Abstract. Knowledge base question answering aims to answer natural language questions by querying external knowledge base, which has been widely applied to many real-world systems. Most existing methods are template-based or training BiLSTMs or CNNs on the task-specific dataset. However, the hand-crafted templates are time-consuming to design as well as highly formalist without generalization ability. At the same time, BiLSTMs and CNNs require large-scale training data which is unpractical in most cases. To solve these problems, we utilize the prevailing pre-trained BERT model which leverages prior linguistic knowledge to obtain deep contextualized representations. Experimental results demonstrate that our model can achieve the state-of-the-art performance on the NLPCC- ICCPOL 2016 KBQA dataset, with an 84.12% averaged F1 score(1.65% absolute improvement).

Keywords: Chinese knowledge base question answering · Entity linking · Predicate mapping · BERT

1 Introduction

Recently, open domain knowledge base question answering (KBQA) has emerged as large-scale knowledge bases develop rapidly, such as DBpedia, Freebase, Yago2 and NLPCC Chinese Knowledge Base [1,2]. The goal of knowledge base question answering is to generate a related answer given a natural language question, which is challenging since it requires a high level of semantic understanding of questions. The mainstream methods can be divided into two paradigms. One line of research is built on semantic parsing-based methods [3–5] and the other utilizes information extraction-based methods [6–8]. In more detail, the former first converts the natural language question into a structured representation, such as logical forms or SPARQL [9,10], then query the knowledge base to obtain the answer. The latter, which is information extraction-based, first retrieves a set of candidate triples and then extracts features to rank these candidates. In this paper, we focus on the semantic-parsing method since it is more popular and general.

© Springer Nature Switzerland AG 2019
M. Sun et al. (Eds.): CCL 2019, LNAI 11856, pp. 81–92, 2019.
https://doi.org/10.1007/978-3-030-32381-3_7

In semantic parsing-based methods, the basic framework of KBQA [11–17] consists of three modules. The first one is *entity linking*, which recognizes all entity mentions in a question (*mention detection*) and links each mention to an entity in KB (*entity disambiguation*). Normally, there are several candidate entities of a single mention, so entity disambiguation is needed. The second one is *predicate mapping*, which finds candidate predicates in KB for the question. The last one is *answer selection*, which ranks the candidate entity-predicate pairs, converts the top one into a query statement and queries the knowledge base to obtain the answer. For example, it first detects the mention "天堂鸟 || *Bird of Paradise* " in the question "天堂鸟是什么界的动物呀? || *Which kingdom does the animal Bird of Paradise belong to?*", then a candidate entity set {天堂鸟 (2001 年李幼斌主演电视剧) || *Bird of Paradise(Teleplay starring Li Youbin in 2001),* 天堂鸟 (迷你专辑) || *Bird of Paradise(ep),* 天堂鸟 (动物) || *Bird of Paradise(animal),* ...} and candidate predicate set {别名 || *Alias,* 中文学名 || *Chinese scientific name,* 界 || *Kingdom,* 门 || *Phylum,* 亚门 *Subphylum,* 纲 || *Class,* 亚属 || *Subgenus,* 种 || *Species*} are obtained from the knowledge base. Finally, it ranks the candidate entity-predicate pairs and selects the top one "天堂鸟 (动物)-界 || *Bird of Paradise(animal)-Kingdom*" to retrieve the factoid triple "< 天堂鸟 (动物),界,动物界 > || < *Bird of Paradise(animal), Kingdom, Kingdom Animalia>*" from the knowledge base, therefore the answer is "动物界 || *Kingdom Animalia*".

In previous studies of entity linking module, Xie et al. [11] regards mention detection as a sequence labeling task with CNN model. Lai et al. [12], Yang et al. [13] and Zhou et al. [14] find all possible candidate entities of a question according to a pre-constructed alias dictionary. To disambiguate the candidate entities, Lai et al. [12] proposes a template-based algorithm which requires considerable hand-crafted templates, Yang et al. [13] utilizes the GBDT model and Zhou et al. [14] adopts a language model. In terms of predicate mapping module, Wang et al. [15] proposes the CGRU model and Yang et al. [13] combines the NBSVM model with CNN to rank candidate entities. Lai et al. [12] measures the token-level similarity between the question and each candidate predicate through a variety of hand-crafted extraction rules and gets the correct predicate. Xie et al. [11] introduces the CNN-DSSM [18] and BiLSTM-DSSM [19] which are variants of the deep semantic matching model (DSSM) [20] to calculate the semantic similarity between the question and each candidate predicate. However, above methods have two drawbacks: On the one hand, although prior linguistic knowledge can be combined directly into hand-crafted templates, the design of templates is time consuming. Meanwhile, hand-crafted templates are often with large granularity which prone to cause exceptions, which damage the generalization ability of models. On the other hand, the performances of BiLSTMs and CNNs are heavily dependent on large scale of training data which is often not available in practice. Recent years, pre-training [21–24] on large-scale unsupervised corpus, which is easy to collect, has shown its advantages on mining prior linguistic knowledge automatically, it indicates a possible way to deal with above two problems.

This paper focuses on exploiting pre-trained language models to ease the problems described above. BERT [23] is effectively combined into the semantic parsing-based framework for KBQA. Two different combining models for different subtasks in KBQA are designed. A BERT-CRF (Conditional Random Field [25]) model is proposed for mention detection, while a BERT-Softmax model is proposed for entity disambiguation and predicate mapping. In the end, we build a **BERT-B**ased **KBQA** model, which achieves the state-of-the-art performance on the NLPCC-ICCPOL 2016 KBQA dataset with averaged F1 score of 84.12%.

Our contributions can be summarized as follows:

1. We propose the BERT-CRF model which integrates both advantages of BERT and CRF to train a efficient mention detection model. Furthermore, BB-KBQA model based on BERT-CRF and BERT-Softmax is proposed which leverages external knowledge and produces deep semantic representations of questions, entities and predicates.
2. Experimental results show that our method can achieve the state-of-the-art on NLPCC-ICCPOL 2016 KBQA dataset. Credit to the powerful feature extraction ability of BERT, our approach can produce more precise and related answer given the question.

2   BB-KBQA Model

As shown in Fig. 1, the KBQA framework consists of three modules: *entity linking* (including *mention detection* and *entity disambiguation*), *predicate mapping* and *answer selection*.

We propose a **BERT-B**ased **KBQA** model based on this framework, where BERT-CRF is adopted for mention detection, and BERT-Softmax is adopted for entity disambiguation and predicate mapping. In the following, we first elaborate on the BERT-based models we design and then introduce these models in KBQA modules.

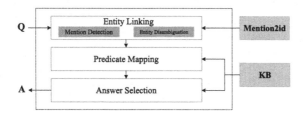

Fig. 1. KBQA framework.

2.1 Models

BERT. BERT is a multi-layer bidirectional Transformer [26] encoder. The input is a character-level token sequence, which is able to unambiguously represent either a single sentence or a pair of sentences separated with a special token [SEP]. For each token of the input sequence, the input representation is a sum of the corresponding token embedding, segment embedding and position embedding. The first token of every sequence is always the special classification symbol ([CLS]), and the final hidden state corresponding to this token can be used for classification tasks. BERT is pre-trained by two unsupervised prediction tasks: masked language model task and next sentence prediction task. After fine-tuning, the pre-trained BERT representations can be used in a wide range of natural language processing tasks. Readers can refer to [23] for more details.

BERT-Softmax. As shown in Fig. 2(a), following [23] fine-tuning procedure, the input sequence of BERT is $\mathbf{x} = \{x_1, \cdots, x_N\}$, and the final hidden state sequence is $\mathbf{H} = \{\mathbf{h}_1, \cdots, \mathbf{h}_N\}$,

$$\mathbf{H} = BERT(\mathbf{x}), \tag{1}$$

where $\mathbf{H} \in \mathbb{R}^{d \times N}$, $\mathbf{h}_i \in \mathbb{R}^d$ and $BERT(\cdot)$ denotes the network defined in [23]. Each hidden state $\mathbf{h}_i$ is followed by a softmax classification layer which outputs the label probability distribution $\mathbf{p}_i$,

$$\mathbf{p}_i = softmax(\mathbf{W}\mathbf{h}_i + \mathbf{b}), \tag{2}$$

here we view BERT-Softmax as a binary sequence classification task, where $\mathbf{W} \in \mathbb{R}^{2 \times d}$, $\mathbf{b} \in \mathbb{R}^2$, $\mathbf{p}_i = \begin{bmatrix} \mathbf{p}_i^{(0)} \\ \mathbf{p}_i^{(1)} \end{bmatrix} \in \mathbb{R}^2$.

For sequence classification task, we only use the final hidden state of the first token (special symbol [CLS]) for softmax classification, which is employed in mention detection and predicate mapping modules; for sequence labeling task, there is a classifier on each hidden state $\mathbf{h}_i$, which is used in mention detection module.

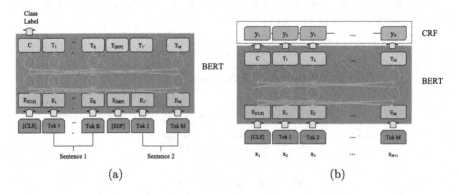

(a)   (b)

Fig. 2. BERT-Softmax model and BERT-CRF model.

BERT-CRF. The model structure is depicted in Fig. 2(b). In the sequence labeling task, the input sequence of the BERT is $\mathbf{x} = \{x_1, \cdots, x_N\}$, and the final hidden state sequence is defined the same as Eq. (1), which is further passed through a CRF [25] layer, then the final output is the predicted labels $\mathbf{Y}$ corresponding to each token,

$$\mathbf{Y} = CRF(\mathbf{WH} + \mathbf{b}), \tag{3}$$

where the label set is { "B" , "I" , "O" }, $W \in \mathbb{R}^{3 \times d}$, $b \in \mathbb{R}^3$, $Y = \{y_1, \ldots, y_N\}$, $y_i \in \{0, 1, 2\}$, $i = 1, \ldots, N$. By employing a CRF layer, we can use past and future labels to predict the current label [27], leading to a state transition matrix of CRF layer, which focuses on sentence level instead of individual positions. Generally speaking, it can obtain higher labeling accuracy with the help of the CRF layer [27].

2.2 Modules

This paper builds each module in Fig. 1 based on the above models.

Entity Linking. Entity linking includes mention detection and entity disambiguation, where the former extracts the mention in a question, and the latter links the mention to its corresponding entity in the knowledge base.

Mention Detection. We treat mention detection as a sequence labeling task, where the BIO format is applied for representing mention labels. We construct a BERT-CRF model with question Q as input sequence $[Q]^1$ to detect the mention m in the question Q,

$$m = BERT\_CRF([Q]). \tag{4}$$

Entity Disambiguation. The candidate entity set $E = \{e_1, \cdots, e_T\}$ is obtained by a mention2id library[2] using the mention m, T is the number of candidate entities. The entity disambiguation can be regarded as a binary sequence classification task. We concatenate the question Q and each candidate entity e_i as input sequence $[Q; e_i]^3$, and feed it into the BERT-Softmax model to output the classification probability distribution $\mathbf{p}_i^e$,

$$\mathbf{p}_i^e = BERT\_Softmax([Q; e_i]), \tag{5}$$

where $\mathbf{p}_i^e = \begin{bmatrix} \mathbf{p}_i^{e(0)} \\ \mathbf{p}_i^{e(1)} \end{bmatrix} \in \mathbb{R}^2$, $i = 1, \ldots, T$. The predicted probability of label "1" is considered as the score $\mathcal{S}^e$ of the candidate entity, $\mathcal{S}^e \in \mathbb{R}^T$,

$$\mathcal{S}^e = \begin{bmatrix} \mathbf{p}_1^{e(1)} & \cdots & \mathbf{p}_T^{e(1)} \end{bmatrix}. \tag{6}$$

---

[1] Insert special symbol [CLS] as the first token of Q. We omit [CLS] from the notation for brevity.

[2] mention2id library "nlpcc-iccpol-2016.kbqa.kb.mention2id" is introduced in [2], which maps the mention to all possible entities.

[3] Insert special symbol [CLS] as the first token of Q. Delimiter [SEP] are added between Q and e_i. We omit [CLS] and [SEP] from the notation for brevity. $[Q; r_{ij}]$ ditto.

Predicate Mapping. Following the entity linking module, we get the candidate predicate set $R_i = \{r_{i1}, \cdots, r_{iL}\}$ from the KB according to the candidate entity e_i, L is the number of candidate predicates. Predicate mapping module scores all candidate predicates according to the semantic similarity between the question and each candidate predicate. The question Q is concatenated with the candidate predicate r_{ij} to form an input sequence $[Q; r_{ij}]$. Similar to entity disambiguation, BERT-Softmax model is employed to produce the score $\mathcal{S}^r$ for candidate predicates,

$$\mathbf{p}_{ij}^r = BERT\_Softmax([Q; r_{ij}]), \tag{7}$$

$$\mathcal{S}^r = \left[\mathbf{p}_{ij}^{r\,(1)} \right]_{T \times L}, \tag{8}$$

where $\mathbf{p}_{ij}^r$ is the label probability distribution, $\mathbf{p}_{ij}^r = \begin{bmatrix} \mathbf{p}_{ij}^{r\,(0)} \\ \mathbf{p}_{ij}^{r\,(1)} \end{bmatrix} \in \mathbb{R}^2, i = 1, \ldots, T,$ $j = 1, \ldots, L, \mathcal{S}^p \in \mathbb{R}^{T \times L}$.

Answer Selection. In answer selection module, we calculate the weighted sum of candidate entity score $\mathcal{S}^e$ and candidate predicate score $\mathcal{S}^r$ as the final score of the candidate "entity-predicate" pair $\mathcal{S}$,

$$\mathcal{S} = \alpha \times \mathcal{S}^e + (1 - \alpha) \times \mathcal{S}^r, \tag{9}$$

where α is a hyper-parameter, $\mathcal{S} \in \mathbb{R}^{T \times L}$. We select the entity-predicate pair with the highest score and query the knowledge base through the query statement to get the answer.

3   Experiments

3.1   Datasets

The NLPCC-ICCPOL 2016 KBQA task [2] provides a training set with 14609 QA pairs, a test set with 9870 QA pairs, a Chinese knowledge base containing approximately 43M triples, and a mention2id library[4] that maps the mention to all possible entities. Since the mention detection, entity linking and predicate mapping modules require respective dataset, we create these three datasets in our own way. Specifically, we obtain the "entity-predicate" pair for the question via the golden answer. For mention detection task, we label the mention in the question manually. For entity disambiguation task, we collect all entities corresponding to the correct mention, and mark the correct entity as a positive example, other entities as negative examples. For predicate mapping dataset, we collect all predicates corresponding to the correct entity from the KB, and mark the correct predicate as a positive example, other predicates as negative examples.

---

[4] Chinese knowledge base "nlpcc-iccpol-2016.kbqa.kb" is introduced in [2].

The provided Chinese KB includes triples crawled from web. Each triple is in the form: <Subject, Predicate, Object>, where 'Subject' denotes a subject entity, 'Predicate' denotes a relation, and 'Object' denotes an object entity. There are about 43M triples in this knowledge base, in which about 6M subjects, 0.6M predicates and 16M objects are mentioned. On average, each subject entity corresponds to 7 triples, and each predicate corresponds to 73 triples. Some examples of triples are shown in Table 1.

Table 1. Triples in knowledge base.

Subject	Predicate	Object
北京 ‖*Beijing*	别名 ‖*Alias*	北京 ‖*Beijing*
北京 ‖*Beijing*	中文名 ‖*Chinese name*	北京市 ‖*Beijing City*
北京 ‖*Beijing*	外文名 ‖*Foreign name*	Municipality of Beijing
北京 ‖*Beijing*	所属地区 ‖*Region*	中国华北 ‖*Northern China*
. . .	. . .	. . .

3.2 Training Details

We use Chinese BERT-Base model[5] pre-trained on Chinese Wikipedia corpus using character level tokenization, which has 12 layers, 768 hidden states, 12 heads and 110M parameters. For fine-tuning, all hyper-parameters are tuned on the development set. The maximum sequence length is set to 60 according to our dataset, the batch size is set to 32. We use Adam [28] for optimization with $\beta_1 = 0.9$ and $\beta_2 = 0.999$. The dropout probability is 0.1. Typically, the initial learning rate is set to 1e-5 for BERT-CRF, 5e-5 for the BERT-Softmax, meanwhile a learning rate warmup strategy [23] is applied. The training epochs of BERT-CRF and BERT-Softmax are 30 and 3, respectively. Hyper-parameter α is set to 0.6 in answer selection module. For all baseline models, the word embedding is pre-trained by word2vec [29] using training set, and the embedding size is set to 300.

3.3 Compare with Baseline Models

We compare our model with the baseline model released in NLPCC-ICCPOL 2016 KBQA task [2], the state-of-the-art model [12] and some other baseline models [11,13–17]. Table 2 demonstrates the experimental results on the NLPCC-ICCPOL 2016 KBQA task. Our model outperforms all other methods. Compared with models using other sophisticated features and hand-craft rules (such as the use of part-of-speech features in the mention detection stage) [11,12], and models using simple LSTM, CNN [13–17], the BB-KBQA model we proposed achieves state-of-the-art result.

---

[5] https://github.com/google-research/bert.

Table 2. NLPCC-ICCPOL 2016 KBQA results (%)

Models	Averaged F1
Baseline model (C-DSSM)	52.47
Wang et al. [15]	79.14
Xie et al. [11]	79.57
Lei et al. [17]	80.97
Zhou et al. [14]	81.06
Yang et al. [13]	81.59
Xie et al. [16]	82.43
Lai et al. [12]	82.47
BB-KBQA	**84.12**

3.4 Module Analysis

Mention Detection. Table 3 summarizes the experimental results of our model and several baselines on mention detection task. BERT-Softmax is BERT model with a linear and softmax classification layer, BERT-BiLSTM-CRF combine BERT and BiLSTM-CRF model [27], BERT-CRF only adds a CRF layer based on BERT. Fine-tuned BERT-Softmax model has obvious improvement compared to traditional BiLSTM-CRF, where F1 score is relatively increased by 6.33%. BERT-BiLSTM-CRF is 0.29% higher than BERT-Softmax, and BERT-CRF which only employes a CRF layer get another 0.44% performance boost. The CRF layer can obtain the global optimal sequence labels instead of the local optimum, and the pre-trained BERT models the word order information and semantic information of the sequence. Adding a BiLSTM layer may disturb the valid information extracted by BERT.

Table 3. Mention detection results (%)

Models	F1
BiLSTM-CRF	90.28
BERT-Softmax	96.61
BERT-BiLSTM-CRF	96.90
BERT-CRF	**97.34**

Entity Disambiguation. It can be seen from Table 4 that the BERT-Softmax model outperforms all baseline models by approximately 2% on average, which shows that the fine-tuned BERT model can extract more comprehensive deep semantic information than other shallow neural network models, such as CNN and BiLSTM models.

Table 4. Entity disambiguation results (%)

Models	Accuracy@1	Accuracy@2	Accuracy@3
BiLSTM-DSSM [19]	85.89	88.50	90.81
Siamese BiLSTM [30]	87.85	92.58	94.59
Siamese CNN [31]	88.04	92.68	94.88
BERT-Softmax	**89.14**	**93.19**	**95.05**

Predicate Mapping. Table 5 demonstrates experimental results on predicate mapping task. The entity mention in the question may bring useful information as well as useless noise. Therefore, a set of comparative experiments are performed according to whether the entity mention in the question is replaced with a special token [ENT] (Siamese BiLSTM(2), Siamese CNN(2) and BERT-Softmax(2) represent models for such replacement operation). The experimental results show that this treatment is effective for the Siamese models, but does not work the same way in the BERT-Softmax model. While training the Siamese models, the entity mentions in the training set are sparse, resulting in insufficient training. However, the BERT model pre-trained with large-scale corpus covers a large amount of general knowledge, and the information of the mention in the question contributes to the predicate mapping task.

Table 5. Predicate mapping results (%)

Models	Accuracy@1	Accuracy@2	Accuracy@3
Siamese BiLSTM	92.54	96.74	98.12
Siamese BiLSTM(2)	93.74	97.46	98.38
Siamese CNN	86.47	93.80	96.16
Siamese CNN(2)	90.61	95.57	97.01
BERT-Softmax	**94.81**	**97.68**	**98.60**
BERT-Softmax(2)	94.66	97.63	98.41

3.5   Case Study

Table 6 gives some examples of our model and other baseline models. By modeling mention detection into a sequence labeling task instead of using hard matching method, our model can detect the mention even there are typos in the question. For example, the correct-written mention in the question "泡泡小兵中文版 的游戏目标是什么? || *What is the goal of the Chinese version of Bubble Soldier?*" is "跑跑小兵中文版 ||*the Chinese version of Run Soldier*". Since there is no "泡泡小兵中文版 ||*the Chinese version of Bubble Soldier*" in the mention2id library, the hard matching method fails to detect it while our model works. By using BERT-CRF, we can detect the correct mentions that baseline models do wrongly. For the question "我要拼是什么国家的啊？ || *Which country is Wo Yao Pin?*", the detection of BERT-CRF is "我要拼 || *Wo Yao Pin*" but the result of baseline models is "我 || *Wo*". Similarly, BERT-Softmax is able to get the right result in some questions that are incorrectly resolved in the baseline models.

We also randomly sample some examples where our model does not generate correct answers. We find that some errors are caused by the dataset itself, which mainly includes: (1) there are unclarified entities of the question. For example, the mention "东山村 || *Dongshan Village*" in the question "有人知道东山村的地理位置吗？ || *Does anyone know the location of Dongshan Village?*" has many corresponding entities in the knowledge base, such as "东山村 (云南省宜良县汤池镇东山村) || *Dongshan Village (Dongshan Village, Tangchi Town, ziliang County, Yunnan Province)*", "东山村 (北京市门头沟军庄镇东山村) || *Dongshan Village (Dongshan Village, Junzhuang Town, Mentougou, Beijing)*" and so on; (2) the question lacks an entity mention, like the question "我想知道官方语言是什么？ || *I want to know what the official language of the is?*".

Table 6. Experiment result examples.

Question	[12]	[11]	BB-KBQA	False analysis
泡泡小兵中文版的游戏目标是什么？ *What is the goal of the Chinese version of Bubble Soldier?*	×	√	√	Mentions are written-wrongly.
我要拼是什么国家的啊？ *Which country is Wo Yao Pin?*	×	×	√	Mention detection fails.
告诉我《兄弟》这本书是几开的书？ *Tell me the size of Brother.*	×	×	√	Entity Disambiguation fails.
沅水的流量有多少？ *How much flow does the Yuanshui River have?*	×	×	√	Predicate mapping fails.
我想知道官方语言是什么 *I want to know what the official language of the is.*	×	×	×	Data error.
有人知道东山村的地理位置吗？ *Does anyone know the location of Dongshan Village?*	×	×	×	Data error.

4   Conclusion

We propose a BERT-based knowledge base question answering model BB-KBQA. Compared to previous models, ours captures deep semantic information of questions, entities and predicates, which achieves a new state-of-the-art result of 84.12% on the NLPCC-ICCPOL 2016 KBQA dataset. In the future we plan to evaluate our model on other datasets and attempt to jointly model entity linking and predicate mapping to further improve the performance.

References

1. Bollacker, K., Evans, C., Paritosh, P., Sturge, T., Taylor, J.: Freebase: a collaboratively created graph database for structuring human knowledge. In: Proceedings of the 2008 ACM SIGMOD International Conference on Management of Data, pp. 1247–1250. ACM (2008)
2. Duan, N.: Overview of the NLPCC-ICCPOL 2016 shared task: open domain Chinese question answering. In: Lin, C.-Y., Xue, N., Zhao, D., Huang, X., Feng, Y. (eds.) ICCPOL/NLPCC-2016. LNCS (LNAI), vol. 10102, pp. 942–948. Springer, Cham (2016). https://doi.org/10.1007/978-3-319-50496-4_89
3. Wang, Y., Berant, J., Liang, P.: Building a semantic parser overnight. In: Proceedings of the 53rd Annual Meeting of the Association for Computational Linguistics and the 7th International Joint Conference on Natural Language Processing (Volume 1: Long Papers), vol. 1, pp. 1332–1342 (2015)
4. Pasupat, P., Liang, P.: Compositional semantic parsing on semi-structured tables. In: Proceedings of the 53rd Annual Meeting of the Association for Computational Linguistics and the 7th International Joint Conference on Natural Language Processing (Volume 1: Long Papers), vol. 1, pp. 1470–1480 (2015)
5. Yang, M.-C., Duan, N., Zhou, M., Rim, H.-C.: Joint relational embeddings for knowledge-based question answering. In: Proceedings of the 2014 Conference on Empirical Methods in Natural Language Processing (EMNLP), pp. 645–650 (2014)
6. Bordes, A., Usunier, N., Chopra, S., Weston, J.: Large-scale simple question answering with memory networks. arXiv preprint arXiv:1506.02075 (2015)
7. Dong, L., Wei, F., Zhou, M., Xu, K.: Question answering over freebase with multi-column convolutional neural networks. In: Proceedings of the 53rd Annual Meeting of the Association for Computational Linguistics and the 7th International Joint Conference on Natural Language Processing (Volume 1: Long Papers), vol. 1, pp. 260–269 (2015)
8. Yih, W., Chang, M.-W., He, X., Gao, J.: Semantic parsing via staged query graph generation: question answering with knowledge base. In: Proceedings of the 53rd Annual Meeting of the Association for Computational Linguistics and the 7th International Joint Conference on Natural Language Processing (Volume 1: Long Papers), vol. 1, pp. 1321–1331 (2015)
9. Berant, J., Chou, A., Frostig, R., Liang, P.: Semantic parsing on freebase from question-answer pairs. In: Proceedings of the 2013 Conference on Empirical Methods in Natural Language Processing, pp. 1533–1544 (2013)
10. Berant, J., Liang, P.: Semantic parsing via paraphrasing. In: Proceedings of the 52nd Annual Meeting of the Association for Computational Linguistics (Volume 1: Long Papers), vol. 1, pp. 1415–1425 (2014)
11. Xie, Z., Zeng, Z., Zhou, G., He, T.: Knowledge base question answering based on deep learning models. In: Lin, C.-Y., Xue, N., Zhao, D., Huang, X., Feng, Y. (eds.) ICCPOL/NLPCC-2016. LNCS (LNAI), vol. 10102, pp. 300–311. Springer, Cham (2016). https://doi.org/10.1007/978-3-319-50496-4_25
12. Lai, Y., Lin, Y., Chen, J., Feng, Y., Zhao, D.: Open domain question answering system based on knowledge base. In: Lin, C.-Y., Xue, N., Zhao, D., Huang, X., Feng, Y. (eds.) ICCPOL/NLPCC-2016. LNCS (LNAI), vol. 10102, pp. 722–733. Springer, Cham (2016). https://doi.org/10.1007/978-3-319-50496-4_65
13. Yang, F., Gan, L., Li, A., Huang, D., Chou, X., Liu, H.: Combining deep learning with information retrieval for question answering. In: Lin, C.-Y., Xue, N., Zhao, D., Huang, X., Feng, Y. (eds.) ICCPOL/NLPCC-2016. LNCS (LNAI), vol. 10102, pp. 917–925. Springer, Cham (2016). https://doi.org/10.1007/978-3-319-50496-4_86

14. Zhou, B., Sun, C., Lin, L., Liu, B.: LSTM based question answering for large scale knowledge base. Beijing Da Xue Xue Bao **54**(2), 286–292 (2018)
15. Wang, L., Zhang, Y., Liu, T.: A deep learning approach for question answering over knowledge base. In: Lin, C.-Y., Xue, N., Zhao, D., Huang, X., Feng, Y. (eds.) ICCPOL/NLPCC-2016. LNCS (LNAI), vol. 10102, pp. 885–892. Springer, Cham (2016). https://doi.org/10.1007/978-3-319-50496-4_82
16. Xie, Z., Zeng, Z., Zhou, G., Wang, W.: Topic enhanced deep structured semantic models for knowledge base question answering. Sci. China Inf. Sci. **60**(11), 110103 (2017)
17. Lei, K., Deng, Y., Zhang, B., Shen, Y.: Open domain question answering with character-level deep learning models. In: 2017 10th International Symposium on Computational Intelligence and Design (ISCID), vol. 2, pp. 30–33. IEEE (2017)
18. Shen, Y., He, X., Gao, J., Deng, L., Mesnil, G.: A latent semantic model with convolutional-pooling structure for information retrieval. In: Proceedings of the 23rd ACM International Conference on Conference on Information and Knowledge Management, pp. 101–110. ACM (2014)
19. Palangi, H., et al.: Semantic modelling with long-short-term memory for information retrieval. arXiv preprint arXiv:1412.6629 (2014)
20. Huang, P.-S., He, X., Gao, J., Deng, L., Acero, A., Heck, L.: Learning deep structured semantic models for web search using clickthrough data. In: Proceedings of the 22nd ACM International Conference on Information & Knowledge Management, pp. 2333–2338. ACM (2013)
21. Peters, M., et al.: Deep contextualized word representations. In: Proceedings of the 2018 Conference of the North American Chapter of the Association for Computational Linguistics: Human Language Technologies, Volume 1 (Long Papers), pp. 2227–2237 (2018)
22. Radford, A., Narasimhan, K., Salimans, T., Sutskever, I.: Improving language understanding by generative pre-training (2018). https://s3-us-west-2.amazonaws.com/openai-assets/research-covers/languageunsupervised/languageunderstanding paper.pdf
23. Devlin, J., Chang, M.-W., Lee, K., Toutanova, K.: BERT: pre-training of deep bidirectional transformers for language understanding. arXiv preprint arXiv:1810.04805 (2018)
24. Radford, A., Wu, J., Child, R., Luan, D., Amodei, D., Sutskever, I.: Language models are unsupervised multitask learners
25. Lafferty, J., McCallum, A., Pereira, F.C.N.: Conditional random fields: probabilistic models for segmenting and labeling sequence data (2001)
26. Vaswani, A., et al.: Attention is all you need. In: Advances in Neural Information Processing Systems, pp. 5998–6008 (2017)
27. Huang, Z., Xu, W., Yu, K.: Bidirectional LSTM-CRF models for sequence tagging. arXiv preprint arXiv:1508.01991 (2015)
28. Kingma, D.P., Ba, J.: Adam: a method for stochastic optimization. arXiv preprint arXiv:1412.6980 (2014)
29. Mikolov, T., Chen, K., Corrado, G., Dean, J.: Efficient estimation of word representations in vector space. arXiv preprint arXiv:1301.3781 (2013)
30. Mueller, J., Thyagarajan, A.: Siamese recurrent architectures for learning sentence similarity. In: Thirtieth AAAI Conference on Artificial Intelligence (2016)
31. Kim, Y.: Convolutional neural networks for sentence classification. arXiv preprint arXiv:1408.5882 (2014)

Reconstructed Option Rereading Network for Opinion Questions Reading Comprehension

Delai Qiu[1], Liang Bao[1], Zhixing Tian[2,3], Yuanzhe Zhang[3], Kang Liu[2,3(✉)],
Jun Zhao[2,3], and Xiangwen Liao[1]

[1] College of Mathematics and Computer Science, Fuzhou University, Fuzhou, China
noneqdl@gmail.com, caldreaming821618@gmail.com
[2] University of Chinese Academy of Sciences, Beijing 100049, China
[3] National Laboratory of Pattern Recognition, Institute of Automation,
Chinese Academy of Sciences, Beijing 100190, China
{zhixing.tian,yzzhang,kliu,jzhao}@nlpr.ia.ac.cn

Abstract. Multiple-choice reading comprehension task has seen a recent surge of popularity, aiming at choosing the correct option from candidate options for the question referring to a related passage. Previous work focuses on factoid-based questions but ignore opinion-based questions. Options of opinion-based questions are usually sentiment phrases, such as "Good" or "Bad". It causes that previous work fail to model the interactive information among passage, question and options, because their approaches are based on the premise that options contain rich semantic information. To this end, we propose a **R**econstructed **O**ption **R**ereading **N**etwork (**RORN**) to tackle it. We first reconstruct the options based on question. Then, the model utilize the reconstructed options to generate the representation of options. Finally, we fed into a max-pooling layer to obtain the ranking score for each opinion. Experiments show that our proposed achieve state-of-art performance on the Chinese opinion questions machine reading comprehension datasets in AI challenger competition.

Keywords: Machine Reading Comprehension · Opinion analysis

1 Introduction

Multiple-choice reading comprehension (MCRC) is a major form of machine reading comprehension (MRC) task, which requires a system to read a given passage and a question for choosing the correct option from the candidate options. Questions of MCRC task are generally divided into factoid-based questions and opinion-based questions. Figure 1 shows two different types of question. As we can see, the options of factoid-based question usually have abundant context information, but the options of opinion-based question usually select a opinion with some short sentiment phrases such as "Yes" and "No". AI challenger 2018[1]

---

[1] https://challenger.ai/competition/oqmrc2018.

M. Sun et al. (Eds.): CCL 2019, LNAI 11856, pp. 93–104, 2019.
https://doi.org/10.1007/978-3-030-32381-3_8

Passage: ... There is no problem if the intake of vitamin no more than the recommended amount ...
Question: Can we take the vitamin C every day? A. Yes   B. No   C. Indeterminacy
Answer: A
Passage: ...... In 1993, New York State ordered stores to charge money on beverage containers. Within a year, consumers had returned millions of aluminum cans and glass and plastic bottles. Plenty of companies were eager to accept the alu-minum and glass as raw material for new products, but because few could figure out ...
Question: What regulation was issued by New York State concerning beverage containers? A. A fee should be charged on used containers for recycling. B. Throwaways should be collected by the state for recycling. C. Consumers had to pay for beverage containers and could get their money back on returning them.
Answer: C

Fig. 1. The above is a factoid-based question, and other is an opinion-based question.

defines a new sub-task called Opinions Question Machine Reading Comprehension, which needs to choose a opinion option from candidates.

Many previous studies have introduced neural-based models on multiple-choice reading comprehension [9, 12, 16, 18, 21, 23, 24], which typically have following pipelines. Firstly, they encode the passage, question and candidate options to generate the contextual representation of them respectively. Secondly, attention mechanisms are employed to acquire the interaction representation among them. Further more, they prove that consider option correlations contribute to the semantic representation. Finally, the final output module computes scores of options based on score function to generate the final predictions.

Though previous work achieve promising results in recent years, they cannot handle the opinion-based questions. When reading the passage, understanding the semantic information of options is a common strategy for human beings, which inspires most of existing models. But the strategy is ineffective in opinion-based questions, because options of opinion-based questions always do not contain the context information. Taking Fig. 1 as a example, we are able to gain more context information from factoid-based options but we cannot access more from candidate options such as "YES" and "NO".

In these paper, we present a novel model to trackle opinion-based questions in multiple choice reading comprehension task. Firstly, Our model introduces a simple but effective method to reconstruct opinion options to acquire the semantic information of options. Then, we employ BERT [2] as our encoder to obtain the context representation of reconstructed options, questions and passages. Then we apply the co-attention mechanism to fuse the information between each options and passage, the rereaded option representation for each option is computed with self-attention mechanism. Finally, we utilize the max pooling layer to make the final prediction.

We conduct experiments on Opinions Question Machine Reading Comprehension dataset in AI challenger 2018. Our experiments show that the validity of option reconstruction and option rereading with passage-aware information. Our contributions can be summarized as:

(1) We reconstruct options of the opinion-based questions, which supplement more semantic information of options.
(2) Attention mechanisms is employed to generate more subtle context-aware representation of the options.
(3) Experiments demonstrate that the model achieve state-of-the-art on Opinion Questions Machine Reading Comprehension datasets.

2 Related Work

Multiple-choice reading comprehension (MCRC) is a major form of machine reading comprehension (MRC) task, aiming to selecting the correct answer from candidate options given a question and a passage. There are some large-scale datasets for this task, such as MCTest [10] and RACE [4]. Differing from extractive machine reading comprehension datasets such as SQuAD [8]and NewsQA [15], the correct answer for most questions in MCRC may not directly appear in the original passage.

With the rapid development of deep learning, various neural networks have been proposed for MCRC in recent years [1,3,5,9,12,14,16,18,22–24]. The Stanford AR [1] and GA Reader [3] variants are used to encode question and passage independent of options, ignoring their correlations. Trischler et al. [16] incorporates hierarchy to compare passage, questions and candidate options. The model [24] observes that leveraging candidate options to boost evidence gathering from the passage play a vital role in this task. So go further, the DCMN [23] model the relationship among passage, question and options bidirectionally, and the OCN [9] incorporate the correlation of options to identify more sublte correlations between options to help reasoning. Their approach is based on the premise that options contain rich semantic information.

Recently, the pre-trained language models such as GPT [7], ELMo [6] and BERT [2] have achieved huge success on various nature language processing datasets, including SQuAD [8]. More and more models treat them as a strong encoder to generate contextual representation or even simple make a finetune on them [9,23].

Almost all the models consider factoid-based questions, in which candidate options can boost the performance of these models. In our RORN model, we focus on opinion-based questions and utilize the strategy that reread options with the information of passage to answer question.

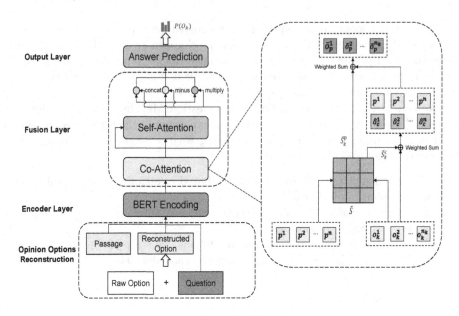

Fig. 2. Framework of our RORN model.

3 Model

The architecture of RORN is show in Fig. 2. It contains four modules: (1) Opinion options reconstruction module, which reconstructs the opinion options based on question; (2) Encoder layer module, which extracts features with BERT for passage, question and option respectively; (3) Fusion layer module, which acquires the representation of context-aware options; (4) Output layer module, which is employed to generate the final answer.

Then, we formally define MCRC task. Thare is a passage $P = \{p_1, p_2, ..., p_n\}$ with n tokens, a question $Q = \{q_1, q_2, ..., q_m\}$ with m tokens and a set of options $O = \{O_1, O_2, ..., O_k\}$ with k options, where each option is $O_k = \{o_{k_1}, o_{k_2}, ...o_{k_{n_o}}\}$ is a option with n_o tokens. Our model aims to compute a probability for each option and take the one with higher probability as the prediction answer.

3.1 Opinion Options Reconstruction

In this module, we reconstruct options according to the question to enhance the contextual information of options. For different type of question, we use templates to rewrite them.

We first divide question to three types, which contain **Normal Question**, **Question with different opinions** and **Question with comparison**. For these types of questions, we do respectively:

Normal Question: This type of question always contain only one view of a specific object, like "**在上海读国际高中好吗**" (How about going to international high school in Shanghai?). We first remove question words such as "吗" and "么", then each opinion option replace the raw opinion word in the question to generate reconstructed options respectively.

Question with Different Opinions: This type of question includes two opposite opinions, like "早上空腹吃芝士威化饼会不会发胖" (Do you get fat if you eat cheese wafers on an empty stomach in the morning?), we transform the opposite opinions such as "会不会" to single opinion "会" or "不会".

Question with Comparison: This type of question compare two entities in the same view, like "学数控技术好还是修车好" (Is it better to learn NC technology or to repair cars in college?), we split it to two options with only one entity.

If the question cannot be overwritten by templates, we remove question words of question and then cat it with each raw option.

There are some examples shown in the Table 1.

Table 1. Some examples of reconstructed options

Question	Raw Option	Reconstructed Option
在上海读国际高中好吗? (How about going to international high school in Shanghai?)	好(Good)	在上海读国际高中好。(It is good at international high school in Shanghai.)
	不好(Bad)	在上海读国际高中不好。(It is bad at international high school in Shanghai.)
早上空腹吃芝士威化饼会不会发胖? (Do you get fat if you eat cheese wafers on an empty stomach in the morning?)	会(Yes)	早上空腹吃芝士威化饼会发胖 (You will get fat if you eat cheese wafers on an empty stomach in the morning.)
	不会(No)	早上空腹吃芝士威化饼不会发胖 (You won't get fat if you eat cheese wafers on an empty stomach in the morning.)
大专学数控技术好还是修车好? (Is it better to learn NC technology or to repair cars in college?)	数控技术 (NC technology)	大专学数控技术好。(It is better to learn NC technology in college.)
	修车(Repair cars)	大专学修车好。(It is better to learn to repair cars in college.)

3.2 Encoder Layer

We encode the tokens with BERT [2]. BERT has become one of the most successful natural language representation models in various NLP tasks. BERT's model architecture is a multi-layer bidirectional Transfomer [17] encoder, which is pretrained on large-scale corpus. We use BERT as an encoder. It takes as input passage P, question Q and each option O_k, then computes the context-aware representation for each token.

Specifically, given passage $P = \{p_i\}_{i=1}^m$, question $Q = \{q_j\}_{j=1}^n$, and the kth option $O_k = \{o_j\}_{j=1}^{n_k}$, we pack them as a sequence of length $m + n + n_k + 4$ as follows:

$$S = [\langle CLS \rangle, P, \langle SEP \rangle, Q, \langle SEP \rangle, O_k, \langle SEP \rangle] \tag{1}$$

where $\langle CLS \rangle$ is a specific classifier token and $\langle SEP \rangle$ is a sentence separator which are defined in BERT.

Then the sequence is fed to BERT to generate the context-aware representation for each token in sequence. The output vectors of BERT are denoted as:

$$[\boldsymbol{P}; \boldsymbol{Q}; \boldsymbol{O}_k] = \text{BERT}(S) \tag{2}$$

where $\boldsymbol{P} \in \mathbb{R}^{d \times n}$, $\boldsymbol{Q} \in \mathbb{R}^{d \times m}$, $\boldsymbol{O}_k \in \mathbb{R}^{d \times n_k}$, and $\text{BERT}(\cdot)$ denotes the network defined in [2].

3.3   Fusion Layer

This module aims to generate the passage-aware representation of each option and reread each option. We utilize the co-attention [20] mechanism to capture the context information of passage to option. Then the self-attention [19] mechanism apply to understand each option deeply.

First, we define our attention weight function. Given input matrices $\boldsymbol{U} = \{u_i\}_{i=1}^N \in \mathbb{R}^{d \times N}$ and $\boldsymbol{V} = \{v_j\}_{i=1}^M \in \mathbb{R}^{d \times M}$, We compute the similarity matrix $S \in \mathbb{R}^{N \times M}$, which contains a similarity score s_{ij} for each pair (u_i, v_j):

$$s_{ij} = \boldsymbol{v}^T [\boldsymbol{u}_i; \boldsymbol{v}_j; \boldsymbol{u}_i \circ \boldsymbol{v}_j] \tag{3}$$

where $\circ$ denotes the element-wise multiplication operation and $[\cdot; \cdot]$ denotes column-wise concatenation, And then the attention weight function $\text{Att}(\cdot)$ is defined as:

$$\bar{\boldsymbol{S}} = \text{Att}(\boldsymbol{U}, \boldsymbol{V}) = \left[\frac{\exp(s_{ij})}{\sum_i \exp(s_{ij})} \right]_{i,j} \tag{4}$$

and $\bar{\boldsymbol{S}} \in \mathbb{R}^{N \times M}$ is the attention weight matrix.

For each option O_k, the co-attention is performed as:

$$\bar{\boldsymbol{S}}_k = \text{Att}(\boldsymbol{O}_k, \boldsymbol{P}) \tag{5}$$

$$\bar{\boldsymbol{S}}_k^p = \text{Att}(\boldsymbol{P}, \boldsymbol{O}_k) \tag{6}$$

$$\hat{\boldsymbol{O}}_k^p = [\boldsymbol{P}; \boldsymbol{O}_k \bar{\boldsymbol{S}}_k] \bar{\boldsymbol{S}}_k^p \tag{7}$$

$$\bar{\boldsymbol{O}}_k^p = \text{ReLU}(\boldsymbol{W}_p \hat{\boldsymbol{O}}_k^p + \boldsymbol{b}_p) \tag{8}$$

where $\bar{\boldsymbol{O}}_k^p \in \mathbb{R}^{d \times n_k}$, $\boldsymbol{W}_p \in \mathbb{R}^{d \times 2d}$ and $\boldsymbol{b} \in \mathbb{R}^d$ are the trainable parameters.

Then, mimicking humans, the options will be reread with passage via self-attention mechanism.

$$\bar{\boldsymbol{O}}_k^s = \text{Att}(\bar{\boldsymbol{O}}_k^p, \bar{\boldsymbol{O}}_k^p) \tag{9}$$

$$\bar{\boldsymbol{O}}_k^f = [\bar{\boldsymbol{O}}_k^p; \bar{\boldsymbol{O}}_k^s; \bar{\boldsymbol{O}}_k^p - \boldsymbol{O}_k^s; \bar{\boldsymbol{O}}_k^p \circ \boldsymbol{O}_k^s;] \tag{10}$$

$$O_k^f = \text{ReLU}(\boldsymbol{W}_f \boldsymbol{O}_k^f + \boldsymbol{b}_f) \tag{11}$$

where $\boldsymbol{W}_f \in \mathbb{R}^{d \times 4d}$ and $\boldsymbol{b}_f \in \mathbb{R}^d$ is the trainable parameter and $\boldsymbol{O}_k^f \in \mathbb{R}^{d \times n_k}$ is the final representation of the kth option.

3.4   Output Layer

To aggregate the final representation for each condidate option, a row-wise max pooling layer is employed to $\bar{\boldsymbol{O}}_k^f$:

$$\bar{\boldsymbol{O}}_k = \text{maxpooling}(\bar{\boldsymbol{O}}_k^f) \tag{12}$$

where $\bar{\boldsymbol{O}}_k \in \mathbb{R}^d$.

And then the score s_k of option O_k to be the correct answer is computed as:

$$s_k = \text{MLP}(\bar{\boldsymbol{O}}_k) \tag{13}$$

where MLP is a 2-layer full connect feed-forward network.

The probability $P(O_k|Q, P)$ of option Q_k to be the correct answer is computed as:

$$P(k|Q, P, O) = \frac{exp(s_k)}{\sum_i exp(s_i)} \tag{14}$$

And our loss function is computed as followed:

$$L(\theta) = -\frac{1}{N} \sum_i log(P(\hat{k}_i|Q_i, P_i, O_i)) \tag{15}$$

where θ denotes all trainable parameters, N is the training example number, and $\hat{k}_i$ is the ground truth for the ith example.

4   Experiments

4.1   Experimental Settings

Dataset
We conduct experiments on the Opinion Questions Machine Reading Comprehension dataset in the AIChallenger competition[2], in which questions are option-based questions. There are 270,000 and 30,000 examples in the training set and development set respectively. We divide the original development set into two parts evenly, one as a split development set for tuning model and the other one as a split test set, which contains 15,000 examples respectively.

---

[2] The dataset can be downloaded in https://challenger.ai/competition/oqmrc2018.

Implementation Details

Our model is implemented with pytorch[3], and uses the framework[4] for BERT model. We use pre-trained BERT on chinese corpus[5] to initialize our encoder. We use Adam optimizer and the learning rate uses the linear schedule to decrease from 3×10^{-5} to 0. Passages, questions and options are trimmed to 300, 30 and 30 tokens respectively. In this work, other hyper-parameter are shown in Table 2.

Table 2. Hyper-parameter of our model

Paramete name	Value
Train epochs	5
Batch size	12
Hidden units	786
Learning rate	0.00003
Dropout	0.8
Max sequence length	384

4.2 Baselines

We choose several baselines:

(1) **MwAN** [13] is a baseline for modeling sentence pair. It proposes the multiway attention network which employ multiple attention function. It is provided as a baseline by the official[6].

(2) **BiDAF** [11] is a strong baseline for MRC tasks. It is a typical neural-based MRC model which utilizes bi-directional attention to obtain query-aware context representation. We compress the representation with max-pooling layer, then feed it into a 2-layer full connect feed-forward network for classification.

(3) **RNET**[7] [19] is one of the top MRC models. It introduces a self-matching attention mechanism to refine the representation by matching the passage against itself. The model is designed for SQuAD-style datasets. So we replace its output layer with a 2-layer full connect feed-forward network.

(4) **BERT** [2] is a powerful pre-trained language model based on Transfomer [17]. We finetune the model with a linear layer on top of the pooled output of BERT.

---

[3] https://github.com/pytorch/pytorch.

[4] https://github.com/huggingface/pytorch-pretrained-BERT.

[5] https://storage.googleapis.com/bert_models/2018_11_03/chinese_L-12_H-768_A-12. zip.

[6] https://github.com/AIChallenger/AI_Challenger_2018.

[7] https://github.com/HKUST-KnowComp/R-Net.

Table 3. The results of different models.

Model	Dev(%acc)	Test(%acc)
MwAN [13]	69.98	69.16
BiDAF [11]	70.48	69.32
RNET [19]	72.84	72.42
BERT finetune [2]	74.18	73.75
RORN (ours)	**79.07**	**78.33**

4.3 Experimental Results

Table 3 shows the results of our RORN model achieve better than other models, which is 4.58% higher in value than BERT finetune model. Our single RORN model achieves 78.33% in term of accuracy.

The modified RNET model is our original baseline for the competition, which can be at the top of the competition leaderboard with complicated data preprocessing. Our RORN model has huge improvement than RNET model, which demonstrates that our model can effectively handle opinion-based questions.

4.4 The Effectiveness of Option Reconstruction

To study the effectiveness of opinion-based option reconstruction, we conduct experiments on the dataset. Table 4 presents their comparison results. We can observe that the RORN model without option reconstruction shows performance drop with 1.81%. The results demonstrate that our reconstruction module can effectively obtain the context of opinion options to improve performance.

4.5 Attention Mechanisms Ablation

In this section, we conduct ablation stduy on attention mechanisms to examine the effectiveness of each attention mechanism. The experimental results are listed in Table 5.

From the best model, if we remove the co-attention mechanism, the accuracy drops by 1.16% on the test set, and if the self-attention mechanism is removed, the accuracy drops by appropriately 1.12% on the test set. The results suggests that rereading the options with passage has more important to guide the model.

When we remove all attention mechanisms, the performance of the model drops 2.25%, which demonstrate that attention mechanisms is indispensable for our model.

4.6 Error Analysis

Based on the analysis of misclassified our instances, we can find some main reasons for misclassification as follows:

Table 4. Effectiveness of option reconstruction.

Model	Dev(%acc)	Test(%acc)
RORN w/o reconstruction	77.25	76.52
RORN	79.07	78.33

Table 5. Influence of different attention mechanisms.

Model	Dev(%acc)	Test(%acc)
RORN w/o attention	76.81	76.08
RORN w/o co-attention	77.24	77.17
RORN w/o self-attention	77.82	77.21
RORN	79.07	78.33

Table 6. Error instances

Instances
柠檬水的正确泡法用凉水还是热水？ (Cold or hot water is the correct way to soak lemonade?) 泡柠檬的水温一般在60到70℃比较合适。 (The water temperature to make soak lemonade is between 60 and 70 degrees centigrade.)
小孩头皮撞破皮了需要剃光头吗? (Does a child need to shave his head when his scalp breaks?) 宝宝头皮撞破最好是别沾水,会引起伤口发炎的。 (Baby scalp breakage is best not to touch water, it will cause wound inflammation.)

(1) Some example need external knowledge or complicated reasoning to infer answer. Taking the fisrt instance in Table 6 as a example, we need to know common sense like "Water of 60 to 70 degrees is hot water." to identify the right answer.

(2) There are some questions whose answers are uncertain according to the passage. The second instance in Table 6 shows a similar situation. We should introduce corresponding solutions to distinguish these questions which can not answer with context.

(3) Some manually annotated example are ambiguous or wrong, which mislead our model to predict wrong answer.

5   Conclusion

In this paper, we propose RORN model for opinion questions reading comprehension task. We use simple but effective method to reconstruct the opinion-based options, which can obtain the context information of options. Then the lastest breakthrough, BERT, is treated as our power encoder in our model. Mimicking humans, two attention mechanisms are employed to fuse semantic information among the passage, question and options. The experimental results demonstrate that our option reconstruction can boost our performance and two type of attention mechanism can influence the context-level fusion.

Acknowledgement. This work was supported by the National Natural Science Foundation of China (No. 61772135, No. U1605251, No.61533018), the Natural Key R&D Program of China (No. 2018YFC0830101). This work was also supported by the Open Project of Key Laboratory of Network Data Science & Technology of Chinese Academy of Sciences (No. CASNDST201708 and No. CASNDST201606), the Open Project of National Laboratory of Pattern Recognition at the Institute of Automation of the Chinese Academy of Sciences (201900041).

References

1. Chen, D., Bolton, J., Manning, C.D.: A thorough examination of the CNN/daily mail reading comprehension task. In: Proceedings of the 54th Annual Meeting of the Association for Computational Linguistics (Volume 1: Long Papers), vol. 1, pp. 2358–2367 (2016)
2. Devlin, J., Chang, M., Lee, K., Toutanova, K.: BERT: pre-training of deep bidirectional transformers for language understanding. CoRR abs/1810.04805 (2018). http://arxiv.org/abs/1810.04805
3. Dhingra, B., Liu, H., Yang, Z., Cohen, W., Salakhutdinov, R.: Gated-attention readers for text comprehension. In: Proceedings of the 55th Annual Meeting of the Association for Computational Linguistics (Volume 1: Long Papers), pp. 1832–1846 (2017)
4. Lai, G., Xie, Q., Liu, H., Yang, Y., Hovy, E.: RACE: large-scale reading comprehension dataset from examinations. In: Proceedings of the 2017 Conference on Empirical Methods in Natural Language Processing, pp. 785–794 (2017)
5. Parikh, S., Sai, A.B., Nema, P., Khapra, M.M.: ElimiNet: a model for eliminating options for reading comprehension with multiple choice questions. In: Proceedings of the 27th International Joint Conference on Artificial Intelligence, pp. 4272–4278. AAAI Press (2018)
6. Peters, M.E., et al.: Deep contextualized word representations. arXiv preprint arXiv:1802.05365 (2018)
7. Radford, A., Narasimhan, K., Salimans, T., Sutskever, I.: Improving language understanding by generative pre-training (2018). https://s3-us-west-2.amazonaws. com/openai-assets/research-covers/languageunsupervised/languageunderstanding paper.pdf
8. Rajpurkar, P., Zhang, J., Lopyrev, K., Liang, P.: SQuAD: 100,000+ questions for machine comprehension of text. In: Proceedings of the 2016 Conference on Empirical Methods in Natural Language Processing, pp. 2383–2392 (2016)
9. Ran, Q., Li, P., Hu, W., Zhou, J.: Option comparison network for multiple-choice reading comprehension. arXiv preprint arXiv:1903.03033 (2019)
10. Richardson, M., Burges, C.J., Renshaw, E.: MCTest: a challenge dataset for the open-domain machine comprehension of text. In: Proceedings of the 2013 Conference on Empirical Methods in Natural Language Processing, pp. 193–203 (2013)
11. Seo, M., Kembhavi, A., Farhadi, A., Hajishirzi, H.: Bidirectional attention flow for machine comprehension. arXiv preprint arXiv:1611.01603 (2016)

12. Sun, K., Yu, D., Yu, D., Cardie, C.: Improving machine reading comprehension with general reading strategies. arXiv preprint arXiv:1810.13441 (2018)
13. Tan, C., Wei, F., Wang, W., Lv, W., Zhou, M.: Multiway attention networks for modeling sentence pairs. In: IJCAI, pp. 4411–4417 (2018)
14. Tay, Y., Tuan, L.A., Hui, S.C.: Multi-range reasoning for machine comprehension. arXiv preprint arXiv:1803.09074 (2018)
15. Trischler, A., et al.: NewsQA: a machine comprehension dataset. arXiv preprint arXiv:1611.09830 (2016)
16. Trischler, A., Ye, Z., Yuan, X., He, J., Bachman, P.: A parallel-hierarchical model for machine comprehension on sparse data. In: Proceedings of the 54th Annual Meeting of the Association for Computational Linguistics (Volume 1: Long Papers), vol. 1, pp. 432–441 (2016)
17. Vaswani, A., et al.: Attention is all you need. In: Advances in Neural Information Processing Systems, pp. 5998–6008 (2017)
18. Wang, S., Yu, M., Jiang, J., Chang, S.: A co-matching model for multi-choice reading comprehension. In: Proceedings of the 56th Annual Meeting of the Association for Computational Linguistics (Volume 2: Short Papers), pp. 746–751 (2018)
19. Wang, W., Yang, N., Wei, F., Chang, B., Zhou, M.: Gated self-matching networks for reading comprehension and question answering. In: Proceedings of the 55th Annual Meeting of the Association for Computational Linguistics (Volume 1: Long Papers), pp. 189–198 (2017)
20. Xiong, C., Zhong, V., Socher, R.: Dynamic coattention networks for question answering. arXiv preprint arXiv:1611.01604 (2016)
21. Xu, Y., Liu, J., Gao, J., Shen, Y., Liu, X.: Towards human-level machine reading comprehension: reasoning and inference with multiple strategies. arXiv preprint arXiv:1711.04964 (2017)
22. Yin, W., Ebert, S., Schütze, H.: Attention-based convolutional neural network for machine comprehension. In: Proceedings of the Workshop on Human-Computer Question Answering, pp. 15–21 (2016)
23. Zhang, S., Zhao, H., Wu, Y., Zhang, Z., Zhou, X., Zhou, X.: Dual co-matching network for multi-choice reading comprehension. arXiv preprint arXiv:1901.09381 (2019)
24. Zhu, H., Wei, F., Qin, B., Liu, T.: Hierarchical attention flow for multiple-choice reading comprehension. In: Thirty-Second AAAI Conference on Artificial Intelligence (2018)

Explore Entity Embedding Effectiveness in Entity Retrieval

Zhenghao Liu[1], Chenyan Xiong[2], Maosong Sun[1(✉)], and Zhiyuan Liu[1]

[1] Department of Computer Science and Technology,
Institute for Artificial Intelligence, State Key Lab on Intelligent Technology
and Systems, Tsinghua University, Beijing, China
sms@tsinghua.edu.cn
[2] Microsoft Research AI, Redmond, USA

Abstract. This paper explores entity embedding effectiveness in ad-hoc entity retrieval, which introduces distributed representation of entities into entity retrieval. The knowledge graph contains lots of knowledge and models entity semantic relations with the well-formed structural representation. Entity embedding learns lots of semantic information from the knowledge graph and represents entities with a low-dimensional representation, which provides an opportunity to establish interactions between query related entities and candidate entities for entity retrieval. Our experiments demonstrate the effectiveness of entity embedding based model, which achieves more than 5% improvement than the previous state-of-the-art learning to rank based entity retrieval model. Our further analysis reveals that the entity semantic match feature effective, especially for the scenario which needs more semantic understanding.

Keywords: Entity retrieval · Entity embedding · Knowledge graph

1 Introduction

In the past decade, large-scale public knowledge bases have emerged, such as DBpedia [3], Freebase [2] and Wikidata [4]. These knowledge bases provide a well-structured knowledge representation and have become one of the most popular resources for many applications, such as web search and question answering. A fundamental process in these systems is ad-hoc entity retrieval, which has encouraged the development of entity retrieval systems. Ad-hoc entity retrieval in the web of data (ERWD) aims to answer user queries through returning entities from publicly available knowledge bases and satisfy some underlying information need.

Knowledge bases represent knowledge with Resource Description Framework (RDF) triples for structural information. Entity related triples contain lots of related information, such as name, alias, category, description and relationship with other entities. Previous entity retrieval works represent an entity by grouping entity related triples into different categories. And the multi-field entity

ⓒ Springer Nature Switzerland AG 2019
M. Sun et al. (Eds.): CCL 2019, LNAI 11856, pp. 105–116, 2019.
https://doi.org/10.1007/978-3-030-32381-3_9

representation provides an opportunity to convert the entity retrieval task to a document retrieval task. Therefore, lots of ranking methods can be leveraged, such as BM25, TF-IDF and Sequential Dependence Models (SDM). Learning to rank (LeToR) models provide an effective way to incorporate different match features and achieve the state-of-the-art for entity retrieval [6]. These entity retrieval systems only leverage text based matches and neglect entity semantics in the knowledge graph. Therefore, field representation shows its limitation with the structural knowledge representation.

Knowledge representation learning provides an effective way to model entity relations with embedding. The relations of entities in a knowledge graph are stored in RDF triples which consist of the head entity, relation and tail entity. Previous works, such as TransE [5], represent both entities and relations as the low-dimensional representation. Then they formalize the entities and relations with different energy functions. Knowledge representation learning helps to learn the structural information of the knowledge graph, which can better help entity retrieval models understand the semantic information from entities.

This work investigates the effectiveness of entity embedding, which contains knowledge graph semantics, for entity retrieval. It utilizes TransE to get the low-dimensional representation for each entity. And then we calculate the soft match feature between query entities and candidate entities. Furthermore, we also follow the previous methods to represent entities textual information with multiple fields and exact match features with different ranking methods for all fields. The learning to rank models is utilized to combine all exact match features and entity soft match feature for the ranking score. Experiments on an entity search test benchmark confirm that entity embedding based soft match feature is critical for entity retrieval and significantly improve the previous state-of-the-art entity retrieval methods by over 5%. Our analyses also indicate that entity embedding based semantic match plays an important role, especially for the scenario which needs more linguistic and semantic understanding. We released all resources of data and codes via github[1].

2   Related Work

The ERWD task, which is first introduced by Pound et al. [14], focuses on how to answer arbitrary keyword queries by finding one or more entities with entity representations. Existing entity retrieval systems concern more about the representation of entities. Early works, especially in the context of expert search, obtain entity representations by considering mentions of the given entity [1,2]. The INEX 2007-2009 Entity Retrieval track (INEX-XER) [7,8] studies entity retrieval in Wikipedia, while the INEX 2012 Linked Data track further considers Wikipedia articles together with RDF properties from the DBpedia and YAGO2 knowledge bases [16]. The recent works usually represent entities as the fielded document [3,17], which divides the entity representation into three or five

---

[1] https://github.com/thunlp/EmbeddingEntityRetrieval.

categories. These entity representation methods provide a possible way to solve the entity retrieval problem with document retrieval methods.

Previous document retrieval models calculate the query and document relevance with bag-of-word representations, such as BM25 and Language Model (LM). Nevertheless, these bag-of-words retrieval models neglect term dependence, which is an important match signal for document retrieval. Markov Random Field (MRF) [12] for document retrieval provides a solid theoretical way to model the dependence among query terms. Sequential Dependence Model (SDM) [12] is a variation of Markov Random Field, which considers unigram, ordered bigram and unordered bigram match features. The SDM provides a good balance between retrieval effectiveness and efficiency.

Entity retrieval models leverage document retrieval models and extend them to multiple fields. They weight all ranking scores from all categories of the entity representation for the ranking score. They mainly leverage the standard bag-of-words framework to calculate the similarity between query and candidate entities. BM25F [15] and Mixture of Language Models (MLM) [13] combine BM25 and Language Model retrieval models to the multi-field entity retrieval. Different from MLM and BM25F, Fielded Sequential Dependence Model (FSDM) [17] considers the sequential dependence and leverages SDM to calculate the relevant score between query and each field of the candidate entity. On the other hand, Probabilistic Retrieval Model for Semistructured Data (PRMS) [10] weights query terms according to document collection statistics for the better retrieval performance. To further leverage the entity based interactions between the query and candidate entities, some works [9] calculate the entity mention based exact match feature between query and candidate entities. State-of-the-art learning to rank models, such as Coordinate Ascent and RankSVM, provide an opportunity to combine features from different models and different fields, which achieves the state-of-the-art for entity retrieval [6].

The knowledge representation learning methods model entity structural information and encode entities into a low-dimensional vector space. TransE [5] is one of the most popular and robust works for knowledge representation learning. TransE interprets knowledge graph triples as a translation: the entity vector plus the relation vector is equal to the tail entity. Moreover, the entity embedding with semantic information has further improved ranking performance [11] for ad-hoc retrieval. Therefore, knowledge embedding may provide a potential way to bring entity semantic information from knowledge graph to entity retrieval.

3 Methodology

In this section, we introduce the text match based retrieval model, entity mention based retrieval model and our entity embedding based model. Given a query $Q = \{q_1, q_2, ..., q_n\}$ and an entity representation E, our aim is to generate a ranking score $f(Q, E)$ to rank candidate entities.

3.1 Text Based Retrieval Model

Existing entity search models leverage term and term dependence based match features to calculate Q and E similarity based on the Markov Random Field (MRF) model. Therefore, we introduce two variations of MRF, the Sequential Dependence Model (SDM) and Fielded Sequential Dependence Model (FSDM) in this part.

Sequential Dependence Model. The Sequential Dependence Model (SDM) considers both unigram and bigram match features for ranking. To calculate the ranking score, it is apparent that computing the following posterior probability is sufficient:

$$P(E|Q) = \frac{P(Q,E)}{P(Q)} \overset{rank}{=} P(Q,E). \tag{1}$$

Based on MRF model, we could get query term and adjacent term cliques. Then we incorporate the term q_i based match feature and the term dependence based match feature to get the SDM ranking function:

$$P(E|Q) \overset{rank}{=} \lambda_T \sum_{q_i \in Q} f_T(q_i, E) + \lambda_O \sum_{q_i, q_{i+1} \in Q} f_O(q_i, q_{i+1}, E) + \lambda_U \sum_{q_i, q_{i+1} \in Q} f_U(q_i, q_{i+1}, E), \tag{2}$$

where λ is the parameter to weight features, which should meet $\lambda_T + \lambda_O + \lambda_U = 1$. $f_T(q_i, E)$ denotes unigram match feature. $f_O(q_i, q_{i+1}, E)$ and $f_U(q_i, q_{i+1}, E)$ represent ordered and unordered bigram match features respectively. Then the specific feature functions are presented as follow:

$$f_T(q_i, E) = \log\left[\frac{tf_{q_i,E} + \mu \frac{cf_{q_i}}{|C|}}{|E| + \mu}\right], \tag{3}$$

$$f_O(q_i, q_{i+1}, E) = \log\left[\frac{tf_{\#1(q_i,q_{i+1}),E} + \mu \frac{cf_{\#1(q_i,q_{i+1})}}{|C|}}{|E| + \mu}\right], \tag{4}$$

$$f_U(q_i, q_{i+1}, E) = \log\left[\frac{tf_{\#uwN(q_i,q_{i+1}),E} + \mu \frac{cf_{\#uwN(q_i,q_{i+1})}}{|C|}}{|E| + \mu}\right], \tag{5}$$

where tf and cf denotes uni-gram or bi-gram term frequency for the candidate entity and entire entity collection respectively. $(\#1(q_i, q_{i+1}), E)$ calculates the exact match for q_i, q_{i+1} and $(\#uwN(q_i, q_{i+1}), E)$ counts the number of co-occurrence times of q_i and q_{i+1} within a N size window. And μ is the Dirichlet prior.

Field Sequential Dependence Model. Field Sequential Dependence Model extends the SDM with a Mixture of Language Model (MLM) for each field of an entity representation. MLM computes each field probability and combines all fields with a linear function. For the field $f \in F$, the match feature functions $f_T(q_i, E)$, $f_O(q_i, q_{i+1}, E)$ and $f_U(q_i, q_{i+1}, E)$ can be extended as follow:

$$f_T(q_i, E) = \log \sum_f w_f^T \frac{tf_{q_i, E_f} + \mu_f \frac{cf_{q_i, f}}{|C_f|}}{|E_f| + \mu_f}, \tag{6}$$

$$f_O(q_i, q_{i+1}, E) = \log \sum_f w_f^O \frac{tf_{\#1(q_i, q_{i+1}), E_f} + \mu_f \frac{cf_{\#1(q_i, q_{i+1}), f}}{|C_f|}}{|E_f| + \mu_f}, \tag{7}$$

$$f_U(q_i, q_{i+1}, E) = \log \sum_f w_f^U \frac{tf_{\#uwN(q_i, q_{i+1}), E_f} + \mu_f \frac{cf_{\#uwN(q_i, q_{i+1}), f}}{|C_f|}}{|E_f| + \mu_f}, \tag{8}$$

where μ_f is the weight for the field f. The FSDM represents entities with a novel five-field schema: The **names** contains entity names, such as the label relation from RDF triples; The **attributes** field involves all text information, such as entity abstract, except entity **names** field; The **categories** field implies entity categories; Then the **SimEn** and **RelEn** denote similar or aggregated entity and related entity respectively. Then FSDM weights all field weight and achieves a further improvement than SDM.

3.2   Entity Mention Based Retrieval Model

In this part, we introduce the entity mention based retrieval model for entity retrieval. The Entity Linking incorporated Retrieval (ELR) model represents the query Q through an annotated entity set $\hat{E}(Q) = \{e_1, e_2, ..., e_m\}$ with the confidence score $s(e_i)$ for each entity. According to MRF graph, ELR involves interactions between $\hat{E}(Q)$ and E. Then ELR extends MRF with a linear combination of correlate entity potential function:

$$P(E|Q) \overset{rank}{=} \sum_{q_i \in Q} \lambda_T f_T(q_i, E) + \sum_{q_i, q_{i+1} \in Q} \lambda_O f_O(q_i, q_{i+1}, E)$$
$$+ \sum_{q_i, q_{i+1} \in Q} \lambda_U f_U(q_i, q_{i+1}, E) + \sum_{e \in \hat{E}(Q)} \lambda_{\hat{E}} f_{\hat{E}}(e, E), \tag{9}$$

Then ELR takes the entity confidence score to weight all entity based matches:

$$P(E|Q) \overset{rank}{=} \lambda_T \sum_{q_i \in Q} \frac{1}{|Q|} f_T(q_i, E) + \lambda_O \sum_{q_i, q_{i+1} \in Q} \frac{1}{|Q| - 1} f_O(q_i, q_{i+1}, E)$$
$$+ \lambda_U \sum_{q_i, q_{i+1} \in Q} \frac{1}{|Q| - 1} f_U(q_i, q_{i+1}, E) + \lambda_E \sum_{e \in \hat{E}(Q)} s(e) f_{\hat{E}}(e, E), \tag{10}$$

where $|Q|$ and $|Q| - 1$ are utilized to smooth TF features according to the sequence length.

3.3   Entity Embedding Based Retrieval Model

Previous entity retrieval models only calculate Q and E correlation with exact matches without considering knowledge based semantic information. For example, given two entities **Ann Dunham** and **Barack Obama**, **Ann Dunham** is a parent of **Barack Obama**. It is inevitable that exact matches will regard **Ann Dunham** and **Barack Obama** as different entities. To solve this problem, we leverage entity embeddings with knowledge graph semantics to improve entity retrieval performance.

To leverage knowledge embedding to calculate our ranking features, we first map both entities in $\hat{E}(Q)$ and E into the same vector space. Then we also consider confidence score $s(e_i)$ for the similarity of i-th entity in the query annotated entity set $\hat{E}(Q)$ and the candidate entity E:

$$f(Q, E) = \sum_{i=1}^{m} s(e_i) \cdot \cos(\boldsymbol{v}_{e_i}, \boldsymbol{v}_E), \tag{11}$$

where $\boldsymbol{v}_{e_i}$ and $\boldsymbol{v}_E$ TansE embedding for entity e_i and E respectively.

Translation based methods successfully model structural knowledge bases for entity representation learning. TransE is a robust and efficient algorithm in translation-based entity embedding model and we use TransE to obtain the entity embedding. For the entity triple in knowledge base S, the tail entity t should be close to the head entity h plus the relationship r. The energy function is demonstrated as follows:

$$J(h, r, t) = ||h + r - t||_+. \tag{12}$$

Then we minimize the pairwise loss function over the training set to optimize both entity and relation embeddings:

$$L = \sum_{(h,r,t) \in S} \sum_{(h',r,t') \in \hat{S}} [\gamma + d(\boldsymbol{v}_h + \boldsymbol{v}_r, \boldsymbol{v}_t) - d(\boldsymbol{v}_{h'} + \boldsymbol{v}_r, \boldsymbol{v}_{t'})], \tag{13}$$

where h' and t' denote negative head entity and tail entity for the relation r. d denotes the distance between two vectors.

4   Experimental Methodology

This section describes our experimental methods and materials, including dataset, baselines and parameters setting.

4.1   Dataset

We use DBpedia version 3.7 as our knowledge base and compare the effectiveness of our knowledge embedding model based on a publicly available benchmark which contains 485 queries and 13090 related entities shown as Table 1. There are four types of queries in this collection: `Entity` (e.g. "NAACP Image Awards"), `Type` (e.g. "circus mammals"), `Attribute` (e.g. "country German language") and `Relation` (e.g. "Which airports are located in California, USA"). Therefore, the four subtasks for entity retrieval evaluate models from different aspect:

- **SemSearch ES:** Queries usually consist of named entity. And queries are oriented to the specific entities, which usually need to be disambiguated. (e.g., "harry potter", "harry potter movie")
- **ListSearch:** A query set combines INEX-XER, SemSearch LS, TREC Entity queries. This subtask aims to a list of entities that matches a certain criteria. (e.g. "Airlines that currently use Boeing 747 planes")
- **INEX-LD:** IR-style keyword queries, including a mixture of `Entity`, `Type`, `Attribute` and `Relation`. (e.g., "bicycle sport races")
- **QALD-2:** Consisting of natural language questions as well as involve four different types. (e.g., "Who wrote the book The pillars of the Earth")

The SemSearch ES, ListSearch and QALD-2 subtasks need more linguistic or semantic understanding. On the other hand, INEX-LD focuses more on keyword matches.

Table 1. Statistic of DBpedia-entity test collection

Query set	#queries	#rel	Query types
SemSearch ES	130	1131	Entity
ListSearch	115	2398	Type
INEX_LD	100	3756	Entity, Type, Attribute, Relation
QALD-2	140	5805	Entity, Type, Attribute, Relation
Total	485	13090	-

Table 2. Traditional baseline features.

Features	Dimension
FSDM	1
SDM on all fields	5
BM25 on all fields	5
Language model on all fields	5
Coordinate match on all fields	5
Cosine similarity on all fields	5

4.2 Baselines

We follow the previous state-of-the-art model [6] and leverage feature based learning to rank methods, RankSVM and Coordinate Ascent, as our ranking algorithms. Our baseline methods also utilize 26 features from different traditional ranking models and different fields, as shown in Table 2.

The entity mention match feature is also incorporated with word based 26 ranking features for our baseline, denoted as "+ELR". The entity mention based match feature only considers exact matches for entity mention. And we further incorporate our entity embedding based semantic match feature with the baseline 26 features and is denoted as "+TransE".

4.3   Implementation Details

We use the Fielded Sequential Dependence Model (FSDM) as the basic retrieval model to generate the candidate entity set with top 100 entities. And all models in our experiments rerank candidate entities. RankSVM implementation is provided by SVMLight toolkit[2]. Coordinate Ascent implementation is provided by RankLib[3]. All models in our experiments are trained and tested using five fold cross validation and all models keep the same partition. Moreover, all parameter settings are kept the same with the previous work [6]. All methods are evaluated by MAP@100, P@10, and P@20. Statistic significances are tested by permutation test with P< 0.05.

For entity embedding, we involve 11,988,202 entities to train our TransE model. The TransE model is implemented with C++ language[4]. We set the entity dimension as 100 dimension. All embeddings are optimized with SGD optimizer and 0.001 learning rate.

5   Evaluation Result

In this section, we present the model performance and the feature weight distribution to demonstrate the effectiveness of our model.

5.1   Overall Performance

In this part, we conduct the overall performance of three models with Coordinate Ascent and RankSVM, as shown in Tables 3 and 4 respectively.

The entity mention based exact match feature is introduced by Entity Linking incorporated Retrieval model (ELR) [9] and shows its effectiveness by improving the baseline with Coordinate Ascent and RankSVM almost 1% and 4% for the whole data. Then +ELR model shows a significant improvement on the ListSearch test scenario. The ListSearch subtask aims to find related entities which share the same type. And +ELR demonstrates that leveraging entity mention based exact match feature can help to enhance the entity retrieval performance.

The entity embedding is a kind of entity semantic representations and models entity relations in the whole knowledge graph. +TransE model overall improves

---

[2] https://www.cs.cornell.edu/people/tj/svm_light/svm_rank.html.
[3] http://sourceforge.net/p/lemur/wiki/RankLib/.
[4] https://github.com/thunlp/Fast-TransX.

Table 3. Entity retrieval performance with Coordinate Ascent. Relative performances compared are in percentages. †, ‡ indicate statistically significant improvements over Baseline† and +ELR‡ respectively.

Models	SemSearch ES						
	MAP		P@10		P@20		W/T/L
Baseline	0.3899	–	0.2908	–	0.2077	–	–/–/–
+ELR	0.3880	−0.49%	**0.3023**†	+3.95%	0.2150†	+3.51%	58/19/53
+TransE	**0.4085**‡	+4.77%	**0.3023**†	+3.95%	**0.2165**†	+4.24%	64/16/50
	ListSearch						
Baseline	0.2334	–	0.3130	–	0.2378	–	–/–/–
+ELR	0.2443†	+4.67%	0.3130	+0.00%	0.2422	+1.85%	54/22/39
+TransE	**0.2507**†	+7.41%	**0.3304**†	+5.56%	**0.2543**†‡	+6.94%	65/20/30
	INDEX-LD						
Baseline	0.1298	–	0.2900	–	0.2285	–	–/–/–
+ELR	0.1275	−1.77%	**0.2920**	+0.69%	**0.2335**	+2.19%	44/10/46
+TransE	**0.1312**	+1.08%	0.2860	−1.38%	0.2255	−1.31%	42/12/46
	QALD-2						
Baseline	0.1998	–	0.1500	–	0.1196	–	–/–/–
+ELR	0.2074	+3.80%	0.1664	+10.93%	0.1282	+7.19%	45/59/36
+TransE	**0.2270**†	+13.61%	**0.1700**†	+13.33%	**0.1371**†‡	+14.63%	48/62/30
	ALL						
Baseline	0.2454	–	0.2540	–	0.1934	–	–/–/–
+ELR	0.2472	+0.73%	0.2544	+0.16%	0.1945	+0.57%	175/145/165
+TransE	**0.2597**†‡	+5.83%	**0.2639**†‡	+3.90%	**0.1970**†	+1.86%	212/122/151

the Coordinate Ascent based baseline and the RankSVM based baseline significantly by over 6% and 5% respectively. +TransE also illustrates its effectiveness with the significant improvement on SemSearch, ListSearch and QALD-2 test scenarios. And the improvement demonstrates the entity semantic match plays an important role in the task which needs more semantic or linguistic understanding. Both +TransE and +ELR improves baseline model with a small margin on the INDEX-LD test scenario, which illustrates that INDEX-LD only needs keyword matches and the multi-field based entity representation can do well on this scenario. The +TransE model also shows its effectiveness by a large margin improvement with +ELR model, especially on the SemSearch scenario, which illustrates entity embeddings can help model better understand the semantic information of entities.

Overall experiments present the entity effectiveness by a significant improvement especially on the scenarios which need more semantic understanding. Nevertheless, the role of entity based matches in entity retrieval is not clear. Therefore, we explore the importance of the entity based match in the following experiments.

Table 4. Entity retrieval performance with RankSVM. Relative performances compared are in percentages. †, ‡ indicate statistically significant improvements over Baseline† and +ELR‡ respectively.

Models	SemSearch ES						
	MAP		P@10		P@20		W/T/L
Baseline	0.3895	–	0.3038	–	0.2169	–	–/–/–
+ELR	0.3881	−0.36%	0.3046	+0.26%	0.2173	+0.18%	48/38/44
+TransE	**0.4061**†‡	+4.26%	**0.3077**	+1.28%	**0.2196**	+1.24%	52/24/54
	ListSearch						
Baseline	0.2323	–	0.3078	–	0.2513	–	–/–/–
+ELR	0.2390†	+2.88%	0.3148	+2.27%	0.2530	+0.68%	60/25/30
+TransE	**0.2439**†	+4.99%	**0.3252**†	+5.65%	**0.2565**	+2.07%	54/23/38
	INDEX-LD						
Baseline	0.1350	–	0.2940	–	0.2345	–	–/–/–
+ELR	0.1390	+2.96%	**0.2980**	+1.36%	**0.2375**	+1.28%	46/19/35
+TransE	**0.1392**	+3.11%	0.2950	+0.34%	0.2365	+0.85%	44/16/40
	QALD-2						
Baseline	0.2229	–	0.1529	–	0.1257	–	–/–/–
+ELR	0.2197	−1.44%	**0.1671**†	+9.29%	0.1286	+2.31%	39/75/26
+TransE	**0.2278**	+2.20%	0.1629	+6.54%	**0.1321**†	+5.09%	44/68/28
	ALL						
Baseline	0.1925	–	0.2245	–	0.1798	–	–/–/–
+ELR	0.2005†	+4.16%	0.2307†	+2.76%	0.1816	+1.00%	200/153/132
+TransE	**0.2054**†	+6.70%	**0.2346**†	+4.50%	**0.1881**†‡	+4.62%	249/129/107

5.2   Feature Weight Distribution

This part presents the weight distribution of ranking features, which are divided into three groups: the FSDM based ranking feature (FSDM), the entity based ranking feature (ENT) and other traditional retrieval model based ranking features (Others), as shown in Fig. 1. Then we calculate the percentage of weight given to each type of features by summing up the absolute weight values according to the feature group. For all 27 features (baseline 26 match features with entity based match features), TransE and FSDM features play the most important roles in our model and other traditional ranking features also show their effectiveness especially for RankSVM. On the other hand, TransE shares more weight than the FSDM based ranking feature, confirming the entity semantic match feature is so important for entity retrieval. Moreover, the different weight distributions between ELR and TransE based entity retrieval models demonstrate the entity embedding based semantic match is more effective and important than the entity mention based entity exact match. The semantic information

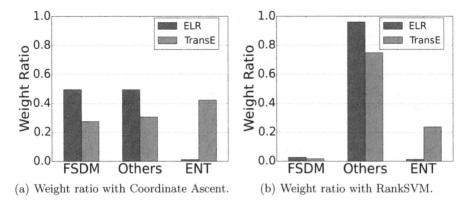

(a) Weight ratio with Coordinate Ascent.   (b) Weight ratio with RankSVM.

Fig. 1. Feature weight distribution for FSDM based ranking feature, entity based ranking feature (ENT) and other traditional retrieval model based ranking features (Others).

from knowledge graph is brought to the entity retrieval system through the entity embedding and helps entity retrieval models achieve further improvement.

6 Conclusion

This paper explores entity embedding effectiveness in entity retrieval with two previous state-of-the-art learning to rank methods which incorporate diverse features extracted from different models and different fields. Entity embedding shows its effectiveness by incorporating entity semantic information from the knowledge graph, which can better model the interaction between query and candidate entities from entity based matches. Experiments on an entity-oriented test collection reveal the power of entity embeddings, especially for the task which needs more semantic and linguistic understanding. Our further analysis reveals that entity embedding based semantic match features plays the same important role as FSDM in entity retrieval and better models query and entity relations than the entity mention based exact match feature. We hope our experiments and models can provide a potential way to better represent entity and leverage semantic information from knowledge graph for entity retrieval systems.

Acknowledgment. This work is supported by National Natural Science Foundation of China (NSFC) grant 61532001.

References

1. Balog, K., Azzopardi, L., De Rijke, M.: Formal models for expert finding in enterprise corpora. In: Proceedings of the 29th Annual International ACM SIGIR Conference on Research and Development in Information Retrieval (2006)
2. Balog, K., Bron, M., De Rijke, M.: Query modeling for entity search based on terms, categories, and examples. ACM Trans. Inf. Syst. **29**, 22 (2011)

3. Balog, K., Neumayer, R.: A test collection for entity search in DBpedia. In: Proceedings of the 36th International ACM SIGIR Conference on Research and Development in Information Retrieval (2013)

4. Bendersky, M., Metzler, D., Croft, W.B.: Learning concept importance using a weighted dependence model. In: Proceedings of the Third ACM International Conference on Web Search and Data Mining (2010)

5. Bordes, A., Usunier, N., Garcia-Duran, A., Weston, J., Yakhnenko, O.: Translating embeddings for modeling multi-relational data. In: Advances in Neural Information Processing Systems (2013)

6. Chen, J., Xiong, C., Callan, J.: An empirical study of learning to rank for entity search. In: Proceedings of the 39th International ACM SIGIR Conference on Research and Development in Information Retrieval (2016)

7. de Vries, A.P., Vercoustre, A.-M., Thom, J.A., Craswell, N., Lalmas, M.: Overview of the INEX 2007 entity ranking track. In: Fuhr, N., Kamps, J., Lalmas, M., Trotman, A. (eds.) INEX 2007. LNCS, vol. 4862, pp. 245–251. Springer, Heidelberg (2008). https://doi.org/10.1007/978-3-540-85902-4_22

8. Demartini, G., Iofciu, T., de Vries, A.P.: Overview of the INEX 2009 entity ranking track. In: Geva, S., Kamps, J., Trotman, A. (eds.) INEX 2009. LNCS, vol. 6203, pp. 254–264. Springer, Heidelberg (2010). https://doi.org/10.1007/978-3-642-14556-8_26

9. Hasibi, F., Balog, K., Bratsberg, S.E.: Exploiting entity linking in queries for entity retrieval. In: Proceedings of the 2016 ACM International Conference on the Theory of Information Retrieval (2016)

10. Kim, J., Xue, X., Croft, W.B.: A probabilistic retrieval model for semistructured data. In: Boughanem, M., Berrut, C., Mothe, J., Soule-Dupuy, C. (eds.) ECIR 2009. LNCS, vol. 5478, pp. 228–239. Springer, Heidelberg (2009). https://doi.org/10.1007/978-3-642-00958-7_22

11. Liu, Z., Xiong, C., Sun, M., Liu, Z.: Entity-duet neural ranking: understanding the role of knowledge graph semantics in neural information retrieval. In: Proceedings of the 56th Annual Meeting of the Association for Computational Linguistics (2018)

12. Metzler, D., Croft, W.B.: A Markov random field model for term dependencies. In: Proceedings of the 28th Annual International ACM SIGIR Conference on Research and Development in Information Retrieval (2005)

13. Ogilvie, P., Callan, J.: Combining document representations for known-item search. In: Proceedings of the 26th Annual International ACM SIGIR Conference on Research and Development in Informaion Retrieval (2003)

14. Pound, J., Mika, P., Zaragoza, H.: Ad-hoc object retrieval in the web of data. In: Proceedings of the 19th International Conference on World Wide Web. ACM (2010)

15. Robertson, S., Zaragoza, H., Taylor, M.: Simple BM25 extension to multiple weighted fields. In: Proceedings of the Thirteenth ACM International Conference on Information and Knowledge Management, pp. 42–49. ACM (2004)

16. Wang, Q., et al.: Overview of the INEX 2012 linked data track. In: CLEF (2012)

17. Zhiltsov, N., Kotov, A., Nikolaev, F.: Fielded sequential dependence model for ad-hoc entity retrieval in the web of data. In: Proceedings of the 38th International ACM SIGIR Conference on Research and Development in Information Retrieval (2015)

Text Classification and Summarization

Part Classification and Segmentation

ERCNN: Enhanced Recurrent Convolutional Neural Networks for Learning Sentence Similarity

Niantao Xie[1], Sujian Li[1,2(✉)], and Jinglin Zhao[3]

[1] MOE Key Laboratory of Computational Linguistics, Peking University,
Beijing, China
{xieniantao,lisujian}@pku.edu.cn
[2] Peng Cheng Laboratory, Shenzhen, China
[3] Faculty of Arts and Social Science, National University of Singapore,
Singapore, Singapore
zhaojinglin92@gmail.com

Abstract. Learning the similarity between sentences is made difficult by the fact that two sentences which are semantically related may not contain any words in common limited to the length. Recently, there have been a variety kind of deep learning models which are used to solve the sentence similarity problem. In this paper we propose a new model which utilizes enhanced recurrent convolutional neural network (ERCNN) to capture more fine-grained features and the interactive effects of keypoints in two sentences to learn sentence similarity. With less computational complexity, our model yields state-of-the-art improvement compared with other baseline models in paraphrase identification task on the Ant Financial competition dataset.

Keywords: Sentence similarity · ERCNN · Soft attention mechanism

1 Introduction

Paraphrase identification is an important NLP task which may be greatly enhanced by modeling the underlying semantic similarity between compared texts. In particular, a good method should not be susceptible to variations of wording or syntax used to express the same idea, and also should be able to measure similarity for both long and short texts like sentences.

Learning such sentences similarity has attracted many research interests [17]. However, it still remains a knotty issue. With the renaissance of deep learning, many text representation methods are combined to take better into account both semantics and structure of sentences for paraphrase identification task. Text representation is the key component in measuring semantic similarity and usually composed of convolutional neural network (CNN) [14] and recurrent neural network (RNN). CNN is supposed to be better at extracting robust and

© Springer Nature Switzerland AG 2019
M. Sun et al. (Eds.): CCL 2019, LNAI 11856, pp. 119–130, 2019.
https://doi.org/10.1007/978-3-030-32381-3_10

abstract features of texts while RNN, especially the Long Short-Term Memory (LSTM) [11] and Gated Recurrent Unit (GRU) [6] have been applied widely in paraphrase identification task because of their naturally suited for variable-length inputs. Some prior work also proposes attention-based CNN or RNN models to focus on the mutual influence of two sentences in text representation [5, 21, 23].

Following common solutions in paraphrase identification task, prior works have demonstrated the effectiveness of two types of deep learning frameworks. The first framework is based on the "Siamese network" [3, 17, 23]. And the second framework is called "matching-aggregation" [5, 22]. The idea behind these two frameworks is to measure semantic similarity from multiple perspectives by capturing interactive features between the text representations of input sentences. And the difference between them is mainly on the way of capturing interactive features. "Siamese network" framework simply concatenates on the text representations of input sentences by calculating their cosine or other distances. Due to its simple way of capturing interactive features, "Siamese network" framework becomes easy to train. Conversely, "matching-aggregation" framework makes more improvements on interactive layer by utilizing a complex way to capture the fine-grained features and therefore performs better than "Siamese network" framework usually.

In this paper, we introduce our model based on one state-of-the-art work named ESIM [5] to measure sentence similarity by jointly leveraging the model predictive effect and parameter size. As a result, the improvements on text representation and optimization of capturing interactive features are main contributions of our model. The details are as follows:

- Different from ESIM, we add CNN layers upon RNN layers to reduce the parameters and capture more fine-grained features during input encoding.
- Moreover, we further simplify ESIM structure by using fully-connected layers to replace the time-consuming BiLSTM layers in overall similarity modeling. Based on the methods above, we achieve additional improvements in paraphrase identification task on Ant Financial competition dataset.

2   Related Work

The deep learning applications for paraphrase identification task have recently received much attention [17], from the availability of high-quality semantic word representations [7, 16, 20] and followed by the seminal papers introduce "Siamese network" and "matching-aggregation" framework for learning sentence similarity. In the past few years, many deep learning models based on "Siamese network" or "matching-aggregation" framework have made progress in learning sentence similarity [5, 17, 22, 23].

For "Siamese network" framework [3, 17, 19, 23], two input sentences are applied by the same neural network encoder like a CNN or a RNN individually, so both of the two sentences are encoded into vectors of the same embedding space. As a result, a matching interaction is calculated solely based on the two sentence vectors. The advantage of this framework is that sharing parameters makes the model simply equipped and therefore effortless to train, and the sentence vectors can be used directly for measuring similarity based on cosine or other type of distances. However, there is no explicit interaction between the two sentences during the encoding procedure, which may lose some crucial information.

To deal with this problem, some prior works also incorporate attention mechanism into "Siamese network" framework [21,23]to address the problem of insufficient interactions. Different from learning each sentence's representation separately, the attention-based models consider the mutual influence between two sentences. One of famous models is ABCNN [23], which is an attention-based CNN for modeling sentence similarity.

For "matching-aggregation" framework [5,22], smaller units such as words or characters of the two sentences are firstly matched, while the prior works usually employ bidirectional RNN to encode variable-length sentences into a fixed-length vector, and then the matching results are usually aggregated by RNN into a vector to make the final decision. This framework captures more interactive features between the two sentences, therefore significant improvements are obtained naturally. However, this framework still remains some defects, and the main shortcoming exists in the time-consuming matching operation in capturing interactive features.

One famous model based on "matching-aggregation" framework is called ESIM [5], which achieves the state-of-the-art performance in many NLP tasks besides sentence similarity by employing attention-based LSTM to capture some high-order interactions between two sentences. Motivated by the idea of ESIM, a recently proposed model named BiMPM (bilateral multi-perspective matching) [22] emerges. BiMPM brings in many complex matching operations to enhance the model ability to extract mutual information between two sentences.

As introduced before, our work is also related to ESIM using attention-based RNN for paraphrase identification task [5]. But unlike other models, we incorporate CNN to process the RNN output and simplify the time-consuming recursive architectures by fully-connected layers.

3   ERCNN Model

Assume we have two sentences $a = (a_1, ..., a_{l_a})$ and $b = (b_1, ..., b_{l_b})$, where a_i or $b_j \in R^k$ is an embedding of k-dimensional vector, which can be initialized with some pre-trained word or character embeddings. The goal is to predict a label $\mathbf{y}$ that indicates the similarity between a and b.

Our ERCNN model is composed of three components: (①) input encoding, (②) local similarity modeling, and (③) overall similarity modeling. These components are explained in details in the following sections. Here the input sentences are preprocessed respectively in word-level and character-level, and they are trained separately and final predictions are the average of two outputs. The left part of Fig. 1 shows a high-level view of the architecture.

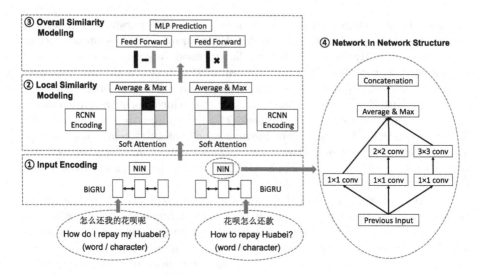

Fig. 1. The architecture of our introduced model.

3.1   Input Encoding

BiGRU. We first employ bidirectional GRU (BiGRU) [6] to encode the two input sentences (Eqs. (1) and (2)). Here BiGRU learns to represent a word or character and its corresponding hidden layer state as follows:

$$\overline{a}_i = BiGRU(a, i), \forall i \in [1, ..., l_a]. \tag{1}$$

$$\overline{b}_j = BiGRU(b, j), \forall j \in [1, ..., l_b]. \tag{2}$$

BiGRU contains two GRUs starting from the left and the right side respectively. The hidden states generated by these two GRUs at each time step are concatenated to represent the status of time step.

Note that we make use of GRU memory blocks in our models instead of LSTMs in original ESIM model. We compared LSTM [11] with GRU in this part and LSTMs are inferior to GRUs on results of experiments for this task.

NIN. The most difference between our model and ESIM is that we apply CNN to aggregate the input encoding after the process of BiGRU. Because CNN and RNN have different mechanism of action in sentence representation: CNN performs better in extracting key words from sentence while RNN behaves preferable in extracting sequence information in the sentence. Therefore, by incorporating both the merits of CNN and RNN, we can capture more fine-grained features during input encoding. Moreover, due to the peculiarity of parameter sharing, CNN can help reduce the number of parameters in the model dramatically. In the next part of experiment, we will prove our improvements by detailed examples.

Here we refer the idea of "Network In Network" (NIN)[10,15] which is illustrated as (④) network in network structure in the right part of Fig. 1 to design our CNN layers. The first layer contains three convolutions of size 1×1 and the second layer consists of two convolutions of sizes 2×2 and 3×3. The 1×1 convolution operations help reduce the dimension of output feature map and increase non-linearity by the active function Relu [18], and the 2×2 and 3×3 convolution operations here further extract robust and abstract features based on the output of BiGRU layer.

Let's take the convolution operation on BiGRU layer for example, the BiGRU output $\overline{\boldsymbol{p}} \in R^{n \cdot d_0}$ which is a sentence of length n (padded where necessary) is represented as:

$$\overline{\boldsymbol{p}}_{1:n} = \overline{\boldsymbol{p}}_1 \oplus \overline{\boldsymbol{p}}_2 \oplus \overline{\boldsymbol{p}}_3 \oplus \dots \oplus \overline{\boldsymbol{p}}_n \tag{3}$$

where $\oplus$ is the concatenation operator and d_0 is the hidden state size. Let $\overline{\boldsymbol{p}}_{i:i+j}$ refer to the concatenation of hidden states $\overline{\boldsymbol{p}}_i, \overline{\boldsymbol{p}}_{i+1}, \dots, \overline{\boldsymbol{p}}_{i+j}$. The convolutional operation involves a filter $\mathbf{W}_f \in R^{w \cdot d_0}$ which is applied to a window of w words to product a new feature. For instance, a feature $\boldsymbol{p}_i$ is generated from a window of words $\overline{\boldsymbol{p}}_{i:i+w-1}$ as the following formula:

$$p_i = ReLU(\mathbf{W}_f \cdot \overline{\boldsymbol{p}}_{i:i+w-1} + b) \tag{4}$$

here $b \in R$ is a bias term. Because the window may be outside of the sentence boundaries when it slides near the boundary, we set special padding tokens for the sentence, which means that we take the out-of-range input vectors $\overline{\boldsymbol{p}}_i$ ($i < 1$ or $i > n$) as zero vectors.

This filter is applied to each possible window of hidden states in the sentence $\{\overline{\boldsymbol{p}}_{1:w}, \overline{\boldsymbol{p}}_{2:w+1}, \dots, \overline{\boldsymbol{p}}_{n-w+1:n}\}$ to produce a feature map:

$$\boldsymbol{p} = [p_1, p_2, \dots, p_{n-w+1}] \tag{5}$$

where $\boldsymbol{p} \in R^{d_1}$ and $d_1 = n - w + 1$.

As a whole, we acquire the NIN encoding in the following way:

$$\widetilde{\boldsymbol{p}} = NIN(\overline{\boldsymbol{p}}) \tag{6}$$

Pooling. Pooling is commonly used to extract features from convolution output. In this part, we employ column-wise average and max pooling, and this architecture supports to stack an arbitrary number of convolution-pooling blocks to extract increasingly abstract features.

3.2 Local Similarity Modeling

Modeling local subsentential similarity between two inputs is the basic component to determine the overall similarity. It needs to employ some forms of alignment [1] to associate the relevant parts between two input sentences. Here we use soft attention to achieve the alignment. Our soft alignment layer computes the attention weights as the similarity of two sentences encoding with Eq. (4).

$$e_{ij} = \overline{\boldsymbol{p}}_{ai}^T \cdot \overline{\boldsymbol{p}}_{bj} \tag{7}$$

where $\overline{\boldsymbol{p}}_{ai}$ and $\overline{\boldsymbol{p}}_{bj}$ are the ith BiGRU output of sentence $\boldsymbol{a}$ and the jth BiGRU output of sentence $\boldsymbol{b}$ separately. Local similarity is measured by the attention weight e_{ij} computed above. For the encoding of a word or character in one sentence, i.e., $\overline{\boldsymbol{p}}_{ai}$, the relevant semantics in another sentence is measured by e_{ij} specifically with Eq. (8).

$$\hat{\boldsymbol{p}}_{ai} = \sum_{j=1}^{l_b} \frac{\exp(e_{ij})}{\sum_{k=1}^{l_b} \exp(e_{ik})} \overline{\boldsymbol{p}}_{bj}, \forall i \in [1, ..., l_a] \tag{8}$$

$$\hat{\boldsymbol{p}}_{bj} = \sum_{i=1}^{l_a} \frac{\exp(e_{ij})}{\sum_{k=1}^{l_a} \exp(e_{kj})} \overline{\boldsymbol{p}}_{ai}, \forall j \in [1, ..., l_b] \tag{9}$$

where $\hat{\boldsymbol{p}}_{ai}$ is a weighted summation of $\{\overline{\boldsymbol{p}}_{bj}\}_{j=1}^{l_b}$. Intuitively, the content in $\{\overline{\boldsymbol{p}}_{bj}\}_{j=1}^{l_b}$ that is relevant to $\overline{\boldsymbol{p}}_{ai}$ will be selected and represented as $\hat{\boldsymbol{p}}_{ai}$. The same process is performed as $\hat{\boldsymbol{p}}_{bj}$ with Eq. (9).

The weighted vectors obtained above are processed with pooling layer. We employ the following strategy: compute both average and max pooling, and concatenate all these vectors with NIN outputs to form the final fixed-length vectors $\boldsymbol{o}_a$ and $\boldsymbol{o}_b$. The process of calculation is as follows:

$$\hat{\boldsymbol{p}}_{a,ave} = \sum_{i=1}^{l_a} \frac{\hat{\boldsymbol{p}}_{a,i}}{l_a}, \ \hat{\boldsymbol{p}}_{a,max} = \overset{l_a}{\underset{i=1}{\max}} \hat{\boldsymbol{p}}_{a,i} \tag{10}$$

$$\hat{\boldsymbol{p}}_{b,ave} = \sum_{j=1}^{l_b} \frac{\hat{\boldsymbol{p}}_{b,j}}{l_b}, \ \hat{\boldsymbol{p}}_{b,max} = \overset{l_b}{\underset{j=1}{\max}} \hat{\boldsymbol{p}}_{b,j} \tag{11}$$

$$\boldsymbol{o}_a = [\hat{\boldsymbol{p}}_{a,ave}; \widetilde{\boldsymbol{p}}_a; \hat{\boldsymbol{p}}_{a,max}] \tag{12}$$

$$\boldsymbol{o}_b = [\hat{\boldsymbol{p}}_{b,ave}; \widetilde{\boldsymbol{p}}_b; \hat{\boldsymbol{p}}_{b,max}] \tag{13}$$

3.3 Overall Similarity Modeling

To determine the overall similarity between two sentences, we design a concatenation layer to perform the similarity modeling. This layer has two type of concatenation. The first concatenation includes the difference and the element-wise product for the interactive sentence representation. And second concatenation further concatenates the output of the first concatenation. The process is specified as follows:

$$\boldsymbol{m}_{out} = [\boldsymbol{o}_a - \boldsymbol{o}_b; \boldsymbol{o}_a \odot \boldsymbol{o}_b] \tag{14}$$

This process could be regarded as a special case of modeling some high-order interaction between the tuple elements. We then put $\boldsymbol{m}_{out}$ into a final multilayer perceptron (MLP) classifier:

$$\mathbf{y} = MLP(\boldsymbol{m}_{out}) \tag{15}$$

This is also a main difference between our ERCNN and ESIM. ESIM here still applies two BiLSTM layers to perform overall similarity modeling but BiLSTM is time-consuming in training process. Based on the feedback of experimental results, here we employ MLP to replace original BiLSTM and achieve the similar results with the number of parameters reduced.

4   Experiments

4.1   Data Descriptions

The experimental dataset[1] comes from AI competition held by Ant Financial Service Group. This competition aims at seeking better question similarity calculation method for customer service. The dataset contains 102477 question pairs with manual labeled and we can take this label as experimental ground truth. The average length of questions is 13.4. During the data preprocessing, we pad each question of same length (20 in word-level and 48 in character-level).

4.2   Evaluation Metrics

The parameter size and Macro F1-score are utilized as the evaluation metrics from two different aspects. The parameter size is the amount of model parameter and used to measure the complexity and time cost of different models, and Macro F1-score aims to evaluate the performance of different models in measuring sentence similarity.

4.3   Training Details

We divide the dataset into development set and test set. The test set has 10000 question pairs to be predicted. We further divide the development set into 10-fold to perform training and use these 10 models for testing. The final results are the average of 10 models.

The optimization method we choose for model training is the Adam [12]. The first momentum is set to be 0.9 and the second 0.999. The initial learning rate is 0.001 and the batch size is 128. The hidden states of GRU have 192 dimensions and linear layers have 384 dimensions. The two CNN layers both have 128 filters with 3 different sizes 1, 2, 3 and word embeddings have 100 dimensions. We set dropout with a rate of 0.5, which is applied to all feedforward connections.

We initialize word and character embeddings with the 100-dimensional pre-trained cw2vec [4] vectors from the competition dataset. For the out-of-vocabulary (OOV) words, we initialize the embeddings randomly. All vectors including word and character embeddings are updated during training.

The entire model is trained end-to-end and take cross-entropy as loss function. For all the experiments, we pick the model which works the best on the development set, and then evaluate it on the test set.

---

[1] https://dc.cloud.alipay.com/index#/topic/data?id=3.

4.4 Experimental Results

In this subsection, we compare our model with state-of-the-art models on the paraphrase identification task. Because Ant Financial competition dataset is a brand-new dataset and no previous results have been published yet, we implement the following baseline models to compare:

Table 1. Experimental results and corresponding parameter sizes on Ant Financial dataset.

Model	Paras	Train	Test
(1) Siamese-CNN [24]	994K	0.5223	0.6021
(2) Siamese-RNN [17]	1847K	0.5462	0.6235
(3) Siamese-RCNN [13]	1674K	0.5668	0.6499
(4) Siamese-Attention-RCNN [8]	1723K	0.5725	0.6526
(5) ABCNN [23]	1536K	0.560	0.6346
(6) ESIM [5]	2118K	0.5870	0.6915
(7) ERCNN	**2077K**	**0.6022**	**0.7014**

- Siamese-CNN [24]: utilizes CNN to encode input sentences into vectors and measures similarity by the cosine distance.
- Siamese-RNN [17]: utilizes BiLSTM to encode input sentences into vectors and measures similarity as same as Siamese-CNN.
- Siamese-RCNN [13]: utilizes RCNN to encode input sentences into vectors and measures similarity by concatenating sentences vectors and performing logistic regression.
- Siamese-Attention-RCNN [8]: utilizes attention-based RCNN to encode input into sentences vectors and measures similarity as same as Siamese-RCNN.
- ABCNN [23]: utilizes attention-based CNN to encode input sentences and capture interactive features and measures similarity as same as Siamese-RCNN.
- ESIM [5]: utilizes "matching-aggregation" framework to encode input sentences and perform multiperspective matching interactions to measure similarity.

We are concerned with both parameter size and Macro F1-score when evaluating different models. We list the parameter sizes when they reached the best performances on test dataset, and Table 1 shows the detailed comparisons.

The experimental results display that our ERCNN has outperformed all the previous models in both development dataset and test dataset. From the results of model (1), (2) and (3), we can see that Siamese-RNN performs better than Siamese-CNN in paraphrase identification task. But with the enhancement of CNN, Siamese-RCNN achieves better results than Siamese-RNN. This finding implies us that the incorporation of both CNN and RNN structures can deepen learning the sentence similarity as CNN is better to extract the robust and abstract features while RNN performs better to capture the fine granularity features. These parts of experimental results also demonstrate that our improvements on ESIM are reasonable and effective. On the other hand, in view of the comparison of parameter size, we show that the model with the combination of CNN have better performance with lower complexity. For example, Siamese-RCNN has less parameters than Siamese-RNN but achieves higher Macro F1-score in experiments.

And from the results of model (4) and (5), we can discover that with the addition of attention layer, the performance of each model performs better indeed as attention layer help extract the mutual information and better measure similarity between two sentences in comparation with other models which neglect these type of information.

Table 2. Partial prediction results of ESIM and ERCNN.

Question Pairs	Label	ESIM	ERCNN
花呗是否需要绑定银行卡 Does Huabei need to be bound with credit card? 花呗可不可以绑定银行卡 Can Huabei be bound with credit card?	1	✓	✓
花呗怎么还款 How to repay Huabei? 花呗可不可以绑定银行卡 How do I repay my Huabei?	1	✗	✓
花呗绑定另一个手机号能解绑吗 Can Huabei be unbound with other mobile phone number? 我解绑花呗被让自己的另一个手机号使用花呗 Can I unbound Huabei to use other phone number?	1	✗	✓
怎么保证花呗安全 How to ensure the security of Huabei? 安全验证没有成功花呗 Security certification doesn't succeed Huabei.	0	✗	✓
花呗显示负的是什么意思 What is the meaning of the Huabei showing negative? 花呗显示负***是总共还欠着***吗 Huabei shows negative *** means that owed a total of ***?	1	✗	✗

At the end of the Table 1, the comparison results of model (6) and (7) shows that our improvements based on ESIM have achieved satisfied results, from aspects of both evaluation metric and model complexity. The most difference between our model and ESIM lies in the supplement of CNN layer after original input encoding. As our former experimental results shown, CNN can help extract more elaborated feature in sentences and reduce the model complexity. Moreover, compared with ESIM, we apply feedforward layer to replace the origin

BiLSTM layer in overall similarity modeling, and this modification of the model assists us further in reducing the model parameter size.

4.5 Further Analysis with Typical Cases

In this subsection, we analyze partial typical prediction results of ESIM and our ERCNN. Table 2 shows these partial prediction results. By comparing the prediction results for different type of question pairs, we can show that our ERCNN performs better than ESIM in learning similarity between two sentences even if two sentences don't have much common words or they have some wrongly written or mispronounced characters.

From the first case in Table 2, we can see that for two semantically similar sentences with much words or characters in common, ESIM and ERCNN are both able to measure the similarity despite the sentences lengths are short.

And for the second case, we choose a question pair with some misspellings ("花被 " is wrongly written and should be "花呗"). In such condition, the prediction result of ESIM is wrong due to the reason that it only makes use of BiLSTM to consider the sequence information and ignore the key word information, but our ERCNN with CNN as a supplement can better identify the relation between "花被" and "花呗", and therefore is not easily to be mislead by misspellings.

Then, for the third and fourth cases, the question pairs to be compared both don't look like grammatical sentences. Because in the scene of intelligent customer service, people post their questions in mobile devices and often full of grammatical mistakes, so a practical model should be able to handle this condition and identify the client's real purpose. The third and fourth cases show comparisons between ESIM and ERCNN in above-mentioned condition. As the result turns out, ESIM is mislead by the semantic confusion and can't identify the similar question pair, but our ERCNN makes out in this condition because the incorporation of CNN improves the ability to capture robust and abstract features. With such enhancement, our model is relatively hard to be puzzled by grammatical mistake in sentences.

In the final case, we're supposed to infer that ESIM and ERCNN both misjudge the similarity of two sentences. We believe the reason is that the two models are mislead by the specific symbol such as "*" in the sentences. With the development of mobile internet, an increasing number of specific symbols or emoticons appear in our daily network communication [9]. These specific symbols are meaningful and full of sentiment information [2]. Our ERCNN and ESIM both don't take these specific symbols into consideration when measuring similarity and we need to attach sufficient weight to them in future work.

5 Conclusions and Future Work

In this paper, we have introduced a new model named ERCNN based on ESIM for paraphrase identification task. ERCNN achieves the state-of-the-art performance in learning sentence similarity on Ant Financial competition dataset. Our model demonstrates the incorporation of CNN in local similarity modeling which helps extract more elaborated information and puts the fully-connected layers to replace the time-consuming BiLSTM layers in overall similarity modeling with lower complexity. In the future work, we plan to extend the ERCNN model to other NLP tasks like question answering and natural language inference, and we will pay more attention to the specific symbols and emoticons in sentences to adapt to the reality.

Acknowledgments. We thank the anonymous reviewers for their helpful comments on this paper. This work was partially supported by National Natural Science Foundation of China (61572049 and 61876009).

References

1. Bahdanau, D., Cho, K., Bengio, Y.: Neural machine translation by jointly learning to align and translate. arXiv preprint arXiv:1409.0473 (2014)
2. Barbieri, F., Ronzano, F., Saggion, H.: What does this emoji mean? A vector space skip-gram model for twitter emojis. In: LREC (2016)
3. Bromley, J., Guyon, I., LeCun, Y., Säckinger, E., Shah, R.: Signature verification using a "siamese" time delay neural network. In: Advances in Neural Information Processing Systems, pp. 737–744 (1994)
4. Cao, S., Lu, W., Zhou, J., Li, X.: cw2vec: Learning Chinese word embeddings with stroke n-gram information. In: Thirty-Second AAAI Conference on Artificial Intelligence (2018)
5. Chen, Q., Zhu, X., Ling, Z., Wei, S., Jiang, H., Inkpen, D.: Enhanced LSTM for natural language inference. arXiv preprint arXiv:1609.06038 (2016)
6. Cho, K., et al.: Learning phrase representations using RNN encoder-decoder for statistical machine translation. arXiv preprint arXiv:1406.1078 (2014)
7. Devlin, J., Chang, M.W., Lee, K., Toutanova, K.: BERT: pre-training of deep bidirectional transformers for language understanding. arXiv preprint arXiv:1810.04805 (2018)
8. Du, C., Huang, L.: Text classification research with attention-based recurrent neural networks. Int. J. Comput. Commun. Control **13**(1), 50–61 (2018)
9. Eisner, B., Rocktäschel, T., Augenstein, I., Bošnjak, M., Riedel, S.: emoji2vec: learning emoji representations from their description. arXiv preprint arXiv:1609.08359 (2016)
10. He, K., Zhang, X., Ren, S., Sun, J.: Deep residual learning for image recognition. In: Proceedings of the IEEE Conference on Computer Vision and Pattern Recognition, pp. 770–778 (2016)
11. Hochreiter, S., Schmidhuber, J.: Long short-term memory. Neural Comput. **9**(8), 1735–1780 (1997)
12. Kingma, D.P., Ba, J.: Adam: a method for stochastic optimization. arXiv preprint arXiv:1412.6980 (2014)
13. Lai, S., Xu, L., Liu, K., Zhao, J.: Recurrent convolutional neural networks for text classification. In: Twenty-Ninth AAAI Conference on Artificial Intelligence (2015)
14. LeCun, Y., Bottou, L., Bengio, Y., Haffner, P., et al.: Gradient-based learning applied to document recognition. Proc. IEEE **86**(11), 2278–2324 (1998)
15. Lin, M., Chen, Q., Yan, S.: Network in network. arXiv preprint arXiv:1312.4400 (2013)

16. Mikolov, T., Sutskever, I., Chen, K., Corrado, G.S., Dean, J.: Distributed representations of words and phrases and their compositionality. In: Advances in Neural Information Processing Systems, pp. 3111–3119 (2013)
17. Mueller, J., Thyagarajan, A.: Siamese recurrent architectures for learning sentence similarity. In: Thirtieth AAAI Conference on Artificial Intelligence (2016)
18. Nair, V., Hinton, G.E.: Rectified linear units improve restricted Boltzmann machines. In: Proceedings of the 27th International Conference on Machine Learning (ICML-10), pp. 807–814 (2010)
19. Neculoiu, P., Versteegh, M., Rotaru, M.: Learning text similarity with siamese recurrent networks. In: Proceedings of the 1st Workshop on Representation Learning for NLP, pp. 148–157 (2016)
20. Peters, M.E., et al.: Deep contextualized word representations. arXiv preprint arXiv:1802.05365 (2018)
21. Sutskever, I., Vinyals, O., Le, Q.V.: Sequence to sequence learning with neural networks. In: Advances in Neural Information Processing Systems, pp. 3104–3112 (2014)
22. Wang, Z., Hamza, W., Florian, R.: Bilateral multi-perspective matching for natural language sentences. arXiv preprint arXiv:1702.03814 (2017)
23. Yin, W., Schütze, H., Xiang, B., Zhou, B.: ABCNN: attention-based convolutional neural network for modeling sentence pairs. Trans. Assoc. Comput. Linguist. **4**, 259–272 (2016)
24. Zhang, X., Zhao, J., LeCun, Y.: Character-level convolutional networks for text classification. In: Advances in Neural Information Processing Systems, pp. 649–657 (2015)

Improving a Syntactic Graph Convolution Network for Sentence Compression

Yifan Wang and Guang Chen[(✉)]

Beijing University of Posts and Telecommunications, Beijing, China
{yifan_wang, chenguang}@bupt.edu.cn

Abstract. Sentence compression is a task of compressing sentences containing redundant information into short semantic expressions, simplifying the text structure and retaining important meanings and information. Neural network-based models are limited by the size of the window and do not perform well when using long-distance dependent information. To solve this problem, we introduce a version of the graph convolutional network (GCNs) to utilize the syntactic dependency relations, and explore a new way to combine GCNs with the Sequence-to-Sequence model (Seq2Seq) to complete the task. The model combines the advantages of both and achieves complementary effects. In addition, in order to reduce the error propagation of the parse tree, we dynamically adjust the dependency arc to optimize the construction process of GCNs. Experiments show that the model combined with the graph convolution network is better than the original model, and the performance in the Google sentence compression dataset has been effectively improved.

Keywords: Sentence compression · Graph convolution network · Sequence-to-Sequence

1 Introduction

Sentence compression is a standard NLP task that simplifies sentences while preserving the most important content. Deletion-based sentence compression treats the task as a word deletion problem: given an input source sentence $\mathbf{x} = (x_0, x_1, \ldots, x_N)$ (each symbol represents a word in the sentence), and the goal is to generate a sentence summarization by deleting the words in the source sentence $\mathbf{x}$. There have been many studies related to syntactic-tree-based methods [1–4] or machine-learning-based methods [5–7] in the past. A common approach is to use the syntactic tree to produce the most readable and informative compression [1, 3]. Meanwhile, Filippova et al. [5] treated the sentence compression task as a sequence labeling problem, using the recurrent neural network (RNN), which has become a main-stream solution. However, the two methods still have the following problems:

1. Automatic generation of compressions by simply pruning the syntactic tree is inevitably vulnerable to error propagation as there is no way to recover from an incorrect parse tree.

© Springer Nature Switzerland AG 2019
M. Sun et al. (Eds.): CCL 2019, LNAI 11856, pp. 131–142, 2019.
https://doi.org/10.1007/978-3-030-32381-3_11

2. The sequence labeling-based recurrent neural network (RNN) approach focuses more on the word itself than on the overall sentence structure; due to the timing characteristics of the RNN, the jumping information with long distance is difficult to capture.

The purpose of our paper is to study how to combine the syntactic structure with the neural network to achieve complementary effects and reduce the error propagation of the parse tree.

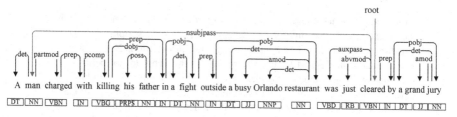

Fig. 1. An example of sentence compression and its syntactic parsing results in Google sentence compression dataset. Words in gray indicate are supposed to be deleted. The tags below the words indicate the corresponding part-of-speech tags in the tree.

Semantic representation is closely related to syntactic features. Intuitively, in a syntactic dependency tree, if the parent node has a high probability to be retained, the child node directly connected to it should also be retained with a certain probability.

For example, Fig. 1 shows an example of compressed sentence and its syntactic dependency tree. The word "*cleared*" connected to the "*root*" is retained with a great probability. The word "*man*" which directly connected to "*cleared*" by dependent arc should also be given a high score, although the two words in the sentence are far apart (15 words apart). Furthermore, following the dependency chain, the phrases "*charged with*" and "*by a grand jury*" have a greater tendency to be retained too. As a result, given the similarities of dependency trees and compressed sentences, syntactic information can be used naturally when generating sentence summaries.

In particular, in this paper we introduce graph convolutional network (GCNs) [8, 9] in the sentence compression task, combining it with the sequence-to-sequence model (Seq2Seq). GCNs is a multi-layer neural network acting on graphs and is widely used in irregular data tasks. By using dependency relations, syntactic context representation is generated for each node in the graph (each word in the text). Further, this representation is incorporated into the encoding structure of the Seq2Seq model in order to obtain more relevant information. In Sect. 3.4 we will discuss two combination methods of GCNs and Seq2Seq: stacked GCNs and parallel GCNs. As we use the parallel method, we achieve the state-of-the-art performance on the widely used dataset for sentence compression.

2 Related Work

The general techniques of text summarization are mainly divided into two categories: extractive methods and abstractive methods. The method of extraction is designed to select valid fragments, sentences or paragraphs from the text, and generate a shorter summary. The abstract summary is closer to natural language expression and intended to be generated using natural language generation techniques. This paper focuses on extractive summaries based on deletions.

Previous work used syntactic-tree-based approach to produce readable and informative compression. Jing [1] used a variety of external resources to determine which part of the sentence can be removed, eventually the component with little relevance to the context was removed. Knight et al. [10] proposed applying the noise channel model to the sentence compression task. Clarke et al. [3] first used the ILP framework for sentence compression. Filippova et al. [4] operated directly on the syntactic parsing tree to get the compressed sentence and constructed a large-scale parallel corpus, which laid the foundation for the subsequent research.

In recent work, researchers have attempted to use neural networks in sentence compression tasks [11, 12]. Filippova et al. [5] treated the sentence compression task as a sequence labeling problem. The model based on the recurrent neural network (RNN) was trained on a large-scale parallel corpus. In terms of introducing the syntactic features, Wang et al. [7] used explicit syntactic features and introduced syntactic constraints through Integer Linear Programming (ILP). In order to solve the problem of long sentence compression, Kamigaito et al. [13] handled dependency features as an attention distribution on LSTM hidden states. Marcheggiani et al. [9] introduced GCNs into the semantic role labeling task and stacked it with LSTMs. On this basis, we explore more ways to combine GCNs and RNN, and the structure of GCNs is further optimized to mitigate the effects of error propagation of the parse tree.

3 Approach

3.1 Problem Definition

Sentence compression based on deletion is regarded as a sequence labeling task. The original sentence is represented as $\mathbf{x} = (x_0, x_1, \ldots, x_N)$ (where N represents the length of the sentence). By retaining or deleting words, a sequence that still retains important information is obtained. The output is a binary decision sequence $\mathbf{y} = (y_0, y_1, \ldots, y_N)$, $y_i \in \{0, 1\}$. Here, $y_i = 1$ represents RETAIN, and $y_i = 0$ represents REMOVE.

3.2 Base Seq2Seq Model

First we introduce our baseline model, an extended model of Seq2Seq. In the deleted based task, LSTM-based tagging model [7, 14] and Seq2Seq model [5, 13, 15] are widely used. Considering that Seq2Seq model can capture the overall sentence information, we improve Seq2Seq model [13] as the basis for experiments.

In the input layer, the input sentence is converted to a dense vector representation. According to the setting of Wang et al. [7], besides pre-training word vectors $\mathbf{w} = (w_0, w_1, \ldots, w_N)$, we also consider part-of-speech tags and dependency relations as additional feature inputs. The part-of-speech tag embedding is denoted as $\mathbf{t} = (t_0, t_1, \ldots, t_N)$ and the dependency relation embedding is denoted as $\mathbf{r} = (r_0, r_1, \ldots, r_N)$. The three are concatenated to form a new word representation as input:

$$
\begin{aligned}
\mathbf{e} &= (e_0, e_1, \ldots, e_N) \\
e_t &= [w_t, t_t, r_t]
\end{aligned}
\tag{1}
$$

Here [] represents the concatenation of vectors, and e_t is the vector input when the time step is t.

In the encoder layer, we use the standard bidirectional LSTM model to encode the input embedding vector into a series of hidden layer vectors $\mathbf{h} = (h_0, h_1, \ldots, h_N)$. The representation of the encoder output h_t is:

$$
h_t = \left[\overrightarrow{h_t}, \overleftarrow{h_t} \right]
\tag{2}
$$

$$
\overrightarrow{h_t} = \overrightarrow{LSTM}\left(\overrightarrow{h_{t-1}}, e_t \right)
\tag{3}
$$

$$
\overleftarrow{h_t} = \overleftarrow{LSTM}\left(\overleftarrow{h_{t-1}}, e_t \right)
\tag{4}
$$

In the decoder layer, LSTM unit is used as the basic unit of the RNN. Unlike Kamigaito et al. [13], in each step i, the decoder state s_i is determined by the previous decoder state s_{i-1}, the previously predicted label y_{i-1}, the aligned encoder hidden layer output h_i, and the context vector c_i:

$$
s_i = f_{LSTM}(s_{i-1}, y_{i-1}, h_i, c_i)
\tag{5}
$$

Where y_i is the predicted label, f_{LSTM} is the state transition operation inside LSTM unit, and the context vector c_i is calculated as the weighted sum of the encoder states [16]:

$$
c_i = \sum_{j=1}^{N} \alpha_{i,j} h_j
\tag{6}
$$

$$
\begin{aligned}
\alpha_{i,j} &= \frac{\exp(e_{i,j})}{\sum_{k=1}^{N} \exp(e_{i,k})} \\
e_{i,k} &= g(s_{i-1}, h_k)
\end{aligned}
\tag{7}
$$

Where h_j is the output of the encoder at different time step j. e_{ij} represents the matching score of the input in the position j and the output in the position i, which is calculated by the hidden layer state s_{i-1} and h_j.

The final output layer calculates the probability of each label:

$$p(y_i|y_{i-1}, \mathbf{x}) = softmax(Ws_i + b) \tag{8}$$

Where W and b are the parameter matrices and bias to be learned.

3.3 Syntactic GCNs

GCNs aims to capture the dependency relations between nodes. Each node in the graph constantly changes its state due to the influence of neighbors and farther points. The closer the neighbor is, the greater its influence is. For a graph $G = (V, E)$, the input $\mathbf{x}$ is a $N*D$ matrix. There is also an adjacency matrix $\mathbf{A}$ of the graph. It is desirable to obtain a $N*F$ feature matrix $\mathbf{Z}$, representing the feature of each node, where F is the dimension of the desired representation.

In Kipf et al. [8], GCNs is proved to be very effective for node classification tasks. In this paper, we introduce it into the sentence compression task. Here G is the parse tree for the input sentence x, $V = (v_1, v_2, \ldots, v_n)$ ($|V| = n$) is the word in x, and $K = (v_i, v_j) \in E$ is the directed dependent arc between the words in x. In addition to the forward arc, we also consider the reverse flow of information: the reverse arc, and the self-circulating arc [9]. As shown in Fig. 1, there is an arc $K("man", "charged") = partmod$ (verb modification relationship) between "man" and "charged". Naturally, there is a modified relationship $K("charged", "man") = partmod'$ between the two nodes, with the apostrophe indicating that the information flows in the opposite direction of the arc.

For one layer of GCNs, the relationship of direct neighbor nodes can be captured; for a k-layers stacked GCNs, the relationship of up to k-order chains can be learned. In the k-th syntactic convolutional network layer, we calculate the graph convolution vector of the node $v \in V$ by the following method:

$$h_v^{(k+1)} = f\left(\sum_{u \in N(v)} (W_{K(u,v)}^{(k)} h_u^{(k)} + b_{L(u,v)}^{(k)}) \right) \tag{9}$$

Where $K(u, v)$ and $L(u, v)$ denotes the type of dependency relations; $W_{K(u,v)}^{(k)}$ and $b_{L(u,v)}^{(k)}$ are weight matrices and bias; $N(v)$ is the set of neighbors of v, including v (because it contains a self-loop); f is the activation function.

Since the parameters W are specific to the dependency tags, assuming that the number is N, then there will be (2N + 1) parameter matrices for one layer of GCNs to be trained. Specifically, for the Google sentence compression dataset used in this paper, there are 54 dependency relations, and 109 parameter matrices will be learned. The training costs make model fall into the predicament of over-parameterization. Following the settings of Marcheggiani et al. [9], we use different dependency relations on the bias terms, but simplify W, only to show three edges (1) forward; (2) reverse;

(3) self-looping. The parameters W are shared in the same direction side of the different tags, and the dependency is expressed as:

$$K(w_i, w_j) = \begin{cases} along, & (w_i, w_j) \in E \\ rev, & i! = j \,\&\, (w_j, w_i) \in E \\ loop, & i == j \end{cases} \tag{10}$$

However, considering the sentence summary task, on the one hand, the structure of GCNs is completely determined by the syntactic parsing result, which is unchangeable during the training process. In order to reduce the influence of the error propagation, we set a dynamic adjustment of the inputs' dependency arc. On the other hand, in the compression process, there is no need to rely equally on the edges contained in the graph. It is necessary to add different weights to the edges. In response to these problems, we make the following two changes:

1. According to Eq. (9), the input of node v is determined by its parent u. For a given relationship, the structure of GCNs is determined. In this paper, the dependent parent of the input parsing is dynamically adjusted to make it structurally variable during the training process. Define $p = (u_0, u_1, \ldots u_N)$ as parent set of sentence x, u_t is the parent node of v_t. The one-hot representation of p is denoted by p_v, W and b are trainable parameter matrices, and σ is the logistic sigmoid function. The adjusted dependent parent p' is represented as:

$$p' = argmax(\sigma(Wp_v + b)) \tag{11}$$

2. Inspired by Marcheggiani et al. [9] and Dauphin et al. [17], a gating unit is added for each node output, using weights to get different importance distributions:.

$$g_{u,v}^{(k)} = \sigma\left(h_u^{(k)} \cdot V_{K(u,v)}^{(k)} + b_{L(u,v)}^{(k)} \right) \tag{12}$$

$V_{K(u,v)}^{(k)}$ and $b_{L(u,v)}^{(k)}$ are the parameter matrix and bias of the gating unit. After adding the gate, the syntactic GCNs's final output is:

$$h_v^{(k+1)} = ReLU\left(\sum_{u \in N(v)} g_{u,v}^{(k)}(W_{K(u,v)}^{(k)} h_u^{(k)} + b_{L(u,v)}^{(k)}) \right) \tag{13}$$

3.4   Combining GCNs and Seq2Seq

GCNs can capture the dependency relations between associated arcs, which indicates that GCNs has the ability to learn semantic information over long distance, but may not perform well on local sequential information mining. Studies have shown that, the LSTMs model has outstanding performance in processing context information in this task [6, 7].

At this point, it can be considered that LSTMs and GCNs play complementary roles. LSTMs can handle timing-related context information. GCNs establishes the connection of semantically related words, although they may be far apart in the same sentence.

In this paper, GCNs structure is integrated into the encoder of Seq 2Seq, and the different connection methods of GCNs and LSTMs are compared.

Stacked Model

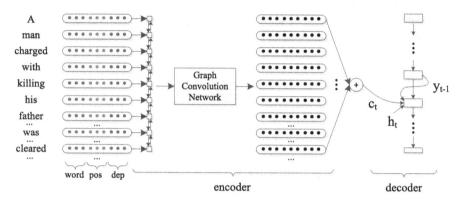

Fig. 2. Stacked model, including input layer, LSTMs layer, GCNs layer, and decoder portion.

As shown in Fig. 2, similar to Marcheggiani et al. [9], LSTMs model and GCNs model are stacked. The output of bidirectional LSTMs layer is obtained from Eq. (2), as an input to GCNs. The local context information is adaptively acquired by LSTMs.

Parallel Model

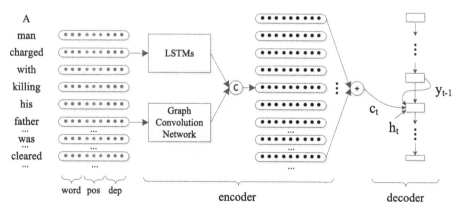

Fig. 3. Parallel model, parallelizing LSTMs layer and GCNs layer in the encoder section. The function C is introduced by formulas 15 and 16.

The above model represents the encoder as a sequential superposition of Bi-LSTMs layer and GCNs layer. However, previous work did not explicitly explore why LSTMs and GCNs should be stacked: this paper attempts to use a parallel form to dynamically control the output of the encoder by calculating the probability of using LSTMs or GCNs. As shown in Fig. 3, the final output of the encoder is determined jointly by the output of LSTMs and GCNs. We demonstrate the validity of this idea in later experiments. The previous model can be considered as stacked, and the model proposed next can be considered as parallel.

After the input layer, the same feature vector is input to Bi-LSTMs layer and GCNs layer, which means using the original input sentence instead of the output of LSTMs as input to GCNs:

$$e = (e_0, e_1, \cdots, e_N) \tag{14}$$

Calculate the probability of using LSTMs or GCNs for each word:

$$g_t^{\alpha} = \sigma(h_t^{\alpha} w_{\alpha} + b_{\alpha}) \tag{15}$$

α represents LSTMs layer or GCNs layer, h_t^{α} is the hidden layer output, W and b are trainable parameter matrices and bias, and σ is the logistics activation function.

The output of the encoder contains the information of LSTMs hidden layer h^{lstm} and GCNs hidden layer h^{gcn}, calculated by the following formula:

$$h_t = [g_t^{lstm} \cdot h_t^{lstm}, g_t^{gcn} \cdot h_t^{gcn}] \tag{16}$$

4   Experiment

4.1   Dataset and Parameters

We use Google sentence compression dataset [5], as shown in Table 1, containing 200,000 sentence compression pairs[1]. The dataset contains each sentence's compressed tags, part-of-speech (POS) tags, dependency parents and dependency tags. For comparison with previous studies [5, 7, 13, 15, 18], the first 1000 and last 1000 sentences of comp-data.eval.json are used as the test dataset and validation dataset, and the rest is used as training dataset.

Table 1. Google sentence compression dataset

	Google dataset
Size	200,000
Compression ratio	0.40
Vocab size	42621

---

[1] https://github.com/google-research-datasets/sentence-compression.

In the experiment, for the input layer, the embedding vector size of pre-training word, the part-of-speech tag and the dependency is 100. The word vector is initialized by GloVe-100 dimension pre-training embedding, and the part of speech and dependency embedding is randomly initialized. The bidirectional LSTMs in the Seq2Seq model encoder is set to two layers, and the depth of decoder is also set to two. As a result, the total number of layers is six, which is comparable to Wang et al. [7] and Kamigaito et al. [13]. Dropout [19] rates are 0.3. The batchsize is set to 16. All experiments use the Adam [20] optimization algorithm, with the initial learning rate being 0.0001.

4.2   Evaluation

We use Google sentence compression dataset with true-value tags for automated evaluation. Word-level accuracy (Accuracy) and F1-score are used to measure the validity of the model. Accuracy is defined as the correct percentage of each predicted label yi in the sentence. The F1 score is calculated based on the precision (Precision) and recall rate (Recall) in terms of tokens kept in the golden and the generated compressions. In addition, we measure the difference between truth and prediction by calculating the Compression Ratio (CR). To compare with the previous work, Table 2 shows the performance of the experiments in our work.

Table 2. Results in Google dataset

Google dataset	F1	Accuracy	CR
&1 Traditional ILP [3][a]	54.0	56.0	62.0
&2 LSTMs [5]	82.0	–	38.0
&3 Seq2Seq [13]	82.7	–	37.4
&4 HiSAN [13]	83.2	–	38.1
&5 Seq2Seq (Our implementation)	82.6	86.2	38.0
&6 Seq2Seq+stacked GCNs (k = 1)	83.0	86.4	38.3
&7 Seq2Seq+stacked GCNs (k = 1), without adjusted	82.9	86.2	38.0
&8 Seq2Seq+stacked GCNs (k = 2)	82.6	86.2	38.1
&9 Seq2Seq+parallel GCNs (k = 1)	**83.3**	**86.5**	**39.0**
&10 Seq2Seq+parallel GCNs (k = 1), without adjusted	82.7	86.1	38.6
&11 Seq2Seq+parallel GCNs (k = 2)	83.1	86.4	38.5
&12 Seq2Seq+GCNs(no LSTMs) (k = 1)	82.0	85.4	38.0
&13 Seq2Seq+GCNs(no LSTMs) (k = 2)	81.9	85.4	39.0

[a]Note that we quote the results from Wang et al. [7].

1. First, we implement a powerful baseline model &5 that is comparable to other research results. This version of the Seq2Seq model, which does not add GCNs, is 0.6% better than LSTMs model &2 with larger datasets in F1 score.
2. In order to verify the validity of GCNs layer, we compare our GCNs model with baseline under the same settings. It was observed that the stacked model &6 and the

parallelized model &9 respectively increased the F1 score by 0.4% and 0.7%, compared to the baseline. The performance of parallel model is better than the best results [13] on the current dataset. The compression ratio of our model is also closer to the original dataset. From model &7 and model &10, which don't adjust the dependent parents, we observe the effectiveness of this improvement.

3. It can be seen from experiments &8 and &11 that increasing the number of layers of GCNs does not improve the experimental effect. After adding a layer of GCNs (k = 2) to the stacked model and the parallel model, the F1 score decreased by 0.4% and 0.2%, respectively. This may be because the combination of LSTMs and one layer of GCNs has obtained most of the information. The multi GCNs layer adds redundant information, which causes interference to the prediction results.

4. Experiments &6 and &9 compare stacked GCNs and parallel GCNs models. The F1 score, accuracy, and compression ratio of the parallel model all performed better.

5. Finally, in experiments &12 and &13, we compare the versions that GCNs are employed alone in the encoder. The results are incomparable with the model combined with LSTMs, which prove the irreplaceability of LSTMs.

4.3 Results and Discussion

To further analyze the experimental results, Table 3 shows examples of the original sentence, the compressed sentence, and its predicted compression in this paper.

Table 3. Examples of original sentence, compressed sentence, and its predicted compression

Source:	A man charged with killing his father in a fight outside a busy Orlando restaurant was just cleared by a grand jury
Ground truth:	A man charged with killing his father was cleared by a grand jury
Seq2Seq:	A man charged with killing his father in a fight was cleared
Stacked model (k = 1):	A man charged with killing his father was cleared by a grand jury
Parallel model (k = 1):	A man charged with killing his father was cleared by a jury

Seen from the example sentences, although the output of the base Seq2Seq model is syntactically coherent, it lacks key information such as "*by a grand jury*". In the model with GCNs, either the stacked model or the parallel model contains this information. As observed in Fig. 1, the word "*cleared*" directly connected to the root is retained with high probability, and the phrase "*by a grand jury*" directly connected by the dependent arc also has a greater retention propensity, retaining in the compressed sentence. This is in line with the experimental expectations of this paper and is confirmed during the evaluation phase.

From the results of the overall evaluation, the performance of parallel model is better than stacked model. We think this may be because that the direct utilization of GCNs leads to an increase in compression ratio and fits the real data better.

5   Conclusion

In this paper, we introduce and improve the graph convolutional neural network in the sentence compression task, and explore a variety of its combinations with the sequence-to-sequence model. The experiment results show that the model based on the graph convolutional neural network can obtain better compression effect, have different degrees of improvement in F1 and accuracy score, and further improve the compression ratio.

References

1. Jing, H.: Sentence reduction for automatic text summarization. In: Sixth Applied Natural Language Processing Conference (2000)
2. McDonald, R.: Discriminative sentence compression with soft syntactic evidence. In: 11th Conference of the European Chapter of the Association for Computational Linguistics (2006)
3. Clarke, J., Lapata, M.: Global inference for sentence compression: An integer linear programming approach. J. Artif. Intell. Res. **31**, 399–429 (2008)
4. Filippova, K., Altun, Y.: Overcoming the lack of parallel data in sentence compression (2013)
5. Filippova, K., Alfonseca, E., Colmenares, C.A., et al.: Sentence compression by deletion with LSTMs. In: Proceedings of the 2015 Conference on Empirical Methods in Natural Language Processing, pp. 360–368 (2015)
6. Zhao, Y., Senuma, H., Shen, X., Aizawa, A.: Gated neural network for sentence compression using linguistic knowledge. In: Frasincar, F., Ittoo, A., Nguyen, L.M., Métais, E. (eds.) NLDB 2017. LNCS, vol. 10260, pp. 480–491. Springer, Cham (2017). https://doi.org/10.1007/978-3-319-59569-6_56
7. Wang, L., Jiang, J., Chieu, H.L., et al.: Can syntax help? Improving an LSTM-based sentence compression model for new domains. In: Proceedings of the 55th Annual Meeting of the Association for Computational Linguistics, vol. 1: Long Papers, pp. 1385–1393 (2017)
8. Kipf, T.N., Welling, M.: Semi-supervised classification with graph convolutional networks. arXiv preprint arXiv:1609.02907 (2016)
9. Marcheggiani, D., Titov, I.: Encoding sentences with graph convolutional networks for semantic role labeling. arXiv preprint arXiv:1703.04826 (2017)
10. Knight, K., Marcu, D.: Statistics-based summarization-step one: Sentence compression. In: AAAI/IAAI, 2000, pp. 703–710 (2000)
11. Zhao, Y., Luo, Z., Aizawa, A.: A language model based evaluator for sentence compression. In: Proceedings of the 56th Annual Meeting of the Association for Computational Linguistics, vol. 2: Short Papers, pp. 170–175 (2018)
12. Fevry, T., Phang, J.: Unsupervised sentence compression using denoising auto-encoders. arXiv preprint arXiv:1809.02669 (2018)
13. Kamigaito, H., Hayashi, K., Hirao, T., et al.: Higher-order syntactic attention network for longer sentence compression. In: Proceedings of the 2018 Conference of the North American Chapter of the Association for Computational Linguistics: Human Language Technologies, vol. 1 (Long Papers), pp. 1716–1726 (2018)

14. Chen, Y., Pan, R.: Automatic emphatic information extraction from aligned acoustic data and its application on sentence compression. In: Thirty-First AAAI Conference on Artificial Intelligence (2017)
15. Tran, N.T., Luong, V.T., Nguyen, N.L.T., et al.: Effective attention-based neural architectures for sentence compression with bidirectional long short-term memory. In: Proceedings of the Seventh Symposium on Information and Communication Technology, pp. 123–130. ACM (2016)
16. Bahdanau, D., Cho, K., Bengio, Y.: Neural machine translation by jointly learning to align and translate. arXiv preprint arXiv:1409.0473 (2014)
17. Dauphin, Y.N., Fan, A., Auli, M., et al.: Language modeling with gated convolutional networks. In: Proceedings of the 34th International Conference on Machine Learning, vol. 70, pp. 933–941. JMLR.org (2017)
18. Klerke, S., Goldberg, Y., Søgaard, A.: Improving sentence compression by learning to predict gaze. arXiv preprint arXiv:1604.03357 (2016)
19. Srivastava, N., Hinton, G., Krizhevsky, A., et al.: Dropout: a simple way to prevent neural networks from overfitting. J. Mach. Learn. Res. 15(1), 1929–1958 (2014)
20. Kingma, D.P., Ba, J.: Adam: a method for stochastic optimization. arXiv preprint arXiv: 1412.6980 (2014)

Comparative Investigation of Deep Learning Components for End-to-end Implicit Discourse Relationship Parser

Dejian Li, Man Lan$^{(\boxtimes)}$, and Yuanbin Wu$^{(\boxtimes)}$

School of Computer Science, East China Normal University,
Shanghai, People's Republic of China
51194506071@stu.ecnu.edu.cn, {mlan,ybwu}@cs.ecnu.edu.cn

Abstract. The neural components in deep learning framework are crucial for the performance of many natural language processing tasks. So far there is no systematic work to investigate the influence of neural components on the performance of implicit discourse relation recognition. To address it, in this work we compare many different components and build two implicit discourse parsers base on the sequence and structure of sentence respectively. Experimental results show due to different linguistic features, the neural components have different effects in English and Chinese. Besides, our models achieve state-of-the-art performance on CoNLL-2016 English and Chinese datasets.

Keywords: Deep learning · Implicit discourse relation classification · Word embedding · Neural network

1 Introduction

Discourse consists of a series of consecutive text units, such as clauses, sentences or paragraphs. They are coherent both in form and content, conveying the complete information together. Discourse relation (e.g., *Contrast, Conjunction*) is the semantic logic relationship between two text units. Discourse relation recognition benefits many downstream NLP tasks such as Sentiment Analysis [28], Machine Translation [7] and Summarization [24], etc.

Discourse relation can be divided into explicit discourse relation and implicit discourse relation according to whether the arguments contain discourse connectives. The recognition of explicit discourse relationship reaches 93% accuracy by using only discourse connectives [14], but the performance of implicit discourse relationship recognition is always poor due to the lack of discourse connectives, which is the bottleneck of the whole discourse parser. In order to fix this problem, early researchers designed many complex features with expert knowledge. [1] used an aggregated approach to word pair features and [18] employed Brown cluster pairs to represent discourse relation. However, this method performs badly in generalization.

© Springer Nature Switzerland AG 2019
M. Sun et al. (Eds.): CCL 2019, LNAI 11856, pp. 143–155, 2019.
https://doi.org/10.1007/978-3-030-32381-3_12

With the development of deep learning in the NLP field, researchers begin to use this method to recognize implicit discourse relation. Their methods can be divided into two lines in general. One research line is to learn from explicit discourse relation. [2,6,19] tried to expand the implicit training dataset with the help of the discourse connectives. [26] learned discourse-specific word embedding from massive explicit data. [16] presented their implicit network to learn from another neural network which has access to connectives. The other line focuses on the expression of words and the structure of the model. [15,23] used word2vec word embedding and Convolutional Neural Network (CNN) to determine the senses. [8] used CNN to model argument pairs with GloVe word embedding and multi-task learning system. [3] used BiLSTM to model the sentences and adopted gated relevance network to calculate the relevance score between two arguments. [25] employed new network structure TreeLSTM to model the sentences.

However, with the emergence of the new word embeddings (e.g., ELMo, BERT) and neural network models, there is no systematic research work to deeply analyze the influence of each component on the performance of the parser in the deep learning method. To fill this gap we conduct comparative experiments from three aspects which are word embedding layer, sentence modeling layer and sentence interaction layer. Specifically, we construct two parsers which are different in sentence modeling. One parser focuses on sentence order and the other focuses on the sentence grammar. Besides, we select four word embeddings and two ways of sentence interactions. We conduct our experiment both in English and Chinese corpora to verify the semantic expression of components in different languages.

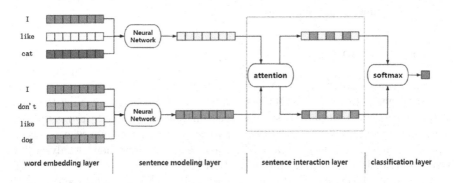

Fig. 1. Architecture of our implicit discourse relation parser system.

2   Implicit Discourse Relation Parser

Our comparative study is based on deep learning framework. We aim to compare the different components to find the key influencing factors. The proposed implicit discourse parser contains four independent components as shown in Fig. 1. First, the word embedding layer converts each single word into word vector. Then the sentence modeling layer makes semantic modeling and obtains the semantic vector expression of the sentence. Later, two sentences interact with each other to obtain semantic information in the sentence interaction layer. Finally, the predictions are obtained in classification layer by using *softmax* function.

2.1 Word Embedding Layer

In deep learning framework, the pre-trained models play an important role because the exciting performance of deep learning relies on the training in large corpus. Word embedding is the first and crucial step in deep learning framework, which transforms the natural language into word vector as the input of the neural network. Different pre-trained word vector models are chosen to verify whether there is any loss or misinterpretation between the conversion.

We convert each word w in the argument into word vector $x \in \mathbb{R}^{d_w}$, where d_w is the dimension of the word vector. Let x_i^1 (x_i^2) be the i-th word vector in Arg-1(Arg-2), then the two discourse arguments are represented as:

$$Arg - 1 : \left[x_1^1, x_2^1, \cdots, x_{L_1}^1 \right] \tag{1}$$

$$Arg - 2 : \left[x_1^2, x_2^2, \cdots, x_{L_2}^2 \right] \tag{2}$$

where Arg-1(Arg-2) has L_1(L_2)words. Generally, the word embeddings are pre-trained on large corpus and supposed to contain latent semantic and syntactic information. In recent years several supreme word embeddings have been presented by researchers. To examine their different effectiveness in word conversion, we choose two types of pre-trained word vector models, i.e., context-free models and contextual models.

Context-free models generate a "word embedding" representation for each word in the vocabulary. This means the vector representation of the word in argument has no relation with the specific context of this argument. Here we choose word2vec [11] and GloVe [12] models which are widely used. The word2vec uses local text which is controlled by window size from large corpus to train the word vector. While GloVe is trained on aggregated global word-word co-occurrence statistics from the corpus.

Contextual models generate a representation of each word which is based on the other words in the sentence. It is usually pre-trained on large corpus by learning the language model rather than the "word embedding". Thus for specific sentence, contextual model generates the word representation base on its context. Here we choose ELMo [13] and BERT [5] models.

2.2 Sentence Modeling Layer

After the word embedding layer, we get the sentence representation in a word vector matrix. This only represents the information of source corpus rather than the target recognition task. Therefore, in order to better fit the task, we use the neural network models to convert the vector matrix into the semantic representation of the arguments.

Considering that both sentence order and grammar are important in understanding the semantics of sentences, so we choose two types of sentence modeling methods, i.e., sequential relation modeling focusing on sentence order and structural relation modeling focusing on sentence grammar.

- **Sequential relation modeling**: We select three representative sequential models: Long short Term Memory (LSTM), Bi-directional Long Short Term Memory (BiLSTM) and Convolutional Neural Network (CNN).
- **Structural relation modeling**: We use Tree-LSTM [22] to capture the structure relation information, which is the combination of LSTM and tree structured neural networks. Here we choose Child-Sum-Tree-LSTM and Binary-Tree-LSTM. The former has flexible number of child nodes in its tree while the latter has only two child nodes for each father node. The above two tree structures come from the sentence constituency parse tree which is obtained by the Stanford CoreNLP toolkit [9].

Given the two arguments representations as Formulas (1) and (2), the LSTM computes the state sequence $[h_1, h_2, \cdots, h_L]$ for each time step i using the following formulate:

$$i_i = \sigma(W_i[x_i, h_{i-1}] + b_i) \tag{3}$$

$$f_i = \sigma(W_f[x_i, h_{i-1}] + b_f) \tag{4}$$

$$o_i = \sigma(W_o[x_i, h_{i-1}] + b_o) \tag{5}$$

$$\tilde{c}_i = tanh(W_c[x_i, h_{i-1}] + b_c) \tag{6}$$

$$c_i = i_i \odot \tilde{c}_i + f_i \odot c_{i-1} \tag{7}$$

$$h_i = o_i \odot tanh(c_i) \tag{8}$$

where σ denotes the *sigmoid* function and $\odot$ denotes element-wise multiplication. On this basis, BiLSTM get the information from both past and future rather than only from the past in LSTM. Therefore, at each position i of the sequence, we obtain two states $\overrightarrow{h}_i$ and $\overleftarrow{h}_i$, where $\overrightarrow{h}_i, \overleftarrow{h}_i \in \mathbb{R}^{d_h}$. Then we concatenate them to get the intermediate state, i.e. $h_i = [\overrightarrow{h}_i, \overleftarrow{h}_i]$. After that, we sum up the sequence states $[h_1, h_2, \cdots, h_L]$ to get the representations of *Arg*-1 and *Arg*-2 respectively as follows:

$$R_{Arg_1} = \sum_{i=1}^{L_1} h_i^1 \tag{9}$$

$$R_{Arg_2} = \sum_{i=1}^{L_2} h_i^2 \tag{10}$$

As for CNN model, we use $Arg[i : j]$ to represent the sub-matrix of Arg from row i to row j. A convolution involves a filter $\mathbf{w} \in \mathbb{R}^{h \times d}$ (h is the height of filter and d is the dimensionality of the word vector). The output sequence o_i of the convolution operator is obtained by repeatedly applying the filter on sub-matrices of $\boldsymbol{Arg}$:

$$o_i = \mathbf{w} \cdot \mathbf{Arg}[i : i + h - 1] \tag{11}$$

where $i = 1...s - h + 1$. A bias term $b \in \mathbb{R}$ and an activation function f are added to each o_i to compute the feature map c_i for this filter:

$$c_i = f\left(o_i + b\right) \tag{12}$$

Then we use *max pooling* operation to get the representation of the argument:

$$\boldsymbol{R_{arg}} = max\{c_i\} \tag{13}$$

2.3 Sentence Interaction Layer

After the sentence modeling layer, the representation of two arguments is still isolate. This is not what we expected. Since discourse relationship is annotated by the two arguments rather than one single argument, we suppose that the interaction relationship between two arguments is helpful to the discourse relation recognition. In order to obtain argument interaction representation, in this work we choose the following interaction mechanisms:

– Attention: Two sentence vectors do Attention operation, then concatenated together
– Con-self-Attention: Two vectors concatenated, then do self-Attention operation
– Self-Attention-con: Two sentence vectors do self-Attention operation respectively, then concatenated together
– Attention-mlp: Two sentence vectors do Attention interact operation, then put into Multi-Layer Perceptron (MLP)

Through above interactions, the two separate representations, i.e., $\boldsymbol{R_{Arg1}}$, $\boldsymbol{R_{Arg2}}$, become joint pair representation $\boldsymbol{R_{pair}}$ which contains the overall information of the two arguments. Finally we feed the $\boldsymbol{R_{pair}}$ into a full-connected *softmax* layer to make sense prediction.

3 Experiment

3.1 Dataset

The Penn Discourse Treebank (PDTB) and the Chinese Discourse Treebank (CDTB) are the most widely used discourse datasets in English and Chinese, respectively. To make comparison reasonable, we use the adapted version of the data provided by CoNLL-2016 Shared Task. Table 1 show the distributions of the two datasets.

Table 1. The distribution of discourse relation types in the English and Chinese data.

	English		Chinese	
	Amount	Percentage(%)	Amount	Percentage(%)
Explicit	18,459	45.5	2,398	21.75
Implicit	16,053	39.5	7,238	65.66
EntRel	5,210	12.8	1,219	11.06
AltLex	624	1.5	223	2.02
NoRel	254	0.6	0	0
Total	40,600	100	11,023	100

Table 2. Statistics of CoNLL-2016 English implicit discourse sense.

Sense label	Train set	Development set	Test set
Comparison	2,035	96	134
Contingency	3,720	134	221
Expansion	7,378	286	434
Temporal	849	51	20
Total	13,982	567	809

Except Explicit and EntRel, we extract remaining relations as our experimental implicit dataset in English. Our experiments focus on multi-class classification on the four top-level classes. The statistics of these four labels are shown in Table 2. As for Chinese, we follow previous research in [17,23] and select the non-Explicit (i.e., Implicit, EntRel and AltLex) samples as our dataset. The statistics of Chinese data is shown in Table 3.

3.2 Experiment Setup

We choose cross-entropy loss function and Adam with a learning rate of 0.001 and a mini-batch size of 64 to train the model. Follow previous work, we use macro-F_1 to evaluate the performance in English and accuracy to evaluate performance in Chinese since the Chinese corpus is skewed distributed and the macro-F_1 is prone to be affected by the uneven samples. In CNN model, we choose filter window size (1, 3, 5) to represent the *unigram, trigram* and *5-gram* features in sentence. And we set hidden size as 50 in LSTM, BiLSTM and TreeLSTM models. We applied dropout before the classification layer and set the dropout rate as 0.5.

In context-free word embedding, we select different models for English and Chinese. For English experiment, we train two word2vec models as follows:

- BLLIP-50d: The 50-dimensional word vector trained on BLLIP [10]
- Google-300d: The 300-dimensional word vector trained on 100 billion words from Google News

Table 3. Statistics of CoNLL-2016 Chinese non-Explicit discourse sense.

Sense label	Training	Development	Test
Conjunction	5,196	189	228
Expansion	1,228	49	40
EntRel	1,098	50	71
Causation	260	12	11
Purpose	79	2	6
Contrast	72	3	1
Temporal	36	0	1
Conditional	32	1	1
Progression	14	0	0
Total	8,015	306	359

and select four GloVe models[1] which are different in vocabulary size and training corpus as follows:

– 6B-50d/100d/300d: 6B tokens, 400K vocab, uncased, 50/100/300 dimensions
– 840B-300d: 840B tokens, 2.2M vocab, cased, 300 dimensions

For Chinese, we use the Tagged Chinese Gigaword[2] to train 300-dimension word2vec and GloVe word embeddings.

In the contextual model, we use ELMo tool provided by allennlp[3] and the three layers respectively as word representations both for English and Chinese experiment. As for BERT, we choose four models for English:

– Base-single/pairs: 12-layer, 768-hidden, encode single sentence/sentence pairs
– Large-single/pairs: 24-layer, 1024-hidden, encode single sentence/sentence pairs

and one pre-trained Chinese model provided by Google with 12-layer, 768-hidden, 12-heads and 110M parameters.

3.3   Results

We perform a series of comparison experiments to explore the influence of each component on the performance of the parser.

First, we design experiments to evaluate the impact of word embeddings. Tables 4 and 5 show the performance comparison of context-free models and contextual models for English. Here we use the sequence sentence modeling without sentence interaction layer. From the two tables, we see that contextual word embeddings perform better than context-free embeddings in English.

---

[1] https://nlp.stanford.edu/projects/GloVe/.
[2] https://catalog.ldc.upenn.edu/LDC2007T03.
[3] https://github.com/allenai/allennlp.

The best performance of contextual models (BERT, 51.18%) is 7.64% higher than context-free models' best performance (GloVe, 43.54%).

Table 4. Comparisons of F_1 scores (%) for English context-free word embedding with sequence sentence modeling.

	word2vec		GloVe			
	BLLIP-50d	Google-300d	6B-50d	6B-100d	6B-300d	840B-300d
LSTM	41.75	39.41	31.61	38.19	43.01	43.08
BiLSTM	40.12	40.29	33.22	38.88	41.54	**43.54**
CNN	39.29	36.44	37.09	40.35	38.11	40.06

Table 5. Comparisons of F_1 scores (%) for English contextual word embedding with sequence sentence modeling.

	ELMo			BERT			
	1	2	3	Base-single	Base-pairs	Large-single	Large-pairs
LSTM	42.19	45.62	44.97	45.24	**51.18**	44.01	50.24
BiLSTM	41.1	48.09	45.97	45.17	49.5	46.32	47.59
CNN	42.59	46.28	46.28	44.27	46.31	45.22	44.14

The Chinese result are shown in Table 6. It is clear that the best performance of contextual models outperform the context-free model. Due to the language difference and the ELMo and BERT generate the "single Chinese character" embeddings rather than the "word" embeddings, the performance gap between contextual models and context-free models in Chinese is not as much as that in English. From Tables 4 and 5, we state that context information in embeddings is helpful to the implicit relation recognition both in English and Chinese.

Next, we evaluate the effects of sentence modeling in sequence and structure on parser performance. Table 7 show the comparison of best sequential models from previous experiments and structural models in English, where GloVe is 840B-300d, ELMo is the 2nd layers, and BERT is Base-single. We see that the Binary-Tree-LSTM outperforms sequential models. This proves that after adding the grammatical parsing information of the sentence, the tree model is able to capture the semantics of the sentence more effectively. In Chinese, we choose ELMo and BERT as embeddings and list the results in Table 6. It is surprising that CNN outperforms BiLSTM and Tree-LSTM, which is conflict with the finding in English. This may result from the different characteristic in languages.

Further, we select several representative models to examine the sentence interaction layer for experiments. For English experiment, we find that the model

Table 6. Comparisons of Accuracy(%) for Chinese word embeddings with sequence sentence modeling.

	word2vec	GloVe	ELMo			BERT	
			1	2	3	Single	Pairs
LSTM	67.68	70.47	70.31	71.59	67.41	67.69	66.57
BiLSTM	70.19	71.30	71.03	72.70	70.47	66.30	68.24
CNN	70.75	70.20	69.92	71.59	72.42	**74.09**	70.47
Child-Sum-Tree-LSTM	-	-	68.24	70.75	70.75	72.98	-
Binary-Tree-LSTM	-	-	70.19	70.19	72.14	73.53	-

Table 7. Comparisons of F_1 scores (%) for English sequence and structural sentence modeling.

	GloVe	ELMo	BERT
LSTM	43.08	45.62	45.24
BiLSTM	43.54	48.09	45.17
Child-Sum-Tree-LSTM	41.14	46.45	47.01
Binary-Tree-LSTM	44.10	48.53	**50.44**

Table 8. Comparisons of F_1 scores (%) for English sentence interaction layer.

			Dropout				
			0.5	0.4	0.3	0.2	0.1
BiLSTM	BERT (Large-single)	without Attention	46.32	45.54	45.69	46.76	44.80
		Attention	47.24	51.93	50.75	48.98	48.28
		Con-self-Attention	21.22	41.10	44.05	47.33	49.77
		Self-Attention-con	38.28	42.52	46.93	49.12	50.06
		Attention-mlp	49.32	**53.05**	48.38	50.54	50.97
	BERT (Base-pairs)	without-Attention	49.5	50.37	48.54	47.57	50.64
		Self-Attention	27.82	36.84	49.58	**52.33**	50.95
		Attention-mlp	28.42	34.90	44.49	45.14	49.64
	ELMo (second)	without Attention	48.09	45.45	43.48	45.12	43.46
		Attention	30.97	32.55	35.80	35.87	33.54
		Con-self-Attention	43.35	45.31	46.68	46.57	47.89
		Self-Attention-con	45.65	47.65	48.87	48.28	45.26
		Attention-mlp	**49.72**	45.63	45.58	47.25	46.92

is unable to fit the training data due to complexity increasing of the model after adding the interaction layer. So we adjusted the dropout values to fit the training data as shown in Table 8. Note that although the BERT-pairs model the two arguments simultaneously and get some interaction information at the level of word representation, we still add the interaction layer to this model to make comparison. The results show that the sentence interaction level is helpful to identify discourse relations.

Table 9. Comparisons of Accuracy(%) for Chinese sentence interaction layer.

	$word2vec$ + BiLSTM	$ELMo_2$ + BiLSTM	$BERT_{single}$ + CNN
without Attention	70.19	72.70	**74.09**
Attention	68.80	68.52	70.75
Con-self-Attention	70.75	65.74	71.30
Self-Attention-con	72.98	70.20	70.75
Attention-mlp	70.47	64.90	69.63

Table 9 lists the results of interaction in Chinese. We find that the interaction layer does not help even after the dropout adjusting. This may be caused by the language characteristics. The interaction layer aims to amplify the corresponding parts of the two arguments and pay less attention on the noise of the sentences to promote the classification performance. But most Chinese sentences are short due to the omission of sentence elements. Thus the attention operation may not effectively capture the interaction information between two arguments, leading to bad performance in Chinese.

Finally, Tables 10 and 11 show the comparison of our best model with recent systems for multi-class classification for English and Chinese result. Our models achieve state-of-the-art performance on English and Chinese datasets.

Table 10. Comparisons of our model with recent systems for English implicit dataset.

	$P(\%)$	$R(\%)$	$F_1(\%)$
Wang and Lan [23]	46.51	46.33	46.42
Xu et al. [27]	60.63	-	44.48
Dai et al. [4]	59.75	-	51.84
Ours (BERT+BiLSTM+Attention-mlp)	58.22	51.14	**53.05**

Table 11. Comparisons of our model with recent systems for Chinese non-Explicit dataset, accuracy(%).

	Development set	Test set
Wang and Lan [23]	**73.53**	72.42
Rutherford and Xue [20]	71.57	67.41
Schenk et al. [21]	70.59	71.87
Rönnqvist et al. [17]	-	73.01
Ours (BERT-single+CNN)	72.54	**74.09**

4   Conclusion

In this paper, we study the influence of each component on the recognition of English and Chinese implicit discourse relation with deep learning method. The contextual word embeddings outperform context-free embeddings for both English and Chinese. But structural sentence modeling and attention interaction have positive impact on English data rather than on Chinese data. Due to different linguistic features, the neural components have different effects on implicit discourse relationship recognition. Besides, our models achieve state-of-the-art performance on English and Chinese discourse benchmark corpora.

References

1. Biran, O., McKeown, K.: Aggregated word pair features for implicit discourse relation disambiguation (2013)
2. Braud, C., Denis, P.: Combining natural and artificial examples to improve implicit discourse relation identification. In: Proceedings of COLING 2014, the 25th International Conference on Computational Linguistics: Technical Papers, pp. 1694–1705 (2014)
3. Chen, J., Zhang, Q., Liu, P., Qiu, X., Huang, X.: Implicit discourse relation detection via a deep architecture with gated relevance network. In: Proceedings of the 54th Annual Meeting of the Association for Computational Linguistics (Volume 1: Long Papers), vol. 1, pp. 1726–1735 (2016)
4. Dai, Z., Huang, R.: Improving implicit discourse relation classification by modeling inter-dependencies of discourse units in a paragraph. In: Proceedings of the 2018 Conference of the North American Chapter of the Association for Computational Linguistics: Human Language Technologies, Volume 1 (Long Papers), pp. 141–151 (2018)
5. Devlin, J., Chang, M.W., Lee, K., Toutanova, K.: Bert: pre-training of deep bidirectional transformers for language understanding. In: Proceedings of the 2019 Conference of the North American Chapter of the Association for Computational Linguistics: Human Language Technologies, Volume 1 (Long and Short Papers), pp. 4171–4186 (2019)
6. Ji, Y., Zhang, G., Eisenstein, J.: Closing the gap: domain adaptation from explicit to implicit discourse relations. In: Proceedings of the 2015 Conference on Empirical Methods in Natural Language Processing, pp. 2219–2224 (2015)
7. Li, J.J., Carpuat, M., Nenkova, A.: Assessing the discourse factors that influence the quality of machine translation. In: Proceedings of the 52nd Annual Meeting of the Association for Computational Linguistics (Volume 2: Short Papers), vol. 2, pp. 283–288 (2014)
8. Liu, Y., Li, S., Zhang, X., Sui, Z.: Implicit discourse relation classification via multi-task neural networks. In: Thirtieth AAAI Conference on Artificial Intelligence (2016)
9. Manning, C., Surdeanu, M., Bauer, J., Finkel, J., Bethard, S., McClosky, D.: The Stanford CoreNLP natural language processing toolkit. In: Proceedings of 52nd Annual Meeting of the Association for Computational Linguistics: System Demonstrations, pp. 55–60 (2014)
10. McClosky, D., Charniak, E., Johnson, M.: Bllip North American news text, complete. In: Linguistic Data Consortium, p. 4:3 (2008)

11. Mikolov, T., Sutskever, I., Chen, K., Corrado, S.G.: Distributed representations of words and phrases and their compositionality. In: Advances in Neural Information Processing Systems (2013)
12. Pennington, J., Socher, R., Manning, C.: Glove: global vectors for word representation. In: Proceedings of the 2014 Conference on Empirical Methods in Natural Language Processing (EMNLP) (2014)
13. Peters, M.E., et al.: Deep contextualized word representations. In: Proceedings of NAACL (2018)
14. Pitler, E., Nenkova, A.: Using syntax to disambiguate explicit discourse connectives in text. In: Proceedings of the ACL-IJCNLP 2009 Conference Short Papers, pp. 13–16. Association for Computational Linguistics (2009)
15. Qin, L., Zhang, Z., Zhao, H.: Shallow discourse parsing using convolutional neural network. In: Proceedings of the CoNLL-16 Shared Task, pp. 70–77 (2016)
16. Qin, L., Zhang, Z., Zhao, H., Hu, Z., Xing, E.: Adversarial connective-exploiting networks for implicit discourse relation classification. In: Proceedings of the 55th Annual Meeting of the Association for Computational Linguistics (Volume 1: Long Papers), pp. 1006–1017 (2017)
17. Rönnqvist, S., Schenk, N., Chiarcos, C.: A recurrent neural model with attention for the recognition of Chinese implicit discourse relations. In: Proceedings of the 55th Annual Meeting of the Association for Computational Linguistics, pp. 256–262. Association for Computational Linguistics, Vancouver (2017)
18. Rutherford, A., Xue, N.: Discovering implicit discourse relations through brown cluster pair representation and coreference patterns. In: Proceedings of the 14th Conference of the European Chapter of the Association for Computational Linguistics, pp. 645–654 (2014)
19. Rutherford, A., Xue, N.: Improving the inference of implicit discourse relations via classifying explicit discourse connectives. In: Proceedings of the 2015 Conference of the North American Chapter of the Association for Computational Linguistics: Human Language Technologies, pp. 799–808 (2015)
20. Rutherford, A., Xue, N.: Robust non-explicit neural discourse parser in English and Chinese. In: Proceedings of the CoNLL-16 Shared Task, pp. 55–59 (2016)
21. Schenk, N., Chiarcos, C., Donandt, K., Rönnqvist, S., Stepanov, E., Riccardi, G.: Do we really need all those rich linguistic features? A neural network-based approach to implicit sense labeling. In: Proceedings of the CoNLL-16 Shared Task, pp. 41–49 (2016)
22. Tai, K.S., Socher, R., Manning, C.D.: Improved semantic representations from tree-structured long short-term memory networks. In: Proceedings of the 53rd Annual Meeting of the Association for Computational Linguistics and the 7th International Joint Conference on Natural Language Processing (Volume 1: Long Papers), pp. 1556–1566 (2015)
23. Wang, J., Lan, M.: Two end-to-end shallow discourse parsers for English and Chinese in CoNLL-2016 shared task. In: Proceedings of the CoNLL-16 Shared Task, pp. 33–40 (2016)
24. Wang, X., Yoshida, Y., Hirao, T., Sudoh, K., Nagata, M.: Summarization based on task-oriented discourse parsing. IEEE Trans. Audio Speech Lang. Process. 23(8), 1358–1367 (2015)
25. Wang, Y., Li, S., Yang, J., Sun, X., Wang, H.: Tag-enhanced tree-structured neural networks for implicit discourse relation classification. In: Proceedings of the Eighth International Joint Conference on Natural Language Processing (Volume 1: Long Papers), pp. 496–505 (2017)

26. Wu, C., Shi, X., Chen, Y., Su, J., Wang, B.: Improving implicit discourse relation recognition with discourse-specific word embeddings. In: Proceedings of the 55th Annual Meeting of the Association for Computational Linguistics (Volume 2: Short Papers), pp. 269–274 (2017)
27. Xu, Y., Hong, Y., Ruan, H., Yao, J., Zhang, M., Zhou, G.: Using active learning to expand training data for implicit discourse relation recognition. In: Proceedings of the 2018 Conference on Empirical Methods in Natural Language Processing, pp. 725–731 (2018)
28. Yang, B., Cardie, C.: Context-aware learning for sentence-level sentiment analysis with posterior regularization. In: Proceedings of the 52nd Annual Meeting of the Association for Computational Linguistics (Volume 1: Long Papers), vol. 1, pp. 325–335 (2014)

Syntax-Aware Attention for Natural Language Inference with Phrase-Level Matching

Mingtong Liu, Yasong Wang, Yujie Zhang$^{(\boxtimes)}$, Jinan Xu, and Yufeng Chen

School of Computer and Information Technology, Beijing Jiaotong University,
Beijing, China
{16112075,18120467,yjzhang,jaxu,chenyf}@bjtu.edu.cn

Abstract. Natural language inference (NLI) aims to predict whether a premise sentence can infer another hypothesis sentence. Models based on tree structures have shown promising results on this task, but the performance still falls below that of sequential models. In this paper, we present a syntax-aware attention model for NLI, by which phrase-level matching between two sentences is allowed. We design tree-structured semantic composition function that builds phrase representations according to syntactic trees. We then introduce cross sentence attention to learn interaction information based on phrase-level representations between two sentences. Moreover, we additionally explore a self-attention mechanism to enhance semantic representations by capturing the context from syntactic tree. Experimental results on SNLI and SciTail datasets demonstrate that our model has the ability to model NLI more precisely and significantly improves the performance.

Keywords: Natural Language Inference · Syntax-aware attention · Tree-structured semantic composition · Phrase-level matching

1 Introduction

Natural Language Inference (NLI) is a core challenge for natural language understanding [18]. More specifically, the goal of NLI is to identify the logical relationship (*entailment, neutral,* or *contradiction*) between a premise and a corresponding hypothesis. Recently, neural network-based models for NLI have attracted more attention for their powerful ability to learn sentence representation [1,25]. There are mainly two class of models: sequential models [7,9,13,20,24–26] and tree-structured models [2,3,14,19,27].

For the first class of models, sentences are regarded as sequences, in which word-level representation is usually used to model interaction between the premise and hypothesis with attention mechanism [7,20,25]. These models make no consideration of the syntax, but syntax has been proved to important for natural language sentence understanding [4,5]. Since the compositional nature of sentence, the same words may produce different semantics because of different

© Springer Nature Switzerland AG 2019
M. Sun et al. (Eds.): CCL 2019, LNAI 11856, pp. 156–168, 2019.
https://doi.org/10.1007/978-3-030-32381-3_13

Show me the fights from **New York** to **Florida**.   Show me the fights from **Florida** to **New York**.

(a) Sentences with same words but different word orders.

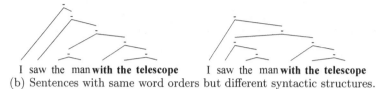

I saw the man **with the telescope**   I saw the man **with the telescope**

(b) Sentences with same word orders but different syntactic structures.

Fig. 1. The examples that are difficult for sequential structure to understand.

word orders or syntactic structures, as shown in Fig. 1. The sentences in Fig. 1(a) have same words but different word orders, and express different meaning. The sentences in Fig. 1(b) have same word orders but different syntactic structures. On the left, "with a telescope" is combined with man, and express that "I saw the man who had a telescope". On the right, "with a telescope" provides additional information about the action "saw the man", and express that "I used the telescope to view the man". Thus, for these language expressions with subtle semantic changes, the sequential models can not always work better than tree-structured models, and syntax is still worth of a further exploration.

For the second class of models, tree structures are used to learn semantic composition [2,3,19,27], in which leaf node is word representation and non-leaf node is phrase representation. The final representation of root node is regarded as sentence representation. Recent evidence [3,8,19,23] reveals that tree-structured models with attention can achieve higher accuracy than sequential models on several tasks. However, the potential of the tree-structured network has not been well exploited for NLI, and the performance of tree-structured models still falls below complex sequential models with deeper networks.

To further explore the potential of tree structure for improving semantic computation, we propose a syntax-aware attention model for NLI, as shown in Fig. 2. It mainly consists of three sub-components: (1) tree-structured composition; (2) cross attention; and (3) self-attention. The tree-structured composition uses syntactic tree to generate phrase representations. Then, we design cross attention to model phrase-level matching that learns interaction between two sentences. A self-attention mechanism is also introduced to enhance semantic representations, which captures the context from syntactic tree within each sentence.

In summary, our contributions are as follows:

- We propose a syntax-aware attention model for NLI. It learns phrase representations by tree-structured composition based on syntactic structure.
- We introduce phrase-level matching with cross attention and self-attention mechanism. The cross attention makes interaction between two sentences, and the self-attention enhances semantic representations by capturing the context from syntactic tree within each sentence.
- We evaluate the proposed model on SNLI and SciTail datasets and the results show that our model has the ability to model NLI more precisely than the previous sequential and tree-structured models.

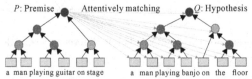

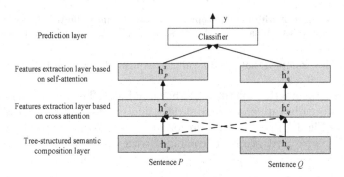

(a) Tree-structured semantic composition and attention-based phrase-level matching.

(b) An overview architecture of our proposed model.

Fig. 2. (a): It learns tree-structured semantic composition in which non-leaf nodes are composed following syntactic tree and represent phrase representations. Then syntax-aware attention is performed for phrase-level matching between two sentences. (b): Based on phrase representations composed syntactically, cross attention and self-attention mechanism are performed to extract features. Finally, a classifier is used to predict the semantic relation between the two sentences.

2   Related Works

Previous work [17,22,23] reveals that models using syntactic trees may be more powerful than sequential models in several tasks, such as sentiment classification [23], neural machine translation [8]. For NLI task, Bowman et al. [2] use constituency parser tree, and explore tree-structured Tree-LSTM to improve sequential LSTM. This method is simple and effective, but ignores the interaction between two sentences. Munkhdalai and Yu [19] use full binary tree, and introduce attention mechanism to model the interaction between two sentences by using node-by-node matching. More recently, Chen et al. [3] design enhanced Tree-LSTM. It shows that incorporating tree-structured information can further improve model performance, and the constituency parser tree is more effective than full binary tree. The latent tree structure [27] is also used to improve semantic computation. However, the existing tree-structured models still fall below complex sequential models [3,7,9,24,25].

In this paper, we focus on how to use syntactic structure to improve semantic computation for complex language understanding. We propose a syntax-aware attention model for NLI, which explores tree-structured semantic composition and implements attention-based phrase-level matching between premise and hypothesis. Experimental results demonstrate the effect of the proposed model.

3 Approach

The model takes two sentences P and Q with syntactic trees as input. Let $P = [p_1, \cdots, p_i, \cdots, p_m]$ with m words and $Q = [q_1, \cdots, q_j, \cdots, q_n]$ with n words. The goal is to predict label y that indicates the logic relation between P and Q. In this paper, we focus on learning semantic composition over constituency tree. We give an example of binarized constituency parser tree in Fig. 2(a).

3.1 Tree-Structured Composition

We apply tree-structured composition for P and Q. In our model, each non-leaf node has two children nodes: leaf child l and right child r. We initiate leaf nodes with BiLSTM [12]. For non-leaf nodes, we adopt S-LSTM [28] as composition function. Each S-LSTM unit has two vectors: hidden state h and memory cell c.

Let (h_t^l, c_t^l) and (h_t^r, c_t^r) represent the left child node l and the right child node r, respectively. We compute a parent node hidden state h_{t+1}^p and memory cell c_{t+1}^p as following equations.

$$i_{t+1} = \sigma(W_{hi}^l h_t^l + W_{hi}^r h_t^r + W_{ci}^l c_t^l + W_{ci}^r c_t^r + b_i) \tag{1}$$

$$f_{t+1}^l = \sigma(W_{hf_l}^l h_t^l + W_{hf_l}^r h_t^r + W_{cf_l}^l c_t^l + W_{cf_l}^r c_t^r + b_{f_l}) \tag{2}$$

$$f_{t+1}^r = \sigma(W_{hf_r}^l h_t^l + W_{hf_r}^r h_t^r + W_{cf_r}^l c_t^l + W_{cf_r}^r c_t^r + b_{f_r}) \tag{3}$$

$$u_{t+1} = \tanh(W_{hu}^l h_t^l + W_{hu}^r h_t^r + b_u) \tag{4}$$

$$c_{t+1} = f_{t+1}^l \odot c_t^l + f_{t+1}^r \odot c_t^r + i_{t+1} \odot u_{t+1} \tag{5}$$

$$h_{t+1} = o_{t+1} \odot \tanh(c_{t+1}) \tag{6}$$

where σ denotes the logistic sigmoid function and $\odot$ denotes element-wise multiplication of two vectors; f_l and f_r are the left and right forget gate; i, o are the input gate and output gate; W and b are learnable parameters, respectively.

We use the hidden state of node as phrase representation. Then, two sentences are represented by $h_p = [h_{p_1}, \cdots, h_{p_i}, \cdots, h_{p_{2m-1}}]$ and $h_q = [h_{q_1}, \cdots, h_{q_j}, \cdots, h_{q_{2n-1}}]$. It noted that there are m-1/n-1 non-leaf nodes composed from the tree for phrase representations and m/n leaf nodes for word representations for P/Q.

3.2 Cross Attention

Cross attention is utilized to capture the phrase-level relevance between two sentences. Give two composed representations based on syntactic trees h_p and h_q for P and Q, we first compute unnormalized attention weights A for any pair of nodes between P and Q with biaffine attention function [6] as follows:

$$A_{ij} = h_{p_i}{}^T W h_{q_j} + \langle U_l, h_{p_i} \rangle + \langle U_r, h_{q_j} \rangle \tag{7}$$

where $W \in \mathbb{R}^{h \times h}$, $U_l \in \mathbb{R}^h$, $U_r \in \mathbb{R}^h$ are learnable parameters, and $\langle \cdot, \cdot \rangle$ denotes the inner production operation. p_i and q_j are the i-th and j-th node in P and Q, respectively. Next, the relevant semantic information for nodes p_i and q_j in another sentence is extracted as follows:

$$\widetilde{h}_{p_i} = \sum_{j=1}^{2n-1} \frac{exp(A_{ij})}{\sum_{k=1}^{2n-1} exp(A_{ik})} h_{q_j} \tag{8}$$

$$\widetilde{h}_{q_j} = \sum_{i=1}^{2m-1} \frac{exp(A_{ij})}{\sum_{k=1}^{2m-1} exp(A_{kj})} h_{p_i} \tag{9}$$

Intuitively, the interaction representation $\widetilde{h}_{p_i}$ is a weighted summation of $\{h_{q_j}\}_{j=1}^{2n-1}$ that is softly aligned to h_{p_i}, and the semantics of h_{q_j} is more probably selected if it is more related to h_{p_i}.

To further enrich the interaction, we use a local comparison function ReLU [10].

$$\overline{h}_{p_i} = [h_{p_i}; \widetilde{h}_{p_i}; |h_{p_i} - \widetilde{h}_{p_i}|; h_{p_i} \odot \widetilde{h}_{p_i}] \tag{10}$$

$$h_{p_i}^c = \mathrm{ReLU}(W_p \overline{h}_{p_i} + b_p) \tag{11}$$

$$\overline{h}_{q_j} = [h_{q_j}; \widetilde{h}_{q_j}; |h_{q_j} - \widetilde{h}_{q_j}|; h_{q_j} \odot \widetilde{h}_{q_j}] \tag{12}$$

$$h_{q_j}^c = \mathrm{ReLU}(W_q \overline{h}_{q_j} + b_q) \tag{13}$$

where W, b are learnable parameters. This operation helps the model to further fuse the matching information, and also reduce the dimension of vector representations for less model complexity.

After that, nodes p_i and q_j in P and Q are newly represented by $h_{p_i}^c$ and $h_{q_j}^c$, respectively.

3.3   Self-attention

We introduce a self-attention layer after cross attention. It captures context from syntactic tree for each sentence and enhances node semantic representations.

For sentence P, we first compute self-attention weights S as Eq. (7).

$$S_{ij} = \langle h_{p_i}^c, h_{p_j}^c \rangle \tag{14}$$

where, S_{ij} indicates the relevance between the i-th node and j-th node in P. Then, we compute the self-attention vector for each node in P as follows:

$$\widetilde{h}_{p_i}^c = \sum_{j=1}^{2m-1} \frac{exp(S_{ij})}{\sum_{k=1}^{2m-1} exp(S_{ik})} h_{p_j}^c \tag{15}$$

Intuitively, $\widetilde{h}_{p_i}^c$ augments each node representation with global syntactic context from P also from Q.

Similarly, we compute self-attention vector $\widetilde{h}_{q_j}^c$ for each node q_j in Q. Then a comparison function is used to $(h_{p_i}^c, \widetilde{h}_{p_i}^c)$ and $(h_{p_j}^c, \widetilde{h}_{p_j}^c)$ to get enhanced representations $h_{p_i}^s$ and $h_{q_j}^s$ as Eqs. (10)–(13).

Table 1. Statistics of datasets: SNLI and SciTail. Avg.L refers to average length of a pair of sentences.

	Train	Dev	Test	Avg.L		Vocab
SNLI	549K	9.8K	9.8K	14	8	36K
SciTail	23K	1.3K	2.1K	17	12	24K

Finally, we further fuse the above cross attention and the self-attention information as follows:

$$\widehat{h}_{p_i} = h_{p_i}^c + h_{p_i}^s \tag{16}$$

$$\widehat{h}_{q_j} = h_{q_j}^c + h_{q_j}^s \tag{17}$$

The representations $\widehat{h}_{p_i}$ and $\widehat{h}_{q_j}$ are learned from cross attention between two syntactically composed trees and then are augmented by self-attention. We then pass them into prediction layer.

3.4 Prediction Layer

We perform mean and max pooling on each sentence as Chen et al. [3], and use two-layer 1024-dimensional MLP with ReLU activation as classifier.

For model training, the object is to minimize the objective function $\mathcal{J}(\Theta)$:

$$\mathcal{J}(\Theta) = -\frac{1}{N} \sum_{i=1}^{N} \log P(y^{(i)} | p^{(i)}, q^{(i)}; \Theta) + \frac{1}{2} \lambda \|\Theta\|_2^2 \tag{18}$$

where Θ denotes all the learnable parameters, N is the number of instances in the training set, $(p^{(i)}, q^{(i)})$ are the sentence pairs, and $y^{(i)}$ denotes the annotated label for the i-th instance.

4 Experiments

4.1 Dataset

We evaluate our model on two datasets: the Stanford Natural Language Inference (SNLI) dataset [1] and the SciTail dataset [14]. The syntactic trees used in this paper are produced by the Stanford PCFG Parser 3.5.3 [16] and they are provided in these datasets.

The detailed statistical information of the two datasets is shown in Table 1.

4.2 Implementation Details

Following Tay et al. [24], we learn word embedding by concatenating pre-trained word vector, learnable word vector and POS vector. Then we use a ReLU layer to

Table 2. Comparison results on SNLI dataset.

Models	Train	Test
SAN [13]	89.6	86.3
BiMPM [25]	90.9	87.5
ESIM [3]	92.6	88.0
DIIN [11]	91.2	88.0
CAFE [24]	89.8	88.5
DR-BiLSTM [9]	94.1	88.5
AF-DMN [7]	94.5	88.6
SPINN [2]	89.2	83.2
NTI [19]	88.5	87.3
syn TreeLSTM [3]	92.9	87.8
Our model (single)	92.0	**88.8**
BiMPM (ensemble) [25]	93.2	88.8
ESIM (ensemble) [3]	93.5	88.6
DIIN (ensemble) [11]	92.3	88.9
CAFE (ensemble) [24]	92.5	89.3
DR-BiLSTM (ensemble) [9]	94.8	89.3
AF-DMN (ensemble) [7]	94.9	89.0
Our model (ensemble)	93.2	**89.5**

the concatenated vector. We set word embeddings, the hidden states of S-LSTM and ReLU to 300 dimensions. The pre-trained word vectors are 300-dimensional *Glove* 840*B* [21] and fixed during training. The learnable word vectors and POS vectors have 30 dimensions. The batch size is set to 64 for SNLI and 32 for SciTail. We use the Adam method [15] for training, and set the initial learning rate to 5e−4 and l_2 regularizer strength to 6e−5. For ensemble model, we average the probability distributions from three single models as in Duan et al. [7].

4.3　Comparison Results on SNLI

The comparison results on SNLI dataset is shown in Table 2.

The first group are sequential models that adopt attention to model word-level matching. SAN [13] is a distance-based self-attention network. BiMPM [25] design a bilateral multi-perspective matching model from both directions. ESIM [3] incorporate the chain LSTM and tree LSTM. CAFE [24] use novel factorization layers compress alignment vectors into scalar valued features. DR-BiLSTM [9] process the hypothesis conditioned on the premise results. DIIN [11] hierarchically extract semantic features using CNN. AF-DMN [7] adopt attention-fused deep matching network.

The second group are tree-structured models, of which SPINN [2] use Tree-LSTM with constituency parser tree, without attention. NTI [19] and syn Tree-LSTM [3] adopt attention for node matching. NTI use full binary tree while syn

Table 3. Comparison results on SciTail dataset.

Models	Dev	Test
Majority class	63.3	60.3
Ngram	65.0	70.6
DecompAtt	75.4	72.3
ESIM	70.5	70.6
DGEM	79.6	77.3
DEISTE	82.4	82.1
CAFE	-	83.3
Our model	**88.1**	**85.8**

Table 4. Ablation study on SNLI dev and test sets.

Models	Dev	Test
Only root node	86.2	85.9
+ Cross attention	88.9	88.2
+ Self-attention	89.4	88.8

Tree-LSTM use constituency parser tree. Compared to Chen et al. [3], we use same parser tree but different tree composition function and attention mechanism.

In Table 2, our single and ensemble models achieve 88.8% and 89.5% test accuracy. The comparison results show that our model outperforms not only the existing tree-structured models, but also state-of-the-art achieved by sequential models on SNLI dataset.

4.4   Comparison Results on SciTail

The comparison results on SciTail dataset is shown in Table 3. SciTail is known to be a more difficult dataset for NLI. The first five models in Table 2 are all implemented in Knot et al. [14]. DGEM is a graph based attention model using syntactic structures. CAFE [24] adopt LSTM and attention for word-level matching. DEISTE [26] propose deep explorations of inter-sentence interaction.

On this dataset, our single model significantly outperforms these previous models, and achieves 85.8% test accuracy.

4.5   Ablation Study

We conduct an ablation study to examine the effect of each key component of our model. As illustrated in Table 4, the first row is the model that uses the representation of root node to represent sentence, without attention. By adding cross attention and self-attention, the model performance is further improved. This proves the effect of our tree-structured composition and matching model.

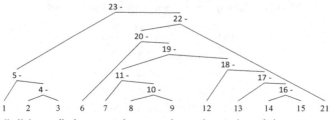

(a) The syntactic tree of premise sentence P.

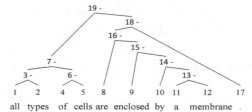

(b) The syntactic tree of hypothesis sentence Q.

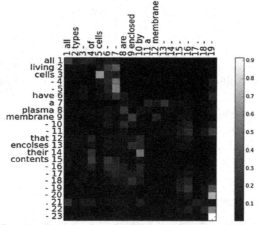

(c) The attention results of two syntactic trees of P and Q.

Fig. 3. The syntactic trees and attention result for sentences P and Q.

4.6   Investigation on Attention

In this section, we investigate what information is captured by the attention, and visualize the cross attention results, as shown in Fig. 3. This is an instance from the test set of SciTail dataset: {P: all living cells have a plasma membrane that encloses their contents. Q: all types of cells are enclosed by a membrane. The label y: entailment.}. From the results, it shows that our syntax-based model can semantically aligns word-level expressions (node 13 "encloses" and node 9 "enclosed") and phrase-level expressions (node 5 "all living cells" and node 7

Table 5. Some complex examples and the classification. The E indicates entailment and the N indicates neutral between P and Q.

ID	Sentence1(P)	Sentence2(Q)	Bleu	Gold	Ours	CAFE
A	Early morning sprinkling reduces water loss from evaporation	Watering plants and grass in the early morning is a way to conserve water because smaller amounts of water evaporate in the cool morning	0.05	E	E	N
B	Slow, deep watering allows plant roots to grow deep, prevents blow-over of your trees, and also minimizes salt buildup	Deep roots will best prevent a tree from being blown over by high winds during a storm	0.10	E	E	N
C	They are among the most primitive of dicotyledonous angiosperm plants	Angiosperms are the most successful phylum of plants	0.50	N	N	E
D	As the wheel turns, the arc causes the body to lift up.	The turning driveshaft causes the wheels of the car to turn	0.42	N	N	E
E	Multiple tissue types compose organs and body structures	a(n) organ is a structure that is composed of one or more types of tissues	0.05	E	N	N
F	Both take approximately one year to orbit the sun	It takes about one year for earth to orbit the sun	0.53	N	E	E

"all types of cells") in the P and Q, respectively. We also observe that attention degree for the phrase expressions is more obvious than the single word that composes the phrase, such as node 17 in the P and node 16 in the Q. An intuitive explantation is that the syntax-based model can capture more rich semantics by using tree-structured composition. Finally, the syntax-based model attends over higher level tree nodes with rich semantics when considering longer phrase or full sentence, such as, the larger sub-trees 20, 22 and 23 in the P is aligned to the root node 19 that represents the whole semantics of the Q. It also indicates the composition function can effectively compute phrase representations.

4.7   Case Study

We show some examples from SciTail test dataset, as shown in Table 5. We compare the proposed syntax-based model with sequential model. For sequential model, we use the representative CAFE model [24]. We compute the Bleu score of the P with referenced to the Q and use 1-gram length. The Bleu score assumes the more overlapped words between two sentences, the closer the semantics are.

Examples A-B are entailment cases, but each of which has low Bleu score. Thus, it is more difficult to recognize the entailment relation between them. Our syntax-based approach correctly favors entailment in these cases. It indicates that the low lexical similarity challenges the sequential model to extract the related semantics, but it maybe solved by introducing syntactic structures.

The second set of examples C-D are neutral cases. Each of them has high Bleu score, where sequential model trends to misidentify the semantic relation to entailment, but our syntax-based model have the ability to correctly recognize the neutral relation. It indicates syntactic structure is more superior to solve semantic understanding involving structurally complex expressions.

Finally, examples E-F are cases that sequential and syntactic models get wrong. Examples E are entailment relation, but it have low Bleu score. Meanwhile, the word orders and structures ("compose" and "is composed of") of two sentences are also quite different. It causes models to failure recognizing the entailment relation between them. Example F is neutral relation where two sentences have high lexical overlap and also the similar word orders, which confuses models to misclassify a entailment class. For the difficult cases, sentence semantics suffer more the issues, such as polysemy, ambiguity, as well as fuzziness, in which the model may need more inference information to distinguish these relations and make the correct decision, such as incorporating external knowledge to help model better understanding the lexical and phrasal semantics.

5   Conclusions and Future Work

In this paper, we explore the potential of syntactic tree for semantic computation and present a syntax-aware attention model for NLI. It leverages tree-structured composition and phrase-level matching. Experimental results on SNLI and Sci-Tail datasets show that our model significantly improves the performance, and that the syntactic structure is important for modeling complex semantic relationship. In the future, we will explore the combination of syntax and pre-trained language model technology, to further improve the performance.

Acknowledgments. The authors are supported by the National Nature Science Foundation of China (Nos. 61876198, 61370130 and 61473294), the Fundamental Research Funds for the Central Universities (Nos. 2018YJS025 and 2015JBM033), and the International Science and Technology Cooperation Program of China (No. K11F100010).

References

1. Bowman, S.R., Angeli, G., Potts, C., Manning, C.D.: A large annotated corpus for learning natural language inference. arXiv preprint arXiv:1508.05326 (2015)
2. Bowman, S.R., Gauthier, J., Rastogi, A., Gupta, R., Manning, C.D., Potts, C.: A fast unified model for parsing and sentence understanding. arXiv preprint arXiv:1603.06021 (2016)
3. Chen, Q., Zhu, X., Ling, Z.H., Wei, S., Jiang, H., Inkpen, D.: Enhanced LSTM for natural language inference. In: Proceedings of ACL (2017)
4. Chomsky, N.: Syntactic Structures. Mouton, The Hague (1965). Aspects of the Theory of Syntax. MIT Press, Cambridge (1981). Lectures on Government and Binding. Foris, Dordrecht (1982). Some concepts and consequences of the theory of government and binding. LI Monographs **6**, 1–52 (1957)

5. Dowty, D.: Compositionality as an empirical problem. In: Direct Compositionality, vol. 14, pp. 23–101 (2007)
6. Dozat, T., Manning, C.D.: Deep biaffine attention for neural dependency parsing (2016)
7. Duan, C., Cui, L., Chen, X., Wei, F., Zhu, C., Zhao, T.: Attention-fused deep matching network for natural language inference. In: IJCAI, pp. 4033–4040 (2018)
8. Eriguchi, A., Hashimoto, K., Tsuruoka, Y.: Tree-to-sequence attentional neural machine translation. arXiv preprint arXiv:1603.06075 (2016)
9. Ghaeini, R., et al.: DR-BiLSTM: dependent reading bidirectional LSTM for natural language inference (2018)
10. Glorot, X., Bordes, A., Bengio, Y.: Deep sparse rectifier neural networks. In: Proceedings of the Fourteenth International Conference on Artificial Intelligence and Statistics, pp. 315–323 (2011)
11. Gong, Y., Luo, H., Zhang, J.: Natural language inference over interaction space. arXiv preprint arXiv:1709.04348 (2017)
12. Hochreiter, S., Schmidhuber, J.: Long short-term memory. Neural Comput. **9**(8), 1735–1780 (1997)
13. Im, J., Cho, S.: Distance-based self-attention network for natural language inference (2017)
14. Khot, T., Sabharwal, A., Clark, P.: SciTail: a textual entailment dataset from science question answering. In: Proceedings of AAAI (2018)
15. Kingma, D., Ba, J.: Adam: A method for stochastic optimization. arXiv preprint arXiv:1412.6980 (2014)
16. Klein, D., Manning, C.D.: Accurate unlexicalized parsing. In: Proceedings of the 41st Annual Meeting on Association for Computational Linguistics, vol. 1, pp. 423–430. Association for Computational Linguistics (2003)
17. Li, J., Luong, T., Jurafsky, D., Hovy, E.: When are tree structures necessary for deep learning of representations? In: Proceedings of the 2015 Conference on Empirical Methods in Natural Language Processing, pp. 2304–2314 (2015)
18. MacCartney, B., Manning, C.D.: Modeling semantic containment and exclusion in natural language inference. In: Proceedings of the 22nd International Conference on Computational Linguistics, vol. 1, pp. 521–528. Association for Computational Linguistics (2008)
19. Munkhdalai, T., Yu, H.: Neural tree indexers for text understanding. In: Proceedings of the Conference. Association for Computational Linguistics. Meeting, vol. 1, p. 11. NIH Public Access (2017)
20. Parikh, A.P., Täckström, O., Das, D., Uszkoreit, J.: A decomposable attention model for natural language inference. arXiv preprint arXiv:1606.01933 (2016)
21. Pennington, J., Socher, R., Manning, C.: Glove: global vectors for word representation. In: Proceedings of the 2014 Conference on Empirical Methods in Natural Language Processing (EMNLP), pp. 1532–1543 (2014)
22. Socher, R., Huval, B., Manning, C.D., Ng, A.Y.: Semantic compositionality through recursive matrix-vector spaces. In: Proceedings of the 2012 Joint Conference on Empirical Methods in Natural Language Processing and Computational Natural Language Learning, pp. 1201–1211. Association for Computational Linguistics (2012)
23. Tai, K.S., Socher, R., Manning, C.D.: Improved semantic representations from tree-structured long short-term memory networks. arXiv preprint arXiv:1503.00075 (2015)

24. Tay, Y., Tuan, L.A., Hui, S.C.: A compare-propagate architecture with alignment factorization for natural language inference. arXiv preprint arXiv:1801.00102 (2017)
25. Wang, Z., Hamza, W., Florian, R.: Bilateral multi-perspective matching for natural language sentences. Corr abs/1702.03814 (2017)
26. Yin, W., Dan, R., Schütze, H.: End-task oriented textual entailment via deep exploring inter-sentence interactions (2018)
27. Yogatama, D., Blunsom, P., Dyer, C., Grefenstette, E., Ling, W.: Learning to compose words into sentences with reinforcement learning. arXiv preprint arXiv:1611.09100 (2016)
28. Zhu, X., Sobhani, P., Guo, H.: Long short-term memory over recursive structures. In: International Conference on International Conference on Machine Learning (2015)

Sharing Pre-trained BERT Decoder
for a Hybrid Summarization

Ran Wei[1,2], Heyan Huang[1,2(✉)], and Yang Gao[1,2]

[1] School of Computer Science and Technology, Beijing Institute of Technology,
Beijing 100081, China
`weiranbit@gmail.com`, {`hhy63,gyang`}`@bit.edu.cn`
[2] Beijing Engineering Research Center of High Volume Language Information
Processing and Cloud Computing Applications, Beijing 100081, China

Abstract. Sentence selection and summary generation are two main
steps to generate informative and readable summaries. However, most
previous works treat them as two separated subtasks. In this paper,
we propose a novel extractive-and-abstractive hybrid framework for single document summarization task by jointly learning to select sentence
and rewrite summary. It first selects sentences by an extractive decoder
and then generate summary according to each selected sentence by an
abstractive decoder. Moreover, we apply the BERT pre-trained model as
document encoder, sharing the context representations to both decoders.
Experiments on the CNN/DailyMail dataset show that the proposed
framework outperforms both state-of-the-art extractive and abstractive
models.

Keywords: Text summarization · Extractive and abstractive ·
Pre-trained based

1 Introduction

Automatic text summarization has played an important role in a variety of natural language processing (NLP) applications, such as question answering [13,21],
report generation [14], and opinion mining [8]. Single document summarization,
the task of generate short, representative and readable summaries from the original text while retaining the main ideas of source articles, has received much
attention in recent years [5,16,18,20].

Current methods for single document summarization using neural network
architectures have primarily focused on two strategies: extractive and abstractive. Extractive summarization forms summaries by selecting originally important segments of the input documents [15,17]. Abstractive summarization potentially generates new sentence or reorganizes their orders to form fluent summaries [6,18,20]. Both methods suffer from obvious drawbacks: extractive systems are sometimes redundant since they cannot trim the original sentences to
fit into the summary, and they lack a mechanism to ensure overall coherence. In

© Springer Nature Switzerland AG 2019
M. Sun et al. (Eds.): CCL 2019, LNAI 11856, pp. 169–180, 2019.
https://doi.org/10.1007/978-3-030-32381-3_14

contrast, abstractive systems require natural language generation and semantic representation, which are complex and can hardly meet the demands of generating correct facts with proper word relations.

In this paper, we present a novel architecture that attempts to combine the extractive and abstractive methods. Our model first decides whether to choose a sentence based on its probability generated by an extractive decoder, and then rewrite the selected sentence by an abstractive decoder, which can remove the meaningless words, reorganize words orders and generate coherent contents. In this way, our model can extract informative contents and generate coherent summaries. Moreover, we choose Bidirectional Encoder Representations from Transformers (BERT [3]) pre-trained language model as basic document encoder, which can provide powerful pre-trained context representations. Both decoders share the same representations, so that our model can be trained simultaneously in one end-to-end framework. Our contributions in this paper are two-folds:

1. We propose a novel extractive-and-abstractive hybrid neural architecture combining the extractive and abstractive decoders, taking advantage of both summarization approaches.
2. We explore a new way that applies pre-trained language model into summarization task, making good use of the pre-trained context representations in the sharing encoder process.

Extensive experiments are conducted on CNN/DailyMail dataset [9,16]. The results show that our model outperforms both extractive and abstractive state-of-the-arts models.

The rest of this paper is organized as follows. We present the related work in Sect. 2. In Sect. 3, we introduce our extractive-and-abstractive hybrid model in details. In Sect. 4, we describe the experiments setup and implementation details. We present the results of our experiments and analysis the performance in Sect. 5. We conclude our work in Sect. 6.

2   Related Work

2.1   Extractive Summarization

Kageback et al. [10] and Yin and Pei [22] use neural networks to map sentences into vectors and select sentences based on those vectors. Cheng and Lapata [2] select sentences based on an LSTM classifier that predicts a binary label for each sentence. Nallapati et al. [15] adopt a similar approach, they differ in their neural architecture for sentence encoding and features used during label prediction, while Narayan et al. [17] equip the same architecture with a training algorithm based on reinforcement learning. While some extractive summarization methods obtain high ROUGE scores, most of them lack a mechanism to ensure overall coherence and suffer from low readability.

2.2   Abstractive Summarization

Rush et al. [19] first bring up the abstractive summarization task and use attention-based encoder to read the input text and generate the summary. Nallapati et al. [16] apply a more powerful sequence-to-sequence model. See et al. [20] combine pointer networks into their models to deal with the out-of-vocabulary (OOV) words. Paulus et al. [18] use policy gradient on summarization and state out the fact that high ROUGE scores might still lead to low human evaluation scores. However, most of them underperform or are on par with the baseline of simply selecting the leading sentences in the document as summaries, the best results for abstractive summarization have been achieved with models that are more extractive in nature than abstractive, since most of the words in the summary are copied from the document (Gehrmann et al. [6]).

2.3   Pre-trained Model Summarization

Edunov et al. [4] combine the pre-trained embedding to the encoder network to enhance the text representations. Zhang et al. [23] propose a BERT based encoder-decoder framework, which use BERT as encoder and refine every word in the draft summary. All these methods demonstrate a pre-trained model on vast corpora can provide improvements for summarization task.

In this paper, we propose a novel hybrid end-to-end architecture combining extractive and abstractive model by using the extractive decoder to select informative sentences and rewrite these selected contents by a abstractive decoder. Both decoders share the same pre-trained representations provided by a BERT encoder.

3   Our Model

Our model extracts sentences from a given document and further rewrites these sentences by a sequence-to-sequence architecture. We denote a document $D = (s_1, ..., s_M)$ as a sequence of M sentences, and a sentence $s_i = (w_{i1}, ..., w_{iN})$ as a sequence of N words. A extractive decoder are used assign a label $z_i \in \{0, 1\}$ to each sentence s_i. $z_i = 1$ indicates s_i is selected and $z_i = 0$ means s_i is skipped. On the other hand, the abstractive decoder generates the summary text $y_j = (w_{j1}, ..., w_{jK})$, where y_j is the summary rewritten by the jth selected sentence. Only the sentences with $z_i = 1$ will go through the abstractive decoder. Figure 1 demonstrates an overview of our proposed model. In the following, we introduce each of its components in details.

3.1   Document Encoder

We use BERT as the basic document encoder, because extractive and abstractive decoder share the same pre-trained document encoder. We require it to output representations in sentence level and word level. To get the representation for each sentence, we adopt similar modifications [12] to the input sequence

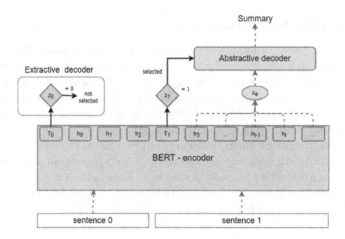

Fig. 1. Framework of our summarization system. The model extracts the most relevant sentences by taking into account the sentence representation T_i. If a sentence is selected ($z_i = 1$), its word level representations are fed into the abstractive decoder to generate final summary.

and embedding of BERT, we insert a [CLS] token before each sentence and use interval segment embeddings to distinguish multiple sentences within a document. As illustrated in Fig. 2, the vector T_i which is the ith [CLS] symbol from the top BERT layer will be used as the representation for sentence s_i, the vector h_i is the representation for each word.

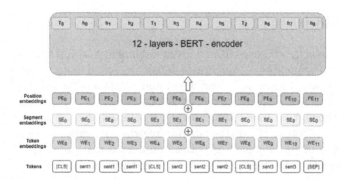

Fig. 2. Document encoder architecture. We insert a [CLS] token before each sentence and use interval segment embeddings to distinguish different sentences.

3.2   Extractive Decoder

The extractive decoder selects the most informative sentences, which cover necessary information that is belonged to the gold summary. For each sentence s_i, the extractive decoder takes a decision based on the encoder representation T_i which is the [CLS] symbol representation from the top BERT encoder layer for s_i. Benefit from self-attention mechanism in the multi-layer transformer based BERT encoder, the sentence representation T_i not only contains the semantic information of s_i but also represent the relationship between s_i and other sentences in the document. The extractive decoder adds a linear layer into the encoder outputs T_i and compute the probability of action $z_i \in \{0, 1\}$ to sentence s_i as:

$$p(z_i|T_i) = \sigma(W_0 T_i + b_0) \tag{1}$$

where W_0 and b_0 are the model parameters, σ is the Sigmoid function. To optimize the extractive decoder, We use a Cross Entropy Loss:

$$L_{ext} = -\frac{1}{N} \sum_{i=1}^{N} (l_i \ln p(z_i = 1|T_i) + (1 - l_i) \ln(1 - p(z_i = 0|T_i)) \tag{2}$$

where $l_i \in \{0, 1\}$ is the ground-truth label for sentence s_i and N is the number of sentences. When $l_i = 1$, it indicates that sentence s_i should be extracted and be attended to the abstractive decoder.

3.3   Abstractive Decoder

After the sentence s_i is selected in the extractive decoder, we input s_i into the abstractive decoder to rewrite the original sentence in a more abstractive way. In practice, we introduce a 4 layer Transformer decoder to learn the conditional probability $P(y_i|T_i)$, where T_i is the hidden representation for s_i from the BERT encoder, y_i is the gold summary sentence.

Transformer Decoder. Transformer decoder takes a shifted sequence of target summary word embeddings as input and produces contextualized representations $o_1, ..., o_n$, from which the target tokens are predicted by a softmax layer. As shown in Eq. (3), at step t, the decoder predicts output probability $P_t^{vocab}(w)$ conditioned on previous outputs $c_{<t}$ and encoder outputs c_e as follow:

$$P_t^{vocab}(w) = softmax(W_1[c_{<t}, c_e] + b_1) \tag{3}$$

$$c_{<t} = \sum_{j=1}^{t-1} \alpha_j^{<t} o_j \tag{4}$$

$$c_e = \sum_{j=1}^{e} \alpha_j^e h_j \tag{5}$$

$$\alpha_j^{<t} = Multihead - attention(o_j, o_1, ..., o_{t-1}) \tag{6}$$

$$\alpha_j^e = Multihead - attention(h_j, h_1, ..., h_m) \tag{7}$$

The transformer decoder calculates source attention α_j^e over the representations $h_1, ..., h_m$ of selected sentence, while the target side self attention $\alpha_j^{\leq t}$ should not be able to look at the representations of later positions. The decoder's learning objective is to minimize negative likelihood of conditional probability as follow:

$$L_{abs} = -\frac{1}{C} \sum_{t=1}^{C} \ln(P(w_t = w_t^* | c_{<t}, c_e)) \tag{8}$$

where w_t^* is the word in gold summary at step t.

Copy Mechanism. As some summary tokens are out-of-vocabulary (OOV) words yet occur in the selected sentences, we incorporate copy mechanism (Gu et al. [7]) into the Transformer decoder. At decoder time step t, we first calculate the attention probability distribution over the selected sentence s_i using dot product of the last layer transformer decoder output o_t and the encoder output h_j as the following:

$$u_t^j = o_t W_c h_j \tag{9}$$

$$\alpha_t^j = \frac{\exp(u_t^j)}{\sum_{k=1}^{N} \exp(u_t^k)} \tag{10}$$

We then calculate copying gate $g_t \in [0, 1]$, which represents the probability of selecting words from source sentence:

$$g_t = sigmoid(W_g[o_t, h] + b_g) \tag{11}$$

$$h = \sum_{j=1} \alpha_t^j h_j \tag{12}$$

Finally we use g_t to calculate the final probability at time step p, which is a weighted sum of copy probability and generation probability:

$$P_t(w) = (1 - g_t) P_t^{vocab}(w) + g_t \sum_{i:w_i=w} a_t^i \tag{13}$$

3.4   Learning and Inference

During training stage, the goal is the combination of extractive and abstractive loss. We use both sentence level label and ground-truth summary jointly training our model and minimizing the following objective:

$$L_{model} = L_{ext} + L_{abs} \tag{14}$$

In the inference stage, we calculate a decision variable $z_i = 1$ if $p(z_i|T_i) > 0.5$ for selecting the sentence s_i, and $z_i = 0$ if $p(z_i|T_i) < 0.5$ for skipping this sentence. To control the length of summaries, we only keep the top 4 selected sentences with the max $p(z_i|T_i)$ if too many sentences selected by our extractive decoder. We use beam search to generate the abstractive summaries.

4   Experiments Setup

In this section, we introduce the summarization dataset, implementation details and evaluation protocol in our experiments.

4.1   Dataset and Preprocess

Experiments are performed on the CNN/DailyMail dataset (Hermann et al. [9]; Nallapati et al. [16]; See et al. [20]) which contains news stories in CNN and Daily Mail websites. We used the standard splits of Hermann et al. [9] for training, validation, and testing (287,113 training pairs, 13,368 validation pairs and 11,490 test pairs). We follow See et al. [20] and use the non-anonymized version data.

Data Preprocess. To train the extractive decoder in our model, sentence level labels are needed, which indicate the sentence should be selected or not. Besides we need to align each selected sentence to one ground-truth summary, so that we can use the abstractive decoder to learn this kind of corresponding relationship. However, CNN/DailyMail dataset only contains abstractive gold summaries, we apply a greedy preprocess algorithm to generate these labels. For each sentence in the gold summary, we calculate ROUGE-1, ROUGE-2 and ROUGE-L to every sentence in the document and assign label 1 to the sentence with maximum sum of these three ROUGE scores, here all the ROUGE scores are the recall value because we want these selected sentences contain complete information of the gold summaries. In order to keep the origin context order of the selected sentences, we also apply a greedy algorithm, for example we have a document $D = (s_1, ..., s_M)$, a gold summary $S = (y_1, y_2)$ and y_1 aligns to the s_i, we will find the selected sentence aligns to y_2 only in the sentences set $(s_{i+1}, ..., s_M)$.

4.2   Implementation Details

In this work, we use the 'bert-base-uncased' version BERT as document encoder. We use the same WordPiece vocabulary (30522 words) for both encoder and abstractive decoder and set the transformer layer to 4, and set the attention heads number to 12. We train the model using an Adam optimizer with learning rate of $5e - 4$, $\beta_1 = 0.9$, $\beta_2 = 0.999$ and use dynamic learning rate. BERT encoder, extractive and abstractive decoder are jointly trained for 10 epochs on 4 GPUs (GTX 1080 Ti with 11 GB memory) with a batch size to 24 (6 in each GPU). Model checkpoints are saved and evaluated on the validation set every 5000 steps.

4.3   Model Evaluation

We evaluated summarization quality using ROUGE F1 (Lin and Hovy [11]), which is the standard evaluation metric for summarization. We report results in terms of unigram and bigram overlap (ROUGE-1) and (ROUGE-2) as a means of assessing informativeness, and the longest common subsequence (ROUGE-L) as a means of assessing fluency.

5   Results and Analysis

In this section, we first compare our model with both extractive and abstractive baselines on benchmark datasets. We then conduct ablation experiments to study the effect of the hybrid decoders. Finally, we present some examples output from our model.

5.1   Evaluation Results

The experimental results on CNN/Dailymail dataset are shown in Table 1. Our model aims to take advantage of both extractive and abstractive approaches, so we compare our model with several previously proposed extractive and abstractive systems.

Table 1. Testing results on the CNN/DailyMail dataset using ROUGE F1.

Model	ROUGE-1	ROUGE-2	ROUGE-L
PGN	39.53	17.28	37.98
BOTTOM-UP	41.22	18.68	38.34
DCA	41.69	19.47	37.92
LEAD	40.42	17.62	36.67
REFRESH	41.0	18.80	37.70
NEUSUM	41.59	19.01	37.98
SRC-ELMO + SHDEMB	41.56	18.94	38.47
TWO-STAGE + RL	41.71	**19.49**	38.79
Our model	**41.76**	19.31	**38.86**

- **LEAD** is an extractive baseline which uses the first-3 sentences of the document as a summary.
- **REFRESH** (Narayan et al. [17]) is an extractive summarization system trained by globally optimizing the ROUGE metric with reinforcement learning.
- **NEUSUM** (Zhou et al. [24]) is the state-of-the-art extractive system that jointly score and select sentences.
- **PGN** (See et al. [20]), is the Pointer Generator Network, an abstractive summarization system based on an encoder-decoder architecture.
- **BOTTOM-UP** (Gehrmann et al. [6]), is a state-of-the-art abstractive summarization system using a bottom-up attention.
- **DCA** (Celikyilmaz et al. [1]) is the Deep Communicating Agents, a state-of-the-art abstractive summarization system with multiple agents to represent the document as well as hierarchical attention mechanism over the agents for decoding.

We also compare our model with two pre-trained based summarization approaches.

- **SRC-ELMO + SHDEMB** (Edunov et al. [4]) is an abstractive model using pre-trained embedding to enhance text representations.
- **TWO-STAGE + RL** (Zhang et al. [23]) is a two stage encoder-decoder model, which applies BERT to the encoder and draft-refine part.

As illustrated in the Table 1, our model outperforms both extractive and abstractive previous baselines. We get a state-of-the-art result in ROUGE-1 and ROUGE-L, which shows that our model can generates informative and coherent summaries. On the ROUGE-2 metric, our model is also comparable with most baselines. Only the DCA and TWO-STAGE + RL outperform our model, in which both apply reinforcement learning for optimizing objective directly derived by ROUGE metrics. Compared to the SRC-ELMO + SHDEMB and TWO-STAGE + RL model, which use pre-trained representations, our model makes a better use of pre-trained language model and achieves better performance.

5.2 Ablation Study

Additionally, in order to study the effect of our abstractive decoder, we perform two extensive baselines.

- **Extractive-1** just uses the sentence level label to train the extractive decoder and select summaries from document.
- **Extractive-2** uses both sentence level label and ground-truth summaries to train complete model, but generates summaries only according to the extractive decoder.

Table 2. Test set results for ablation study.

Model	ROUGE-1	ROUGE-2	ROUGE-L
Extractive-1	40.54	17.97	37.04
Extractive-2	41.15	18.74	37.51
Complete model	41.76	19.31	38.86

The ablation results are shown in Table 2. The results of Extractive-2 performs better than Extractive-1, suggesting that jointly training of extractive and abstractive decoder can improve the model's ability of extracting informative contents. This improvement mainly benefits from that we use both sentence level label and ground-truth summary to train our model. Compared to the Extractive-2, the complete model achieves significant improvements which demonstrates the abstractive decoder is helpful to generate informative and readable summaries.

Article

as the countdown continues to floyd mayweather 's mega-fight with manny pacquiao in las vegas on may 2 , the money man 's daughter iyanna mayweather has shared her thoughts about life in training with her champion father . mayweather vs pacquiao will generate revenue upwards of $ 300 million in what is being billed as the most lucrative bout in boxing history and , ahead of the may showdown , iyanna mayweather offered some insight into her dad 's intense training regime . ` when i watch my dad train , it 's inspiring to me , ' she said . iyanna mayweather has been spending time in her father floyd 's training camp , iyanna watches on as her champion dad gets through another gruelling training session , iyanna says she is amazed by her dad 's work ethic in the gym and is amazed by his jump rump skills ` to work at hard not only at working out , but to work hard at everything , ' i think my dad fighting pacquiao ... it 's just another fight in my opinion . ' floyd mayweather and pacquiao have been keeping boxing fans updated daily on social media with their training schedules and iyanna mayweather explained how impressed she was with her father 's work ethic in the gym . ' i like watching my dad jump rope because i 've never seen anyone jump rope like that before , ' she added . mayweather posted an update to his instagram account on friday as he embarked on another shopping trip ` it 's fun coming to the gym because when dad 's not in training camp , the money team does n't see each other often so when my dad gets back in training camp , we get back to seeing each other . ` we hang out a lot , we play around , we just have fun outside of the gym . my dad is my best friend . '

Extractive-2

- ~~as the countdown continues to~~ floyd mayweather 's mega-fight with manny pacquiao in las vegas on may 2 , ~~the money man 's daughter iyanna mayweather has shared her thoughts about life in training with her champion father~~ .
- mayweather vs pacquiao will generate revenue upwards of $ 300 million ~~in what is being billed~~ as the most lucrative bout in boxing history ~~and , ahead of the may showdown , iyanna mayweather offered some insight into her dad 's intense training regime~~ .
- iyanna mayweather has been spending time ~~in her father~~ floyd 's training camp , iyanna watches on as her champion dad gets through another gruelling training session , ~~iyanna says she is amazed by her dad 's work ethic in the gym and is amazed by his jump rump skills ` to work at hard not only at working out , but to work hard at everything~~ .

Gold

- floyd mayweather will fight manny pacquiao in las vegas on may 2.
- the bout is expected to generate $ 300 million in revenue.
- iyanna mayweather has been in training camp with her father floyd.

Ours

- floyd mayweather fight with manny pacquiao in las vegas on may 2 .
- mayweather vs pacquiao will generate revenue upwards of 300 million as most lucrative in boxing history .
- iyanna mayweather has been spending time in floyd 's training camp and watches her champion dad training session .

Fig. 3. Example output summaries, article, Extractive-2 result and gold summary. Our entractive decoder selects informative sentences of the article (yellow highlight). The abstractive decoder rewrites sentence by removing unrelevant words (blue part) and generating coherent expressions (green highlight, here "spending" and "watches" are the actions of same subject, our model use "and" to keep the sentence coherence instead of using two separated subject "iyanna"). (Color figure online)

5.3   Case Study

We investigate an example of generated output in Fig. 3. Our extractive decoder first extracts 3 informative sentences from document (the sentences in Extractive-2). Then, our abstractive decoder rewrites these sentences and generates the final summaries. More specifically, our abstractive decoder is able to remove the meaningless and unrelevant words in the selected sentences, reorganizes words orders and generates coherent expressions.

6   Conclusion

In this work, we propose a novel extractive-and-abstractive hybrid model combining extractive and abstractive decoder for text summarization task. Experimental results shows that the ability of our model to extract informative contents and generate coherent and readable abstractive summary. Our model outperforms both state-of-the-art extractive and abstractive systems on the CNN/DailyMail dataset.

Acknowledgments. This work is supported by Ministry of Education - China Mobile Research Foundation NO. MCM20170302.

References

1. Celikyilmaz, A., Bosselut, A., He, X., Choi, Y.: Deep communicating agents for abstractive summarization. arXiv preprint arXiv:1803.10357 (2018)
2. Cheng, J., Lapata, M.: Neural summarization by extracting sentences and words. arXiv preprint arXiv:1603.07252 (2016)
3. Devlin, J., Chang, M.W., Lee, K., Toutanova, K.: Bert: Pre-training of deep bidirectional transformers for language understanding. arXiv preprint arXiv:1810.04805 (2018)
4. Edunov, S., Baevski, A., Auli, M.: Pre-trained language model representations for language generation. arXiv preprint arXiv:1903.09722 (2019)
5. Fan, A., Grangier, D., Auli, M.: Controllable abstractive summarization. arXiv preprint arXiv:1711.05217 (2017)
6. Gehrmann, S., Deng, Y., Rush, A.M.: Bottom-up abstractive summarization. arXiv preprint arXiv:1808.10792 (2018)
7. Gu, J., Lu, Z., Li, H., Li, V.O.: Incorporating copying mechanism in sequence-to-sequence learning. arXiv preprint arXiv:1603.06393 (2016)
8. Hariharan, S., Srimathi, R., Sivasubramanian, M., Pavithra, S.: Opinion mining and summarization of reviews in web forums. In: Proceedings of the Third Annual ACM Bangalore Conference, p. 24. ACM (2010)
9. Hermann, K.M., et al.: Teaching machines to read and comprehend. In: Advances in Neural Information Processing Systems, pp. 1693–1701 (2015)
10. Kågebäck, M., Mogren, O., Tahmasebi, N., Dubhashi, D.: Extractive summarization using continuous vector space models. In: Proceedings of the 2nd Workshop on Continuous Vector Space Models and their Compositionality (CVSC), pp. 31–39 (2014)
11. Lin, C.Y., Hovy, E.: Automatic evaluation of summaries using n-gram co-occurrence statistics. In: Proceedings of the 2003 Human Language Technology Conference of the North American Chapter of the Association for Computational Linguistics (2003)
12. Liu, Y.: Fine-tune bert for extractive summarization. arXiv preprint arXiv:1903.10318 (2019)

13. Liu, Y., Li, S., Cao, Y., Lin, C.Y., Han, D., Yu, Y.: Understanding and summarizing answers in community-based question answering services. In: Proceedings of the 22nd International Conference on Computational Linguistics, vol. 1, pp. 497–504. Association for Computational Linguistics (2008)
14. Mani, S., Catherine, R., Sinha, V.S., Dubey, A.: AUSUM: approach for unsupervised bug report summarization. In: Proceedings of the ACM SIGSOFT 20th International Symposium on the Foundations of Software Engineering, p. 11. ACM (2012)
15. Nallapati, R., Zhai, F., Zhou, B.: SummaRuNNer: a recurrent neural network based sequence model for extractive summarization of documents. In: Thirty-First AAAI Conference on Artificial Intelligence (2017)
16. Nallapati, R., Zhou, B., Gulcehre, C., Xiang, B., et al.: Abstractive text summarization using sequence-to-sequence RNNs and beyond. arXiv preprint arXiv:1602.06023 (2016)
17. Narayan, S., Cohen, S.B., Lapata, M.: Ranking sentences for extractive summarization with reinforcement learning. arXiv preprint arXiv:1802.08636 (2018)
18. Paulus, R., Xiong, C., Socher, R.: A deep reinforced model for abstractive summarization. arXiv preprint arXiv:1705.04304 (2017)
19. Rush, A.M., Chopra, S., Weston, J.: A neural attention model for abstractive sentence summarization. arXiv preprint arXiv:1509.00685 (2015)
20. See, A., Liu, P.J., Manning, C.D.: Get to the point: Summarization with pointer-generator networks. arXiv preprint arXiv:1704.04368 (2017)
21. Shi, Z.: Question answering summarization of multiple biomedical documents. In: Kobti, Z., Wu, D. (eds.) AI 2007. LNCS (LNAI), vol. 4509, pp. 284–295. Springer, Heidelberg (2007). https://doi.org/10.1007/978-3-540-72665-4_25
22. Yin, W., Pei, Y.: Optimizing sentence modeling and selection for document summarization. In: Twenty-Fourth International Joint Conference on Artificial Intelligence (2015)
23. Zhang, H., et al.: Pretraining-based natural language generation for text summarization. arXiv preprint arXiv:1902.09243 (2019)
24. Zhou, Q., Yang, N., Wei, F., Huang, S., Zhou, M., Zhao, T.: Neural document summarization by jointly learning to score and select sentences. arXiv preprint arXiv:1807.02305 (2018)

Title-Aware Neural News
Topic Prediction

Chuhan Wu[1][(✉)], Fangzhao Wu[2], Tao Qi[1], Yongfeng Huang[1], and Xing Xie[2]

[1] Department of Electronic Engineering, Tsinghua University, Beijing 100084, China
{wu-ch19,qit16,yfhuang}@mails.tsinghua.edu.cn
[2] Microsoft Research Asia, Beijing 100080, China
{fangzwu,xing.xie}@microsoft.com

Abstract. Online news platforms have gained huge popularity for online news reading. The topic categories of news are very important for these platforms to target user interests and make personalized recommendations. However, massive news articles are generated everyday, and it too expensive and time-consuming to manually categorize all news. The news bodies usually convey the detailed information of news, and the news titles usually contain summarized and complementary information of news. However, existing news topic prediction methods usually simply aggregate news titles and bodies together and ignore the differences of their characteristics. In this paper, we propose a title-aware neural news topic prediction approach to classify the topic categories of online news articles. In our approach, we propose a multi-view learning framework to incorporate news titles and bodies as different views of news to learn unified news representations. In the title view, we learn title representations from words via a long-short term memory (LSTM) network, and use attention mechanism to select important words according to their contextual representations. In the body view, we propose to use a hierarchical LSTM network to first learn sentence representations from words, and then learn body representations from sentences. In addition, we apply attention networks at both word and sentence levels to recognize important words and sentences. Besides, we use the representation vector of news title to initialize the hidden states of the LSTM networks for news body to capture the summarized news information condensed by news titles. Extensive experiments on a real-world dataset validate that our approach can achieve good performance in news topic prediction and consistently outperform many baseline methods.

Keywords: News topic prediction · Multi-view learning · Attention mechanism

1 Introduction

Online news platforms such as Bing News and Google News have gained huge popularity and attracted millions of users to read digital news online [5]. On

© Springer Nature Switzerland AG 2019
M. Sun et al. (Eds.): CCL 2019, LNAI 11856, pp. 181–193, 2019.
https://doi.org/10.1007/978-3-030-32381-3_15

Title	James Harden's incredible heroics lift **Rockets** over **Warriors** in overtime	The 5 best **movies screening** around Orange this week
Body	James Harden was blanketed with nowhere to go and no time to get there. It did not matter. The hottest **player** in the **NBA** would not be stopped, not by a 20-point deficit, not by all the defensive clamps the tag team of Klay Thompson and Draymond Green could wrap around him...	**Spider-Man**: Into the Spider-Verse Miles Morales is juggling his life between being a high school student and being **Spider-Man**. However, when Wilson "Kingpin" Fisk uses a super collider, another **Spider-Man** from another dimension, Peter Parker, accidentally winds up in Miles' dimension...
Category	sports	movies

Fig. 1. Two illustrative example news with different topics. Several important words in news title and body are highlighted.

many news platforms, news articles are classified into different topic categories such as sports and finance to target user interests and make news recommendations [20]. However, hundreds of thousands of news articles emerge everyday, and it is too expensive and time-consuming to manually categorize all news articles [6]. Thus, automatically predicting the topic categories of news is very important for online news platforms to provide personalized news services [23].

Learning accurate representations of news is critical for news topic prediction. Many of existing news topic prediction methods build news representations via manual feature engineering [6,25,28]. For example, Dilrukshi et al. [6] proposed to use support vector machine (SVM) to classify Twitter news, and use word unigram features to represent news. However, the bag-of-word features used in these methods cannot capture the contexts and orders of words, both of which are important for the prediction of news topics. In recent years, several deep learning based text classification methods are applied to news topic prediction. For example, Zhang et al. [29] proposed to use a character-level convolutional neural network (CNN) to learn news representations from original characters. Lai et al. [13] proposed to use a convolutional recurrent neural network to learn news representations from news bodies. Conneau et al. [4] proposed to apply a deep CNN network with shortcut connections to learn hidden news representations. However, these methods usually simply formulate the news topic prediction task as a document classification problem, i.e., aggregating the title and body of news as a single document, while the differences in characteristics between news title and news body are not taken into consideration.

Our work is motivated by the observation that both news titles and news bodies are useful for learning news representations. The bodies usually convey the detailed information of the news, and the titles usually convey summarized and supplementary information of news. For example, in Fig. 1 the title of the first news shows that this news is about an NBA event, and the body introduces its details. In addition, in the second news, the news body only introduces the details of movies, but the title clearly summarizes the news topic. Thus, incorporating the information of both news titles and bodies has the potential to enhance the learning of news representations for topic prediction. In addition, the characteristics of news titles and news bodies usually have some differences, since titles are usually short and concise sentences, while bodies are usually long documents with rich details. Thus, they should be handled differently. Besides, different words in the same news title or body usually have different informativeness for learning news representations. For example, in Fig. 1 the word "NBA" is more informative than "would". Moreover, different sentences in the same news body may also have different importance. For instance, the first sentence in the body of the first news in Fig. 1 is more informative than the second sentence in inferring the topic of this news.

In this paper, we propose a title-aware neural news topic prediction approach which can utilize both news title and news body information. In our approach, we propose to use a multi-view learning framework to incorporate both titles and bodies as different views of news for learning unified news representations. In the title view, we learn title representations from words via an LSTM network, and use attention mechanism to recognize important words. In the body view, we use a hierarchical LSTM network to first learn sentence representations from words, and then learn body representations from sentences. In addition, we apply attention mechanism to select important words and sentences for learning informative news representations. Moreover, we propose to use the representations vector of news title to initialized the hidden states of the LSTM networks in the body view. Thus, our approach can learn more accurate representations of news bodies with the help of summarized information provided by news titles. Extensive experiments are conducted on a real-world dataset, and the results show that our approach can effectively improve the performance of news topic prediction and consistently outperform many baseline methods.

2   Related Work

News topic prediction is an important task in the natural language processing field and has been extensively explored over years [23]. News topic prediction is usually formulated as a text classification problem, and learning accurate news representations is a core step. Many of existing methods build news representations via manual feature engineering [1–3, 6, 10, 15, 19, 24–26, 28]. For example, Yang et al. [26] proposed to use logistic regression (LR) to classify news topics and they used the TF-IDF features to build representations of news documents. Dilrukshi et al. [6] proposed to use support vector machine (SVM) to classify Twitter news, and they used word unigrams to represent news. Antonellis et al. [1] proposed to build news representations via the addition of the sentence representations learn by the cosine similarities between news representations and category term representations. Joulin et al. [9] proposed to use a bag of n-grams as the features to represent news documents. In addition, they used a hashing strategy to reduce the memory cost. However, these methods cannot effectively utilize contextual information and word orders. In addition, they cannot distinguish informative contexts from uninformative ones.

In recent years, several deep learning based methods have been proposed to automatically learn news representations from their content [4, 13, 16, 18, 21, 29, 31]. For example, Zhang et al. [29] proposed to apply a CNN network at character-level to learn representations of news from their original characters.

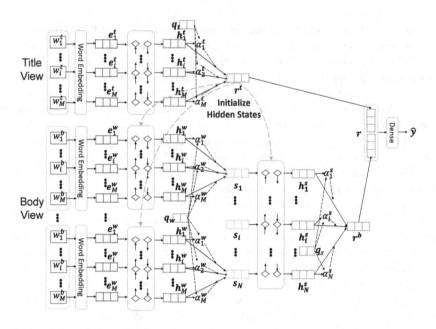

Fig. 2. The framework of our *TAP* approach.

Lai et al. [13] proposed a convolutional recurrent neural network to learn representations of news from their bodies by capturing both local and global contexts. Zhu et al. [31] proposed a voting method to learn representations of news from their titles based on an ensemble of a CNN network, a gated recurrent units (GRU) network, and an SVM trained using TF-IDF features. Lu et al. [18] proposed to use a Bi-LSTM network to learn news representations from news headlines, and they used attention mechanism to select important words. However, these methods usually only consider either the title or the body of news, which is usually insufficient for representing news accurately. Different from these methods, our approach can learn title-aware news representations by utilizing the information in both news titles and news bodies.

3   Our Approach

In this section, we will introduce our Title-Aware neural news topic Prediction approach (TAP). There are two main components in our model, i.e., a *title-aware news representation* module to learn news representations from both news titles and bodies, and a *topic classification* module to classify news into different topic categories. The overall framework of our approach is shown in Fig. 2. We will introduce the details of our *TAP* approach as follows.

3.1   Title-Aware News Representation

The *title-aware news representation* aims to learn news representations by incorporating the information of both news title and news body. In our approach, we propose to use a multi-view learning framework to learn unified news representations by regarding titles and bodies as different views of news.

Title Representation. The first view is *title representation*. It contains three layers. The first one is word embedding. It aims to convert a news title with M words from a word sequence into a low-dimensional vector sequence. Denote the input news title as $[w_1^t, w_2^t, ..., w_M^t]$, the output of this layer is an embedding sequence $[e_1^t, e_2^t, ..., e_M^t]$.

The second layer is a bi-directional long-short term memory network (Bi-LSTM) [11]. Global contexts in a news title are usually necessary for understanding this news. For example, in the title of the first news of Fig. 1, the contexts of "Rockets" such as "lift" and "James Harden's" are very useful for representing this news. In addition, the contexts in the future are also useful for learning informative word representations. For example, the word "Warriors" is also important for understanding "Rockets". Thus, we use a Bi-LSTM network to learn word representations by summarizing the past and future contextual information in both directions. It scans the input embedding sequence forward and backward, and outputs a hidden word representation sequence $[\mathbf{h}_1^t, \mathbf{h}_2^t, ..., \mathbf{h}_M^t]$.

The third layer is a word-level attention network. Usually, different words have different contributions to the representation learning of news title. For example, in the second news title of Fig. 1, the word "movies" is very informative for inferring news topic, while the word "around" is uninformative. Thus, we apply an attention network to select important words to learn informative title representations for topic prediction. Denote the attention weight of the i-th word in the same news title as α_i^t, which is computed as:

$$a_i^t = \mathbf{q}_t \times \tanh(\mathbf{W}_t \times \mathbf{h}_i^t + \mathbf{w}_t), \quad \alpha_i^t = \frac{\exp(a_i^t)}{\sum_{j=1}^{M} \exp(a_j^t)}, \tag{1}$$

where $\mathbf{W}_t$ and $\mathbf{w}_t$ are projection parameters, $\mathbf{q}_t$ is the word attention query. The contextual representation of the news title is the summation of the hidden word representations weighted by their attention weights, i.e., $\mathbf{r}^t = \sum_{j=1}^{M} \alpha_j^t \mathbf{h}_j^t$.

Body Representation. The second view is *body representation*, which is used to learn news representations from news bodies. It contains two major modules. The first module is *sentence representation*. Similar with the title view, there are also three layers in this module.

The first one is a shared word embedding layer. Denote a sentence in a news body as $[w_1^b, w_2^b, ..., w_M^b]$, it is converted from a word sequence into a vector sequence $[e_1^b, e_2^b, ..., e_M^b]$.

The second one is a title-aware Bi-LSTM network. Global contexts within the same sentence in a news body are also important for learning body representations. For example, in the first news of Fig. 1, the contexts of the word "player" such as "hottest" and "NBA" are important for inferring the topic of this news. However, it may be difficult to capture important topic information with limited contexts. For example, the word "20-point" may not be topic sensitive in many news, but it is very useful in inferring the topic of this news. Fortunately, the summarized information provided by news titles has the potential to help learn more topic discriminative body representations. Thus, we propose to use a title-aware Bi-LSTM network by using the title representation vector to initialize the hidden states of the Bi-LSTM. In addition, we need to apply a linear transformation to the title representation vector to align its dimension with the size of LSTM hidden states. After initialization, the Bi-LSTM scans the word embedding sequence $[\mathbf{e}_1^w, \mathbf{e}_2^w, ..., \mathbf{e}_M^w]$ in both directions, and outputs the hidden word representation sequence $[\mathbf{h}_1^w, \mathbf{h}_2^w, ..., \mathbf{h}_M^w]$.

The third one is a word attention network. Different words in the same sentence usually have different importance in learning sentence representations. For example, in the first news of Fig. 1, the word "NBA" is much more informative than "would" in representing its third sentence "The hottest...". Thus, we use an attention network to select important words according to their hidden representations. Denote the attention weight of the i-th word in a sentence as α_i^w, which is computed as:

$$a_i^w = \mathbf{q}_w \times \tanh(\mathbf{W}_w \times \mathbf{h}_i^w + \mathbf{w}_w), \quad \alpha_i^w = \frac{\exp(a_i^w)}{\sum_{j=1}^M \exp(a_j^w)}, \tag{2}$$

where $\mathbf{W}_w$, $\mathbf{w}_w$ and $\mathbf{q}_w$ are projection parameters. The contextual sentence representation is the summation of the word representations weighted by their attention weight, i.e., $\mathbf{s} = \sum_{i=1}^M \alpha_i^w \mathbf{h}_i^w$. The representations of each sentence in the news body are computed in a similar way. We denote the sentence representation sequence as $[\mathbf{s}_1, \mathbf{s}_2, ..., \mathbf{s}_N]$, where N is the number of sentences.

The second module is *document representation*. It contains two layers. Global contexts of sentences are also important for learning body representations. For example, in the first news of Fig. 1, the relatedness between the first and the third sentence is important for representing the entire document. In addition, since news titles can provide the global topic information, incorporating the summarized information condensed by news titles may help improve the sentence representation learning. Thus, we use a title-aware Bi-LSTM network at sentence-level to learn contextual sentence representations. It takes the sentence representation sequence $[\mathbf{s}_1, \mathbf{s}_2, ..., \mathbf{s}_N]$ as input, and outputs the contextual sentence representation sequence $[\mathbf{h}_1^s, \mathbf{h}_2^s, ..., \mathbf{h}_N^s]$.

The second one is a sentence-level attention network. Different sentences in the news body may also have different informativeness in learning body representations. For example, in the first news of Fig. 1, the first sentence is more informative than the second sentence for learning body representations. Thus, we propose to use a sentence attention network to select important sentences. Denote the attention weight of the i-th sentence in the news body as α_i^s, which is computed as:

$$a_i^s = \mathbf{q}_s \times \tanh(\mathbf{W}_s \times \mathbf{h}_i^s + \mathbf{w}_s), \quad \alpha_i^s = \frac{\exp(a_i^s)}{\sum_{j=1}^N \exp(a_j^s)}, \tag{3}$$

where $\mathbf{W}_s$, $\mathbf{w}_s$ and $\mathbf{q}_s$ are attention parameters. The contextual body representation is computed as: $\mathbf{r}^b = \sum_{i=1}^N \alpha_i^s \mathbf{h}_i^s$. The final unified news representation $\mathbf{r}$ is the concatenation of the new representations learned from the title view and the body view, i.e., $\mathbf{r} = [\mathbf{r}^t; \mathbf{r}^b]$.

3.2 Topic Classification

The *topic classification* module is used to classify the topic category of a news by predicting the probabilities $\hat{\mathbf{y}}$ of a news in different topic category as follows:

$$\hat{\mathbf{y}} = softmax(\mathbf{W}_y \times \mathbf{r} + \mathbf{w}_y), \tag{4}$$

where $\mathbf{W}_y$ and $\mathbf{w}_y$ are parameters. In the model training stage, cross entropy is used as the loss function, which is computed as follows:

$$\mathcal{L} = -\sum_{i=1}^{S}\sum_{t=1}^{T} y_{i,t} \log(\hat{y}_{i,t}), \tag{5}$$

where S is the number of training samples, T is the number of topic categories, $y_{i,k}$ and $\hat{y}_{i,k}$ represent the gold and predicted probability of the i-th news in the k-th topic category, respectively.

4 Experiments

4.1 Datasets and Experimental Settings

Since existing news topic classification datasets such as 20 News [14], AG News[1], and UCI News [17] usually only contain either news title or news body, we constructed a new news topic classification dataset which contains both news title and news body information by crawling news articles from the MSN News[2] platform during 12/13/2018 and 01/12/2019. The final dataset has 31,908 news articles in 14 topic categories. The topic distributions of these news articles are shown in Fig. 3. The average number of words per title is 11.16, and the number for news body is 761.86. We use 80% of the news for training, 10% for validation, and the remaining 10% for test.

In our experiments, the word embeddings were 300-dimensional and we used the pre-trained Glove embeddings [22] to initialize them. The dimension of the hidden states in the Bi-LSTM networks was 2×200. The attention query vectors were also 200-dimensional. We used the Adam [12] algorithm to optimize the loss function. To mitigate overfitting, we added 30% dropout to each layer. The size of a training minibatch was 64. These hyperparameters were tuned on the validation set. Each experiment was repeated 10 times and we reported the average accuracy and macro F1score.

4.2 Performance Evaluation

First, we want to evaluate the performance of our approach by comparing it with several baseline methods for news topic prediction, including: (1) *SVM* [6], support vector machine; (2) *LR* [26], logistic regression; (3) *FastText* [9], a famous

---

[1] http://www.di.unipi.it/~gulli/AG_corpus_of_news_articles.html.

[2] https://www.msn.com/en-us/news.

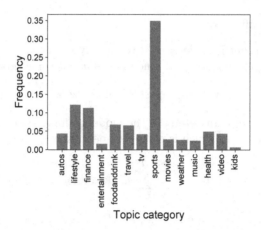

Fig. 3. Distributions of different news topics.

text classification method based on n-grams; (4) *CNN* [11], convolutional neural network. (5) *LSTM* [8], long short-term memory network. (6) *CLSTM* [16,30], a combination of both CNN and LSTM network. (7) *CNN-Att* [7], CNN with attention mechanism. (8) *LSTM-Att* [18], LSTM with attention mechanism. (9) *HAN* [27], a hierarchical attention network for document classification. (10) *TAP*, our title-aware neural news topic prediction approach. In traditional methods including *SVM* and *LR*, we used the TF-IDF features extracted from news titles and bodies as the input. In other baseline methods (3–9), we used the combination of news titles and bodies by simply regarding them as a single document. The performance of these methods under different ratios of training data are summarized in Table 1. We have several observations from the results.

First, the methods which learn news representations from news content (e.g., *CNN*, *LSTM* and *TAP*) outperform the methods which build news representations via feature engineering (*SVM*, *LR* and *Linreg*). This is probably because the latter methods rely on bag-of-words features, while the contextual information and word orders cannot be effectively modeled. Second, the methods with attention mechanism (e.g., *CNN-Att*) outperform the methods without mechanism (e.g., *CNN*). This is probably because different contexts within a news usually have different informativeness in representing this news, and distinguishing important contexts from unimportant ones can benefit news representation learning. Third, the methods based on hierarchical neural architectures (*HAN* and *TAP*) outperform the methods based on flatten models (e.g., *CNN-Att* and *LSTM-Att*). This is because news are usually long documents, and it may be more effective to model news in a hierarchical manner to exploit the document structures of news. Fourth, our approach can outperform all other baselines. This is probably because the characteristics of news titles and news bodies are quite different, and they should be processed differently. Thus, the performance of other baselines such as *HAN* is sub-optimal. Different from these methods, our approach employs a multi-view learning framework to learn unified news

Table 1. The performance of different methods under different ratios of training data. *The advantage of our approach over all baselines is significant at $p < 0.005$.

Method	25%		50%		100%	
	Accuracy	Fscore	Accuracy	Fscore	Accuracy	Fscore
SVM [6]	73.59 ± 0.65	61.11 ± 0.69	74.45 ± 0.71	62.36 ± 0.75	74.88 ± 0.77	63.11 ± 0.79
LR [26]	73.18 ± 0.68	60.24 ± 0.72	74.47 ± 0.72	61.88 ± 0.76	75.35 ± 0.74	62.44 ± 0.78
FastText [9]	74.08 ± 0.72	61.88 ± 0.78	76.10 ± 0.69	63.22 ± 0.73	78.20 ± 0.62	66.34 ± 0.65
CNN [11]	74.79 ± 0.64	62.40 ± 0.66	76.55 ± 0.53	64.31 ± 0.58	78.44 ± 0.40	67.61 ± 0.43
LSTM [8]	74.55 ± 0.48	61.96 ± 0.52	76.66 ± 0.44	64.48 ± 0.47	78.33 ± 0.36	67.55 ± 0.38
CLSTM [30]	75.04 ± 0.60	62.33 ± 0.63	77.11 ± 0.58	65.00 ± 0.62	78.86 ± 0.43	67.94 ± 0.47
CNN-Att [7]	75.22 ± 0.56	62.59 ± 0.60	77.33 ± 0.51	65.29 ± 0.53	78.98 ± 0.44	68.15 ± 0.47
LSTM-Att [18]	74.89 ± 0.51	62.47 ± 0.54	77.59 ± 0.43	65.54 ± 0.46	79.12 ± 0.39	68.23 ± 0.42
HAN [27]	76.22 ± 0.44	63.15 ± 0.47	78.22 ± 0.35	66.06 ± 0.40	79.96 ± 0.35	68.88 ± 0.39
TAP*	$\mathbf{77.63 \pm 0.38}$	$\mathbf{64.34 \pm 0.42}$	$\mathbf{79.50 \pm 0.31}$	$\mathbf{67.44 \pm 0.34}$	$\mathbf{81.49 \pm 0.26}$	$\mathbf{70.22 \pm 0.30}$

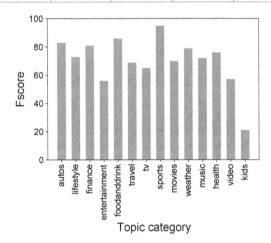

Fig. 4. Performance of the recognition of different topic categories.

representations by regarding news titles and bodies as different views of news, which is useful for learning better news representations.

Then, we want to explore the performance of our approach in the classification of different topic categories. The performance in Fscore of recognizing different topic categories is shown in Fig. 4. From Fig. 4, we find our approach can achieve satisfactory performance in recognizing most topics, even the sports news are dominant in our dataset. However, it is very difficult to classify the news in the category "kids", since the training samples in this category are too scarce. These results show that our approach is effective in news topic prediction.

4.3 Effectiveness of Learning Title-Aware News Representations

In this section, we conducted several experiments to validate the effectiveness of learning title-aware news representations in our approach. We compare the

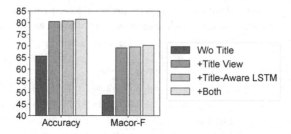

Fig. 5. Effectiveness of learning title-aware news representations.

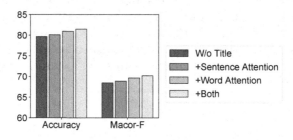

Fig. 6. Effectiveness of different attention networks.

performance of our approach and its three variants, i.e., without title information, incorporating titles via multi-view learning only, and using title-aware
LSTMs only. The results are shown in Fig. 5. According to Fig. 5, we have several observations. First, we find the performance of our approach will decline
seriously if title information is not considered. This is probably because news
titles are usually synthesis of news topics and highlights, which are critical for
topic prediction. Thus, it is necessary to incorporate the information of news
titles. Second, incorporating news title and body via multi-view learning can
effectively improve the performance of our approach. This may be because news
titles can provide useful complementary information to news bodies In addition,
using title-aware LSTM networks is also useful. This is probably because news
titles can provide summarized information of news, and can help learn more
informative body representations. Third, combining both techniques can further
improve the performance of our approach. These results validate the effectiveness
of learning title-aware news representations in our approach.

4.4   Effectiveness of Attention Mechanism

In this section, we conducted experiments to verify the effectiveness of the word-
level and sentence-level attention networks in our approach. The performance
of our approach and its variant with different combinations of attention networks is shown in Fig. 6. The results in Fig. 6 lead to several observations. First,
the word-level attention networks are very important for our approach. This

is probably because different words in the same news title and body usually have different informativeness for learning news representations. Thus, selecting important words in news titles and bodies can help learn more informative news representations. Second, the sentence-level attention network is also useful for our approach. This is probably because different sentences also have different informativeness in representing news bodies, and selecting important sentences is beneficial for body representation learning. Third, combining both kinds of attention networks can further improve our approach. These results validate the effectiveness of the hierarchical attention networks in our approach.

5 Conclusion

In this paper, we propose a title-aware neural news topic prediction approach which can incorporate both news title and news body to learn informative representation of news. In our approach, we propose a multi-view learning framework to exploit the titles and the bodies as different views of news articles. In the title view, we learn the representations of news titles via a Bi-LSTM network, and use an attention network to select important words. In the body view, we learn the representations of news bodies in a hierarchical way. We first learn sentence representations from words, and then learn body representations from sentences. We apply a hierarchical attention network to select important words and sentences. In addition, we propose a title-aware LSTM network by using the representation of news titles to initialize the hidden states of the LSTM networks for body representation learning. Extensive experiments on a real-world dataset validate the effectiveness of our approach.

Acknowledgments. The authors would like to thank Microsoft News for providing technical support and data in the experiments, and Jiun-Hung Chen (Microsoft News) and Ying Qiao (Microsoft News) for their support and discussions. This work was supported by the National Key Research and Development Program of China under Grant number 2018YFC1604002, the National Natural Science Foundation of China under Grant numbers U1836204, U1705261, U1636113, U1536201, and U1536207, and the Tsinghua University Initiative Scientific Research Program.

References

1. Adi, A.O., Çelebi, E.: Classification of 20 news group with Naïve Bayes classifier. In: SIU, pp. 2150–2153. IEEE (2014)
2. Bracewell, D.B., Yan, J., Ren, F., Kuroiwa, S.: Category classification and topic discovery of Japanese and English news articles. ENTCS **225**, 51–65 (2009)
3. Cecchini, D., Na, L.: Chinese news classification. In: BigComp, pp. 681–684 (2018)
4. Conneau, A., Schwenk, H., Barrault, L., Lecun, Y.: Very deep convolutional networks for text classification. In: EACL, pp. 1107–1116 (2017)
5. Das, A.S., Datar, M., Garg, A., Rajaram, S.: Google news personalization: scalable online collaborative filtering. In: WWW, pp. 271–280. ACM (2007)

6. Dilrukshi, I., De Zoysa, K., Caldera, A.: Twitter news classification using SVM. In: ICCSE, pp. 287–291. IEEE (2013)
7. Du, J., Gui, L., Xu, R., He, Y.: A convolutional attention model for text classification. In: Huang, X., Jiang, J., Zhao, D., Feng, Y., Hong, Y. (eds.) NLPCC 2017. LNCS (LNAI), vol. 10619, pp. 183–195. Springer, Cham (2018). https://doi.org/10.1007/978-3-319-73618-1_16
8. Hochreiter, S., Schmidhuber, J.: Long short-term memory. Neural Comput. **9**(8), 1735–1780 (1997)
9. Joulin, A., Grave, E., Bojanowski, P., Mikolov, T.: Bag of tricks for efficient text classification. In: EACL, vol. 2, pp. 427–431 (2017)
10. Kaur, G., Bajaj, K.: News classification using neural networks. Commun. Appl. Electron **5**(1), 42–45 (2016)
11. Kim, Y.: Convolutional neural networks for sentence classification. In: EMNLP, pp. 1746–1751 (2014)
12. Kingma, D.P., Ba, J.: Adam: A method for stochastic optimization. arXiv preprint arXiv:1412.6980 (2014)
13. Lai, S., Xu, L., Liu, K., Zhao, J.: Recurrent convolutional neural networks for text classification. In: AAAI, pp. 2267–2273. AAAI Press (2015)
14. Lang, K.: Newsweeder: learning to filter netnews. In: Machine Learning Proceedings, pp. 331–339. Elsevier (1995)
15. Lange, L., Alonso, O., Strötgen, J.: The power of temporal features for classifying news articles. In: WWW, pp. 1159–1160. ACM (2019)
16. Li, C., Zhan, G., Li, Z.: News text classification based on improved Bi-LSTM-CNN. In: ITME, pp. 890–893. IEEE (2018)
17. Lichman, M., et al.: UCI machine learning repository (2013)
18. Lu, Z., Liu, W., Zhou, Y., Hu, X., Wang, B.: An effective approach for Chinese news headline classification based on multi-representation mixed model with attention and ensemble learning. In: Huang, X., Jiang, J., Zhao, D., Feng, Y., Hong, Y. (eds.) NLPCC 2017. LNCS (LNAI), vol. 10619, pp. 339–350. Springer, Cham (2018). https://doi.org/10.1007/978-3-319-73618-1_29
19. Majeed, F., Asif, M.W., Hassan, M.A., Abbas, S.A., Lali, M.I.: Social media news classification in healthcare communication. J. Med. Imaging Health Inform. **9**(6), 1215–1223 (2019)
20. Okura, S., Tagami, Y., Ono, S., Tajima, A.: Embedding-based news recommendation for millions of users. In: KDD, pp. 1933–1942. ACM (2017)
21. Peng, H., et al.: Large-scale hierarchical text classification with recursively regularized deep graph-CNN. In: WWW, pp. 1063–1072 (2018)
22. Pennington, J., Socher, R., Manning, C.: Glove: global vectors for word representation. In: EMNLP, pp. 1532–1543 (2014)
23. Qiu, X., Gong, J., Huang, X.: Overview of the NLPCC 2017 shared task: Chinese news headline categorization. In: Huang, X., Jiang, J., Zhao, D., Feng, Y., Hong, Y. (eds.) NLPCC 2017. LNCS (LNAI), vol. 10619, pp. 948–953. Springer, Cham (2018). https://doi.org/10.1007/978-3-319-73618-1_85
24. Sawaf, H., Zaplo, J., Ney, H.: Statistical classification methods for Arabic news articles. In: Arabic Natural Language Processing in ACL2001. Citeseer (2001)
25. Tenenboim, L., Shapira, B., Shoval, P.: Ontology-based classification of news in an electronic newspaper. In: Advanced Research in Artificial Intelligence, p. 89 (2008)
26. Yang, B., Sun, J.T., Wang, T., Chen, Z.: Effective multi-label active learning for text classification. In: KDD, pp. 917–926. ACM (2009)
27. Yang, Z., Yang, D., Dyer, C., He, X., Smola, A., Hovy, E.: Hierarchical attention networks for document classification. In: NAACL, pp. 1480–1489 (2016)

28. Yin, Z., Tang, J., Ru, C., Luo, W., Luo, Z., Ma, X.: A semantic representation enhancement method for Chinese news headline classification. In: Huang, X., Jiang, J., Zhao, D., Feng, Y., Hong, Y. (eds.) NLPCC 2017. LNCS (LNAI), vol. 10619, pp. 318–328. Springer, Cham (2018). https://doi.org/10.1007/978-3-319-73618-1_27
29. Zhang, X., Zhao, J., LeCun, Y.: Character-level convolutional networks for text classification. In: NIPS, pp. 649–657 (2015)
30. Zhou, C., Sun, C., Liu, Z., Lau, F.: A C-LSTM neural network for text classification. arXiv preprint arXiv:1511.08630 (2015)
31. Zhu, F., Dong, X., Song, R., Hong, Y., Zhu, Q.: A multiple learning model based voting system for news headline classification. In: Huang, X., Jiang, J., Zhao, D., Feng, Y., Hong, Y. (eds.) NLPCC 2017. LNCS (LNAI), vol. 10619, pp. 797–806. Springer, Cham (2018). https://doi.org/10.1007/978-3-319-73618-1_69

How to Fine-Tune BERT
for Text Classification?

Chi Sun, Xipeng Qiu$^{(\boxtimes)}$, Yige Xu, and Xuanjing Huang

Shanghai Key Laboratory of Intelligent Information Processing,
School of Computer Science, Fudan University,
825 Zhangheng Road, Shanghai, China
{sunc17,xpqiu,ygxu18,xjhuang}@fudan.edu.cn

Abstract. Language model pre-training has proven to be useful in learning universal language representations. As a state-of-the-art language model pre-training model, BERT (Bidirectional Encoder Representations from Transformers) has achieved amazing results in many language understanding tasks. In this paper, we conduct exhaustive experiments to investigate different fine-tuning methods of BERT on text classification task and provide a general solution for BERT fine-tuning. Finally, the proposed solution obtains new state-of-the-art results on eight widely-studied text classification datasets.

Keywords: Transfer learning · BERT · Text classification

1 Introduction

Text classification is a classic problem in Natural Language Processing (NLP). The task is to assign predefined categories to a given text sequence. An important intermediate step is the text representation. Previous work uses various neural models to learn text representation, including convolution models, recurrent models, and attention mechanisms.

Recently, pre-trained language models have shown to be useful in learning common language representations by utilizing a large amount of unlabeled data: e.g., ELMo [20], OpenAI GPT [22] and BERT [6]. Among them, BERT is based on a multi-layer bidirectional Transformer [24] and is trained on plain text for masked word prediction and next sentence prediction tasks.

Although BERT has achieved amazing results in many natural language understanding (NLU) tasks, its potential has yet to be fully explored. There is little research to enhance BERT to improve the performance on target tasks further.

In this paper, we investigate how to maximize the utilization of BERT for the text classification task. We explore several ways of fine-tuning BERT to enhance its performance on text classification task. We design exhaustive experiments to make a detailed analysis of BERT.

© Springer Nature Switzerland AG 2019
M. Sun et al. (Eds.): CCL 2019, LNAI 11856, pp. 194–206, 2019.
https://doi.org/10.1007/978-3-030-32381-3_16

The contributions of our paper are as follows:

- We propose a general solution to fine-tune the pre-trained BERT model, which includes three steps: (1) further pre-train BERT on within-task training data or in-domain data; (2) optional fine-tuning BERT with multi-task learning if several related tasks are available; (3) fine-tune BERT for the target task.
- We also investigate the fine-tuning methods for BERT on target task, including layer-wise learning rate, catastrophic forgetting, and few-shot learning problems.
- We achieve the new state-of-the-art results on eight widely-studied text classification datasets.

2   Related Work

Borrowing the learned knowledge from the other tasks has a rising interest in the field of NLP. We briefly review two related approaches: language model pre-training and multi-task Learning.

2.1   Language Model Pre-training

Pre-trained word embeddings [18,19], as an important component of modern NLP systems can offer significant improvements over embeddings learned from scratch. The generalization of word embeddings, such as sentence embeddings [8,14] or paragraph embeddings [9], are also used as features in downstream models.

Peters et al. [20] concatenate embeddings derived from language model as additional features for the main task and advance the state-of-the-art for several major NLP benchmarks. In addition to pre-training with unsupervised data, transfer learning with a large amount of supervised data can also achieve good performance, such as natural language inference [4] and machine translation [16].

More recently, the method of pre-training language models on a large network with a large amount of unlabeled data and fine-tuning in downstream tasks has made a breakthrough in several natural language understanding tasks, such as OpenAI GPT [22] and BERT [6]. Dai and Le [5] use language model fine-tuning but overfit with 10k labeled examples while Howard and Ruder [7] propose ULMFiT and achieve state-of-the-art results in the text classification task. BERT is pre-trained on *Masked Language Model Task* and *Next Sentence Prediction Task* via a large cross-domain corpus. Unlike previous bidirectional language models (biLM) limited to a combination of two unidirectional language models (i.e., left-to-right and right-to-left), BERT uses a Masked Language Model to predict words which are randomly masked or replaced. BERT is the first fine-tuning based representation model that achieves state-of-the-art results for a range of NLP tasks, demonstrating the enormous potential of the fine-tuning method. In this paper, we have further explored the BERT fine-tuning method for text classification.

2.2   Multi-task Learning

Multi-task learning [1,3] is another relevant direction. Rei [23] and Liu et al. [11] use this method to train the language model and the main task model jointly. Liu et al. [13] extend the MT-DNN model originally proposed in [12] by incorporating BERT as its shared text encoding layers. MTL requires training tasks from scratch every time, which makes it inefficient and it usually requires careful weighing of task-specific objective functions [2]. However, we can use multi-task BERT fine-tuning to avoid this problem by making full use of the shared pre-trained model.

3   BERT for Text Classification

BERT-base model contains an encoder with 12 Transformer blocks, 12 self-attention heads, and the hidden size of 768. BERT takes an input of a sequence of no more than 512 tokens and outputs the representation of the sequence. The sequence has one or two segments that the first token of the sequence is always [CLS] which contains the special classification embedding and another special token [SEP] is used for separating segments.

For text classification tasks, BERT takes the final hidden state $\mathbf{h}$ of the first token [CLS] as the representation of the whole sequence. A simple softmax classifier is added to the top of BERT to predict the probability of label c:

$$p(c|\mathbf{h}) = \text{softmax}(W\mathbf{h}), \tag{1}$$

where W is the task-specific parameter matrix. We fine-tune all the parameters from BERT as well as W jointly by maximizing the log-probability of the correct label.

4   Methodology

When we adapt BERT to NLP tasks in a target domain, a proper fine-tuning strategy is desired. In this paper, we look for the proper fine-tuning methods in the following three ways.

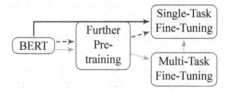

Fig. 1. Three general ways for fine-tuning BERT, shown with different colors. (Color figure online)

(1) **Fine-Tuning Strategies:** When we fine-tune BERT for a target task, there are many ways to utilize BERT. For example, the different layers of BERT capture different levels of semantic and syntactic information, which layer is better for a target task? How we choose a better optimization algorithm and learning rate?

(2) **Further Pre-training:** BERT is trained in the general domain, which has a different data distribution from the target domain. A natural idea is to further pre-train BERT with target domain data.

(3) **Multi-task Fine-Tuning:** Without pre-trained LM models, multi-task learning has shown its effectiveness of exploiting the shared knowledge among the multiple tasks. When there are several available tasks in a target domain, an interesting question is whether it still bring benefits to fine-tune BERT on all the tasks simultaneously.

Our general methodology of fine-tuning BERT is shown in Fig. 1.

4.1 Fine-Tuning Strategies

Different layers of a neural network can capture different levels of syntactic and semantic information [7,27].

To adapt BERT to a target task, we need to consider the overfitting problem. A better optimizer with an appropriate learning rate is desired. Intuitively, the lower layer of the BERT model may contain more general information. We can fine-tune them with different learning rates.

Following [7], we split the parameters θ into $\{\theta^1, \cdots, \theta^L\}$ where θ^l contains the parameters of the l-th layer of BERT. Then the parameters are updated as follows:

$$\theta_t^l = \theta_{t-1}^l - \eta^l \cdot \nabla_{\theta^l} J(\theta), \tag{2}$$

where η^l represents the learning rate of the l-th layer.

We set the base learning rate to η^L and use $\eta^{k-1} = \xi \cdot \eta^k$, where ξ is a decay factor and less than or equal to 1. When $\xi < 1$, the lower layer has a lower learning rate than the higher layer. When $\xi = 1$, all layers have the same learning rate, which is equivalent to the regular stochastic gradient descent (SGD). We will investigate these factors in Sect. 5.3.

4.2 Further Pre-training

The BERT model is pre-trained in the general-domain corpus. For a text classification task in a specific domain, such as movie reviews, its data distribution may be different from BERT. Therefore, we can further pre-train BERT with masked language model and next sentence prediction tasks on the domain-specific data. Three further pre-training approaches are performed:

(1) Within-task pre-training, in which BERT is further pre-trained on the training data of a target task.
(2) In-domain pre-training, in which the pre-training data is obtained from the same domain of a target task. For example, there are several different sentiment classification tasks, which have a similar data distribution. We can further pre-train BERT on the combined training data from these tasks.
(3) Cross-domain pre-training, in which the pre-training data is obtained from both the same and other different domains to a target task.

We will investigate these different approaches to further pre-training in Sect. 5.4.

4.3   Multi-task Fine-Tuning

Multi-task Learning is also an effective approach to share the knowledge obtained from several related supervised tasks. Similar to [13], we also use fine-tune BERT in multi-task learning framework for text classification.

All the tasks share the BERT layers and the embedding layer. The only layer that does not share is the final classification layer, which means that each task has a private classifier layer. The experimental analysis is in Sect. 5.5.

Table 1. Statistics of eight text classification datasets.

Dataset	Classes	Type	Average lengths	Max lengths	Train samples	Test samples
IMDb [15]	2	Sentiment	292	3,045	25,000	25,000
Yelp P. [28]	2	Sentiment	177	2,066	560,000	38,000
Yelp F. [28]	5	Sentiment	179	2,342	650,000	50,000
TREC [25]	6	Question	11	39	5,452	500
Yahoo! Answers [28]	10	Question	131	4,018	1,400,000	60,000
AG's News [28]	4	Topic	44	221	120,000	7,600
DBPedia [28]	14	Topic	67	3,841	560,000	70,000
Sogou News [28]	6	Topic	737	47,988	54,000	6,000

5   Experiments

We investigate the different fine-tuning methods for seven English and one Chinese text classification tasks. We use the base BERT models: the uncased BERT-base model[1] and the Chinese BERT-base model[2] respectively.

---

[1] https://storage.googleapis.com/bert_models/2018_10_18/uncased_L-12_H-768_A-12.zip.
[2] https://storage.googleapis.com/bert_models/2018_11_03/chinese_L-12_H-768_A-12.zip.

5.1 Datasets

We evaluate our approach on eight widely-studied datasets. These datasets have varying numbers of documents and varying document lengths, covering three common text classification tasks: sentiment analysis, question classification, and topic classification. We show the statistics for each dataset in Table 1.

Data Pre-processing. Following [6], we use WordPiece embeddings [26] with a 30,000 token vocabulary and denote split word pieces with ##. So the statistics of the length of the documents in the datasets are based on the word pieces. For further pre-training with BERT, we use spaCy[3] to perform sentence segmentation in English datasets and we use "∘", "?" and "!" as separators when dealing with the Chinese Sogou News dataset.

5.2 Hyperparameters

We use the BERT-base model [6] with a hidden size of 768, 12 Transformer blocks [24] and 12 self-attention heads. We further pre-train with BERT on 1 TITAN Xp GPU, with a batch size of 32, max squence length of 128, learning rate of 5e−5, train steps of 100,000 and warm-up steps of 10,000.

We fine-tune the BERT model on 4 TITAN Xp GPUs and set the batch size to 24 to ensure that the GPU memory is fully utilized. The dropout probability is always kept at 0.1. We use Adam with $\beta_1 = 0.9$ and $\beta_2 = 0.999$. We use *slanted triangular learning rates* [7], the base learning rate is 2e−5, and the warm-up proportion is 0.1. We empirically set the max number of the epoch to 4 and save the best model on the validation set for testing.

5.3 Exp-I: Investigating Different Fine-Tuning Strategies

In this subsection, we use the IMDb dataset to investigate the different fine-tuning strategies. The official pre-trained model is set as the initial encoder[4].

Layer-Wise Decreasing Layer Rate. Table 2 show the performance of different base learning rate and decay factors (see Eq. (2)) on IMDb dataset. We find that assign a lower learning rate to the lower layer is effective to fine-tuning BERT, and an appropriate setting is $\xi = 0.95$ and lr $= 2.0$e−5.

Catastrophic Forgetting. Catastrophic forgetting [17] is usually a common problem in transfer learning, which means the pre-trained knowledge is erased during learning of new knowledge. Therefore, we also investigate whether BERT suffers from the catastrophic forgetting problem.

We fine-tune BERT with different learning rates, and the learning curves of error rates on IMDb are shown in Fig. 2.

---

[3] https://spacy.io/.
[4] https://github.com/google-research/bert.

Table 2. Decreasing layer-wise layer rate.

Learning rate	Decay factor ξ	Test error rates (%)
2.5e−5	1.00	5.52
2.5e−5	0.95	5.46
2.5e−5	0.90	**5.44**
2.0e−5	1.00	5.42
2.0e−5	0.95	**5.40**
2.0e−5	0.90	5.52

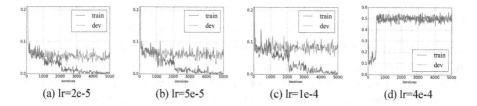

(a) lr=2e-5   (b) lr=5e-5   (c) lr=1e-4   (d) lr=4e-4

Fig. 2. Catastrophic forgetting

Table 3. Performance of in-domain and cross-domain further pre-training on seven datasets. Each was further pre-trained for 100k steps. The first column indicates the different further pre-training dataset. "all sentiment" means the dataset consists of all the training datasets in sentiment domain. "all" means the dataset consists of all the seven training datasets.

Domain	Sentiment			Question		Topic	
Dataset	IMDb	Yelp P.	Yelp F.	TREC	Yah. A.	AG's News	DBPedia
IMDb	**4.37**	2.18	29.60	2.60	22.39	5.24	0.68
Yelp P.	5.24	1.92	29.37	2.00	22.38	5.14	**0.65**
Yelp F.	5.18	1.94	29.42	2.40	22.33	5.43	**0.65**
All sentiment	4.88	**1.87**	29.25	3.00	22.35	5.34	0.67
TREC	5.65	2.09	29.35	3.20	22.17	5.12	0.66
Yah. A.	5.52	2.08	29.31	**1.80**	22.38	5.16	0.67
All question	5.68	2.14	29.52	2.20	**21.86**	5.21	0.68
AG's News	5.97	2.15	29.38	2.00	22.32	**4.80**	0.68
DBPedia	5.80	2.13	29.47	2.60	22.30	5.13	0.68
All topic	5.85	2.20	29.68	2.60	22.28	4.88	**0.65**
All	5.18	1.97	**29.20**	2.80	21.94	5.08	0.67
W/o pretrain	5.40	2.28	30.06	2.80	22.42	5.25	0.71

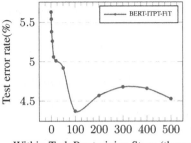

Fig. 3. Benefit of different further pre-training steps on IMDb datasets. BERT-ITPT-FiT means "BERT + withIn-Task Pre-Training + Fine-Tuning".

We find that a lower learning rate, such as $2e-5$, is necessary to make BERT overcome the catastrophic forgetting problem. With an aggressive learn rate of $4e-4$, the training set fails to converge.

5.4 Exp-II: Investigating the Further Pre-training

Besides, fine-tune BERT with supervised learning, we can further pre-train BERT on the training data by unsupervised masked language model and next sentence prediction tasks. In this section, we investigate the effectiveness of further pre-training. In the following experiments, we use the best strategies in Exp-I during the fine-tuning phase.

Within-Task Further Pre-training. Therefore, we first investigate the effectiveness of within-task further pre-training. We take further pre-trained models with different steps and then fine-tune them with text classification task.

As shown in Fig. 3, the further pre-training is useful to improve the performance of BERT for a target task, which achieves the best performance after 100 K training steps.

In-domain and Cross-Domain Further Pre-training. Besides the training data of a target task, we can further pre-train BERT on the data from the same domain. In this subsection, we investigate whether further pre-training BERT with in-domain and cross-domain data can continue to improve the performance of BERT.

We partition the seven English datasets into three domains: topic, sentiment, and question. The partition way is not strictly correct. Therefore we also conduct extensive experiments for cross-task pre-training, in which each task is regarded as a different domain.

The results is shown in Table 3. We find that almost all further pre-training models perform better on all seven datasets than the original BERT-base model (row 'w/o pretrain' in Table 3). Generally, in-domain pretraining can bring better performance than within-task pretraining. On the small sentence-level TREC dataset, within-task pre-training do harm to the performance while in-domain pre-training which utilizes Yah. A. corpus can achieve better results on TREC.

Comparisons to Previous Models. We compare our model with the feature-based transfer learning methods such as rigion embedding [21] and CoVe [16] and the language model fine-tuning method (ULMFiT) [7], which is the current state-of-the-art for text classification.

We implement BERT-Feat through using the feature from BERT model as the input embedding of the biLSTM with self-attention [10]. The result of BERT-IDPT-FiT corresponds to the row of 'all sentiment', 'all question', and 'all topic' in Table 3, and the result of BERT-CDPT-FiT corresponds to the row of 'all' in it. As is shown in Table 4, BERT-Feat performs better than all other baselines except for ULMFiT. In addition to being slightly worse than BERT-Feat on DBpedia dataset, BERT-FiT outperforms BERT-Feat on the other seven datasets. Moreover, all of the three further pre-training models are better than BERT-FiT model. Using BERT-Feat as a reference, we calculate the average percentage increase of other BERT-FiT models on each dataset. BERT-IDPT-FiT performs best, with an average error rate reduce by 18.57%.

Table 4. Test error rates (%) on eight text classification datasets. BERT-Feat means "BERT as features". BERT-FiT means "BERT + Fine-Tuning". BERT-ITPT-FiT means "BERT + withIn-Task Pre-Training + Fine-Tuning". BERT-IDPT-FiT means "BERT + In-Domain Pre-Training + Fine-Tuning". BERT-CDPT-FiT means "BERT + Cross-Domain Pre-Training + Fine-Tuning".

Model	IMDb	Yelp P.	Yelp F.	TREC	Yah. A.	AG	DBP	Sogou	Avg. Δ
Region Emb. [21]	/	3.60	35.10	/	26.30	7.20	1.10	2.40	/
CoVe [16]	8.20	/	/	4.20	/	/	/	/	/
ULMFiT [7]	4.60	2.16	29.98	3.60	/	5.01	0.80	/	/
BERT-Feat	6.79	2.39	30.47	4.20	22.72	5.92	0.70	2.50	-
BERT-FiT	5.40	2.28	30.06	2.80	22.42	5.25	0.71	2.43	9.22%
BERT-ITPT-FiT	**4.37**	1.92	29.42	3.20	22.38	**4.80**	0.68	**1.93**	16.07%
BERT-IDPT-FiT	4.88	**1.87**	29.25	**2.20**	**21.86**	4.88	**0.65**	/	**18.57%**
BERT-CDPT-FiT	5.18	1.97	**29.20**	2.80	21.94	5.08	0.67	/	14.38%

5.5   Exp-III: Multi-task Fine-Tuning

When there are several datasets for the text classification task, to take full advantage of these available data, we further consider a fine-tuning step with multi-task learning. We use four English text classification datasets (IMDb, Yelp P., AG, and DBP). The dataset Yelp F. is excluded since there is overlap between the test set of Yelp F. and the training set of Yelp P., and two datasets of question domain are also excluded.

Table 5. Test error rates (%) with multi-task fine-tuning.

Method	IMDb	Yelp P.	AG	DBP
BERT-FiT	5.40	2.28	5.25	0.71
BERT-MFiT-FiT	5.36	2.19	5.20	0.68
BERT-CDPT-FiT	5.18	**1.97**	**5.08**	**0.67**
BERT-CDPT-MFiT-FiT	**4.96**	2.06	5.13	**0.67**

Table 5 shows that for multi-task fine-tuning based on BERT, the effect is improved. However, multi-task fine-tuning does not seem to be helpful to BERT-CDPT in Yelp P. and AG. Multi-task fine-tuning and cross-domain pre-training may be alternative methods since the BERT-CDPT model already contains rich domain-specific information, and multi-task learning may not be necessary to improve generalization on related text classification sub-tasks.

5.6   Exp-IV: Few-Shot Text Classification

One of the benefits of the pre-trained model is being able to train a model for downstream tasks within small training data. We evaluate BERT-FiT and BERT-ITPT-FiT on different numbers of training examples. We select a subset of IMDb training data and feed them into BERT-FiT and BERT-ITPT-FiT. We show the result in Fig. 4.

This experiment result demonstrates that further pre-training brings a significant improvement for few-shot text classification. On IMDb dataset, when there are only 100 labeled data for each class, the accuracy of BERT-ITPT-FiT can reach 92%.

5.7   Exp-V: Further Pre-training on BERT Large

In this subsection, we investigate whether the $BERT_{LARGE}$ model has similar findings to $BERT_{BASE}$. We further pre-train Google's pre-trained $BERT_{LARGE}$ model[5] on 1 Tesla-V100-PCIE 32G GPU with a batch size of 24, the max sequence length of 128 and 120 K training steps. For target task classifier BERT fine-tuning, we set the batch size to 24 and fine-tune $BERT_{LARGE}$ on 4 Tesla-V100-PCIE 32G GPUs with the max sequence length of 512.

As shown in Table 6, ULMFiT performs better on almost all of the tasks compared to $BERT_{BASE}$ but not $BERT_{LARGE}$. This changes however with the task-specific further pre-training where even $BERT_{BASE}$ outperforms ULMFiT on all tasks. $BERT_{LARGE}$ fine-tuning with task-specific further pre-training achieves state-of-the-art results.

---

[5] https://storage.googleapis.com/bert_models/2018_10_18/uncased_L-24_H-1024_A-16.zip.

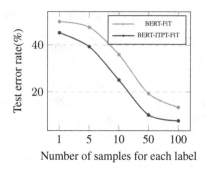
Number of samples for each label

Fig. 4. Few-shot text classification on IMDb dataset

Table 6. Test error rates (%) on five text classification datasets.

Model	IMDb	Yelp P.	Yelp F.	AG	DBP
ULMFiT	4.60	2.16	29.98	5.01	0.80
BERT$_{\text{BASE}}$	5.40	2.28	30.06	5.25	0.71
+ ITPT	4.37	1.92	29.42	4.80	0.68
BERT$_{\text{LARGE}}$	4.86	2.04	29.25	4.86	0.62
+ ITPT	**4.21**	**1.81**	**28.62**	**4.66**	**0.61**

6   Conclusion

In this paper, we conduct extensive experiments to investigate the different approaches to fine-tuning BERT for the text classification task. There are some experimental findings: (1) With an appropriate layer-wise decreasing learning rate, BERT can overcome the catastrophic forgetting problem; (2) Within-task and in-domain further pre-training can significantly boost its performance; (3) A preceding multi-task fine-tuning is also helpful to the single-task fine-tuning, but its benefit is smaller than further pre-training; (4) BERT with further pre-training performs well in few-shot text classification.

With the above findings, we achieve state-of-the-art performances on eight widely studied text classification datasets. In the future, we will probe more insight of BERT on how it works.

Acknowledgments. The research work is supported by China National Key R&D Program No. 2018YFC0831103.

References

1. Caruana, R.: Multitask learning: a knowledge-based source of inductive bias. In: Proceedings of the Tenth International Conference on Machine Learning (1993)
2. Chen, Z., Badrinarayanan, V., Lee, C.Y., Rabinovich, A.: GradNorm: gradient normalization for adaptive loss balancing in deep multitask networks. arXiv preprint arXiv:1711.02257 (2017)

3. Collobert, R., Weston, J.: A unified architecture for natural language processing: deep neural networks with multitask learning. In: Proceedings of the 25th International Conference on Machine Learning, pp. 160–167. ACM (2008)
4. Conneau, A., Kiela, D., Schwenk, H., Barrault, L., Bordes, A.: Supervised learning of universal sentence representations from natural language inference data. arXiv preprint arXiv:1705.02364 (2017)
5. Dai, A.M., Le, Q.V.: Semi-supervised sequence learning. In: Advances in Neural Information Processing Systems, pp. 3079–3087 (2015)
6. Devlin, J., Chang, M.W., Lee, K., Toutanova, K.: BERT: pre-training of deep bidirectional transformers for language understanding. arXiv preprint arXiv:1810.04805 (2018)
7. Howard, J., Ruder, S.: Universal language model fine-tuning for text classification. arXiv preprint arXiv:1801.06146 (2018)
8. Kiros, R., et al.: Skip-thought vectors. In: Advances in Neural Information Processing Systems, pp. 3294–3302 (2015)
9. Le, Q., Mikolov, T.: Distributed representations of sentences and documents. In: International Conference on Machine Learning, pp. 1188–1196 (2014)
10. Lin, Z., et al.: A structured self-attentive sentence embedding. arXiv preprint arXiv:1703.03130 (2017)
11. Liu, L., et al.: Empower sequence labeling with task-aware neural language model. In: Thirty-Second AAAI Conference on Artificial Intelligence (2018)
12. Liu, X., Gao, J., He, X., Deng, L., Duh, K., Wang, Y.Y.: Representation learning using multi-task deep neural networks for semantic classification and information retrieval (2015)
13. Liu, X., He, P., Chen, W., Gao, J.: Multi-task deep neural networks for natural language understanding. arXiv preprint arXiv:1901.11504 (2019)
14. Logeswaran, L., Lee, H.: An efficient framework for learning sentence representations. arXiv preprint arXiv:1803.02893 (2018)
15. Maas, A.L., Daly, R.E., Pham, P.T., Huang, D., Ng, A.Y., Potts, C.: Learning word vectors for sentiment analysis. In: Proceedings of the 49th Annual Meeting of the Association for Computational Linguistics: Human Language Technologies-Volume 1, pp. 142–150. Association for Computational Linguistics (2011)
16. McCann, B., Bradbury, J., Xiong, C., Socher, R.: Learned in translation: contextualized word vectors. In: Advances in Neural Information Processing Systems, pp. 6294–6305 (2017)
17. McCloskey, M., Cohen, N.J.: Catastrophic interference in connectionist networks: the sequential learning problem. In: Psychology of Learning and Motivation, vol. 24, pp. 109–165. Elsevier (1989)
18. Mikolov, T., Sutskever, I., Chen, K., Corrado, G.S., Dean, J.: Distributed representations of words and phrases and their compositionality. In: Advances in Neural Information Processing Systems, pp. 3111–3119 (2013)
19. Pennington, J., Socher, R., Manning, C.: Glove: global vectors for word representation. In: Proceedings of the 2014 Conference on Empirical Methods in Natural Language Processing (EMNLP), pp. 1532–1543 (2014)
20. Peters, M.E., et al.: Deep contextualized word representations. arXiv preprint arXiv:1802.05365 (2018)
21. Qiao, C., et al.: A new method of region embedding for text classification. In: International Conference on Learning Representations (2018)

22. Radford, A., Narasimhan, K., Salimans, T., Sutskever, I.: Improving language understanding by generative pre-training (2018). https://s3-us-west-2.amazonaws. com/openai-assets/research-covers/languageunsupervised/languageunderstanding paper.pdf
23. Rei, M.: Semi-supervised multitask learning for sequence labeling. arXiv preprint arXiv:1704.07156 (2017)
24. Vaswani, A., et al.: Attention is all you need. In: Advances in Neural Information Processing Systems, pp. 5998–6008 (2017)
25. Voorhees, E.M., Tice, D.M.: The TREC-8 question answering track evaluation. In: TREC, vol. 1999, p. 82. Citeseer (1999)
26. Wu, Y., et al.: Google's neural machine translation system: Bridging the gap between human and machine translation. arXiv preprint arXiv:1609.08144 (2016)
27. Yosinski, J., Clune, J., Bengio, Y., Lipson, H.: How transferable are features in deep neural networks? In: Advances in Neural Information Processing Systems, pp. 3320–3328 (2014)
28. Zhang, X., Zhao, J., LeCun, Y.: Character-level convolutional networks for text classification. In: Advances in Neural Information Processing Systems, pp. 649–657 (2015)

A Comprehensive Verification of Transformer in Text Classification

Xiuyuan Yang, Liang Yang$^{(\boxtimes)}$, Ran Bi, and Hongfei Lin

Dalian University of Technology, Dalian 116023, Liaoning, China
yxy815754134@mail.dlut.edu.cn,
{liang,biran,hflin}@dlut.edu.cn

Abstract. Recently, a self-attention based model, named Transformer, is proposed in Neural Machine Translation (NMT) domain, and outperforms the RNNs based seq 2seq model in most cases, hence it becomes the state-of-the-art model for NMT task. However, some studies find that the RNNs based model integrated with the Transformer structures could achieve almost the same experiment effect as the Transformer on the NMT task. In this paper, following the previous researches, we intend to further verify the performance of Transformer structures on the text classification task. Based on RNNs model, we gradually add each part of the Transformer block and evaluate their influence on the text classification task. We carry out the experiments on NLPCC2014 and dmsc_v2 datasets, and the experiment results show that multi-head attention mechanism and multiple attention layers could improve the performance of the model on the text classification task. Furthermore, the visualization of the attention weights also illustrates that multi-head attention outperforms the traditional attention mechanism.

Keywords: RNNs · Transformer · Text classification

1 Introduction

Since Deep Learning methods are widely implemented in Natural Language Processing (NLP) tasks, great improvements have been achieved in recent years. Initially, Recurrent Neural Networks (RNNs) [1], such as Long Short-Term Memory (LSTM) [2] networks and the Gated Rectified Unit (GRU) [3], which can better capture the long-range context information, are commonly used in various research domains.

For example, in the text classification task, RNNs are used for encoding text information. Bi-RNNs aggregate information from both directions of sentence to get a comment on the word and merge it into the text vectors. For Lexical analysis, including word segmentation, part-of-speech tagging (POS) [4], named entity recognition (NER) [5], etc., RNNs can carry more deep information such as syntax structure and dependencies. In addition, RNNs are skilled in dealing with variable-length input sentences and can be used as a language model to generate sentences, thus it is a natural choice for using them as encoder and decoder in the neural machine translation (NMT) systems.

© Springer Nature Switzerland AG 2019
M. Sun et al. (Eds.): CCL 2019, LNAI 11856, pp. 207–218, 2019.
https://doi.org/10.1007/978-3-030-32381-3_17

Recently, Vaswani et al. [6] propose Transformer as a novel feature extractor for NMT task and it outperforms RNNs in most cases. Tang et al. [7] hypothesize that the reasons, why the Transformer has a strong performance, are the abilities of semantic feature extraction and capturing long-range dependencies. Their experimental results show that Transformer outperforms distinctly better than RNNs on word sense disambiguation. Meanwhile, Transformer model and RNNs have a close performance in modeling subject-verb agreement over long distance. Unlike the above mentioned works, Radford et al. [8] evaluate the performance of models by introducing different feature extractors into different tasks. The experiments show that on eight different NLP tasks with large datasets, the feature extractor is only changed from Transformer to LSTM and the average of eight tasks scores drop points under the same conditions.

Due to these analyses, we can find that the performance of Transformer has obvious advantages over native RNNs. Does it mean that we can just simply replace RNNs with Transformer in most tasks? Actually, it is not the case. As we know, the Transformer block makes Transformer work well, which is a small system composed of several components such as multi-head attention, layer norm, and feed-forward network. Domhan et al. [9] find that RNNs can also make close achievements to Transformer by gradually adding each part of Transformer block on NMT task.

Inspired by this, following the researches from Domhan et al. [9], we intend to further verify the performance of Transformer block on the text classification task, by gradually adding each part of Transformer block to RNNs and evaluate their impact in this paper.

Our observations and analysis conclusions in this paper are as follows:

- Transform not always performs well on text classification.
- Unlike NMT, adding feed-forward to RNN based models has no obvious effect on the text classification task.
- The contributions of multi-head attention and multiple attention layer are crucial in different classification tasks.
- Attention visualization demonstrates the strong effects of multi-head attention.

2   Related Work

Before the Transformer is proposed, RNNs with attention are the state-of-the-art models for most cases in NLP. Bahdanau et al. [10] firstly introduce the single layer attention-like mechanism for RNNs based NMT models, to solve the problem of long-term gradient disappearance in the RNNs. Luong et al. [11] further perform different single layer attention mechanisms for RNNs based NMT models. Yang et al. [12] propose a hierarchical attention mechanism for RNNs, which is a two levels of attention structure, capturing the word and sentence representation on the document classification.

Recently, Vaswani et al. [6] propose Transformer as a novel feature extractor for NMT task and it outperforms RNNs in most cases. The Transformer is based solely on attention mechanisms, dispensing with recurrence entirely. Their experiments show that the Transformer models to be superior in quality while being more parallelizable and requiring significantly less time to train.

However, Domhan et al. [9] find that RNNs can make close achievements to Transformer by applying some parts of the Transformer block into RNNs on NMT tasks. Chen et al. [13] demonstrate that several modeling improvements (including multi-head attention and layer normalization) as well as optimization techniques are applicable across different model architectures. Moreover, they further propose new model architectures that combine components from the RNMT+ (an enhanced version of RNNs on NMT) and the Transformer model, and achieve better results than both individual architectures.

Different from the above research on NMT, this paper proposes new improvements to RNNs on text classification, based on the strategy adopted by Domhan et al. [9]. The effect for each part of Transformer block is different, which complements the relationship between RNNs and Transformer on text classification.

3 Model Architecture

The basic architecture of the model is based on RNNs. To get better results, we choose to use GRU and LSTM alternately instead of using a single RNNs model. The baseline model consists of several parts as shown in the solid line in Fig. 1, which contain an input embedding layer, an encoder layer, an attention-pooling layer and an output layer.

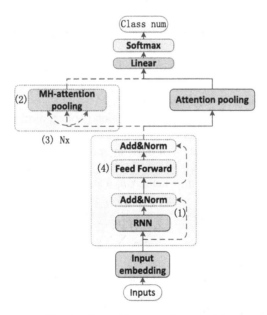

Fig. 1. The architecture of our model

In order to clearly explain how we add structural components to the basic model, we describe each structural component as shown in the dotted line in Fig. 1: (1) A residual [14] connection around each of the two sub-layers, followed by layer normalization. (2) MH-attention pooling. (3) Multiple-attention layer. (4) A feed-forward layer.

3.1 RNN Encoder Layer

Through word embedding layer, we get the input text with words $x_t, t \in [1, T]$. After that, the text representations are constructed by using Bi-Recurrent Neural Networks (Bi-RNNs), which combine a Long Short-Term Memory (LSTM) network [15] and a Gated Rectified Unit (GRU) [16], showed as following:

$$h_t, ce_t = \overset{\rightarrow\leftarrow}{LSTM}(x_t, h_{t-1}, ce_t)$$
$$h_t^* = \overset{\rightarrow\leftarrow}{GRU}(h_{t-1}, h_{t-1}^*) \tag{1}$$

Where h_t is the hidden state and ce_t is cell state for Bi-LSTM, and the h_t^* is the output hidden state for Bi-GRU.

3.2 Attention Mechanism

All attention mechanisms take a set of query vectors $Q \in \mathbb{R}^{M \times d}$, key vectors $K \in \mathbb{R}^{N \times d}$, and value vectors $V \in \mathbb{R}^{N \times d}$, in order to produce one context vector $C \in \mathbb{R}^{M \times d}$, which is a linear combination of the value vectors.

$$\text{Attention}(Q, K, V) = \text{softmax}(QWK^T)V \tag{2}$$

Attention-Pooling for Text Classification
In the baseline model, we construct the attention-pooling mechanism like [11], A context vector $u_w \in \mathbb{R}^{1 \times d}$ to extract such words that are important to the meaning of the text and aggregate the representation of those informative words to form a text vector.

$$u_i = \tanh(W_s h_t^* + b_s)$$
$$a_i = \frac{\exp(u_i^T, u_w)}{\sum_i \exp(u_i^T, u_w)} \tag{3}$$
$$c = \sum_i a_i u_i$$

where the h_t^* is the output state of RNNs encoder, the $c \in \mathbb{R}^{1 \times d}$ is the text vector. It is noteworthy that the context vector $u_w \in \mathbb{R}^{1 \times d}$ can be randomly initialized and jointly learned during the training process.

Multiple Head Attention-Pooling for Text Classification

Vaswani et al. [6] propose multiple head attention, as follows:

$$head_i(Q_i, K_i, V_i) = \text{softmax}(\frac{Q_i K_i^T}{\sqrt{d_k}})V_i$$

$$\text{MutltiHead}(Q, K, V) = \text{Concat}(head_1, \ldots, head_h) \tag{4}$$

In the multi-head attention, the query, key, and value will be separated into parts $(Q_i \in \mathbb{R}^{M \times d_i}, K_i \in \mathbb{R}^{N \times d_i}, V_i \in \mathbb{R}^{N \times d_i})$. Then, model conducts dot product attention for each part ($head_i$) independently. Finally, the multiple head attention vector will be produced by contacting $head_i$.

For text classification, we will formulate the query, key, and value as following:

$$Q = W_q u_w, K = W_k h_t^*, V = W_v h_t^*$$

$$c = MultiHead(Q, K, V) \tag{5}$$

Where the W_q, W_k, W_v are the learnable weights and $u_w \in \mathbb{R}^{1 \times d}, c \in \mathbb{R}^{1 \times d}$ is the context vector and text vector. In this paper, we employ $h = 8$ parallel attention heads, which means that we will separate them into 8 parts. For each part, $d_i^q = d_i^k = d_i^v = d_{model}/h$.

3.3 Layer Normalization

Layer normalization [17] fixes the mean and the variance of the accumulated inputs within each layer. It computes as:

$$norm(h_t) = \frac{g}{\sigma_t} \otimes (h_t - \mu_t) + b$$

$$\mu_j = \frac{1}{d}\sum_{i=1}^{d} h_{t,j}, \sigma_t = \sqrt{\frac{1}{d}\sum_{i=1}^{d}(h_{t,j} - \mu_j)^2} \tag{6}$$

Where g, b are learned scale and shift parameters with the same dimension as h_t. In this paper, we will apply this structure to each output of sub RNNs layers.

3.4 Feed-Forward Layer

The feed-forward layer can be taken as a 1-D convolution layer, which includes two linear transformations in between with a ReLU activation.

$$ff(h_t, d_o) = dropout(\max(0, Wh_t + b))) \tag{7}$$

Where h_t is the input, and $W \in \mathbb{R}^{d_o \times d_{model}}$ is the learnable weight. In this paper, we will apply it between each sub RNNs layers.

3.5   Classification Layer

The classification layer includes a linear layer and a softmax layer. The text vector c can be used as features for text classification.

$$p = \text{softmax}(W_c c + b_c) \tag{8}$$

We use the negative log likelihood of the correct labels as training loss.

$$L = -\sum_d \log p_{dj}, \tag{9}$$

Where j is the label of the text d.

4   Experiments and Analysis

The following is an extensive empirical analysis of models with different parts of the Transformer block, and how different certain parts affect the performance.

4.1   Experiment Setup

We conduct experiments on corpus with different sizes to get more typical results. For this goal, a small size dataset of NLPCC2014 and dmsc_v2 dataset with large samples are selected. The former small dataset is collected from Chinese product review web site. This task of NLPCC2014 aims to predict the polarity of each review in the data set and the polarity of each review is binary, either positive or negative. We use 10 thousand training sentences and 2500 sentences as test data. Detailed data statistics are shown in Table 1.

Table 1. NLPCC2014 dataset related information

Dataset	Positive	Negative
NLPCC2014-train	5000	5000
NLPCC2014-test	1250	1250

The latter is a big dataset, with more than 2 million ratings from Douban movies. This task of dmsc_v2 aims to analyze the level of each rating and the scores are divided into five levels. We use 1.68 million training examples and 0.42 million examples as test data. Detailed data statistics are shown in Table 2:

Table 2. Dmsc_v2 dataset related information

Dataset	Total_num	Level-1	Level-2	Level-3	Level-4	Level-5
Train	1.68 million	505 K	151 K	142 K	376 K	508 K
Test	0.42 million	126 K	38 K	35 K	93 K	127 K

For both datasets, we process them using the Spacy tokenizer and collect the vocabulary by filtering the word with a frequency less than 5. We also use pre-trained word embedding [18], learning upon Baidu Encyclopedia with SGNS (skip-gram model with negative sampling) [19].

In this paper, the same set of hyperparameters is shared across models. We make a 300 dimensions word embedding layer, and in order to facilitate adding residual connections, all sub-layers in the model produce outputs with 512 dimensions. Adam optimizer with default setting is used for training [20], except that learning rate is 0.0001. Negative Log Likelihood Loss is selected as the loss function, and clip normal in this paper is range from −5 to 5. Each training experiment runs three more times, and then the output is recorded, producing the mean of the Accuracy, Precision, Recall, and F1-score.

4.2   Results and Analysis

Different from the RNNs, the Transformer is more than just self-attention. The differences between RNNs and Transformer also include multi-head attention, multiple attention layers and norm layer. That is why we study the contribution of each part of the Transformer block and try to find which one plays an important role. The details of the models are as follows:

(1) SVM: Traditional SVM is used to classify datasets.

(2) CNN+1 h: Using CNN for text classification, which kernel sizes are the set of [3–5] + attention mechanism.

(3) Transformer: Transformer has the same hyperparameters of transformer block with the follow models.

(4) RNNs + 1 h: Basic model (RNNs + single-attention pooling + classifiaction layer).

(5) RNNs + mh: Replace single-attention pooling with multi-head attention pooling based on (4).

(6) RNNs + mh + norm: Adding a residual connection around each of RNNs sub layers, followed by layer normalization, based on (5).

(7) RNNs + multi-att-1 h/mh + norm: Appling attention pooling (1 h/mh) at each output of RNNs sub layers and contacting multiple attention states together as text vector.

(8) RNNs + multi-att-1 h/mh + norm + ffn: Adding a feed-forward layer based on (7).

Results and Analysis on the NLPCC2014 Dataset

The results of different models on the small dataset are shown in Table 3.

Table 3. Transforming RNNs into a Transformer-style architecture on the small dataset.

Models	Accuracy	F1-Score	Precision	Recall
(1) SVM	0.7316	0.7365	0.7249	0.7539
(2) CNN + 1 h	0.7576	0.7638	0.7447	0.7840
(3) Transformer	0.7584	0.7626	0.7496	0.7760
(4) RNN + 1 h	0.7815	0.8037	0.7312	0.8997
(5) RNN + mh	0.7940	0.8101	0.7524	0.8822
(6) RNN + mh + norm	0.7811	0.8079	0.7202	0.9222
(7) RNN + multi-att-1 h + norm	0.7848	0.8000	0.7502	0.8687
RNN + multi-att-mh + norm	**0.7948**	**0.8249**	0.7196	0.9689
(8) RNN + multi-att-1 h + norm + ffn	0.7856	0.8041	0.7400	0.8807
RNN + multi-att-mh + norm + ffn	0.7892	0.8084	0.7427	0.8959

By comparing (1) with (2, 3, 4), we can find that the basic model is significantly better than the traditional model, which proves that deep networks can reduce artificial feature extraction and make better use of text feature.

By comparing (2, 3) with (4), we know that RNN models are better than CNN or Transformer on this task, which demonstrate that RNNs can better capture the long-range context information.

By comparing (3) with other, we may find that Transformer not always work well on text classification task. Meanwhile, adding each part of Transformer block to RNNs model is an effective way to improve model on text classification work.

By comparing (4) with (5), we can see that the model benefits from the multi-head attention mechanism, which illustrates that Multi-head attention allows the model to jointly attend information from different representation sub spaces at different positions.

By comparing (5) with (6), we can demonstrate that layer normalization on residual inputs blocks has a small drop on evaluation metrics.

By comparing (4) and (5) with (7), we can confirm that the model gains better performance when adding multiple attention layers to both model (4) and model (5) (1 h/mh). It means that multiple attention layers can get more context information.

Compared (7) with (8), the experiments show that adding a feed-forward layer has a slight effect on the models. Adding a feed-forward layer leads to large and consistent performance boost on NMT task [9] while it is not strictly necessary on the text classification task, which confirms that text classification task is different from the NMT task, so we need to make different structures of the model on them.

In the end, we can easily find that the biggest benefits for model on the small dataset come from **multi-head attention** and **multiple attention layers.**

Results and Analysis on the Dmsc_v2 Dataset

The results of different models on the large dataset are shown in Table 4.

Firstly, Most of comparisons are consistent with that on the small dataset, such as (1) with (2, 3, 4) (the traditional model and the neural network), (2, 3) with (4) (the comparison between CNN, Transformer and RNN), (3) with the rest (the comparison between Transformer and enhanced RNNs), (4) with (5) (the effect of adding mh-head attention pooling), and (5) with (6) (the influence of adding layer-norm).

Table 4. Transforming RNNs into a Transformer-style architecture on the large dataset.

Models	Accuracy	F1-Score	Precision	Recall
(1) SVM	0.4986	0.4619	0.4887	0.4594
(2) CNN + 1 h	0.5643	0.5337	0.5547	0.5268
(3) Transformer	0.5437	0.5068	0.5253	0.5030
(4) RNN + 1 h	0.5786	0.5442	0.5701	0.5382
(5) RNN + mh	**0.5845**	0.5526	0.5704	0.5462
(6) RNN + mh + norm	0.5831	0.5455	0.5723	0.5372
(7) RNN + multi-att-1 h + norm	0.5780	0.5447	0.5738	0.5318
RNN + multi-att-mh + norm	0.5837	**0.5584**	0.5776	0.5460
(8) RNN + multi-att-1 h + norm + ffn	0.5795	0.5466	0.5702	0.5355
RNN + multi-att-mh + norm + ffn	0.5839	0.5500	0.5743	0.5397

In the rest of Table 4, by comparing (4), (5) with (7), we can see that adding multiple attention layers has a slight effect on (4) (1 h-attention pooling), but it improves F1-Score on (5) (mh-attention pooling). In addition, adding a feed-forward layer improve slightly performance on both single attention pooling and the multi-head attention pooling. It means that whether the size of the dataset is big or not, adding a feed forward layer has no obvious effect on text classification tasks.

In the end, we can clearly find that the main benefits on the large dataset come from multi-head attention and multiple attention layers like the small dataset.

4.3   Visualization of Different Attention Mechanisms

We visualize the attention weights of examples from test set in Fig. 2. Each part is the same example, which uses both the single-head attention and the multi-head attention. The color shade denotes the weight of words from attention pooling. The higher the weight, the darker the color. For visualization purposes, we normalize the word weight and display the weight higher than threshold 0.1 to make sure that only important words are emphasized.

From Fig. 2, the results show that the single-head attention model can select the words carrying some strong sentiment like "很好 (very good)", "不 (no)". However, we can see the words just like the first sentence, whether they are the keywords or not, marked in single-head attention have almost the same color shadow. It shows that single-head attention is not effective all the time. Fortunately, we observe that for multi-head attention model, that is to say, eight attention operations, most of them can find the keywords in the same example. After integrating this information, it will recognize all keywords in the sentence, which outperform than the single-head attention mechanism. That is the reason why the multi-head attention mechanism has greatly improved the experimental results in the above experiments.

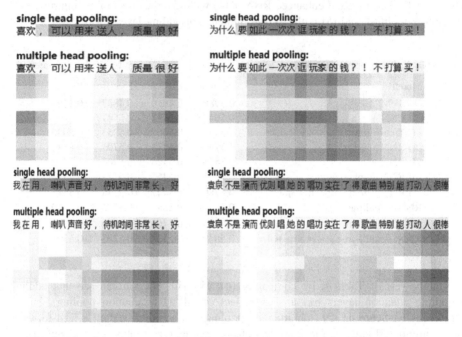

Fig. 2. Visual the same examples with single-head attention and multi-head attention.

5   Conclusion

In this paper, we aim to explore and evaluate how much specific parts of the Transformer block can successfully improve RNNs' performance, especially on the text classification task. Hence, we conduct extensive evaluation experiments on different datasets, such as NLPCC2014 and dmsc_v2 datasets. The scales of there two datasets are different, and the former is a small one, while the other is large.

Based on the experiments and results, we find that RNNs based model in both the small dataset and the large dataset on classification task can benefit from multi-head attention mechanisms and multiple attention layers. In addition, layer normalization on residual inputs blocks drop the performance on both datasets. Moreover, adding a feedforward layer leads to large and consistent performance boost on NMT task [9] while it is not strictly necessary on the text classification task, which confirms that text classification task is different from the NMT task, so we need to make different structures of the model on them. On the other hand, by comparing the attention visualization results between single-head attention and multi-head attention, we obtain some conclusions which confirm that the multiple head attention mechanism could optimize attention weights, and improve the performance of models. In a word, the RNNs basic model can greatly improve the performance by continuously adding each part of the Transformer block.

In the future work, considering the strong performance of Transformer block, we will replace the self-attention from Transformer block with more feature extractors on the text classification task, like TextRCNN [21], to further discover the influence of Transformer block.

Acknowledgments. This work is partially supported by a grant from the National Key Research and Development Program of China (No. 2018YFC0832101), the Natural Science Foundation of China (No. 61702080 and 61632011), the Fundamental Research Funds for the Central Universities (No. DUT19RC(4)016), and Postdoctoral Science Foundation of China (2018M631788).

References

1. Elman, J.L.: Finding structure in time. Cogn. Sci. **14**(2), 179–211 (1990)
2. Hochreiter, S., Schmidhuber, J.: Long short-term memory. Neural Comput. **9**(8), 1735–1780 (1997)
3. Cho, K., et al.: Learning Phrase Representations using RNNs encoder–decoder for statistical machine translation. In: EMNLP, pp. 1724–1734. ACL, Doha (2014)
4. Huang, Z., Xu, W., Yu, K.: Bidirectional LSTM-CRF models for sequence tagging. arXiv preprint arXiv:1508.01991 (2015)
5. Lample, G., Ballesteros, M., Subramanian, S., Kawakami, K., Dyer, C.: Neural architectures for named entity recognition. In: NAACL-HLT, pp. 260–270. NAACL, San Diego (2016)
6. Vaswani, A., et al.: Attention is all you need. In: NIPS, pp. 5998–6008. Curran Associates, Inc., Long Beach (2017)
7. Tang, G., Müller, M., Rios, A., Sennrich, R.: Why self-attention? A targeted evaluation of neural machine translation architectures. In: EMNLP, pp. 4263–4272. ACL, Brussels (2018)
8. Radford, A., Narasimhan, K., Salimans, T., Sutskever, I.: Improving language understanding by generative pre-training (2018). https://s3-us-west-2.amazonaws.com/openai-assets/ research-covers/languageunsupervised/languageunderstandingpaper.pdf
9. Domhan, T.: How much attention do you need? A granular analysis of neural machine translation architectures. In: ACL, vol. 1, pp. 1799–1808. ACL, Melbourne (2018)
10. Bahdanau, D., Cho, K., Bengio, Y.: Neural machine translation by jointly learning to align and translate. In: ICLR, San Diego, USA (2015)
11. Luong, M.T., Pham, H., Manning, C.D.: Effective approaches to attention-based neural machine translation. In: EMNLP. ACL, Lisbon (2015)
12. Yang, Z., Yang, D., Dyer, C., He, X., Smola, A., Hovy, E.: Hierarchical attention networks for document classification. In: NAACL-HLT, pp. 1480–1489. NAACL, San Diego (2016)
13. Chen, M.X., et al.: The best of both worlds: combining recent advances in neural machine translation. In: ACL. ACL, Melbourne (2018)
14. He, K., Zhang, X., Ren, S., Sun, J.: Deep residual learning for image recognition. In: CVPR, pp. 770–778. IEEE (2016)
15. Graves, A., Mohamed, A.R., Hinton, G.: Speech recognition with deep recurrent neural networks. In: ICASSP, pp. 6645–6649. IEEE, Vancouver (2013)
16. Chung, J., Gulcehre, C., Cho, K., Bengio, Y.: Empirical evaluation of gated recurrent neural networks on sequence modeling. In: NIPS Workshop on Deep Learning (2014)

17. Ba, J.L., Kiros, J.R., Hinton, G.E.: Layer normalization. arXiv preprint arXiv:1607.06450 (2016)
18. Li, S., Zhao, Z., Hu, R., Li, W., Liu, T., Du, X.: Analogical reasoning on chinese morphological and semantic relations. In: ACL, vol. 2, pp. 138–143. ACL, Melbourne (2018)
19. Soutner, D., Müller, L.: Continuous distributed representations of words as input of LSTM network language model. In: Sojka, P., Horák, A., Kopeček, I., Pala, K. (eds.) TSD 2014. LNCS (LNAI), vol. 8655, pp. 150–157. Springer, Cham (2014). https://doi.org/10.1007/978-3-319-10816-2_19
20. Kingma, D.P., Ba, J.: Adam: a method for stochastic optimization. In: ICLR. Scottsdale, Arizona, USA (2015)
21. Lai, S., Xu, L., Liu, K., Zhao, J.: Recurrent convolutional neural networks for text classification. In: AAAI, pp. 2267–2273. AAAI, Austin (2015)

Knowledge Graph and Information Extraction

Next News Recommendation
via Knowledge-Aware Sequential Model

Qianfeng Chu[1], Gongshen Liu[1(✉)], Huanrong Sun[2], and Cheng Zhou[2]

[1] Shanghai Jiaotong University, Shanghai, China
lgshen@sjtu.edu.cn
[2] Shanghai Songheng Network Technology Co., Ltd., Shanghai, China

Abstract. A news recommendation system aims to predict the next news based on users' interaction histories. In general, the clicking sequences from the interaction histories indicate users' latent preference, which plays an important role in predicting their future interest. Besides, news articles consist of considerable knowledge entities which have deep connections from common sense of human. In this paper, we propose a Self-Attention Sequential Knowledge-aware Recommendation (*Saskr*) system consisting of sequential-aware and knowledge-aware modelling. We use the self-attention mechanism to uncover sequential patterns in the sequential-aware modelling. The knowledge-aware modelling leverage the knowledge graph as side information to mine deep connections between news, thus improving diversity and extensibility of recommendation. Content-based news embeddings help to address the item cold-start problem. Through extensive experiments on the real-world news dataset, we demonstrate that the proposed model outperforms state-of-the-art deep neural sequential recommendation systems.

Keywords: News recommendation · Sequential recommendation · Knowledge-aware modelling

1 Introduction

The news recommendation system makes recommendations based on users' historical behavior and characteristics. Generally, three major features distinguish the news recommendation from other general systems. First, expired news articles are substituted by later ones within a shorter period of time compared with other type of items such as movies and books, which makes traditional collaborative filtering (CF) based methods suffer from severe item cold start problem. Second, users' latent preference hide in the sequential interaction histories. The clicked news history should be considered as time-queued sequence. While traditional methods fail in characterizing sequence data, the sequential recommendation can predict the successive item(s) for a user given her interaction sequence. Third, news articles consist of considerable knowledge entities which may seem irrelevant but have deep connections from common sense of human.

ⓒ Springer Nature Switzerland AG 2019
M. Sun et al. (Eds.): CCL 2019, LNAI 11856, pp. 221–232, 2019.
https://doi.org/10.1007/978-3-030-32381-3_18

To overcome the above challenges in news recommendation, the powerful Transformer [15] mechanism and the knowledge graph are considered. Our model is a content-based model which generates recommendation score for each candidate news given the users' sequential click history.

Empirically, we apply our model to a real-world dataset collected from East-day Toutiao news website. Extensive experiments show that our model out-performs state-of-the-art recommendation models on Hit Rate@10(HR@10) and Mean Reciprocal Rank (MRR). The main contributions of this work are summarized as follows:

1. We propose an encoder-decoder recommendation model with multi-head self-attention to generate recommendation score for candidate news.
2. News articles are represented as content-based embeddings to address item cold start problem.
3. We leverage the knowledge graph as side information to mine deep connections between news, thus improving reasonable diversity and extensibility of recommendation.

2   Related Work

2.1   Deep Neural Network for Sequential Recommendation

Sequential Recommendation models leveraging deep neural network encode can translate user interaction history into latent state vectors. GRU4REC [3] uses GRU for sequence prediction. [13] and [1] improve the GRU4REC by considering the dwell time and applying four additional training strategies, respectively. Quadrana et al. [10] proposes a hierarchical RNN model utilizing session-level and user-level GRU to describe preference. NARM [8] adds the attention mechanism to the user interaction history encoder to characterize the user sequence behavior. In order to extract union-level features, Caser [14] uses a Convolutional Neural Network (CNN) to encode user interaction history. [20] learns the user's short-term interest from the user's interaction history based on the self-attention [15] mechanism, while retaining the user's long-term interest as well.

2.2   Knowledge Graph in Recommendation

A knowledge graph (KG) is a type of directed heterogeneous graph, typically consisting of entity-relation-entity triples (h, r, t). Fruitful facts and connections about items in KG give great contribution on recommendation. DKN [17] utilizes TransE [2] to generate the entity embedding and context embedding, then feeds them into a CNN framework to recommend. CKE [19] combines a CF module with knowledge&text&image embedding of items in a unified Bayesian framework. KSR [4] integrates the RNN-based networks with Key-Value Memory Network(KV-MN) and incorporate KG information to enhance the semantic representation of KV-MN. RippleNet [16] simulates the propagation of user interest on the knowledge graph. An end-to-end model is constructed to ultimately get the click probability.

3   The Proposed Model

In this paper, we propose the Self-Attention Sequential Knowledge-aware Recommendation (*Saskr*) system to accomplish the sequential recommendation task. Laterally, our model incorporates sequential-aware modelling and knowledge-aware modelling. The framework of *Saskr* is illustrated in Fig. 1.

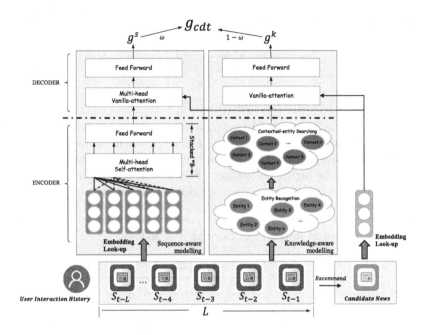

Fig. 1. The framework of *Saskr*

3.1   Problem Formulation

The problem assumes a set of users $U = \{u_1, u_2, ..., u_M\}$ and a set of items $I = \{i_1, i_2, ..., i_N\}$. Each user $u \in U$ interacts with a sequence of items $S = (S_1, S_2, ..., S_{|S|})$, where $S_k \in I$ and S is sorted in chronological order. The sequential recommendation aims to predict the next item most likely to be clicked given the interaction history S.

The task specializes in news sequential recommendation. Suppose user's intent can be derived from L news articles in the interaction sequence. Every L successive news articles are extracted from S, denoted as $S_t = (S_{t-L}, S_{t-L+1}, ..., S_{t-1})$, where t is the time step. Then given a candidate news $i_{cdt} \in I$, the recommendation score g_{cdt} is generated from g^s of the sequence-aware modelling and g^k of the knowledge-aware modelling with the factor ω:

$$g_{cdt} = \omega \cdot g^s + (1 - \omega) \cdot g^k \tag{1}$$

3.2 Sequential-Aware Modelling

Users' interaction histories reflect the reading preference from a sequential perspective.

A. Embedding Layer. The embedding layer projects each news article into a d-dimensional latent space. As for news i in S_t, the embedding $Q_i \in \mathbb{R}^d$ is generated through to form the matrix $E \in \mathbb{R}^{L \times d}$ at time step t:

$$
E = \begin{bmatrix} Q_{S_{t-1}} + P_1 \\ Q_{S_{t-2}} + P_2 \\ \dots \\ Q_{S_{t-L}} + P_L \end{bmatrix}
\tag{2}
$$

Each news article is embeded based on its content to deal with the item cold start problem. Content-based methods [6] aim to represent items descriptively.

Specific for Chinese news, we focus on the news body because the titles in Chinese news could be highly misleading with useless eye-catching terms. Based on words in the news body, we try two strategies to generate the news embedding:

– **TF-IDF:** Term Frequency-Inverse Document Frequency gets the weight $w(m, i)$ for a term m in news i is calculated as follows:

$$
w(m, i) = \frac{tf_{m,i} \log \frac{N}{df_m}}{\sqrt{\sum_{m_j \in i}(tf_{m_j,i})^2 \log(\frac{N}{df_{m_j}})^2}}
\tag{3}
$$

where $tf_{m,i}$ is the frequency of m in the news i, df_m is the number of news containing the term m in the whole news set I, and N is the length of I. Then the news is embedded as:

$$
Q_i = \sum_m w(m, i) \cdot emb(m)
\tag{4}
$$

where $emb(m)$ is the pre-trained d-dimensional word embedding for m.
– **Entity-stacking:** Named entity recognition [5] identify entities with specific meanings in the text. The stacking of entities can be regarded as a direct representation of news. The news i is embedded as:

$$
Q_i = \frac{1}{|entity(i)|} \sum_{e_i \in entity(i)} emb(e_i)
\tag{5}
$$

where $entity(i)$ is the set of entities in news i, filtering out those occurring only once because of meaningless or irrelevance.

Position Embedding. The learnable positional embeddings $P \in \mathbb{R}^{L \times d}$ give the relative position information.

B. Multi-head Self-attention Module. Being proposed in Transformer [15], the multi-head self-attention module is a special attention mechanism extracting the latent representation of a sequence by uncovering sequential patterns in between:

$$Attention(Q, K, V) = softmax(\frac{QK^\top}{\sqrt{d}})V \tag{6}$$

$$MultiheadAtt(Q, K, V) = Concat(head_1, \cdots, head_n)W^O$$
$$where\ head_i = Attention(QW_i^Q, KW_i^K, VW_i^V) \tag{7}$$

where Q, K, V are the query, key and value respectively. The scaling factor d is the number of dimension of queries and keys. Q, K, V are then into n subspaces with the projection matrixes W_i^Q, W_i^K, W_i^V, $i \in [1, n]$.

As for self-attention, the Q, K, V here are the embedding E:

$$M = MultiheadAtt_{encoder}^s(E, E, E) \tag{8}$$

C. Feed Forward Module. Two fully-connected layers are applied after the multi-head self-attention module:

$$C^s = FFM_{encoder}^s(M) = ReLU(MW_1 + b_1)W_2 + b_2 \tag{9}$$

where W_1, W_2, b_1, b_2 is the trainable parameters. Note that these two modules can be stacked as B (3 in our model) blocks for deeper understanding of the input sequence.

D. Multi-head Vanilla-Attention Module. The candidate news i_{cdt} is embedded to Q_{cdt} through the same embedding layer as in section A. Then Q_{cdt} and the output of the encoder C^s are fed into the decoder for recommendation score:

$$O^s = MultiheadAtt_{decoder}^s(Q_{cdt}, C^s, C^s) \tag{10}$$

$$g^s = FFM_{decoder}^s(O^s) \tag{11}$$

here g^s is the recommendation score predicted from the sequence-aware modelling.

3.3 Knowledge-Aware Modelling

Knowledge-aware modelling leverages the knowledge graph as side information, aiming to mine deep connections among news. The modelling is built via a knowledge-searching encoder and a preference-interpreting decoder.

A. Knowledge-Searching Encoder. A Chinese news-specific knowledge graph G is proposed for the recommendation task. It contains fruitful facts and connections among entities, which provides us a wider perspective to explore uses' preferences in the field of news.

To leverage knowledge in G, we encode the interaction history in four steps: (1) The technique of Named Entity Recognition [18] is used to search the interaction history for entities, forming the set $\{Entity_i\}$. They represent the user's intuitive preference. (2) Based on these identified entities, we use the technique of Entity Linking [9] to associate them to the predefined entities in G. This procedure is to disambiguate mentions in news. (3) We search the contextual-entities of $\{Entity_i\}$: the adjacent entities connected to $Entity_i$ within one hop in G, denoted as $\{ContextualEntity_j\}$. These entities are of potential interest to the user. (4) $\{ContextualEntity_j\}$ are embedded to $\{e_j\}$ through pre-trained word embedding [12], and finally formed as the contextual-entity embedding matrix C^k.

B. Preference-Interpreting Decoder. The decoder takes C^k and Q_{cdt} as input:

$$O^k = MultiheadAtt^k_{decoder}(Q_{cdt}, C^k, C^k) \tag{12}$$

$$g^k = FFM^k_{decoder}(O^k) \tag{13}$$

here g^k is the recommendation score predicted from the knowledge-aware modelling. Note that here we set the number of head as 1, since we assume the key and value here are not latent vectors but meaningful word embeddings. Multihead operation will take away the useful information embedded.

3.4 Model Training

Objective Function. Instead of just one news article, we consider the next T news articles $D_t = \{S_t, S_{t+1}, ..., S_{t+T-1}\}$ as the target news that the user clicks in ground truth. Thus the goal of model training is to rank the target news $j \in D_t$ higher than all other news ($j' \in I \setminus D_t$). To this end, we adopt a pairwise ranking method Sequential Bayesian Personalized Ranking(S-BPR) [11] to train the model:

$$\mathcal{L}(\theta) = \sum_{u \in U} \sum_t \sum_{j \in D_t} \sum_{j' \in D_t^-} \ln \sigma(g_j - g_{j'}) + \lambda \|\theta\|_2^2 \tag{14}$$

here we randomly sample several (5 in our experiments) negative instances and denote the set by D_t^-.

4 Experiments

In this section, we present our experiments and the results. Then *Saskr* is compared with several baseline methods.

4.1   Dataset

We use a real-world Chinese news dataset from Eastday Toutiao[1], containing nearly a million serve logs from 6960 users during the time period of October 30, 2018 to November 13, 2018. Each piece of log consists of user id, news id, news title&text, click time and category.

Given a great many key words from the real-world news, we build up the news-specific knowledge graph G by collecting knowledge of the key words from the web. Then G is finely grinded for recommendation tasks from two aspects: (1) We exclude out the triples such as $(Yaoming, Height, 7.4\ ft)$. They are leaf nodes which have no outgoing links and are regarded as useless in recommendation tasks. (2) We assume the entities of **one** hop are the most valuable knowledge because entities too distant from a user's interaction history may bring severe noise. Table 1 shows the detailed statistics of the news dataset and the knowledge graph.

Table 1. Detailed statistics of the news dataset and the knowledge graph. ("#" denotes "the number of")

# users	6960	avg # entities per news	12.0
# news	108684		
# logs	861996	avg # contextual-entities per entity	3.95
# entities	146267		

4.2   Evaluation Metrics

In the validation and test stage, following [7], we randomly sample 100 news articles that are not interacted by the user in the next time step and rank the ground-truth news among them. To evaluate the model, we adopt the evaluation metrics Hit Ratio@10 (HR@10) and Mean Reciprocal Rank (MRR).

4.3   Baselines

We compare our model Saskr with the following baselines:

- **GRU4Rec** [3] is a session-based model using the GRU network for predictions. It utilizes the session-parallel mini-batch strategy and the ranking loss function for training.
- **Narm** [8] is an encoder-decoder attentive network which utilizes the global and local encoder to capture both the user's sequential behavior and main purpose in the session.

---

[1] http://mini.eastday.com/.

- **RA-DSSM** [7] uses an attention-based bidirectional RNN to tackle the problem of changing interests of users. Then it adapts DSSM to capture the similarity between users and items.
- **Caser** [14] models the user's historical interactions with horizontal and vertical CNN networks.
- **Attrec** [20] utilizes self-attention mechanism to infer the user's transient interests. The model is trained in a metric learning framework taking short-term and long-term intentions into consideration.

Note that all the baselines compared here are deep neural networks which model users' interactions from the sequence or session level. In order to show our model's ability to address item cold start problem, two separate experiments are conducted by replacing the embedding layers as ID-based Embedding and Content-based Embedding. ID-based Embedding randomly initializes the item embeddings according to the item IDs and train them with the loss functions. Content-based Embedding embeds the items (news) following Sect. 3.3(A) and sets them untrainable when training.

4.4   Result

We report the result of each model under its optimal hyperparameter settings in Table 2. Firstly, when adopting the ID-based Embedding, we observe that compared to the RNN-based models (GRU4Rec, Narm, RA-DSSM), the self-attention based models (Attrec) can deeper extract users' preference, thus achieving higher performance. The CNN-based model (Caser) also performs well owing to its feature extraction on both horizontal and vertical level. Saskr achieves the best result owing to the power multi-head self-attention module and the knowledge graph as side information.

Table 2. Performance comparison of *Saskr* with baselines

Models	ID-based		Content-based		ID vs Content	
Metrics	HR@10	MRR	HR@10	MRR	HR@10	MRR
GRU4Rec	0.7991	0.4866	0.6967	0.2942	−12.8%	−39.5%
Narm	0.8453	0.5013	0.7567	0.3118	−10.4%	**−37.8%**
RA-DSSM	0.7764	0.5157	0.6973	0.2894	**−10.2%**	−43.9%
Caser	0.8428	0.5770	0.6635	0.2805	−21.3%	−51.4%
Attrec	0.8653	0.5969	0.7270	0.3520	−13.8%	−41.0%
Saskr	**0.9035**	**0.6578**	**0.8113**	**0.4069**	**−10.2%**	−38.1%

However, the ID-based embedding method suffers from item cold start problem. To illustrate, we segregate the target news from test data set which have never been seen before, i,e, these news' embeddings have never been tuned since initialized. There are 60 such news articles. And out of these news articles, the metric of HR@10 only reaches 13.33%. To tackle this problem, we adopt the content-based method here.

The performance of each model drops when utilizing the content-based embedding. The last column is the drop rate of each model using content-based embedding relative to the ID-based embedding, defined as $\frac{IdBased-ContentBased}{IdBased}$. Owing to the powerful modules in Saskr, the drop rate of HR@10 is 10.2%, lower than the baselines. It indicates that the performance of the baselines is closely connected with the information stored in news embeddings. Once facing the item cold start problem, these models may show poor performance. As a final recapitulation, Saskr outperforms all baselines.

4.5 Discussion

In the following studies, we dive into an in-depth analysis of Saskr, aiming to examine the influence of hyperparameters. We focus on HR@10 metric.

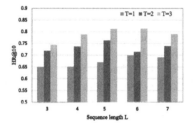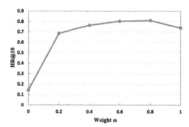

(a) HR@10 w.r.t sequence length L (b) HR@10 w.r.t weight factor ω
and number of targets T

Fig. 2. Influence of hyperparameters

TF-IDF Embedding vs Entity-Stacking Embedding. News articles embeddings are generated following the TF-IDF and Entity-stacking strategy. It is believed that an effective news embedding method can cluster similar news articles in accordance with their categories. To this end, we generate embeddings for 4,000 news articles randomly selected from 6 categories. Then we apply the T-distributed Stochastic Neighbor Embedding (t-SNE) technique to reduce the dimension and visualize them in different color according to its category. Figure 3 present the 2D embeddings respectively. As shown, the boundaries of each category in Fig. 3(a) are more clear than in Fig. 3(b). The Entity-stacking news embedding provides a better clustering and is adopted in the final Saskr model.

Influence of Sequence Length L and Number of Targets T. We vary the sequence length L while keeping other hyperparameters unchanged. The longer the sequence is, the more information it will carry, at the same time the more likely it will bring noises. The setting of the number of targets T is related to the

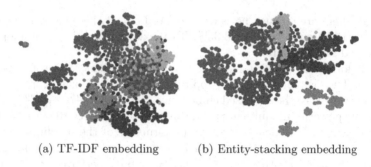

(a) TF-IDF embedding   (b) Entity-stacking embedding

Fig. 3. Visualization of two embedding methods in 2D using t-SNE

skip behavior [14], that is, the impact from past behaviors may skip a few steps and still have strength. Additionally, users may click news articles casually or mistakenly. The results are shown in Fig. 2(a). On our dataset, Saskr performs best with L of 5 and T of 3.

Influence of Weight Factor ω. The weight factor ω controls the contribution of the sequence-aware and knowledge-aware modelling. We can observe from Fig. 2(b) that enabling only knowledge-aware modelling gets poor performance, which indicates the sequence-aware modelling plays a more important role in resolving the user preference from the interaction history. Saskr preforms best at ω equaling 0.8, which proves the knowledge-aware modelling does help the recommendation as the side information.

Table 3. Illustration of an instance from the test set.

	News ID	Date	Category	Entities	Contextual-entities
interaction history	20453	2018/11/7 13:44	健康	自助餐；牛排；牛肉	生活；食品
	6411	2018/11/7 13:49	历史	中国；后宫	君主；政治人物
	1321	2018/11/7 13:54	历史	封建社会；女性	（无）
	5345	2018/11/7 13:57	科技	互联网；阿里巴巴集团；马云	社会；企业家；技术；马化腾
	74	2018/11/8 14:18	历史	朱元璋；明朝	明朝；政治人物；朱棣
target news	61	2018/11/8 14:25	社会	小康社会	-
	5905	2018/11/8 14:26	健康	癌细胞；红薯；花生；红枣	-
	21646	2018/11/8 14:28	历史	朱元璋	-

Network Visualization. We look into the attention module by visualizing them. To illustrate, an instance from the test set is shown in Table 3. The interaction history ($L = 5$) and target news ($T = 3$) are listed chronologically with entities from each news articles and contextual-entities from the knowledge graph. Figure 4(a) and (b) shows the value of two weight matrixes of head 4 and head 6 respectively from the first block in the sequence-aware modelling's encoder. Intuitively, we can inspect the latent relation of the 5 input news articles against themselves, in two different subspaces corresponding to head 4 and head 6, and tell which news article contributes more in the specific subspace.

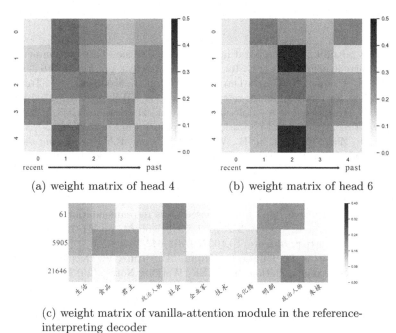

(a) weight matrix of head 4   (b) weight matrix of head 6

(c) weight matrix of vanilla-attention module in the reference-interpreting decoder

Fig. 4. Attention visualization for an instance from the test set.

Figure 4(c) visualizes the weight matrix of the vanilla-attention module in the reference-interpreting decoder. It shows the attention weight of 3 target news articles towards the contextual-entities of the input sequence, which intuitively present how much a contextual-entity contributes to the recommendation score of the knowledge-aware modelling.

5 Conclusion

Saskr is a self-attention sequential knowledge-aware recommendation system that takes advantage of the powerful self-attention mechanism and the knowledge graph. The self-attention module is used to uncover sequential patterns of the interaction history, and the knowledge graph as side information to improve diversity and extensibility of recommendation. Content-based news embeddings help to address the item cold-start problem. Extensive experiments on a news dataset from Eastday Toutiao demonstrate that Saskr outperforms state-of-the-art deep neural sequential recommendation systems.

Acknowledgments. This research work has been funded by the National Natural Science Foundation of China (Grant No. 61772337, U1736207), and the National Key Research and Development Program of China NO. 2016QY03D0604.

References

1. Bogina, V., Kuflik, T.: Incorporating dwell time in session-based recommendations with recurrent neural networks. In: RecTemp@ RecSys, pp. 57–59 (2017)
2. Bordes, A., Usunier, N., Garcia-Duran, A., Weston, J., Yakhnenko, O.: Translating embeddings for modeling multi-relational data. In: NIPS, pp. 2787–2795 (2013)
3. Hidasi, B., Karatzoglou, A., Baltrunas, L., Tikk, D.: Session-based recommendations with recurrent neural networks. arXiv preprint arXiv:1511.06939 (2015)
4. Huang, J., Zhao, W.X., Dou, H., Wen, J.R., Chang, E.Y.: Improving sequential recommendation with knowledge-enhanced memory networks. In: SIGIR, pp. 505–514. ACM (2018)
5. Huang, Z., Xu, W., Yu, K.: Bidirectional LSTM-CRF models for sequence tagging. arXiv preprint arXiv:1508.01991 (2015)
6. Kompan, M., Bieliková, M.: Content-based news recommendation. In: Buccafurri, F., Semeraro, G. (eds.) EC-Web 2010. LNBIP, vol. 61, pp. 61–72. Springer, Heidelberg (2010). https://doi.org/10.1007/978-3-642-15208-5_6
7. Kumar, V., Khattar, D., Gupta, S., Gupta, M., Varma, V.: Deep neural architecture for news recommendation. In: CLEF (Working Notes) (2017)
8. Li, J., Ren, P., Chen, Z., Ren, Z., Lian, T., Ma, J.: Neural attentive session-based recommendation. In: CIKM, pp. 1419–1428. ACM (2017)
9. Milne, D., Witten, I.H.: Learning to link with Wikipedia. In: CIKM, pp. 509–518. ACM (2008)
10. Quadrana, M., Karatzoglou, A., Hidasi, B., Cremonesi, P.: Personalizing session-based recommendations with hierarchical recurrent neural networks. In: RecSys, pp. 130–137. ACM (2017)
11. Rendle, S., Freudenthaler, C., Schmidt-Thieme, L.: Factorizing personalized Markov chains for next-basket recommendation. In: WWW, pp. 811–820. ACM (2010)
12. Song, Y., Shi, S., Li, J., Zhang, H.: Directional skip-gram: explicitly distinguishing left and right context for word embeddings. In: NAACL, pp. 175–180 (2018)
13. Tan, Y.K., Xu, X., Liu, Y.: Improved recurrent neural networks for session-based recommendations. In: DLRS, pp. 17–22. ACM (2016)
14. Tang, J., Wang, K.: Personalized top-n sequential recommendation via convolutional sequence embedding. In: WSDM, pp. 565–573. ACM (2018)
15. Vaswani, A., et al.: Attention is all you need. In: NIPS, pp. 5998–6008 (2017)
16. Wang, H., et al.: RippleNet: propagating user preferences on the knowledge graph for recommender systems. In: CIKM, pp. 417–426. ACM (2018)
17. Wang, H., Zhang, F., Xie, X., Guo, M.: DKN: deep knowledge-aware network for news recommendation. In: WWW, pp. 1835–1844. International World Wide Web Conferences Steering Committee (2018)
18. Xu, B., et al.: CN-DBpedia: a never-ending Chinese knowledge extraction system. In: Benferhat, S., Tabia, K., Ali, M. (eds.) IEA/AIE 2017. LNCS (LNAI), vol. 10351, pp. 428–438. Springer, Cham (2017). https://doi.org/10.1007/978-3-319-60045-1_44
19. Zhang, F., Yuan, N.J., Lian, D., Xie, X., Ma, W.Y.: Collaborative knowledge base embedding for recommender systems. In: KDD, pp. 353–362. ACM (2016)
20. Zhang, S., Tay, Y., Yao, L., Sun, A.: Next item recommendation with self-attention. arXiv preprint arXiv:1808.06414 (2018)

Improving Relation Extraction with Relation-Based Gated Convolutional Selector

Qian Yi[1,3(✉)], Guixuan Zhang[1,2], Shuwu Zhang[1,2], and Jie Liu[1,2]

[1] Beijing Engineering Research Center of Digital Content Technology, Institute of Automation, Chinese Academy of Sciences, Beijing, China
yiqian2016@ia.ac.cn
[2] Advanced Innovation Center for Future Visual Entertainment, Beijing, China
[3] University of Chinese Academy of Sciences, Beijing, China

Abstract. Distant supervision is an effective way to collect large-scale training data for relation extraction. To better solve the wrong labeling problem accompanied by distant supervision, some methods have been proposed to remove noise sentences directly. However, these methods seldom consider the relation label when removing noise sentences, neglecting the fact that a sentence is regarded as noise because the relation it expresses is inconsistent with the relation label. In this paper, we propose a novel method to improve the performance of bag-level relation extractor via removing noise data with a relation-based sentence selector. Specifically, the relation-based gated convolutional unit of the sentence selector can selectively output features related to the given relation, and these features will be used to judge whether a sentence expresses the given relation. The sentence selector is trained with the data automatically labeled by the relation extractor, and the relation extractor improves its performance with the high-quality data selected by the sentence selector. These two modules are trained alternately, and both of them have achieved better performance. Experimental results show that our model significantly improves the performance of the relation extractor and outperforms competitive baseline methods.

1 Introduction

Relation extraction aims to obtain the relationship between two entities from unstructured text. For example, given a sentence *'Donald Trump was born in America.'* and two entities *'Donald Trump'* and *'America'*, relation extraction intends to get the relation *'place of birth'* from them. Earlier works use manually labeled data to train the classifier in a supervised manner and have achieved good performance [2,10,14,16,17]. However, the performances of these models are limited by the scale of the training data, and constructing a large-scale manually labeled dataset is labor consuming. In order to build large-scale dataset automatically, Mintz et al. [9] proposed distant supervision. Distant supervision is based on the idea that if an entity pair (h, t) is contained by a triple (h, t, r) of

© Springer Nature Switzerland AG 2019
M. Sun et al. (Eds.): CCL 2019, LNAI 11856, pp. 233–245, 2019.
https://doi.org/10.1007/978-3-030-32381-3_19

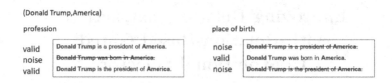

(Donald Trump,America)

profession		place of birth	
valid	Donald Trump is a president of America.	noise	Donald Trump is a president of America.
noise	Donald Trump was born in America.	valid	Donald Trump was born in America.
valid	Donald Trump is the president of America.	noise	Donald Trump is the president of America.

Fig. 1. An example of the noise problem for bag-level relation extraction.

a given knowledge base, all sentences that contain the entity pair (h, t) will be labeled as the relation r. The h, t, r represent head entity, tail entity and relation, respectively. However, due to the existence of the multi-relational entity pairs, distant supervision suffers from the wrong labeling problem.

Various methods have been proposed to alleviate this issue. One common way among these studies is to employ Multi-Instance Learning (MIL) schema [3,13], in which sentences containing the same entity pair are divided into the same bag and the classification proceeds on bag-level. Zeng et al. [18] selected the most important sentence to represent the bag and trained the model with these selected sentences. Lin et al. [6] applied attention mechanism to give the important sentences lager weights and combined all sentences to obtain the bag representation. Jiang et al. [5] used cross-sentence max-pooling to find the most prominent features among all sentence representations. Recently, some researchers suggested that it was not enough to attenuate the effects of noise data through 'soft' means like attention mechanism. They tended to remove the noise data directly. Feng et al. [1] and Qin et al. [11] trained a sentence selector to distinguish between noise sentences and valid sentences through reinforcement learning (RL). Qin et al. [12] trained a generative adversarial network (GAN) and used the classifier to remove the noise data.

However, these 'hard' methods neglect the fact that when we consider a sentence as noise, it means that this sentence expresses a relation inconsistent with its label. Just as Fig. 1 shows, for the three sentences 'Donald Trump is a presedent of America.', 'Donsld Trump was born in America.', 'Donsld Trump is the presedent of America.' and the entity pair 'Donald Trump' and 'America', if the labeling relation is 'place of birth', the first and the third sentence are noise sentences. But when the labeling relation is 'profession', the second sentence becomes the noise data. Therefore, we think it is crucial to consider the labeling relation when identifying noise data.

In this paper, we propose a novel method to improve the performance of bag-level relation extractor via removing noise data with a relation-based sentence selector. We design a relation-based gated convolutional network for the sentence selector. The relation-based gated convolutional network has two convolutional components. One acts as a feature extractor to extract the semantic features of the sentence. The other is a relation-based gate, which can select the features related to the given relation. Like the previous models, we encounter the problem of lacking training data for the sentence selector. To deal with this problem, we adopt an easy and reasonable method. We treat each sentence as a bag with only one sentence and use the pre-trained bag-level relation classifier to classify it. A sentence will be labeled as a positive sample if the classification result is identical to its label. Otherwise, the sentence is labeled as a negative sample. This labeling method is consistent with the idea that the label of valid data is the same as the relation it conveys. As for bag-level relation extractor, we adopt two widely used architectures: piecewise convolutional neural networks (PCNN) [18] with attention mechanism and the convolutional neural networks (CNN) with attention mechanism. Moreover, because our model is a generic framework, the relation extractor here can be replaced by any other bag-level relation extractor with different structures. Then we train the bag-level relation extractor and the sentence selector alternately so that their performance can be improved jointly.

The main contributions of this paper can be summarized as follows:

- We design a novel relation-based gated convolutional sentence selector to select valid sentences for distantly supervised relation extraction.
- We propose a framework which can train the sentence selector without manually labeled data and jointly improve the performance of the bag-level relation extractor and the sentence selector.
- Experimental results show that our model significantly improves the performance of the relation extractor and outperforms competitive baseline methods.

2   Related Work

The purpose of relation extraction is to obtain the relationship between two entities from unstructured text. Traditional methods leveraged syntactic information and adopted kernel-based classifier to build multi-class relation classifier [10,16]. Recently, more attention has been paid to neural networks methods.

In order to extract relation features, previous neural networks models employed various structures to encode the sentence. Zeng et al. [17] adopted CNN to extract the semantic information of the sentence. Xu et al. [15] encoded sentence with Long Short-Term Memory (LSTM) along the shortest dependency path. Zhou et al. [22] combined attention mechanism and LSTM to encode the sentence. Zeng et al. [18] proposed PCNN to extract features from different parts of the sentence separately. Zhang et al. [21] adopted graph convolution over pruned dependency trees to improve the performance of relation extraction. Zhang et al. [19] used attention-based capsule networks to encode the sentence.

In order to solve the problem of lacking for manual annotation data, distant supervision was proposed [9]. To deal with the accompanying wrong labeling problem, researchers have proposed various methods. Zeng et al. [18] adopted the MIL framework. They collected all the sentences containing the same entity-pair as a bag and selected the most important sentence in each bag to train the network. Lin et al. [6] used attention mechanism to give each sentence an importance weight and combined all the sentences to represent the bag. Jiang et al. [5] used cross-sentence max-pooling to extract the features of a sentence bag. Liu et al. [7] softly revised incorrect bag labels with the posterior probability

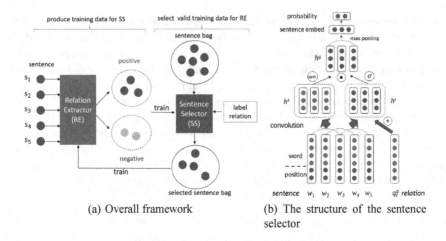

(a) Overall framework

(b) The structure of the sentence selector

Fig. 2. The architecture of our model.

constraint. The above works focused on highlighting the valid sentences of the sentence bag and reduce the effects of noise.

However, some researchers suggested that it was not enough to only weaken the effects of noise data by giving them a small weight, they tended to remove the noise data directly. Feng et al. [1] and Qin et al. [11] trained a sentence classifier to distinguish between noise sentences and other sentences through RL. Qin et al. [12] trained a generative adversarial network and used the classifier to remove the noise data.

3   Model

An overview of our framework is shown in Fig. 2(a). The model consists of two parts: a relation extractor and a sentence selector. The sentence selector is trained with data automatically labeled by the relation extractor. And the relation-based selector selects data from each sentence bag according to the labeling relation. The relation extractor is then further trained with the selected high-quality sentence bags. These two modules help each other to obtain better training data and finally achieve better performance. In this section, we will first describe these two parts in detail, and then introduce the specific details of training and test.

3.1   Input Layer

Given a sentence s, the input layer transforms the sentence into an embedding matrix, which contains both semantic information and positional information of each word, and feed it to the subsequent networks.

Embedding. Word embeddings are low dimensional, continuous and real-valued vectors, which can capture semantic meanings of words. Each word in the vocabulary corresponds to a word embedding vector $v_w \in \mathcal{R}^{d_w}$. In this paper, we use word embeddings pre-trained on the New York Times (NYT) corpus with Skip-Gram [8].

Position Embedding. Position embeddings are vectors that embed the relative distances of each token to the two target entities. For example, in the sentence "*SteveJobs* was the co-founder and CEO of *Apple* and...", the relative position from token *co − founder* to entity *SteveJobs* and *Apple* is 3 and −4, respectively. Each relative position value corresponds to a position embedding vector $v_p \in \mathcal{R}^{d_p}$.

For each word w, we concatenate its word embedding v_w and two position embeddings (each corresponds to the relative distance from one entity) $v_{p_{en1}}$ and $v_{p_{en2}}$ as its representation $v \in \mathcal{R}^{d_w + d_p * 2}$. Then for each sentence with n words $s = \{w_1, ..., w_n\}$, we obtain an embedding matrix $S = \{v_1; ...; v_n\}$ by concatenating all words representations.

3.2   Relation Extractor

Given an entity pair (h, t) and its sentence bag $S_{h,t} = \{s_1, s_2, ...\}$, the relation extractor intends to obtain the relation of the bag. Because our model is a generic framework, the relation extractor module can use any bag-level relation extractor. In this paper, we adopt two widely used models: CNN with attention mechanism and PCNN with attention mechanism.

Sentence Encoder. We use CNN and PCNN as sentence encoder to encode the sentence embedding matrix into a representation vector.

Convolution. A filter $W_r \in \mathcal{R}^{k_h \times d_v \times m}$ is applied to extract local features of a sentence, where $d_v = d_w + 2 * d_p$ is the dimension of the word vector, m is the width of the filter and k_h is the dimension of the output channel. By sliding W_r along the sentence embedding matrix S_i, we can get the k_h-dimensional feature vector:

$$h_i = ReLU([v_{i-(m-1)/2}, ..., v_{i+(m-1)/2}] \otimes W_r + b_r) \tag{1}$$

where $b_r \in \mathcal{R}^{k_h}$ is a bias. Then all the feature vectors are concatenated to form a feature map $H = \{h_1, ..., h_n\}$.

Max-pooling and Piece-Wise Max-pooling. Max-pooling operation is then applied over the feature map H to get the final sentence representation:

$$q_j = \max_{1 \leqslant i \leqslant n} \{h_{i,j}\} \tag{2}$$

Piece-wise Max-pooling is an extension of Max-pooling:

$$q^{(1)}{}_j = \max_{1 \leqslant i \leqslant i_{en1}} \{h_{i,j}\}$$

$$q^{(2)}{}_j = \max_{i_{en1} < i \leqslant i_{en2}} \{h_{i,j}\} \tag{3}$$

$$q^{(3)}{}_j = \max_{i_{en2} < i \leqslant n} \{h_{i,j}\}$$

where the subscript j represents the j-th value of a vector, i_{en1} and i_{en2} are the positions of two entities. Then we concatenate the three pooling vectors to get the final sentence representation:

$$q = \{q^{(1)}, q^{(2)}, q^{(3)}\}. \tag{4}$$

Sentence Selective Attention. After obtaining sentence representation, we apply selective attention to compute the attention score α_i for each sentence. Then the bag embedding u is computed as a weighted sum of sentence representations:

$$u = \sum_i^{|S_{h,t}|} \alpha_i q_i \tag{5}$$

where the weight α_i indicates the degree of correlation between sentence and the relation, and $|S_{h,t}|$ is the number of sentences in the bag. We assign a query vector $q_r \in \mathcal{R}^{d_q}$ for each relation r, where d_q is the dimension of sentence representation q. The attention score is computed as:

$$e_i = q_r^\mathrm{T} W_a q_i$$

$$\alpha_i = \frac{\exp(e_i)}{\sum_j^N \exp(e_j)} \tag{6}$$

where $W_a \in \mathcal{R}^{d_q \times d_q}$ is the weight matrix and N is the number of relations.

Loss Function. Finally, we obtain the conditional probability $p(r|S_{h,t}, \theta)$ through feeding the bag representation u to a fully connected layer:

$$p(r|S_{h,t}, \theta) = \frac{\exp(o_r)}{\sum_k \exp(o_k)} \tag{7}$$

$$o = W_c u + b_c$$

where $W_c \in \mathcal{R}^{N \times d_q}$ is a weight matrix and $b_c \in \mathcal{R}^N$ is a bias vector.

Given the collection of sentence bags $\Omega = \{S_{h_1,t_1}, S_{h_2,t_2}, ...\}$ and corresponding labeling relation $\{r_i, r_2, ...\}$, the loss function is defined as follows:

$$J_R = -\frac{1}{|\Omega|} \sum_{i=1}^{|\Omega|} \log p(r_i|S_{h_i,t_i}, \theta) \tag{8}$$

where $|\Omega|$ is the number of bags.

3.3   Sentence Selector

The sentence selector is a binary classifier which can judge whether a sentence expresses the given relation. Its detailed structure is shown in Fig. 2(b).

Obtaining Training Data. Since there is no training data for the sentence selector, we propose a method to label data automatically. We transform the relation extractor introduced in Sect. 3.2 into a sentence-level relation extractor by regarding each sentence as a bag with only one sentence. Those sentences whose classification result is consistent with their labeling relation are labeled as positive. Otherwise, they are labeled as negative.

After that, we get a data set $D = \{s, e, y, r\}$, in which s represents the sentence, e is the entity pair, y is the two-category label and r is the labeling relation obtained by distant supervision.

Relation-Based Gated Convolutional Network. Given a sentence s_i and its corresponding relation r_i, the input to the relation-based gated convolutional network is the same as the input embedding matrix in Sect. 3.1. Specifically, each token is embedded into a word embedding and two position embeddings, so that we get the input embedding matrix $S_i = \{v_1; ...; v_n\}$.

Then we feed the input embedding matrix S_i to the relation-based gated convolutional unit. The relation-based gated convolutional unit contains two convolutional components. One is a plain convolution operation mentioned in Sect. 3.2:

$$h_i^s = tanh([v_{i-(m-1)/2}, ..., v_{i+(m-1)/2}] \otimes W_s + b_s) \tag{9}$$

where $W_s \in \mathcal{R}^{k_h \times d_v \times m}$ is the filter, and $b_s \in \mathcal{R}^{k_h}$ is a bias. The other convolution operation integrates relation information when extracting the local features:

$$h_i^r = \sigma([v_{i-(m-1)/2}, ..., v_{i+(m-1)/2}] \otimes W_g + W_q q_r^g + b_g) \tag{10}$$

where σ is the *sigmoid* function, $W_g \in \mathcal{R}^{k_h \times d_v \times m}$ is the filter and $b_g \in \mathcal{R}^{k_h}$ is a bias, $W_q \in \mathcal{R}^{d_v \times d_q}$ is the weight matrix, q_r^g is the query vector which contains information related to relation r_i. It is worth noting that we initialize the query vector q_r^g with the value of q_r trained in the relation extractor when training the sentence selector. Then we compute an element-wise multiplication between the feature vector h^s and the relation gate vector h^r:

$$h_i^g = h_i^s \odot h_i^r \tag{11}$$

where the symbol $\odot$ represents the element-wise multiplication. Through the element-wise multiplication, the relation gate can selectively output the sentence features related to the given relation. The max pooling procedure is then performed over the feature maps to obtain the sentence embedding:

$$q_{\ j}^g = \max_{1 \leqslant i \leqslant n} \{h_{i,j}^g\} \tag{12}$$

Algorithm 1. Overall Training Procedure

Input: Episode number L, collection of sentence bags $B = \{S_1, S_2, ..., S_n\}$, collection of sentences $S = \{s_1, s_2, ..., s_m\}$

Initialize the relation extractor and the sentence selector randomly.

1: **for** $l = 1$ to L **do**
2:   **if** Not converge **then**
3:     Train the relation extractor with B.
4:     Label S with the relation extractor to obtain S_{label}.
5:     **if** $l = 1$ **then**
6:       Initalize the relation query matrix $Q^g = \{q_{r_1}^g; q_{r_2}^g; ...; q_{r_N}^g\}$ with the relation query matrix $Q = \{q_{r_1}; q_{r_2}; ...; q_{r_N}\}$ of the relation extractor.
7:     **end if**
8:     Train the sentence selector with S_{label}.
9:     Select the sentences in each sentence bags with the sentence selector to obtained $B = B_{selected}$.
10:   **else**
11:     End training.
12:   **end if**
13: **end for**

Loss Function. We feed the sentence embedding q_i^g to a fully connected layer to compute the posterior probability $p(y'|s, r, \phi)$:

$$p(y'|s, r, \phi) = \frac{\exp(o_{y'}^g)}{\sum_k \exp(o_k^g)}$$

$$o^g = W_o q^g + b_o$$

(13)

where $W_o \in \mathcal{R}^{2 \times d_q}$ is a weight matrix, $b_o \in \mathcal{R}^2$ is a bias vector and y' is the two-category label.

Given the collection of sentences $\Lambda = \{s_1, s_2, ...\}$, its label relation $\{r_i, r_2, ...\}$ and the corresponding label $\{y_i, y_2, ...\}$, the loss function is defined as follows:

$$J_S = -\frac{1}{|\Lambda|} \sum_{i=1}^{|\Lambda|} \log p(y_i | s_i, r_i, \phi)$$

(14)

where $|\Lambda|$ is the number of bags.

3.4   Training and Test

Because the performance of the relation extractor and the sentence selector influence each other. We train the two modules alternately.

During training, we first adopt Adam algorithm to minimize the loss function Eq. 8 with the original dataset. After using relation extractor to generate training data for the sentence selector, we then optimize the sentence selector by minimizing the Eq. 14 with Adam. Next, we utilize the sentence selector to select the sentences to further train the relation extractor. The relation extractor and the instance selector are trained alternately as described above until convergence. The complete training process is described in Algorithm 1.

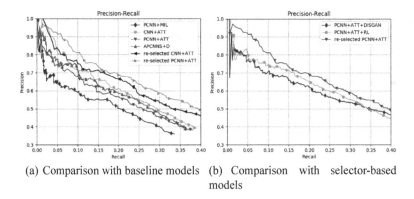

(a) Comparison with baseline models   (b) Comparison with selector-based models

Fig. 3. Comparison with previous methods.

Table 1. Comparison of P@N between our model and the original bag-level relation extractor

P@N	100	200	500	Mean
CNN+ATT	0.73	0.68	0.56	0.657
re-selected CNN+ATT	**0.79**	**0.76**	**0.60**	**0.717**
PCNN+ATT	0.80	0.71	0.59	0.700
re-selected PCNN+ATT	**0.86**	**0.76**	**0.64**	**0.753**

During testing, we first select the test data with the sentence selector. Because there is no label for sentence bags during the test procedure, for each sentence bag $S_{h,t}$, we generate one relation-based sentence bag $S_{h,t}^r$ for each relation r by selecting valid sentences from $S_{h,t}$ given relation r. We will obtain N relation-based sentence bags for each sentence bag, where N is the number of relations. Then we feed all N relation-based sentence bags into the relation extractor, and the relation r whose bag obtains the highest score $p(r|S_{h,t}^r, \theta)$ is chosen as the predicted relation for the entity pair (h, t).

What needs to be mentioned is that, in order to increase the recall, we set a threshold $u < 0.5$ when selecting data, and only sentences whose selecting probability is smaller than u will be removed. This operation increases the tolerance for classification errors of sentence selector and enhances the recall of the relation extractor.

4 Experiments

4.1 Dataset and Evaluation

In this paper, we evaluate our model on the widely used New York Times (NYT) dataset developed by [13]. This dataset is constructed by aligning Freebase with New York Times (NYT) corpus through distant supervision. There are 522611 sentences in the training set and 172448 sentences in the test set, and these sentences are labeled by 53 candidate relations. Among the 53 relations, there is a label NA, which represents there is no relation between the two entities in a sentence. During training, We randomly extract ten percent of the sentences from the training data as the validation data and the rest as the training data.

We evaluate all methods with the held-out evaluation. The held-out evaluation compares the relational facts extracted from the test set by models with all the facts existing in the test sentences (which is labeled by Freebase through distant supervision). For evaluation, we present precision-recall curves for all models.

4.2   Implementation Detail

In our experiment, our parameter settings are as follows: the dimension of word embedding d_w and position embedding d_p are 50 and 5; the width of the convolution kernel m is 3 and the dimension of the output channel k_h of the convolution filter is 230; the max sentence length is 120; the batch size is fixed to 50 and dropout probability is fixed to 0.5. We adopt Adam to update the parameters, and the learning rate for training relation extractor and sentence selector are set to 0.001 and 0.0005. As for the threshold u for the sentence selector during selecting, we tune it on the validation dataset and pick $u = 0.3$ in the candidate set $\{0.1, 0.2, 0.3, 0.4, 0.5\}$.

4.3   Comparison with Previous Methods

To evaluate the performance of our proposed model, we compare our model with various baseline models. **PCNN+MIL** [18] proposed piecewise CNN to encode the sentence and adopted the MIL framework. **CNN + ATT** and **PCNN+ATT**[6] employed attention mechanism to reduce the influence of noise data and used CNN and PCNN as sentence encoder respectively. **APC-NNS+D** [4] used external entity descriptions and attention mechanism to obtain better bag representation. We also compare the performance of our model with other selector-based models to further assess the sentence selector. **PCNN+ATT+DSGAN** [12] trained a generative adversarial network and used the classifier to remove the noise data. **PCNN+ATT+RL** [1] trained a sentence selector through reinforcement learning.

As shown in Fig. 3(a), our models, which are denoted as re-selected+PCNN+ATT and re-selected+CNN+ATT, have a significant improvement on all baseline models. For a more detailed comparison, we show the precision@N(P@N) of our models (re-selected+PCNN+ATT and re-selected+CNN+ATT) and the corresponding baseline models (CNN+ATT and PCNN+ATT) in Table 1. The results demonstrate the effectiveness of the sentence selector for distant supervision relation extraction. The re-selected CNN+ATT model and the re-selected PCNN+ATT both achieve higher values for P@100, P@200, P@500 compared to the original baseline models. Moreover,

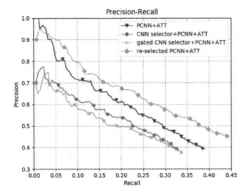

Fig. 4. Comparison between CNN selector, plain gated-CNN selector and relation-based gated-CNN selector.

the mean value of re-selected+CNN+ATT is 6% higher than CNN+ATT, and re-selected+PCNN+ATT is 5.3% higher than PCNN+ATT.

As Fig. 3(b) shows, when compared to other selector-based models, PCNN+ATT with relation-based gated selector also achieves better performance on both precision and recall. Moreover, compared to the RL and GAN, our model is more stable and easier to converge when training the sentence selector.

4.4 Effect of Relation-Based Convolutional Gate Unit

To further show the effectiveness of the relation-based gated convolutional unit and the necessity of relation information, we compare the performance of three sentence selector architectures: the CNN selector which removes the gate component h_i^r, the plain gated-CNN selector which removes the relation information component $W_q q_r^g$ from Eq. 10 and the relation-based gated-CNN selector.

As Fig. 4 shows, the performance of the relation extractors decline drastically when trained with the sentences selected by the plain gated-CNN selector and the CNN selector. The experimental results indicate that the sentence selector is unable to identify the noise sentences with only the semantic features extracted by the CNN or the gated CNN. The good performance of the sentence selector was brought by the relation-based gate which can integrate relation information when extracting the sentence features.

5 Conclusion

In this paper, we propose a novel method to improve the performance of bag-level relation extractor via removing noise data with a relation-based sentence selector for neural relation extraction under the distant supervision scenario. The whole model contains a relation extractor and a sentence selector composed of a well-designed relation-based gated convolutional network. We train the sentence selector without manually labeled data and employ the selector to select high-quality data for training the relation extractor. We conduct experiments on a widely used dataset. The experimental results confirm the effectiveness of the relation-based gated convolutional unit and our framework significantly improves the performance of the original bag-level relation extractor.

References

1. Feng, J., Huang, M., Zhao, L., Yang, Y., Zhu, X.: Reinforcement learning for relation classification from noisy data. In: Proceedings of AAAI (2018)
2. GuoDong, Z., Jian, S., Jie, Z., Min, Z.: Exploring various knowledge in relation extraction. In: Proceedings of the 43rd Annual Meeting of the Association for Computational Linguistics, pp. 427–434 (2005)
3. Hoffmann, R., Zhang, C., Ling, X., Zettlemoyer, L., Weld, D.: Knowledge-based weak supervision for information extraction of overlapping relations. In: Proceedings of the 49th Annual Meeting of the Association for Computational Linguistics, pp. 541–550 (2011)
4. Ji, G., Liu, K., He, S., Zhao, J.: Distant supervision for relation extraction with sentence-level attention and entity descriptions. In: Proceedings of AAAI (2017)
5. Jiang, X., Wang, Q., Li, P., Wang, B.: Relation extraction with multi-instance multi-label convolutional neural networks. In: Proceedings of COLING 2016, the 26th International Conference on Computational Linguistics, pp. 1471–1480(2016)
6. Lin, Y., Shen, S., Liu, Z., Luan, H., Sun, M.: Neural relation extraction with selective attention over instances. In: Proceedings of the 54th Annual Meeting of the Association for Computational Linguistics, pp. 2124–2133(2016)
7. Liu, T., Wang, K., Chang, B., Sui, Z.: A soft-label method for noise-tolerant distantly supervised relation extraction. In: Proceedings of the 2017 Conference on Empirical Methods in Natural Language Processing, pp. 1790–1795 (2017)
8. Mikolov, T., Sutskever, I., Chen, K., Corrado, G., Dean, J.: Distributed representations of words and phrases and their compositionality. In: Advances in Neural Information Processing Systems, pp. 3111–3119 (2013)
9. Mintz, M., Bills, S., Snow, R., Jurafsky, D.: Distant supervision for relation extraction without labeled data. In: Proceedings of the Joint Conference of the 47th Annual Meeting of the ACL and the 4th International Joint Conference on Natural Language Processing of the AFNLP, pp. 1003–1011 (2009)
10. Mooney, R., Bunescu, R.: Subsequence kernels for relation extraction. In: Advances in Neural Information Processing Systems, pp. 171–178 (2006)
11. Qin, P., Xu, W., Wang, William Y.: Robust distant supervision relation extraction via deep reinforcement learning. In: Proceedings of the 56th Annual Meeting of the Association for Computational Linguistics, pp. 2137–2147 (2018)
12. Qin, P., Xu, W., Wang, William Y.: DSGAN: generative adversarial training for distant supervision relation extraction. In: Proceedings of the 56th Annual Meeting of the Association for Computational Linguistics, pp. 496–505(2018)
13. Riedel, S., Yao, L., McCallum, A.: Modeling relations and their mentions without labeled text. In: Joint European Conference on Machine Learning and Knowledge Discovery in Databases, pp. 148–163 (2010)
14. Santos, C., Xiang, B., Zhou, B.: Classifying relations by ranking with convolutional neural networks. In: Proceedings of the 53rd Annual Meeting of the Association for Computational Linguistics and the 7th International Joint Conference on Natural Language Processing, pp. 626–634(2015)

15. Xu, Y., Mou, L., Li, G., Chen, Y., Peng, H., Jin, Z.: Classifying relations via long short term memory networks along shortest dependency paths. In: Proceedings of the 2015 Conference on Empirical Methods in Natural Language Processing, pp 1785–1794 (2015)

16. Zelenko, D., Aone, C., Richardella, A.: Kernel methods for relation extraction. J. Mach. Learn. Res. **3**, 1083–1106 (2003)

17. Zeng, D., Liu, K., Lai, S., Zhou, G., Zhao, J.: Relation classification via convolutional deep neural network. In: Proceedings of COLING 2014, the 25th International Conference on Computational Linguistics, pp. 2335–2344 (2014)

18. Zeng, D., Liu, K., Chen, Y., Zhao, J.: Distant supervision for relation extraction via piecewise convolutional neural networks. In: Proceedings of the 2015 Conference on Empirical Methods in Natural Language Processing, pp. 1753–1762 (2015)

19. Zhang, N., Deng, S., Sun, Z., Chen, X., Zhang, W., Chen, H.: Attention-based capsule networks with dynamic routing for relation extraction. In: Proceedings of the 2018 Conference on Empirical Methods in Natural Language Processing, pp. 985–992 (2018)

20. Zhang, Y., Zhong, V., Chen, D., Angeli, G., Manning, C.: Position-aware attention and supervised data improve slot filling. In: Proceedings of the 2017 Conference on Empirical Methods in Natural Language Processing, pp. 35–45(2017)

21. Zhang, Y., Qi, P., Manning, C.: Graph convolution over pruned dependency trees improves relation extraction. In: Proceedings of the 2018 Conference on Empirical Methods in Natural Language Processing, pp. 2205–2215 (2018)

22. Zhou, P., et al.: Attention-based bidirectional long short-term memory networks for relation classification. In: Proceedings of the 54th Annual Meeting of the Association for Computational Linguistics, pp. 207–212 (2016)

Attention-Based Gated Convolutional Neural Networks for Distant Supervised Relation Extraction

Xingya Li[(⊠)], Yufeng Chen, Jinan Xu, and Yujie Zhang

Beijing Jiaotong University, Beijing, China
{17120394, chenyf, jaxu, yjzhang}@bjtu.edu.com

Abstract. Distant supervision is an effective method to generate large-scale labeled data for relation extraction without expensive manual annotation, but it inevitably suffers from the wrong labeling problem, which would make the corpus much noisy. However, the existing research work mainly focuses on sentence-level noise filtering, without considering noisy words which widely exist inside sentences. In this paper, we propose an attention-based gated piecewise convolutional neural networks (AGPCNNs) for distant supervised relation extraction, which can effectively reduce word-level noise by selecting the inner-sentence features. On the one hand, we construct a piecewise convolutional neural network with gate mechanism to extract features that are related to relations. On the other hand, we employ a soft-label strategy to enable model to select important features automatically. Furthermore, we adopt an attention mechanism after the piecewise pooling layer to obtain high-level positive features for relation predicting. Experimental results show that our method can effectively filter word-level noise and outperforms all baseline systems significantly.

Keywords: Relation extraction · Distant supervision · Gate mechanism · Attention mechanism

1 Introduction

Relation extraction (RE) aims to identify the semantic relationship between two entities from natural language texts, and it is an important part of information extraction. One of the major challenges faced by RE is that its training requires large-scale labeled corpus, while manual annotation is too time-consuming and laborious. Thus, Mintz et al. (2009) proposed a distant supervised method for RE, which could annotate large-scale data automatically and heuristically with the existing knowledge bases. The labeling process is as follows: given a triplet in a knowledge base, also known as a relation fact, (*Syracuse, contains, Lake Onondaga*), all sentences containing the above two named entities will be labeled as relation *contains*.

Distant supervision is an effective method of automatically labeling training data, nevertheless, it is plagued by the wrong labeling problem (Riedel et al. 2010), since a sentence that mentions two entities does not necessarily express the relation contained in a known knowledge base. For example, in the sentence *[The Onondaga nation is an*

M. Sun et al. (Eds.): CCL 2019, LNAI 11856, pp. 246–257, 2019.
https://doi.org/10.1007/978-3-030-32381-3_20

11-square-mile parcel in a valley south of Syracuse and about eight miles from Onondaga Lake]. There is no *contains* relation between the entities *Syracuse* and *Onondaga Lake*, but it will still be regarded as a positive instance. In response, some research work adopted the Multi-instance Learning (MIL) method (Dietterich et al. 1997), and used a probabilistic graphical model to select sentences (Hoffmann et al. 2011; Suideanu et al. 2012). Zeng et al. (2015) combined MIL and piecewise convolutional neural networks (PCNNs) to select the most likely positive sentence. Lin et al. (2016) introduced a sentence-level attention mechanism on the basis of PCNNs, and extracted effective features of all sentences by assigning different attention weights to sentences. Moreover, Ji et al. (2017) proposed a model architecture called APCNNs, which also used attention mechanism and added entity description information, and achieved the best performance in such methods.

Although the above researches have achieved good results in distant supervised relation extraction, there still exists two problems. Specifically, (1) Not all words in a sentence contribute to judging relation labels. For example, considering a sentence *[The cultural appreciation was driven by literacy theories like Roland Barthes, not by Jean Baudrillard]*, where *Roland Barthes* and *France* are two corresponding entities. Obviously, the sentence describes the *national* relation between *Roland Barthes* and *France*, but the sub-sentence *[not by Jean Baudrillard]* has little effect on judging the relation *national*, regarded as noisy words or word-level noise, of which the features will cut down the precision of the relation extraction model. (2) The Hard-label method, in which the labels of entity pairs are immutable during model training, would enlarge the impact of wrong labels in distant supervision.

In this paper, we propose a novel word-level distant supervised method for relation extraction, named AGPCNNs (Attention-based Gated Piecewise Convolutional Neural Networks), which reduces word-level noise by selecting important features inside sentences. To settle the above first problem, a gate mechanism (Hochreiter et al. 1997) is used to screen out the features extracted by convolution layer. Besides, an attention mechanism is applied after piecewise pooling layer to gain the high-level positive features. To tackle the second problem, we introduce a soft-label strategy (Wang et al. 2018) in our model, by which the same instances may have different labels in different epochs of training. Finally, combined with a sentence-level noise filtering module, more positive correlation features are obtained. We tested on the public data sets, and the experimental results show that the performance of our proposed model is significantly better than that of all baseline systems, which verifies the effectiveness of the proposed approach. Our contributions are summarized as follows:

- To handle the problem of noisy words, a gate mechanism is proposed to filter inner-sentence features uncorrelated to relation labels and an attention mechanism after piecewise pooling layer to obtain high-level features related to relation labels.
- A soft-label strategy is utilized to weaken the impact of hard labels on feature selection during training. Specifically, we use the bilinear transformation between entity pairs to conduct the feature selection process in the gate mechanism, which makes it more precise and suitable for reducing word-level noise.

- The proposed model achieves significant results for distant supervised relation extraction. Furthermore, the gate mechanism could be adopted by other neural networks and enhance the performance of the corresponding tasks.

2   Methodology

Given a bag (a set of sentences containing the same entity pair) $B = \{s_1, s_2, \cdots, s_n\}$ and the corresponding two entities, our model will predict the probability of each relation label on B. The overall architecture of our proposed model is illustrated in Fig. 1.

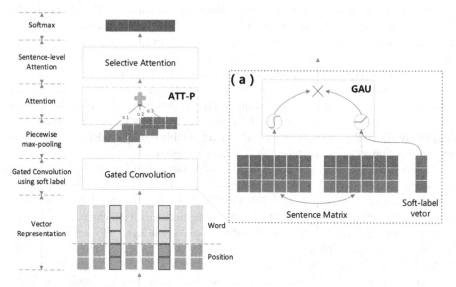

The appreciation in [France] was driven by [Roland Barthes]

Fig. 1. The architecture of our model (AGPCNNs) used for distant supervised relation extraction. The right part (a) describes the Gated Convolution layer, which takes soft labels as the supervised information.

Compared with the traditional PCNNs model, we improve the convolution layer by adding a gate mechanism and the piecewise pooling layer by adding an attention mechanism to select important inner-sentence features. Furthermore, we adopt a soft-label strategy for the above two mechanisms, so that they can work better. We will describe the gate mechanism, the soft-label strategy and the attention mechanism in detail in Sects. 2.2–2.4.

2.1 Vector Representation

The input of AGPCNNs is represented by embeddings, which are composed of two parts: word embeddings and position embeddings.

Word Embeddings. Word Embeddings are distributed representation of words, aiming at mapping each word into a low-dimensional vector. The vector is obtained by looking up a pre-trained vector matrix $\mathbf{V}$ (or lookup table), where $\mathbf{V} \in \mathbb{R}^{|\mathbf{V}| \times d^w}$, $|\mathbf{V}|$ is the size of the matrix $\mathbf{V}$ and d^w is the dimension of word embeddings.

Position Embeddings. Following Zeng et al. (2015), we employ position features to track the relative distances of the current word to the head entity e_1 and the tail entity e_2. Figure 2 shows an example of the relative distances. The relative distances from word *theorists* to *France* (e_1) and *Roland Barthes* (e_2) are 4 and -2 respectively. With two position matrices $\mathbf{PF}_1$ and $\mathbf{PF}_2$, which are initialized randomly, we can transfer relative distances to real-value vectors $\mathbf{E}_p \in \mathbb{R}^{d^p}$, where d^p is the dimension.

$$4 \qquad\qquad -2$$

The appreciation in [France] was driven by theorists like [Roland Barthes].

Fig. 2. An example of relative distances

Finally, the input representation of a word is a vector concatenated by word embeddings and position embeddings. With the vector representation of words, we transfer the sentence s into a matrix $\mathbf{S} \in \mathbb{R}^{|s| \times d}$, where $|s|$ is the length of s, $d = d^w + d^p$.

2.2 Convolution with Gate Mechanism

Convolution. Convolutional neural networks (CNNs) can effectively extract all local features of the input and perform global predictions.

Convolution is an operation between a weight matrix (also known as filter) w and the vector matrix $\mathbf{S}$ of a sentence s. The result of convolution operation is $c = \{c_1, c_2, \cdots, c_{|s|-w+1}\}$, where w is the window size, and $c_j = f(w \otimes \mathbf{S}_{(j-w+1):j} + b_c)$, where $1 \leq j \leq |s| - w + 1$, f is a nonlinear activation function and $b_c \in \mathbb{R}$ is bias.

Gate Mechanism. Considering the impact of inner-sentence noise on the model performance, we use a gate mechanism to select positive features at word level. Gate mechanism have shown effectiveness of gate mechanisms in language modeling (Kalchbrenner et al. 2016; Gehring et al. 2017). We improve the gate mechanism based on GTU (Gated Tanh Units) and name it as GAU (Gated Activation Units, as shown in Fig. 1 (a)), which is represented by:

$$c_{\text{GAU}} = \tanh(w_c \otimes S + b_1) \times \text{relu}(v \otimes S + b_2) \tag{1}$$

The relu gates control features extracted by the tanh units according to its own outputs to achieve the purpose of selecting the important word-level features.

2.3 Soft-Label Strategy

Generally, the relation labels of entity pairs are unchangeable during training, no matter whether they are correct or not, which would enlarge the negative impact of the wrong labeling problem on the feature selection process. For this, we introduce a soft-label strategy into GAU to weaken the impact of wrong labels on the model performance, i.e., we replace hard labels with soft labels generated from the entity pairs to guide feature selection and cut down inner-sentence noise during training.

As shown in Fig. 1, GAU is connected to two convolutional networks (one is the original CNN and the other has label features). We use the bilinear transformation $l_{\text{relation}} = e_1 W_B e_2$ as the soft label between the two entities (e_1, e_2) to help model to select important features. Specifically, we obtain the feature c_j by:

$$m_j = \text{relu}(w_m \otimes S_{(j-w+1):j} + l_{\text{relation}} + b_m) \tag{2}$$

$$n_j = \tanh(w_n \otimes S_{(j-w+1):j} + b_n) \tag{3}$$

$$c_j = m_j \times n_j \tag{4}$$

The ability to capture different features typically requires the use of multiple filters in the convolution, so we use n filters $W = \{w_1, w_2, \cdots, w_n\}$. The convolution result is:

$$c_{ij} = m_{i,j} \times n_{i,j} \tag{5}$$

where $1 \leq i \leq n$, $1 \leq j \leq |s| - w + 1$. The overall output result of the gated convolution layer is $C = \{c_1, c_2, \cdots, c_n\}$

2.4 Pooling

Piecewise Max Pooling. Max pooling operation is usually used to extract the most dominant features in feature maps, but ignores the structure information and fine-grained information. Thus, PCNNs divides an instance into three segments according to the given entity pair and does max pooling operation on each segment. For each convolutional result C_i, it can be divided into $C_i = \{c_{i,1}, c_{i,2}, c_{i,3}\}$, then the piecewise max pooling process is defined as $p_{ij} = \max(c_{i,j})$, where $1 \leq i \leq n, j = 1, 2, 3$. We concatenate all vectors $p_i = [p_{i,1}, p_{i,2}, p_{i,3}]$ as $P \in \mathbb{R}^{3n}$.

Attention Mechanism. To select the positive feature more precisely, we propose an attention mechanism to help our model focus on positive features at high level. Inspired by Wu et al. (2018), we adopt the attention mechanism after the piecewise max pooling layer (denoted as ATT-P) to obtain more positive high-level features. For getting more positive related features, we also adopt the soft-label strategy here:

$$\alpha_H^j = \frac{\exp(\eta^j)}{\sum_k \exp(\eta^k)} \tag{6}$$

$$\eta^j = \boldsymbol{p}_i^j \mathbf{A}(e_1 \mathbf{W}_B e_2) \tag{7}$$

where $\mathbf{A}$ and $\mathbf{W}_B$ is weighted matrices. Then the representation $\gamma \in \mathbb{R}^{3n}$ of the sentence s is $\gamma = \sum_{j=1}^{3} \alpha_H^j \boldsymbol{p}_i^j$.

Finally, we obtain the feature vector $\boldsymbol{b}_s \in \mathbb{R}^{3n}$ of the sentence s from:

$$\boldsymbol{b}_s = \tanh(\gamma) \tag{8}$$

2.5   Sentence-Level Attention and Output

In previous studies, bilinear and nonlinear attention mechanisms have been proved helpful to model performance. Considering the computational efficiency and effectiveness, we adopt the non-linear form in our method.

We also use $l_{relation} = e_1 \mathbf{W}_B e_2$ as relation labels between two entities. For the feature vector $\boldsymbol{b}_i^j$ of the j-th sentence in the i-th bag B_i, the corresponding attention weight α^j is calculated as follows:

$$\alpha^j = \frac{\exp(\omega^j)}{\sum_k \exp(\omega^k)} \tag{9}$$

$$\omega^j = \mathbf{W}_a^T(\tanh[\boldsymbol{b}_i^j; l_{relation}]) + b_a \tag{10}$$

The final feature vector of B_i is expressed as $\boldsymbol{r}_i = \sum_{j=1}^{|B_i|} \alpha^j \boldsymbol{b}_i^j$.

The vector representation of the bag is then fed to the softmax classifier to predict the final relation labels and calculate the cross-entropy objective function on all training bags (T):

$$J(\theta) = -\sum_{i=1}^{T} \log p(y_i | \boldsymbol{r}_i; \theta) \tag{11}$$

3 Experiments

3.1 Datasets and Evaluation Metrics

We evaluate our approach on a widely used dataset which is developed by Riedel et al. (2010). This dataset is generated by aligning the relations in Freebase with the New York Times corpus (NYT). We use aligned sentences from 2005 to 2006 as training data and sentences from 2007 as testing data. The dataset has 53 kinds of relation labels including label NA which means that there is no relation between entity pairs. The training data includes 522,611 sentences, 281,270 entity pairs and 18,252 relation facts. The testing data includes 172,448 sentences, 96,678 entity pairs and 1,950 relational facts.

We use the held-out evaluation to evaluate our model. It provides an approximate precision measurement method without time-consuming manual evaluation by comparing the relation instances extracted from bags against Freebase relations data automatically. We will show the aggregated P-R curve (Precision/recall Curve) and Precision@N (precision at top n predictions) in the experiments.

3.2 Parameter Settings

In our experiments, we use the word2vec tool (Mikolov et al. 2013) to pre-train the word embeddings on NYT corpus. We tune all of the models using three-fold validation on the training set. We select the dimension of word embeddings d^w among {50, 100, 200, 300}, the dimension of position embeddings d^p among {5, 10, 20}, the window size w among {3, 5, 7}, the number of filters n among {50, 100, 230, 300}, batch size among {50, 100, 160}, the learning rate λ among {0.001, 0.01, 0.1, 0.5}. The best configurations are: $d^w = 50$, $d^p = 5$, $w = 3$, n = 300, $\lambda = 0.1$, the batch size is 50. We use dropout strategy and Adadelta to train our models. According to experience, the dropout rate is fixed to 0.5.

3.3 Performance Evaluation

We compare our method with three previous works: PCNNs + MIL (Zeng et al. 2015) selects the sentence with the highest score as the representation of a bag; PCNNs + ATT (Lin et al. 2016) and APCNNs (Ji et al. 2017) use sentence-level bilinear and nonlinear form attention to synthesize all sentences' information in a bag as it's representation. Moreover, we add GAU module to PCNNs + MIL (denoted as PCNNs + MIL + GAU) to verify the effectiveness of GAU. In addition, we remove ATT-P module (denoted as GPCNNs) to testify the contribution of ATT-P to model. At the same time, we replace GAU with GTU (denoted as PCNNs + GTU) to prove the validity of our soft-label strategy. Figure 3 shows the aggregated P-R curves, and Table 1 shows the Precision@N with N = {100, 200, 300} of our approach and all the baselines.

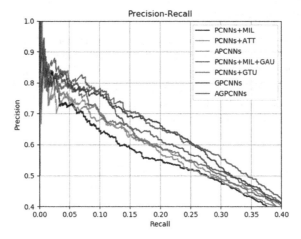

Fig. 3. Aggregate precision/recall curves for AGPCNNs and all the baseline models. For the sake of clarity, we show all the curves with different colors and bold lines.

Table 1. Precision@N of our model and all the baseline models.

Precision@N (%)	Top 100	Top 200	Top 300	Average
PCNNs + MIL	72.89	69.23	64.05	68.72
PCNNs + ATT	74.26	72.14	68.44	72.61
APCNNs	76.24	74.13	69.44	73.27
PCNNs + MIL + GAU	81.19	77.61	74.42	77.74
PCNNs + GTU	78.22	76.62	68.77	74.54
GPCNNs	83.16	77.11	74.09	78.12
AGPCNNs	**83.17**	**79.10**	**75.08**	**79.12**

From Fig. 3 and Table 1 we have the following observations:

(1) AGPCNNs achieves the best P-R curve over baselines. And its Precision@N values are the highest, which are about 5% higher than baselines on average. This indicates that AGPCNNs is effective because GAU and ATT-P module can select more important inner-sentence features at fine-grained level.

(2) All the models with GAU outperform that without GAU. It demonstrates that sentence-level noise filtering methods combined with the gate mechanism can obtain more positive features than those denoising at sentence level only, which verifies the effectiveness of the GAU module.

(3) AGPCNNs performs much better than PCNNs + GTU. It means that the gate mechanism can select important features more precisely with the help of the soft-label strategy, and also verifies the bilinear transformation between entity pairs can map their relation effectively.

(4) Integrated with ATT-P module, AGPCNNs achieves more improvements than GPCNNs. It shows that ATT-P can obtain helpful high-level global features and resist noise further.

3.4 Effectiveness of GAU Module

In order to verify the effectiveness of the gate mechanism in word-level feature selection more intuitively, we (1) remove all the attention mechanisms of AGPCNNs (denoted as PCNNs + GAU); (2) apply GAU to CNNs (we name it as CNNs + GAU). CNNs and PCNNs are used as baselines in this experiment. Note that these methods are sentence-level extraction, different from bag-level extraction in Sect. 3.1. The experimental results are shown in Fig. 4 below:

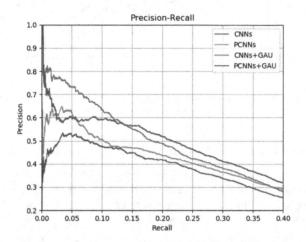

Fig. 4. Aggregate precision/recall curves for a variety of models on sentence-level extraction.

From Fig. 4 we can see:

(1) Compared with PCNNs and CNNs, PCNNs + GAU and CNNs + GAU have achieved significant improvement, indicating that GAU module can effectively improve the performance of the model with different pooling mechanisms, verifying the effectiveness of the gate mechanism and reflecting the robustness of the GAU module.
(2) The performance of PCNNs + GAU is significantly improved compared with PCNNs, indicating that the word-level noise has a great impact on the performance of the RE models, and also demonstrating the GAU module can effectively filter word-level noise by obtaining more important features.

3.5 Case Study

To explicitly illustrate the adverse impact of word-level noise on feature selection and the effectiveness of our proposed model, we show an example of attention weights in a bag during testing. As shown in Table 2, all sentences contain the phrase *[president emeritus of]*, which clearly indicates the/*company* relation between the entities *John Brademas* and *New York University*, so they are all positive instances.

Table 2. An example of attention weights. The bold strings are head/tail entities and the underlined strings are keywords to predict the relation. The relation/*company* corresponds the/ *business/person/company* in Freebase.

Triplet	Instances	APCNNs	AGPCNNs
(John Brademas, New York University/company)	1. Thirty-five years ago, President vetoed the legislation, refusing to encourage "the family-centered child rearing". "I don't think we've ever recovered from that veto message." said **John Brademas**, <u>president emeritus of</u> **New York University**, and, as a former Democratic congressman from Illinois, a sponsor of that legislation	0.085	0.354
	2. An article on Jan.11 about a conference in New York misidentified the home state of **John Brademas**, <u>president emeritus of</u> **New York University**	0.665	0.324
	3. Correction: January 25, 2006, Wednesday An article on Jan.11 about a conference in New York misidentified the home state of **John Brademas**, <u>president emeritus of</u> **New York University**	0.249	0.322

It can be seen that, except the phrase *[president emeritus of]*, other words do not have direct or indirect connection with the entity pair relation/*company*, which means these words are noisy. Compared with the other two sentences, the first sentence contains the most amount of noisy information, the third sentence contains the second, and the second sentence the least. Due to the lack of effective inner-sentence feature selection mechanisms, the APCNNs model assigns the higher weight to the second sentence (0.665) and the lesser weight to the first sentence and the third sentence (0.085, 0.249). However, AGPCNNs model is barely affected by noisy information, and assigns similar weights to the three sentences (0.354, 0.324, 0.322). The attention weights verify that our proposed model can effectively select more important word-level features no matter how much the noisy information sentences contains and make full use of the supervision information in a bag.

4   Related Work

Due to the high costs of manual annotation, distant supervision plays an increasingly important role in RE. However, this method faces the challenge brought by the wrong labeling problem, resulting in being prone to generating lots of noise instances. For this, Riedel et al. (2010) modeled distant supervised RE as a single labeling problem by using multi-instance learning. The following research work (Hoffmann et al. 2011; Surdeanu et al. 2012) adopted multi-instance multi-label learning and used a proba-bilistic graphical model to select sentences. However, all of the above methods rely heavily on the quality of features generated by NLP tools and are deeply troubled by the problem of error propagation.

As neural networks have been widely used and achieved good results in many tasks, Zeng et al. (2015) proposed PCNNs with MIL to select the most likely positive sentences. Lin et al. (2016) use selective attention over instance with PCNNs to select valid sentences. Ji et al. (2017) assign more precise attention weights by making use of entity descriptions. Focused on the imbalance of datasets, a label-free method has been proposed by Wang et al. (2018). Besides, reinforcement learning has been used to select the valid instances before training for relation extraction (Feng et al. 2018; Qin et al. 2018). However, all the above approaches filter noise at the sentence level, ignoring the word-level (inner-sentence) noise, resulting in the insufficient use of the supervision information in a bag. On the other hand, the fixed relation labels (hard labels) of entity pairs during training also enlarge the influence of the wrong labels.

Different from the existing researches, we propose an AGPCNNs model, which uses a gate mechanism in convolution layer and an attention mechanism after piecewise pooling layer to reduce word-level noise by selecting more important inner-sentence features. Furthermore, we also introduce a soft-label strategy (Wang et al. 2018; Kalchbrenner et al. 2016; Gehring et al. 2017) in our model by adopting bilinear transformation results of entity pairs as relation labels, making feature selection more precise. Experimental results verify the effectiveness of the AGPCNNs model. In addition, Liu et al. (2018) also proposed a word-level noise filtering method. The differences between our model and theirs lie in: (1) Liu et al. (2018) used NLP tools to build dependency subtrees, and introduced external knowledge to filter word-level noise through transfer learning. While our model only uses two mechanisms (gate mechanism and attention mechanism), and does not use external tools and knowledge; (2) The soft-label strategy is introduced into our model, which weakens the impact of wrong labels on feature selection.

5   Conclusions and Future Work

Aiming at tackling the low-quality corpus problem, we propose a novel distant supervised approach for relation extraction, named AGPCNNs, which uses gate mechanism, attention mechanism and soft-label strategy to cut down word-level noise by valid inner-sentence feature selection. The Gate mechanism can effectively select word-level features extracted by convolution layer. The Attention mechanism is adopted after piecewise pooling layer to obtain high-level global feature relevant to relation labels. The Soft-label strategy is introduced to improve the accuracy of feature selection by using bilinear transformation results between entity pairs to conduct the selection process. The experimental results show that our model is superior to all baseline systems and achieves the best results.

In the future, we will incorporate reinforcement learning to filter noise from different aspects. Meanwhile, we will introduce the external prior knowledge to explore ways to improve the performance of relation extraction further.

Acknowledgments. The authors are supported by the National Nature Science Foundation of China (Nos. 61473294, 61370130 and 61876198), the Fundamental Research Funds for the Central Universities (Nos. 2015JBM033), and the International Science and Technology Cooperation Program of China (No. K11F100010).

References

Mintz, M., Bills, S., Snow, R., Jurafsky, D.: Distant supervision for relation extraction without labeled data. In: Proceedings of the Joint Conference of the 47th Annual Meeting of the ACL and the 4th International Joint Conference on Natural Language Processing of the AFNLP, pp. 1003–1011 (2009)

Riedel, S., Yao, L., McCallum, A.: Modeling relations and their mentions without labeled text. In: Machine Learning and Knowledge Discovery in Databases, pp. 148–163 (2010)

Dietterich, T.G., Lathrop, R.H., Lozano-Pérez, T.: Solving the multiple instance problem with axis-parallel rectangles. Artif. Intell. **89**(1–2), 31–71 (1997)

Hoffmann, R., Zhang, C., Ling, X., Zettlemoyer, L., Weld, D. S.: Knowledge-based weak supervision for information extraction of overlapping relations. In: Proceedings of ACL, pp. 541–550. Association for Computational Linguistics (2011)

Surdeanu, M., Tibshirani, J., Nallapati, R., Manning, C.D.: Multi-instance multi-label learning for relation extraction. In: Proceedings of the 2012 Joint Conference on Empirical Methods in Natural Language Processing and Computational Natural Language Learning, pp. 455–465. Association for Computational Linguistics (2012)

Zeng, D., Liu, K., Chen, Y., Zhao, J.: Distant supervision for relation extraction via piecewise convolutional neural networks. In: Proceedings of EMNLP, pp. 1753–1762 (2015)

Lin, Y., Shen, S., Liu, Z., Luan, H., Sun, M.: Neural relation extraction with selective attention over instances. In: Proceedings of the 54th Annual Meeting of the Association for Computational Linguistics (Volume 1: Long Papers), pp. 2124–2133 (2016)

Ji, G., Liu, K., He, S., Xu, L., Zhao, J.: Distant supervision for relation extraction with sentence-level attention and entity descriptions. In: AAAI, pp. 3060–3066 (2017)

Hochreiter, S., Schmidhuber, J.: Long short-term memory. Neural Comput. **9**(8), 1735–1780 (1997)

Wu, W., Chen, Y., Xu, J., Zhang, Y.: Attention-Based Convolutional Neural Networks for Chinese Relation Extraction. In: Chinese Computational Linguistics and Natural Language Processing Based on Naturally Annotated Big Data, pp. 147–158. Springer, Cham (2018)

Wang, G., Zhang, W., Wang, R., Zhou, Y., Chen, X., Zhang, W., Zhu, H., Chen, H.: Label-free distant supervision for relation extraction via knowledge graph embedding. In: Proceedings of EMNLP, pp. 2246–2255 (2018)

Kalchbrenner, N., Espeholt, L., Simonyan, K., Oord, A., Graves, A., Kavukcuoglu, K.: Neural machine translation in linear time. arXiv preprint arXiv, 1610.10099 (2016)

Gehring, J., Auli, M., Grangier, D., Yarats, D., Dauphin, Y.: Convolutional sequence to sequence learning. In: Proceedings of the 34th International Conference on Machine Learning-Volume 70, pp. 1243–1252. JMLR. org (2017)

Mikolov, T., Chen, K., Corrado, G., Dean, J.: Efficient estimation of word representations in vector space. arXiv preprint arXiv, 1301.3781 (2013)

Liu, T., Zhang, X., Zhou, W., Jia, W.: Neural relation extraction via inner-sentence noise reduction and transfer learning. In: Proceedings of EMNLP, pp. 2195–2204 (2018)

Feng, J., Huang, M., Zhao, L., Yang, Y., Zhu X.: Reinforcement learning for relation classification from noisy data. In: AAAI, pp. 5779–5786 (2018)

Qin, P., Xu, W., Wang, W.Y.: Robust distant supervision relation extraction via deep reinforcement learning. In: Proceedings of ACL, pp. 2137–2147 (2018)\

Relation and Fact Type Supervised Knowledge Graph Embedding via Weighted Scores

Bo Zhou[1,2], Yubo Chen[1], Kang Liu[1,2(✉)], and Jun Zhao[1,2]

[1] National Laboratory of Pattern Recognition, Institute of Automation, Chinese Academy of Sciences, Beijing 100190, China
{bo.zhou,yubo.chen,kliu,jzhao}@nlpr.ia.ac.cn
[2] University of Chinese Academy of Sciences, Beijing 100049, China

Abstract. Knowledge graph embedding aims at learning low-dimensional representations for entities and relations in knowledge graph. Previous knowledge graph embedding methods use just one score to measure the plausibility of a fact, which can't fully utilize the latent semantics of entities and relations. Meanwhile, they ignore the type of relations in knowledge graph and don't use fact type explicitly. We instead propose a model to fuse different scores of a fact and utilize relation and fact type information to supervise the training process. Specifically, scores by inner product of a fact and scores by neural network are fused with different weights to measure the plausibility of a fact. For each fact, besides modeling the plausibility, the model learns to classify different relations and differentiate positive facts from negative ones which can be seen as a muti-task method. Experiments show that our model achieves better link prediction performance than multiple strong baselines on two benchmark datasets WN18 and FB15k.

Keywords: Knowledge graph embedding · Relation supervised · Fact type supervised · Weighted scores

1 Introduction

Knowledge graphs can be regarded as large knowledge bases (KBs) which consist of structured triples in the form (entity, relation, entity). There are many KBs, such as DBpedia [14], YAGO [24] and Freebase [1] which can offer great help in many natural language processing applications such as relation extraction [8,21,28], question answering [2,4,6] and machine reading comprehension [30]. However, these KBs are far from complete, that is to say, many valid facts aren't contained in the KBs. Therefore, many researches have been focused on the task *knowledge base completion* which aims to predict the tail entity when given the head entity and relation, or vice versa.

In order to conduct the *knowledge base completion* task, different models have been proposed in recent years. Roughly, these can be divided into two categories

© Springer Nature Switzerland AG 2019
M. Sun et al. (Eds.): CCL 2019, LNAI 11856, pp. 258–267, 2019.
https://doi.org/10.1007/978-3-030-32381-3_21

[26], one is *Translational Distance Models* and they measure the plausibility of a fact as the distance between the two entities, usually after a translation carried out by the relation. The other is *Semantic Matching Models* and they measure plausibility of facts by matching latent semantics of entities and relations embodied in their vector space representations.

These models can learn good representation for entities and relations in KBs and perform well in knowledge base completion task. But there are two problems in existing methods, one is that they just use one score to measure the plausibility for each fact triple. Take the classic model TransE [3] for example, for each fact triple (h, r, t), the score is represented as $f_r(h, t) = -\|h + r - t\|_2$ by looking up the embedding table and the score is used for subsequent training. We argue that the only one score is too simple to make full use of the latent semantics of entities and relations encoded in the low-dimensional vector representation. The other problem is that they ignore the relation type and don't explicitly use the fact type, most models just minimize a loss like the pairwise ranking loss to encourage positive triples to get high scores and negative ones to get low scores.

To solve the above two problems, we propose a model to fuse different scores of a fact and utilize relation and fact type information. First, scores by inner product of a fact and scores by neural network are fused with different weights. Then for each fact, besides obtaining a score, we use multi-task learning to learn a classifier to differentiate positive facts from negative ones. At last, we add a relation classification loss and a fact type classification loss to the multi-task learning loss to jointly train the model.

In summary, our contributions in this paper are as follows:

- We propose a model to fuse different scores of a fact which can fully utilize the latent semantics of entities and relations.
- In order to make better use of relation and fact type information, we use muti-task learning to simultaneously train a relation classifier and fact classifier besides modeling the scores of a triple.
- We evaluate our model on two benchmark datasets and our model achieves better link prediction performance than multiple strong baselines. Besides, we conduct ablation study which shows the effectiveness of both the relation and fact type information and the weighted scores.

2   Related Work

Translational Distance Models use additive functions over embeddings to obtain a score. TransE [3] is the first model to introduce translation-based embedding, which represents both entities and relations as real vectors of same length. It assumes $h + r \approx t$ and minimizes $f_r(h, t) = \|h + r - t\|_{1/2}$. TransH [27] introduces relation-specific hyperplanes in order to better model the 1-to-N, N-to-1 and N-to-N relations which can't be well dealt with in TransE. TransR [15] is similar to TransH, but it introduces relation-specific spaces instead of hyperplanes. In order to reduce the number of parameters, TransD [9] decomposes the relation-specific matrix in TransR into product of two vectors. TranSparse [10] also reduces the parameters in TransR by utilizing sparse relation-specific matrix. There are also some works considering the uncertainty in knowledge graph and modeling entities and relations as random vectors [7,29].

Semantic Matching Models use product-based functions over embeddings to obtain a score. RESCAL [20] represents each entity as a vector and each ralation as a matrix. It defines the score of a triple by $f_r(h,t) = \boldsymbol{h}^\top \boldsymbol{M}_r \boldsymbol{t}$ rather than the translational distance in TransE. In order to reduce the number of parameters, DistMult [31] replace the relation matrix in RESCAL with a diagonal matrix. ComplEx [25] extends DistMult by introducing complex-valued embeddings which can better model asymmetric relations. There are also some works using neural netwok to calculate the score for each triple. MLP [5] first maps embeddings of entities and relations into hidden representations which will be added up, then the score is obtained by dot product. ConvKB [17] uses CNN to produce feature maps and then calculated the score by dot product.

3   The Proposed Model

A knowledge graph $\mathcal{G}$ can be seen as a set which contains valid tiples (head entity, relation, tail entity) denoted as (h, r, t) such that $h, t \in \mathcal{E}$ and $r \in \mathcal{R}$ where $\mathcal{E}$ is a set of entities and $\mathcal{R}$ is a set of relations. Each embedding model aims to define a score for a triple such that valid triples get higher scores than invalid ones. Table 1 gives score functions of some previous SOTA models.

Table 1. The score functions in previous SOTA models. $\langle \boldsymbol{v}_h, \boldsymbol{v}_r, \boldsymbol{v}_t \rangle = \sum_i \boldsymbol{v}_{h_i} \boldsymbol{v}_{r_i} \boldsymbol{v}_{t_i}$ denotes a tri-linear dot product. *Re* is an operation to take the real part of a complex number.

Model	The score function $f(h, r, t)$
TransE	$-\left\| \boldsymbol{v}_h + \boldsymbol{v}_r - \boldsymbol{v}_t \right\|_{1/2}$
ComplEx	$Re\left(\langle \boldsymbol{v}_h, \boldsymbol{v}_r, \overline{\boldsymbol{v}}_t \rangle \right)$
DistMult	$\langle \boldsymbol{v}_h, \boldsymbol{v}_r, \boldsymbol{v}_t \rangle$
MLP	$\boldsymbol{w}^\top \tanh\left(\boldsymbol{M}^1 \boldsymbol{h} + \boldsymbol{M}^2 \boldsymbol{r} + \boldsymbol{M}^3 \boldsymbol{t} \right)$

We have two lookup tables for entities in $\mathbb{R}^{|\mathcal{E}| \times d}$ named head table and tail table respectively and two lookup tables for relations in $\mathbb{R}^{|\mathcal{R}| \times d}$ named relation table and reverse relation table respectively, where d denotes the embbeding size.

Given a triple (h, r, t), we first look up the head table, relation table and tail table to get their corresponding embeddings denoted as a matrix $\boldsymbol{E} = [\boldsymbol{v}_h, \boldsymbol{v}_r, \boldsymbol{v}_t] \in \mathbb{R}^{d \times 3}$. Then $\boldsymbol{v}_r$ is input to a MLP to conduct relation classification:

$$\boldsymbol{p}_r = \text{softmax}\left(f\left(\boldsymbol{v}_r^T \boldsymbol{W}_r + \boldsymbol{b}_r \right) \right) \tag{1}$$

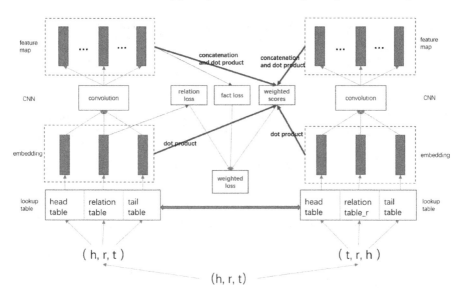

Fig. 1. Illustration of proposed model with relation and fact type supervision via weighted scores.

here $\boldsymbol{W}_r \in \mathbb{R}^{d \times n_r}$ is a weight matrix and $b_r \in \mathbb{R}^{n_r}$ is a weight vector, n_r is the number of relation type and $\boldsymbol{p}_r \in \mathbb{R}^{n_r}$ is a relation probability distribution for $\boldsymbol{v}_r$. Meanwhile, we get score $s_1 = \langle \boldsymbol{v}_h, \boldsymbol{v}_r, \boldsymbol{v}_t \rangle$.

Then the embedding matrix $\boldsymbol{E}$ is input to a convolutional neural network [13] to get different feature maps which will be concatenated to form a vector $\boldsymbol{m}$. Specifically, we have k filters and each is of shape $[f_h, 3]$. Thus each filter will produce a feature map of size $[d - f_h + 1, 1]$. After concatenation of these k feature maps we have $\boldsymbol{m}$ of shape $[k(d - f_h + 1), 1]$ and get score $s_2 = \langle \boldsymbol{m}, \boldsymbol{w} \rangle$ where $\boldsymbol{w}$ is a weight parameter. Meanwhile, another task to classify a fact is conducted with $\boldsymbol{m}$:

$$\boldsymbol{p}_f = \mathrm{softmax}\left(f\left(\boldsymbol{m}^T \boldsymbol{W}_f + \boldsymbol{b}_f \right) \right) \tag{2}$$

here $\boldsymbol{W}_f \in \mathbb{R}^{v_m \times n_f}$ is a weight matrix and $b_f \in \mathbb{R}^{n_f}$ is a weight vector, v_m(i.e., $k(d - f_h + 1)$) is the dimension of vector $\boldsymbol{m}$, n_f is the number of fact type(In our experiment n_f equals 2, namely the positive fact and the negative fact.) and $\boldsymbol{p}_f \in \mathbb{R}^{n_f}$ is a fact probability distribution for $\boldsymbol{m}$.

Similar to [11], for the triple (h, r, t) we also reverse it to get a triple (t, r, h), then we look up the head table, reverse relation table and tail table to get their corresponding embeddings denoted as a matrix $\boldsymbol{E}' = [\boldsymbol{v}'_t, \boldsymbol{v}'_r, \boldsymbol{v}'_h] \in \mathbb{R}^{d \times 3}$ and in the meantime we get score $s'_1 = \langle \boldsymbol{v}'_t, \boldsymbol{v}'_r, \boldsymbol{v}'_h \rangle$. The matrix $\boldsymbol{E}'$ will be input to another CNN of the same structure as described above to get score $s'_2 = \langle \boldsymbol{m}', \boldsymbol{w}' \rangle$. It should be noted that we don't do relation or fact type classification this time. Figure 1 illustrates the structure of our model.

With score s_1, s_2, s_1' and s_2', we define the weighted score function f of our model as follows:

$$f(h, r, t) = \lambda_1 s_1 + \lambda_2 s_2 + \lambda_3 s_1' + \lambda_4 s_2' \tag{3}$$

where λ_1, λ_2, λ_3 and λ_4 are weight parameters tuned on validation set.

The loss function $\mathcal{L}_1$ generated by the weighted scores with L_2 regularization on all the embedding vector e of the model is:

$$\mathcal{L}_1 = \sum_{(h,r,t) \in \{\mathcal{G} \cup \mathcal{G}'\}} \log\left(1 + \exp\left(l_{(h,r,t)} \cdot f(h, r, t)\right)\right) + \frac{\lambda}{2} \|e\|_2^2 \tag{4}$$

$$l_{(h,r,t)} = \begin{cases} 1 \text{ for } (h, r, t) \in \mathcal{G} \\ -1 \text{ for } (h, r, t) \in \mathcal{G}' \end{cases} \tag{5}$$

where $\mathcal{G}'$ is a set of triples generated by corrupting valid triples in $\mathcal{G}$.

We use cross entorpy loss $\mathcal{L}_2$ as our relation classification loss:

$$\mathcal{L}_2 = -\sum_{i=1}^{N} \sum_{j=1}^{n_r} \left(l_{ri}^j \log\left(p_{ri}^j\right)\right) \tag{6}$$

where l_r is the label vector of relation and N is the number of training instances. Similarly, the fact classification loss $\mathcal{L}_3$ is:

$$\mathcal{L}_3 = -\sum_{i=1}^{N} \sum_{j=1}^{n_f} \left(l_{fi}^j \log\left(p_{fi}^j\right)\right) \tag{7}$$

where l_f is the label vector of fact type and N is the number of training instances. With $\mathcal{L}_1$, $\mathcal{L}_2$ and $\mathcal{L}_3$, the total weighted loss $\mathcal{L}$ is:

$$\mathcal{L} = \alpha_1 \mathcal{L}_1 + \alpha_2 \mathcal{L}_2 + \alpha_3 \mathcal{L}_3 \tag{8}$$

where α_1, α_2 and α_3 are weight parameters tuned on validation set. We optimize our model by minimizing the total loss $\mathcal{L}$.

4   Experiments

4.1   Datasets

Our experiments are conducted on two benchmark datasets: WN18 and FB15k. WN18 is a subset of Wordnet [16] which is a KB whose entities (termed synsets) correspond to senses of words, and relationships between entities define lexical relations. FB15k is a subset of Freebase [1] which is a huge and growing KB for common facts with around 1.9 billion triplets. In our experiments, we use the same train/valid/test sets split as in [3]. The detailed statistics of WN18 and FB15k are showed in Table 2.

Table 2. Statistics of the experimental datasets.

Dataset	$|\mathcal{E}|$	$|\mathcal{R}|$	#Triples in train/valid/test		
WN18	40943	18	141442	5000	5000
FB15k	14951	1345	483142	50000	59071

4.2 Baselines

We compare our model with several previous methods. Our baselines include TransE, TransR, STransE [19], NTN [22], DistMult, ComplEx and SimpleE, some of them are strong baselines. We report the results of TransE, STransE, DistMult, and ComplEx from [25]. The results of TransR and NTN are reported from [18], and SimpleE is reported from [11].

4.3 Evaluation Metrics

The purpose of link prediction or KB completion task [3] is to predict a missing entity given the relation and another entity in the valid triple, i.e, predicting h giving (r, t) or predicting t giving (h, r). Then the results are evaluated based on the rankings of the scores calculated by the score function.

Specifically, for each valid triple, we first replace the head entity or tail entity randomly by all the other entity in $\mathcal{G}$ to produce a set of corrupted triples, i.e., the negative triple sets for the valid triple. Then the score function will be used to calculate all the scores of corrupted triples together with the valid triple and we rank the results based on the scores. We employ common metrics to evaluate the ranking list: mean reciprocal rank (MRR) (i.e., the reciprocal mean rank of the correct test triples), Hits@10 (i.e., the proportion of the correct test triples ranked in top 10 predictions), Hits@3 and Hits@1.

As pointed out in [3], the above corrupted triple set for each test triple may contain some valid triples in $\mathcal{G}$ and these valid triples may be ranked above the test triple. To avoid this, we follow [3] to remove from the corrupted triple set all the triples that appear in $\mathcal{G}$. The former is called Raw setting and the latter is called Filter setting. We then evaluate results with MR, MRR, Hit@10, Hit@3 and Hit@1 on the new ranking list.

4.4 Training Protocol

We use the common Bernoulli trick [15,27] to generate the head or tail entities when producing negative triples, i.e.,with probably p the head entity of a test triple is replaced and with probabily $1 - p$ the tail entity is replaced. We calculate MRR in both Raw and Filter setting and Hit@1, Hit@3 and Hit@1 in Filter setting.

In our experiment, the dimension d of entities and relations are all 200. We initialize all the lookup tables by uniform distribution between $\left[-\frac{6.0}{\sqrt{200}}, \frac{6.0}{\sqrt{200}}\right]$.

The filters of CNN are initialized by a uniform distribution $\left[-\frac{1}{\sqrt{303}}, \frac{1}{\sqrt{303}}\right]$. The number of filters for CNN is 3 and the filter size is $[3, 101]$. We also use Dropout [23] after the CNN and the dropout rate is 0.5. We choose ReLU as the activation function f. The L_2 regularizer λ is fixed at 0.03. The weight parameters λ_1, λ_2, λ_3, λ_4 α_1, α_2 and α_3 are set to 0.4, 0.2, 0.3, 0.1, 0.6, 0.2 and 0.2 respectively. Besides, we also try attention mechanism to dynamically tune these weight parameters in our experiment. For the WN18 dataset, the batchsize is 1415 and for the FB15k dataset the batchsize is 4832. For both datasets, we sample 1 negative triple for each positive triple. The Adam optimizer [12] is used to train our model with initial learning rate 0.1. We run our model on both datesets to 1000 epochs and the validation set is used to select the best model in these epochs to do test set evaluation.

4.5   Experimental Results

Table 3 compares the performance of our model with results of previous models, from which we can see that our model outperforms TransE, TransR, STransE, NTN and DistMult on both datasets for all the valuation metrics. This shows the effectiveness of our model.

Compared with ComplEx and SimplE, on WN18 dataset our model achieves best performance as for MRR and Hit@1. Hit@3 is slightly lower than ComlpEx and we obtain the same Hit@10 as ComlpEx and SimplE. On FB15k dataset, our model obtains lower MRR in the Filter setting and Hit@3 but we achieve best performance as for MRR in the Raw setting, Hit@1 and Hit@10. Besides, our model achieves significant improvement compared with all the baselines as for the MRR in the Raw setting on both datasets. This also shows the effectiveness of our model, in fact ComplEx and SimplE are strong baselines which are previous SOTA models.

Table 3. Results on WN18 and FB15k. Best results are in bold.

Model	WN18					FB15k				
	MRR		Hit@			MRR		Hit@		
	Filter	Raw	1	3	10	Filter	Raw	1	3	10
TransE	0.454	0.335	0.089	0.823	0.934	0.380	0.221	0.231	0.472	0.641
TransR	0.605	0.427	0.335	0.876	0.940	0.346	0.198	0.218	0.404	0.582
STransE	0.657	0.469	-	-	0.934	0.543	0.252	-	-	0.797
NTN	0.530	-	-	-	0.661	0.250	-	-	-	0.414
DistMult	0.822	0.532	0.728	0.914	0.936	0.654	0.242	0.546	0.733	0.824
ComplEx	0.941	0.587	0.936	**0.945**	**0.947**	0.692	0.242	0.599	0.759	0.840
SimplE	0.942	0.588	0.939	0.944	**0.947**	**0.727**	0.239	0.660	**0.773**	0.838
Ours	**0.943**	**0.596**	**0.940**	0.944	**0.947**	0.715	**0.258**	**0.661**	0.770	**0.843**

Ablation Study. Table 4 shows the results of our ablation study on WN18, from which we can see that the full model achieves the best performance as for MRR in both Filter and Raw setting. If we ablate Relation classifier or Fact classifier, the performance will degrade. After ablation of both classifier, the model obtains the lowest performance. This demonstrates the effectiveness of both the relation and fact type information for knowledge graph embedding task. Besides, after ablating the Relation classifier the performance degrades more than ablating the Fact classifier, which indicates that relation information is more important than fact type information for knowledge graph embedding task.

Table 4. Ablation study for WN18.

Ablation	MRR(Filter)	MRR(Raw)
Full model	0.943	0.596
Relation classifier	0.940	0.590
Fact classifier	0.940	0.594
Both	0.939	0.580

5   Conclusion

In this paper, we propose a novel model for the knowledge graph embedding task. The model fuses scores by inner product of a fact and scores by neural network with different weights to measure the plausibility of a fact. The model also learns a classifier to classify different relations and differentiate positive facts from negative ones which can be seen as a muti-task method when modeling the plausibility for each fact. Both of these can help to obtain better representations for entities and relations. Experiments show that our model achieves better performance in link prediction task than multiple strong baselines on two benchmark datasets WN18 and FB15k. In the future, we plan to use more complicated fusion strategy and further utilize relation and fact type information to better model the knowledge embedding task.

Acknowledgments. This work is supported by the National Natural Science Foundation of China (No. 61533018), the Natural Key R&D Program of China (No. 2017YFB1002101), the National Natural Science Foundation of China (No. 61806201, No. 61702512) and the independent research project of National Laboratory of Pattern Recognition. This work was also supported by CCF-Tencent Open Research Fund.

References

1. Bollacker, K., Evans, C., Paritosh, P., Sturge, T., Taylor, J.: Freebase: a collaboratively created graph database for structuring human knowledge. In: Proceedings of the 2008 ACM SIGMOD International Conference on Management of Data, pp. 1247–1250. ACM (2008)
2. Bordes, A., Chopra, S., Weston, J.: Question answering with subgraph embeddings. arXiv preprint arXiv:1406.3676 (2014)

3. Bordes, A., Usunier, N., Garcia-Duran, A., Weston, J., Yakhnenko, O.: Translating embeddings for modeling multi-relational data. In: Advances in Neural Information Processing Systems, pp. 2787–2795 (2013)

4. Bordes, A., Weston, J., Usunier, N.: Open question answering with weakly supervised embedding models. In: Calders, T., Esposito, F., Hüllermeier, E., Meo, R. (eds.) ECML PKDD 2014. LNCS (LNAI), vol. 8724, pp. 165–180. Springer, Heidelberg (2014). https://doi.org/10.1007/978-3-662-44848-9_11

5. Dong, X., et al.: Knowledge vault: a web-scale approach to probabilistic knowledge fusion. In: Proceedings of the 20th ACM SIGKDD International Conference on Knowledge Discovery and Data Mining, pp. 601–610. ACM (2014)

6. Hao, Y., et al.: An end-to-end model for question answering over knowledge base with cross-attention combining global knowledge. In: Proceedings of the 55th Annual Meeting of the Association for Computational Linguistics (Volume 1: Long Papers), pp. 221–231 (2017)

7. He, S., Liu, K., Ji, G., Zhao, J.: Learning to represent knowledge graphs with Gaussian embedding. In: Proceedings of the 24th ACM International on Conference on Information and Knowledge Management, pp. 623–632. ACM (2015)

8. Hoffmann, R., Zhang, C., Ling, X., Zettlemoyer, L., Weld, D.S.: Knowledge-based weak supervision for information extraction of overlapping relations. In: Proceedings of the 49th Annual Meeting of the Association for Computational Linguistics: Human Language Technologies-Volume 1, pp. 541–550. Association for Computational Linguistics (2011)

9. Ji, G., He, S., Xu, L., Liu, K., Zhao, J.: Knowledge graph embedding via dynamic mapping matrix. In: Proceedings of the 53rd Annual Meeting of the Association for Computational Linguistics and the 7th International Joint Conference on Natural Language Processing (Volume 1: Long Papers), vol. 1, pp. 687–696 (2015)

10. Ji, G., Liu, K., He, S., Zhao, J.: Knowledge graph completion with adaptive sparse transfer matrix. In: Thirtieth AAAI Conference on Artificial Intelligence (2016)

11. Kazemi, S.M., Poole, D.: Simple embedding for link prediction in knowledge graphs. In: Advances in Neural Information Processing Systems, pp. 4284–4295 (2018)

12. Kingma, D.P., Ba, J.: Adam: a method for stochastic optimization. arXiv preprint arXiv:1412.6980 (2014)

13. LeCun, Y., Bottou, L., Bengio, Y., Haffner, P., et al.: Gradient-based learning applied to document recognition. Proc. IEEE 86(11), 2278–2324 (1998)

14. Lehmann, J., et al.: Dbpedia-a large-scale, multilingual knowledge base extracted from Wikipedia. Semant. Web 6(2), 167–195 (2015)

15. Lin, Y., Liu, Z., Sun, M., Liu, Y., Zhu, X.: Learning entity and relation embeddings for knowledge graph completion. In: Twenty-Ninth AAAI Conference on Artificial Intelligence (2015)

16. Miller, G.A.: WordNet: a lexical database for English. Commun. ACM 38(11), 39–41 (1995)

17. Nguyen, D.Q., Nguyen, T.D., Nguyen, D.Q., Phung, D.: A novel embedding model for knowledge base completion based on convolutional neural network. arXiv preprint arXiv:1712.02121 (2017)

18. Nguyen, D.Q.: An overview of embedding models of entities and relationships for knowledge base completion. arXiv preprint arXiv:1703.08098 (2017)

19. Nguyen, D.Q., Sirts, K., Qu, L., Johnson, M.: STransE: a novel embedding model of entities and relationships in knowledge bases. arXiv preprint arXiv:1606.08140 (2016)

20. Nickel, M., Tresp, V., Kriegel, H.P.: A three-way model for collective learning on multi-relational data. ICML **11**, 809–816 (2011)
21. Riedel, S., Yao, L., McCallum, A., Marlin, B.M.: Relation extraction with matrix factorization and universal schemas. In: Proceedings of the 2013 Conference of the North American Chapter of the Association for Computational Linguistics: Human Language Technologies, pp. 74–84 (2013)
22. Socher, R., Chen, D., Manning, C.D., Ng, A.: Reasoning with neural tensor networks for knowledge base completion. In: Advances in Neural Information Processing Systems, pp. 926–934 (2013)
23. Srivastava, N., Hinton, G., Krizhevsky, A., Sutskever, I., Salakhutdinov, R.: Dropout: a simple way to prevent neural networks from overfitting. J. Mach. Learn. Res. **15**(1), 1929–1958 (2014)
24. Suchanek, F.M., Kasneci, G., Weikum, G.: YAGO: a core of semantic knowledge. In: Proceedings of the 16th International Conference on World Wide Web, pp. 697–706. ACM (2007)
25. Trouillon, T., Welbl, J., Riedel, S., Gaussier, É., Bouchard, G.: Complex embeddings for simple link prediction. In: International Conference on Machine Learning, pp. 2071–2080 (2016)
26. Wang, Q., Mao, Z., Wang, B., Guo, L.: Knowledge graph embedding: a survey of approaches and applications. IEEE Trans. Knowl. Data Eng. **29**(12), 2724–2743 (2017)
27. Wang, Z., Zhang, J., Feng, J., Chen, Z.: Knowledge graph embedding by translating on hyperplanes. In: Twenty-Eighth AAAI Conference on Artificial Intelligence (2014)
28. Weston, J., Bordes, A., Yakhnenko, O., Usunier, N.: Connecting language and knowledge bases with embedding models for relation extraction. arXiv preprint arXiv:1307.7973 (2013)
29. Xiao, H., Huang, M., Zhu, X.: TransG: a generative model for knowledge graph embedding. In: Proceedings of the 54th Annual Meeting of the Association for Computational Linguistics (Volume 1: Long Papers), vol. 1, pp. 2316–2325 (2016)
30. Yang, B., Mitchell, T.: Leveraging knowledge bases in LSTMS for improving machine reading. arXiv preprint arXiv:1902.09091 (2019)
31. Yang, B., Yih, W.T., He, X., Gao, J., Deng, L.: Embedding entities and relations for learning and inference in knowledge bases. arXiv preprint arXiv:1412.6575 (2014)

Leveraging Multi-head Attention Mechanism to Improve Event Detection

Meihan Tong[1,2,3], Bin Xu[1,2,3(✉)], Lei Hou[1,2,3], Juanzi Li[1,2,3], and Shuai Wang[1,2,3]

[1] DCST, Tsinghua University, Beijing 100084, China
{tongmh17,shuai-wa16}@mails.tsinghua.edu.cn,
{xubin,houlei,lijuanzi}@tsinghua.edu.cn
[2] KIRC, Institute for Artificial Intelligence, Tsinghua University, Beijing, China
[3] Beijing National Research Center for Information Science and Technology, Beijing, China

Abstract. Event detection (ED) task aims to automatically identify trigger words from unstructured text. In recent years, neural models with attention mechanism have achieved great success on this task. However, existing attention methods tend to focus on meaningless context words and ignore the semantically rich words, which weakens their ability to recognize trigger words. In this paper, we propose MANN, a multi-head attention mechanism model enhanced by argument knowledge to address the above issues. The multi-head mechanism gives MANN the ability to detect a variety of information in a sentence while argument knowledge acts as a supervisor to further improve the quality of attention. Experimental results show that our approach is significantly superior to existing attention-based models.

Keywords: Event detection · Multi-head attention mechanism · Knowledge enhancement

1 Introduction

Event Detection (ED) intends to detect and categorize event trigger at the sentence level, and often serves as a prerequisite of Event Extraction (EE) which needs to identify its corresponding arguments simultaneously [1]. It has a wide range of applications in information retrieval, text recommendation and text summarization.

Trigger ambiguity is the major challenge in Event Detection task. For example, in the sentence "Davies is leaving to **become** chairman of the London school of economics", "become" is a trigger word indicating a "Start_Position" event. However, in the sentence "BEGALA **become** honorable by winning an election of some sort or having a high post", "become" only acts as a normal verb and does not trigger any event. If the context information of the triggers cannot be effectively perceived, it is difficult to distinguish between the two situations and simply categorizing "become" as NEGATIVE classes.

M. Sun et al. (Eds.): CCL 2019, LNAI 11856, pp. 268–280, 2019.
https://doi.org/10.1007/978-3-030-32381-3_22

There are two branches of related researches addressing the problem, feature-based and representation-based models. Feature-based models [5,6,10] design syntactic, entity-related and trigger-related features to reduce semantic confusion and leverage various classifiers for prediction. Representation-based models [1, 3,11] determine the trigger-relevant information via dynamic-CNN, skip-CNN and memory-LSTM model and utilize full-connected neural networks to make prediction. Attention mechanism [7,9] is an effective technique that is often incorporated into the representation-based models to improve their performance in handling trigger ambiguity. However, existing attention models still suffer from two problems, mono-attention and knowledge absence.

Mono-Attention. Existing attention models [7,9] only learn one attention weight for each context word. We call this type of model as mono-attention model. The disadvantage of mono-attention models is that they fail to absorb various information contained in a sentence. As shown in Fig. 1, without capture all of the useful words("leaving", "chairman"), both of the mono-attention models mislabel "become" into NEGATIVE class. We argue that multi-head attention model is more suitable for Event Detection task. By learning multiple attention weights for each context word, multi-head attention model can capture all of the useful context words.

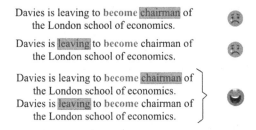

Fig. 1. Mono v.s. multiple attention. We intend to detect "become" as "Start_Position" event trigger. The first two attention weights are derived from the mono-attention model. The following attention weight is derived from our model, which can capture a wide range of information.

Knowledge Absence. Existing attention models [7] lack human knowledge to directly supervise the attention weight. They usually implicitly optimize the attention weight with the final event classification target, resulting in most proportion of the attention words are meaningless words. Figure 2 shows the difference between human attention and machine attention. Some of the researchers [9] utilize argument knowledge to guide the attention. If the model pays more attention on "chairman" (i.e., the *position* argument of the "Start-position" event), it can safely classify the trigger "become" into "Start-Position" type. We follow them to incorporate argument knowledge into our model, but the difference is that we need to supervise multiple attentions, they only need to supervise single attention.

Davies is leaving to become chairman of
the London school of economics.

Davies is leaving to become chairman of
the London school of economics.

Fig. 2. Knowledge absence example. The above sentence represents the attention of human, and the following sentence represents what the machine notices.

In this paper, we propose a novel multi-head attention model enhancing by arguments information named MANN to address the above issues. Multi-head attention mechanism allows the model to capture the context information from different aspects, and the arguments information guides the model to concentrate on human-concern words and filter out meaningless words.

MANN consists of three modules: query encoder obtains the semantic meaning of the candidate trigger and its directly adjacent words, context encoder models the sentence representation, and event detector concatenates them to feed the final classifier. Specifically, we model the multi-head attention using a scale-dot product attention network as the basic unit with each unit capturing one aspect of the sentence. Moreover, we propose an efficient unilateral supervision approach to integrate the external knowledge.

Finally, we conduct extensive experiments on the widely-used ACE2005 datasets, and the results demonstrate the effectiveness of MANN. Detailed investigation validates the effectiveness of the multi-head attention and argument knowledge integration mechanism. It is worth noting that the multi-head mechanism is more helpful for the promotion of long sentence trigger word detection.

Our contributions in this paper can be summarized as follows:

- We develop a novel multi-head attention mechanisms to capture multi-faced semantics within the sentence, including local information, action information and subject information. To the best of our knowledge, this is the first work that introduces the multi-head attention mechanism to event detection task.
- We incorporate external knowledge into the multi-head attention model via an efficient unilateral supervision approach, which significantly improves the quality of the attention.
- Extensive experiments on the ACE2005 corpus demonstrate the effectiveness of the proposed method, which achieves the best recall rate among the existing attention-based model.

2   Problem Definition

In this section, we formalize the event detection task. Before that, we first review several related basic concepts.

Definition 1 (Event). *As defined in ACE (Automatic Context Extraction) event extraction program, an event is composed of two elements: event trigger w and event arguments A. Trigger w is often a single verb or noun, most clearly* *expressing the event mention. Argument $a \in A$ could be an entity mention, temporal expression or value that serves as a participant or attribute with a specific role in an event mention.*

As shown in Fig. 2, "leaving" and "become" are triggers of the **End_Position** and **Start_Position** events respectively. "Davies" (Role = "People") is the argument of both events, and **Start_Position** event has two additional arguments: "chairman" (Role = "Position") and "school" (Role = "Entity").

Different from event extraction that is expected to extract both event trigger and arguments, event detection only detects and categorizes the event trigger, which is the focus of this paper. We first introduce the notations briefly, and then formally define the detection task.

Notations. *Given the corpus D with T sentences, each sentence $s \in D$ is denoted as a word sequence of length n, i.e., $s = \langle w_1, w_2, \ldots, w_n \rangle$. For the i-th word w_i, e_i, y_i denote its entity type and event type respectively. Because n varies in size, a window of size $L = 2l$ is introduced to fix the length. Without loss of generality, assuming w_i is the candidate trigger, then its word context and entity context are defined as $C_{w_i} = \{w_k|_{k=i-l}^{k=i+l}\}$ and $E_{e_i} = \{e_k|_{k=i-l}^{k=i+l}\}$. For convenience, we use w, C, E to represent the current candidate trigger, its word context and its entity context.*

Definition 2 (Event Detection). *Given the trigger candidate w, its word context C and entity context E, the goal is to figure out the event type y that w belongs to. Essentially, the model need to estimate the probability $p_\Phi(y|w, C, E)$.*

In this paper, we regard the detection task as a classification problem. To be specific, we treat all words in the sentence as trigger candidates. If the word is initially not an event trigger (without any event type label), we will classify it into NEGATIVE class.

3 Methodology

In this section, we present our proposed multi-head attention neural network (MANN). Figure 3 shows its architecture, it consists of three modules: Query Encoder (QE) is a fully-connected network, aiming to encode the candidate trigger and its surrounding words. Context Encoder (CE) is the core module. It has multiple attention layers. Each attention layer is designed to focus on one aspect of information. At each layer, the candidate trigger is regarded as the query word, and the rest words are regarded as the queried words. Then they are multiplied together to obtain the final contextual representation. Event Detector (ED) absorbs the representation learned in QE and CE and assigns an event label (including NEGATIVE) to the candidate trigger via a three-layer MLP model.

Next, we will introduce module implementation details, and then highlight the approach to integrate external knowledge.

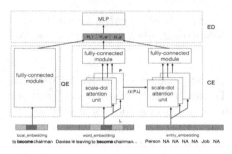 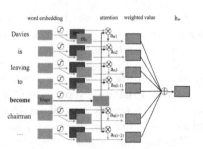

Fig. 3. The architecture of MANN   **Fig. 4.** The scale-dot product unit

3.1 Pre-trained Embedding

Representing word with distribution embedding has been proved to be useful in many NLP tasks. Compared with the one-hot representing, distributed representation is capable of capturing rich semantics between words. In this section, we discuss the way to obtain the word and entity embedding.

Word Embedding. We choose the New York Times (NYK) corpus for word embedding training because the articles in NYK is very similar to those in the ACE2005 corpus. Following [5], we obtain the pre-trained word embedding using Skip-Gram model with the window size as 5 and the minimum word frequency as 10. The final vocabulary size is 201,370, which is large enough to cover all the words in the ACE corpus. In order to preserve the semantic information in the pre-trained embedding, we fixed the word embedding in the following steps.

Entity Embedding. Entity information is quite useful for trigger disambiguation, e.g., the "End-Position" event cannot have a subject of "Country". To incorporate entity information into MANN, we adopt a uniform distribution to initialize the entity embedding and gradually optimize it during the training process.

3.2 Query Encoder

Query Encoder module aims to capture the local semantics of the candidate trigger. For the candidate trigger w, we concatenate its embedding with those of its left word w_{i-1} and right word w_{i+1}, and then feed the hidden representation into a one-layer fully-connected neural network.

$$H_l = W_q[w_{i-1}; w; w_{i+1}] + b_q \tag{1}$$

where; means the concatenation operation, H_l stands for the hidden representation of the candidate trigger w, W_q and b_q are parameters which need to be optimized during the training process.

3.3   Context Encoder

In this section, we leverage the multi-head attention mechanism to represent context words and entity information. As defined in Sect. 2, C and E represent the contextual words and entities of the candidate trigger w with fixed length L. In our model, we use w and C to calculate the word contextual representation h_w, and use w and E to calculate the entity contextual representation h_e.

Word Contextual. Given the candidate trigger w and its contextual words C, we exploit scale-dot attention network (SDA) to calculate the weight distribution of each attention head. To enhance the model flexibility, SDA adopts the key-value split technique. As illustrated in Fig. 4, the contextual words C are copied and assigned to C^k and C^v respectively. C^k is used in the process of calculating attention weights, as opposed to C^v in the process of calculating the final representation.

Here is the specific process of SDA. We first feed w, C^k, C^v into a linear network and activate them explicitly with the activation function. (We denote the dimensions of q, k, v as d_q, d_k, d_v respectively.)

$$
\begin{aligned}
q &= \sigma(W_q w + b_q) \\
\mathbf{k} &= \sigma(W_k \mathbf{C^k} + b_k) \\
\mathbf{v} &= \sigma(W_v \mathbf{C^v} + b_v)
\end{aligned}
\tag{2}
$$

Then, we multiply q and k to obtain the attention weight α_i for the i-th word in k. It is worth noting that we add a compression operation here and the scale of the compression is proportional to d_k. This operation intends to avoid the dot-product attention being too large [16], which might push the softmax function into regions with extremely small gradients.

$$
\begin{aligned}
\mathbf{s} &= \frac{q \cdot \mathbf{k}}{\sqrt{d_k}} \\
\alpha_i &= \frac{s_i}{\sum_{i=1}^{L} s_i}
\end{aligned}
\tag{3}
$$

Finally, we can obtain the word contextual representation by sum v with regard of α.

$$
h_w = \alpha \mathbf{v}^\mathrm{T}
\tag{4}
$$

Now, we have obtained the word contextual representation h_w from one scale-dot attention (SDA) unit. By repeating the process multiple times, we can obtain multiple word contextual representations from multiple scale-dot attention (SDA) units. Assuming we have P SDA units, then we will get P word contextual representations for the current candidate trigger w. We concatenate them into one embedding vector and feed the vector into a fully-connected network to obtain the final word contextual representation H_w.

$$
H_w = W_o[h_{w_1}; h_{w_2}; \ldots; h_{w_P}] + b_o
\tag{5}
$$

Entity Contextual. Similarly, we first transform the entity contextual E of current candidate trigger w into the hidden representation.

$$\mathbf{u} = \sigma(W_u \mathbf{E} + b_u) \tag{6}$$

Note that we do not use w's entity type as keyword to calculate the attention weight once again. Instead, we share the attention weight α between the word and entity context.

$$h_e = \alpha \mathbf{u}^{\mathrm{T}} \tag{7}$$

We do this for two reasons. On the one hand, most entity types are NEGATIVE, which cannot reflect the semantics of the query. On the other hand, entity embedding is randomly initialized and does not include any semantic information while word embedding contains a lot.

Finally, we feed the hidden representation h_e into multiple SDA units to obtain different entity contextual representation and integrate them via the linear network.

$$H_e = W_o[h_{e_1}; h_{e_2}; \ldots; h_{e_P}] + b_o \tag{8}$$

3.4   Event Detector

In this section, we aim to illustrate the event detector (ED) module. We will first introduce the basic ED model, then we analyze the characteristics of multi-head attentions and exploit external knowledge to supervise the attention.

Basic MLP Model. As illustrated in Fig. 3, we first splice the local representation H_l, word representation H_w and entity representation H_e, i.e., $H = \sigma(H_l \oplus H_c \oplus H_e)$, and then feed H into a three-layer perception model [4], which has proven to be very effective for event detection task [1,8].

Let $x = \langle w, C, E \rangle$ denote a training sample, where w, C, E represent the candidate trigger as well as its word and entity context, H denote its hidden representation (i.e., the input of the MLP model), MLP will output a result vector O, where the k-th entry of O represents the probability that x belongs to the corresponding event class. Specifically, the conditional probability is calculated by softmax function.

$$P(y_t | x_t) = \frac{exp(o_{tk})}{\sum_{k=1}^{K} exp(o_{tk})} \tag{9}$$

Given the input corpus $D = \{x_t, y_t\}|_{t=1}^{T}$, the negative loss function is defined as:

$$J(\theta) = -\sum_{t=1}^{T} \log p(y_{(t)} | x_{(t)}, \theta) \tag{10}$$

The model can be optimized using SGD with Adadelta rule. To prevent overfitting, we adopt L_2 norm and dropout mechanism.

3.5 Argument-Enhanced Event Detector

In this section, we introduce how to leverage the argument knowledge to guide the attention.

Gold Attention Generation. Liu [9] proved that argument information is very useful to supervise the attention. We follow them to use a Gaussian distribution to generate the gold attention.

Multi-head Attention Integration. Existing attention models normally have one attention layer, while MANN leverages multi-head attention mechanism. To facilitate external knowledge incorporation, we need to transform multiple attentions into one representation. Thus we develop the following two methods:

I1: Sum Supervision assumes that the arguments knowledge is generated after people comprehensively analyze all the information in the sentence. Under this assumption, we integrate multi-head attention by adding all the attention together.

I2: Unilateral Supervision assumes that argument words attention is an aspect of the multi-head attention. Without loss of generality, we choose the first head attention as argument words attention.

Jointly Detection Model. Given the gold attention α^* and the integrated attention α', we adopt Mean Square Error (MSR) to define the attention loss function.

$$A = \sum_{t=1}^{T}(\alpha_t^* - \alpha_t') \tag{11}$$

Combining the classification loss and the attention loss, we rewrite the final loss function as follows:

$$J'(\theta) = J(\theta) + \lambda A \tag{12}$$

where λ controls the weight of the supervised attention loss.

4 Experiment

In this section, we evaluate the proposed MANN using widely-used benchmark ACE2005. We will first introduce the experimental settings, then present the comparison results, and finally investigate some method details.

4.1 Experimental Settings

Datasets. As mentioned previously, ACE2005 is chosen as the evaluation benchmark and New York Times corpus is used for pre-trained word embedding training. ACE2005 task defines 8 superclasses and 33 subclasses of the event. In the experiment, we ignore the hierarchical information and regard it as a 34 multiple classification problem (33 subclasses and 1 NEGATIVE class). Note that we exclude the NEGATIVE class when computing the model precision.

Following [6], we split the dataset into training, validate and test sets with the size of 529/30/40.

Baselines. We denote the proposed method as MANN, and the argument knowledge integrated version as MANN-Aug. To validate the effectiveness, we compare our models to three kinds of mainstream models to demonstrate the advancement of our model. **CrossEvent**: a Max-Entropy model adopting document-level information [6]. **Combined-PSL**: a probabilistic soft logic model aiming to exploit global information, the best reported feature-based system [10]. **Skip-CNN**: a CNN model with non-continue n-grams as input [13]. **DLRNN**: a LSTM-based model extracting cross-sentence clues to improve sentence-level event detection [2]. **ANN-Aug**: a attention model with additional arguments information [9]. **GMLATT**: a gated multilingual attention approach. It is the best reported sentence-level attention approaches [7].

Training Details. The dim of input word embedding, word attention and output word attention are 200/300/400, and those for entity are all 100. The number of attention head P is 5 for MANN model and 10 for MANN-Aug, and the context window size L is set to 60. The training is conducted on a TitanX GPU with learning rate as $1e - 6$ and batch size as 100, the training time of each epoch is 80 ms in average and the optimal epoch number is around 100.

4.2   Overall Performance

Compared with feature-based methods, representation-based approaches perform better and attention-based model surpasses the representation-based methods and achieve the best F score in recent years. This is caused by the powerful representation ability of the attention mechanism.

Among the attention-based models, our approaches achieve the best F value. Besides, comparing with the model without external knowledge, incorporating the argument knowledge can improve the F score by 0.3%. Compared with the existing attention models like ANN, GMLATT, our model significantly improves the recall ($\geq$4%), which proves that our model can handle the mono-attention problem effectively (Table 1).

Table 1. Overall comparison results

Methods	P	R	F1
CrossEvent	68.7	68.9	68.8
Combined-PSL	75.3	64.4	69.4
DLRNN	77.2	64.9	70.5
Skip-CNN	n/a	n/a	71.3
ANN-Aug	78.0	66.3	71.7
GMLATT	78.9	66.9	72.4
MANN (our)	73.2	**72.3**	72.7
MANN-Aug (our)	76.3	69.8	**73.0**

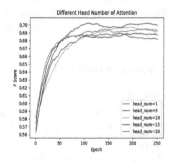

Fig. 5. The number of attention heads

4.3 The Effectiveness of Multi-head Attention

In this section, we discuss the effectiveness of multi-head attention in terms of (1) the length of sentence and (2) the number of attention heads.

The Length of Sentence. We first analyze how different sentence lengths influence the performance. We define short sentences as sentences of less than 10 words and long sentences as sentences of more than 10 words. Table 2 presents the results of our proposed MANN and a mono-attention model ANN.

Table 2. Results with different sentence lengths

Methods		Precision	Recall	F1
ANN	Short	100.0	66.7	80.0
	Long	71.5	67.3	69.3
MANN	Short	80.0	80.0	80.0
	Long	73.2	72.3	72.8

As we can see, both of the models perform better on short sentences. The reason is that the semantics of the short sentences are more uniform, reducing the ambiguity of the trigger words. Comparing with the short sentence, our model significantly improve the F1-value of long sentences(80%–80% v.s 69%–73%), which proves that the multi-head attention mechanism can better capture the multi-level and multi-angle information in long sentences.

The Number of Attention Heads. In order to illustrate the effect of the number of attention heads on the experimental results, we plot the training process of MANN with different heads in Fig. 5. For a more rigorous comparison, we use the MANN model without external knowledge. Experiment results show that the number of attention heads greatly influence the performance. As the number of attention heads increases, the optimal F1-value is gradually improved and the model can converge faster. However, it is worth noting that the model is more likely to encounter the over-fitting problem with a larger head number.

4.4 The Effectiveness of Argument Knowledge

We demonstrate the effectiveness of argument knowledge by comparing the performance of argument-enhanced MANN with basic MANN (MANN without argument knowledge). In order to ensure the rigor of the experiment, we set the head number as 10 and guarantee other hyper parameters are completely identical. Table 3 presents the comparison results.

Table 3. MANN with/without argument knowledge

Model	Precision	Recall	F1
MANN	74.3	69.0	71.6
MANN-Aug	**76.3**	**69.8**	**73.0**
MANN-Aug (I1)	69.8	**72.2**	71.0
MANN-Aug (I2)	**76.3**	69.8	**73.0**

As we can see from the Table 3, argument-enhanced MANN significantly outperform the basic MANN (71.6 v.s 73.0), which demonstrate the attention guide coming from argument knowledge is quite helpful. We further analyze the performance of argument-enhanced MANN by validating the effectiveness of I1/I2. Reviewing Sect. 3.5, I1/I2 are two approaches to integrate multi-head attention. As shown in Table 3, I2 consistently outperforms I1 (73.0 v.s 71.0). This indicates that argument information is just an aspect of attention that MANN takes into account when making decisions. If we sum all the attention and supervise them with arguments information (what I1 exactly does), MANN will lose the ability to discover other useful information.

5   Related Work

Early **Feature-based method** designs various features [6] by domain experts, and later researches exploit the maximum entropy classifier or SVM [15] to automatically distinguish important features. **Representation-based methods** have yielded impressive results in event detection. DMCNN [1] is the pioneering work that employs convolution neural network (CNN) in this task, followed by various CNN [12,13] and LSTM models [3,11]. However, these methods still have the problem of unexplainable and high computational costs. **Attention mechanism** is a widely-used technique, which has been successfully applied to machine translation [16] and text classification [14]. In the field of event detection, [9] and [7] leverage semantic and multilingual attention mechanism to address the trigger ambiguity issue. However, these methods still suffer from mono-attention and knowledge absence problem.

6   Conclusion

We propose a multi-head attention neural network (MANN-Aug) incorporating argument knowledge to address the mono-attention and knowledge absence problem. Extensive experiments on ACE2005 dataset show that our method significant outperforms existing attention-based methods.

Acknowledgements. This work is supported by the National Key Research and Development Program of China (2018YFB1005100 and 2018YFB1005101) and NSFC key projects (U1736204, 61533018, 61661146007), Ministry of Education and China Mobile Joint Fund (MCM20170301), a research fund supported by Alibaba Group, and THUNUS NExT Co-Lab. It also got partial support from National Engineering Laboratory for Cyberlearning and Intelligent Technology, and Beijing Key Lab of Networked Multimedia.

References

1. Chen, Y., Xu, L., Liu, K., Zeng, D., Zhao, J.: Event extraction via dynamic multi-pooling convolutional neural networks. In: The Meeting of the Association for Computational Linguistics, pp. 167–176 (2015)
2. Duan, S., He, R., Zhao, W.: Exploiting document level information to improve event detection via recurrent neural networks. In: Proceedings of the Eighth International Joint Conference on Natural Language Processing, pp. 352–361 (2017)
3. Feng, X., Qin, B., Liu, T.: A language-independent neural network for event detection. Sci. Chin. Inf. Sci. **61**(9), 092106 (2018)
4. Hagan, M.T., Demuth, H.B., Beale, M.H., De Jesús, O.: Neural Network Design, vol. 20. PWs Pub, Boston (1996)
5. Li, Q., Ji, H., Huang, L.: Joint event extraction via structured prediction with global features. In: Meeting of the Association for Computational Linguistics, pp. 73–82 (2013)
6. Liao, S., Grishman, R.: Using document level cross-event inference to improve event extraction. In: Proceedings of the 48th Annual Meeting of the Association for Computational Linguistics, pp. 789–797 (2010)
7. Liu, J., Chen, Y., Liu, K., Zhao, J.: Event detection via gated multilingual attention mechanism. Statistics **1000**, 1250 (2018)
8. Liu, S., Chen, Y., He, S., Liu, K., Zhao, J.: Leveraging FrameNet to improve automatic event detection. In: Proceedings of the 54th Annual Meeting of the Association for Computational Linguistics, pp. 2134–2143 (2016)
9. Liu, S., Chen, Y., Liu, K., Zhao, J.: Exploiting argument information to improve event detection via supervised attention mechanisms. In: Proceedings of the 55th Annual Meeting of the Association for Computational Linguistics, pp. 1789–1798 (2017)
10. Liu, S., Liu, K., He, S., Zhao, J.: A probabilistic soft logic based approach to exploiting latent and global information in event classification. In: AAAI, pp. 2993–2999 (2016)
11. Nguyen, T.H., Cho, K., Grishman, R.: Joint event extraction via recurrent neural networks. In: Proceedings of the 2016 Conference of the North American Chapter of the Association for Computational Linguistics: Human Language Technologies, pp. 300–309 (2016)
12. Nguyen, T.H., Grishman, R.: Event detection and domain adaptation with convolutional neural networks. In: Proceedings of the 53rd Annual Meeting of the Association for Computational Linguistics and the 7th International Joint Conference on Natural Language Processing, pp. 365–371 (2015)
13. Nguyen, T.H., Grishman, R.: Modeling skip-grams for event detection with convolutional neural networks. In: Proceedings of the 2016 Conference on Empirical Methods in Natural Language Processing, pp. 886–891 (2016)

14. Pappas, N., Popescu-Belis, A.: Multilingual hierarchical attention networks for document classification. CoRR (2017). http://arxiv.org/abs/1707.00896
15. Suykens, J.A.K., Vandewalle, J.: Least squares support vector machine classifiers. Neural Process. Lett. **9**(3), 293–300 (1999)
16. Vaswani, A., et al.: Attention is all you need. In: Advances in Neural Information Processing Systems, vol. 30, pp. 5998–6008 (2017)

Short-Text Conceptualization Based on a Co-ranking Framework via Lexical Knowledge Base

Yashen Wang[✉]

China Academy of Electronics and Information Technology of CETC, Beijing, China
yashen_wang@126.com

Abstract. The problem of short-text conceptualization is important, and has attracted increasing attention. Recent probabilistic algorithms have demonstrated remarkable successes. However most of them are limited to the assumption that all the observed terms in given short-text are conditionally independent, ignoring the interaction among terms (and concepts), as well as the beneficial reactions from concepts to terms. To overcome these problems, recently some co-rank paradigms are proposed, unfortunately neither they fails to integrate the co-occurrence feature nor they fails to utilize the semantic similarity implicit in the lexical knowledge base. Therefore, previous works could not release robust concept representation. Faced with this problem, this paper proposes a novel framework based on both statistic information (e.g., co-occurrence feature in large-scale corpus) and semantic information (e.g., semantic similarity in lexical knowledge base), for co-ranking terms and their corresponding concepts simultaneously, This co-ranking framework utilizes several graphs: the concept graph, the term graph and the subordination graph. The experimental results show that our method achieves higher accuracy and efficiency in short-text conceptualization than the state-of-the-art algorithms.

Keywords: Conceptualization · Co-ranking · Lexical knowledge base

1 Introduction

Shot-text conceptualization, is an interesting task to infer the most likely concepts for terms in the short-text, which could help better make sense of text data, and extend the texts with categorical or topical information [1,9,23,26,27,29]. It is a task to map a piece of short-text to a set of open domain concepts with different granularities [8,10,12,25]. Recent works on short-text understanding have put more emphasis on using signals from lexical knowledge bases to assist short-text conceptualization [9,27,31], and achieve great success. Many probabilistic (graph-based) algorithms have been proposed [10,21,23,29]. Generally, these kind of algorithms is closely integrated with the knowledge base, and the

© Springer Nature Switzerland AG 2019
M. Sun et al. (Eds.): CCL 2019, LNAI 11856, pp. 281–293, 2019.
https://doi.org/10.1007/978-3-030-32381-3_23

knowledge base has been demonstrated to be used to helping short-text understanding [27,28,31]. Given a short-text as input, we map each term to the corresponding candidate concepts defined in lexical knowledge base (e.g., Probase), and therefore a semantic graph is constructed based on the terms, concepts and the links among them. Note that, This semantic graph is heterogeneous [5,19,20], including three sub-graph: (i) the concept graph G_C connecting concepts (defined in the lexical knowledge base); (ii) the term graph G_T connecting terms (embedded in the short-text), and (iii) the subordination graph G_{TC} that ties the two previous graphs together.

As concluded in [10], after mapping the terms $T = \{t_i | i = 1, \ldots, n_T\}$ in given short-text to some candidate concepts in lexical knowledge base (e.g., Probase), previous works aim at estimating optimal set of concepts $C = \{c_j | j = 1, \ldots, k_C\}$, which maximizes conditional probabilities $P(c_j | T) \propto P(c_j) \prod_{i=1}^{n_T} P(t_i | c_j)^1$ based on the co-occurrence of terms and concepts under Naive Bayes assumption [21, 29]. In fact, they fail in mining the holistic concept-set for the entire short-text. The reasons are discussed as follows: (i) They assume that all the observed terms (and concepts) are conditionally independent, ignoring the beneficial reactions from concept to terms, which could reflect the global concepts, and simply regard the multiplication of conditional probabilities from each term as the likelihood of concept c_j [21,23]. (ii) Recently some co-rank paradigms are proposed to investigate the beneficial reactions among terms and concepts, unfortunately neither they fails to integrate the co-occurrence feature [17] nor they fails to utilize the semantic similarity implicit in the lexical knowledge base [10].

So as to overcome these problems, we must: (i) devise a framework that enables the signals (i.e., terms and concepts) to fully interplay to derive solid conceptualization for short-text; and (ii) combines *global* statistic information (e.g., co-occurrence feature from large corpus), *local* information (heuristic information implicit in context, i.e., correlation function) and *manual-defined knowledge* (semantic similarity in lexical knowledge base). Therefore, we propose a framework to co-rank terms and their concepts simultaneously in the concept graph (G_C), the term graph (G_T) and the subordination graph (G_{TC}). As a result, improved rankings of terms and their concepts depend on each other in a mutually reinforcing way, thus taking advantage of the additional information implicit in such heterogeneous graph of terms and concepts. The main intuition behind the co-ranking strategy is that, there is a mutually reinforcing relationship among concepts and terms that could be reflected in the rankings.

2   Preliminary

2.1   Problem Definition

Following [25], we define the notation "concept" as a set or class of "entities" or "things" within a domain, such that words belonging to similar classes get similar

---

[1] Notation n_T indicates the number of the terms occurring in this heterogeneous semantic graph, which will be discussed later.

representations. Probase [31] is used in our study as lexical knowledge base. Probase is widely used in research about short-text understanding [22,23,30] and text representation [10,27]. Probase uses an automatic and iterative procedure to extract concept knowledge from 1.68 billion Web pages. It contains 2.36 millions of open domain terms. Each term is a concept, an instance, or both. Meanwhile, it provides around 14 millions relationships with two kinds of important knowledge related to concepts: concept-attribute co-occurrence (isAttrbuteOf) and concept-instance co-occurrence (isA). Moreover, Probase provides huge number of high-quality and robust concepts without builds.

Given a short-text $S = \{t_i | i = 1, \ldots, n_T\}$, wherein t_i denotes a term, we could obtain the following results via short-text conceptualization: (i) concept distribution $\phi_C = \{\langle c_i, RS_C(i) \rangle | i = 1, \cdots, k_C\}$ from lexical knowledge base, wherein $RS_C(i)$ indicates the ranking score of concept c_i representing the importance of concept c_i contributing to model the entire semantic of the given short-text S (details in Sect. 3); and (ii) key-term distribution $\phi_T = \{\langle t_j, RS_T(j) \rangle | j = 1, \cdots, k_T\}$, wherein $RS_T(j)$ indicates the ranking score of term t_j representing the importance of term t_j contributing to model the entire semantic of the given short-text S. Through the above-mentioned definition, the essence of short text conceptualization is to map a given short-text to a concept space. This mapping process could filter out the incorrect concepts that are not suitable for the current given context, and then achieve the semantic disambiguation of polysemy.

2.2 Heterogeneous Semantic Graph

As discussed in the begining section, the proposed co-ranking framework operates on a heterogeneous semantic graph, which consists of three sub-graphs. Overall, we denote the heterogeneous semantic graph as $G = (V_C \cup V_T, E_{CC} \cup E_{TC} \cup E_{TT})$. Wherein, V_C is the set of candidate concepts with size of $n_C = |V_C|$, and V_T is the set of terms with size of $n_T = |V_T|$. E_{CC} is the set of links representing correlation ties among concepts, E_{TT} is the set of links among terms established by their co-occurrence relations, and E_{TC} is the set of links representing the subordination relations among terms and concepts. The overall heterogeneous semantic graph G is composed of three sub-graphs: (i) the Concept Graph $G_C = (V_C, E_C)$ respecting concepts, (ii) the term-correlation graph $G_T = (V_T, E_T)$ respecting terms, and (iii) the bipartite subordination graph $G_{TC} = (V_{TC}, E_{TC})$ that ties concepts (in G_C) and terms (in G_T) together.

2.3 Affinity Matrix

Overall, the proposed co-ranking framework is controlled by four affinity matrices. Note that, the affinity matrix is also reviewed as the transition matrix in Markov chain and is a stochastic matrix prescribing the transition probabilities from one vertex (concept or term) to the next, as discussed in the following Sect. 3. The affinity matrix $\mathbf{M}$ of G is defined as follows:

$$\mathbf{M} = \begin{bmatrix} \mathbf{M_{CC}} & \mathbf{M_{CT}} \\ \mathbf{M_{TC}} & \mathbf{M_{TT}} \end{bmatrix} \tag{1}$$

Wherein $\mathbf{M}_{CC}$ represents the correlation relationship among concepts (in G_C, defined in Sect. 2.2), and $\mathbf{M}_{TT}$ represents the co-occurrence relationship among terms (in G_W, defined in Sect. 2.2). $\mathbf{M}_{CT}$ and $\mathbf{M}_{TC}$ denote bipartite subordination (in G_{TC}, defined in Sect. 2.2), measuring how likely the given term is assigned with some concepts and vice versa.

Concept Graph (G_C): The Concept Graph $G_C = (V_C, E_C)$, representing the relatedness among candidate concepts associated with the given short-text, is a weighted undirected graph. The individual concept is denoted as $\{c_i | c_i \in V_C, i = 1, 2, \ldots, n_C\}$. $M_{CC}[i][j]$, the element of $\mathbf{M}_{CC}$, is derived by aggregating the co-occurrences between all instances of the two concepts c_i and c_j and the semantic similarity between the two concepts c_i and c_j. To achieve this goal, we firstly define the semantic similarity between concept c_i and concept c_j, as follows:

$$\text{sim}(c_i, c_j) = \frac{|T_{c_i} \cap T_{c_j}|}{|T_{c_i}|} \tag{2}$$

Wherein, T_{c_i} indicate the set of terms belong to concept c_i defined in lexical knowledge base Probase, and T_{c_j} could be defined in the same way. With efforts above, we could utilize the following equation to define $M_{CC}[i][j]$:

$$M_{CC}[i][j] = \frac{\eta_{CC} \cdot \sum_{t_p \in c_i, t_q \in c_j} n(t_p, t_q) + (1 - \eta_{CC}) \cdot \text{sim}(c_i, c_j)}{\sum_{l=1}^{n_C} [\eta_{CC} \cdot \sum_{t_p \in c_i, t_q \in c_l} n(t_p, t_q) + (1 - \eta_{CC}) \cdot \text{sim}(c_i, c_l)]} \tag{3}$$

Wherein, η_{CC} is the parameter controlling the weights about the co-occurrence feature and the semantic similarity feature. t_p and t_q are terms in vocabulary, and $n(t_p, t_q)$ represents the co-occurrence frequency between them through statistics. Furthermore, we could define correlation function for each pair of c_i and c_j resulting from their co-participation of term t_k:

$$\tau(c_i, c_j, t_k) = \frac{\amalg(M_{TC}[k][i] \neq 0, M_{TC}[k][j] \neq 0)}{|t_k|(|t_k| - 1)/2} \tag{4}$$

Wherein $\amalg(M_{TC}[k][i] \neq 0, M_{TC}[k][j] \neq 0)$ is the indicator function of whether term t_k could be mapped to concepts c_i and c_j simultaneously, and $|t_k|$ denotes the number of all the concepts related to t_k. Hence, adding up correlation function from all terms, we obtain:

$$M_{CC}[i][j] = \frac{[\eta_{CC} \cdot \sum_{t_p \in c_i, t_q \in c_j} n(t_p, t_q) + (1 - \eta_{CC}) \cdot \text{sim}(c_i, c_j)] \cdot \sum_{k=1}^{n_T} \tau(c_i, c_j, t_k)}{\sum_{l=1}^{n_C} [\eta_{CC} \cdot \sum_{t_p \in c_i, t_q \in c_l} n(t_p, t_q) + (1 - \eta_{CC}) \cdot \text{sim}(c_i, c_l)] \cdot \sum_{k=1}^{n_T} \tau(c_i, c_l, t_k)} \tag{5}$$

Term Graph (G_T): We segment the given short-text into a set of terms $\{t_i | t_i \in V_T, i = 1, 2, \ldots, n_T\}$, by utilizing the Probase [29] as our lexicon. The Term Graph $G_T = (V_T, E_T)$ is an weighted undirected graph representing co-occurrence relations among terms in given short-text. Similarly, $M_{TT}[i][j]$, the element of $\mathbf{M}_{TT}$, is derived by aggregating the co-occurrences between the two term t_i and t_j and the semantic similarity between them. For given term t_i, we denote its concept set as C_{t_i}, consisting the corresponding concepts deriving from Probase by leveraging single instance conceptualization algorithm [10,27,30]. Therefore, we define the semantic similarity between term t_i and concept t_j, as follows:

$$\text{sim}(t_i, t_j) = \frac{|C_{t_i} \cap C_{t_j}|}{|C_{t_i}|} \tag{6}$$

With efforts above, $M_{TT}[i][j]$ could be defined as follows:

$$M_{TT}[i][j] = \frac{\eta_{TT} \cdot n(t_i, t_j) + (1 - \eta_{TT}) \cdot \text{sim}(t_i, t_j)}{\sum_{k=1}^{n_T} [\eta_{TT} \cdot n(t_i, t_k) + (1 - \eta_{TT}) \cdot \text{sim}(t_i, t_k)]} \tag{7}$$

Wherein, η_{TT} is the parameter controlling the weights about the co-occurrence feature and the semantic similarity feature. Moreover, we could also take local information implicit in this context into consideration. Therefore, we also introduce a correlation function for each pair of t_i and t_j (given concept c_k) to differentiate different attention:

$$\sigma(t_i, t_j, c_k) = \frac{\text{II}(M_{TC}[i][k] \neq 0, M_{TC}[j][k] \neq 0)}{|c_k|(|c_k| - 1)/2} \tag{8}$$

Wherein $|c_k|$ denotes the number of all the terms related to concept c_k. Adding up correlation function from all concepts, we obtain

$$M_{TT}[i][j] = \frac{[\eta_{TT} \cdot n(t_i, t_j) + (1 - \eta_{TT}) \cdot \text{sim}(t_i, t_j)] \cdot \sum_{k=1}^{n_C} \sigma(t_i, t_j, c_k)}{\sum_{l=1}^{n_T} [\eta_{TT} \cdot n(t_i, t_l) + (1 - \eta_{TT}) \cdot \text{sim}(t_i, t_l)] \cdot \sum_{k=1}^{n_C} \sigma(t_i, t_l, c_k)} \tag{9}$$

Subordination Graph (G_{TC}): $G_{TC} = (V_{TC}, E_{TC})$ is a weighted bipartite graph representing relationship among terms all of their corresponding concepts and leveraging the previous graphs, wherein $V_{TC} = V_T \cup V_C$. $M_{TC}[i][j]$ represents the link from a term t_i to a concept c_j. We formulate the subordinate degree of term t_i and concept c_j:

$$\text{sub}(t_i, c_j) = \frac{n_{ins}(t_i, c_j)}{\sum_{k=1}^{n_C} n_{ins}(t_i, c_k)} \tag{10}$$

Wherein, $n_{ins}(t_i, c_j)$ is the frequency that term t_i is an instance of concept c_j. Moreover, we also takes local information embedded in the current context into consideration, by introducing an correlation function for pair of term t_i and term t_j (given concept c_k):

$$\varphi(t_i, t_j, c_k) = \frac{\text{II}(M_{TC}[i][k] \neq 0, M_{TC}[j][k] \neq 0)}{|t_j|(|t_j| - 1)/2} \tag{11}$$

Furthermore, add up this correlation function from all terms:

$$M_{TC}[i][j] = \frac{\text{sub}(t_i, c_j) * \sum_{k=1}^{n_T} \varphi(t_i, t_k, c_j)}{\sum_{l=1}^{n_C} \text{sub}(t_i, c_l) * \sum_{k=1}^{n_T} \varphi(t_i, t_k, c_l)} \tag{12}$$

As demonstrated in [10], we also assign the normalization of subordinate degree to $M_{CT}[i][j]$ straightly, rather than introducing the correlation function above:

$$M_{CT}[i][j] = \frac{\text{sub}(t_j, c_i)}{\sum_{l=1}^{n_T} \text{sub}(t_l, c_i)} \tag{13}$$

3   The Proposed Co-ranking Framework

Based on the construction of the heterogeneous semantic graph (Sect. 2.2) and the affinity matrixes (Sect. 2.3), the proposed co-ranking framework operates the following iteration procedure, consisting for step in each iteration, on G_C, G_T and G_{TC} mutually to mine the most expressive concepts similar to [10], and when this iteration procedure converges, we choose the top-k_C concepts and top-k_T terms according to descending ranking-scores as final results. The algorithm typically converges when difference between the ranking-scores computed at two successive iterations falls below a presupposed threshold. As shown in Fig. 1, to simultaneously tune intra-class rankings (among homogenesis elements) and inter-class rankings (among heterogeneous elements), a set of asymmetric parameters $\gamma_{CC}, \gamma_{CT}, \gamma_{TC}, \gamma_{TT} \in [0,1]$ is used to determining the weights of different random walk procedure in different sub-graphs, with the following constraints: $\gamma_{CC} + \gamma_{CT} + \gamma_{TC} + \gamma_{TT} = 1$ (as demonstrated in [10]).

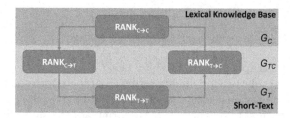

Fig. 1. The proposed co-ranking framework for short-text conceptualization.

Term Ranks Concept (RANK$_{T \to C}$ in Fig. 1). The ranking-scores of terms are used to reinforcing the scores of concepts, and these values are initially set as TF-IDF in corpus.

$$RS_C^{(z+1)}(i) = \gamma_{TC} \Sigma_{k=1}^{n_T} M_{TC}[k][i] * RS_T^{(z)}(k) \tag{14}$$

$$\mathbf{RS}_\mathbf{C}^{(\mathbf{z+1})} = \mathbf{RS}_\mathbf{C}^{(\mathbf{z+1})} / \|\mathbf{RS}_\mathbf{C}^{(\mathbf{z+1})}\| \tag{15}$$

Wherein, $\mathbf{RS}_\mathbf{C}^{(z+1)}$ and $\mathbf{RS}_\mathbf{T}^{(z+1)}$ denote ranking-score vector for concepts and terms in $z+1$-th iteration, and $RS_C^{(z+1)}(i)$ and $RS_T^{(z)}(k)$ denote the ranking-scores of concept c_i and term t_k. To guarantee convergence, $\mathbf{RS}_\mathbf{C}^{(z)}$ and $\mathbf{RS}_\mathbf{T}^{(z)}$ are normalized after each iteration [7, 29, 33].

Concept Ranks Concept (RANK$_{C \to C}$ in Fig. 1). The ranking-scores of concepts are used to reinforcing the scores of other concepts based on their relevance. Note that, We rank the concept graph G_C following the PageRank paradigm [3], which is somewhat similar to weighted PageRank [6]. Consider a random walk on G_C, and the affinity matrix $\mathbf{M_{CC}}$ could be reviewed as the transition matrix. Note taht, a random walk on a graph is a Markov chain [15], its states being the vertices of the graph

$$RS_C^{(z+1)}(i) = \gamma_{CC}(1 - \beta_{CC} + \beta_{CC} * \Sigma_{k=1}^{n_C} M_{CC}[k][i] * RS_C^{(z)}(k)) \tag{16}$$

$$\mathbf{RS_C^{(z+1)}} = \mathbf{RS_C^{(z+1)}} / \|\mathbf{RS_C^{(z+1)}}\| \tag{17}$$

Wherein β_{CC} is the damping factor as used in PageRank, and at each time step with probability $(1 - \beta_{CC})$ we stick to random walking and with probability β_{CC} we do not make a usual random walk step, but instead jump to any vertex, chosen uniformly at random.

Concept Ranks Term (RANK$_{C \to T}$ in Fig. 1). The ranking-scores of concepts are used to reinforcing the scores of terms.

$$RS_T^{(z+1)}(j) = \gamma_{CT} \Sigma_{k=1}^{n_C} M_{CT}[k][j] * RS_C^{(z)}(k) \tag{18}$$

$$\mathbf{RS_T^{(z+1)}} = \mathbf{RS_T^{(z+1)}} / \|\mathbf{RS_T^{(z+1)}}\| \tag{19}$$

Term Ranks Term (RANK$_{T \to T}$ in Fig. 1). The ranking-scores of terms are used to reinforcing the scores of other terms based on their relevance.

$$RS_T^{(z+1)}(j) = \gamma_{TT}(1 - \beta + \beta_{TT} * \Sigma_{k=1}^{n_T} M_{TT}[k][j] * RS_T^{(z)}(k)) \tag{20}$$

$$\mathbf{RS_T^{(z+1)}} = \mathbf{RS_T^{(z+1)}} / \|\mathbf{RS_T^{(z+1)}}\| \tag{21}$$

Wherein β_{TT} is also the damping factor as used in PageRank. The fact that there exists a unique solution to Eq. (21) follows from the random walk $\mathbf{M_{TT}}$ being ergodic[2].

4 Experiments and Results

Since there exists no concept-annotated corpus for short-texts, to validate the performance of our co-ranking framework and other state-of-the-art algorithms, we conduct experiments on text clustering task, which is widely used for evaluating text conceptualization [21,23], to evaluate the results.

---

[2] $\beta_{TT} > 0$ guarantees irreducibility [17], because we can jump to any vertex.

4.1   Datasets

Following [10], we preprocess the Wikipedia articles to construct corpus **Wiki** for construction of the affinity matrix **M** of the heterogeneous semantic graph G, which contains 3.74 million Wikipedia articles. For text clustering task, we use three datasets: **NewsTitle, Twitter, WikiFirst** and **TREC**, as follows:

NewsTitle: We extract news titles from a news corpus containing 3.62 million articles searched from Reuters and New York Time. The news articles are classified into six categories: *company, religion, science, traffic, politician,* and *sport.* We randomly select 5,000 news titles in each category. The average word count of titles is 9.37.

Twitter: We utilize the official tweet collections used in TREC Microblog Task 2013/2014 to construct this dataset. By manually labeling, the dataset contains 41,536 tweets which are in four categories: *food, sport, entertainment,* and *device/IT company.* We remove the URLs and stop-words. The average length of the tweets is 12.95 words. Because of noise and sparsity, this dataset is more challenging.

WikiFirst: this dataset includes 330,000 Wikipedia articles, which are divided into 110 categories based on the mapping relationship between Wikipedia articles and Freebase topics. For example, Wikipedia articles titled The "Big Bang Theory" are categorized into *Tv_program* in Freebase. Each category contains 3,000 Wikipedia articles. We extract the first sentence of each Wikipedia article to construct this dataset, and the average length of the first sentence is 12.67 words. Note that, this dataset is a challenging data set because of its large number of categories, strong diversity of categories and strong correlation among many categories.

TREC: It is the corpus for question clustering on TREC [14], which is widely used as benchmark. The entire dataset of 5,952 sentences are classified into the six categories: *person, entity, abbreviation, description, location* and *numeric.*

4.2   Alternative Algorithms and Experiment Settings

We compare the proposed framework with the following short-text conceptualization algorithms:

BOW: It represents short-text as bag-of-words with the TF-IDF scores [18].

LDA: It represents short-text as its inferred topic distribution [2], and the dimensions of the short-text vector of is number of topics as we presuppose.

IJCAI$_{11}$: [21] proposed a probabilistic framework, which performed a simple co-clustering of concepts and terms by identifying the disjoint cliques, and then derived the most likely concepts using Bayesian inference.

IJCAI$_{11}$+CL: By introducing the clustering strategy, [21] extends **IJCAI$_{11}$**. [21] first mines dense k-exclusive clusters that maximize conditional probability $P(t_i|c_j)$, where the words in the same cluster are considered to belong to the same semantic cluster. Then the algorithm **IJCAI$_{11}$** is implemented on each semantic cluster to complete short-text conceptualization.

IJCAI$_{15}$: Taking verbs and adjectives into consideration, [29] conceptualized terms using a random-walk based iterative algorithm.

RW: It is a pure random walk variant of **IJCAI$_{15}$**, without adjusting the weights on links during whole procedure.

Co-Rank$_{IP}$: [10] ranks the concepts and terms simultaneously in an iterative procedure based on a co-ranking framework.

Co-Rank (Ours): By leveraging concept-based similarity among terms, the proposed co-ranking framework boosts [10], and achieves the goal of combining global statistic information (e.g., co-occurrence feature from large corpus), local information (heuristic information implicit in context, i.e., correlation function) and manual-defined knowledge (semantic similarity in lexical knowledge base).

Co-Rank$_{AD}$: A co-ranking framework by simply coupling two random walks [33], which separately rank different type of vertices under PageRank [3].

With the limitation of space, we briefly describe the experimental settings here. The dimensions of vector in **BOW** is 25,000, which releases the optimal experimental results. For **Co-Rank, Co-Rank$_{AD}$, Co-Rank$_{HITS}$, IJCAI$_{11}$, IJCAI$_{11}$+CL, IJCAI$_{15}$** and **RW**, we select 5,000 concepts as features in text clustering task, which is like the number of concept clusters in Probase [10, 25, 27, 29]. In the proposed **Co-Rank**: (i) we set the damping factor β_{CC} in Eq. (17) and β_{TT} in Eq. (21) and to 0.15 following the standard PageRank paradigm; (ii) we set the set of asymmetric parameters $\{\gamma_{CC}, \gamma_{CT}, \gamma_{TC}, \gamma_{TT}\}$ as 0.2, 0.25, 0.35, 0.2, which yields the best results. (iii) we test the convergence threshold from $\{10^{-3}, 10^{-4}, 10^{-5}, 10^{-6}, 10^{-7}\}$, and analysis results indicate that when the threshold tails off to below 10^{-5}, the co-ranking framework yields the best results in the short-text clustering task.

4.3 Experiments on Text Clustering

Because news title data, tweet data, Wiki first-sentence data and question data have no ground-truth labels, to evaluate the effectiveness of conceptualizing, we design text clustering task. Totally, we first generate "concepts"[3] of each short-text (in the aforementioned datasets) based on different algorithms, and then use these concepts as features to construct short-text vector, and run spherical K-means clustering [13, 16] to evaluate each algorithm. This paper uses Purity [32], Adjusted Rand Index (ARI) [11] and Normalized Mutual Information (NMI) [4, 24] to measure the quality of the short-text clustering task. The larger the Purity (ARI or NMI) is, the better the clustering result and the better the performance of the corresponding algorithm achieves.

We discuss these measurements as follows. Let $X = \{x_1, x_2, \cdots, x_{|X|}\}$ denote the set of short-text clusters after short-text clustering, wherein x_i indicates the i-th short-text cluster. Similarly, Let $Y = \{y_1, y_2, \cdots, y_{|Y|}\}$ denote the ground-truth set of short-text clusters. Besides, N denotes the total count of the short-text. Therefore, Purity could be measured as follows:

---

[3] Except for algorithm **LDA** generating topic as "concept", all the other algorithms generate concepts which are defined by lexical knowledge base Probase.

$$\text{Purity} = \frac{1}{N} \sum_{i=1}^{|X|} \sum_{j=1}^{|Y|} \max |x_i \cap y_i| \tag{22}$$

Let n_{ij} denotes the count of short-texts which occurs in cluster x_i and cluster y_j simultaneously, and ARI could be defined as follows:

$$\text{ARI} = \frac{\sum_{i=1}^{|X|} \sum_{j=1}^{|Y|} C_2^{n_{ij}} - \frac{\sum_{i=1}^{|X|} C_2^{|x_i|} \cdot \sum_{j=1}^{|Y|} C_2^{|y_j|}}{C_2^N}}{\frac{\sum_{i=1}^{|X|} C_2^{|x_i|} + \sum_{j=1}^{|Y|} C_2^{|y_j|}}{2} - \frac{\sum_{i=1}^{|X|} C_2^{|x_i|} \cdot \sum_{j=1}^{|Y|} C_2^{|y_j|}}{C_2^N}} \tag{23}$$

Moreover, the measurement of NMI is discussed as follows. Let $H(X)$ denote the Information Entropy, defined as follows:

$$H(X) = - \sum_{i=1}^{|X|} P(x_i) \cdot \log P(x_i) \tag{24}$$

Wherein, $P(x_i)$ indicates the probability that the short-text occurs in cluster x_i, and $P(y_j)$ indicates the probability that the short-text occurs in cluster y_j. Let $I(X;Y)$ denotes the mutual information of set X and set Y, as follows:

$$I(X;Y) = \sum_{i=1}^{|X|} \sum_{j=1}^{|Y|} [P(x_i \cap y_j) \cdot \log \frac{P(x_i \cap y_j)}{P(x_i) \cdot P(y_j)}] \tag{25}$$

Therefore, we could utilize the following equation to define NMI:

$$\text{NMI} = \frac{2 \cdot I(X;Y)}{H(X) + H(Y)} \tag{26}$$

Table 1. Evaluation results of short-text clustering task.

	NewsTitle			Twitter			WikiFirst			TREC		
	Purity	ARI	NMI	Purity	ARI	NMI	Purity	ARI	NMI	Purity	ARI	NMI
BOW	0.617	0.569	0.781	0.212	0.211	0.250	0.297	0.419	0.531	0.712	0.663	0.863
LDA	0.619	0.575	0.683	0.319	0.323	0.341	0.274	0.387	0.490	0.719	0.672	0.760
IJCAI$_{11}$	0.681	0.635	0.807	0.365	0.353	0.354	0.311	0.439	0.556	0.757	0.752	0.875
IJCAI$_{11}$+CL	0.711	0.651	0.809	0.378	0.351	0.387	0.326	0.460	0.583	0.812	0.748	0.881
IJCAI$_{15}$	0.737	0.675	0.822	0.419	0.381	0.416	0.343	0.484	0.613	0.832	0.770	0.882
RW	0.760	0.695	0.847	0.413	0.395	0.443	0.346	0.488	0.617	0.862	0.791	0.904
Co-Rank$_{AD}$	0.731	0.677	0.806	0.423	0.387	0.421	0.343	0.483	0.612	0.833	0.782	0.874
Co-Rank$_{IP}$	0.785	0.738	**0.879**	**0.461**	0.428	0.478	0.369	0.521	0.659	0.854	0.831	0.942
Co-Rank (Ours)	**0.801**	**0.753**	0.876	0.456	0.441	**0.482**	**0.380**	**0.537**	**0.679**	**0.871**	**0.848**	**0.961**

Experimental results are shown in Table 1. The results show the proposed co-ranking framework improves the baseline models in most cases: (i) **Co-Rank**

(**Ours**) is superior to **Co-Rank$_{IP}$**, which ignores the manual-defined knowledge (e.g., semantic similarity in lexical knowledge base); (ii) Taking measurement NMI as an example, **Co-Rank (Ours)** exceeds the recognized baseline model **IJCAI$_{15}$** by 6.57% and **IJCAI$_{11}$** by 8.55% on dataset **NewsTitle**, exceeds **IJCAI$_{15}$** by 15.87% and **IJCAI$_{11}$** by 36.16% on dataset **Twitter** (Note that, dataset **Twitter** is challenging because of its noise), and exceeds **IJCAI$_{15}$** by 8.94% and **IJCAI$_{11}$** by 22.08% on dataset **WikiFirst**, indicating that it is essential to utilize the beneficial interactions among terms and concepts.

5 Conclusion

Short-text conceptualization plays an increasingly vital role in text understanding and other applications. This paper proposes a novel co-ranking framework to address the problem of short-text conceptualization, which operates an iterative procedure over a heterogeneous semantic graph and reinforces the terms and corresponding concepts simultaneously. Furthermore, this framework is found to automatically detect the contextual salient key-terms in the short-text. Experiments on real-world datasets suggest that the proposed co-ranking framework is effective.

Acknowledgements. The authors are very grateful to the editors and reviewers for their helpful comments. This work is funded by: (i) the China Postdoctoral Science Foundation (No. 2018M641436); (ii) the Joint Advanced Research Foundation of China Electronics Technology Group Corporation (CETC) (No. 6141B08010102); (iii) 2018 Culture and tourism think tank project (No. 18ZK01); (iv) the New Generation of Artificial Intelligence Special Action Project (18116001); (v) the Joint Advanced Research Foundation of China Electronics Technology Group Corporation (CETC) (No. 6141B0801010a); and (vi) the Financial Support from Beijing Science and Technology Plan (Z181100009818020).

References

1. Agrawal, R., Gollapudi, S., Kannan, A., Kenthapadi, K.: Similarity search using concept graphs. In: 23rd ACM International Conference on Conference on Information and Knowledge Management, pp. 719–728 (2014)
2. Blei, D.M., Ng, A.Y., Jordan, M.I.: Latent Dirichlet allocation. J. Mach. Learn. Res. **3**, 993–1022 (2003)
3. Brin, S., Page, L.: The anatomy of a large-scale hypertextual web search engine. In: International Conference on World Wide Web, pp. 107–117 (1998)
4. Cho, D., Lee, B.: Optimized automatic sleep stage classification using the normalized mutual information feature selection (NMIFS) method. In: Conference of the IEEE Engineering in Medicine and Biology Society 2017, pp. 3094–3097 (2017)
5. Dathathri, R., et al.: Gluon: a communication-optimizing substrate for distributed heterogeneous graph analytics. In: ACM SIGPLAN Conference on Programming Language Design & Implementation (2018)
6. Ding, Y.: Applying weighted PageRank to author citation networks. J. Assoc. Inf. Sci. Technol. **62**(2), 236–245 (2011)

7. Fujiwara, Y., Nakatsuji, M., Onizuka, M., Kitsuregawa, M.: Fast and exact top-k search for random walk with restart. Proc. VLDB Endowment **5**(5), 442–453 (2012)
8. Gabrilovich, E., Markovitch, S.: Wikipedia-based semantic interpretation for natural language processing. J. Artif. Intell. Res. **34**(4), 443–498 (2014)
9. Hua, W., Wang, Z., Wang, H., Zheng, K., Zhou, X.: Short text understanding through lexical-semantic analysis. In: IEEE International Conference on Data Engineering, pp. 495–506 (2015)
10. Huang, H., Wang, Y., Feng, C., Liu, Z., Zhou, Q.: Leveraging conceptualization for short-text embedding. IEEE Trans. Knowl. Data Eng. **30**(7), 1282–1295 (2018)
11. Hubert, L., Arabie, P.: Comparing partitions. J. Classif. **2**(1), 193–218 (1985)
12. Kim, D., Wang, H., Oh, A.: Context-dependent conceptualization. In: International Joint Conference on Artificial Intelligence, pp. 2654–2661 (2013)
13. Li, M., Xu, D., Zhang, D., Zou, J.: The seeding algorithms for spherical k -means clustering. J. Glob. Optim. **3**(1), 1–14 (2019)
14. Li, X., Roth, D.: Learning question classifiers. In: 19th International Conference on Computational Linguistics, pp. 1–7 (2002)
15. Gasparini, M.: Markov chain monte Carlo in practice. Technometrics **39**(3), 338–338 (1999)
16. Peterson, A.D., Ghosh, A.P., Maitra, R.: Merging k-means with hierarchical clustering for identifying general-shaped groups. Stat **7**(1), e172 (2018)
17. Rui, Y., Lapata, M., Li, X.: Tweet recommendation with graph co-ranking. In: Meeting of the Association for Computational Linguistics: Long Papers (2012)
18. Salton, G., Mcgill, M.J.: Introduction to Modern Information Retrieval. McGraw-Hill, New York (1983)
19. Sebastian, Y., Eu-Gene, S., Orimaye, S.O.: Learning the heterogeneous bibliographic information network for literature-based discovery. Knowl.-Based Syst. **115**, 66–79 (2016)
20. Shi, C., Li, Y., Zhang, J., Sun, Y., Yu, P.S.: A survey of heterogeneous information network analysis. IEEE Trans. Knowl. Data Eng. **29**, 17–37 (2017)
21. Song, Y., Wang, H., Wang, Z., Li, H., Chen, W.: Short text conceptualization using a probabilistic knowledgebase. In: International Joint Conference on Artificial Intelligence, pp. 2330–2336 (2011)
22. Song, Y., Wang, H., Wang, Z., Li, H., Chen, W.: Short text conceptualization using a probabilistic knowledgebase. In: Proceedings of the Twenty-Second International Joint Conference on Artificial Intelligence, vol. 3, pp. 2330–2336 (2011)
23. Song, Y., Wang, S., Wang, H.: Open domain short text conceptualization: a generative+ descriptive modeling approach. In: Proceedings of the 24th International Conference on Artificial Intelligence, pp. 3820–3826 (2015)
24. Strehl, A., Ghosh, J.: Cluster ensembles – a knowledge reuse framework for combining multiple partitions (2003)
25. Wang, F., Wang, Z., Li, Z., Wen, J.R.: Concept-based short text classification and ranking. In: The ACM International Conference, pp. 1069–1078 (2014)
26. Wang, Y., Huang, H., Feng, C.: Query expansion based on a feedback concept model for microblog retrieval. In: International Conference on World Wide Web, pp. 559–568 (2017)
27. Wang, Y., Huang, H., Feng, C., Zhou, Q., Gu, J., Gao, X.: CSE: conceptual sentence embeddings based on attention model. In: 54th Annual Meeting of the Association for Computational Linguistics, pp. 505–515 (2016)
28. Wang, Z., Cheng, J., Wang, H., Wen, J.: Short text understanding: a survey. J. Comput. Res. Dev. **53**(2), 262–269 (2016)

29. Wang, Z., Zhao, K., Wang, H., Meng, X., Wen, J.R.: Query understanding through knowledge-based conceptualization. In: International Conference on Artificial Intelligence, pp. 3264–3270 (2015)
30. Wu, W., Li, H., Wang, H., Zhu, K.Q.: Probase: a probabilistic taxonomy for text understanding. In: SIGMOD Conference (2012)
31. Wu, W., Li, H., Wang, H., Zhu, K.Q.: Probase: a probabilistic taxonomy for text understanding. In: ACM SIGMOD International Conference on Management of Data, pp. 481–492 (2012)
32. Ying, Z., Karypis, G.: Empirical and theoretical comparisons of selected criterion functions for document clustering. Mach. Learn. $55(3)$, 311–331 (2004)
33. Zhou, D., Orshanskiy, S.A., Zha, H., Giles, C.L.: Co-ranking authors and documents in a heterogeneous network. In: 7th IEEE International Conference on Data Mining, pp. 739–744 (2007)

Denoising Distant Supervision for Relation Extraction with Entropy Weight Method

Mengyi Lu and Pengyuan Liu$^{(\boxtimes)}$

Beijing Language and Culture University, Beijing, China
`lmy0722@foxmail.com`, `liupengyuan@pku.edu.cn`

Abstract. Distant supervision for relation extraction has been widely used to construct training set by aligning the triples of the knowledge base, which is an efficient method to reduce human efforts. However, this method inevitably suffers from wrong labeling problems leading too much noise that will severely hurt the performance of relation extraction. To tackle this problem, in this paper, we propose a denoising model based on Entropy Weight Method (EWM) to filter the noise and select most relevant sentences. First, in a pretraining stage, we develop a sentence-level relation aware attention mechanism to distinguish several most relevant sentence, increasing the attention weights for those critical sentences. Second, we filter the noisy sentences by calculating the entropy weight using the above attention matrix, and then we employ intra-bag and inter-bag attentions to aggregate these selected sentence representations. Experiments on the NYT dataset show that our method can significantly reduce the noisy instance and achieve the state-of-the-art model performance.

Keywords: Relation extraction · Distant supervision · Noise filtering

1 Introduction

Relation Extraction (RE) is defined as a task of generating relation triple facts from plain texts, which is widely used to facilitate a lot of Natural Language Processing (NLP) tasks including knowledge base construction [1] and question answering [2]. As the fully supervised RE approaches are limited by the consuming and labour intensive labeled training set, distant supervision strategy [1] is proposed as a promising approach to automatically create training data via aligning Knowledge Bases (KBs) with texts. The basic assumption of distant supervision is that if two entities e_1 and e_2 have a relation r in KBs, then all the sentences in corpus that contain these two entities will express this specific relation and will be labeled as the training instances of r. Although distant supervision is effective to label data automatically, it suffers from the noisy labeling problem.

© Springer Nature Switzerland AG 2019
M. Sun et al. (Eds.): CCL 2019, LNAI 11856, pp. 294–305, 2019.
https://doi.org/10.1007/978-3-030-32381-3_24

To address the issue of noisy labeling, previous studies adopt multi-instance learning to consider the noises of instances (Riedel et al. [4]; Hoffmann et al. [17]; Surdeanu et al. [19]; Zeng et al. [11]; Lin et al. [5]; Ji et al. [12]). For example, Zeng et al. propose to combine multi-instance learning [11] with Piecewise Convolutional Neural Networks (PCNNs) to choose the most likely valid sentence and predict relations. In these studies, the training and test process is proceeded at the bag level, where a bag contains noisy sentences mentioning the same entity pair but possibly not describing the same relation. However, these methods unable to handle the sentence-level prediction and are sensitive to the bags with all noisy sentences which do not describe a relation at all.

In this paper, we proposed a novel approach of filtering sentences called a entropy weighted method (EWM) to distinguish the relevant sentence and alleviate negative effect of noisy labeling problem. The overall idea of the model is as follows. First, We use a pretrain strategy, In this stage, we extracts all sentence features using PCNNs and learns the weights of sentences by the relation aware attention module. We hope that the attention mechanism is able to selectively focus on the relevant sentences through assigning higher weights for valid sentences and lower weights for the invalid ones. In this way, the attention matrix can recognize multiple valid sentences in a bag, and we retrain the model to filter the noisy sentences by calculating the entropy weight using the above attention matrix (we will show more calculation details later). More specifically, For a bag, we first use PCNNs to extract each sentence's feature vector, then compute the attention weight for each sentence through multiplying each possible relation which is utilized as the query, and according the attention weight we calculate the entropy weight setting a threshold to filter the noisy sentence, and then the bag's representation compute by the weighted sum of the selected valid sentence feature vectors, Furthermore, the representation of a group of bags in the training set which share the same relation label is calculated by weighting bag representations using a similarity-based inter-bag attention module. Finally, a bag group is utilized as a training sample when building our relation extractor.

Our contributions of this paper include:

- We introduce a denoise approach named the Entropy Weight Method which can select the most relevant sentences and filter those wrong labeling sentences, this strategy can effectively reduce the noise and improve the model performance.
- We propose a two-step model to better capture different levels of structural information and fuse them for classification.
- Our method achieve the-state-of-art performance on the widely used New York Times (NYT) dataset [4].

2   Related Work

Distant supervision for relation extraction, first introduced by Mintz et al. [3], automatically generates training data through heuristic alignment between a knowledge base and plain texts. Although distant supervision is an efficient way to scale relation extraction to a large number of relations, the basic assumption used in the alignment is so strong that it will inevitably bring wrong labeling problem.

To alleviate noise, Hoffmann et al. [26], Riedel et al. [4] and Surdeanu et al. [19] build multi-instance learning paradigms. Specifically, Riedel et al. [4] uses at-least-one assumption to resolve the problem. Hoffmann et al. [26] builds a probabilistic graphic model and intends to resolve multi-instance with over-lapping relations in distant supervision. Surdeanu et al. [19] trains a Bayesian framework by expectation maximization (EM) algorithm. In addition, researchers notice that the incompleteness of the knowledge base (i.e., Freebase) will result in the false negative problem and design a latent-variable approach (Ritter et al. [18]). Later, considering automatic feature engineering, Lin et al. [5] and Zeng et al. [11] integrate multi-instance learning model with PCNNs to extract relations on distantly supervised data. Although proved effective, MIL suffers from information loss problem because it ignored the presence of more than one valid instances in most bags. Recently attention mechanism attracted a lot of interests of researchers [10,24,25]. Considering the flaw of MIL, Lin et al. [5] and Ji et al. [12] introduced bilinear and non-linear attention respectively into this task to make full use of supervision information by assigning higher weights to valid instances and lower weights to invalid ones. The two attention models significantly outperform MIL method. However, they suffer from noise residue problem because noisy sentences have harmful information but still have positive weights. The residue weights of noisy data mean that attention methods cannot fully eliminate the negative effects of noise.

3   Methodology

In this section, we present an overview of our model for distant supervised RE, as illustrated in Fig. 1. Our model follows three steps. First, We train the attention matrix in pretraining stage, then we use the entropy weight method to filter sentences, Finally, we adopt the Ye et al. [22] intra-bag and inter-bag attention mechanism to classify relationships.

3.1   Sentence Encoder

Sentence encoder transforms the sentence into its distributed representation. First, words in a sentence are transformed into dense real-valued vectors. For word token w, we use pre-trained word embeddings as low dimension vector representation. Following Zeng et al. [8], we use position embeddings as extra position feature. We compute the relative distances between each word and two entity words, and transform them to real-valued vectors by looking up randomly-initialized embedding matrices. We denote the word embedding of word w by $w_w \in \mathbb{R}^{d^w}$ and two position embeddings by p_{w_1}, p_{w_2}. The word representation is then composed by horizontal concatenating word embeddings and position embeddings:

$$s_w = [w_w; p_w^1; p_w^2].s_w \in \mathbb{R}^{(d^w + 2 \times d_p)} \tag{1}$$

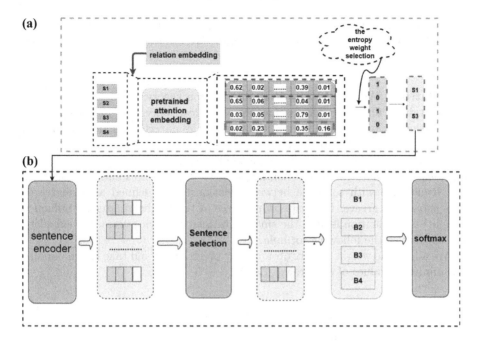

Fig. 1. The architecture of our model. It has two parts: (a) pretraining module and (b) overall framework.

Then, given a sentence and corresponding entity pair, we apply PCNN to construct a distributed representation of the sentence.

3.2 Pre-trained Attention Embedding

In our implementation, a pre-training strategy is adopted. We first train the model with relation-aware attention mechanism until convergence. We use PCNNs to extract each sentence's feature vector, then compute the attention weight for each sentence through multiplying each possible relation which is utilized as the query. Therefore we pre-train the attention embedding in relation extraction, and use the pre-trained attention embedding to calculate the Entropy Weight. In this way, the Entropy Weight will more accurately reflect the information contained in a sentence bag for its relations.

3.3 Sentence Selection

Although previous work yields high performance, there still exists some drawbacks. MIL suffers from information loss problem because it ignored multiple valid sentences and used only one sentence for representing a bag and training. Attention-based methods have noise residue problem because they assigned small but still positive weights to harmful noisy sentences, which means noise effects weren't completely removed.

In our work, we narrow our sentence space and focus on select the sentences that are likely to be relevant to the relation r. We use the following formulas as our selection mechanism.

The Entropy Weight Method: In information theory, entropy is a measure of uncertainty. The more information, the less uncertainty, the less entropy; The smaller the amount of information, the greater the uncertainty, and the greater the entropy. According to the characteristics of entropy, the randomness and dispersion degree of sentences can be judged by calculating entropy value. We use this idea to simulate the correlation degree between sentences and relationships. The irrelevance with sentence between relationship is defined as a degree of dispersion. The greater the dispersion degree of the is, the greater its influence (weight) on evaluation is, and the smaller its entropy is vice versa. We calculate it according to the following formula.

We adopt the weight matrix obtained by pre-training and take the maximum and minimum values of each row:

$$s_{ij} = \frac{a_{ij} - min(a_i)}{max(a_i) - min(a_i)} \tag{2}$$

where s_{ij} is the embedding normalization by attention value.

$$E_i = -\frac{1}{ln(n)} \sum_{i=1}^{n} p_{ij} ln(p_{ij}) \tag{3}$$

where $p_{ij} = \frac{s_{ij}}{\sum_{i=1}^{n} s_{ij}}$. E_i means the sentence information entropy.

$$M = E_w = \frac{e^{E_i}}{\sum_{i=1}^{n} e^{E_i}} \tag{4}$$

we take a threshold M, if the entropy value is greater than the threshold meaning that the sentence dispersion degree is smaller, the degree of irrelevance of sentences to relationships is smaller, and it will be reserved. If it is less than the threshold, the sentence dispersion degree is greater, it is considered as noise to discard it. We alleviate noise residue problem by only assigning attention weights to selected sentences. The unselected noisy data will not be assigned weights, and will not participate in training process.

3.4   Intra-bag Attention

We hope that the attention model can learn higher weights for valid instances and lower weights for the invalid ones. In experiments, we will show the weights of an example. We use $S_i \in \mathbb{R}^{m_i \times 3d_c}$ represent the sentences representations within bag b_i, where $R \in \mathbb{R}^h \times 3d_c$ denote relation embedding matrix where h is the number of relations. $S_1, S_2, ..., S_q$ are feature vectors (computed by PCNNs) of all instances in a bag, we propose the following formulas to compute the attention weight. For each sentence j in the bag i with respect to each relation k, and aggregates to bag representation as b_k^i, by the following equations:

$$b_k^i = \sum_{j=1}^{m} a_{kj}^i s_j^i \qquad (5)$$

where k is the relation index and a_{kj}^i is the attention weight between the k-th relation and the j-th sentence in bag b_i. And the a_{kj}^i can be calculated as:

$$a_{kj}^i = \frac{exp(e_{kj}^i)}{\sum_{j=1}^{m_i} exp(e_{kj}^i)} \qquad (6)$$

where e_{kj}^i is the matching degree between the k-th relation query and the j-th sentence in bag b_i, and it is defined as:

$$e_{kj}^i = r_k s_j^i \top \qquad (7)$$

where r_k is the k-th row of the relation embedding matrix R.

3.5   Inter-bag Attention

Inspired by the works of Ye et al. [22] and Yuan et al. [23] which utilizes bag-level attention mechanism to deal with the noisy bag problem. At the sentence level, noise cannot be completely removed because it assumes that at least one correct sentence exists. However, in the process of distant supervising the construction of the data set, there may be not exist a correct sentence in a bag, furthermore, the knowledge base cannot be covered all the relationships. The entity expressing the truly relationships may not exist in a given corpus, so there will still exist noise. To solve the problem, we combine several sentence bags of the same relation type and to get more attention to the more relevant bags. We obtain the superbag representation as follow equation:

$$f = \sum_{i=1}^{m} \gamma_{ik} b_k^i \qquad (8)$$

$$\gamma_i = \frac{e^{(S(r_k, b_{ik})}}{\sum_{j=1}^{m} e^{(S(r_k, b_{ik}))}} \qquad (9)$$

$$S_{i,j,k} = \frac{x_{i,j} r_k}{|x_{i,j}| |r_k|} \qquad (10)$$

where b_k^i is the bag representation w.r.t. B_i for the k-th relation and r_k is the attention parameter corresponding to the j-th relation.

4   Experiment

Our experiments are intended to show that our model can capture high weight sentences and take full advantage of informative sentences for distant supervised relation extraction. In the experiments, we first introduce the dataset and evaluation metrics used. Next, we determine some parameters of our model. And then we evaluate the effects of our model performance, and we also compare our method to some classical methods. Finally, we do some experimental analysis and case study.

4.1   Dataset and Evaluation Metrics

We evaluate our model on a widely used dataset which is developed by Riedel et al. [4]. This dataset was generated by aligning Freebase with the New York Times (NYT) corpus. The dataset contains 53 relations (including no relation "NA") and 39,528 entities. The training data contains 522,611 sentences, 281,270 entity pairs and 18,252 relational facts. The test dataset contains 172,448 sentences, 96,678 entity pairs and 1,950 relational facts. We use word2vec to train word embedding on the NYT corpus and use the embeddings as initial values. Following previous work [1,2,5,8], we evaluate our model in the held-out evaluation, which evaluates our model by comparing the extracted relation facts with those in Freebase, and report both the precision/recall curves and Precision@N (P@N) of the experiments.

4.2   Experimental Settings

In this section, we study the influence of one parameter on our model: the threshold value M is defined in Eq. (4). We tune our models using three-fold validation on the training set. We use a grid search to determine the optional parameter: $M \in \{0, 0.3, 0.5, 0.7, 1\}$. Most of them followed the hyperparameter settings in Ye et al. [22].

4.3   Performance Evaluation

We compare our method with these previous works: PCNN+ONE [11] selects the sentence with the highest right probability as bag representation; PCNN+ATT [5] use non-linear attention to assign weights to all sentences in a bag. Ye et al. [22] adopt the intra-bag and inter-bag attention modules achieve the state-of-the-art performance denoted PCNN+ATT_RA+BAG_ATT.ATT_RA means the relation-aware intra-bag attention method, and BAG_ATT means the inter-bag attention method.

In order to prove the superiority of our module, we propose a more intuitive and simpler way for instance selection: we set a threshold on attention weights calculated by entropy weight and filter sentences with lower weights than threshold. We denote this method as Entropy Weight Method (EWM). We

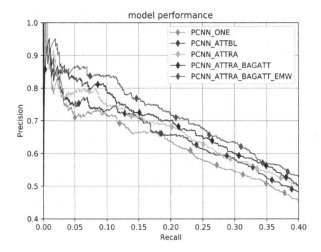

Fig. 2. The precision/recall curves for the combined model and the baselines.

adopt EWM to PCNN+ATT_RA+BAG_ATT to demonstrate the effectiveness of instance selectors, denoted as PCNN+ATT_RA+BAG_ATT+EWM. Figure 2 shows the aggregated precision/recall curves, and Table 1 shows the Precision@N with N = {100, 200, 300, ALL} of our approaches and all the baselines.

We can see our proposed methods achieved highest P@N values than all previous work. Furthermore, we have the following observations: (1) Similar to the results of Ye et al. [22], ATT_RA outperformed ATT_BL. It can be attributed to that the ATT _BL method only considered the target relation when deriving bag representations at training time, while the ATT_RA method calculated intra-bag attention weights using all relation embeddings as queries, which improved the flexibility of bag representations. (2) The BAG_ATT method performs better than the ones without BAG_ATT, it verified the effectiveness of the method. (3) The EWM method is efficient for this task. Especially when the test data set is all, the effect is more obvious. The results show that the EWM method can capture high weight sentences and take full advantage of informative sentences.

4.4   Analysis of Entropy Weight Threshold

In this section, we will investigate the effectiveness of the entropy weight threshold as denoted EWM. We fine-tune the hyperparameter threshold to achieve its best performance. Higher thresholds bring back information loss problem because more informative sentences are neglected. Lower thresholds bring back noise residue problem because more noisy sentences are selected and assigned weights. We conduct experiments on EWM with different thresholds. For clarify, we use a histogram to approximate precision/recall curves of different thresholds, shown in Fig. 2. In our experiment, we found EWM value is set by 0.3 achieving the best performance, with higher threshold the performance will decline,

Table 1. Top-N precision (P@N) for relation extraction in the entity pairs with different number of sentences. Following (Lin et al. [5]), One, Two and All test settings random select one/two/all sentences on the bags of entity pairs from the testing set which have more than one sentence to predict relation.

Method	One				Two				All			
	100	200	300	Mean	100	200	300	Mean	100	200	300	Mean
PCNN+ONE (Zeng [11])	66.7	62.8	54.8	61.4	71.3	68.3	62.2	67.2	70.2	68.1	61.4	66.5
PCNN+ATT (Lin [5])	73.3	69.2	60.8	67.8	77.2	71.6	66.1	71.6	76.2	73.1	67.4	72.2
PCNN+ATT_BL	78.6	73.5	68.1	73.4	77.8	75.1	70.3	74.4	80.8	77.5	72.3	76.9
PCNN+ATT_RA	79.2	73.9	68.3	73.8	81.5	77.5	72.7	77.5	83.6	79.9	72.3	78.6
PCNN+ATT_BL+BAG_ATT	84.8	**78.9**	70.7	78.1	84.5	79.5	74.2	79.4	88.8	83.9	77.3	83.3
PCNN+ATT_RA+BAG_ATT	86.8	77.6	73.8	**79.4**	90.8	**79.1**	74.4	81.4	91.8	83.9	77.6	84.4
PCNN+ATT_BL+EWM	78.8	72.6	67.9	73.1	77.2	72.4	66.9	72.1	81.2	77.0	71.7	76.6
PCNN+ATT_RA+EWM	79.2	73.6	68.2	73.6	81.6	77.7	70.3	76.5	83.8	84.0	78.7	84.8
PCNN+ATT_RA+BAG_ATT+EWM	**86.9**	77.6	**73.9**	79.3	**91.1**	78.9	**74.5**	81.5	**91.8**	**84.6**	**78.9**	**85.1**

In this way, some relevant sentences are filtered out and there is less effective information to be utilized. EWM(1.0) selects only the sentence with maximum attention weight to train. Similar to MIL select strategy, EWM(1.0) also has similar performance to PCNN+ONE. When the threshold set lower, more noisy

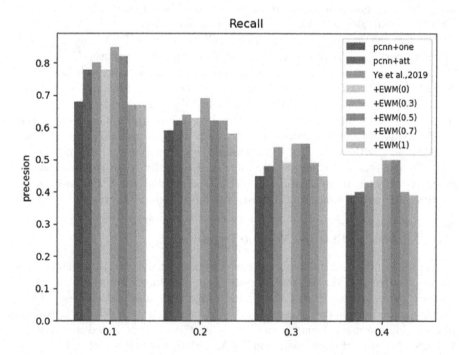

Fig. 3. Aggregate precision/recall histogram of EWM with different thresholds.

sentences get involved, so the performance behave dissatisfactory. The result also close to PCNN+ATT model (equivalent to EWM(0)) (Fig. 3).

4.5 Case Study

Table 2 shows an example of our method selection result. The bag contains 4 sentences which the 4-th instance are invalid sentence. With the help of EWM, attention mechanism only assigns high weights to selected sentences. The fourth sentence was a noisy sentence because the sentence in this bag didn't express the relation */location/location/contains* between the two entities *NewOrleans*, and *DillardUniversity*. Therefore, the attention mechanism can select the valid instances and is useful in our task.

Table 2. An example of Entropy Weight.

Tuple	Instance	Select	EMW
/location/location/contains (New Orleans, Dillard University)	1. She graduated from [Dillard University] in [New Orleans] and received a master's degree in marine science from the College of William and Mary	1	0.23
	2. Jinx Broussard, a communications professor at [Dillard University] in [New Orleans], said four members of her family had lost their houses to the hurricanes	1	0.37
	3. I was grieving from the death when I graduated from high school, but I decided to go to [Dillard University] in [New Orleans]	1	0.31
	4.4. When he came here in May 2003 to pick up an honorary degree from [Dillard University], his dense schedule didn't stop him from calling Dooky Chase's, the Creole restaurant he sang about in "Early in the Morning Blues", "where he'd eaten his favorite dish ever since he lived in [New Orleans] In the 1950's"	0	0.09

5   Conclusion

In this paper, we proposed a novel approach of filtering sentences called a entropy weighted method (EWM) to distinguish the relevant sentence and alleviate negative effect of noisy labeling problem in distant supervision relation extraction.

Experimental results show our method is able to selectively focus on the relevant sentences through assigning higher weights for valid sentences and lower weights for the invalid ones.

Acknowledgements. This work is supported by Beijing Natural Science Foundation (4192057).

References

1. Han, X., Liu, Z., Sun, M.: Neural knowledge acquisition via mutual attention between knowledge graph and text (2018)
2. Lee, C., Hwang, Y.G., Jang, M.G.: Fine-grained named entity recognition and relation extraction for question answering. In: Proceedings of the 30th Annual International ACM SIGIR Conference on Research and Development in Information Retrieval, pp. 799–800. ACM (2007)
3. Mintz, M., Bills, S., Snow, R., Jurafsky, D.: Distant supervision for relation extraction without labeled data. In: Proceedings of Joint Conference 47th Annual Meeting ACL 4th International Joint Conference Natural Language Processing (AFNLP), pp. 1003–1011. Association for Computational Linguistics, August 2009
4. Riedel, S., Yao, L., McCallum, A.: Modeling relations and their mentions without labeled text. In: Balcázar, J.L., Bonchi, F., Gionis, A., Sebag, M. (eds.) ECML PKDD 2010. LNCS (LNAI), vol. 6323, pp. 148–163. Springer, Heidelberg (2010). https://doi.org/10.1007/978-3-642-15939-8_10
5. Lin, Y., Shen, S., Liu, Z., Luan, H., Sun, M.: Neural relation extraction with selective attention over instances. In: Proceedings of the 54th Annual Meeting of the Association for Computational Linguistics (Volume 1: Long Papers), vol. 1, pp. 2124–2133 (2016)
6. Mikolov, T., Chen, K., Corrado, G., Dean, J.: Efficient estimation of word representations in vector space. arXiv preprint arXiv:1301.3781 (2013)
7. Santos, C.N.D., Xiang, B., Zhou, B.: Classifying relations by ranking with convolutional neural networks. arXiv preprint arXiv:1504.06580 (2015)
8. Zeng, D., Liu, K., Lai, S., Zhou, G., Zhao, J.: Relation classification via convolutional deep neural network. In: Proceedings of COLING 2014, the 25th International Conference on Computational Linguistics: Technical Papers, pp. 2335–2344 (2014)
9. Nguyen, T.H., Grishman, R.: Relation extraction: perspective from convolutional neural networks. In: Proceedings of the 1st Workshop on Vector Space Modeling for Natural Language Processing, pp. 39–48 (2015)
10. Luong, M.T., Pham, H., Manning, C.D.: Effective approaches to attention-based neural machine translation. arXiv preprint arXiv:1508.04025 (2015)
11. Zeng, D., Liu, K., Chen, Y., Zhao, J.: Distant supervision for relation extraction via piecewise convolutional neural networks. In: Proceedings of the 2015 Conference on Empirical Methods in Natural Language Processing, pp. 1753–1762 (2015)
12. Ji, G., Liu, K., He, S., Zhao, J., et al.: Distant supervision for relation extraction with sentence-level attention and entity descriptions. In: AAAI, pp. 3060–3066 (2017)
13. Zhou, P., et al.: Attention-based bidirectional long short-term memory networks for relation classification. In: Proceedings of the 54th Annual Meeting of the Association for Computational Linguistics (Volume 2: Short Papers), vol. 2, pp. 207–212 (2016)

14. Vaswani, A., et al.: Attention is all you need. In: Advances in Neural Information Processing Systems, pp. 5998–6008 (2017)
15. Dietterich, T.G., Lathrop, R.H., Lozano-Perez, T.: Solving the multiple instance problem with axis-parallel rectangles. Artif. Intell. **89**(1–2), 31–71 (1997)
16. Riedel, S., Yao, L., McCallum, A.: Modeling relations and their mentions without labeled text. In: Balcázar, J.L., Bonchi, F., Gionis, A., Sebag, M. (eds.) ECML PKDD 2010. LNCS (LNAI), vol. 6323, pp. 148–163. Springer, Heidelberg (2010). https://doi.org/10.1007/978-3-642-15939-8_10
17. Hoffmann, R., Zhang, C., Ling, X., Zettlemoyer, L., Weld, D.S.: Knowledge-based weak supervision for information extraction of overlapping relations. In: Proceedings of 49th Annual Meeting Association Computer Linguistics Human Language Technologies, pp. 541–550, June 2011
18. Ritter, A., Zettlemoyer, L., Etzioni, O.: Modeling missing data in distant supervision for information extraction. Trans. Assoc. Comput. Linguist. **1**, 367–378 (2013)
19. Surdeanu, M., Tibshirani, J., Nallapati, R., Manning, C.D.: Multi-instance multi-label learning for relation extraction. In: Proceedings Joint Conference Empirical Methods Natural Language Processing Computer Natural Language Learning Association Computer Linguistics, pp. 455–465, July 2012
20. Socher, R., Huval, B., Manning, C.D., Ng, A.Y.: Semantic compositionality through recursive matrix-vector spaces. In: Proceedings of the 2012 Joint Conference on Empirical Methods in Natural Language Processing and Computational Natural Language Learning, pp. 1201–1211. Association for Computational Linguistics (2012)
21. Sorokin, D., Gurevych, I.: Context-aware representations for knowledge base relation extraction. In: Proceedings of the 2017 Conference on Empirical Methods in Natural Language Processing, pp. 1784–1789 (2017)
22. Ye, Z.X., Ling, Z.H.: Distant supervision relation extraction with intra-bag and inter-bag attentions. Comput. Lang. arXiv:1904.00143 (2019)
23. Yuan, Y., et al.: Cross-relation cross-bag attention for distantly-supervised relation extraction. In: National Conference on Artificial Intelligence (2019)
24. Bahdanau, D., Cho, K., Bengio, Y.: Neural machine translation by jointly learning to align and translate. arXiv preprint arXiv:1409.0473 (2014)
25. Mnih, V., Heess, N., Graves, A., et al.: Recurrent models of visual attention. In: Advances in Neural Information Processing Systems, pp. 2204–2212 (2014)
26. Hoffmann, R., Zhang, C., Ling, X., Zettlemoyer, L., Weld, D.S.: Knowledge-based weak supervision for information extraction of overlapping relations. In: Proceedings of ACL, pp. 541–550 (2011)

Cross-View Adaptation Network
for Cross-Domain Relation Extraction

Bo Yan[1(✉)], Dongmei Zhang[1], Huadong Wang[2], and Chunhua Wu[3]

[1] School of Computer Science, Beijing University of Posts and Telecommunications,
Beijing, China
{mjkbyb,zhangdm}@bupt.edu.cn
[2] Samsung Research China-Beijing (SRC-B), Beijing, China
huadong.wang@samsung.com
[3] School of Cyberspace Security,
Beijing University of Posts and Telecommunications, Beijing, China
wuchunhua@bupt.edu.cn

Abstract. In relation extraction, directly adopting a model trained in
the source domain to the target domain will suffer greatly performance
decrease. Existing studies extract the shared features between domains
in a coarse-grained way, which inevitably introduce some domain-specific
features or suffer from information loss. Inspired by human beings often
using different views to find connection between domains, we argue that,
there exist some fine-grained features which can be shared across different
views of origin data. In this paper, we proposed a cross-view adaptation
network, which use adversarial method to extract shared features and
introduce cross-view training to fine-turn it. Besides, we construct some
novel views of input data for cross-domain relation extraction. Through
experiments we demonstrated that the different views of data we construct can effectively avoid introducing some domain-specific features
into unified feature space and help the model learn a fine-grained shared
features of different domain. On the three different domains of ACE 2005
dataset, Our method achieved the state-of-the-art results in F1-score.

Keywords: Relation extraction · Domain adaption · Cross-view
training · Adversarial training

1 Introduction

Relation extraction refers to extracting the relation between entities within a
sentence. For example, given the sentence "*His hometown is Beijing*", we
can extract Located relation between "*hometown*" and "*Beijing*". Relation
extraction is often seen as a supervised classification task, and many methods have show great performance on it. However, for the relation extraction
across domains, due to the different data distribution, the models learned in
one domain directly applied to another domain often have poor performance, so
recent research pay more attention to cross-domain relation extraction.

© Springer Nature Switzerland AG 2019
M. Sun et al. (Eds.): CCL 2019, LNAI 11856, pp. 306–317, 2019.
https://doi.org/10.1007/978-3-030-32381-3_25

To solve the problem of domain adaptation, a simple method is training a model on the source domain and using target domain data to fine-tune it [1,10]. However, this method requires expensive labeling costs and expects a high quality data distribution on the target domain. [9,13–15,19] use some manually crafting features such as word clustering and dependency path to adapt models. This method effectively captures some domain invariant features, but will loss information due to the limited human knowledge. With the development of Adversarial Training in recent years, some studies have used adversarial method to automatically extract domain invariant features. [7] used CNN with multiple kernels as shared feature extractor, to extract domain invariant features through adversarial training, and jointly optimizes with relation classifier. [16] train the model on the source domain, and use another model to extract target domain features to match the source feature distribution.

Most of the above studies project the shared features and private features into one unified space, to make the model learn the shared features of different domains, but these methods inevitably introduce some domain-specific features. [17] using the domain separation network [3] to extract the domain-specific features and domain-general features separately, so as to limit domain-specific features into shared feature space. Though it introduce reconstruction loss, but still suffer from information loss due to the dimension reduction of intermediate representation and at the same time the model becomes more complicated.

To address these problems, we proposed a cross-view adaptation network which adopted cross-view training [6] on the target domain. Source labeled data and target unlabeled data is fed into a shared feature extractor to learn a common representation and produce relation prediction using these features. Besides these full-view data, we construct some restrict-view data which loss some contextual information on the target domain, such as masking the entity word. These restrict-view data also be fed into shared feature extractor to produce prediction. Then full-view data's prediction will act as a teacher to teach the different restrict-view data learn the same prediction. By matching the predict distribution, the model can learn some contextual information that don't relevant to target-specific features such as entity word.

The major contributions of this paper are as follows:

– A novel domain adaptation method for relation extraction is proposed, which uses cross-view training on the target domain to fine-tune the shared feature space. This method can make the model learn some fine-grained shared features and more effectively adapt to the target domain.
– We construct different views of target domain data for relation extraction. Experiments on ACE 2005 dataset shows our model can significant improve the cross-domain relation extraction performance.

2   Cross-View Adaptation Network

In this section, we present a adaptation method for the cross-domain relation extraction task. This task can be formulated as follows: given a labeled

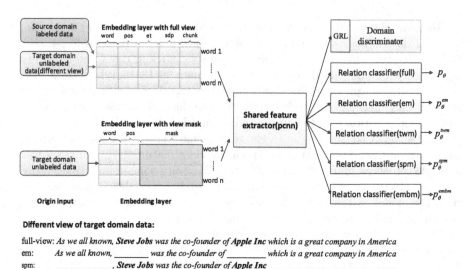

Fig. 1. The overall of cross-view domain adaptation network. We construct different views on target domain unlabeled data. In the origin input layer, we mask some words such as entity pairs. In the embedding layer, we mask some embeddings such as entity type. During training, source domain labeled data and target domain unlabeled data are fed into the network to extract shared features though domain discriminator and relation classifier (full). Different views of target domain data (origin input mask and embedding mask) will be used to fine-tune the shared feature extractor through matching the distribution of relation classifier (full)'s output. In the test stage, we use relation classifier (full) to get the predict of the model. The particular example shows the different views of target data.

source domain corpus $\mathcal{S} = \{(s_1, e_{11}, e_{12}, r_1), \ldots, (s_S, e_{S1}, e_{S2}, r_S)\}$, where $s_i = [w_1, \ldots, w_m]$ denotes a word sentence, e_{i1} and e_{i2} represent the candidate entity pairs and r_i denotes their relation type. The goal of this task is to build a relation extraction model on $\mathcal{S}$ and apply it to an unlabeled target domain corpus $\mathcal{T} = \{(\hat{s}_1, \hat{e}_{11}, \hat{e}_{12}), \ldots, (\hat{s}_T, \hat{e}_{T1}, \hat{e}_{T2})\}$. In other words, for the source domain, the data is labeled, but without any labels in the target domain. Using $\mathcal{S}$ and $\mathcal{T}$ as training data to train a model, so that given a sentence of the target domain and two candidate entities, the model can correctly extract the relation between the entity pairs. The overall of our model is shown in Fig. 1.

2.1   Embedding Layer

From the previous work on relation extraction, we know that embedding features improve relation extraction a lot. Motivated by the work [7,9,13], the embedding layer consists of the following parts:

Word Embedding. We use pre-trained 300 dimensions word embedding from word2vec [12], and every word w_i is converted to the corresponding vector by looking up word embedding table W.

Position Embedding. The position of a word refers to the relative distance between the word and two entities respectively. For example, i and j are the position index of the two entities in a sentence. For the word with index k, its position is $k - i$ and $k - j$. After getting the position of the word, we can get the corresponding position embedding p_{i1} and p_{i2} by looking up position embedding table P.

Entity Type Embedding. Each entity has a type to which it belongs. For each entity in the sentence, we get its entity type embedding t_i through entity type embedding table E. Non-entity word will have the same entity type embedding t_i. For every word in the sentence, we will get two entity type embedding t_{i1} and t_{i2} because we have two candidate entities in each sentence.

Chunks Embedding. Chunks is regard as a phrase that has a specific structure and a relatively stable meaning. We use the method of sequence labeling to get the chunks of the sentence, so that each word has a chunks representation, and then use the chunks embedding table C to get the chunks embedding c_i.

Dependency Path Embedding. We use a binary number to indicate whether a word is on the shortest dependency tree path between two entities, and use dependency path table D to get the dependency path embedding d_i for each word.

At last, we concatenate all above types of embedding to get the representation of one word: $v_i = [e_i; p_{i1}; p_{i2}; t_{i1}; t_{i2}; c_i; d_i]$, all the embeddings are randomly initialized and optimized during training except for the pre-trained word embedding.

2.2 Shared Feature Extractor

We use Piecewise Convolutional Neural Networks (PCNN) [20] as the shared feature extractor. PCNN pays attention to the distance and position of entities, and the context information near the entities, which are the most important features in the relation extraction. It divides a sentence into three parts according to the entity position, such as "*As we all known, **Steve Jobs** was the co-founder of **Apple Inc.** which is a great company in America*", which will be divide into: (1) "*As we all known, **Steve Jobs**"*, (2) "***Steve Jobs** was the co-founder of **Apple Inc.***", (3) "***Apple Inc.** which is a great company in America*".

Applying the multiple convolution kernels on embedding layer, we get a convoluted context matrix $C = [\hat{c}_1, \hat{c}_2, \cdots, \hat{c}_n]$. Let i_1, i_2 denote the index of the last token of first entity e_1 and the second entity e_i respectively, the context matrix can be segmented into three parts $C_1 = [\hat{c}_i]_{1 \leq i \leq i_1}$, $C_2 = [\hat{c}_i]_{i_1 < i \leq i_2}$ and $C_3 = [\hat{c}_i]_{i_2 < i \leq n}$. Using the piecewise max pooling (max-over-time) procedure on them, we generate the fixed size shared feature representation

$$f(V; \theta_s) = [max(C_1), max(C_2), max(C_3)] \tag{1}$$

where V is the input embeddings described in Sect. 2.1, θ_s denotes the parameters of shared feature extractor.

2.3   Relation Classify Loss

The source labeled data together with target unlabeled data are fed into PCNN to get the shared features $f(V; \theta_s)$, but only the shared features $f(V_s; \theta_s)$ from source domain are used to predict the relation type, there V_s denote the source domain data embedding. We use one hidden layer with tanh activation function followed by a softmax to produce the relation distribution:

$$p_{rel} = softmax(R(f(V_s; \theta_s); \theta_y)) \tag{2}$$

where θ_y denotes the parameters of the relation classify layer, R is the relation classify layer. The relation classify loss L_{rel} is defined as:

$$L_{rel} = -\frac{1}{S} \sum_{i=1}^{S} \sum_{j=1}^{M} y_{ij} \log p_{ij} \tag{3}$$

where S is the number of source domain data, M is the total number of relation types, y_{ij} is a binary number to indicate whether the example i has the relation j and p_{ij} is the probability of example i has the relation j.

2.4   Domain Discriminator Loss

Following previous work [7,8], we use adversarial training to learn the shared features of the source and target domain. All the shared features $f(V; \theta_s)$ will be feed into the domain discriminant layer to predict the domain to which the sample belongs. The domain discriminant layer includes one hidden layer with tanh activation function followed by a softmax:

$$p_{dom} = softmax(D(f(V; \theta_s); \theta_d)) \tag{4}$$

where θ_d is the parameters of domain discriminator layer. We use L_{dom} to represent the loss of domain classification:

$$L_{dom} = -\frac{1}{S+T} \sum_{i}^{S+T} (1 - y_i) \log(1 - p_i) + y_i \log p_i \tag{5}$$

where T is the number of target domain data, y_i indicates which domain the sample belongs, p_i is the probability of the sample belonging to source domain.

To make the shared feature extractor extract shared features between domains, following previous work [7], we use the gradient reversed layer (GRL) to reverse the gradient of the parameters before the domain discriminator, then the forward and back propagation are formulated as follows:

$$GRL(x) = x \qquad (6)$$

$$\frac{dGRL(x)}{dx} = -I \qquad (7)$$

where I is an identity matrix. Using GRL, the parameters θ_s are optimized to maximize the L_{dom}, i.e., the domain discriminator can't distinguish which domain the features come from. Meanwhile, the parameters θ_d are trained to minimize the L_{dom}, which tends to correctly distinguish the domain, thus through adversarial training, the shared feature extractor will learn some shared features of source and target domain.

2.5   Cross-View Adaptation Loss

The model we described above can extract shared features of source and target domain, but it inevitable introduce some domain-specific features. In addition, we use some external features, such as chunks information, dependency parsing, and so on. These external features are obtained by other model, and may exist some errors. In order to solve the above problems, we apply cross-view training on the target domain.

First, we construct four restricted views of target domain data for cross-domain relation extraction task:

Entity Pair Mask (em). The effect of relation extraction depends on the context information near the entity pair, and the similar context information often has the same relation. For example, in the phrase "***Basra*** *is a port city in* ***Iraq***", "***Iraq***" and "***Basra***" have a relation of PART-WHOLE, and in the sentence "***Paris*** *is the most prosperous city in* ***France***", "***Paris***" and "***France***" are also have PART-WHOLE relation. In cross-domain relation extraction task, entities in different domains are different, but contexts often similar across domains. These different entities are domain-specific features that hurt the performance of relation extraction on the target domain. Based on this idea, we construct a restricted view of input data by masking the target domain entity pairs, leaving only the context information near entities. By feeding data from this view we get the relation distribution $p_\theta^{em}(y|x_i)$:

$$p_\theta^{em}(y|x_i^{em}) = softmax(R(f(x_i^{em}; \theta_s); \theta_{em})) \qquad (8)$$

where x_i^{em} is the entity pair mask on the target domain data. By matching the distribution of full-view data, the model learns some fixed context patterns that do not depend on entity pairs, this avoid introduce some domain-specific features to some extent. In the test stage, the model tends to predict relation as same as context-like example on the training set.

Target Specific Word Mask (twm). Similar to the idea of entity pair mask, a more violent way is directly masking the words that only appeared in the target domain:

$$p_\theta^{twm}(y|x_i^{twm}) = softmax(R(f(x_i^{twm}; \theta_s); \theta_{twm})) \tag{9}$$

where x_i^{twm} is the target specific word mask on target domain data. Experiments show that this view of input data also improved the cross-domain relation extraction.

Shortest Dependency Path Mask (spm). Shortest dependency path is the shortest path between two entities in the dependency tree. In a sentence, the information needed for extracting the relation between two entities is usually determined by the shortest path between the two entities in the dependency graph [4]. So we mask the words outside the shortest path, thus not only preserving the information needed for the relation extraction, but also removing some domain-specific information. The shortest dependency path mask can be formulated as:

$$p_\theta^{spm}(y|x_i^{spm}) = softmax(R(f(x_i^{spm}; \theta_s); \theta_{spm})) \tag{10}$$

where x_i^{spm} is the shortest dependency path mask on the target domain data.

Embedding Mask (embm). It seems less data efficient that masking word directly. So except for constructing the word level mask, we also explored the mask of the embedding layer. Among the features of our embedding layer, chunk and dependency path features are obtained from other trained models and therefore inevitable have some errors, entity type features are often domain-specific features. So we remove these embedding, leaving only word embedding and position embedding as input to the shared features extractor:

$$p_\theta^{embm}(y|x_i^{embm}) = softmax(R(f(x_i^{embm}; \theta_s); \theta_{embm})) \tag{11}$$

where x_i^{embm} is the embedding mask on target domain data. Through the construction of this input view, the model will be more domain-general and have stronger Fault-tolerance.

Then we feed these restricted views of target domain data together with full views to the network. We use $p_\theta(y|x_i)$ to represent the relation distribution of full views described in (2), where x_i is the origin input sentence on the target domain. During training, we minimize the difference between $p_\theta(y|x_i)$ and $p_\theta^j(y|x_i)$. Specifically, we first get $p_\theta(y|x_i)$ by feeding full views of target domain data and fix it (i.e., do not perform back-propagation) in every training batch, then use the relation distribution $p_\theta^j(y|x_i)$ of restricted views data to match it. We use KL divergence to measure the difference in data distribution, and let the relation distribution of restricted view data fit the distribution of the full-view data by minimizing L_{cv}:

$$L_{cv} = \frac{1}{T} \sum_{i=1}^{T} \sum_{j \in K} D(p_\theta(y|x_i), p_\theta^j(y|x_i^j)) \tag{12}$$

where D is the KL-divergence, $K = \{em, spm, twm, embm\}$ is the set of all input views.

During training, we jointly optimize all the loss function by minimize L_{loss}:

$$L_{loss} = L_{rel} + \alpha L_{dom} + \beta L_{cv} \tag{13}$$

We set $\alpha = 0.1$ and $\beta = 0.01$ through development set.

3   Experiments

3.1   Dataset

We use the English part of ACE 2005, which is a widely used dataset for cross-domain relational extraction. It covers six domains: Newswire (nw), Broadcast Conversation (bc), Broadcast News (bn), Telephone Speech (cts), Usenet Newsgroups (un), and Weblogs (wl). Following previous work [9,17,19], we use nw and bn as the training set, half of bc as the development set, and the remaining half of bc as the test set. After processing, 43497 entity pairs were generated for training. Table 1 show the detailed statistics.

Table 1. ACE 2005 dataset statistics.

Domain	Total	Number of entities	No relation rate
bn+nw	43497	5442	91.6%
bc dev	7004	936	91.2%
bc test	8083	1107	91.1%
cts	15803	769	96.1%
wl	13882	2150	94.9%

3.2   Implementation Details

Following Fu and Grishman [7], we use the same settings for the embeddding layer. The pre-trained word embedding is 300 dimensions generated by word2vec. The embedding size of position/chunk/dependency path is 50. For each convolution layer of PCNN, the convolution step is set to 2, 3, 4, 5, and total filter numbers are 150. Our fixed sentence length is 155 which is the maximum of all sentence and the dropout rate is 0.5. All of the input views shared the same PCNN parameters, but the fully connected layer are different respectively, and we only use full-view fully connected layer for testing, the hidden size of fully connected layer and domain discrimination layer is 300. We use Adam for optimizing the model, the learning is 0.001 and will be halved every two epochs.

3.3   Baseline Models

We compare our proposed model to the following baseline models:

- **Log-linear** [13]. This paper explored some neural network method for relation extraction, include CNN, bidirectional RNN, forward RNN, backward RNN and proposed a combined model called Log-linear. We compare with the single models rather than the combined models they proposed.
- **FCM & Hybrid FCM** [9]. Feature-rich Compositional Embedding Model (FCM) is a model that combines both hand-crafted features with learned word embeddings. The Hybrid model combines the FCM with an existing log-linear model.
- **LRFCM** [19]. Low-rank FCM (LRFCM) dramatically reduced the number of FCM's parameters and can scale to more features and more labels.
- **DANN** [7]. This model is similar with our proposed model but only use domain adversarial training to extract shared features.
- **GSN** [17]. This paper demonstrates that traditional method for cross-domain relation extraction inevitable introduced some domain-specific features. So they extract shared features and private features separately and maximize the difference in their distribution. They also use shared feature to perform relation extraction.

Table 2. Relation extraction performance on different models (Macro F1-score %). +em, +spm, +twm, +embm mean adding entity type mask, shortest dependency path mask, target word mask and embedding mask respectively to the PCNN+DANN model. "Combine" is our final model, taking all input views into account.

Model	bc	cts	wl
Forward RNN	61.44	54.93	55.10
Backward RNN	60.82	56.03	51.78
Bidirectional RNN	63.07	56.47	53.65
CNN	63.26	55.63	53.91
FCM	61.9	52.93	50.36
Hybrid FCM	63.48	56.12	55.17
LRFCM	59.4	-	-
GSN	66.38	57.92	56.84
CNN+DANN	65.16	-	-
PCNN+DANN	65.78	58.56	56.62
+em	66.78	59.28	57.45
+spm	65.53	59.11	57.12
+twm	66.71	58.42	57.21
+embm	66.65	58.77	57.34
Combine	**66.81**	**60.88**	**57.62**

3.4   Evaluation and Analysis

We use macro F1-score to measure relation extraction performance. The result shows that our proposed baseline model PCNN+DANN achieved comparable performance to state-of-the-art models (Table 2). After using cross-view training with different restricted views we construct, our combine model outperform all the previous methods in the three domains of ACE2005 dataset. Specifically, on bc domain, our model obtained 1% gains compared with PCNN+DANN model, and achieved comparable results to GSN model, but GSN is more complex and requires more time training. On cts domain, our model increased the F1-score by 3%, which is a significant improvement.

We also performed ablation experiments to explore the effects of different input views (Table 2). Among them, the entity pair mask has the greatest impact on the results, which also verifies our previous analysis. Some entities are domain-specific and have a negative effect on the results. Using the entity pair mask, to some extent, the shared feature extractor is prevented from introducing some domain-specific features so that the model can better transfer from source domain to target domain. Target specific word mask have a very different effect on different domains. We calculated the proportion of shared words in different domains and found that when the number of these words decrease, the model gets less gain. We analyze that this may be due to more context information is lost when there are fewer shared words, which affects the relation extraction. This also explains that the data utilization efficiency is lower by directly masking the origin word. The embedding mask view compensates for this shortcoming. It used all words in a sentence as input, but only masked some features in the embedding layer. We also tried to mask some intermediate representations of the model, such as masking one output feature of PCNN's piece, but the experiments show that this will make the result worse, which is obvious because it masked most of the context near entity. Like Clark and Manning [6], we also use LSTM as a relation extractor and construct forward and backward restricted views, but the results are also worse. We suppose that sequence tagging task relies more on sequential information but relation extraction task does not.

4   Related Work

Most existing cross-domain relation extraction studies focus on learning a shared feature representation between source and target domains. [15] use manually constructed features such as word clustering and tree kernels to improve the ability of domain adaptation. Some neural network based methods such as CNN, RNN, or compositional models also significantly advance the performance of cross-domain relation extraction [9,13,14,19]. Recently some methods based on adversarial training have also been applied to cross-domain relation extraction. [7] and [16] used adversarial training to project the common features of source and target domains into one feature space, and then use these features to identify relations. [5,11,17] respectively extract private and shared features based on the domain separation network [3] to avoid introduce private features into shared feature space. These methods all find the connection between different domains from a full view, and intuitively, if observing the data from different part views, we can more clearly discover the shared features of different domains.

Cross-view training [6] is a semi-supervised learning algorithm. It constructs some restricted data input views and use them to improve the intermediate representation of the model. A very similar algorithm is multi-view learning, which divides features into sub-features [18], and uses co-training algorithm [2] to train two separate models. On unlabeled data, each one acts as a "teacher" for the other model. While in cross-view training, there is only one model and use different views of unlabeled data to improve the shared model. Cross-view training has achieved good results in some tasks such as sequence labeling and machine translation, but it has not been applied to domain adaptation. We used similar training methods and constructed some novel input views for relation extraction. Experiments show this method is suitable for cross-domain relation extraction.

5 Conclusion

We proposed a cross-view training based method for cross-domain relation extraction, and not required any labeled data in target domain. Specifically, We innovatively constructed some data input views for cross-domain relation extraction, and experiments demonstrated that by changing the data input views, such as masking some domain-specific information, the model can learn the shared features more effectively. To the best of our knowledge, this is the first study to apply cross-view training to cross-domain relation extraction. We hope that this method can be extended to other domain adaptation tasks in future research.

Acknowledgements. This work was supported by National Key R&D Program of China (2017YFB0802703) and National Natural Science Foundation of China (61602052).

References

1. Blitzer, J., McDonald, R., Pereira, F.: Domain adaptation with structural correspondence learning. In: Proceedings of the 2006 Conference on Empirical Methods in Natural Language Processing, EMNLP 2006, Stroudsburg, pp. 120–128. Association for Computational Linguistics (2006)
2. Blum, A., Mitchell, T.: Combining labeled and unlabeled data with co-training. In: Proceedings of the Eleventh Annual Conference on Computational Learning Theory, COLT 1998, pp. 92–100. ACM, New York (1998). https://doi.org/10.1145/279943.279962
3. Bousmalis, K., Trigeorgis, G., Silberman, N., Krishnan, D., Erhan, D.: Domain separation networks. In: Lee, D.D., Sugiyama, M., Luxburg, U.V., Guyon, I., Garnett, R. (eds.) Advances in Neural Information Processing Systems 29, pp. 343–351. Curran Associates Inc., New York (2016)
4. Bunescu, R.C., Mooney, R.J.: A shortest path dependency kernel for relation extraction, January 2005. https://doi.org/10.3115/1220575.1220666

5. Chen, X., Shi, Z., Qiu, X., Huang, X.: Adversarial multi-criteria learning for Chinese word segmentation. In: Proceedings of the 55th Annual Meeting of the Association for Computational Linguistics (Volume 1: Long Papers), Vancouver, July 2017, pp. 1193–1203. Association for Computational Linguistics (2017). https://doi.org/10.18653/v1/P17-1110
6. Clark, K., Luong, T., Manning, C.D., Le, Q.V.: Semi-supervised sequence modeling with cross-view training (2018)
7. Fu, L., Nguyen, T.H., Min, B., Grishman, R.: Domain adaptation for relation extraction with domain adversarial neural network. In: Proceedings of the Eighth International Joint Conference on Natural Language Processing (Volume 2: Short Papers), Taipei, November 2017, pp. 425–429. Asian Federation of Natural Language Processing (2017)
8. Ganin, Y., Lempitsky, V.: Unsupervised domain adaptation by backpropagation. In: Proceedings of the 32nd International Conference on Machine Learning, ICML 2015, vol. 37, pp. 1180–1189. JMLR.org (2015)
9. Gormley, M.R., Yu, M., Dredze, M.: Improved relation extraction with feature-rich compositional embedding models. In: Proceedings of the 2015 Conference on Empirical Methods in Natural Language Processing, Lisbon, September 2015, pp. 1774–1784. Association for Computational Linguistics (2015). https://doi.org/10.18653/v1/D15-1205
10. Jiang, J., Zhai, C.: Instance weighting for domain adaptation in NLP, January 2007
11. Liu, P., Qiu, X., Huang, X.: Adversarial multi-task learning for text classification. arXiv preprint arXiv:1704.05742 (2017)
12. Mikolov, T., Chen, K., Corrado, G., Dean, J.: Efficient estimation of word representations in vector space. In: Proceedings of Workshop at ICLR 2013, January 2013
13. Nguyen, T.H., Grishman, R.: Combining neural networks and log-linear models to improve relation extraction, November 2015. arXiv e-prints
14. Nguyen, T.H., Grishman, R.: Employing word representations and regularization for domain adaptation of relation extraction. In: Proceedings of the 52nd Annual Meeting of the Association for Computational Linguistics (Volume 2: Short Papers), Baltimore, June 2014, pp. 68–74. Association for Computational Linguistics (2014). https://doi.org/10.3115/v1/P14-2012
15. Plank, B., Moschitti, A.: Embedding semantic similarity in tree kernels for domain adaptation of relation extraction, vol. 1, pp. 1498–1507, August 2013
16. Rios, A., Kavuluru, R., Lu, Z.: Generalizing biomedical relation classification with neural adversarial domain adaptation. Bioinformatics 34(17), 2973–2981 (2018)
17. Shi, G., et al.: Genre separation network with adversarial training for cross-genre relation extraction. In: Proceedings of the 2018 Conference on Empirical Methods in Natural Language Processing, Brussels, October–November 2018, pp. 1018–1023. Association for Computational Linguistics (2018)
18. Xu, C., Tao, D., Xu, C.: A survey on multi-view learning. CoRR abs/1304.5634 (2013)
19. Yu, M., Gormley, M.R., Dredze, M.: Combining word embeddings and feature embeddings for fine-grained relation extraction, pp. 1374–1379, January 2015. https://doi.org/10.3115/v1/N15-1155
20. Zeng, D., Liu, K., Chen, Y., Zhao, J.: Distant supervision for relation extraction via piecewise convolutional neural networks. In: Proceedings of the 2015 Conference on Empirical Methods in Natural Language Processing, Lisbon, September 2015, pp. 1753–1762. Association for Computational Linguistics (2015). https://doi.org/10.18653/v1/D15-1203

Machine Translation and Multilingual Information Processing

Character-Aware Low-Resource Neural Machine Translation with Weight Sharing and Pre-training

Yichao Cao[1,2], Miao Li[1(✉)], Tao Feng[1,2], and Rujing Wang[1,2]

[1] Institute of Intelligent Machines, Chinese Academy of Sciences,
Hefei 230031, People's Republic of China
{mli,rjwang}@iim.ac.cn
[2] University of Science and Technology of China,
Hefei 230026, People's Republic of China
{cycao,ft2016}@mail.ustc.edu.cn

Abstract. Neural Machine Translation (NMT) has recently achieved the state-of-the-art in many machine translation tasks, but one of the challenges that NMT faces is the lack of parallel corpora, especially for low-resource language pairs. And the result is that the performance of NMT is much less effective for low-resource languages. To address this specific problem, in this paper, we describe a novel NMT model that is based on encoder-decoder architecture and relies on character-level inputs. Our proposed model employs Convolutional Neural Networks (CNN) and highway networks over character inputs, whose outputs are given to an encoder-decoder neural machine translation network. Besides, we also present two other approaches to improve the performance of the low-resource NMT system much further. First, we use language modeling implemented by denoising autoencoding to pre-train and initialize the full model. Second, we share the weights of the front few layers of two encoders between two languages to strengthen the encoding ability of the model. We demonstrate our model on two low-resource language pairs. On the IWSLT2015 English-Vietnamese translation task, our proposed model obtains improvements up to 2.5 BLEU points compared to the baseline. We also outperform the baseline approach more than 3 BLEU points on the CWMT2018 Chinese-Mongolian translation task.

Keywords: Low-resource neural machine translation ·
Character-level · Weight sharing · Pre-training

1 Introduction

Machine Translation (MT) is a challenging task in the field of natural language processing, which is aimed at transforming a source language into a target language automatically. Thanks to recent advances in deep learning, NMT has reached large improvements in some standard benchmarks and even has achieved

© Springer Nature Switzerland AG 2019
M. Sun et al. (Eds.): CCL 2019, LNAI 11856, pp. 321–333, 2019.
https://doi.org/10.1007/978-3-030-32381-3_26

near human-level performance on several language pairs [1,2]. Unfortunately, the performance of NMT depends heavily on numerous high-quality parallel corpora, which is only available for a few language pairs. In most cases, the NMT model has better performance in high-resource settings because of the help of significant amounts of training data. Some recent advances have reported poor performance of NMT systems in low-resource settings [3,4]. Therefore, improving the performance of low-resource NMT is a valuable study.

In this work, we propose a novel NMT model that exploits character information by a character-level CNN, whose output is used as an input to an encoder-decoder neural machine translation network module. Unlike previous works [5] that combine input word embeddings with map features from a character-level CNN module, our model only utilizes feature representations from CNN. Hence, we can take advantages of intra-word information and subword information to handle rare or out-of-vocabulary words of low-resource languages, especially when processing morphologically rich languages. Besides, our model no longer requires word vocabulary in source side by replacing word embedding layer with character-level CNN. The translation system of our model is based on the Transformer model [1], which follows an encoder-decoder architecture. It is based on attention mechanisms solely and dispenses with recurrence and convolutions entirely.

In order to improve the performance of our model even further, we present two approaches for low-resource neural machine translation. The first one is sharing weights between languages. More concretely, we employ the multi-task learning framework to build our NMT model, which is based on encoder-decoder architecture. We first build two NMT models in two opposite directions for a given low-resource language pair such as English-Vietnamese (e.g. English-to-Vietnamese and Vietnamese-to-English). Inspired by [6], we then share the weights of the front few layers of two encoders that are responsible for mapping input sentences into high-level representation space. Note that, two decoders are not shared. Intuitively, the shared encoder can make better use of the similarity and complementarity between different languages, especially for low-resource language pairs. The second approach utilizes strong language modeling to pre-train and initialize our full model, which can enhance the quality of translation results for low-resource NMT. Similar to [7], we apply language modeling pre-training via training the encoder-decoder system as a denoising autoencoder [8]. Furthermore, denoising autoencoding is still in progress during translation training. To summarize, our contributions are as follows:

- We propose a character-aware and weight-sharing constraint for low-resource NMT, and at the same time apply language modeling which is implemented by denoising autoencoding to pre-train and initialize the entire model. To maintain the internal characteristics of each language for the model, we also keep denoising autoencoding throughout the training.
- We conduct extensive experiments on English-Vietnamese and Chinese-Mongolian low-resource translation tasks. The experiment results demonstrate that the proposed approach is effective for low-resource NMT and significantly outperforms the baseline model.

2   Background and Related Work

In recent years, a neural machine translation model is usually implemented by an encoder-decoder architecture [9], and the encoder and decoder of the NMT model are often based on recurrent neural networks (RNN) or Transformer. The encoder reads a source sentence $X = (x_1, \ldots, x_{T_x})$ as an input and encodes it into a high-level representation $H = (h_1, \ldots, h_{T_x})$. Then the decoder generates corresponding translation $Y = (y_1, \ldots, y_{T_y})$ based on the encoded sequence of hidden states H, where x_t, h_t and y_t are the symbols of source language, hidden states and target language, respectively. Generally, the NMT model is trained to maximize the conditional log-probability of a target sentence Y given a source sentence X as:

$$p\left(Y \mid X\right) = \prod_{t=1}^{T} \left(Y_t \mid Y_{0:t-1}, X; \theta\right) \tag{1}$$

where θ is the parameters of the model, and T represents the length of Y. In this work, our model is based on the Transformer model, which uses only attention mechanisms. Attention mechanism can be described as mapping queries (Q) and a set of key-value (K-V) pairs to outputs, where Q, K and V are all vectors. In practice, we compute the attention function as follows:

$$R_a = softmax\left(QK^T\right)V \tag{2}$$

where R_a denotes the attention vector.

In the past, many studies have produced competitive results on low-resource NMT. [3] introduced a transfer learning approach to improve the performance of low-resource NMT, and [10] took advantage of a model-agnostic meta-learning algorithm to train the model. Other attempts exploited the availability of monolingual corpora to obtain better results on low-resource NMT [11,12]. However, most of the NMT models belong to a family of word-level systems, whose large vocabulary is often filled with many similar words. Besides, many words are out-of-vocabulary (OOV) because of the paucity of data for low-resource language pairs. In our approach, character-level NMT and language modeling pre-training are employed to address the above problems. There have been numerous great efforts at character-level neural machine translation. [13] made attempt to compose words from individual characters with the help of Long Short Term Memory (LSTM). [14] exploited character-aware word vectors to replace the standard word representations on the source side and demonstrated it is much effective. There is also a line of work on character-level NMT that processes words at the character level [15,16]. For language modeling pre-training, the Generative Pre-trained Transformer (OpenAI GPT) [17] and the Bidirectional Encoder Representations from Transformers (BERT) [7] are trained on the downstream tasks by fine-tuning the pre-trained language modeling parameters. Our work is most aligned with the language modeling proposed by [18], which made use of denoising autoencoding to train models on both source and target languages.

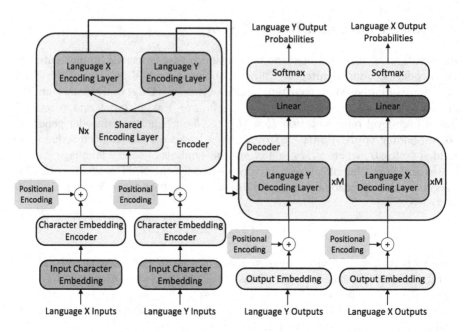

Fig. 1. The architecture of the proposed model. The model uses the multi-task learning framework for two languages X, Y of a given low-resource language pair. In source side, the standard word input embedding is replaced by a combination of character-level CNN and highway network which is called Character Embedding Encoder. In target side, we still use the standard word embedding lookup table. We share N layers of the encoder but do not share any layers of the decoder. The feed forward layers and the softmax layers are employed to get output probabilities of two languages.

A recent trend in multi-task neural machine translation is to share the weights of the model between different languages. [6,19] investigated how to leverage weight sharing in unsupervised and supervised NMT to improve the quality of the translation results. Our proposed model also belongs to this scenario, which can exploit the similarity and complementarity between different languages and mitigate the overfitting issues for low-resource NMT.

3   Model Architecture

It is well known that low-resource NMT systems usually suffer from the limited amount of parallel data, OOV words and model overfitting. To address these specific problems, we propose a character-aware low-resource NMT model that follows the encoder-decoder architecture. As shown in Fig. 1, we introduce the multi-task learning framework to build the whole model and use the data of one low-resource language pair as the model inputs. The character-level inputs of different languages are fed into the combination of CNN and highway network that is called Character Embedding Encoder (CEE). The machine translation encoder-decoder module of our model is based on the Transformer model, and the input of encoder is obtained from CEE. Note that we provide the standard word-level embeddings as the decoder inputs rather than character-level, and predictions are also still at the word-level.

3.1 Character Embedding Encoder

In recent years, NMT has benefited from character-aware representations on the source side or target side [14,20]. Inspired by [21], we replace the word embedding layer with a combination of CNN and highway network [22] which called CEE to acquire the embedding of a word. The structure of CEE is illustrated in Fig. 2, and the input of the model encoder for each word is the output from CEE. Let $X = [x_1, \ldots, x_n]$ be the input sentence of our model, where x_i is i-th word of sentence. And assume that word x_i consists of a sequence of characters $[x_c^1, \ldots, x_c^m]$, where m is the length of word x_i. Let $\mathbf{c_i}$ be the embedding of a character, then character embeddings are concatenated together as input $\mathbf{C_{x_i}}$ of CNN for each word x_i. A narrow 1D convolution is applied between concatenation of character embeddings $\mathbf{C_{x_i}}$ and a kernel $K \in \mathbb{R}^{d_c \times w}$ with width w, where d_c denotes the dimensionality of character embeddings. Then we utilize max-over-time pooling to keep only the maximum value of the output for each convolutional filter, and concatenate these feature maps as the input Z_f of the highway network.

To acquire a better embedding for word x_i, we employ the highway networks and a fully connected layer on Z_f. A highway network allows some character n-grams to be combined to build new features, and other character n-grams to remain primitive features. We can compute a new set of features Z with Z_f by the following formula,

$$Z = u \odot g\left(W_H Z_f + b_H\right) + u' \odot Z_f \tag{3}$$

where $u = \sigma\left(W_u Z_f + b_u\right)$ is called the *transform* gate, $u' = 1 - u$ is called the *carry* gate, g denotes a nonlinearity, and σ, W_H, W_u, b_H, b_u are all trainable parameters. Besides, a fully connected feed-forward layer with a *ReLU* activation is applied to seek a more satisfying word representation for each word, and calculated as follows:

$$E_w = ReLU\left(Z W_{ff}\right) + b_{ff} \tag{4}$$

where E_w is the final word embedding of the word x_i, W_{ff} and b_{ff} denote the transformation matrix and bias respectively. The dimensionality of E_w is d_m, which matches with the dimension of model encoder and decoder hidden layers.

3.2 Weight Sharing Encoder

It is well known that there are the similarity and complementarity between different languages, [23] has demonstrated that multilingual translation is indeed

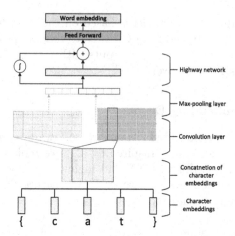

Fig. 2. The architecture of Character Embedding Encoder (CEE). Firstly, combining character embeddings of a word into a representation **C**. Then convolution operations are applied between **C** and multiple filter matrices. We only show 2 convolution filters in the above illustration. A max-over-time pooling operation is employed to get a high-level representation **Z** of the word, and **Z** will be used as an input to the highway network. Finally, a fully connected feed-forward layer is applied over the output of the highway network to achieve the embedding of this word.

strongly beneficial for resource-scarce language pairs. Our proposed model takes advantage of the multi-task learning framework, and employs jointly training for two languages in a given low-resource language pair.

Specifically, for example, we build two encoder-decoder translation systems for English-Vietnamese whose parallel corpus is not sufficient. One of them is trained to translate from English to Vietnamese, and the other is trained in the opposite direction (Vietnamese to English). And then we share the weights of the front few layers of two encoders, but not share the weights of two decoders. Intuitively, encoder weight sharing can make the best of the similarity and complementarity between two languages to extracting the high-level representation of each input. Furthermore, the shared encoder may keep the balance between two languages of a given low-resource language pair, and do not cause overfitting caused by the data scarcity problem. Note that we do not share the last layer of two encoders and all layers of two decoders in our approach, because the separate encoder or decoder preserves the uniqueness and internal characteristics of each language. Another reason we do not share decoders of two translation systems is that the decoder utilizes not only self-attention mechanism but also dot-product attention [1]. The alignment matrix calculated from the dot-product attention mechanism is not same for different source and target languages, therefore we utilize two independent decoders to building the NMT model.

Algorithm 1. Training for our proposed model

Input: Monolingual data D'_x, D'_y and parallel data D_x, D_y
Output: An NMT model: $M_{x \to y, y \to x}$
1: **for** i=1 to N **do**
2:     Sample data $X'(i) \in D'_x$ and $Y'(i) \in D'_y$
3:     Add noise to $X'(i)$, $Y'(i)$ and obtain noised inputs $X_c'(i)$, $Y_c'(i)$
4:     Train character-aware language modeling L_x and L_y via denoising autoencoding using $X_c'(i)$ and $Y_c'(i)$
5: Initialize the translation model $M^{(0)}_{x \to y, y \to x}$ using L_x and L_y
6: **for** t=1 to T **do**
7:     Train $M^{(t)}_{x \to y, y \to x}$ using D_x and D_y
8:     Fine-tune $M^{(t)}_{x \to y, y \to x}$ via denoising autoencoding using D'_x and D'_y
9: **return** $M^{(T)}_{x \to y, y \to x}$

3.3 Denoising Autoencoding and Training

Language modeling pre-training has achieved great success for most natural language processing tasks, and [24] also showed the effectiveness of cross-lingual pre-training on unsupervised machine translation. In our method, language modeling is accomplished via denoising autoencoding, that can reconstruct the original sentences from the noised inputs and learn some useful structures in the data. To be specific, let $X' = [x'_1, \ldots, x'_n]$ be the original sentence sampled from monolingual data D'_x, we add three different types of noise to X' and get the noised input X'_c. Firstly, we shuffle the input sentence with a random permutation γ and verify the condition:

$$|\gamma(i) - i| \leq W', \forall i \in \{1, n\} \tag{5}$$

where W' is a tunable parameter. Secondly, we delete some word of X' with a probability P_d. In practice, we consider $P_d = 0.1$. Last but not least, we sample randomly 10% of the word from X', and replace them by a special token '$<mask>$' 80% of the time and keep them unchanged 20% of the time similar to [7]. This way, the encoder is forced to keep a distributional contextual representation of X', and the decoder is trained to learn the internal structure of the sentence to generate the original inputs.

In our work, we first pre-train the language modeling for N step by minimizing:

$$\mathcal{L}_{ae} = \mathbb{E}_{X' \sim D'_x}[-log P_{x \to x}(X' \mid X'_c)] + \mathbb{E}_{Y' \sim D'_y}[-log P_{y \to y}(Y' \mid Y'_c)] \tag{6}$$

where X'_c and Y'_c denote noised inputs, D'_x and D'_y are monolingual data, $P_{x \to x}$ and $P_{y \to y}$ are the denoising autoencoder both operating on two languages, respectively.

As we described, the proposed model is based on the multi-task learning framework. Hence, joint training is applied to a given low-resource language pair translation, and the NMT model is initialized using the pre-trained model after

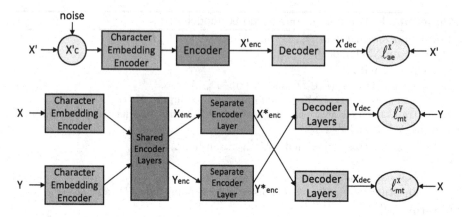

Fig. 3. Illustration of denoising autoencoding and translation procedure. Top (denoising autoencoding): the model learns to reconstruct the original sentence X' from the noised sentence X'_c. X'_{enc} and X'_{dec} are the output of the encoder and decoder, respectively. Bottom (translation): as described before, we jointly train the translation model for a given low-resource language pair. X and Y are source inputs, X_{enc}, Y_{enc} and X^*_{enc}, Y^*_{enc} are outputs of shared encoder layers and separate encoder layer respectively. X_{dec} and Y_{dec} are decoder layers outputs. The gray ellipses indicate terms in the loss function.

finishing N step pre-training. Then we can train our NMT model in two opposite directions for one language pair, and fine-tune it through denoising autoencoding at every iteration. In practice, it is worth noting that the proportion of denoising autoencoding loss in the total loss is decreased along with the training. The overview of our algorithm is given in Algorithm 1. Let X and Y be sentences sampled from parallel data D_x and D_y, the translation loss can be written as:

$$\mathcal{L}_{mt} = \mathbb{E}_{X \sim D_x, Y \sim D_y} \left[(-log P_{x \to y}(Y \mid X)) + (-log P_{y \to x}(X \mid Y)) \right] \quad (7)$$

where $P_{x \to y}$ or $P_{y \to x}$ denotes the translation model.

In summary, as shown in Fig. 3, the final objective function at one iteration of our method is thus:

$$\mathcal{L}_{model} = \lambda_{ae} \mathcal{L}_{ae} + \lambda_{mt} \mathcal{L}_{mt} \quad (8)$$

where λ_{ae} and λ_{mt} are hyper-parameters weighting the importance of the denoising autoencoding and translation loss. λ_{ae} is decreased along with the training.

4 Experiments

In this section, we first describe the datasets and experimental protocol that we used. Then, we compare the proposed approach with the baseline method and other methods. The results and related discussions will be reported in the end.

4.1 Datasets

In our experiments, we consider two low-resource language pairs: English-Vietnamese and Chinese-Mongolian. For English-Vietnamese, we utilize available parallel corpus from IWSLT 2015 which is composed of 131 K sentence pairs. For Chinese-Mongolian, we use CWMT 2018 Chinese-Mongolian dataset consisting of about 260 K sentence pairs. We separate all parallel sentences as the monolingual corpora. We measure the performance of all methods by BLEU score, which is often used in translation tasks. We report BLEU scores on tst2012 and tst2013 for English-Vietnamese, and report results on test2017 for Chinese-Mongolian.

Moreover, we apply word segmentation to the Chinese sentences using THU Lexical Analyzer for Chinese (THULAC) [25]. Because our model is a character-to-word (*char2word*) translation model, we only need the character vocabulary for the *source-side* inputs. In practice, we give a source character vocabulary of size 200–300 for each language except Chinese. For Chinese character vocabulary, we take the most frequent five thousand Chinese characters and replace the rest with '$<unk>$' token. Like most word-level NMT systems, the standard word vocabulary is employed on the target side.

4.2 Training Details

In this study, we built the proposed model upon the character-level CNN, highway networks and Transformer cells. Following [21], the CNN has filters of width [1, 2, 3, 4, 5, 6, 7] of size [50, 100, 150, 200, 200, 200, 200] for a total of 1100 filters, and the highway network has 2 layers with a *ReLU* activation. Besides, the dimensionality of character embeddings is set to 50. Our encoder-decoder translation architecture is based on Transformer cells with 2048 hidden units and 8 heads. We use 6 layers both in the encoder and decoder, and share the weights of the front 5 layers of the encoder between two source languages. The dimensionality of the word embeddings and the hidden layers is set to 512. The model is trained using the Adam optimizer [26] with $\beta_1 = 0.9, \beta_2 = 0.98$ and $\epsilon = 10^{-9}$, and the initial learning rate is set to 2. The learning rate varies according to the formula:

$$learning\_rate = d_m^{-0.5} \cdot min\left(gs \cdot ws^{-1.5}, gs^{-0.5}\right) \tag{9}$$

where d_m is the dimensionality of the hidden layers, gs and ws denote the global step and warmup step, respectively. We also apply dropout [27] with a rate of 0.1 to the output of each sub-layer, the embeddings and the positional encodings.

To evaluate the effectiveness of the proposed method and the importance of different components, we compare our model with the baseline method Transformer and other models. We utilize the symbol *CWSP* to denote our proposed model, and the following methods will be compared:

- Transformer (word2word): This approach is our baseline method similar to [1].

Table 1. The translation performance on IWSLT2015 English-Vietnamese tst2012 set, tst2013 set and CWMT2018 Chinese-Mongolian test2017 set.

Models	en-vi (tst2012)	vi-en (tst2012)	en-vi (tst2013)	vi-en (tst2013)	zh-mn (test2017)	mn-zh (test2017)
Transformer (word2word)	24.69	22.36	26.94	24.42	29.14	15.87
Transformer (char2word)	24.77	22.75	27.71	24.65	30.66	16.33
CWSP (without LM)	24.95	23.60	27.84	25.58	31.20	16.55
CWSP (4 shared layers)	25.29	24.23	27.78	26.46	30.94	16.37
CWSP (6 shared layers)	25.08	24.12	27.79	26.59	31.25	18.69
CWSP (our full model)	**25.41**	**24.40**	**27.93**	**26.93**	**31.92**	**18.86**

- Transformer (char2word): This model is the same as the baseline method except for the source inputs, which is character-level.
- CWSP (without LM): This is our model without denoising autoencoding.
- CWSP (4 shared layers): This is our approach which only shares four layers of the encoder.
- CWSP (6 shared layers): This is our model which shares all six layers of the encoder.
- CWSP (our full model): This is our full model which shares 5 layers of the encoder and does not share any layers of the decoder.

4.3   Results and Analysis

Table 1 reports the translation BLEU scores of different systems on English-Vietnamese and Chinese-Mongolian tasks. The results show that our proposed approach obtains significant improvements compared with the baseline method. Our model can achieve at least +0.7 BLEU points improvement in English-to-Vietnamese translation, and up to +2.5 BLEU points in Vietnamese-to-English translation. We also can reach 31.92 and 18.86 BLEU scores in Chinese-Mongolian test2017, outperforming the baseline method by +2.78 and +2.99 points respectively. As can be seen, the proposed approach is effective for low-resource machine translation.

Besides, compared to the baseline model, the Transformer (char2word) model leads to improvements of up to +1.52 BLEU on two translation tasks. This confirms the CEE module of our model is able to encode semantically meaningful features, and then generates better word embeddings. It has a beneficial effect on low-resource NMT. The approach named *CWSP*(without LM) gets improvements compared with the Transformer (char2word) model, with up to +0.93 BLEU points on English-Vietnamese translation task. It reveals that the model can encode the inputs better via sharing the weights of layers of encoders between two languages. And the encoder of our model can utilize the similarity and complementarity of two different languages to enhance the low-resource neural machine translation performance. We also find that the shared encoder is less prone to overfitting in practice.

Our full model obtains improvements up to +1.35 BLEU points compared to the proposed model without denoising autoencoding (*CWSP*(without LM)) on English-Vietnamese test sets, and at least +0.72 BLEU points on Chinese-Mongolian test set. It shows that pre-training allows the NMT model to learn the semantic features of the language, and guides the decoder to generate more satisfactory sentences. We also investigate how the number of weight sharing layers of encoders affects translation performance. From Table 1, we find that better performance is achieved when 5 layers are shared in our model. Therefore, we set the number of the encoder layers as 5 in our full model. Our model with 4 shared layers obtains up to -0.47 points decline than our full model in English-Vietnamese translation, and at least -0.98 points in Chinese-Mongolian translation. It verifies less shared layers of two encoders do not take full advantage of the relationship between two languages of a given resource-scarce language pair. The BLEU score is worse than the result of our full model when we share all of the six layers of two encoders. And we also share some layers of two decoders, the results are even worse than the baseline model. We explain these as that the shared encoder or the shared decoder are weak in keeping the unique characteristics of each language of given low-resource language pairs. The separate encoder and decoder can extract the distinctive characteristics for different languages, thus we share the front few layers of two encoders rather than all layers, and do not share any decoders.

5   Conclusion and Future Work

In this work, we introduce a character-aware encoder-decoder architecture for low-resource neural machine translation, and two strategies are employed to improve the translation performance for resource-scarce language pairs. On the one hand, we make use of weight sharing to enhance the encoder of the NMT model, which can also reduce overfitting. On the other hand, we initialize our model using the pre-trained model which is implemented by denoising autoencoding. Meanwhile, the model is fine-tuned via denoising autoencoding during training. The experiments show that the proposed model outperforms the baseline method and is effective for low-resource language pairs such as English-Vietnamese and Chinese-Mongolian.

In the next stage of the study, we may investigate how to apply character aware to the unsupervised NMT model. Furthermore, we will also research the performance of our model on some large-scale parallel corpora.

Acknowledgements. This work is supported by the National Natural Science Foundation of China under Grant No. 61572462 and the 13th Five-year Informatization Plan of Chinese Academy of Science, Grant No. XXH13505-03-203.

References

1. Vaswani, A., Shazeer, N.: Attention is all you need. In: Advances in Neural Information Processing Systems, pp. 5998–6008 (2017)
2. Hassan, H., Aue, A., et al.: Achieving human parity on automatic Chinese to English news translation. arXiv preprint arXiv:1803.05567 (2018)
3. Zoph, B., Yuret, D.: Transfer learning for low-resource neural machine translation. arXiv preprint arXiv:1604.02201 (2016)
4. Koehn, P., Knowles, R.: Six challenges for neural machine translation. arXiv preprint arXiv:1706.03872 (2017)
5. Santos, C.N.D., Guimaraes, V.: Boosting named entity recognition with neural character embeddings. arXiv preprint arXiv:1505.05008 (2015)
6. Yang, Z., Chen, W.: Unsupervised neural machine translation with weight sharing. arXiv preprint arXiv:1804.09057 (2018)
7. Devlin, J., Chang, M.W.: BERT: pre-training of deep bidirectional transformers for language understanding. arXiv preprint arXiv:1810.04805 (2018)
8. Vincent, P., Larochelle, H.: Extracting and composing robust features with denoising autoencoders. In: Proceedings of the 25th International Conference on Machine learning, pp. 1096–1103. ACM (2008)
9. Bahdanau, D., Cho, K.: Neural machine translation by jointly learning to align and translate. arXiv preprint arXiv:1409.0473 (2014)
10. Gu, J., Wang, Y.: Meta-learning for low-resource neural machine translation. arXiv preprint arXiv:1808.08437 (2018)
11. Sennrich, R., Haddow, B.: Improving neural machine translation models with monolingual data. arXiv preprint arXiv:1511.06709 (2015)
12. Currey, A., Barone, A.V.M.: Copied monolingual data improves low-resource neural machine translation. In: Proceedings of the Second Conference on Machine Translation, pp. 148–156 (2017)
13. Ling, W., Trancoso, I.: Character-based neural machine translation. arXiv preprint arXiv:1511.04586 (2015)
14. Ruiz Costa-Jussà, M., Rodríguez Fonollosa, J.A.: Character-based neural machine translation. In: Proceedings of the 54th Annual Meeting of the Association for Computational Linguistics, pp. 357–361 (2016)
15. Luong, M.T., Manning, C.D.: Achieving open vocabulary neural machine translation with hybrid word-character models. arXiv preprint arXiv:1604.00788 (2016)
16. Passban, P., Liu, Q., Way, A.: Improving character-based decoding using targetside morphological information for neural machine translation. arXiv preprint arXiv:1804.06506 (2018)
17. Radford, A., Narasimhan, K.: Improving language understanding with unsupervised learning. Technical report, OpenAI (2018)
18. Lample, G., Ott, M.: Phrase-based & neural unsupervised machine translation. arXiv preprint arXiv:1804.07755 (2018)
19. Firat, O., Cho, K.: Multi-way, multilingual neural machine translation with a shared attention mechanism. arXiv preprint arXiv:1601.01073 (2016)
20. Renduchintala, A., Shapiro, P.: Character-aware decoder for neural machine translation. arXiv preprint arXiv:1809.02223 (2018)

21. Kim, Y., Jernite, Y.: Character-aware neural language models. In: Thirtieth AAAI Conference on Artificial Intelligence (2016)
22. Srivastava, R.K., Greff, K.: Training very deep networks. In: Advances in Neural Information Processing Systems, pp. 2377–2385 (2015)
23. Lee, J., Cho, K.: Fully character-level neural machine translation without explicit segmentation. Trans. Assoc. Comput. Linguist. **5**, 365–378 (2017)
24. Lample, G., Conneau, A.: Cross-lingual language model pretraining (2019)
25. Sun, M., Chen, X.: THULAC: an efficient lexical analyzer for Chinese. Technical report (2016)
26. Kingma, D.P., Ba, J.: Adam: a method for stochastic optimization. arXiv preprint arXiv:1412.6980 (2014)
27. Srivastava, N., Hinton, G.: Dropout: a simple way to prevent neural networks from overfitting. J. Mach. Learn. Res. **15**(1), 1929–1958 (2014)

Mongolian-Chinese Unsupervised Neural Machine Translation with Lexical Feature

Ziyu Wu, Hongxu Hou$^{(\boxtimes)}$, Ziyue Guo, Xuejiao Wang, and Shuo Sun

Department of Computer Science, Inner Mongolia University, Hohhot, China
cshhx@imu.edu.cn

Abstract. Machine translation has achieved impressive performance with the advances in deep learning and rely on large scale parallel corpora. There have been a large number of attempts to extend these successes to low-resource language, yet requiring large parallel sentences. In this study, we build the Mongolian-Chinese neural machine translation model based on unsupervised methods. Cross-lingual word embedding training plays a crucial role in unsupervised machine translation which generative adversarial networks (GANs) training methods only perform well between two closely-related languages, yet the self-learning method can learn high-quality bilingual embedding mappings without any parallel corpora in low-source language. In this work, apply the self-learning method is better than using GANs to improve the BLEU score of 1.0. On this basis, we analyze the Mongolian word lexical features and use stem-affixes segmentation in Mongolian to replace the Bytes-Pair-Encoding (BPE) operation, so that the cross-lingual word embedding training is more accurate, and obtain higher quality bilingual words embedding to enhance translation performance. We reporting BLEU score of 15.2 on the CWMT2017 Mongolian-Chinese dataset, without using any parallel corpora during training.

Keywords: Mongolian-Chinese · Neural machine translation · Unsupervised method · Stem-affix segmentation

1 Introduction

With the progress of deep learning (Sutskever et al. [1]; Bahdanau et al. [2]) and the availability of large parallel corpora, neural machine translation (NMT) have achieved excellent performance on some language pairs (Wu et al. [3]). However, these models can only perform well when they have large parallel corpora. Unfortunately, build parallel corpora is expensive because they require specialized expertise. In contrast, the monolingual corpora are easier to obtain than parallel corpora. Unsupervised NMT models that aim to train a model without using any labeled data have performed well in recent machine translation researches. Recent work has attempted to learn cross-lingual word embedding without parallel data by mapping monolingual embedding to the shared space using adversarial learning methods (Lample et al. [4]; Artetxe et al. [5]). However, these methods based on generative adversarial networks (GANs) is only applicable to bilingual dictionaries trained between two closely-related languages, but apply this method to build bilingual vocabulary between Mongolian and Chinese is

© Springer Nature Switzerland AG 2019
M. Sun et al. (Eds.): CCL 2019, LNAI 11856, pp. 334–345, 2019.
https://doi.org/10.1007/978-3-030-32381-3_27

poor. Previous works have shown that self-learning can learn high-quality bilingual embedding mappings without any parallel corpora in low-resource language. In this work, we use a fully unsupervised initialization based on self-learning methods to improve the performance of cross-lingual word embedding training. Different pre-processing methods in language corpora will also affect the training effect of cross-lingual word embedding. The BPE algorithm is mainstream methods, but the BPE based on the number of co-occurrences, this segmentation method can't consider the semantic features of Mongolian. It leads to the decline of cross-lingual word embedding training effects in low-resource language pairs. Therefore, we perform word stem-affixes segmentation operations on Mongolian in the corpora preprocessing stage.

In summary, this paper makes the following main works:

a. Constructing a Mongolian-Chinese NMT model based on unsupervised. The unsupervised method is used to first realize the translation of Mongolian-Chinese word-by-word, and then through the large-scale language model and back-translation to guide the optimization of model parameters until the model converge.
b. In the Mongolian language, we use the stem-affix segmentation instead of the BPE to preserve the semantic information of Mongolian as much as possible while ensuring the granularity of the segmentation.
c. Use a cross-lingual training method based on self-learning combined with stem-affix segmentation for the original unsupervised translation model to improve the accuracy of the bilingual dictionary.

2   Related Work

Machine translation task is divided into supervised machine translation, unsupervised machine translation and semi-supervised machine translation, which is depending on whether supervision is performed.

2.1   Supervised Methods

There is a rich body of supervised methods for Mongolian-Chinese machine translation based on a large number of Mongolian-Chinese bilingual parallel corpora. Wu et al. [6] introduced the NMT model into the Mongolian Chinese machine translation task, and the machine translation model of cyclic neural network based on the attention mechanism is realized. Fan et al. [7] proposes the method of using similar words instead of low-frequency words. Li et al. [8] proposes a method of introducing a dictionary improves the translation effect of low-frequency words; Wang et al. [9] proposes the method to control the segmentation granularity to improve the translation effect.

2.2   Unsupervised Methods

NMT for scarce resources has become a hotspot in recent years' research, and unsupervised method is one of them. The unsupervised method enables the NMT task to train well-behaved machine translation models without bilingual parallel corpora

(Artetxe et al. [10]). There are three key steps in an unsupervised machine translation system, including translation model initialization, language models, and back-translation. Facebook proposed an unsupervised machine translation method (Lample et al. [4]), which achieved good translation results in English-French machine translation. Based on this model, this paper proposes an unsupervised neural machine translation model from Mongolian to Chinese.

2.3   Semi-supervised Methods

He et al. [12] proposes a semi-supervised neural network model based on dual-learning, which can translate low-resource languages using some monolingual corpora and small parallel corpora. The result shows that semi-supervised neural machine translation can achieve reasonable results with parallel corpora which are insufficient to train a common neural model.

3   Unsupervised Mongolian-Chinese Neural Machine Translation

The Mongolian-Chinese machine translation task is limited by the lack of bilingual parallel corpora. The translation model often cannot be fully trained, which leads to the translation performance not being improved. Recent researches have shown that unsupervised methods enable machine translation tasks to train well-performing machine translation models without bilingual parallel corpora. In the unsupervised machine translation system, the three key steps are translation model initialization, training of language models and back translation. The overall architecture of the system is shown (see Fig. 1).

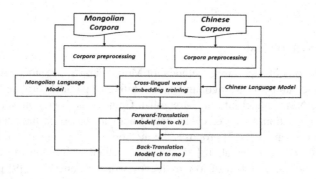

Fig. 1. The overall architecture of the unsupervised Mongolian-Chinese neural machine translation system.

3.1 Word Stem-Affix Segmentation

In recent years, the BPE Segmentation technology has been used to corpora prepro-cessing in machine translation models and obtain good performance. However, this method only relies on word frequency merging and does not take into account the semantic characteristics of any language itself. This makes the effect of Mongo-lian BPE operation less than that of stem-affix segmentation. In this paper, we use the discriminant stem-affix segmentation based on a directed graph morphology analyzer. The method uses the idea of discriminant classification to model the stem affixation of words into the labeling problem of the letters in the word. This method owns gener-alization ability and can deal with the problem that the word contains unregistered stems.

In the Mongolian sentence $S = W_1 W_2 \cdots W_r, W_i (1 \leq i \leq r)$, for the Mongolian word $W_i = C_{i\_1} C_{i\_2} \cdots C_{i\_n}$. $C_{i\_j} (1 \leq j \leq n)$ is the jth char of W_i. n represents the length of word. The problem of stem-affix segmentation becomes the division of the alphabetic sequence: $C_{i\_1} C_{i\_2} \cdots C_{i\_n} \to C_{i\_1:e_1} C_{i\_e_1+1:e_2} \cdots C_{i\_e_{m-1}+1:e_m}, e_m = i\_n$. The letter sequence $C_{i\_1:n}$ is divided into m subsequences, the first subsequence is the stem, and others are affixes. Jiang et al. [13] based on the Chinese word segmentation, divided each Mongolian letter C_{i_j} into four categories: b is the letter of the beginning of the stem or the affix; m in the middle of the stem or affixes is the end of the stem or the affix; s indicates that the word is a single stem or affix. The marked corpora are trained by using the maximum entropy toolkit to obtain the final segmentation result. As shown (see Fig. 2).

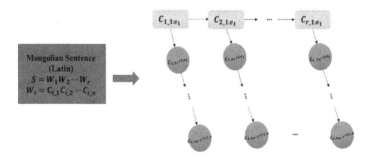

Fig. 2. The process of word stem-affix segmentation. (Color figure online)

Orange squares indicate stem, and the number of the stem is one. A green circle indicates an affix, and the affix can be 0 or more. In this paper, the experiment of the main system uses the stem-affix segmentation for the Mongolian corpora. For Mongolian, the number of affixes is limited, and the data of the training corpora can easily cover all affixes. The situation of stems is much more complicated, and new words will continue to emerge with social development. When the stem of the word to be analyzed does not exist in the training material, the simple enumeration method cannot find the correct candidate for the analysis result. However, the discriminative stem-affix strategy may have a good generalization ability, just like the situation in Chinese word segmentation.

The segmentation results based on the BPE algorithm and the stem-affix segmentation method are shown in Table 1.

Table 1. Mongolian sentences with different segmentation methods

Method	Example
Source	᠊ᠠᠮᠠᠷ ᠊ᠤᠷᠤᠮᠪᠤ ᠊ᠤᠷᠤᠬᠤ ᠊ᠮᠤᠷᠠᠯ ᠊ᠮᠤᠭ ᠊ᠤᠷ ᠊ᠤᠷᠤᠪ ᠊ᠠᠷᠤᠪᠤ ''
BPE(35000)	᠊ᠠᠷ ᠊ᠠᠪ ᠊ᠤᠷᠤᠮᠪᠤ ᠊ᠤᠷᠤᠬᠤ ᠊ᠮᠤᠷᠠᠯ ᠊ᠮᠤᠭ ᠊ᠤᠷ ᠊ᠠᠪ ᠊ᠠᠷᠤ ᠊ᠠᠷᠤᠪᠤ ''
Stem-Affix	᠊ᠠᠷᠤ	᠊ᠠᠮᠠᠷ ᠊ᠤᠷᠤᠮᠪᠤ ᠊ᠤᠷᠤᠬᠤ ᠊ᠮᠤᠷᠠᠯ ᠊ᠮᠤᠭ ᠊ᠤᠷ ᠊ᠠᠪ	᠊ᠠᠷᠤᠪᠤ ''

Table 1 illustrated that the granularity of sentences after BPE segmentation is similar to that after Stem-Affix segmentation, such as "᠊ᠮᠤᠷᠤᠪ" is divided into "᠊ᠠᠪ" and "᠊ᠤᠪ" in both methods. While the stem-affix segmentation method contains more semantic information, like "᠊ᠠᠮᠠᠷ", according to the result of BPE segmentation, the word attribute is changed into a verb, and the meaning of the stem-affix segmentation is consistent with the meaning of the BPE, and the part of speech still no change. So, this method more helpful to our model training.

3.2   Cross-Lingual Word Embedding

In the early training of the unsupervised machine translation, we need to construct a mapping relationship between Mongolian and Chinese, this mapping is called cross-lingual word embedding. Facebook proposed in their unsupervised machine translation system to use GANs (Gouws et al. [14]) training cross-lingual word embedding. However, their evaluation has focused on closed-related languages, while in cross-lingual learning from Mongolian to Chinese, they are often failing.

There are many ways to calculate the distance between the source language word embedding and the target language word embedding, including maximum mean difference, cosine similarity, and Cross-domain Similarity Local Scaling (CSLS) method (Lample et al. [11]). In this work, we adopt the CSLS as a criterion for training word embedding pairs in cross-lingual word embedding training. It's used to represent the average cosine similar measure that word embedding from source X to target Y. This part is an important part of the unsupervised Mongolian-Chinese machine translation model. The quality of cross-language word vector training will directly affect the quality of the Mongolian Chinese bilingual dictionary. The training methods can be based on GANs (Yang et al. [17]; Carone et al. [18]), self-learning method (Artetxe et al. [19]) and so on. However, experiments show that the method based on GANs is not suitable for translation tasks between two languages with low similarity, but the self-learning method is more suitable for tasks like Mongolian Chinese machine translation, so we adopt the self-learning method.

The process of cross-lingual word embedding training by self-learning is shown (see Fig. 3).

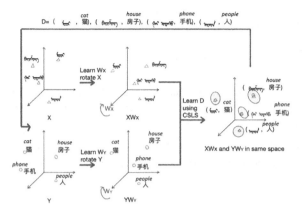

Fig. 3. A process sketch of bilingual dictionary generation by cross-lingual word embedding. The bilingual dictionary is constructed by learning the mapping matrix between X and Y.

Let X and Y be the word embedding matrices in Mongolian and Chinese, their ith row X_{i*} and Y_{i*} denote the embedding of the ith word in their respective vocabularies. Our goal is to learn the linear transformation matrices W_X so the mapped embedding XW_X and YW_Y are in the same cross-lingual space. We build a dictionary between Mongolian and Chinese, encoded as a matrix D where $D_{ij} = 1$ if the jth word in Chinese is a translation of the ith word in Mongolian. It is divided into three parts: word embedding normalization, dictionary initialization and self-learning, and symmetric reweighting.

Word Embedding Normalization. The implementation of this part requires two steps: the first normalize according to the length of word embedding, then average the center of each dimension and normalize again according to the length. The advantage of this operation is that we can guarantee that the final embedding has a unit length. In other words, for any two word embedding, their dot product is their cosine similar distances.

Dictionary Initialization. The difficulty of initializing a bilingual dictionary in this paper is that the Mongolian word embedding X and the Chinese word embedding Y are not aligned (no matter which dimension is not aligned). Therefore, we construct two aligned the Mongolian word embedding matrices X_1 and the Chinese word embedding matrices Y_1 as the initial dictionary. There are many methods for initializing a dictionary, including random dictionary induction, word frequency-based lexical cut-off, nearest neighbor search, and Cross-Lingual Similarity Local Scaling (CSLS). We adopt the CSLS method for dictionary initialization.

Given two map embedding matrices X_1 and Y_1, respectively calculate $r_T(x)$ and $r_S(y)$, $r_T(x)$ is expressed as the average cosine similarity of the k nearest neighbors of the Mongolian word embedding x in the Chinese word embedding matrix Y_1, $r_S(y)$ is expressed as the average cosine similarity of the k nearest neighbors of the Chinese word embedding y in the Mongolian word embedding matrix X_1. The calculation method of CSLS is shown in formula (1)

$$CSLS(x, y) = 2 \cos(x, y) - r_T(x) - r_S(y) \tag{1}$$

The process uses a self-learning method. After calculating the initial dictionary, X_1 and Y_1 are discarded, and the remaining self-learning iterations are performed on the original X and Y.

Self-learning Iterative Improvement. Corresponding rotation matrix W_X and W_Y are obtained by singular value decomposition. As shown in Eqs. (2)–(4):

$$USV^T = X^T DY \tag{2}$$

$$W_X = U \tag{3}$$

$$W_Y = VS \tag{4}$$

The Mongolian-Chinese machine translation model based on neural under unsupervised method mainly includes the following four parts: the Mongolian-Chinese bilingual dictionary training; the Mongolian language model and the Chinese language model training; the translation model initialization from Mongolian to Chinese; back-translation. Next, we will introduce in detail.

Mongolian-Chinese Bilingual Dictionary. In the Mongolian-Chinese NMT with the supervised method, the bilingual dictionary consists of word pairs in the parallel corpora, but in the unsupervised method, the Mongolian corpora and the Chinese corpora are not aligned, so it's impossible to find the one-to-one correspondence of Mongolian and Chinese through the traditional method of supervising machine translation. So before building the bilingual dictionary, we use the fasttext to train the word embedding in Mongolian and Chinese monolingual corpora, then build a Mongolian-Chinese bilingual dictionary by aligning monolingual word embedding spaces in an unsupervised way, which is CSLS method.

3.3 Language Model

In this work, language modeling is accomplished via de-noising auto-encoding (Lample et al. [4]), it's loss function as formula (5), our goal is minimizing L^{lm}:

$$L^{lm} = E_{x \sim S}[-\log P_{s \to s}(x|N(x))] + E_{y \sim T}[-\log P_{t \to t}(y|N(y))] \tag{5}$$

where $N(\cdot)$ is a noise function with some words dropped in Lample et al. [4]. $P_{s \to s}$ and $P_{t \to t}$ are combinations of encoder and decoder operating on the Mongolian side and the Chinese side, respectively.

3.4   Translation Model Initialization

According to the already trained Mongolian-Chinese bilingual dictionary and two language models, through the word-by-word method to initialize the translation model, get the initial translation results from Mongolian to Chinese. The Chinese language model is used to adjust the sequence of translation results. At the same time get the first translation model.

3.5   Back Translation

To train the new system in a real translation environment without violating the limitations of using only monolingual corpora, we introduce the back translation method proposed by Sennrich et al. [15]. Specifically, this method is an input sentence for a given language, and the system uses greedy decoding to translate it into another language in an inferred mode (using a shared encoder and a decoder of another language). Using this method, we can get pseudo-parallel corpora and then train the system to predict the original text based on the translation.

The translation results in those previous Mongolian-Chinese translation model is translated into the Chinese-Mongolian translation model (still through word-by-word). The Mongolian language model is used to correct the translated results and the source Mongolian after back translation. Repeat the previous two translation processes until the model converges. The loss of the back translation model is shown as formula (6)

$$L^{back} = E_{y \sim T}[- \log P_{s \to t}(y|u^*(y))] + E_{x \sim S}[- \log P_{t \to s}(x|v^*(x))] \qquad (6)$$

$u^*(y) = argmax P_{t \to s}(u|y)$ is the back translation result from Chinese to Mongolian, $v^*(x) = argmax P_{s \to t}(v|x)$ is the back translation result from Mongolian to Chinese, and $(u^*(y), y)$, $(x, v^*(x))$ are pseudo-parallel sentences.

In the process of translation, the final objective function is the weighting of the loss of language models and the loss of the back translation. As shown in formula (7):

$$L = \alpha L^{lm} + (1 - \alpha)L^{back} \qquad (7)$$

4   Experiment

4.1   Dataset

Monolingual Data. All the methods being evaluated in all tasks (except for supervised translation systems) take monolingual word embedding in each language as the input data. Use CWMT2017 Mongolian-Chinese parallel corpora for 0.26 M as a training set. Randomly disrupted the corpora sentences to ensure the model run in unsupervised. We use BPE to segment Mongolian corpora according to the number of word combinations. In both baseline systems we choose the BPE method. In the main system performs the technology of stem-affix segmentation in Mongolian. Remove the noise sentences and keep sentences from 1 to 100 in length. Our unsupervised Mongolian-Chinese machine translation tasks based on the transformer. Experiments in NVIDIA TITAN X.

Bilingual Data. Use the 1001 sentence pair test set of the CWMT2017 Mongolian-Chinese daily language translation as the test set for our experiments. The corresponding dataset statistics are summarized in Table 2.

Table 2. Corresponding dataset statistics

Method	Language	vocab.size
WBW	mo/ch	69413/5288
Unsupervised	mo/ch	31738/20754
Ours	mo/ch	41909/5288

Election of Various Parameters. The number of BPE codes is 35000 in Mongolian. The number of encoder layers is 4 and the number decoder layers is 4. The number of share encoder and decoder layers both are 3. The dimensionality of the word embedding is 100. The hidden units are 100, dropout is 0.1, blank is 0.2, the learning rate is 0.0001, the batch size is 32, the epoch size is 500000. We take α is 0.5 to train the model in turn. At the decoding time, we generate greedily.

4.2   Baselines

We used two baseline systems as a comparison of the experiments.

Word-by-word translation (WBW) (Lample et al. [11]): The first baseline system is that it performs word-by-word translation using an initialized Mongolian-Chinese bilingual dictionary. Simultaneously, this model is also our initial translation model.

Unsupervised training (Lample et al. [16]): Unsupervised Mongolian Chinese neural machine translation model, in which the corpora preprocessing part uses the BPE segmentation technique to segments. The parameters of this model are also the same as the parameters of our experiment. The training time is one week.

4.3   Experimental Results and Analysis

Through several comparative experiments, we made the following analysis. As shown in Tables 3, 4 and 5.

Table 3. Comparison of three unsupervised methods

Models	BLEU
WBW (Word-by-word)	5.4
Unsupervised (BPE & GAN)	13.5
Ours_model1 (BPE & Self-learning)	14.5
Ours_model2 (Stem-Affix Segmentation & GAN)	14.3
Ours_model3 (Stem-Affix Segmentation & Self-learning)	15.2

Table 3 shows the BLEU scores in different unsupervised Mongolian-Chinese neural machine translation models. Compare to WBW (baseline system 1), the BLEU score increase 9.8. Compare to Unsupervised (baseline system 2), the BLEU score increase 1.7. We analyze the reason for this situation is the affixation of the Mongolian corpora can preserve as much as possible reducing the size of the dictionary, which can further effectively reduce the out of the vocabulary problem and the unknown word problem. According to the second and the third line, we verify the effectiveness of cross-lingual word embedding training based on self-learning. To compare the second and fourth line, we verify the advantage of the stem-affix segmentation method.

Table 4. Translation results (short sentences) for different models

	Example
Source	᠊ᠣᠪᠠᠠ᠂ ᠊ᠣᠯᠠᠠᠠᠠ᠂ ᠊ᠣᠯᠠᠠᠠᠠ ᠂ ᠊ᠣ᠂ ᠊ᠣᠯᠠᠠ᠂ ᠊ᠣᠯᠠᠠ᠂ ᠊ᠣᠯᠠᠠ᠂ ᠊ᠣᠯᠠᠠ᠂ ᠂ ᠊ᠣᠯᠠᠠ ᠊ᠣᠯᠠ ᠃ ᠊ᠣ ᠵ ᠊ᠣᠯ ᠂᠂
WBW	但是以为说着。
Unsupervised	但是的话蒙古。
Ours_model3	但听过蒙古的。
Reference	但是听说过很多关于蒙古地区的传说。
English	But I have heard a lot about the legends of Mongolia.

Table 5. Translation results (long sentences) for different models

	Example
Source	[Mongolian script text] ... 4 57 ·· 3 ...
WBW	乌力吉、呼锦德政权、残疾人发展475，托拉、康复、补助、西塔格玛、工作措施。
Unsupervised	呼和残疾人事业发展资助资金4个。下拨3万元，残疾人主要用于多取、康复、托养、西塔高加、工作补贴措施，为残疾人正镜，推动残疾人发展事业。
Ours_model3	呼和浩特共有475.3万元残疾人事业资金。主要用于残疾人技术、康复、托养。为残疾人正镜，推动残疾人发展事业。
Reference	近日，呼和浩特市财政下达残疾人事业发展补助资金457.3万元，主要用于落实残疾人在技能培训、康复救助、托养服务补贴、燃油补贴、机构补贴、工作补贴等方面的保障措施，提升服务机构对残疾人的服务水平，促进残疾人事业的全面发展。
English	Recently, Hohhot issued a subsidy of 4.573 million yuan for the development of the disabled, mainly for the implementation of safeguards for skills training, rehabilitation assistance, care support subsidies, fuel subsidies, institutional subsidies, work subsidies, etc. The level of service provided by the institution to the disabled and the overall development of the cause of the disabled.

Table 4 shows the translation result of the three models in short sentences (sentence length less than 20), where WBW uses simple word-to-word translation to obtain a correct translation result. The second baseline system received the correct translation of two words ("But" and "Mongolia"), but the location of "Mongolia" in the target language was incorrect. Through analysis we find is due to the BPE segmentation causes the Mongolian words part of speech changed, so it's position in the decoding process change. The results of the third line confirm our analysis. Our model does not fully translate the correct results, but the result is best in these unsupervised models.

Table 5 explains the translation effects of the three models in long sentences (sentence length of more than 50). The results are similar to those obtained in Table 4. In WBW, two target words ("disabled" and "rehabilitation") were translated; five words were translated on the Unsupervised model (baseline system 2), but there are still cases where the translation results do not match the target end position; better translation performance is still achieved in our model than baseline systems. Regarding the phenomenon of lack of translation, we analyze the reason that the training corpus is still small, resulting in insufficient training of the model.

5 Conclusion

In this work, we build the Mongolian-Chinese NMT model based on the unsupervised method, which is greatly alleviated the problem of the sparse corpus. At the same time, we solve the solution that the previous cross-domain word embedding training performed poorly in low-resource language by self-learning and stem-affix segmentation. Laid a good foundation for the study of translation models between Mongolian-Chinese machine translation and another low-resource language. In future researches, we will consider using higher dimension word embedding size, deeper networks or some new unsupervised methods to improve the quality of translation.

References

1. Sutskever, I., Vinyals, O., Le, Q.V.: Sequence to sequence learning with neural networks. In: Neural Information Processing Systems (NIPS), pp. 3104–3112 (2014)
2. Bahdanau, D., Cho, K., Bengio, Y., et al.: Neural machine translation by jointly learning to align and translate. In: International Conference on Learning Representations (ICLR). arXiv preprint arXiv:1409(0473) (2015)
3. Wu, Y., Schuster, M., Chen, Z., et al.: Google's Neural Machine Translation System: Bridging the Gap between Human and Machine Translation. arXiv: Computation and Language (2016)
4. Lample, G., Conneau, A., Denoyer, L., et al.: Unsupervised machine translation using monolingual corpora only. In: International Conference on Learning Representations (ICLR). arXiv preprint arXiv:1711(00043) (2018)
5. Artetxe, M., Labaka, G., Agirre, E., et al.: Unsupervised neural machine translation. In: International Conference on Learning Representations (ICLR). arXiv preprint arXiv:1710 (11041) (2018)

6. Wu, J., Hou, H., Shen, Z., et al.: Adapting attention-based neural network to low-resource Mongolian-Chinese machine translation. In: International Conference on the Computer Processing of Oriental Languages (ICCPOL), pp. 470–480 (2016)
7. Fan, W., Hou, H., Wang, H., et al.: Machine translation model of Mongolian-Chinese neural network fusing priori information. Chin. J. Inf. Sci. **32**(06), 36–43 (2018)
8. Jinting, L., Hongxu, H., Jing, W., et al.: Combining discrete lexicon probabilities with NMT for low-resource Mongolian-Chinese translation. In: Parallel and Distributed Computing: Applications and Technologies (PDCAT), pp. 104–111 (2017)
9. Wang, H.: Multi-granularity Mongolian Chinese neural network machine translation research. In: Inner Mongolia University, pp. 15–35 (2018)
10. Artetxe, M., Labaka, G., Agirre, E., et al.: Learning bilingual word embeddings with (almost) no bilingual data. In: Meeting of the Association for Computational Linguistics (MACL), pp. 451–462 (2017)
11. Lample, G., Conneau, A., Ranzato, M., et al.: Word translation without parallel data. In: International Conference on Learning Representations (ICLR). arXiv preprint arXiv:1710 (04087) (2018)
12. He, D., Xia, Y., Qin, T., et al.: Dual learning for machine translation. In: Neural Information Processing Systems (NICS), pp. 820–828 (2016)
13. Jiang, W., Wu, J., Wu, R., et al.: Discriminant stem affixation of Mongolian directed graph morphology analyzer. J. Chin. Inf. Process. **25**(04), 30–34 (2011)
14. Gouws, S., Bengio, Y., et al.: BilBOWA: fast bilingual distributed representations without word alignments. In: International Conference on Machine Learning (ICML), pp. 748–756 (2015)
15. Sennrich, R., Haddow, B., Birch, A., et al.: Improving neural machine translation models with monolingual data. In: Meeting of the Association for Computational Linguistics (MACL), pp. 86–96 (2016)
16. Lample, G., Ott, M., Conneau, A., et al.: Phrase-based & neural unsupervised machine translation. In: Empirical Methods in Natural Language Processing (EMNLP), pp. 5039–5049 (2018)
17. Yang, Z., Chen, W., Wang, F., et al.: Improving neural machine translation with conditional sequence generative adversarial nets. In: North American chapter of the association for computational linguistics (NAACL), pp. 1346–1355 (2018)
18. Barone, A.: Towards cross-lingual distributed representations without parallel text trained with adversarial autoencoders. In: Meeting of the Association for Computational Linguistics (ACL), pp. 121–126 (2016)
19. Artetxe, M., Labaka, G., Agirre, E.: A robust self-learning method for fully unsupervised cross-lingual mappings of word embeddings. In: Proceedings of the 56th Annual Meeting of the Association for Computational Linguistics (ACL), pp. 789–798 (2018)

Learning Multilingual Sentence Embeddings from Monolingual Corpus

Shuai Wang[1,2,3], Lei Hou[1,2,3(✉)], Juanzi Li[1,2,3], Meihan Tong[1,2,3], and Jiabo Jiang[4]

[1] DCST, Tsinghua University, Beijing 100084, China
{shuai-wa16,tongmh17}@mails.tsinghua.edu.cn,
{houlei,lijuanzi}@tsinghua.edu.cn
[2] KIRC, Institute for Artificial Intelligence, Tsinghua University, Beijing, China
[3] Beijing National Research Center for Information Science and Technology, Beijing, China
[4] Daqing Oilfield Information Technology Company, Beijing 100043, China
jiangjb@cnpc.com.cn

Abstract. Learning multi-lingual sentence embeddings usually requires large scale of parallel sentences which are difficult to obtain. We propose a novel self-learning approach which is capable of learning multi-lingual sentence embeddings from monolingual corpora. Our assumption is that, irrelevant to languages, sentences appearing in similar contexts are similar. Thus, we first train monolingual sentence embeddings of different languages with shared parameters as initialization. Then we iteratively extract similar sentence pairs and exchange their positions regardless of languages. Through their relations to their new contexts we predict the similarities between a similar sentence pair. Our experiments show that the proposed approach outperforms existing unsupervised approaches and is competitive to supervised approaches.

Keywords: Sentence representation · Multilingual · Unsupervised learning

1 Introduction

Pre-training language representation from unlabelled data is effective in many natural language processing tasks. Recently many works start to focus on sentence representation instead of word representation [10,12,17]. However, most of them only consider monolingual situation and fail to generalize to multi-lingual settings, but many tasks involve dealing more than one languages. Besides, many low-resource languages lack labelled data, and a unified multilingual sentence representation is helpful to deal with those languages. Hence, learning multilingual sentence embeddings is a significant research in language representation.

Currently the best performance is achieved by LASER [3,14]. It utilizes sentence level parallel data to train machine translation model and takes the encoder

© Springer Nature Switzerland AG 2019
M. Sun et al. (Eds.): CCL 2019, LNAI 11856, pp. 346–357, 2019.
https://doi.org/10.1007/978-3-030-32381-3_28

as sentence features extractor. Although it achieves satisfying performance on 93 languages, the need of large scale of parallel data still limits its applications. What's more, the direct usage of machine translation model is also unable to utilize information of adjacent sentences which is important in many tasks [7].

Since unlabelled corpora are almost infinite and easy to obtain, fully unsupervised methods have strong potential. Currently multi-lingual BERT model [7] attracts much attention, which is a simple generalization of monolingual BERT model. It takes as input unlabelled texts from different languages with shared parameters. Different languages are actually trained independently, so there is few interaction among languages. Although, with the strong capacity of BERT, it achieves good results on many cross-lingual tasks, its performance on cross-lingual tasks is strikingly worse than that on monolingual tasks.

The wide gap between multi-lingual BERT and LASER reveals the significance of interactions among languages in training multi-lingual sentence embeddings. Given only monolingual corpora, to increase connections among languages, an intuitive approach is to extract parallel sentences as seeds from unlabelled corpora and connect different languages by parallel sentences. In word level, [2] propose an iterative approach to learn cross-lingual word embeddings in an unsupervised manner. Their self-learning approach iteratively extracts a bilingual dictionary as seeds and trains the cross-lingual word embeddings according to the extracted seeds. However, the following three problems make it challenging to implement such an idea in sentence level:

- **Large number of sentences:** The number of sentences is far larger than that of words, so it is nearly impossible to traverse all sentence pairs and extract enough parallel sentences.
- **Existence of parallel sentences:** Due to the diversity of sentences, strictly parallel sentences do not necessarily exist in two corpora.
- **Generalization to multi-languages:** The approach only considers learning a mapping between two languages, but in reality an encoder which is capable of encoding multiple languages, like LASER, is far more useful.

In this paper, we propose a novel iterative approach to learn sentence embeddings from monolingual corpora. Utilizing two hypotheses: sentence-level distributional hypothesis [13] and language isomorphism [16], we assume that similar sentences appear in similar contexts even across different languages. As illustrated in Fig. 1, two pieces of texts in different languages are still coherent in semantics after two similar sentences are exchanged.

We model sentences by transformer [18] and take the mean-max pooling [22] of output hidden states as sentence representation. We build a shared word piece vocabulary of all languages and all parameters are shared among languages so that a single encoder is capable of encoding multiple languages.

The proposed approach consists of two parts, i.e., sentence-context classification and sentence coherence regression. As illustrated in the first and second row of Fig. 1, the positive instance of classification is constructed by concatenating a sentence with its context. For negative instances, we replace some sentences with random ones and also concatenate to its adjacent sentences. Then we design a classifier to distinguish them. The classification is taken as an initialization of

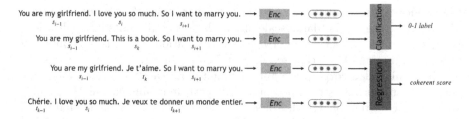

Fig. 1. The overall architecture of the proposed approach. The top part is the construction of our monolingual sentence-context classification and the bottom part is our cross-lingual coherence regression.

the multi-lingual coherence regression. In regression task, we iteratively extract similar sentence pairs and exchange their positions in the original corpora and concatenate them with their new contexts. Since similar sentences are not parallel, we label concatenated sentences by a **coherence score**, which is defined as the similarity between similar sentences, as shown in the third and fourth line of Fig. 1, rather than 0–1 labels.

The experiments show that the proposed approach captures the most cross-lingual sentence information among all unsupervised approaches. Furthermore, it is capable of learning meaningful cross-lingual sentence embeddings under fully disjoint monolingual corpora.

2   Preliminaries and Framework

Our model only requires unlabelled multi-lingual sentences, and sentences are not necessarily paralleled, but should be in document-level because we need context information. Then we concatenate all documents regardless of languages altogether as our training materials.

In this paper, we are mainly dealing with sentences, so we first give a definition of a sentence.

Definition 1 (Sentence). *A sentence is defined as a sequence. The i-th sentence in a corpus is defined as $s_i = \langle w_i^1, w_i^2, \ldots, w_i^{l_i} \rangle$, where l_i is the length of the i-th sentence, and w_i^k is the k-th unit of the sentence i. The basic unit could be words, characters or word pieces depending on the language we are dealing.*

Now we have corpora in different languages, we concatenate them all as our training corpus D, and record the start and end index of each language. Since sentences in different languages are processed in an exactly same way, we will not mention a specific language in the following part.

We denote the concatenation of k sentences as $concat([s_1, s_2, \ldots, s_k])$, i.e., we merge them as one sentence. Our sentence encoder requires a fixed length input, so we normalize the lengths of all sentences to a fixed length $maxlen$.

Our task is to learn a multi-lingual sentence encoder which is capable of encoding similar sentences into nearby vectors. We define it as follows.

Definition 2 (Multi-lingual Sentence Encoder). *Given the training materials D, we learn a multi-lingual sentence encoder $Enc : S \rightarrow R^d$, which satisfies that given two sentences s_i and s_j, the distance between $Enc(s_i)$ and $Enc(s_j)$ reflects their similarity, where d is the dimension of sentence embeddings, S is the collection of sentences.*

Our assumption is, sentences appearing in similar contexts are similar in semantics even across different languages. To fully utilizing the isomorphism among languages, we divide the proposed approach into two stages, **Monolingual Sentence-Context Classification**, which utilizes the sentence-context relations in monolingual situation and provides an initialization for the second stages, and **Multi-lingual Coherence Regression**, as an interactive process, which generalizes the sentence-context relations to multi-lingual circumstance.

3 The Proposed Approach

In this section, we will illustrate the proposed approach, including the encoder and architecture, monolingual sentence-context classification and multi-lingual coherent regression.

Encoder and Architecture: Our sentence encoder Enc is a multi-layer transformer [7,18], which is based on multi-layer self-attention. We employ exactly the same structure as in the original paper, so we omit the details. Given a sentence $s_i = \langle w_i^1, w_i^2, ..., w_i^{l_i} \rangle$, the multi-layer transformer outputs hidden states $H_i = \langle \mathbf{h}_{i1}, \mathbf{h}_{i2}, ..., \mathbf{h}_{ilcn} \rangle$, and we apply mean-max pooling all hidden states as final representation of the sentence.

$$mean(H_i) = \frac{1}{len} \sum_{j=1} \mathbf{h}_{ij}; \quad max(H_i) = \max_j \mathbf{h}_{ij}$$

$$Enc(s_i) = [mean(H_i), max(H_i)]$$

The max operation selects the most salient features and the mean operation captures the general situation of the sequence, so we combine them together as our sentence representation and it is proved to be useful in [22].

3.1 Monolingual Sentence-Context Classification

As mentioned before, our approach is mainly based on sentence-level distributional hypothesis [13]. The contexts of similar sentences are similar, so in this section we train a sentence encoder utilizing the sentence-context information. We define the sentence context as follows.

Definition 3. *The context of a sentence s_i is denoted as $C_k(s_i)$, which is the collection of nearby sentences with distance less than k, i.e.,*

$$C_k(s_i) = \{s_j | |i - j| \leq k\}$$

Specifically, we denote the previous sentences in the context of s_i as $C_k^-(s_i)$, and $C_k^+(s_i)$ for subsequent sentences.

We concatenate sentences with their contexts in the original corpus as positive instances. Then we replace each sentence s_i with s_{p_i}, where p_i is a random index from the whole corpus, and these random sentences are also concatenated with their current contexts as negative instances. We denote the dataset as $E = \{(x_i, y_i)\}$, where $x_i = concat([C_k^-(s_i), s_i, C_k^+(s_i)])$ or $concat([C_k^-(s_i), s_{p_i}, C_k^+(s_i)])$ and $y_i = 1$ or 0 indicate x_i is a positive or negative instance.

After concatenation these sentences are padded or clipped to a same length and then taken as input of our sentence encoder Enc. The encoded sentences are passed to a classifier denoted as M which predicts labels mentioned above. Note that the classifier here is a linear classifier because we want our sentence encoder capture more semantics. Given a sentence and its label in $(x_i, y_i) \in D$, the probability of s_i being consistent in semantics is given by

$$p(y_i = 1|x_i) = M(Enc(x_i)).$$

Our loss function is to maximize the probability of ground truth labels, i.e.,

$$L_{ml} = \sum_{(x_i,y_i) \in E} (p(y_i = 1)y_i + (1 - p(y_i = 1))(1 - y_i))$$

The intuition behind the proposed approach is that if a linear classifier can discriminate through sentence embedding whether a sentence is semantic inconsistent, then the sentence embedding should contain enough semantics.

3.2 Multi-lingual Coherent Regression

The above method is capable of capturing cross-lingual sentence information even though different languages have literally no connections between each other like multi-lingual BERT. To increase the interactions among languages, we generalize sentence level distributional hypothesis to cross-lingual setting. The linguistic isomorphism [16] assumes that, if two sentences in different languages are similar, then they should also be in similar contexts.

However, our training corpora are not assumed to contain parallel sentences, so we cannot replace a sentence with a parallel sentence in another language. In such a large corpora two arbitrary sentences are almost impossible to be similar, and they are useless for our training. Hence, we search some similar sentence pairs as follows. For the i-th place in training corpus, i.e., s_i, we randomly sample b sentences $\{s_{t_{i1}}, s_{t_{i2}}, ..., s_{t_{ib}}\}$, find the one with the largest cross-domain local scaling (CDLS for short) [6] with s_i

$$r_i = \arg \max_{l=1}^{b} CDLS(Enc(s_i), Enc(s_{t_{il}})),$$

and replace s_i with $s_{t_{ir_i}}$. Then we still concatenate new sentence and its contexts, and this time we predict a **coherent score**, which is defined as the similarity between the original sentence and the replaced sentence.

$$c_i = concat([C_k^-(s_i), s_{t_{ir_i}}, C_k^+(s_i)])$$

$$score(c_i) = dist(Enc(s_{t_{ir_i}}), Enc(s_i)),$$

where $dist$ is a similarity measurement in vector space. Now we achieve our regression dataset $F = \{(c_i, score(c_i))\}$, and we also adopt a linear mapping R for the task. Mean Squared Error (MSE) is applied to optimize the regression, i.e., the loss function is

$$L_{cl} = \sum_{(c_i, score(c_i)) \in F} (score(c_i) - R(Enc(c_i)))^2$$

The retrieval process is expensive in computation, so we cannot traverse all possible sentence pairs. We randomly sample $m \times m$ sentences and choose $\min\{m \times m, 1000\}$ similar sentence pairs with the highest similarities. We repeat such selection until we obtain enough sentences pairs to ensure the interactions among languages (we use 100,000 in experiment). Intuitively, larger m corresponds higher quality training sentence pairs, but costs more computation, and we will discuss this setting in experiment.

4 Experiments

4.1 Experiment Setup

Dataset. We use Wikipedia as our training corpus and conduct experiments on English, Chinese, French, Spanish and German. We create word pieces by BPE [15] for languages other than Chinese, and we use characters for Chinese. All sentences are padded or clipped to 150.

Settings. The proposed model contains 3 layers of transformer with 8 heads in multi-head attention and the hidden dimension is 512. The learning rate is 1e-4 with linear decay and dropout rate is 0.1. The model is trained on the classification task for 1 epoch, and then is iterated on regression task for 3 times with 1 epoch for each iteration.

Baselines. We compare our proposed method with three most recent methods, multi-lingual BERT, LASER [3] and vecmap [2].

Multi-lingual BERT model is not mentioned in their original paper, but they give a brief introduction of the model in their Github repository and provide a pre-trained model. With limited computing resources, we fail to train a model as large as BERT, and the results of fine-tuning on a specific task are strikingly influenced by the capacity of model. What's more, the purpose of our experiments is to compare how much cross-lingual information the sentence encoder can capture. Hence, to decrease the influence of model capacity and make a fair comparison, we do not tune any model on specific task, and we extract features of sentences by their provided pre-trained model.

LASER is a sentence encoder supporting 93 languages trained by parallel sentences. The latest version translates every language to English and Spanish respectively and takes the encoder as their sentence encoder. It requires large scale of parallel data, so we just list its results to show the distance of our method with state-of-the-art supervised method. It is not taken into our comparison.

Vecmap is a state-of-the-art unsupervised method to learn cross-lingual word embeddings. It is robust to training corpus and even competitive to supervised methods. We take the mean and sum of cross-lingual word embeddings respectively as sentence embeddings.

We list two versions of our model. One is the model trained after the initial classification task, and the other is the final model after iteration in order to show the effect of iteration.

4.2   XNLI: Cross-Lingual Natural Language Inference

Natural language inference [5,19] is a typical task to evaluate the performance of sentence embeddings. Given two sentences, one called premise, denoted as p, and another is called hypothesis, denoted as h, this task is to predict the relation between them, including entailment, contradiction and neutral. XNLI is a multi-lingual version of natural language inference dataset, which contains 2500 development sentences and 5000 test sentences translated from English.

The training set is not translated to other languages, so in this task we train the model on English and evaluate it on other languages. We extract features by the pretrained models mentioned above and the features of each sentence pair is the concatenation of $Enc(p)$, $Enc(h)$, $Enc(p) * Enc(h)$ and $|Enc(p) - Enc(h)|$. We train a three-layer fully connected neural network, with hidden dimensions 512 and 384 respectively, on the extracted features. We do not use any regularization here and we just early stop the training on English development set. Note that BERT and LASER have statistics on this dataset, but here we use our own task-specific model to make a fair comparison.

As shown in Table 1, in monolingual evaluation, BERT obviously performs best even without fine-tune. In multi-lingual evaluation, as supervised method, LASER learns the most cross-lingual sentence information. Before the iteration, the performance of the proposed approach is much worse than multi-lingual BERT. But after iteration, the proposed approach improves strikingly especially in cross-lingual evaluations.

Table 1. Results on XNLI dataset.

Method		en	zh	fr	es	de	Overall
Supervised	LASER	63.0	59.1	60.5	56.4	55.7	57.9
Unsupervised	Mean	49.2	44.3	46.1	46.9	45.2	45.6
	Sum	48.1	41.0	45.5	46.1	45.2	44.5
	Multi-lingual bert	**67.0**	47.0	49.1	48.5	**49.8**	48.6
Our approach	−iteration	59.1	45.4	46.7	44.8	43.2	45.0
	+iteration	59.6	**50.0**	**49.2**	49.9	48.7	**49.5**

Another evaluation metric here is the distance of the performance of multilingual task to that of monolingual task. With the huge capacity, BERT achieves striking results on monolingual evaluation, but the performance of multi-lingual evaluation is 20% lower. LASER still performs best on this aspect. The proposed approach decreases the distance by 5% after the interaction which proves the effectiveness and significance of the interactions among languages.

4.3   RCV2: Cross-Lingual Text Classification

Cross-lingual text classification is another important task for the evaluation of cross lingual sentence embeddings. In this task we train a classification model on one language and test it on other languages. RCV2 [11] is a dataset containing 487,000 articles in 13 languages. There are no parallel sentences or documents among different languages, and it has four classes. However, articles in this dataset is too long for sentence encoders, so we only take headlines of articles as the input of sentence encoders and extract features only from these titles.

Here we still follow the above settings. We first extract features by sentence encoders and then train a feed-forward neural network with one hidden layer on the features extracted. The L2 regularization of LR is tuned on the development set that is in the same language as training set. The results of this experiment are shown in Table 2. We can find that, the proposed method achieves the best performance on all datasets except English to French and German to English. Different sentence encoders perform alike in this evaluation which proves the importance of interactions among sentences.

Table 2. Results on RCV2 title dataset.

Method		en-zh	en-de	en-fr	en-es	zh-en	de-en	es-en	fr-en
Supervised	LASER	73.2	70.3	68.1	75.4	68.2	66.4	69.0	71.1
Unsupervised	Mean	41.1	60.2	61.1	46.3	55.2	51.3	54.9	56.6
	Sum	52.7	59.1	58.4	69.4	52.0	62.1	39.1	58.3
	Multi-lingual bert	58.3	51.4	**72.3**	52.6	47.2	**57.6**	54.6	58.4
Our approach	−iteration	58.6	50.4	56.1	51.0	48.6	54.3	55.4	57.6
	+iteration	**64.5**	**65.1**	67.6	**60.2**	**59.2**	56.7	**59.8**	**61.8**

4.4   Parameter Analysis

In this section, we investigate the impact of some important factors on the performance of sentence embeddings. The task we use here is RCV2 dataset on cross-lingual sentence classification because it is an easier task so that we can observe the influences clearly. The language pair we select is English and Chinese because they are remote and some languages, like English and German, naturally share some common word pieces under BPE.

Training Corpora. Our experiments are conducted on Wikipedia, and Wikipedia in different languages share many common contents, in which we can dig many parallel sentences. To prove that the proposed approach is robust to training corpora, we conduct an experiment on Toronto Book Corpus [23] and Chinese Wikipedia. The two corpora are from exactly different domains and they are also in remote languages, so it is nearly impossible for a parallel signal to exist. The results of the experiment are shown in Table 2.

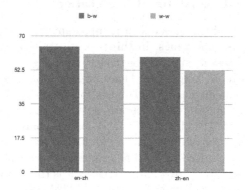

Fig. 2. Results on different Training Corpora. w-w means that two languages are both trained on Wikipedia, and b-w means that English sentences are trained on Toronto Book Corpus and Chinese sentences are trained on Wikipedia. The y-axis represents accuracy of classification.

As shown in Fig. 2, the performance of the proposed approach decreases about 6% on disjoint training corpora. Although the decrease is striking, the sentence encoder can still learn meaningful cross-lingual signals, which means the iteration process does not rely on the existence of parallel sentences, and training can work only if we can retrieve some similar sentences. Hence, our training objective is robust to the domain of training corpora.

Size of Samples. In the multi-lingual coherent regression, we retrieve some sentences to extract semi-parallel sentences. In the above experiments we set the number of sentences $m = 10,000$ for fully utilizing GPU. If the number of samples is small, we will fail to generate meaningful semi-parallel sentences and the training is completely meaningless. Otherwise, if the number is large, the model will be expensive in computation. We try some different number in the experiment, including 1, 100, 1000, 10000, 20000.

As shown in Table 3, the proposed approach is sensitive to the size of samples when the size of samples is small, but when the size of samples is large enough, increasing the size of samples does not improve the performance obviously but causes extra troubles for GPU computation. When the size of samples is too small, the training is even harmful to the performance because the cosine similarity of disjoint sentences is meaningless. Actually the training is useful only if part of sampled semi-parallel sentences are similar.

Table 3. The influence of the size of samples to the overall performance.

Size	1	100	1000	10000	20000
en-zh	50.3	51.7	59.1	64.5	64.9
zh-en	51.6	50.2	52.6	59.2	59.2

Number of Iterations. We repeat the similar sentence extraction and coherence regression process for several times. The number of iterations is an influential factor for the performance of the model. We analyze the accuracy on English to Chinese and Chinese to English cross lingual sentence classification task.

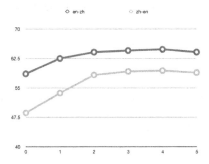

Fig. 3. The x-axis represents the number of iterations and the y-axis means the accuracy on the two datasets.

As shown in Fig. 3, the performance increases obvious in the first and the second iteration, but it almost stays fixed after 3 iterations, so we stop the training after 3 iterations. After 5 iterations the performance even starts to decrease, so too many iterations is not necessarily beneficial for the performance of sentence embeddings.

5 Related Works

This paper involves two research directions including unsupervised methods aligning languages and general purpose sentence representation. In this section, we will briefly introduce the recent progress of them.

Unsupervised Alignment of Words: GAN-based methods regard word embedding of different languages as different probability distributions, and directly align two distributions as a whole. such as [4,6,20,21]. [21] employ Wasserstein-GAN to train the model and minimize earth mover's distance to refine the vectors after training. Although GAN-based methods work well in their original paper, [2] point out that they lack robustness. Iteration-based methods are more robust by this way. [1] firstly propose the iteration approach to learn across language mapping from a small seed dictionary (as small as 25 parallel words). After that, [2] create a new fully unsupervised method to generate the initial dictionary and proceed the above iteration. Their experiments show that their approach is competitive and more robust than GAN-based methods.

General-Purpose Sentence Representation: General-purpose sentence representation methods, based on sentence level distributional hypothesis [13], usually learn a sentence encoder from a large amount of unlabelled data. The encoders can be used as sentence feature extractors to initialize other tasks. Most of such works are designed only for monolingual situations. The most simple approach is to train a sentence level log-linear model, like [8–10,17]. In spite of the good performance, training of seq2seq is time-consuming on large datasets. [12] transform the task to a classification problem. They abandon the decoding process and convert the problem to a simple classification task (to classify if a sentence is in the context of another sentence).

6 Conclusion

We propose a novel approach to learn multi-lingual sentence embeddings from mono-lingual corpora by utilizing Language Isomorphism and sentence-level Distributional Hypothesis. Although the performance is still not competitive to supervised methods, we provide a new view of extracting and utilizing similar sentences as supervised signal.

Acknowledgement. The work is supported by NSFC projects (U1736204, 61533018, 61661146007), Ministry of Education and China Mobile Joint Fund (MCM20170301), a research fund supported by Alibaba Group, and THUNUS NExT Co-Lab.

References

1. Artetxe, M., Labaka, G., Agirre, E.: Learning bilingual word embeddings with (almost) no bilingual data. In: Proceedings of the 55th Annual Meeting of the Association for Computational Linguistics, pp. 451–462 (2017)
2. Artetxe, M., Labaka, G., Agirre, E.: A robust self-learning method for fully unsupervised cross-lingual mappings of word embeddings. In: Proceedings of the 56th Annual Meeting of the Association for Computational Linguistics, pp. 789–798 (2018)
3. Artetxe, M., Schwenk, H.: Massively multilingual sentence embeddings for zero-shot cross-lingual transfer and beyond. arXiv preprint arXiv:1812.10464 (2018)
4. Barone, A.V.M.: Towards cross-lingual distributed representations without parallel text trained with adversarial autoencoders. In: Proceedings of the 1st Workshop on Representation Learning for NLP, pp. 121–126 (2016)
5. Bowman, S.R., Angeli, G., Potts, C., Manning, C.D.: A large annotated corpus for learning natural language inference. In: Proceedings of the 2015 Conference on Empirical Methods in Natural Language Processing, pp. 632–642 (2015)
6. Conneau, A., Lample, G., Ranzato, M., Denoyer, L., Jégou, H.: Word translation without parallel data. arXiv preprint arXiv:1710.04087 (2017)

7. Devlin, J., Chang, M.W., Lee, K., Toutanova, K.: BERT: pre-training of deep bidirectional transformers for language understanding. In: Proceedings of NAACL-HLT 2019, pp. 4171–4186 (2018)
8. Gan, Z., Pu, Y., Henao, R., Li, C., He, X., Carin, L.: Learning generic sentence representations using convolutional neural networks. In: Proceedings of the 2017 Conference on Empirical Methods in Natural Language Processing, pp. 2390–2400 (2017)
9. Hill, F., Cho, K., Korhonen, A.: Learning distributed representations of sentences from unlabelled data. In: Proceedings of NAACL-HLT 2016, pp. 1367–1377 (2016)
10. Kiros, R., et al.: Skip-thought vectors. In: Advances in Neural Information Processing Systems, pp. 3294–3302 (2015)
11. Lewis, D.D., Yang, Y., Rose, T.G., Li, F.: RCV1: a new benchmark collection for text categorization research. J. Mach. Learn. Res. **5**, 361–397 (2004)
12. Logeswaran, L., Lee, H.: An efficient framework for learning sentence representations. arXiv preprint arXiv:1803.02893 (2018)
13. Sahlgren, M.: The distributional hypothesis. Ital. J. Disabil. Stud. **20**, 33–53 (2008)
14. Schwenk, H., Douze, M.: Learning joint multilingual sentence representations with neural machine translation. In: Proceedings of the 2nd Workshop on Representation Learning for NLP, pp. 157–167 (2017)
15. Sennrich, R., Haddow, B., Birch, A.: Neural machine translation of rare words with subword units. In: Proceedings of the 54th Annual Meeting of the Association for Computational Linguistics, pp. 1715–1725 (2015)
16. Storer, T.: Linguistic isomorphisms. Univ. Chic. Press Behalf Philos. Sci. Assoc. **19**(1), 77–85 (1952)
17. Tang, S., Jin, H., Fang, C., Wang, Z., de Sa, V.R.: Rethinking skip-thought: a neighborhood based approach. In: Proceedings of the 2nd Workshop on Representation Learning for NLP, pp. 211–218 (2017)
18. Vaswani, A., et al.: Attention is all you need. In: Advances in Neural Information Processing Systems, pp. 5998–6008 (2017)
19. Williams, A., Nangia, N., Bowman, S.R.: A broad-coverage challenge corpus for sentence understanding through inference. In: Proceedings of NAACL-HLT 2018, pp. 1112–1122 (2017)
20. Zhang, M., Liu, Y., Luan, H., Sun, M.: Adversarial training for unsupervised bilingual lexicon induction. In: Proceedings of the 55th Annual Meeting of the Association for Computational Linguistics, pp. 1959–1970 (2017)
21. Zhang, M., Liu, Y., Luan, H., Sun, M.: Earth mover's distance minimization for unsupervised bilingual Lexicon induction. In: Proceedings of the 2017 Conference on Empirical Methods in Natural Language Processing, pp. 1934–1945 (2017)
22. Zhang, M., Wu, Y., Li, W., Li, W.: Learning universal sentence representations with mean-max attention autoencoder. In: Proceedings of the 2018 Conference on Empirical Methods in Natural Language Processing, pp. 4514–4523 (2018)
23. Zhu, Y., et al.: Aligning books and movies: towards story-like visual explanations by watching movies and reading books. arXiv preprint arXiv:1506.06724 (2015)

Chinese Historical Term Translation Pairs Extraction Using Modern Chinese as a Pivot Language

Xiaoting Wu[2] , Hanyu Zhao[1] , Lei Jing[3], and Chao Che[1(✉)]

[1] Key Laboratory of Advanced Design and Intelligent Computing,
Ministry of Education, Dalian University, Dalian, China
{hanyuzhao7, chechao101}@163.com
[2] State Grid Info & Telecom Group Beijing China-Power Information
Technology Co., LTD., Beijing, China
wuxiaoting2017@163.com
[3] Chiping Vocational Education School, Liaocheng, China
jinglei.lei@163.com

Abstract. Term translation of Chinese historical classics is very difficult and time-consuming work, and using term alignment methods to extract term translation pairs is of great help for historical term translation. However, the limited bilingual corpora resources of historical classics and special morphology of the ancient Chinese result in poor performance of term alignment. To this end, this paper proposes a historical term alignment method using modern Chinese as a pivot language. The method first identifies English terms by rules, then aligns them from English to modern Chinese and then from modern Chinese to ancient Chinese. The use of English-modern Chinese corpus and modern-ancient Chinese corpus instead of English-ancient Chinese corpus solves the shortage problem of the parallel corpus. Moreover, using modern Chinese as a pivot language effectively reduces the alignment errors caused by the abbreviations and the interchangeable characters of ancient Chinese. In the term alignment experiment on *Shiji*, our method outperformed the direct alignment method significantly, which proves the validity of our method.

Keywords: Chinese historical classics · Term alignment · Pivot language

1 Introduction

Translating Chinese classics into English is an important way for the world to understand the history and culture of China. However, most Chinese classic books remain untranslated. At present, only about 0.2% of the 35,000 Chinese classic books have been translated [1]. One main reason for this is the translation difficulty of Chinese classic books due to the dynamic development of history and the cultural differences between China and the West. Term translation is one of the most difficult and time-consuming works in classics translation, and sometimes translators spend more than 60% of their time on searching the proper term translation. Therefore, it will reduce the translation difficulty and save a lot of translation time to perform term

© Springer Nature Switzerland AG 2019
M. Sun et al. (Eds.): CCL 2019, LNAI 11856, pp. 358–367, 2019.
https://doi.org/10.1007/978-3-030-32381-3_29

alignment to extract different term translations from the bilingual corpora as the reference for the translator, which will also accelerate the translation of history books.

To the best of our knowledge, very few researches were conducted on the historical term alignment. Co-occurrence frequency [2] and maximum entropy model [3] were explored for term alignment but did not achieve satisfactory performance. Since historical terms refers to ancient official titles, numbers, institutions, apparatus, systems, events, etiquette, customs names, etc., which are similar to the named entity (NE). We investigate the NE alignment method for the term alignment research. At present, there are three main lines of bilingual NE alignment methods: (1) The symmetric method, which identifies the NEs in two languages, respectively, and then uses the alignment model to align the NEs in two sides [4]; (2) The asymmetric method, which recognizes the NEs in one language, then find its corresponding translation in another language [5–7]; (3) The integration method, which jointly perform bilingual NE alignment with other NLP tasks such as NE recognition [8, 9] or word alignment [10]. However, most approaches rely on word alignment relationship to map the NEs in both languages and the performance of word alignment in historical classics are very poor due to the shortage of the parallel corpus. In addition, the special morphology of ancient Chinese further introduces many word alignment errors.

To the end, this paper proposes a pivot-based method to perform term alignment, which is an effective solution to overcome the scarceness of parallel corpus by introducing a third language that has parallel corpus with both source and target languages [11–14]. We first recognize the English terms using rule-based method, then align them from English to modern Chinese and from modern Chinese to ancient Chinese. Finally, we get the historical terms translation pairs by combing the two alignment results. We employ modern Chinese as the pivot language because most historical classic book are explained by modern Chinese and many English-modern Chinese corpora are available in public. Using modern Chinese as the pivot language can also reduce the alignment error caused by the special wording of ancient Chinese.

2 Motivation

The ancient Chinese has many lexical and syntactic differences with modern Chinese, two of which can result in many alignment errors. First, some ancient Chinese words such as people names and official names are often abbreviated while they are translated into the full name in English. Thus, it is difficult to find the corresponding English translation of the abbreviated words. For example, in the example of Fig. 1, "阳成延" is abbreviated as "延" in ancient Chinese and is difficult to align with its full translation "Yang-ch'eng Yen". Second, it is very difficult to align the words containing the interchangeable characters. The interchangeable character is a character that is used to replace another character with same or similar pronunciation. Since the words containing interchangeable character usually appears in another form, it is almost impossible to find the correct translation.

Because modern Chinese and ancient Chinese share the same language system, we can address the above problems by using the corresponding modern Chinese terms. For the first problem, modern Chinese can complement the shortened terms in ancient Chinese, which make it easy to find a corresponding English translation. For example, as shown in the example in Fig. 1, "延" in ancient Chinese is easily aligned with the English translation "Yang-ch'eng Yen" after mapping to the full name of "阳成延" in modern Chinese. For the second problem, modern Chinese can help identify the character that interchangeable character replaces. It is much easier to align the word containing an interchangeable character with its familiar form.

In addition, the accuracy of the word alignment between ancient and modern Chinese is very high since many words co-occur in both languages. Therefore, we can still have good alignment performance after combining two alignment results.

3   Pivot Term Alignment Method

3.1   Term Alignment Steps

In many English translations of historical books, the first letter of the words in terms are capitalized. In the same time, the ancient Chinese terms need to be identified by term recognizer, which is usually trained on tagged corpora. The English terms are much easier to identify than the ancient Chinese terms. Therefore, instead of identifying the ancient Chinese terms directly, we extract them using the term alignment. Specifically, we first identify the English term, and then perform term alignment between English and modern Chinese, the alignment between modern Chinese and ancient Chinese, respectively, and finally get the ancient Chinese terms by mapping the English terms to ancient Chinese term via modern Chinese. This procedure can be illustrated by the example in Fig. 1.

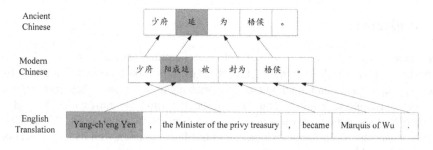

Fig. 1. An example of the term alignment using modern Chinese as the pivot language.

Given English-modern Chinese corpus $EM = \{(E_1, M_1), (E_2, M_2), \cdots, (E_N, M_N)\}$, modern-ancient Chinese corpus $MA = \{(M_1, A_1), (M_2, A_2), \cdots, (M_N, A_N)\}$, wherein E, M, A is English, modern Chinese and ancient Chinese sentence, respectively, the steps of our term alignment method can be described as Fig. 2.

3.2 English Term Recognition

We make use of the capitalization rule to identify English terms since most English translations of historical books capitalize the first letter of words in the term. But the capitalization extraction rule has two problems: (1) The first word of sentence is extracted as the wrong term. (2) Some articles and conjunctions that are not capitalized in terms are missed. Thus, we make some supplementary rules for the above problems.

For the first problem, we do not treat it as a term when the extracted term is at the beginning of the sentence and contains only one word of following part of speech: numeral, preposition, adverb, a conjunction, etc. For the second problem, if "the" is followed by a capital word, or "of" is sandwiched between two capital words, they are added to the extracted terms.

Input: English-modern Chinese corpus EM and modern-ancient Chinese corpus MA
(1) Perform word segmentation for ancient Chinese and modern Chinese.
(2) Perform word alignment between English and modern Chinese and obtain English-modern Chinese word alignment matrix A_{em}. Perform word alignment between modern Chinese and ancient Chinese and obtain modern-ancient Chinese word alignment matrix A_{ma}
(3) Recognize the English terms in each English sentence E_i and get the English terms set $T_e = \{e_1, e_2, \cdots, e_n\}$
(4) Extract the corresponding modern Chinese term m_i of English term e_i according to word alignment matrix A_{em} and obtain the term translation pair set between English and modern Chinese $T_{em} = \{(e_1, m_1), (e_2, m_2), \cdots, (e_n, m_n)\}$.
(5) Extract the corresponding ancient Chinese term a_i of modern Chinese term m_i according to word alignment matrix A_{ma} and obtain the term translation pair set between modern Chinese and ancient Chinese $T_{ma} = \{(m_1, a_1), (m_2, a_2), \cdots, (m_n, a_n)\}$.
(6) Obtain the term translation pair set between English and ancient Chinese $T_{ea} = \{(e_1, m_1), (e_2, m_2), \cdots, (e_n, m_n)\}$ by combing the term translation pair set T_{em} and T_{ma}.
Output: English-ancient Chinese term translation pairs set T_{ea}

Fig. 2. The step of the term alignment method using modern Chinese as the pivot language.

3.3 Word Alignment

We employ IBM-4 model [15] to perform word alignment of English-modern Chinese, modern-ancient Chinese, respectively. Since the performance of word alignment model with limited scale of parallel corpora are rather unsatisfactory, there are many errors in the word alignment result. Two measures are adopted to reduce the word alignment errors:

(1) Words co-occurrence is used to improve the word alignment of modern Chinese and ancient Chinese. Specifically, if more than one character of a modern Chinese word is the same with another ancient Chinese word, they should be aligned to each other. For example, modern Chinese word "齐威王" and ancient Chinese word "齐威". If the co-occurring word pair does not appear in the word alignment matrix, they will be added. If one word of the word pair aligns to other words, we will remove the wrong word alignment and add the co-occurring word pairs.

(2) The term integrity is utilized to complement the word alignment after recognizing the English term. Specifically, all words in an English term should correspond to the same Chinese term. If some words in the English term do not map to a term, we will add the words to the alignment. For example, English term "the Lord of Hao Lake" matches Chinese term "滈池君", while "Lord" does not align to "滈池君" in the word alignment matrix. Thus, we think the algorithm miss the alignment and add ("Lord", "滈池君") to the word alignment matrix.

3.4   The Bilingual Term Pairs Extraction

We conduct two kinds of term alignment to extract English-ancient Chinese term pairs, i.e., English-modern Chinese term alignment and modern Chinese-ancient Chinese alignment. Two kinds of term alignment both share the similar steps. For each term in the source sentence, we search the term in target sentence fulfilling the following conditions as the corresponding target term: (1) words of target term must be consecutive in target sentence; (2) word alignment in bilingual term pairs must be compatible with the alignment matrix, that is, the target words should align to the words in the source term or align to NULL according to the alignment matrix.

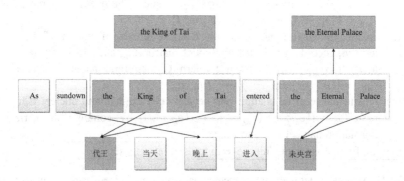

Fig. 3. A term alignment example between English and modern Chinese

We take the example shown in Fig. 3 to illustrate the term alignment process between English and modern Chinese. To match the modern Chinese term for English term "the King of Tai", we first look up the word alignment table to find the corresponding word in Chinese sentence for "King", which is "代王". Then we find the next word "of" is a preposition. We skip it and continue to match next word "Tai", which also aligned to "代王". Therefore, words "of", "Tai" and "the King" align to the same term "代王". Term translation pair ("the King of Tai", "代王") is added to the English-modern Chinese term pair set.

4 Experiment

4.1 Experimental Setup

We built three parallel corpora for the term alignment experiment, namely, English-modern Chinese corpus, modern-ancient Chinese corpus and English-ancient Chinese corpus. Each corpus comprised of 4064 sentence pairs, which are from the five basic annals of *Shiji* and the corresponding translation. The English translation was extracted from *the Records of the Grand Historian of China* [16, 17]. The number of term translation pairs is 1170.

In our method, word segmentation (WS) of modern Chinese was implemented by Jieba[1] and ancient Chinese was segmented by the word segmentation method based on word alignment (WSWA)[2]. The word alignment model, IBM 4 model, was implemented by GIZA++[3][18].

We employ precision (P), recall (R), and F-1 score as the evaluation metrics, which is defined as follows.

$$R = \frac{N_{correct}}{N_{gold}} \times 100\% \qquad\qquad . (1)$$

$$P = \frac{N_{correct}}{N_{segment}} \times 100\% \qquad\qquad (2)$$

$$F1 = \frac{2PR}{P+R} \qquad\qquad (3)$$

Wherein, N_{gold} is the number of words in the gold standard, N_{align} is the number of words aligned by our method; N_{corret} is the number of correctly aligned words by our method.

4.2 Experimental Results and Analysis

To justify the use of modern Chinese as the pivot language, we compared the proposed method with the direct alignment method, which ran IBM 4 model to align directly from English to ancient Chinese. Table 1 shows the comparison results of different term alignment methods.

---

[1] https://github.com/fxsjy/jieba.

[2] https://github.com/supercar101/Word-Segmentation-Method-of-Ancient-Chinese/tree/master.

[3] https://codeload.github.com/moses-smt/giza-pp/zip/master.

Table 1. The comparison results of different term alignment methods.

Term Alignment method	P	R	F-1
Direct alignment method	66.6%	57.8%	61.8%
Pivot based method	**81.2%**	**69.3%**	**74.8%**

In Table 1, our method shows an obvious advantage over the direct alignment method in precision, recall, and F-1 score. This contributes to the following two reasons:

(1) Using modern Chinese as the pivot language can reduce alignment errors caused by the abbreviation of ancient Chinese. The abbreviation can easily find the corresponding English translation by mapping to the full word in modern Chinese. This can be clearly illustrated by the example in Table 2.

Table 2. An example of aligning abbreviation in ancient Chinese.

Ancient Chinese	与齐威、楚宣、魏惠、燕悼、韩哀、赵成侯并。
Modern Chinese	秦孝公与齐威王、楚宣王、魏惠王、燕悼王、韩哀侯、赵成侯并称。
English	The ruler King Wei of Qi, King Xuan of Chu, King Hui of Wei, Duke Dao of Yan, Duke Ai of Hann, and Duke Cheng of Zhao being ranged side by side.

In the ancient Chinese of Table 2, terms "齐威", "楚宣", "魏惠", "燕悼" and "韩哀" all omit the titles, the full official name should be the terms in modern Chinese, i.e. "齐威王", "楚宣王", "魏惠王", "燕悼王" and "韩哀侯". When aligning ancient Chinese and English term directly, all the terms can not find the correct translation due the omission of ancient Chinese. "齐威", "楚宣" did not match English term and "魏惠", "燕悼" and "韩哀" align incorrectly to "Duke Dao", "Duke Ai" and "Duke Cheng". In the pivot alignment, the term alignment of ancient Chinese and modern Chinese was first carried out. After finding the full name of ancient Chinese term, the term alignment between modern Chinese and English was used to find the correct translation of English terms.

(2) Modern Chinese can also help align the ancient Chinese words containing interchangeable characters to the correct translation. Interchangeable character called "通假字" in Chinese refers to a special use of ancient Chinese characters. The interchangeable characters are used to replace the characters with the same or similar pronunciation. In the direct alignment, those words containing interchangeable characters barely align to the right translation due to very low frequency. For example, ancient Chinese term "甯昌" did not find the correct

translation in the direct alignment because "甯" is an interchangeable character of "宁" and is seldom used. By aligning "甯昌" to "宁昌" in modern Chinese, we know they refer to the same name and find correct translation "Ning Ch'ang" via "宁昌".

4.3 The Influence of Ancient Chinese WS

To test the influence of the ancient Chinese WS method on the term alignment, we perform term alignment using ancient Chinese WS results obtained by WSWA and the ground truth, respectively. The comparison results are shown in Table 3.

Table 3. The term alignment results using different WS results.

Methods		P	R	F-1
WSWA	Ancient Chinese WS	89.3%	83.6%	86.3%
	Term Alignment	81.2%	69.3%	74.8%
Ground Truth	Ancient Chinese WS	100%	100%	100%
	Term Alignment	**80.2%**	**78.8%**	**79.5%**

From Table 3, we can see Chinese WS results have a very direct influence on term alignment. The term alignment is subject to the performance limit of ancient Chinese. The recall of term alignment using the WS result is also very low. Using the ground truth can increase 9 point of recall since English terms cannot find the correct ancient Chinese term when the ancient Chinese is not segmented correctly.

Besides ancient Chinese WS, modern Chinese WS also have significant impact on the term alignment results. If the modern Chinese is segmented wrongly, English term aligns to wrong modern Chinese term and it is very hard to find the right ancient term. In the example shown in Table 4, due to the wrongly segmented modern Chinese word "缪公对", "Duke Mu" can not find the corresponding ancient Chinese term "缪公".

Table 4. An example of wrong modern Chinese word segmentation.

Ancient Chinese	缪公/之/怨/此/三人/入於/骨髓
Modern Chinese	缪公对/这/三个人/恨之入骨
English	The hatred Duke Mu bears these men eats into his very bones and marrow!

5 Conclusion

In this paper, we proposed a term alignment method using modern Chinese as a pivot. The method aligned the historical term from English to modern Chinese, then to ancient Chinese. Using modern Chinese as the pivot language not only solves the shortage problem of parallel corpus but also reduces the alignment error caused by abbreviation and the interchangeable characters of the ancient Chinese.

Our method only explores word alignment to perform term alignment. Currently, the performance of word alignment is far from satisfactory. This limits the performance improvement of the term alignment. In the future, we will investigate the method using more information to extract term translation pairs.

Acknowledgements. This work is supported by the National Natural Science Foundation of China (No. 61402068).

References

1. Huang, Z.: English translation of cultural classics and postgraduate teaching of translation in Suzhou University. Shanghai J. Translators **1**, 56–58 (2007). (In Chinese)
2. Li, X., Che, C., Liu, X., Lin, H., Wang, R.: Corpus-based extraction of Chinese historical term translation equivalents. Int. J. Asian Lang. Process. **20**(2), 63–74 (2010)
3. Che, C., Zheng, X.: The extraction of term translation pairs for chinese historical classics based on sub-words. J. Chin. Inf. Process. **30**(3), 46–51 (2016). (In Chinese)
4. Yout, G.-W., Hwangt, S.-W., Song, Y.-I., et al.: Mining name translations from entity graph mapping. In: Proceedings of the 2010 Conference on Empirical Methods in Natural Language Processing (EMNLP), pp. 430–439. ACL, Stroudsburg (2010)
5. Feng, D., Lv, Y., Zhou, M.: A new approach for English-Chinese named entity alignment. In: Proceedings of the Conference on Empirical Methods in Natural Language Processing (EMNLP), pp. 372–379. ACL, Stroudsburg (2004)
6. Lee, C.-J., Chang, J.S., Jang, J.-S.R.: Alignment of bilingual named entities in parallel corpora using statistical models and multiple knowledge sources. ACM Trans. Asian Lang. Inf. Process. (TALIP) **5**(2), 121–145 (2006)
7. Zhang, Y., Wang, Y., Cen, L., et al.: Fusion of multiple features and ranking SVM for web-based English-Chinese OOV term translation. In: 23rd International Conference on Computational Linguistics (COLING), pp. 1435–1443. ACM, New York (2010)
8. Chen, Y., Zong, C., Su, K.-Y.: On jointly recognizing and aligning bilingual named entities. In: Proceedings of the 48th Annual Meeting of the Association for Computational Linguistics, pp. 631–639. ACL, Stroudsburg (2010)
9. Chen, Y., Zong, C., Su, K.-Y.: A joint model to identify and align bilingual named entities. Comput. Linguist. **39**(2), 229–266 (2013)
10. Wang, M., Che, W., Manning, C.D.: Joint word alignment and bilin-gual named entity recognition using dual de-composition. In: Proceedings of the 51st Annual Meeting of the Association for Computational Linguistics, pp. 1073–1082. ACL, Stroudsburg (2013)
11. Wu, H., Wang, H.: Pivot language approach for phrase-based statistical machine translation. Mach. Transl. **21**(3), 165–181 (2007)

12. Wu, H., Wang, H.: Revisiting pivot language approach for machine translation. In: Proceedings of the Meeting of the Association for Computational Linguistics and the International Joint Conference on Natural Language Processing of the AFNLP, pp. 154–162. ACL, Stroudsburg (2009)

13. Durrani, N., Koehn, P.: Improving machine translation via triangulation and transliteration. In: Conference of the European Association for Machine Translation, Dubrovnik, Croatia, pp. 71–78 (2014)

14. Zhu, X., He, Z., Wu, H., Zhu, C., Wang, H., Zhao, T.: Improving pivot-based statistical machine translation by pivoting the co-occurrence count of phrase pairs. In: Proceedings of the 2014 Conference on Empirical Methods in Natural Language Processing (EMNLP), pp. 1665–1675. ACL, Stroudsburg (2014)

15. Brown, P.E., Pietra, S.A.D., Pietra, V.J.D., Mercer, R.L.: The mathematics of statistical machine translation: parameter estimation. Comput. Linguis. **19**(2), 263–311 (1993)

16. Watson, B.: Records of the Grand Historian of China. Columbia University Press, New York (1961)

17. Watson, B.: Records of the Grand Historian: Qin Dynasty. Chinese University of Hong Kong/Columbia University Press, Hong Kong/New York (1993)

18. Och, F.J., Ney, H.: A systematic comparison of various statistical alignment models. Comput. Linguis. **29**(1), 19–51 (2003)

Minority Language Processing

Point the Point: Uyghur Morphological Segmentation Using PointerNetwork with GRU

Yaofei Yang[1], Shupin Li[1], Yangsen Zhang[1], and Hua-Ping Zhang[2(✉)]

[1] Beijing Information Science and Technology University, Beijing, China
yangyaofei@gmail.com, susanli200808@163.com, zhangyangsen@bistu.edu.cn
[2] Beijing Institute of Technology, Beijing, China
kevinzhang@bit.edu.cn

Abstract. Uyghur is an agglutinative language that has many morphemes. It is necessary for processing Uyghur to segment words into morphemes. This work is called morphological segmentation. Previous works treat morphological segmentation as a tagging task and classify each character as one of four classes, which are $\{b, m, e, s\}$. However, these labels are not independent from each other, which makes the models easily overfitted. We propose a new method for the segmentation task. Instead of using these labels, we use only segmentation points for modeling. The model used in our method is more robust and easier to train than previous methods. Applying our model to Uyghur morphological segmentation, it achieves high accuracy and higher recall and f1 score than previous models.

Keywords: Morphological segmentation · Uyghur · Linguist · Agglutinative language · PointerNetwork · NLP

1 Introduction

Morpheme is the smallest grammatical unit in a language. There are two classes of morphemes: (1) free morphemes, which have meaning, and (2) bound morphemes, which have no meaning and need free morphemes to construct words. Language can be classified as two major types according to the ratio of morphemes per word: synthetic language and analytic language. Synthetic language has a high morpheme-per-word ratio. On the contrary, most words are free morphemes in analytic language. More specifically, synthetic language can be classified into inflected language, which combines morphemes by inflection, and agglutinative language, which combines morphemes by concatenation. For example, the English word *unbreakable* is inflected from *un + break + able*. Morphological segmentation is a task that segments words into morphemes, which is a basic natural-language-processing (NLP) task. The superiorities of morphological segmentation before further processing are as follows:

© Springer Nature Switzerland AG 2019
M. Sun et al. (Eds.): CCL 2019, LNAI 11856, pp. 371–381, 2019.
https://doi.org/10.1007/978-3-030-32381-3_30

(i) Reduction of vocabulary size and alleviation of the sparsity problem because words can the share same morpheme;

(ii) Alleviation of the out-of-vocabulary (OOV) problem because new words can be constructed with known morphemes.

Uyghur is a typical agglutinative language with numbers of morphemes, and most Uyghur words are combined from numbers of morphemes. It is almost like a phrase in English that leads to it having very low frequency. Moreover, it is difficult to further process the Uyghur language without morphological segmentation, so it is necessary for Uyghur to undergo morphological segmentation.

Parts of previous works use linguists to build a rules-based system (Orhun et al. 2009), which is complex and low-performing. In recent years, a statistics method has been used in this field, in which segmentation is treated as a sequence of classification tasks, using $\{b, m, e, s\}$ to label the sequence (Wang et al. 2016). However, there is a problem with this labeling method, specifically the tags $\{b, m, e, s\}$ are not independent of each other, which may make the model overfitted, and, in turn, cause high accuracy, low recall, and eventually a low f1 score.

The motivation of our work is to find a new method for tagging data in which the tags are independent of each other, and to use this method to build a better model.

In our model, we do not use the aforementioned set of tags; instead, segmentation points that are independent of each other will be used. Further, to fit the shape of data using the proposed method, we use PointerNetwork (Vinyals et al. 2015) as the main modeling framework. Eventually, the results are more improved than those using all previous reported methods of Uyghur morphological segmentation.

2   Related Work

The famous unsupervised tools for morphological segmentation is Morfessor (Creutz and Lagus (2002)), which uses a minimum description length (MDL) algorithm. MDL-based unsupervised methods were an important breakthrough and have been applied to several languages (Goldsmith 2001). However, this method cannot achieve good performance without manual human retrofitting. Poon et al. (2009) built a log-linear model and used an expectation maximization (EM) algorithm for training, which is a classic model framework for unsupervised learning. Based on this model, Bergmanis and Goldwater (2017) used linguistic information from pre-trained word-vectors as the main feature for their log-linear model and realized better performance.

For supervised machine learning, morphological segmentation has been treated as a classification task for each character of the sequence that classifies each character as one of four classes, $\{b, m, e, s\}$. With that approach, the original method is called a conditional random-field- (CRF-) based model. Cotterell et al. (2016) used a hidden Markov model (HMM) and the CRF model for segmentation. For part-of-speech tagging, Plank et al. (2016) used a bidirectional LSTM (Hochreiter and Schmidhuber 1997) with the CRF model, which also works for morphological segmentation tasks. Similarly, Wang et al. (2016) used a bidirectional window LSTM for modeling and achieved outstanding performance.

For the Uyghur language, Osman et al. (2019) used CRF to build a model, and, in addition, they enlarged the tagging set for orthography. Abudukelimu et al. (2017) used a bidirectional GRU (Cho et al. 2014) model for the Uyghur language and achieved the best performance, to date, in Uyghur morphological segmentation.

All of the aforementioned good-performance methods approached the morphological segmentation task as a tagging task. This method can be expressed as shown in Fig. 1, where, in a $\{b, m, e, s\}$ tagging method, the sequence is tagged and each character is assigned one of the following pre-determined classes:

(i) **B** represents the beginning of a multi-character segmentation;
(ii) **M** represents the middle of a multi-character segmentation;
(iii) **E** denotes the end of a multi-character segmentation;
(iv) **S** denotes a single-character segmentation.

When we reviewed this tagging method, we found that if we remove the symbols M and E, segmentation can still proceed with the remaining symbols. Thus, the total information in all symbols is equal to the information in symbols B and S; that is to say, those tags are not independent of each other based on information theory. When training the model with the above four classes, the model will learn what the tag is and how to use it in the correct order, which is slightly too much effort for a model to just segment a sequence. In other words, there are rules in those tags that the model must learn, and if it focuses on these rules, overfitting may result.

It is therefore necessary to make the model concentrate on segmentation, and a new method is needed to facilitate that.

Fig. 1. Tagging sequence with $\{b, m, e, s\}$

3   Segmentation Models

3.1   Labeling Corpus

In the preceding section, we found that B and S have all of the information, which is the start of each segmented sequence. These tags can be expressed as a pointer to point to where to segment the sequence. Therefore, we treat this task as a generation task that generate a set of pointers pointing to the first character of each segmented sub-sequence, as shown in Fig. 2.

This method, which only uses segmentation points, clearly instructs the model how to segment a word and it is thus easier for a model to understand what it should learn from data. It is difficult for the model to make a mistake and this method makes the model more robust.

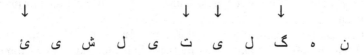

Fig. 2. Tagging sequence with pointer

The two aforementioned labeling methods are formalized in the following. For the $\{b, m, e, s\}$ method,

$$\theta = \arg\max_{\theta} \sum_{i=1}^{n} \log p(b_i | L : \theta),$$

where $L = (l_1, l_2 ... l_n)$ is a sequence that must be segmented, $B = (b_1, b_2, ... b_n)$ is the $\{b, m, e, s\}$ label, and θ is the model's parameter.

In the proposed method,

$$\theta_p = \arg\max_{\theta_p} \sum_{i=1}^{m} \log p(p_i | L : \theta_p),$$

where $P = (p_1, p_2 ... p_m)$ is the pointer sequence for which $m < n$, and θ_p is the model's parameter.

3.2   Model Selection

Because the length of P is not equal to the length of L, we cannot use a recent method like the CRF-based model. To deal with non-aligned data, the best method is to use a sequence to sequence (Seq2Seq) model (Sutskever et al. 2014). A Seq2Seq model is also known as an encoder-decoder (Cho et al. 2014) model. It was proposed to model machine translation with a deep neural network and obtain state-of-the-art performance. A Seq2Seq model has an encoder that can obtain the information of input data and push that information to the decoder, which decodes outputs.

However, a weakness is revealed upon reviewing the difference between the output of the Seq2Seq model and our proposed model. A Seq2Seq model can actually contain our outputs, but there is no strong relation between the input set and output set in the original Seq2Seq model. However, in this segmentation task, our output is a pointer pointing to the input that has a strong relation to the input.

In Vinyals et al. (2015)'s work, their PointerNetwork is a special Seq2Seq network. Each of the outputs of the PointerNetwork is a pointer that points to an input sequence that is usually used on text summarization and Q&A tasks that need the model to point out something from the input.

3.3   PointerNetwork with Scaled Attention

In our work, we use a modified PointerNetwork with scaled attention for the proposed method. There are three parts in our scaled attention PointerNetwork: encoder, decoder, and attention. In the encoder and decoder, we replace the LSTM (Hochreiter and Schmidhuber 1997) with GRU (Cho et al. 2014) to reduce the size of the model. After passing through the encoder and decoder, the data flow into the attention sub-model and the probability of pointing to each input item is calculated. The entire process is detailed below.

Figure 3 shows a sample procedure for the modified PointerNetwork that we propose. Suppose the input of the PointerNetwork is $X = (x_1, x_2, x_3, x_4, x_5)$, in which x_5 is an end-of-sequence symbol <eos>. In addition, the sequence X is to be segmented into (x_1, x_2) and (x_3, x_4).

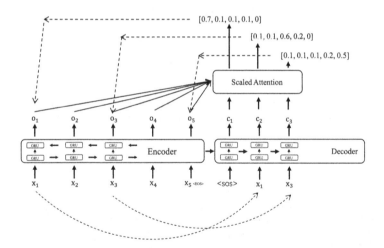

Fig. 3. Proposed modified PointerNetwork with scaled attention with example input $(x_1, x_2, x_3, x_4, x_5)$ in which x_5 is an end-of-sequence symbol <eos>

First, X goes through the encoder to obtain the output O and a hidden state h that will be sent to the decoder. The first input of the decoder is a symbol <sos>, which means the start of the sequence.

The decoder then calculates an output c_1 that will be sent to the scaled attention sub-model, and the probability vector with O will be calculated. The probability vector's dimension is equal to the length of the input X, which is 5 in this case. The largest probability in this vector will be the pointer pointing to the input, which is x_1 in this first step. The character that is pointed out by the output of the attention mechanism is sent into the decoder, which is output c_2 in the second step. Through use of the attention model, the output in this case is a pointer pointing to x_3.

Finally, in the third step, the decoder and attention model with the x_3 output pointer points to x_5, which is an end-of-sequence symbol. Then, the entire decoding stops, and the output 3 pointer segments the sequence $(x_1, x_2, x_3, x_4, x_5)$ into (x_1, x_2) and (x_3, x_4).

The entire procedure is demonstrated over, and the three parts of the proposed model are detailed below.

Encoder and Decoder. The encoder and decoder have a similar structure, which lets data go through them in each time step. Both also have a stacked GRU. Details are as follows.

Encoder:
In the encoder, we define $X = (x_1, x_2, x_3...x_n)$ as an input sequence and, through the encoder $f_{encoder}$, an output sequence $O = (o_1, o_2, o_3...o_n)$ and a hidden state h_n. The encoder can be represented as

$$O, h_n = f_{encoder}(X),$$

and for each time step,

$$o_i, h_i = f_{encoder_{i-1}}(x_{i-i}, h_{i-1}).$$

Decoder:
The input of the decoder in each time step is a character that is pointed out in the last time step. Because the encoder and decoder use the same charset for input, we share char embedding between encoder and decoder.

Each time step can be represented as follows. In time step j, we assume the output of time step $j - 1$ points to x_{i-1}, and the hidden state is h_{j-1},

$$c_j = f_{decoder_j}(x_{i-1}, h_{j-1}),$$

where the output of the decoder is c_j, which will be sent to the attention model to calculate the probability of each input pointer.

Scaled Attention. The attention model uses one of the decoder's outputs and all of the encoder's output to calculate the probability of pointing to each input item. Suppose that the length of input is n, and then the attention model's output vector has n dimension. Each dimension represents a probability of an input item:

$$y_i = f_{atten}(O, c_i),$$

where y_i has n dimension.

In the original PointerNetwork, Bahdanau-style attention (Bahdanau et al. 2014) is used. This attention mechanism is calculated as follows:

$$f_{atten}(o_i, c_j) = V_a^\top tanh(W_1 o_i + W_2 c_j),$$

where V is a learned vector parameter and W_1 and W_2 are two learned matrix parameters. o_i and c_j are the outputs of the encoder and decoder, respectively.

In our work, we use dot-product attention instead, which is much faster and more space-efficient in practice, since it can be implemented using highly optimized matrix multiplication code. In addition, we follow Vaswani et al. (2017)'s scaled-attention mechanism. According to the latter paper, for large dimension d_K, dot-product attention obtains small gradients when using a softmax function in this part without scaling. The calculation is expressed as

$$f_{atten} = softmax(\frac{QK^\top}{\sqrt{d_K}}),$$
$$Q = W_Q C,$$
$$K = W_K O,$$

where C is the encoder's output and O the decoder's output; W_Q and W_K are two independent learnable matrices.

In practice, we mask out (set to ∞) all values larger than the length of input after the softmax function with a batch because we need to pad the batch sequence to the same length for batch input.

3.4 Decoding

Our model is different when inference. For the previous model, the input goes through the model and gets output for each character. But our model is a Seq2Seq model, the length of outputs are not equal to the input. And the decoder will emit each output at each timestamp. The important is in each timestamp decoder need previous output to calculate the output in current timestamp. In another words, the decoder needs a decoding procedure which is not needed in previous model.

There is a flaw in our model if we use previous decoding method. In decoding, the previous output which the decoder need is the previous pointer points to the previous segmentation point. And the output in this timestamp should be a pointer after the previous pointer, which can be a pointer before the previous one. It is an error in that situation.

To fix that flaw, we mask the output which let the probability to point before the previous one be zero, when decoding.

4   Experimentation

4.1 Dataset

For Uyghur morphological segmentation, we use the THUUyMorph (Halidanmu et al. 2018) dataset. For the comparison experiment, we process this corpus into two files: one is a source-words list and the other is a label list that uses $\{b, m, e, s\}$ symbols.

In our model experiment, the corpus is processed into three files: a source-words list, a pointer list, and a character list that are pointers pointing to the source words.

The dataset was split into two parts, one for training (70%) and one for testing (30%).

4.2   Training

For the sequence classification method, we trained the CRF model as the baseline model and the bidirectional GRU model as represented by Abudukelimu et al. (2017) as the state of the art. For the CRF model, we use CRF++. In contrast, for the bidirectional GRU model, we follow the setting in Abudukelimu et al. (2017), which has a 1024 GRU dimension and a 300-char embedding size.

In the original PointerNetwork, Vinyals et al. (2015) did not use the output of the hidden state from the encoder. We double it because the hidden state has rich linguistic features that should help performance. Therefore, we train our model using the two aforementioned models separately. In addition, we use the scaled-attention mechanism. To determine whether our procedure works, the attention-without-scaling mechanism is also trained. Moreover, to realize a better hyperparameter for our model, we trained various sizes of embeddings and GRUs. For the char embedding, the encoder and decoder share the same embedding weight because they use the same char vocabulary. For optimization, all neural network models are optimized by stochastic gradient descent (SGD) (Kiefer and Wolfowitz 1952) with learning rate 0.1 and decay to 0.9 per 10000 steps. For normalization, all neural network models include bi-GRU model are used dropout with rate 0.3.

Table 1. Results of comparisons to baseline

Method	Accuracy	Precision	Recall	f1
CRF	94.20	**98.42**	95.23	96.80
Bi-GRU	**98.5**	96.05	96.15	96.15
Small pointer*	97.80	96.39	**98.70**	97.50
Large pointer**	98.03	96.52	98.56	**97.53**

*Uses size-256 GRU and size-32 embedding.
**Uses size-1024 GRU and size-64 embedding.

4.3   Analysis

The results are shown in Tables 1 and 2.

Table 2. Results using various hyperparameters

GRU size	Embedding	Hidden state*	Scaled**	Accuracy	Precision	Recall	f1
256	32	F	F	97.63	95.61	97.48	96.35
256	32	F	T	97.60	95.76	98.24	96.98
256	32	T	F	97.69	95.99	97.27	96.63
256	32	T	T	97.70	96.24	98.70	96.80
512	32	F	F	97.92	96.29	97.66	96.97
512	32	F	T	97.96	96.29	97.78	97.03
512	32	T	F	97.79	96.38	97.21	96.79
512	32	T	T	97.80	96.39	**98.70**	97.50
1024	32	F	F	97.84	96.58	97.17	96.87
1024	32	F	T	97.98	96.35	97.75	97.05
1024	32	T	F	97.72	96.04	97.32	96.67
1024	32	T	T	97.91	**96.89**	98.48	**97.68**
1024	64	F	F	97.30	95.48	97.76	96.61
1024	64	F	T	97.72	96.15	98.16	97.15
1024	64	T	F	97.41	95.86	97.69	96.76
1024	64	T	T	**98.03**	96.52	98.56	97.53
1024	128	T	T	97.93	96.43	98.42	97.41
2048	128	T	T	97.92	96.36	98.44	97.39

*Uses encoder's hidden state output or not that used by the decoder as initiation state.
**Uses scaled attention in attention mechanism or not.

In Table 1, we apply two models to facilitate a comparison with the baseline and SoTA. As we expected, we obtain a higher recall and f1 score. Because we use a more independent label to train our model, it is difficult to overfit the data and the former output would not affect the subsequent output. It makes our model less error-prone and we eventually obtain a low recall and f1 score, which makes our model more robust.

In Table 2, we compare different hyperparameters of our models. We train several models by using hidden state and scaled attention in various hyperparameters. In all hyperparameters, we can get the same conclusion. First, as mentioned above, using the encoder's hidden state is definitely helpful because there is useful information in it. The scaled-attention mechanism used in our model is more important than the hidden state, since the performance is significantly reduced without scaled attention. We believe that scaled attention can make the attention model more clear when the lengths of input are varied and the pointers need to point to them. In other words, it is a kind of dimension-wise normalization. Regarding the size of GRU and embedding, bigger is not better. A GRU size of 1024 and an embedding size of 64 obtains the best result; larger values of either of these two hyperparameters will degrade the performance.

5   Conclusions

In previous work, morphological segmentation was viewed as a classification problem for each item in a sequence using the set $\{b, m, e, s\}$ to label the sequence. We found this method to have a label-dependent problem that is damaging to training and performance. To solve this problem, we propose using fewer independent labels to model morphological segmentation tasks. We first apply PointerNetwork, modify it with a scaled attention mechanism, and use the hidden state of the encoder, in contrast to the original PointerNetwork. We thus obtain a higher recall and f1 score compared with the previous baseline and the SoTA Uyghur morphological segmentation model.

Results show that our new treatment of the segmentation task works and can achieve a more robust model. Our new method can be ported to other segmentation or tagging tasks, e.g., Chinese segmentation or morphological segmentation of the English, Urdu, and Turkish languages. Combining previous methods and our method, a higher performance and more robust ensemble model can be realized.

Acknowledgements. This work was supported by National Science Foundation of China (Grant No. 61772075), National Science Foundation of China (Grant No. 61772081), Scientific Research Project of Beijing Educational Committee (Grant No. KM201711232022), Beijing Municipal Education Committee (Grant No. SZ20171123228), Beijing Institute of Computer Technology and Application (Grant by Extensible Knowledge Graph Construction Technique Project).

References

Abudukelimu, H., Cheng, Y., Liu, Y., Sun, M.: Uyghur morphological segmentation with bidirectional GRU neural networks. J. Tsinghua Univ. (Sci. Technol.) **57**(1), 1–6 (2017)

Bahdanau, D., Cho, K., Bengio, Y.: Neural Machine Translation by Jointly Learning to Align and Translate. arXiv:1409.0473 [cs, stat], September 2014

Bergmanis, T., Goldwater, S.: From segmentation to analyses: a probabilistic model for unsupervised morphology induction. In: Proceedings of the 15th Conference of the European Chapter of the Association for Computational Linguistics: Volume 1, Long Papers, pp. 337–346. Association for Computational Linguistics, Valencia, April 2017

Cho, K., et al.: Learning phrase representations using RNN encoder-decoder for statistical machine translation. In: Proceedings of the 2014 Conference on Empirical Methods in Natural Language Processing (EMNLP), pp. 1724–1734, October 2014

Cotterell, R., Vieira, T., Schütze, H.: A joint model of orthography and morphological segmentation. In: Proceedings of the 2016 Conference of the North American Chapter of the Association for Computational Linguistics: Human Language Technologies, pp. 664–669. Association for Computational Linguistics, San Diego (2016)

Creutz, M., Lagus, K.: Unsupervised discovery of morphemes. In: Proceedings of the ACL-2002 Workshop on Morphological and Phonological Learning (2002)

Goldsmith, J.: Unsupervised learning of the morphology of a natural language. Comput. Linguis. **27**(2), 153–198 (2001)

Halidanmu, A., Abudukelimu, A., Sun, M., Liu, Y.: THUUyMorph: an uyghur morpheme segmentation corpus. J. Chin. Inf. Process. **32**(2), 81 (2018)

Hochreiter, S., Schmidhuber, J.: Long short-term memory. Neural Comput. **9**(8), 1735–1780 (1997)

Kiefer, J., Wolfowitz, J.: Stochastic estimation of the maximum of a regression function. Ann. Math. Stat. **23**(3), 462–466 (1952)

Orhun, M., Tantug, A.C., Adali, E.: Rule based analysis of the uyghur nouns. Int. J. Asian Lang. Proc. **19**(1), 33–44 (2009)

Osman, T., Yang, Y., Tursun, E., Cheng, L.: Collaborative analysis of uyghur morphology based on character level. Beijing Daxue Xuebao (Ziran Kexue Ban)/Acta Scientiarum Naturalium Universitatis Pekinensis **55**, 47–54 (2019)

Plank, B., Søgaard, A., Goldberg, Y.: Multilingual part-of-speech tagging with bidirectional long short-term memory models and auxiliary loss. In: Proceedings of the 54th Annual Meeting of the Association for Computational Linguistics (Volume 2: Short Papers), pp. 412–418, August 2016

Poon, H., Cherry, C., Toutanova, K.: Unsupervised morphological segmentation with log-linear models. In: Proceedings of Human Language Technologies: The 2009 Annual Conference of the North American Chapter of the Association for Computational Linguistics, pp. 209–217. Association for Computational Linguistics, Boulder (2009)

Sutskever, I., Vinyals, O., Le, Q.V.: Sequence to sequence learning with neural networks. In: Ghahramani, Z., Welling, M., Cortes, C., Lawrence, N.D., Weinberger, K.Q. (eds.) Advances in Neural Information Processing Systems, vol. 27, pp. 3104–3112. Curran Associates, Inc. (2014)

Vaswani, A., et al.: Attention is all you need. In: Guyon, I., et al. (eds.) Advances in Neural Information Processing Systems, vol. 30, pp. 5998–6008. Curran Associates, Inc. (2017)

Vinyals, O., Fortunato, M., Jaitly, N.: Pointer networks. In: Cortes, C., Lawrence, N.D., Lee, D.D., Sugiyama, M., Garnett, R. (eds.) Advances in Neural Information Processing Systems, vol. 28, pp. 2692–2700. Curran Associates, Inc. (2015)

Wang, L., Cao, Z., Xia, Y., de Melo, G.: Morphological segmentation with window LSTM neural networks. In: Thirtieth AAAI Conference on Artificial Intelligence, March 2016

Construction of an English-Uyghur WordNet Dataset

Kahaerjiang Abiderexiti and Maosong Sun[(✉)]

Department of Computer Science and Technology, Institute for Artificial Intelligence,
State Key Lab on Intelligent Technology and Systems, Tsinghua University,
Beijing, China
khejabdr15@mails.tsinghua.edu.cn, sms@tsinghua.edu.cn

Abstract. Automatically building semantic resources is essential to low resource-languages like Uyghur. However, Uyghur suffers from a lack of publicly available evaluation dataset for automatically building semantic resources like WordNet. To cope with this problem, first, we build the largest Uyghur-English and English-Uyghur dictionaries by exploiting many possible online and offline resources. Then by using Princeton WordNet (PWN) 3.0 and Contemporary Uyghur Detailed Dictionary (CUDD), we construct an English-Uyghur WordNet evaluation dataset which is publicly available (https://github.com/kaharjan/uywordnet). In this dataset, more than 73,000 English synsets are mapped Uyghur automatically, in which over 20,000 are annotated manually. And the corresponding Uyghur words include definition and examples in Uyghur language context. We also propose a Synset Mapping based on Word Embeddings (SMWE) method. The experimental results on the dataset are promising.

Keywords: Uyghur · WordNet · Dataset · Synset mapping

1 Introduction

Since the introduction of Princeton WordNet (PWN) by professor George A. Miller [10], the construction of WordNet in other languages has begun. So far, WordNets in more than 50 languages have been constructed[1]. However, little research has been done on Uyghur WordNet construction or evaluation. Although BabelNet[2] contains Uyghur, its Uyghur parts are based on Wikipedia in Uyghur which contains around 2000 articles, and most of them are not aligned to other languages. So aligned concepts in Uyghur in the knowledge graph are scarce, and there are few definitions of these concepts in Uyghur. English-Uyghur WordNet constructed merely by translating PWN would suffer from lacking definitions and example sentences in Uyghur context. This would narrow the value of Uyghur WordNet to a certain extent and help downstream applications little. Also,

---

[1] http://globalwordnet.org/resources/wordnets-in-the-world/.
[2] http://babelnet.org/.

M. Sun et al. (Eds.): CCL 2019, LNAI 11856, pp. 382–393, 2019.
https://doi.org/10.1007/978-3-030-32381-3_31

there are no off-the-shelf English-Uyghur or Uyghur-English dictionaries. Further, there is no evaluation dataset for automatic construction of English-Uyghur WordNet. As a result, there is little research about the automatic construction of English-Uyghur WordNet. Fortunately, Contemporary Uyghur Detailed Dictionary (CUDD) is available. The dictionary contains more than 60,000 words. Each sense of each word is explained by a definition and some example sentences. So far, only Aierken *et al.* [6] attempted to construct a semantic lexicon using CUDD. Since then, the dictionary has not been used for WordNet construction.

To address those problems, first, we preprocess CUDD to get all words' POS tags, definitions, and example sentences. We also build the largest English-Uyghur and Uyghur-English dictionary from online and offline using most of the available resources. Then, we map PWN 3.0 synsets to CUDD nouns using the dictionary to get a preliminary mapping dataset. And we manually annotate more than 20,000 English synsets to Uyghur to get high-quality development and test dataset. Finally, by coping with the common problems of Uyghur language processing like stemming and word representation, we propose a synset mapping method based on word embeddings using our dataset. The experimental results are promising. We make our dataset available online. In our dataset, every Uyghur word has a definition in Uyghur and some example sentences in Uyghur language context which would improve the value of this dataset to the downstream applications.

2   Related Work

In more than 30 years of the WordNet construction in various languages, many construction methods are proposed. The construction methods of WordNet can be divided into the following three kinds: construction by humans, external resource linking, and automatically.

PWN is built manually [10]. Wang *et al.* [25] merge the Southeast University WordNet (SEW) [26], Sinica Bilingual Ontological WordNet (Sinica BOW) [12], China Taiwan University WordNet (CWN) [13] and the Chinese part of the open multilingual WordNet (OMW) [8], then manually proofread to construct Chinese Open WordNet (COW). For the external resource linking methods, through a data integration approach Bond *et al.* [8] combine open-licensed WordNet, Wiktionary[3] and Unicode Common Locale Data Repository CLDR[4] to link and extend the Open Multilingual WordNet (OMW). For the automatic methods, Montazery *et al.* [17] build a Persian WordNet relying on bilingual dictionaries as well as Persian and English monolingual corpus. Lam *et al.* [15] build Arabic, Assamese, Dimasa, Karbi, and Vietnam WordNet using the vocabulary of other languages and Microsoft's online machine translation system. They use the direct translation, intermediate WordNets, intermediate WordNets, and a dictionary (IWND) to create the multilingual WordNet synsets. Tarouti *et al.* [23] improve the above-mentioned methods by calculating word similarity based

---

[3] https://www.wiktionary.org.
[4] http://cldr.unicode.org.

on word vectors, which improves the accuracy of the automatically constructed Arabic WordNet. Arcan et al. [7] use a multilingual parallel corpus for sense disambiguation, extend four European languages WordNets. Khodak *et al.* [14] automatically build WordNets through an unsupervised method based on word embedding and word induction, which utilizes machine translation and large-scale unlabeled monolingual corpus. Ercan and Haziyev [9] propose a graph clustering method in sense detection and apply it to multilingual WordNet construction.

In Uyghur language processing, there are some works about semantic resource construction. Aierken *et al.* [6] by using CUDD to construct a Uyghur language semantic lexicon based on PWN. The semantic lexicon contains 1300 pairs of synonyms and 1059 pairs of antonyms. Yilahun *et al.* [27] investigate Uyghur ontology construction methods and attempt to construct Uyghur ontology in computer science and mathematics. Qiu *et al.* [21] review construction of Uyghur knowledge graph. They also design and implement Uyghur-Chinese-English online dictionary based on knowledge graph [19] then apply the K-means method to semantic search [20]. Maimaiti *et al.* [16] construct named entity corpus from parallel corpus. Abiderexiti *et al.* [1,2] propose named entity relation annotation specification and construct a corresponding corpus. However, there are no reports about publicly available English-Uyghur WordNet datasets. As a result, it is hard to evaluate the accuracy and precision of automatically constructed English-Uyghur WordNet.

3   Dataset

3.1   Data Processing

CUDD mainly uses dictionary form of a word (lemma) as the description object and explains each sense of the word. Every sense has an explanation, and most senses have example sentences as shown in Fig. 1. However, the granularity of each sense in CUDD is rougher than PWN. In general, a Uyghur word-sense pair corresponds to one or more than one word-sense pairs in PWN 3.0. To construct high-quality Uyghur-English WordNet, we preprocess CUDD and extract all the one sense Uyghur nouns from CUDD, whose total number is 10,973.

To get strong initial mapping relations between PWN 3.0 and CUDD, we need Uyghur-English and English-Uyghur dictionaries. But there are no off-the-shelf Uyghur-English or English-Uyghur dictionaries. We crawl Uyghur-English and English-Uyghur online dictionary from two websites (http://dict.izda.com, https://panlex.org). We have used 1s time interval for every word to not affect their server and to avoid our IP bing blocked. However, in some dictionary entries, there are irregular patterns and errors. Some Uyghur words are written in Latin and even Cyrillic script. Therefor, we first, eliminate these irregular patterns and errors manually. Then we convert all Latin and Cyrillic scripts into Uyghur scripts based on the Arabic alphabet which is the current writing system in the Xinjiang Uyghur Autonomous Region. We also use some offline dictionaries which are typed manually. Finally, we combine these dictionaries to

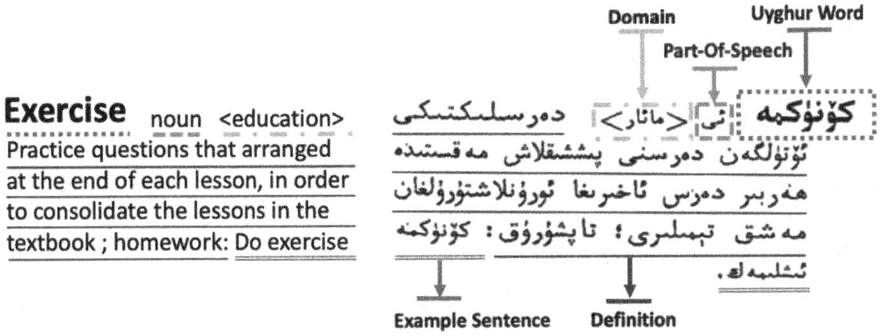

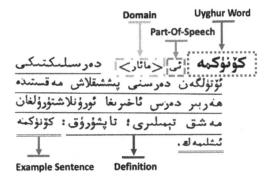

Fig. 1. An example from Contemporary Uyghur Detailed Dictionary (CUDD). The word in the red box with round dot is an Uyghur word, the character in the green box with dash line denote Part-Of-Speech of the word, the words in the orange box with dash dot line denote the domain of the word, the sentence marked with the blue line is a definition of the word, The sentence marked with the brown double line is an example sentence of the word. (Color figure online)

{"ئەۈرىشكە": ["specimen.n.01", "specimen.n.02",   "convention.n.02", "pattern.n.05",

"radiation_pattern.n.01", "sample.n.01", "sample_distribution.n.01", "sample.n.03",

"model.n.02", "model.n.04", "exemplar.n.01", "model.n.07", "model.n.09",

"design.n.06"]}}

{"ئەبنۈس": ["persimmon.n.01", "persimmon.n.02", "japanese_persimmon.n.01",

"kaki.n.02"]}}

_____

{"ئەۈرىشكە": [

    {"def":"ئۈلگە قىلىنىدىغان بويۇم ؛ ئۆرنەك ، نۈسخا",

    "egs":"مال ئەۈرىشكىسى"}]}

{"ئەبنۈس": [

    {"def":"يەر شارىنىڭ ئىسسىق بەلۇاغ رايونلىرىدا ئۈسىدىغان ، ياغىچى قاتتىق ، قارامتۇل بىر خىل دەرەخ",

    "egs":"NULL"}]}

Fig. 2. The JSON file format of the dataset: Arabic script in blue is an Uyghur word written in the current Uyghur writing system which is based on the Arabic alphabet. The upper part of the Figure indicates the relation between words in Contemporary Uyghur Detailed Dictionary (CUDD) and synsets in PWN 3.0. The lower part of the Figure indicates the definition (denoted by "def") and example sentences (denoted by "egs") of the corresponding Uyghur word.

get the largest English-Uyghur and Uyghur-English dictionaries by exploiting many possible online and offline resources. The vocabulary sizes are 222, 842 and 230, 287, respectively.

3.2 Building the Dataset

We denote Uyghur nouns as $W = \{w_1, w_2, \cdots, w_n\}$, where w_i to represent a word. T represents the English translation set of W, i.e. $T = \{t_1, t_2, \cdots, t_m\}$, Uyghur-English dictionary mapping function f_{ue} is:

$$f_{ue}(w_i) = T_i \tag{1}$$

where $T_i \subset T$, $T_i = \{t_{i1}, t_{i2}, \cdots, t_{io}\}$, $T_i \neq \emptyset$.

We denote $S = \{s_1, s_2, \cdots, s_p\}$ to represent the synsets in PWN. The relationship between English words and its synset is based on the following Eq. (2):

$$P_{syn}(T_i) = S_i \tag{2}$$

where $S_i \subset S$, $S_i = \{s_{i1}, s_{i2}, \cdots, s_{io}\}$.

For every word $w_i \in W$, we get the corresponding English synsets S_i by Eqs. (1) and (2). In this way, we build first English-Uyghur WordNet raw mapping data, and the synsets are 73,491. We split the data into three parts for training, developing, and testing. We annotate the 20,702 English synsets manually according to the definitions and example sentences of both Uyghur and English. It is hard to find someone who is proficient in two languages and also have linguistics and computer science background. So one person who satisfies above condition annotate the synsets two times and then rechecked it. Because one person annotates the synsets, so we do not calculate the annotation agreements between two times annotations. Finally we get 6,642 English synsets and corresponding 1,750 Uyghur by checking three times. The statistics of the training set, development set, and test set is shown Table 1.

Table 1. Evaluation dataset for automatic English-Uyghur WordNet construction

	# of training set	# of development set	# of test set
English	52,789	5,493	1,149
Uyghur	4,376	1,450	300

We make our automatically and manually labeled dataset publicly available (https://github.com/kaharjan/uywordnet). The Uyghur words include the meaning of words and examples, the format of the file is JSON which is shown by Fig. 2. Each of the training, development, and test set contains two files. One is CUDD and PWN 3.0 mapping relations shown in the upper part of Fig. 2. Another is the definition and examples of Uyghur word in CUDD shown in the lower part of Fig. 2.

4 Mapping Method

4.1 Uyghur Word Representation

To map corresponding English synsets to Uyghur words, We take advantage of the idea of word embeddings, which allows words with similar meaning to have a similar word representation. We have used the monolingual corpus, including more than three million sentences, which are the largest corpus in Uyghur. However, for the low resource language and agglutinative language like Uyghur, beside the corpus size, getting reasonable word representation also faces challenges. In Uyghur, one word could have several suffixes to form phrases, even sentences. For example, " بېيجىڭدىكىلەرنىڭكىمۇ؟ " (Does it belongs to people who are from Beijing?). So getting word representation in forms of the dictionary would be difficult. Fortunately, morpheme-enhanced continuous bag-of-words model (mCBOW) [3] cope with this problem. The idea of mCBOW is to treat morphemes rather than words as the basic unit of representation learning. To use mCBOW, we need to stem or lemmatize Uyghur words. Although there are some works about Uyghur word morphological segmentation [5,18] and annotated corpus [4], no off-the-shelf stemming, lemmatization tool or large scale annotated corpus for practical use. So we have applied two existing solutions. First is probabilistic generative models that use sparse priors inspired by the Minimum Description Length (MDL) principle [11,24]. The second is the subword model based on the byte pair encoding compression algorithm [22]. We conduct several experiments to compare and tune the models, then use the best solution to train the mCBOW model to get optimal word representation. The comparing experiment results will discuss in Sect. 5.

4.2 Synset Mapping Based on Word Embeddings

After getting optimal results for word representation, we use the training data described in Sect. 3 to get the corresponding lemmas in English. More formally, We denote P_{lemm} to represent the mapping function from synsets to lemmas in Eq. (3):

$$P_{lemm}(S_i) = L_i \tag{3}$$

where L_i represents a subset of the lemma set L, $L_i = \{l_{i1}, l_{i2}, \cdots, l_{io}\}$. L_i is a true subset of L $L_i \subset L$.

We derive the Uyghur word set W_i from the English-Uyghur dictionary mapping function f_{eu} according to Eq. (4):

$$f_{eu}(L_i) = W_i \tag{4}$$

where $W_i = \left\{ W_{i1}', W_{i2}', \cdots, W_{io}' \right\}, W_{ij}' \subset W$.

The word embeddings of Uyghur word $\boldsymbol{v}_{w_i}$ is calculated by Eq. (5):

$$\boldsymbol{v}_{w_i} = \boldsymbol{u}_{w_i}' + \frac{1}{m} \sum_{k=1}^{m} \boldsymbol{u}_k \tag{5}$$

where $\boldsymbol{u}'_{w_i}$ is representation of surface form of the word w_i, $\boldsymbol{u}_k$ is representation of the kth unit of the word w_i. In other words, w_i consists of m units.

We calculate the cosine similarity between each Uyghur word in the set W'_i and the word vector corresponding to w_i in Eq. (6):

$$Sim(w_i, W'_{ij}) = \frac{1}{|W'_{ij}|} \sum_{w'_{ijk} \in W'_{ij}} \boldsymbol{v}_{w'_{ijk}} \cdot \boldsymbol{v}_{w_i} \tag{6}$$

where $\boldsymbol{v}_{w_i}$ and $\boldsymbol{v}_{w'_{ijk}}$ are the vectors of the Uyghur words w_i and w'_{ijk}. $|W'_{ij}|$ the number of the words in the Uyghur word set W'_{ij} corresponding to Uyghur word w_i. Finally, we get $S'_i = \{s_{i1}, s_{i2}, \cdots, s_{ip}\}$, $0 \leq p \leq o$ as the mapping result for w_i. S'_i is English synsets corresponding to the Uyghur words in $match(w_i)$ obtained by Eq. (7). The lower and upper bound of synset number p is between 0 and o.

$$match(w_i) = \arg \max_{W'_{ij} \in W_i} Sim(w_i, W'_{ij}) \tag{7}$$

5   Experiments

5.1   Word Representation

As discussed above section, first we evaluate stemming for mCBOW. We separately train the models based on Morfessor 2.0 [24] and FlatCat [11] on the same corpus. We use two different corpora for training. The one is Uyghur text in parallel Chinese-Uyghur Corpus[5], the Uyghur is the translation of Chinese text that most of them are news and laws. The other one is the part of our Uyghur monolingual multi-domain corpus. To compare fairly, we set the size of the monolingual corpus equal to the previous corpus. For semi-supervision, we use the annotated corpus in [4]. We conduct several experiments about the different parameters of Morfessor 2.0 and FlatCat. In the experiments, we use test data in [5] and the standard precision, recall, and F1 score as the primary performance indicators to evaluate stemming. Note that, we do not compare the models of Morfessor 2.0 and FlatCat with the models of BPE. Because BPE controls the granularity of the segmented unit by operation number, so we directly use this model in our word representation. The Table 2 is comparison of the various models of Morfessor 2.0 and FlatCat.

The Table 2 shows that although the FlatCat model based on semi supervision (Mono_fc_semi) has achieved the highest precision, the recall is much lower than other models of Morfessor 2.0. Initially, the models based on Morfessor 2.0 are chosen. Then we compare models based on Morfessor 2.0. In the Morfessor 2.0 models, the semi-supervised methods are better than the unsupervised method (CLDC_mf_unsup). There is not much significant difference between the

---

[5] http://www.chineseldc.org.

Table 2. The Comparison between various stemming models

Models	P	R	F1
CLDC_mf_unsup	0.827	0.436	0.571
CLDC_mf_semi	0.802	0.729	0.764
MONO_mf_semi	0.819	0.733	**0.774**
MONO_fc_semi	0.999	0.207	0.343

model trained on the monolingual corpus (Mono_mf_semi) and the model based on translation corpus (Mono_mf_semi). So, finally, we choose the two models, *Mono_mf_semi* and *Mono_mf_semi*, to train mCBOW.

We evaluate the performance of word embeddings on the *uyWordSim-196* dataset which is a subset *uyWordSim-353* [3]. The *uyWordSim-353* is Uyghur translation of popular English *WordSim-353* dataset. In the translation of the *WordSim-353*, some of the words in English corresponding to two or more Uyghur words which would become OOV words in training data without multiword expression identification. So [3] filtered 196 pairs that corresponding to one Uyghur words in *uyWordSim-353* to form *uyWordSim-196*. They mainly test the performance of the word representation on the *uyWordSim-196*. So we also choose this subset as our benchmark. The result of Spearman (ρ) correlation to human judgment is shown in the Table 3:

Table 3. Spearman (ρ) correlation results for word similarity on the *uyWordSim-196* dataset

Models	*uyWordSim-196*
mCBOW_CLDC_mf_semi	50.99
mCBOW_Mono_mf_semi	51.73
mCBOW_BPE_32K	58.26
mCBOW_BPE_100K	**63.13**
mcBOW_BPE_150K	62.15

As the Table 3 is shown, we segment the text applying two categories of models which are based on Morfessor 2.0 and BPE. We set the operation number of BPE 32K, 100K, and 150K. The result shows that the mCBOW based model is depended on segmentation, the highest score on *uyWordSim-196* is 63.13. The BPE based model is better than the Morfessor 2.0 based model. When the operation number is 32K or 150K, word segmentation granularity is too coarse or too fine, the performance suffers from low precision. When the operation number is 100K, the model performs best. So we choose BPE based mCBOW for the OOV problem in Uyghur.

5.2 Synset Mapping Based on Word Embdeddings

We evaluate our Synset Mapping based on Word Embeddings (SMWE) approach on our dataset, which will be publicly available for other's comparative research. We tune the similarity threshold by using the development set. On the test set our approach 59.24 precision, 73.06 recall, and 65.03 F1-score. It can be seen that the SMWE make it possible to construct Uyghur WordNet automatically using only monolingual corpus and bilingual dictionaries.

5.3 Case Study

To demonstrate the effectiveness of the synset mapping method, we provide instances in our test dataset for example shown by Table 4. The Uyghur word "پەرمانچى" and English synset "herald.n.01", "harbinger.n.01" map each other in raw mapping. SMWE able to map correctly. In the second example, Uyghur word "گۈلزار" is mapped to wrong synsets like "modling.n.03", "border.n.05", but

Table 4. Examples from test dataset for synset mapping method based on word embbeddings (SMWE)

Uyghur Word	Definition	Raw Mapping	SMWE	Gold Standard
پەرمانچى	پەرماننى باشقىلارغا يەتكۈزگۈچى؛ جاكارلىغۇچى	herald.n.01, harbinger.n.01	herald.n.01	herald.n.01
گۈلزار	ئۆستۈرۈۋېلگەن گىياھلار ـ گۈل قاپلانغان بەلەن، گۈللۈك، جاي، چىمەنزار	flowerbed.n.01 , flower_ garden.n.01 , boundary_ line.n.01, margin.n.01 , edge.n.01 , molding.n.03 , border.n.05	flowerbed.n.01 , flower_ garden.n.01	flowerbed.n.01, flower_ garden.n.01
تۈزۈت	ئارتۇقچە، تارتىنچاقلىق بەلەن تەكەللۈپ مۇئامىلە قىلغان	politeness.n.01, politeness.n.02, civility.n.01, cold.n.01, coldness.n.03, cold.n.03, coldness.n.02, frigidity.n.01, chill.n.01, frisson.n.01, chill.n.03, chill.n.04	politeness.n.01, politeness.n.02	politeness.n.01, politeness.n.02, civility.n.01

SMWE still maps correctly. However, in the third example Uyghur word "تۇزۇت" is mapped to "embellishment.n.01", "embroidery.n.02", which is correct. But according to both definition of English and Uyghur, it is missed the correct one "civility.n.01". So our method still some rooms for improvement.

6 Conclusion

The main contribution of this paper is that we construct an English-Uyghur WordNet dataset for automatically evaluation. In the first time make this dataset available publicly (https://github.com/kaharjan/uywordnet). In this dataset, 73,491 English synsets in PWN 3.0 are mapped to Uyghur in Contemporary Uyghur Detailed Dictionary (CUDD), 20,702 are manually annotated to get 6,642 English synsets. The annotated data are divided into developing and test for the evaluation of English-Uyghur WordNet synset mapping methods. In this dataset, every Uyghur word has definitions or examples in the Uyghur language context, which make this dataset more valuable for downstream application. During these works, we build the largest Uyghur-English and English-Uyghur dictionary using all available online and offline resources, which put a considerable amount of work. We get better word representation of Uyghur using the largest monolingual corpus. This word representation alleviates the OOV problem based on previous work. Also, we propose an English-Uyghur synset mapping based on word embeddings (SMWE) method. The experimental results on the dataset are promising. Because there is no machine translation (MT) between English and Uyghur, we do not exploit the definition of Uyghur words and examples in our SMWE. In the future, we will further investigate how to expand our dataset and improve the performance of our algorithm.

Acknowledgments. This work is supported by National Natural Science Foundation of China (NSFC) grant 61532001.

References

1. Abiderexiti, K., Maimaiti, M., Yibulayin, T., Wumaier, A.: Annotation schemes for constructing Uyghur named entity relation corpus. In: The 2016 International Conference on Asian Language Processing (IALP 2016), pp. 103–107 (2016)
2. Abiderexiti, K., Maimaiti, M., Yibulayin, T., Wumaier, A.: Construction of Uyghur named entity relation corpus. Int. J. Asian Lang. Process. **27**(2), 155–172 (2017)
3. Abudukelimu, H., Liu, Y., Chen, X., Sun, M., Abulizi, A.: Learning distributed representations of Uyghur words and morphemes. In: Sun, M., Liu, Z., Zhang, M., Liu, Y. (eds.) CCL 2015. LNCS (LNAI), vol. 9427, pp. 202–211. Springer, Cham (2015). https://doi.org/10.1007/978-3-319-25816-4_17
4. Abudukelimu, H., Sun, M., Liu, Y., Abulizi, A.: THUUyMorph: an Uyghur morpheme segmentation corpus. J. Chin. Inf. Process. **32**(02), 81–86 (2018). (In Chinese)
5. Abudukelimu, H., Cheng, Y., Liu, Y., Sun, M.: Uyghur morphological segmentation with bidirectional GRU neural networks. J. Tsinghua Univ. (Sci. Technol.) **57**(1), 1–5 (2017). (In Chinese)

6. Aierken, R., Xiao, L., Tohti, A., Jiang, Z.M.: Constructing a Uyghur language semantic lexicon based on WordNet. In: 2014 Science and Information Conference, pp. 182–186 (2014)
7. Arcan, M., McCrae, J.P., Buitelaar, P.: Expanding WordNets to new languages with multilingual sense disambiguation. In: Proceedings of COLING 2016: Technical Papers, pp. 97–108 (2016)
8. Bond, F., Foster, R.: Linking and extending an open multilingual WordNet. Proc. ACL **2013**, 1352–1362 (2013)
9. Ercan, G., Haziyev, F.: Synset expansion on translation graph for automatic WordNet construction. Inf. Process. Manag. **56**(1), 130–150 (2019)
10. Fellbaum, C.: WordNet. In: Poli, R., Healy, M., Kameas, A. (eds.) Theory and Applications of Ontology: Computer Applications, pp. 231–243. Springer, Dordrecht (2010). https://doi.org/10.1007/978-90-481-8847-5_10
11. Grönroos, S.A., Virpioja, S., Smit, P., Kurimo, M.: Morfessor FlatCat: an HMM-based method for unsupervised and semi-supervised learning of morphology. In: Proceedings of COLING 2014, the 25th International Conference on Computational Linguistics, pp. 1177–1185 (2014)
12. Huang, C.R., Chang, R.Y., Lee, S.B.: Sinica BOW (Bilingual Ontological WordNet): integration of bilingual WordNet and SUMO. In: Proceedings of the 4th International Conference on Language Resources and Evaluation (LREC 2004), Lisbon, Portugal, pp. 26–28 (2004)
13. Huang, C., et al.: Chinese WordNet: design, implementation, and application of an infrastructure for cross-lingual knowledge processing. J. Chin. Inf. Process. **24**(02), 14–23 (2010). (In Chinese)
14. Khodak, M., Risteski, A., Fellbaum, C., Arora, S.: Automated WordNet construction using word embeddings. In: Proceedings of the 1st Workshop on Sense, Concept and Entity Representations and their Applications, pp. 12–23 (2017)
15. Lam, K.N., Al Tarouti, F., Kalita, J.: Automatically constructing WordNet synsets. In: Proceedings of the 52nd Annual Meeting of the Association for Computational Linguistics (Volume 2: Short Papers), pp. 106–111 (2014)
16. Maimaiti, M., Wumaier, A., Abiderexiti, K., Wang, L., Wu, H., Yibulayin, T.: Construction of Uyghur named entity corpus. In: Yang, E., Sun, L. (eds.) Proceedings of the Eleventh International Conference on Language Resources and Evaluation (LREC 2018). European Language Resources Association (ELRA), Miyazaki (2018)
17. Montazery, M., Faili, H.: Automatic Persian WordNet construction. In: COLING 2010: Poster, Beijing, China, pp. 846–850 (2010)
18. Osman, T., Yang, Y., Tursun, E., Cheng, L.: Collaborative analysis of Uyghur morphology based on character level. Acta Scientiarum Naturalium Universitatis Pekinensis **55**(01), 47–54 (2019). (In Chinese)
19. Qiu, L., Yang, H., Zhou, R.: The design and implementation of Chinese-Uighur-English online dictionary based on knowledge graph. In: 22017 IEEE International Conference on Computational Science and Engineering (CSE) and IEEE International Conference on Embedded and Ubiquitous Computing (EUC), pp. 883–886 (2017)
20. Qiu, L., Yang, N., Maolimamuti, M.: Chinese-Uyghur-English semantic search based on the knowledge graphs. In: 2017 IEEE International Conference on Computational Science and Engineering (CSE) and IEEE International Conference on Embedded and Ubiquitous Computing (EUC), pp. 879–882 (2017)

21. Qiu, L., Zhang, H.: Review of development and construction of Uyghur knowledge graph. In: 2017 IEEE International Conference on Computational Science and Engineering (CSE) and IEEE International Conference on Embedded and Ubiquitous Computing (EUC), pp. 894–897 (2017)
22. Sennrich, R., Haddow, B., Birch, A.: Neural machine translation of rare words with subword units. In: Proceedings of the 54th Annual Meeting of the Association for Computational Linguistics, Berlin, Germany, pp. 1715–1725 (2016)
23. Tarouti, F.A., Kalita, J.: Enhancing automatic WordNet construction using word embeddings. In: Proceedings of the Workshop on Multilingual and Cross-lingual Methods in NLP, pp. 30–34. Association for Computational Linguistics (2016)
24. Virpioja, S., Smit, P., Grönroos, S.A., Kurimo, M.: Morfessor 2.0: Python implementation and extensions for Morfessor baseline. Technical report, Aalto University, School of Electrical Engineering, Department of Signal Processing and Acoustic (2013)
25. Wang, S., Bond, F.: Building the Chinese open WordNet (COW): starting from core synsets. In: International Joint Conference on Natural Language Processing, pp. 10–18. Asian Federation of Natural Language Processing, Nagoya (2013)
26. Xu, R., Gao, Z., Pan, Y., Qu, Y., Huang, Z.: An integrated approach for automatic construction of bilingual Chinese-English Wordnet. In: Domingue, J., Anutariya, C. (eds.) ASWC 2008. LNCS, vol. 5367, pp. 302–314. Springer, Heidelberg (2008). https://doi.org/10.1007/978-3-540-89704-0_21
27. Yilahun, H., Imam, S., Hamdulla, A.: A survey on Uyghur ontology. Int. J. Database Theor. Appl. 8(4), 157–168 (2015)

Endangered Tujia Language Speech Enhancement Research Based on Improved DCGAN

Chongchong Yu$^{(\boxtimes)}$, Meng Kang, Yunbing Chen, Mengxiong Li,
and Tong Dai

College of Computer and Information Engineering,
Beijing Technology and Business University, Beijing 100048, China
yucc@btbu.edu.cn, {1830401006,10011316215,1604010312,
1604010403}@st.btbu.edu.cn

Abstract. As an endangered language, Tujia language only rely on oral communication. There must exist noises in the process of collecting Tujia language corpus. This paper studies an end-to-end speech enhancement model based on improved deep convolutional generative adversarial network (DCGAN) to extract nearly pure Tujia language speech in noisy environment. Due to the low resource nature of Tujia language, using Chinese corpus as an extension of the Tujia language can effectively solve the problem of insufficient data. The speech enhancement function of the Tujia language was realized using the end-to-end method that consists of symmetric encoding and decoding. By modifying the loss function and network hierarchy parameters, adding the spectrum normalization and imbalanced learning rate made the model more stable during the training process. The experimental results show that the speech enhancement method proposed in this paper can achieve better noise reduction effect on the Tujia language dataset than traditional speech enhancement algorithm and neural network enhancement algorithms.

Keywords: Tujia language · Speech enhancement · Deep Convolutional Generative Adversarial Network

1 Introduction

Tujia language contains rich national culture that is passed on from generation to generation by Tujia in China. However, there is no text recording, which has faced the crisis of extinction. In addition, the scope of use of Tujia language is extremely limited, which are usually located in the mountains and valleys with inconvenient traffic. Finding professional recording studios is hard in this environment. The phenomenon that audio files contain noises during the process of recording Tujia language is difficult to avoid [1]. These noises will submerge the useful speech information and impact the subsequent tasks of Tujia language annotation and speech recognition. Removing the noises from the Tujia language speech is a challenge to ensure improving the accuracy of speech recognition and helping phoneticians complete the recording and preservation of endangered languages.

© Springer Nature Switzerland AG 2019
M. Sun et al. (Eds.): CCL 2019, LNAI 11856, pp. 394–404, 2019.
https://doi.org/10.1007/978-3-030-32381-3_32

There are three main methods of speech noise reduction, namely speech enhancement algorithm, using robust speech feature parameters and noise compensation based on model parameter adaptation. Among them, speech enhancement is an effective method to solve noise pollution. Its purpose is mainly two points. The first is to suppress background noise, improve voice quality, and eliminate people's hearing fatigue, which is subjective measurement. The second is to improve the intelligibility of speech, which is an objective measurement [2]. The traditional speech enhancement algorithms have spectral subtraction which has a small amount of calculation and can easily control speech signal distortion and residual noise, but it is easy to exist musical noises [3]. Adaptive filtering, such as Wiener filtering [4] needs to know some features or statistical characteristics of noise. Subspace decomposition based on time domain can also be used for speech enhancement. For example, Chengli et al. proposed a signal subspace speech enhancement method based on joint low-rank sparse matrix decomposition, but it had better effect under the condition of low SNR or white noise [5]. In the 1980 s, four-layer fully-connected BP network was used for extracting signals from various stationary and non-stationary noises [6]. On this basis, the method of using deep neural network for speech enhancement has also received extensive attention, which has obvious advantages in processing non-stationary noises compared with traditional methods [7–9]. However, the deep neural network models are mostly supervised training and rely on a large amount of annotation data. Goodfellow put forward Generative Adversarial Network (GAN), it is not dependent on any priori assumption [10]. At present, GAN has been successfully applied in image processing [11], language text generation [12], audio generation [13] and other aspects. Speech Enhancement GAN (SEGAN) proposed by Pascual et al. obviously reduced noise, but stability and convergence of the model architecture still need further exploration [14]. Alec Radford et al. proposed Deep Convolutional Generative Adversarial Network (DCGAN) for image processing, which used Convolutional Neural Network (CNN) for stabilizing GAN training [15].

In view of the diversity, randomness and non-stationarity of environmental noises in Tujia speech dataset, the speech enhancement model based on improved DCGAN is proposed in this paper, which can carry out rapid enhancement processing, reduce the step of speech feature extraction and realize end-to-end speech enhancement. Because of the low resource nature of Tujia language and the limited amount of data, using Chinese speech dataset as extended dataset can solve the insufficiency of the Tujia language, it is necessary to simplify the network structure and reduce the network depth. The hinge loss function is used for the loss of the model. Moreover, spectral normalization and imbalanced learning rate are added to the model training process. Finally, the PESQ evaluation index is used for evaluating the performance of the model.

The rest of this paper is organized as follows. Section 2 introduces the knowledge of DCGAN. Section 3 demonstrates the proposed approach in detail. The dataset description and experimental results are presented in Sect. 4. Section 5 concludes our work finally.

2 DCGAN

GAN is a hot research direction in the field of artificial intelligence currently. Many variants have been derived, DCGAN is one of them. Whether it is GAN or its variant, the basic model architecture consists of two neural networks the Generator(G) and the Discriminator(D). G generates new samples that look real by learning input samples. D receives samples from real dataset and the output of G to distinguish the data source. They improve their performance by adversarial learning until generated samples are indistinguishable from real samples. The generator and discriminator of DCGAN adopt improved CNN, which can greatly stabilize the training of GAN.

Different from general CNN, the structure of G in DCGAN becomes fully convolutional network because fully connected hidden layer is cancelled. In addition to the output layer which uses Tanh activation, other layers use ReLU activation. Batch normalization (BN) is added to each layer, which helps to solve the training problems caused by bad network initialization and the flow of gradients in deep networks. D uses ordinary convolution, and all layers use LeakyReLU activation. Similar to G, each layer adds BN operation, and finally fully connected hidden layer is removed.

The training process of DCGAN is the same as ordinary GAN, namely the trainings of G and D are carried out alternately. It is shown in Fig. 1. when G is trained, D remains fixed. Then G accepts a random vector z(assuming that z is subject to some distribution p) to simulate the generation of sample $G(z)$. The error is calculated according to the output of D. Finally, error back propagation algorithm is used to update the parameters of G. when D is trained, G remains fixed. Then the output generator $G(z)$ is taken as negative samples and the real dataset X is taken as positive samples. These positive samples and negative samples are input to D, the error is calculated according to the output of D and the sample labels. Error back propagation algorithm is used to update the parameters of D finally.

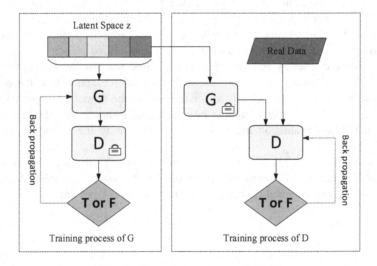

Fig. 1. The training process of GAN

$D(x)$ denotes the probability that D predicts the input data x as real sample. To maximize the expectation $\mathbb{E}_{x \sim p_{data}}$ which the data is from real samples, let D accurately predict $D(x) = 1$ when x obeys the probability density of the real samples. To maximize the expectation $\mathbb{E}_{x \sim p_G}$ which the data source is generated samples, let D accurately predicts $D(x) = 0$. The purpose of D is to maximize the loss as much as possible. However, the sample distribution generated by G should be as close as possible to real sample distribution, namely $p_G(x) = p_{data}(x)$. D cannot make a judgment on the source of input data so that $D(x)$ is equal to 0.5. The discriminant probability of generated samples $D(G(z))$ needs to be maximized, namely $\log(1 - D(G(z)))$ should be minimized. Thus, optimization of the whole network architecture is actually a min-max problem, which is described by the formula as follow:

$$\min_{G} \max_{D} V(D, G) = \mathbb{E}_{x \sim p_{data}(x)}[\log D(x)] + \mathbb{E}_{z \sim p_z(z)}[\log(1 - D(G(z)))] \qquad (1)$$

3 The Proposed Approach

3.1 The Speech Enhancement

In this paper, DCGAN is used for study the speech enhancement of Tujia language speech corpus inspired by SEGAN. The main part is G to complete the speech enhancement function. D is responsible for discriminating the generated data from the real data and feeding the result back to the G so that the output of the G is closer to the real data distribution. Until D is difficult to distinguish the authenticity of the input signal, the purpose of removing the noise signal is achieved.

The characteristic of G is the end-to-end structure with encoder-decoder. Encoder consists of convolution layers and PReLU activation. After encoding we can get a vector and connect it to the latent vector z. Decoder consists of deconvolution layers and PReLU activation. After decoding, we can get the enhanced speech waveform. It can be seen that G is designed as a fully convolutional neural network which eliminates the fully connected layer of feature vector classification. This structure allows the network to focus on the time correlation between the input signal and the processing of each layer. In addition, it also reduces the number of training parameters, thus reducing the training time. D is composed of a one-dimensional two-class convolutional network that only outputs the result of judging true and false. The speech enhancement model is shown in Fig. 2.

Skip connection is the response from a convolutional layer is directly propagated to the corresponding mirrored deconvolutional layer [16]. In this way, the information of the waveform can be better transmitted to the decoding stage, so that the reconstructed speech waveform is more refined and the detailed information is not lost after the multi-layer compression.

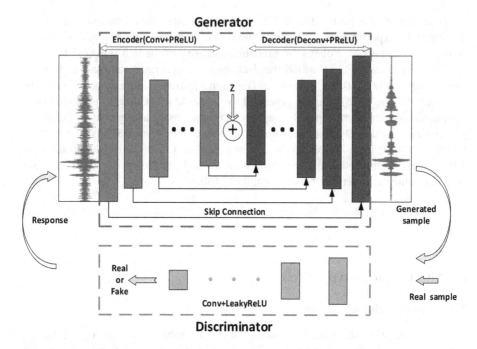

Fig. 2. The speech enhancement model

In order to stabilize the training and improve the quality of the generated samples, the loss function of the model replaces the cross-entropy loss with Hinge loss function [17] which can be better classified. The loss function is defined as follows:

$$L_D = \mathbb{E}_{x \sim p_{data}}[\min(0, -1 + D(x))] + \mathbb{E}_{z \sim p_z}[\min(0, -1 - D(G(z)))] \quad (2)$$

$$L_G = -\mathbb{E}_{z \sim p_z}[D(G(z))] \quad (3)$$

Where L_D is the loss function of D and L_G is the loss function of G.

3.2   Optimizing Train Methods

Spectral normalization and imbalanced learning rate were added to the model training process. spectral normalization of generator and discriminator makes it possible to significantly reducing the computational cost of training [18]. The imbalanced learning rate can often be better solved the problem of stability of G and D.

Spectral Normalization. SN performs Singular Value Decomposition (SVD) for the parameter W of every layer of neural network. Every update of W is divided by the largest singular value of W. So, the maximum stretch factor for each layer input x will not exceed 1. It is supposed that after SN every layer of the neural network satisfies:

$$\frac{\|D(x) - D(y)\|}{\|x - y\|} \leq K \tag{4}$$

Where $\|\bullet\|$ is L_2 regularization. If K is minimal, then K is called the Lipschitz constraint. However, it is difficult to perform SVD each layer of the neural network in each training iteration, especially when the weight dimension is large. Therefore, power iteration method is adopted to get an approximate solution to the singular value. We initialize a random $\hat{u}$, then update $\hat{u}$ and $\hat{v}$ iteratively according to the follows:

$$\hat{v} \leftarrow \frac{W^T \hat{u}}{\|W^T \hat{u}\|_2} \tag{5}$$

$$\hat{u} \leftarrow \frac{W^T \hat{v}}{\|W^T \hat{v}\|_2} \tag{6}$$

the maximum singular value $\sigma(W)$ of matrix W can be calculated:

$$\sigma(W) \approx \hat{u} W^T \hat{v} \tag{7}$$

Every time the network updates parameters, SN is executed:

$$W \leftarrow \frac{W}{\sigma(W)} \tag{8}$$

Imbalanced Learning Rate. It is assumed that ω and θ are the parameter vectors of the D and the G. They learn based on the stochastic gradient $\tilde{g}(\theta, \omega)$ of the discriminator loss function L_D and the stochastic gradient $\tilde{h}(\theta, \omega)$ of the generator loss function L_G. There are the actual gradients $g(\theta, \omega) = \nabla_w L_D$ and $h(\theta, \omega) = \nabla_\theta L_G$. The random approximation of the actual gradient is defined according to the random vector $M^{(\omega)}$ and $M^{(\theta)}$:

$$\tilde{g}(\theta, \omega) = g(\theta, \omega) + M^{(\omega)} \tag{9}$$

$$\tilde{h}(\theta, \omega) = h(\theta, \omega) + M^{(\theta)} \tag{10}$$

According to the Two Time Update Rule (TTUR) [19], we use the learning rates $b(n)$ and $a(n)$ for the discriminator and the generator update respectively:

$$\theta_{n+1} = \theta_n + a(n)(h(\theta_n, w_n) + M_n^{(\theta)}) \tag{11}$$

$$w_{n+1} = w_n + b(n)(g(\theta_n, w_n) + M_n^{(w)}) \tag{12}$$

4   Experiments and Result

4.1   Datasets

Due to the limited Tujia language data, we expand the dataset and two datasets are used in the experiment. The first dataset is the Tujia language corpus, which includes 27 oral corpora. There are a total 7830 sentences, with a duration of 7 h, 8 min and 59 s. The Tujia language corpus is divided into two parts, one part contains noises, called the noisy corpus A. The noise types include rooster crowing, chicken crowing, motor vehicle sounds, electronic equipment noises and other noises. The noise fragments are intercepted manually from noisy corpus A using the Elan tool[1]. The details of all kinds of noises are shown in Table 1. Another part is noise-free corpus called clean corpus B. The second dataset is the thchs30 Chinese corpus [20] recorded by 25 people. There are 13395 sentences in total, the recording time is 30 h, the sampling frequency is 16 kHz, and the sampling size is 16 bits.

Table 1. Type and number of Tujia language noises.

Noise type	Noise number
Rooster crowing	64
Chicken crowing	31
Motor vehicle sounds	6
Electronic equipment noises	3
Other noises	33

First, the noise segments are added to the clean corpus B and the thchs30 Chinese corpus by the sox tool[2]. Noise injection method is to randomly select the starting position at the sampling point and inject different noises into the thchs30 Chinese corpus according to the proportion of each type of noises to the total number of noises. The formula is as follows:

$$m_{ij} = \frac{N_i M_j}{\sum\limits_{i=1}^{5} N_i}, \quad i = 1, \cdots, 5 \ , \ j = 1, \cdots, 25 \tag{13}$$

Where N_i is the number of noise i, M_j is the number of recordings of the j th person in the thchs30 Chinses corpus, m_{ij} is the number of noises injected into the recording of the j th person. The new corpus is called the noisy corpus thchs30. The noisy corpus B is obtained by injecting into the clean corpus B in the same way.

---

[1] https://tla.mpi.nl/tools/tla-tools/elan/.
[2] http://sox.sourceforge.net/.

4.2   Experimental Setup

In this study, the Dell PowerEdge R730 server device is used, in which the processor is Intel(R) Xeon(R) CPU E5-2643 v3 @3.40 GHz, the memory size is 64 G, the GPU is NVIDIA Tesla K40 m × 2, and the memory size is 12 GB × 2. The experimental environment for the deep learning framework installed on the Ubuntu 16.04 system is the GPU version of Tensorflow 0.12.

The experimental scheme of speech enhancement based on improved DCGAN is shown in Fig. 3. Firstly, Improved DCGAN (IDCGAN) model is trained by using noisy thchs30 as input and clean thchs30 as output. Then the IDCGAN model is fine-tuned by using noisy corpus B as input and clean corpus B as output to get Fine-tuning IDCGAN(FIDCGAN) model. Finally, the FIDCGAN model is tested and evaluated with noisy corpus A.

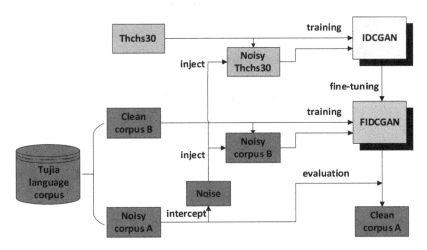

Fig. 3. The experimental scheme of speech enhancement based on improved DCGAN

We extract chunks of waveforms with a sliding window. The window length is about 1 s. The moving distance of the window is 500 ms. The encoding stage of G consists of 11 convolutional layers that are one-dimensional strided convolutional layers of filter width 31 and strides of N = 2. The number of filters per layer is 16, 32, 32, 64, 64, 128, 128, 256, 256, 512, 1024. The decoding stage is symmetric with the encoding stage. The D is also one-dimensional convolutional structure. It has two input channels, The LeakyReLU nonlinear activation function with alpha = 0.3 is used, the last layer is the convolution of 1 × 1, and the output is the result of judging true and false. In IDCGAN model, the learning rate for the D is 0.0003 and the learning rate for the G is 0.0001, use the imbalanced learning rates to train G and D with 1:1 update. The batch size is 24.

Fine-tuning is performed with noisy corpus B and clean corpus B on the basis of IDCGAN model. the learning rate for the D is 0.0002 and the learning rate for the G is 0.00008, the batch size is 16 to obtain the FIDCGAN model. Finally, the FIDCGAN model is tested and evaluated with noisy corpus A.

Except compared with the conventional speech enhancement methods and the speech enhancement methods based on DNN and RNN, Adding BN, SN, BN and SN to each layer of the model in this paper for comparison.

4.3   Experimental Results

Perceptual Evaluation of Speech Quality (PESQ) and Mean Opinion Score of Listening Quality Objective (MOSLQO) are selected as evaluation indexes. PESQ is a typical algorithm in speech quality evaluation. It adopts a linear scoring system with a value between −0.5 and 4.5. The higher the score, the better the quality of speech. PESQ performs level adjustment, input filter filtering, time alignment and compensation, and auditory transformation for he input noisy speech signal and a reference speech signal. Then parameters of the two signals are extracted, and the time-frequency characteristics are integrated to obtain the PESQ score. The MOSLQO value is calculated through the PESQ tool[3]. The evaluation results are shown in Table 2.

Table 2. The results of different speech enhancement methods.

Method	PESQ	MOSLQO
Spectral subtraction	1.423	1.309
Wiener filtering	1.526	1.324
DNN	1.732	1.501
RNN	1.843	1.523
FIDCGAN (BN)	1.921	1.606
FIDCGAN (SN)	2.040	1.664
FIDCGAN (BN + SN)	1.810	1.436

Table 2 shows that the speech enhancement performance proposed in this paper is significant, especially the FIDCGAN model with SN. In addition, the two operations "dividing by the variance" and "multiplying by scaling factor" of BN impact the Lipschitz continuity of the discriminator, the performance will be worse when SN and BN are simultaneously added. Therefore, the BN operation of each layer should be cancelled eventually. The speech spectrum before and after enhancement are shown in Fig. 4.

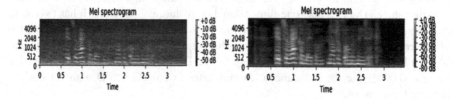

Fig. 4. The speech spectrum before enhancement (left) and after enhancement (right).

---

[3] https://www.itu.int/rec/T-REC-P.862/en.

The experimental results show that the speech enhancement method based on improved DCGAN can effectively remove the environmental noise in the Tujia language dataset.

5 Conclusion

Aiming at the specific scene of the endangered Tujia language, we combine deep CNN and GAN to construct the end-to-end improved DCGAN speech enhancement algorithm. The model preserves phase detail information in the time domain of original speech signal. Spectral normalization and unbalanced learning rates are used to enhance the stability of network training. Compared with the mainstream speech enhancement methods, the experimental results show that the proposed method can obtain better performance. It makes further exploration in the field of speech enhancement and lays a stable foundation for speech feature extraction and recognition. For the low resource Tujia language, we have expanded the data. However, noise type is limited in the specific environment. Therefore, we will attempt to strengthen the generalization of the model and optimize network structure in subsequent work.

Acknowledgment. This research is supported by Ministry of Education Humanities and Social Sciences Research Planning Fund Project, grant number 16YJAZH072, and Major projects of the National Social Science Fund, grant number 14ZDB156.

References

1. Shixuan, X.: On the recording and preservation of endangered language data. J. Guangxi Univ. Natl. (Philos. Soc. Sci. Ed.) **28**(5), 11–15 (2006)
2. Hang, H.: Modern Speech Signal Processing, pp. 351–352. Electronic Industry Press, Beijing (2014)
3. Dailong, X., Guanyu, L., Ning, M.: Speech enhancement research based on spectral subtraction. J. Northwest University (Nat. Sci.) **38**(02), 21–25, 87 (2017)
4. Navneet, U., Rahul, K.: Single channel speech enhancement: using wiener filtering with recursive noise estimation. Procedia Comput. Sci. **84**, 22–30 (2016)
5. Chengli, S., Jianxiao, X., Yan, L.: A signal subspace speech enhancement approach based on joint low-rank and sparse matrix decomposition. Arch. Acoust. **41**(2), 245–254 (2016)
6. Tamura, S., Waibel, A.: Noise reduction using connectionist models. ICASSP **1988**(1), 553–556 (1988)
7. Yong, X., Jun, D., Lirong, D., et al.: An experimental study on speech enhancement based on deep neural networks. IEEE Signal Process. Lett. **21**(1), 65–68 (2014)
8. Shi, W., Zhang, X., Sun, M., et al.: Deep neural network based monaural speech enhancement with sparse and low-rank decomposition. In: IEEE 17th International Conference on Communication Technology (ICCT), pp. 1644–1647 (2017)
9. Huang, Q., Bao, C., Wang, X., et al.: DNN-based speech enhancement using MBE model. In: IWAENC, pp. 196–200 (2018)
10. Goodfellow, I., Pouget-Abadie, M., Mirza, B., et al.: Generative adversarial nets. In Advances in Neural Information Processing Systems (NIPS), pp. 2672–2680 (2014)

11. He, H., Philip S, Y., Changhu, W.: An introduction to image synthesis with generative adversarial nets. arXiv:1803.04469 (2018)
12. Jiaxian, G., Sidi, L., Han, C., et al.: Long text generation via adversarial training with leaked information. arXiv:1709.08624 (2017)
13. Engel, J., Agrawal, K.K., Chen, S., et al.: GANSynth: adversarial neural audio synthesis. In: ICLR (2019)
14. Pascual, S., Bonafonte, A., Serra, J.: SEGAN: speech enhancement generative adversarial network. In: INTERSPEECH (2017)
15. Alec, R., Luke, M.: Unsupervised representation learning with deep convolutional generative adversarial networks. In: ICLR (2016)
16. Xiaojiao, M., Chunhua, S., Yubin, Y.: Image restoration using very deep convolutional encoder-decoder networks with symmetric skip connections. arXiv:1603.09056 (2016)
17. Takeru, M., Toshiki, K., Masanori, K., et al.: Spectral normalization for generative adversarial networks. In: ICLR (2018)
18. Zhang, H., Goodfellow, I., Metaxas, D., et al.: Self-attention generative adversarial networks. arXiv:1805.08318 (2018)
19. Heusel, M., Ramsauer, H., Unterthiner, T., et al.: GANs trained by a two time-scale update rule converge to a local Nash equilibrium. arXiv:1706.08500 (2018)
20. Dong, W., Xuewei, Z.: THCHS-30: a free Chinese speech corpus. Comput. Sci. (2015)

Research for Tibetan-Chinese Name Transliteration Based on Multi-granularity

Chong Shao[1], Peng Sun[1], Xiaobing Zhao[1,2], and Zhijuan Wang[1,2(✉)]

[1] School of Information Engineering, Minzu University of China,
Beijing 100081, China
[2] Minority Languages Branch,
National Language Resource and Monitoring Research Center, Beijing, China
wangzj_muc@126.com

Abstract. In order to solve the problem of data sparseness caused by less training corpus in Tibetan-Chinese transliteration, this paper analyzes the alignment granularity of Tibetan-Chinese names as the research object and uses the pronunciation feature to reduce the corresponding relationships. The method of transliteration of Tibetan and Chinese names and the design of related experiments is comparable with traditional methods and improve the top-1 accuracy of transliteration of Tibetan and Chinese names to 65.72%. The experimental results show that the method can improve the accuracy of Tibetan-Chinese name transliteration.

Keywords: Transliteration · Segmentation granularity ·
Tibetan-Chinese

1 Introduction

In the various tasks of cross-lingual natural language processing, the problem of translation of Out-of-Vocabulary (OOV) is often encountered. Usually, we can use a transliteration method to translate OOV, and the accuracy of OOV transliteration results can directly affect the actual Application [10]. Due to the difference of language feature, when transliterating the person name, the transliteration unit (ie, segmentation) of both source language and the target language should be appropriately adjusted. Therefore, translation granularity has always been one of the key points of transliteration research [12].

Verspoor proposes a forward-transliteration method based on four-step rules from English to Chinese [11]. By artificially establishing a conversion table for English syllables to Chinese Pinyin, and then establishing a conversion table for Chinese Pinyin to Chinese characters, better performance has been obtained. Lin and Chen used the English-Chinese unified phonetic table IPA (International Phonetic Alphabet) to realize the conversion of English to phonetic symbols and phonetic symbols to Chinese [6]. Knight and Graehl used the similarity between English phonemes and Japanese phonemes to achieve English to

© Springer Nature Switzerland AG 2019
M. Sun et al. (Eds.): CCL 2019, LNAI 11856, pp. 405–413, 2019.
https://doi.org/10.1007/978-3-030-32381-3_33

English phonemes, English phonemes to Japanese phonemes, Japanese phonemes to Japanese conversion [4]. Li et al. proposed a joint source-channel model in English-to-Chinese transliteration tasks, directly aligning English and Chinese, and achieved good transliteration result [3,8]. Kunchukuttan and Bhattacharyya finished transliteration of eight languages pairs with three granularities: alphabet, word, mixed alphabet and the word [5]. Zhou and Zhao regard the transliteration of person name as sentence pair in statistical machine translation. Each transliteration unit is regarded as a word in a sentence, and the machine learning method is used to transliterate names and reached a great translation effect [1]. Yu et al. proposed a method of using word graphs to fuse multiple granularities to achieve transliteration of English-Chinese names [10]. Liu et al. used the combination of statistics and rules to analyze the characteristics of English pronunciation, constructed the rules of fine division of transliteration units, and proposed a method of transliteration unit based on glyphs and speech [7]. The above methods have achieved good results in segmentation granularity of transliteration from different views, but each method has its own inadequacies, mainly in the following aspects:

(1) Rule-based methods require a large number of artificially constructed transliteration rules between specific language pairs, these rules are language-dependent and less robust.
(2) The method of transliteration granularity by letter or token is language-independent, so the robustness is better. But because the information of person name's pronunciation is not used, a large number of errors will be generated in the alignment stage, thereby affects the final transliteration effect.
(3) Through the method of phoneme or unified phonetic table, the information of person name's pronunciation can be used to generate the alignment with higher accuracy, but because of more conversion steps, the probability error or errors' accumulation also causes the final transliteration accuracy.
(4) The direct alignment method reduces the error caused by too many conversion steps and improves the accuracy of transliteration. However, this method skips the speech step, information loss is inevitable.

2   Methodology

The main pipeline of person name transliteration consists of four phases: preprocessing, transliteration model, decoding and post-processing, as Fig. 1 displays. Preprocessing is the main topic that this paper focused on.

Step one, both Tibetan and Chinese corpus are sliced into three granularities for the experimental settings, so we get three corpora including same transliteration units. Step two, the transliteration model is trained with the aligned parallel corpus. Step three, a decoding experiment is performed on the source language test corpus that has been divided into transliteration units. Step four, because it doesn't need to reorder the result of transliteration, the output is mainly combined into pinyin, and then converted into corresponding Chinese.

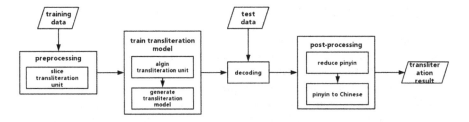

Fig. 1. The pipeline of transliteration.

2.1 Preprocessing

This paper studies the influence of segmentation granularity on the transliteration of Tibetan and Chinese names. Therefore, the same corpus is processed with different segmentation granularity, and three corpora with the same names but different transliteration units are obtained. Three experiments are carried out to verify the research objectives by comparing the experimental results. In the preprocessing part, the emphasis is on the segmentation of transliteration units between source and target language. According to the different granularity of segmentation and alignment units, we divide them into three parts: direct alignment, further transformation, and further segmentation. We will describe the following methods based on the transliteration example offered by Fig. 2.

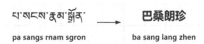

Fig. 2. An example of Tibetan person name and its Chinese transliteration. The following line is transcribed Latin (for Tibetan) or Pinyin (for Chinese).

Direct Alignment. Tibetan has a special symbol for segmenting syllables, each syllable corresponds to a character in Chinese. We call the alignment of Tibetan syllables and Chinese characters as direct alignment.

Transcribed Latin to Chinese Pinyin. We use Latin Transliteration to get representation Tibetan and Chinese person names as Fig. 2 displays.

Fine-Grained Segmentation. Slice Tibetan Latin with rules ([p a] [s angs] [rn am] [sgr on]) and slice Chinese Pinyin by initial and vowel ([b a] [s ang] [l ang] [zh en]).

2.2   Transliteration Model

Alignment. After corpus segmentation, it is necessary to align the segmented corpus so as to align the transliteration units of Tibetan and Chinese bilingual names. For the three different corpora obtained by three different segmentation methods, we call them corpus one, corpus two and corpus three respectively. In machine transliteration, there is no ordering problem of transliteration units, so for the above three corpora, alignment is to combine the Tibetan list elements into the same number of parts according to the number of Chinese names segmentation, and then align the transliteration units of the same number parts.

In the process of segmentation, there are different numbers of transliteration units between the source language strings and the target language strings. Generally, one to zero often occurs in automatic alignment. For corpus three is constructed by fine-grained segmentation, using common automatic alignment method will cause faults, for example, vowels may appear before vowels. The alignment probably increases the error rate of transliteration. Therefore, we adopt the improved automatic alignment algorithm, the specific steps are as follows:

(1) For corpus one and two, if the number of transliteration units of Chinese and Tibetan names is the same after segmentation, they will be aligned directly. That is to say, the segmented corpus will be aligned one by one and form corresponding transliteration pairs.

(2) For corpus one and corpus two, if the number of transliteration units of Chinese and Tibetan names is different after segmentation, The maximum expectation algorithm (EM) algorithm is used to align them. The method of dividing the most probabilistic values is taken as the final dividing method, and then the corresponding transliteration pairs are obtained by aligning them in order.

(3) For corpus three, we add the boundary symbol "|" to the segmented corpus, so that the whole segregated by the two boundary symbols is a whole before further segmenting, that is, "p a | s angs | rn am | sgr on" and "b a | s ang | l ang | zh en", and then align them with EM algorithm at the initial stage, if the boundary is a | s angs | rnam | sgr on". If the symbol is not at the beginning or end of the alignment result, we think that the alignment scheme is wrong. We eliminate them directly. Then we calculate the initial probability of the remaining schemes and iterate. We take the most probabilistic partition as the final partition method, and then align them in order to get the corresponding transliteration pairs of names.

Train. For Tibetan to Chinese person name transliteration, if Chinese name denotes $\alpha = c_1c_2...c_m$, Tibetan name denotes $\beta = b_1b_2...b_m$, c_i represents the minimum segmented unit of Chinese name. So the relationship between Tibetan and Chinese is γ, predicted Tibetan name alignment sequence denotes $\bar{\beta}$:

$$\overline{\beta} = \arg\max_{\beta} P(\beta, \alpha)$$

$$= \arg\max_{\beta} \sum_{\gamma} P(\beta, \alpha, \gamma)$$

$$\approx \arg\max_{\beta} \left(\arg\max_{\beta} P(\beta, \alpha, \gamma) \right) \tag{1}$$

$$= \arg\max_{\beta, \gamma} P(\beta, \alpha, \gamma)$$

Among, $P(\beta, \alpha, \gamma)$ represents the joint probability of α, β and γ, for K aligned transliteration units, we have

$$P(\beta, \alpha, \gamma) \approx \prod_{k=1}^{K} P\left(\langle b, c \rangle_k | \langle b, c \rangle_1^{k-1}\right) \tag{2}$$

2.3 Post-processing

For experiment one, we get the corresponding Chinese results after the decoding step. We just need to format the results for our needs. For experiment two and experiment three, we get the corresponding Chinese Pinyin after decoding, so We need to convert it into the corresponding Chinese.

We use the Conditional Random Field (CRF) model to transform the Chinese of the training corpus into the corresponding Pinyin. Then we input the corresponding pinyin into the CRF as the training corpus and get the corresponding pinyin-to-Chinese model. Then we input the Pinyin from dataset two and dataset three into the CRF. The output of the model is the Chinese result we need.

3   Experiment

This experiment adopts the strategy of determining pronunciation first and then font shape. First, the pronunciation of Chinese names is determined by the transliteration pairs of Tibetan and Chinese names, and then the Chinese names are determined by the Chinese monolingual names. The data used in the experiment include 6405 transliteration pairs of Tibetan and Chinese names. The training set contains 5121 transliteration pairs of Tibetan and Chinese names, and the test set contains 1284 transliteration pairs of Tibetan and Chinese names. The NEWS 2018 Named Entity Transliteration Shared Task [2] has supported a statistical machine transliteration Baselines [9], we followed their work on the Tibetan-Chinese corpus and the result were served as our baseline.

3.1   Evaluate

A precision method is used in this experiment. Only when all parts of a person's name are transliterated correctly, can we think that the result of transliteration is accurate.

$$Precision = \frac{Number\ of\ Correct\ Transliteration}{Total\ Number\ of\ Transliteration} \tag{3}$$

3.2   Result

In order to compare the effects of different segmentation granularity on transliteration results as a whole, we designed three experiments.

Setting One. We divide the Chinese name into characters, divide the Tibetan name into syllables, align the segmented corpus with EM algorithm, and train the corresponding transliteration model, then decode the corpus by Viterbi algorithm and output the best result. The result of transliteration serves as a reference.

Setting Two. In the preprocessing stage, the Tibetan is converted to corresponding Latin transcription, the Chinese are converted to corresponding Pinyin, and then the other steps refer to setting one remain unchanged.

Setting Three. Preprocess data based on setting two, the Latin transcriptions corresponding to Tibetan are separated according to the five letters of "a, e, i, o, u", and the Chinese phonetic alphabets corresponding to Chinese are separated according to the form of initial and vowel, then the other steps refer to setting one unchanged.

After corpus alignment, we extract transliteration parameters from the aligned corpus: logarithmic probability table of binary-order transliteration, probability table of one-order transliteration, and transliteration unit table of target language corresponding to source language transliteration units. In order to better analyze the experimental results, we remove duplication of the transliteration units of the source language and the target language. The statistic of the data set is shown in Table 1.

Table 1. Experimental settings and data statistic.

	Number of Tibetan Transliteration Units	Number of Chinese Transliteration Units
Setting one	538	1314
Setting two	538	323
Setting three	727	75

We can learn that the types of transliteration units in the target language have been significantly reduced from setting one to setting three, and the total number of corresponding relations between the source language and the target language has been significantly reduced. Under the same corpus, it is helpful to obtain the differentiated probability distribution, thus getting closer to the real value and improving the accuracy of transliteration.

In setting two and three, we adopted the strategy of determining pronunciation first and then font shape. We carried out the statistics of pronunciation accuracy before transliterating Pinyin into Chinese characters. That is, if all pronunciations of a person's name are correctly transliterated, then we think that the pronunciation of that person's name is correct. The overall results of the experiment are as shown in Table 2.

Table 2. Results for the Tibetan to Chinese transliteration task.

Setting	Precision of Pronunciation (%)	Precision (%)
Baseline	/	21.31
Setting one	/	32.14
Setting two	84.95	62.95
Setting three	92.36	65.72

Compared with setting one, setting two and setting three have achieved better performance. The analysis is as follows:

(1) Tibetan is a low resource language. It is difficult to obtain enough transliteration pairs of names. In experiment 1, when Tibetan and Chinese are aligned directly and the corresponding relationship is trained, a large number of data sparsity problems arise, which have a great impact on the results, while the other two experiments have almost no problem of data sparsity.

(2) There are a large number of homonyms in Chinese. Because of the limited number of transliteration pairs of names, there are a lot of data sparseness problems in the corresponding matrix we get. Even if we use data smoothing technology, the corresponding relationship obtained is far from the real situation. There is no loss conversion between Tibetan and Latin transcriptions. Converting Latin transcriptions to Pinyin first can get better correspondence because the types of Pinyin are far less than the number of Chinese characters, and then the Pinyin can be converted to Chinese characters. Because of the higher accuracy of pronunciation, the final result of this step is better than that of direct transliteration even if some errors.

(3) Finer-grained segmentation of Latin transcription and Pinyin can further reduce the total number of corresponding relationships, thus increasing the average number of occurrences of each corresponding relationship, making

the number of occurrences of each corresponding relationship more different, thus closer to the real value, and improving the effectiveness of transliteration.

After more fine-grained segmentation of corpus, the accuracy of transliteration has been significantly improved, which verifies that finer-grained segmentation not only reduces the problem of data sparsity caused by the method based on the morphology of character but also transfers the problem of corpus size in less-resourced languages to the problem of the unilateral corpus in Chinese. Thus, a finer-grained method of name segmentation can be obtained. It can improve the effect of transliteration.

4   Conclusion

This paper proposes a more fine-grained method of Latin transcription and Pinyin segmentation through the conversion of Tibetan Latin transcription and Chinese Pinyin based on shape segmentation of character. The experimental result shows that the proposed segmentation method can improve the accuracy of transliteration of Tibetan and Chinese names, and it also performs well in solving the problem of scarcity of names in low resource scenarios. This research has the following innovations:

(1) The idea of removing duplication of correspondence on the basis of pronunciation is put forward. In previous studies, most of them have focused on expanding the size of the corpus and improving the coverage and complexity of correspondence. Such ideas are often not ideal in the transliteration of languages with low resources. In this study, we first determine the pronunciation, then determine the shape of the characters, and effectively removing duplication of the corresponding relationship on the basis of pronunciation, thus reducing the requirements of the model for the size of Tibetan-Chinese names corpus, and better enhancing the effect of Tibetan-Chinese names transliteration.
(2) The problem that bilingual person name pairs in low resource languages are transformed into the problem of Chinese monolingual person-name pairs that can be easily solved. The feasibility of this method is verified by experiment.

However, we only segment Tibetan Latin transcription based on simple rules. In the process of alignment, some wrong segmentation and alignment will occur, which will reduce the accuracy of transliteration. In the future work, we will introduce more rules of language segmentation and alignment to reduce the errors caused by the segmentation and alignment stage, so as to further improve the accuracy of transliteration of Tibetan and Chinese person names.

References

1. Bo, Z., Zhao, J.: Comparison of several English-Chinese name transliteration methods. In: The Fourth National Seminar on Computational Linguistics for Students, pp. 24–30 (2008)
2. Chen, N., Banchs, R.E., Zhang, M., Duan, X., Li, H.: Report of news 2018 named entity transliteration shared task. In: Proceedings of the Seventh Named Entities Workshop, pp. 55–73 (2018)
3. Haizhou, L., Min, Z., Jian, S.: A joint source-channel model for machine transliteration. In: Proceedings of the 42nd Annual Meeting on association for Computational Linguistics, p. 159. Association for Computational Linguistics (2004)
4. Knight, K., Graehl, J.: Machine transliteration. Comput. Linguist. **24**(4), 599–612 (1998)
5. Kunchukuttan, A., Bhattacharyya, P.: Data representation methods and use of mined corpora for Indian language transliteration. In: Proceedings of the Fifth Named Entity Workshop, pp. 78–82 (2015)
6. Lin, W.H., Chen, H.H.: Backward machine transliteration by learning phonetic similarity. In: proceedings of the 6th conference on Natural language learning-Volume 20, pp. 1–7. Association for Computational Linguistics (2002)
7. Liu, B., Xu, J., Yefeng, C., Zhang, Y.: Integrating of grapheme-based and phoneme-based transliteration unit alignment method. Acta Scientiarum Naturalium Universitatis Pekinensis, 75–80 (2016)
8. Min, Z., Haizhou, L., Jian, S.: Direct orthographical mapping for machine transliteration. In: Proceedings of the 20th International Conference on Computational Linguistics, p. 716. Association for Computational Linguistics (2004)
9. Singhania, S., Nguyen, M., Ngo, G.II., Chen, N.: Statistical machine transliteration baselines for news 2018. In: Proceedings of the Seventh Named Entities Workshop, pp. 74–78 (2018)
10. Tingting, L.: Research on Nonparametric Bayesian Based Multi-language Names Transliteration. Ph.D. thesis. Harbin Institute of Technology, Harbin (2013)
11. Wan, S., Verspoor, C.M.: Automatic English-Chinese name transliteration for development of multilingual resources. In: The 17th International Conference on Computational Linguistics COLING 1998, vol. 2 (1998)
12. Yu, H., Tu, Z., Liu, Q., Liu, Y.: Lattice-based multi-granularity name-entity machine transliteration. J. Chin. Inf. Process. **27**(4), 16–22 (2013)

An End-to-End Method for Data Filtering on Tibetan-Chinese Parallel Corpus via Negative Sampling

Sangjie Duanzhu, Cizhen Jiacuo, Rou Te, Sanzhi Jia, and Cairang Jia[✉]

Key Laboratory of Tibetan Information Processing and Machine Translation,
Qinghai Normal University, Xining, China
sangjeedondrub@live.com, czjcaiyaogun@hotmail.com, crpengcuo13@yahoo.com,
samdrubgyal@yeah.net, zwxxzx@163.com

Abstract. In the field of machine translation, parallel corpus serves as the most important prerequisite for learning complex mappings between targeted language pairs. However, in practice, the scale of parallel corpus is not necessarily the only factor to be taken into consideration for improving performance of translation models due to the quality of parallel data itself also has tremendous impact on model capacity. In recent years, neural machine translation systems have become the *de facto* choice of implementation in MT research, but they are more vulnerable to noisy disturbance presented in training data compared with traditional statistical machine translation models. Therefore, data filtering is an indispensable procedure in NMT pre-processing pipeline. Instead of utilizing discrete feature representations of basic language units to build a ranking function of given sentence pairs, in this work, we proposed a fully end-to-end parallel sentence classifier to estimate the probability of given sentence pairs being equivalent translation for each other. Our model was tested in three scenarios, namely, classification, sentence extraction and NMT data filtering tasks. All testing experiments showed promising results, and especially in Tibetan-Chinese NMT experiments, **3.7** BLEU boost was observed after applying our data filtering method, indicating the effectiveness of our model.

Keywords: Tibetan-Chinese · Data filtering · Neural machine translation

1 Introduction

Parallel corpus plays an indispensable role in many multi-lingual nature language processing systems, especially for machine translation (MT) applications,

This work was supported by National Natural Science Foundation of China (grant numbers: 61063033, 61662061) and the National Key Research and Development Program of China (grant number: 2017YFB1402200).

© Springer Nature Switzerland AG 2019
M. Sun et al. (Eds.): CCL 2019, LNAI 11856, pp. 414–423, 2019.
https://doi.org/10.1007/978-3-030-32381-3_34

parallel resources serve as a vital prerequisite for learning the complex mappings between pairs of languages. Recently neural machine translation (NMT) advances draw lots of attention in the fields, resulting in remarkable translation performance boost which surpasses statistical machine translation (SMT) models by a wide margin and becomes the *de facto* implementation in MT research. Transformer [12] was presented as a new powerful sequence transduction framework, hitting many state-of-the-art records in MT and many other machine learning tasks. Researchers even claimed that their model produced near-human translation quality in certain domain [4].

Even though NMT follows a data-thirsty learning paradigm compared with traditional approaches, the scale of parallel data is not necessarily the only factor to be considered for improving model capacity. For MT development and research, the collection and construction of large-scale parallel bi-texts are often crowdsourced, leading to the issue that the data itself is likely to be presented with a considerable portion of noisy disturbance, which usually emerges in forms of misalignment, bad translation, partial or over translation, even semantical irrelevance, etc. NMT models by its nature tend to assign a high probability for low-occurrence language mapping instances and events [4], giving rise to absence of immunity to all these noise emerging in parallel data. In MT practice data filtering and selection process is an essential part in pre-processing pipeline, however, manually selection is nearly impossible in consideration of the scale of bi-texts in NMT model training procedure.

Among SMT literature, some form of difference such as geometric mean of the perplexities [3] or cross-entropy [8] are measured to serve as a ranking function for estimating the probability of the given sentence pairs being semantically equivalent. Like all traditional statistical approaches in NLP fields, all basic language units are encoded in the form of discrete representations, which causes severe semantic information loss and constrains the learning potentials of model.

In this work, we proposed an end-to-end classifier in a simple neural network architecture to estimate the probability of translation equivalence for pair of sentences in Chinese and Tibetan. We used this classifier in sentence extraction and data filtering tasks in NMT pre-processing pipeline, the former indicated that our proposed model gained a promising performance in cross-lingual inference and the latter one indicated that (1) NMT models could benefit from data filtering and selection process, by reducing the training data in scale and in the meanwhile gaining boost in translation performance (we observed a **3.7** increase in BLEU score over baseline in our experiments on Tibetan-Chinese NMT models); (2) Our model could fit into the need of such data filtering during training NMT models.

Even though we took Tibetan-Chinese corpus as researching data, the model presented in this work is language-independent by its architecture design and could be pivoted to any other language pairs if there is availability of moderate scale parallel data for training the classifier.

2 Model

2.1 Negative Sampling

The parallel corpus C consists of N pairs of sentences (S_k^S, S_k^T), $k = 1, 2, 3, \ldots, N$, where S_k^S and S_k^T represent the source and target sentences in the parallel corpus, respectively. Due to (S_k^S, S_k^T) are sampled from parallel corpus, there are all positive data examples. However, in our scenario, we want to train a classifier to distinguish parallel sentence pairs from non-parallel ones in an end-to-end fashion. To enable the effective training of such supervised model, adequate negative examples also need to be presented within training dataset. In this case, we used an automatically generation strategy. In the process of training, for each sentence pair (S_k^S, S_k^T) we automatically extract m pairs of sentences (S_k^S, S_j^T) from the set C on the fly, where $j \neq k$. For each epoch during training, the training data will contains $n(m+1)$ examples in a form of triple (S_i^S, S_i^T, y_i), where $S_i^S = (w_{i,1}^S, w_{i,2}^S, \ldots, w_{i,N}^S)$ denotes the source side sentence containing N words, while $S_i^T = (w_{i,1}^T, w_{i,2}^T, \ldots, w_{i,M}^T)$ denotes the target side sentence containing M words, and label $y_i \in \{0, 1\}$ indicates whether (S_i^S, S_i^T) are parallel in term of translation equivalence.

2.2 Model Architecture

Our basic idea in this work is that learning cross-lingual semantic representation using neural networks model, and estimate the probability $p(y_i = 1 | S_i^S, S_i^T)$ of two given sentences (S_i^S, S_j^T) being parallel. The models we proposed used a bidirectional recurrent neural networks (BiRNN). Among many network architecture alternatives, we experimented with both Long Short-term Memory network [5] and Gated Recurrent Unit (GRU) [1]. As shown in Fig. 1, the model used a shared BiRNN layers to encode the source and target sentences into continuous vector representations.

For the collection of source side sentences S, at timestamp t, the word in i position is defined by its corresponding index k in source side vocabulary V^S, which is represented in the form of one-hot vector $w_k \in \mathbb{R}^{|V^S|}$, where the k index is 1 and all other elements are 0. The one-hot vector is then multiply with the matrix of embedding layer $E^s \in \mathbb{R}^{|V|^S \times D_e}$ to get a vectorized dense representation of the given word $w_{i,t}^S \in \mathbb{R}^{d_e}$. These dense representations are then fed into follow-up BiRNN layers in a sequential manner. The forward RNN layer starts encoding the sentence from its first word to a special symbol $<EOS>$ which indicates the end of sentence, and finally produces a forward fix-width recurrent hidden state of the whole sentence $\overrightarrow{h}_{i,N}^S \in \mathbb{R}^{d_h}$. Likewise, the backward RNN layer encodes the same sentence in a reverse direction, produces backward recurrent hidden state $\overleftarrow{h}_{i,1}^S \in \mathbb{R}^{d_h}$. The final hidden state of the sentence is simply the concatenation of these two hidden states $h_i^S = \left[\overrightarrow{h}_{i,N}^S; \overleftarrow{h}_{i,1}^S \right]$. By following the exact same procedure, the target sentence is encoded to get the

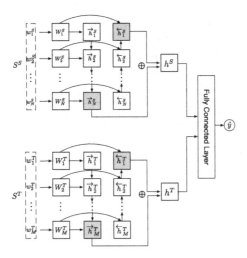

Fig. 1. Model architecture

hidden representation $h_i^T = \left[\overrightarrow{h}_{i,M}^T; \overleftarrow{h}_{i,1}^T\right]$. The encoding process is described in Eqs. 1, 2 and 3.

$$\mathrm{w}_{i,t}^S = E^{S^T} w_k \tag{1}$$

$$\overrightarrow{h}_{i,t}^S = \phi(\overrightarrow{h}_{i,t-1}^S, \mathrm{w}_{i,t}^S) \tag{2}$$

$$\overleftarrow{h}_{i,t}^S = \phi(\overleftarrow{h}_{i,t+1}^S, \mathrm{w}_{i,t}^S) \tag{3}$$

Where ϕ indicating a function a RNN like LSTM or GRU trying to approximate.

After obtaining encoder representations for both source and target side sentences, the point-wise product $h_i^{(1)} = h_i^S \odot h_i^T$ and difference $h_i^{(2)} = |h_i^S - h_i^T|$ of these two hidden states are fed into the subsequent fully-connected layer to capture reliable evidence of features for estimating the probability of the given sentence pair being equivalent translation for each other. As indicated in Eqs. 4, 5.

$$h_i = \tanh(W^{(1)}h_i^{(1)} + W^{(2)}h_i^{(2)} + b) \tag{4}$$

$$p(y_i = 1|h_i) = \sigma(W^{(3)}h_i + c) \tag{5}$$

Where h_i denotes the output of fully-connected layer, σ denotes `softmax` function on h_i and $W^{(1)}, W^{(2)}, b, c$ are weights and biases model need to be learned and optimized.

The model is trained by minimizing the cross-entropy loss of estimation on corrected labeled sentence pairs y_i, as shown in Eq. 6.

$$
\begin{aligned}
\mathcal{L} = - &\sum_{i=1}^{n(1+m)} y_i \log \sigma(W^{(3)}h_i + c) \\
&- (1 - y_i) \log(1 - \sigma(W^{(3)}) + c)
\end{aligned} \tag{6}
$$

Where y_i denotes sentence pair is correctly labeled and $1-y_i$ denotes the sentence pair is not correctly labeled. During model inference, if the probability score of sentence pair being parallel exceeds a predefined threshold ρ then label $\hat{y}_i$ as 1 otherwise label as 0, as shown in Eq. 7.

$$\hat{y}_i = \begin{cases} 1 & if \ p(y_i = 1 \mid h_i) \geqslant \rho \\ 0 & if \ p(y_i = 1 \mid h_i) < \rho \end{cases} \tag{7}$$

3 Experiments

3.1 Dataset

We choose to use in-house Chinese-Tibetan parallel corpus as training data. We took Chinese sentence length as reference to filter out sentences with more than 90 words or less than 5 words in whole corpus. To constrain on parameter space we limited the Chinese and Tibetan vocabulary size to $50K$ and $40K$ respectively. Chinese text was tokenized using *Stanford Word Segmenter* [11][1] and Tibetan text was tokenized using neural networks based approach proposed in [9] (Table 1).

Table 1. Training data scale

	Sentences	Tokens	Syllables[a]	Unique tokens
Tibetan training set	780K	22.3M	123M	500
Chinese training set	780k	34.2M	45M	12K
Tibetan test set	2k	2.4k	3.5k	110
Chinese test set	2k	3.1k	4.5k	2k
Tibetan dev set	2k	2.2k	3.6k	101
Chinese dev set	2k	3.2k	4.6k	1.8k

[a]For Chinese we refer each *hanzi* as syllable

In NMT task, we appended our original $780K$ parallel corpus with additional $100K$ synthesized parallel texts generated by a pre-trained NMT model, to mimic noise might present in the crowdsourced dataset under real world condition.

3.2 Evaluating Tasks

We evaluate the model performance in three scenarios, including

1. Classification task: directly test as a classification task.
2. Sentence extraction task: randomly shuffle the test set and then recover the sentence orders.

---

[1] https://nlp.stanford.edu/software/segmenter.shtml.

3. NMT data filtering task: evaluate BLEU score improvement by applying data filtering in Tibetan-Chinese NMT pre-processing pipeline.

What needs extra attention is that in sentence extraction task, we compute the probability estimation of sentence pairs in a set $\mathcal{J}$, where $\mathcal{J}$ is Cartesian product of source side sentences and shuffled target side sentences in test set as denoted in Eq. 8.

$$\mathcal{J} = S^S \times shuffle(S^T) \tag{8}$$

Where S^S and S^T indicate source and target side sentences from test set, $shuffle(\cdot)$ indicates randomly shuffling operation on target side sentences. After computing the probability evaluation of all items in set $\mathcal{J}$ we try to recover the original order of shuffled sentences by setting a threshold ρ, as shown in Eq. 7. By the definition of Cartesian product, to finishing the test, a total number of $|S^S|^2$ of inference is required, in experiments, we conduct inference via *mini-batch* strategy.

3.3   Experiment Settings

We choose Tensorflow as machine learning framework for model implementation. BiRNN in encoder was composed in one layer RNN for both forward and backward directions. Dimension for input word representation is 512. We experimented both LSTM and GRU as building block of the encoder, and experiments showed that LSTM needed a longer training time and only gain negligible performance improvement in term of classification accuracy, as shown in Table 2. Hidden unit in fully-connected layer was set to 256. During parameter initialization, all weights was initialized with Uniform Scale Distribution and all biases are initialized with *zeros*. During model training, *Adam* [7] was selected as optimizer, initial learning rate was set to 0.001 and *minibatch* size was set to 128 samples.

We trained the model for 30 epochs, to prevent gradient exploration, we applied *gradient clipping* [6] and to prevent model overfitting on training dataset, we applied *dropout* [10], the dropout ratio for BiRNN and fully-connected layer were set to 0.3 and 0.4 respectively.

3.4   Results

In classification scenario, test dataset is a subset of standard parallel corpus. We experimented with different threshold values ρ and noticed that when ρ was set to a smaller value (e.g. 0.5), the high perplexity appears during inference, but when ρ was set to a larger value (e.g. 0.95) all performance metrics improved drastically and became stable. Furthermore, we also experimented with different RNN architectures for encoder, namely LSTM and GRU, and the results indicated that LSTM converged much slowerly than GRU, only producing negligible performance gains over GRU.

Table 2. Model performance in classification scenario

RNN selections	Accuracy (%)	Recall (%)	F_1	ρ
LSTM	68.2	73.1	71.8	0.50
GRU	66.7	76.0	79.3	0.50
LSTM	93.1	84.3	87.2	0.90
GRU	92.6	82.7	85.4	0.90

In parallel sentence extraction task we randomly shuffle the order of Chinese sentences in test dataset, and then try to recover the original order of Chinese sentences in which sentences pairs with same line number is translation equivalence in different threshold value settings. Same findings were observed as experiments in classification task (as shown in Table 2) that bigger ρ produced much better performance in terms of all metrics, including accuracy, recall and F_1, as shown in Table 3.

Table 3. Model performance on parallel sentence extract scenario

Accuracy (%)	Recall (%)	F_1	ρ
89	76	73	0.50
95	87	79	0.99

In NMT task, we experimented two identical translation models in terms of network architecture and hyper-parameter settings as shown in Table 4. The only difference lied in data pre-processing that for one of them data selection technique proposed in this work was applied and for the another experiment data selection wasn't done. As shown in Fig. 2, by applying data filtering, a considerable improvement is observed on validation set during training. Giving experiments are conducted in identical model architecture and hyper-parameter setting, the results indicate the effectiveness of the proposed data filtering technique in NMT.

We use Transformer's official implementation[2] in our experiments which showed that adding our data selection technique to model's pre-processing pipeline significantly boosted performance by surpassing our baseline **3.7** BLEU score, even in the scenario where the training dataset was less than $1M$ in scale. The performance on test set is as shown in Table 5.

---

[2] https://github.com/tensorflow/tensor2tensor/.

Table 4. Hypyer-parameter settings for training Transformer models

Hyper-parameter names	Hyper-parameter settings
Label smoothing	0.1
Optimizer	LazyAdamOptimizer
Learning rate	2.0
β_1	0.9
β_2	0.998
Learning rate decay function	Noam
Hot-loading steps	10000
Length penalty	0.6
Mini-batch type	Token
Mini-batch size	4200
Encoder input dimension	512
Self attention layer number	4
Hidden unit number in self attention layer	512
Head number in multi-head attention layer	8
Hidden unit number in feed-forward layer	2048
Drop out ratio in feed-forward layer	0.1
Dropout rate in self-attention layer	0.1
Dropout in $ReLu$ layer	0.2

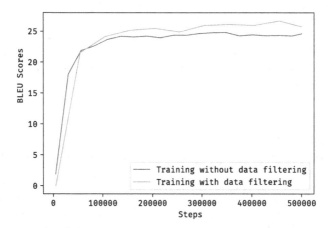

Fig. 2. Validation BLEU score changes over time during training

Table 5. Model performance on NMT tasks

Tasks	NMT model	BLEU
No data filtering applied	Transformer	21.5
Data filtering applied	Transformer	**25.2(+3.7)**

4   Conclusion

In this work we presented a simple neural network architecture which is composed in BiRNN and subsequent fully connected layer to automatically evaluate probability of two given Chinese and Tibetan sentence pair being translation equivalence in a fully end-to-end fashion. For training such classifier we proposed to use to a negative sampling strategy to generating negative samples on the fly. The classifier was experimented and tested in three scenarios, namely, classification, sentence extraction and NMT tasks. The results showed that the proposed model and techniques produced a significantly performance boost in Chinese-Tibetan NMT models.

Like many other languages in the world, Chinese-Tibetan is categorized into low-resourced language pairs in MT research, the availability of parallel bi-texts is very limited. In this work we only have $780K$ parallel sentences to experiment on, and the parallel data is relatively in high quality. However, in other situations where bi-texts is presented with translation noisy then our models can fit into the need of data selection and filtering.

Even though the model presented in this work is promising in term of performance, the architecture itself is fairly simple. Recently pre-training contextual representations such as BERT [2] attracted a lot of attention in the field, resulting in state-of-the-art performance in many NLP tasks. In following work, we are planning to explore such powerful language model pre-training techniques and integrate it with our model to further improve the performance.

References

1. Cho, K., et al.: Learning phrase representations using RNN encoder-decoder for statistical machine translation. In: EMNLP (2014)
2. Devlin, J., Chang, M.W., Lee, K., Toutanova, K.: BERT: pre-training of deep bidirectional transformers for language understanding. arXiv abs/1810.04805 (2018)
3. Foster, G.F., Goutte, C., Kuhn, R.: Discriminative instance weighting for domain adaptation in statistical machine translation. In: EMNLP (2010)
4. Hassan, H., et al.: Achieving human parity on automatic Chinese to English news translation. arXiv abs/1803.05567 (2018)
5. Hochreiter, S., Schmidhuber, J.: Long short-term memory. Neural Comput. **9**(8), 1735–1780 (1997)
6. Kanai, S., Fujiwara, Y., Iwamura, S.: Preventing gradient explosions in gated recurrent units. In: NIPS (2017)

7. Kingma, D.P., Ba, J.: Adam: a method for stochastic optimization. arXiv abs/1412.6980 (2014)
8. Moore, R.C., Lewis, W.D.: Intelligent selection of language model training data. In: ACL (2010)
9. Sangjie, D., Cairang, J.: A study on neural network based tibetan word segmentation method. Qinghai Technol. **25**, 15–21 (2018). (in Chinese)
10. Srivastava, N., Hinton, G., Krizhevsky, A., Sutskever, I., Salakhutdinov, R.: Dropout: a simple way to prevent neural networks from overfitting. J. Mach. Learn. Res. **15**(1), 1929–1958 (2014)
11. Tseng, H., Chang, P., Andrew, G., Jurafsky, D., Manning, C.: A conditional random field word segmenter for SIGHAN bakeoff 2005. In: Proceedings of the Fourth SIGHAN Workshop on Chinese Language Processing (2005)
12. Vaswani, A., et al.: Attention is all you need. In: Advances in Neural Information Processing Systems, pp. 5998–6008 (2017)

An Attention-Based Approach for Mongolian News Named Entity Recognition

Mingyan Tan, Feilong Bao[✉], Guanglai Gao, and Weihua Wang

College of Computer Science, Inner Mongolian Key Laboratory of Mongolian
Information Processing Technology, Inner Mongolia University, Hohhot, China
csfeilong@imu.edu.cn

Abstract. In the field of Natural Language Processing (NLP) of Mongolian, Named Entity Recognition (NER) has great significance. The traditional model is to use the Conditional Random Field (CRF) and Long-Short Term Model (LSTM) method. According to the characteristics of Mongolian, a named entity recognition method based on attention mechanism is proposed in this paper. According to the characteristic of the word-building of the Mongolian language, the suffix of the partial word is divided into morphemes. Based on morphemes, character vectors are trained by LSTM. After that, the word vector is sent to another LSTM to get its context representation. Then the attention mechanism is used to obtain the representation of the full text range of the character vector. Finally, the label sequence of the article is obtained by using CRF. The experimental results show that the Mongolian Named Entity Recognition of attention mechanism is superior to the traditional Bi-LSTM-CRF joint model.

Keywords: Named Entity Recognition · Attention mechanism · Conditional Random Field · Long-Short Term Model

1 Introduction

Named entity recognition is of great significance in the field of natural language processing. The study of named entity recognition dates back to the sixth message Understanding Conference (MUC-6) [1]. The main research topic of MUC conference is to identify and classify important nouns as well as numerical expressions such as quantity, currency, time and so on from unstructured texts. The purpose of this conference is to understand the relationship between text content mining and text extraction. With the public release of evaluation tasks on named entity recognition by conll2002 [2], conll2003 [3] and ace2004 [4], named entity recognition has ushered in a new upsurge.

At present, the research on Mongolian named entity recognition is still relatively few. The reason is that Mongolian corpus is scarce and there is no public Mongolian corpus. Tonglaga of the Minzu University of China [5] analyzes the characteristics of the times, regional characteristics and the changing law of the internal model of human names. Seven groups of different feature templates were constructed by using conditional random field model of Cai [6] in Inner Mongolia University, and the experiment

© Springer Nature Switzerland AG 2019
M. Sun et al. (Eds.): CCL 2019, LNAI 11856, pp. 424–435, 2019.
https://doi.org/10.1007/978-3-030-32381-3_35

of name recognition was carried out. On the basis of analyzing the composition characteristics of Mongolian place names, Wu [7] and others realized the named entity recognition of Mongolian place names by combining conditional random fields with dictionaries. According to the characteristics of Mongolian language, Wang [8] combines character vector and language model to realize Mongolian named entity recognition.

As a characteristic of the national minority in our country, the Mongolian language has a far-reaching significance for its research. At present, the lack of the Mongolian language, the disunity of the coding format, the derivation of the new name, different from the writing characteristics of English and Chinese, the way of word formation and so on, have brought difficulties to the identification of the Mongolian named entity. In the following, the paper will introduce the Mongolian word-formation analysis, and then cut the suffix to get the morpheme.

2 Method

This chapter first introduces the morpheme segmentation based on the characteristics of Mongolian word formation, and then introduces the Mongolian character vector. At the end of this section, we will discuss the specific algorithm and work of BiLSTM-CRF and attention mechanism in Mongolian.

2.1 Mongolian Morpheme Segmentation

Mongolian writing has its own characteristics: according to its position, the form of expression of a word is also different. Mongolian letters change in form at the beginning, end, and word. Therefore, in order to better show the language characteristics, this paper uses Latin form to deal with Mongolian. The contrast between Latin characters and Mongolian letters is shown in Table 1.

Mongolian suffixes can be divided into three types: word formation suffix, configuration suffix and ending suffix. The positions of the three types are relatively fixed. In a Mongolian word, there are one or more word-formation suffixes and configuration suffixes, and only one suffix at the end. However, Mongolian words can have two ending suffixes when they are added with a reverse collar suffix. If you splice all the roots with different suffixes, you can build nearly a million words. In the above three types of suffixes, the ending suffix only represents its grammatical rules and does not change the meaning of the word. Therefore, in the course of training, the ending suffix brings difficulties to the training of CRF. Figure 1 shows the relationship between root, stem, and suffix. Take the word "herdsman" as an example, the red font word "ᠬᠦ" means cow, with a green suffix "ᠴᠢ", it means herdsmen. The black ending suffix only represents the part of speech, and the Mongolian word means "herdsmen's."

Table 1. Comparison between Latin alphabet and Mongolian alphabet

Mongolian alphabet	Latin letter	Mongolian alphabet	Latin letter	Mongolian alphabet	Latin letter
᠕	a	᠌	w	᠖	L
᠍	e	᠎	f	᠏	Z
᠐	i	᠑	k	᠒	Q
᠓	q	᠔	K	᠕	s
᠖	v	᠗	C	᠘	x
᠙	o	᠚	z	᠛	t
᠜	u	᠝	H	᠞	d
᠟	E	ᠠ	R	ᠡ	c
ᠢ	n	ᠣ	g	ᠤ	j
ᠥ	N	ᠦ	m	ᠧ	r
ᠨ	b	ᠩ	l	ᠪ	h
ᠫ	p	ᠬ	y		

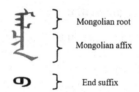

Mongolian root

Mongolian affix

End suffix

Fig. 1. Character embedding with BLSTM (Color figure online)

2.2 Morpheme Vector

According to the characteristics of Mongolian word formation, the strategy of dealing with Mongolian suffix segmentation is to divide the ending suffix of Mongolian word into a new training unit. The reason for choosing this method is that if the suffix is used as a separate training unit, the classifiers can get more context information around the suffixes to help the classifiers work. After the suffix of Mongolian word is segmented, the morpheme can be obtained, and the morpheme vector is used as the training unit for training. There are two training methods of morpheme vector, one is Continuous Bag Of Word (CBOW) [12] model, the other is skip-gram [11] model. In this paper, the skip-gram training morpheme vector is used. The specific training method is to set up a vocabulary with a vocabulary of 10000. After the word-dot code, as the input to the skip-gram model. The output of the model is a probability matrix. The function of probability matrix is to predict contextual morphemes with current morphemes. Giving a morpheme string length T, Its model is represented as:

$$\frac{1}{T}\sum_{t=1}^{T}\sum_{-c\leq j\leq c,j\neq 0}\log p\left(m_{t+j}|m_t\right)$$

(1)

In formula (1), c represents the size of the context window, where the context window size is 8. For the conditional probability $p\left(m_{t+j}|m_t\right)$ in formula (1), its simplest form is:

$$p(o|c) = \frac{\exp\left(u_o^T v_c\right)}{\sum_{m=1}^{M}\exp\left(u_o^T v_c\right)}$$

(2)

In formula (2), o is the ordinal number of the morpheme output and c is the ordinal number of the central morpheme. U is the morpheme output, v is the input morpheme. M is the total number of morphemes.

2.3 Character Vector

Character vectors are different from morpheme vectors. Morpheme vector focuses on the semantics of the Mongolian word, while the character vector focuses on the spelling of Mongolian words, that is, the spelling of morphemes. Character vectors can be used to better describe the attributes of Mongolian words.

In this paper, LSTM is used to learn character vectors. However LSTM can only learn the information in a single direction of the sequence. Therefore, in order to learn all the characteristics of the current time sequence. The forward LSTM and the reverse LSTM need to be spliced together to form the BiLSTM [13]. A forward LSTM, is used to forward processing morphemes; a reverse LSTM, is used to reverse process morphemes. The output $\overleftarrow{h}$ and $\overrightarrow{h}$ of lstm can be expressed as:

$$i_t = \sigma(W_{xi}x_i + W_{hc}h_{t-1} + W_{ci}c_{t-1} + b_i)$$

(3)

$$c_i = (1 - i_t) \odot c_{t-i} + i_t \odot \tanh(W_{xc}x_t + W_{hc}h_{t-1} + b_c)$$

(4)

$$o_t = \sigma(W_{xo}x_t + W_{ho}h_{t-1} + W_{co}c_t + b_o)$$

(5)

$$h_t = o_t \odot \tanh(c_t)$$

(6)

Where σ is the nonlinear activation function softmax, $\{W_{xi}, W_{hc}, W_{ci}, W_{xc}, W_{hc}, W_{xo}, W_{ho}, W_{co}\}$ is the parameter matrix of LSTM. $\{b_i\, b_c\, b_o\}$ is a bias term of the model. In this paper, a 100-dimensional vector is randomly initiated for each character, and then the vector order and inverse order of the characters corresponding to the current word are input into BiLSTM respectively. The final output $[\overrightarrow{h}; \overleftarrow{h}]$ is the character vector of the current word (Fig. 2).

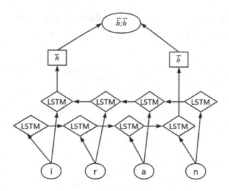

Fig. 2. Character embedding with BLSTM

2.4 BiLSTM-CRF

For a Mongolian, after the above processing, word and character vectors can be obtained. The text representation of each Mongolian word can be obtained by splicing character vector and word vector directly to BiLSTM.

In the task of serial annotation, the marking strategy IOBES is widely used. In this paper, the Mongolian news corpus uses this kind of tagging form. There are three types of entities marked: "PER", "LOC" and "ORG", which represent person name, place name and organization name. In the actual annotation sequence, "I" "B" "O" tags do not appear arbitrarily, and there is a close logical relationship between them, that is the entity tags of a word are affected not only by the context of the word and the meaning of the word itself, but also by the context label of the word. However, this constraint is not taken into account in the ordinary sequence tagging model. In the label judgment of the current word, they only use the context of the current word, and do not use the context of the current word label. Therefore, in some cases, impossible tag sequences will be produced. For example, the I tag appears after the o tag and so on. In order to further improve the accuracy of entity recognition, we draw lessons from the work of Collobert et al. [9] and Huang et al. [10], combined with the advantages of crf model considering label transfer probability, this paper add the tag transfer information of the whole sentence to the original BiLSTM.

Firstly, this paper defines a label transfer matrix A, where A_{ij} represents the score of the transfer from tag i to j, meanwhile parameters are trained with the model. Defines a parameter that the original BiLSTM needs to learn. Then $\theta' = \theta \cup \{A_{ij}, \forall i, j\}$ is all the parameters that the whole model needs to learn. Given a sentence $[x]_l^T$, T is the length of the sentence, Define $[f_\theta]_{it}$ the output score of the ith word, the tth label, Then the formula for calculating the total score of the first sentence in the given label sequence $[i]_l^T$ is:

$$S\left([x]_l^T [i]_l^T, \theta'\right) = \sum_{t=1}^{T} \left(A_{[i]_{t-1}, [i]_t} + [f_\theta]_{[i]_t, t}\right) \tag{7}$$

In this paper, we use softmax to calculate the conditional probability of a sentence $[x]_l^T$ on the real label sequence $[y]_l^T$:

$$p\left([y]_l^T|[x]_l^T,\theta'\right) = \frac{e^{s\left([x]_l^T[y]_l^T,\theta'\right)}}{\sum_j e^{s\left([x]_l^T[y]_l^T,\theta'\right)}} \tag{8}$$

In this formula, $[j]_l^T$ represents all possible label sequences. Finally, the maximum logarithmic likelihood function is used to train the model parameters. The calculation formula is:

$$\ln p\left([y]_l^T|[x]_l^T,\theta'\right) = s\left([y]_l^T|[x]_l^T,\theta'\right) - \ln \sum_{\forall[j]_l^T} e^{s\left([x]_l^T[\varphi]_l^T,\theta'\right)} \tag{9}$$

In this paper, the random gradient drop method is used to optimize the parameters. After the training, at the end of the training, the goal is to find the tag sequence with the highest score as the prediction tag sequence, that is:

$$\arg\max\left(s\left([y]_l^T|[x]_l^T,\theta'\right)\right) \tag{10}$$

In this paper, Viterbi algorithm is used to find the best tag sequence.

2.5   Attention Mechanism

So far, one of the inevitable problems in the methods based on deep learning and traditional machine learning is the inconsistency of the full text of word labels: In one article, the same word, the same entity is often given different entity tags by the model. Obviously, this will reduce the accuracy of the model, and it is not easy to use in the actual project. The main reason for this problem is that today's models usually use sentences as separate processing units. In a separate processing unit, models assign tags according to the context of the word, that is, these models only make use of sentence information and are sentence-level methods. In the same article, if the context of the same entity in different sentences is different, the tags assigned to the sentence level model will also be different, which is the reason for the inconsistency of the full text of the word tags. At the same time, in the same article, for the same entity in its many contexts, if only one or more contexts play a decisive role in judging the label category of that entity, the current sentence-level approach does not deal with the problem well.

The sentence level method also has the problem of word label inconsistency in the specific task of Mongolian news named entity recognition. At the same time, the inconsistency of word labels is reflected in the task of organizational name entity recognition, which is the accuracy of the task. In a news article, the author usually gives the full name only when referring to an organization for the first time, and then gives it in the form of abbreviation or abbreviation. Usually, a general model can label an abbreviation correctly according to the first reference to the abbreviation. when the author refers to the entity by abbreviation, the context relationship is weak, and it is difficult for the ordinary model to assign the correct entity label according to the context information at the statement level. Therefore, in order to solve this problem, only by introducing text-level information can we better solve this problem. For this problem, this paper introduces attention mechanism to solve this problem. The attention mechanism is used to introduce text-level information, and with the help of text information, the problem is alleviated through the continuous training and learning of the model. Attention mechanism was first applied to the field of image recognition. After achieving good results in the field of image recognition, it was later applied to the field of NLP. Attention mechanism is mainly a simulated human attention mechanism [14]: When you look at an image, you don't distract your attention evenly to every part of the image, but most of them focus on specific parts of the image as needed, such as portraits, and usually focus on the face. In this paper, the attention mechanism is used to obtain text-level information for each word, and then the full-text inconsistency of the tags of the same word is improved. Specifically, for an article $[s]_1^N$, N denotes the number of sentences, x_l^T denotes one of the sentences, and t is the length of the sentence. In this paper, **attended** is defined as the word vector or character vector of $[s]_l^N$ and their combination; Define a one of the corresponding one of the $state_i$ for the ith word in the **attended**. Define **source** as the context for each word in the full text, that is, the output of $[s]_l^N$ through BiLSTM. Then it can use the formula to obtain the attention α_i that the ith word should allocate in the full text:

$$energy_i = f(attended, state_i, W) \tag{11}$$

$$\alpha_i = softmax(energy_i) \tag{12}$$

In formula (11), f($\cdot$) is a function used to measure the correlation between **attended** and $state_i$. The **W** in this function is trained with the model. The correlation function used in this article is Manhattan distance. Because the distance between α and oneself is 0, and the weaker the relevance of the meaning of different words, the greater the distance between them in Manhattan:

$$d(a, b, W) = \sum_{t=1}^{N} w_i |a_i - b_i| \tag{13}$$

In reality, we initialize **W** to 1, and we keep it positive during training.

Then, we use the attention weight α to filter and fuse the information in **source** to get the context of the current word in the full text, which we define as **glimpse:**

$$glimpse_i = \alpha_i^T source \tag{14}$$

In order to make the attention model easier to train, and the entity tag of the current word depends not only on the context information within the scope of the full text, but also on the context information adjacent to the current word, this paper combines *glimpse$_i$* and inputs it into the subsequent model structure:

$$context_i = g(\boldsymbol{glimpse}_i, \boldsymbol{source}_i, \boldsymbol{U}) \tag{15}$$

In formula (15), g(•) is a nonlinear function tanh(), U is used as a parameter trained with the model.

By using the attention mechanism and the BiLSTM, described earlier for each article $[s]_l^N$, we can get the $\sum_N \sum_T context$ (in Fig. 3, abbreviated as C), and then through the tanh layer, it can get the score of the model on each label category for each word of the document, which is denoted as $\sum_N \sum_T output$ (abbreviated as O in Fig. 3). Finally, the total score of the article $[s]_l^N$ under the given tag sequence $\sum_M [m]_l^T$ can be calculated:

$$S\left([s]_l^n, \sum_M [m]_l^T, \theta'\right) = \sum_M \sum_{t=1}^T \left(A_{[m]_{t-1}, m_t} + [output]_{[m]_t, t}\right) \tag{16}$$

Then, as in the previous section, the softmax function is used to obtain the probability, and the parameters of the model are trained by maximizing the logarithmic likelihood probability. In the prediction stage, different from the previous section, Viterbi decoding is used for each sentence.

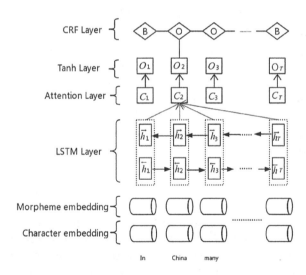

Fig. 3. The model architecture of Attended-BiLSTM-CRF

3 Experimental

3.1 Experimental Setting

The traditional Mongolian language materials used in this paper come from China Mongolian News Network, people's Network (Mongolian version), Chinese

Mongolian Broadcasting Network and other websites from 2013 to 2014. Through the process of mark coding correction, 33292 sentence tagging corpus is obtained, which contains 59562 entities. There are three types of entities that are marked: person name (PER), location name (LOC), organization name (ORG) Table 2 shows the number and proportion of the three entities:

Table 2. Entity type table

Entity type	Number	Proportion
PER	12354	20.74%
LOC	28361	47.62%
ORG	189847	31.64%

In this corpus, 10% are randomly selected as test sets and the rest as development sets. In the evaluation results, this paper uses three commonly used indicators in sequence tagging: the precision (p), recall rate (r), F value as the experimental evaluation index. Table 3 shows the parameters used in this model.

Table 3. The hyper-parameters of model

Parameters	Description	Value
word_embeding_dim	The dimension of word embedding layer	300
Char_embedding_dim	The dimension of char embedding layer	100
Char_for_lstm_dim	The dimension of forward char LSTM layer	100
Char_rev_lstm_dim	The dimension of reverse char LSTM layer	100
For_lstm_dim	The dimension of forward LSTM layer	300
Rev_lstm_dim	The dimension of reverse LSTM layer	300
Learming_rate	Learning rate	0.001

3.2   Result

This article compares the traditional BiLSTM-CRF with the approach of joining the attention mechanism. To explore the effect of Mongolian voxel vector and common word vector on model performance. The experiment is treated in three forms: (1) Which Mongolian word vector or Mongolian morpheme vector can improve the performance of the model. (2) Whether the word vector or morpheme vector combined with the character vector passes through the bilstm layer, that is, whether the morpheme feature or character feature is used for the final classification. (3) Whether morpheme features or character features pass through the attention layer. In other word, in what way does attention align in the full text. The experimental results are shown in Table 4.

Table 4. Expermental results for the feature of attention

Model	Char LSTM	Morpheme LSTM	Morpheme attention	Char attention	P(%)	R(%)	F(%)
BiLSTM-CRF	√	√	×	×	**87.26**	**87.80**	**87.55**
BiLSTM-CRF	√	×	×	×	83.99	83.98	84.48
BiLSTM-CRF	×	√	×	×	84.81	86.36	85.58
Attended-BiLSTM-CRF	×	√	√	×	88.13	87.98	88.05
Attended-BiLSTM-CRF	√	×	×	√	89.58	90.47	90.08
Attended-BiLSTM-CRF	√	√	√	√	*91.24*	*90.17*	*90.67*

Note: "√" indicates that our model uses this feature. "×" indicates that our model does not use this feature.

The results can be obtained from Table 4:

(1) Morpheme segmentation based on Mongolian word formation grammar can really improve the performance of label classification.
(2) In LSTM layer, morpheme features combined with character features are higher than word features combined with character features. In Table 4, this result is marked in bold fonts.
(3) In the attention layer, the use of character features alone is higher than the use of morpheme features alone. Moreover, the simultaneous use of morpheme features and character features in the attention layer will degrade performance. In Table 4, this result is marked in italic font;

The reasons for the above results are as follows:

(1) The attention mechanism can learn the text level information in Mongolian. Thus, the consistency of the full text and the recognition rate of abbreviations can be improved.
(2) When judging the label category, it mainly depends on the meaning of the word, not the meaning of the character, and the meaning of the word and the character are not mutually exclusive. In Mongolian word formation, morphemes can roughly express the meaning of a word. Therefore, in the LSTM layer, morpheme features are better than word features, and the combination of morpheme features and character features is better.
(3) There are a large number of unknown words, so when using morphemes to do attention, the unlogged words will affect the generation of attention weights, resulting in improper weight distribution. There is no such problem with characters. Therefore, the single character feature of attention layer is better than the word feature. When the two are used at the same time, the model cannot completely eliminate the shortcomings of word features, so the performance of the joint use is degraded.

3.3   Performance Comparison Experiment

In order to verify the performance of Attended-BiLSTM-CRF in Mongolian, this paper compares the performance with other methods. The method compared in this paper is the result of Wang's experiment [8] on the corpus used in this paper. The specific experimental data of each entity are given here by the author, as shown in Table 5:

Table 5. Performance comparison experimental table

Model	Entity	P(%)	R(%)	F(%)
BiLSTM-CRF [8]	PER	89.26	87.19	88.21
	LOC	83.57	86.78	85.15
	ORG	88.17	85.64	86.88
Attended-BiLSTM-CRF	PER	93.31	89.16	91.18
	LOC	90.45	90.10	89.82
	ORG	90.40	87.89	89.12

it is not difficult to see from the table that the performance of each entity recognition has been improved. Among them, the recognition performance of geographical names and organizational names is greatly improved. The f value of geographical names increased by 4.67, and the f value of organizational names increased by 2.24.

4   Conclusion

In this paper, the problem of the identification of the Mongolian named entity is solved by using the method of Attended-BiLSTM-CRF on the Mongolian news material. The experimental results show that this method has better results than the existing BiLSTM-CRF method based on Mongolian, which has the following reasons:

1. Based on the characteristics of Mongolian word composition, word-element segmentation is beneficial to the training of the model.
2. Low dimension, dense Morpheme vectors and character vectors have better performance than traditional machine learning, while depth models such as LSTM, can better learn high-level abstract information.
3. Attention mechanism makes use of text-level information to effectively reduce the inconsistency of the full text of word tags, and at the same time, it also improves the accuracy of abbreviation recognition.

However, in the field of Mongolian named entity recognition, attention mechanism still has a lot of room for improvement. From the experiments in this paper, it can be seen that the text-level information can effectively improve the performance, especially in abbreviated entities such as place names and organizational structure names. At the same time, the research of Mongolian named entity recognition provides a good basis for the establishment of Mongolian knowledge base and Mongolian question and answer database in the future.

Acknowledgement. This work was supported by the National Natural Science Foundation of China (Nos. 61563040, 61773224); Natural Science Foundation of Inner Mongolia (Nos. 2018MS06006, 2016ZD06).

References

1. Grishman, R., Sundheim, B.: Message understanding conference-6: a brief history. In: Proceedings of COLING 1996. The 16th International Conference on Computational Linguistics, vol. 1, pp. 1–12 (1996)
2. Sang, E.F.T.K., De Meulder, F.: Introduction to the CoNLL-2002 shared task: language-independent named entity recognition. In: Proceedings of the Sixth Conference on Natural Language Learning, pp. 121–128. Association for Computational Linguistics (2002)
3. Sang, E.F.T.K., De Meulder, F.: Introduction to the CoNLL-2003 shared task: language-independent named entity recognition. In: Proceedings of the Seventh Conference on Natural Language Learning at HLT-NAACL 2003, vol. 4, pp. 142–147. Association for Computational Linguistics (2003)
4. Doddington, G.R., Mitchell, A., Przybocki, M.A., et al.: The Automatic Content Extraction (ACE) program-tasks, data, and evaluation. In: Proceedings of LREC, pp. 21–26, February 2004
5. Tongkala: Automatic name recognition based on Mongolian corpus. Doctor, Central University for Nationalities (2013)
6. Cai, J.: Auotmatic Mongolian personal recognition based on CRF. Doctor, Inner Mongolia University (2016)
7. Wu, J., Li, L., Yang, Z.: Research on Mongolian place name recognition based on CRF and dictionary. Comput. Eng. Sci. **38**(05), 1046–1051 (2016)
8. Wang, W.: Research on Mongolian named entity recognition. Doctor, Inner Mongolia University (2018)
9. Collobert, R., Weston, J., Bottou, L., et al.: Natural language processing (almost) from scratch. J. Mach. Learn. Res. **12**(8), 2493–2537 (2018)
10. Huang, Z., Xu, W., Yu, K.: Bidirectional LSTM-CRF models for sequence tagging. https://arxiv.org/abs/1508.01991. Accessed 04 July 2017
11. Word2Vec (Part 1): NLP with deep learning with Tensorflow (Skip-gram). http://www.thushv.com/natural_language_processing/word2vec-part-1-nlp-with-deep-learning-with-tensorflow-skip-gram/. Accessed 21 Nov 2018
12. Continuous Bag of Words. https://iksinc.online/tag/continuous-bag-of-words-cbow/. Accessed 21 May 2019
13. Graves, A., Schmidhuber, J.: Framewise phoneme classification with bidirectional LSTM and other neural network architectures. Neural Netw. **18**(5), 602–610 (2015)
14. Mnih, V., Heess, N., Graves, A., et al.: Recurrent models of visual attention. In: Advances in Neural Information Processing Systems (2014)
15. Svozil, D., Kvasnicka, V., Pospichal, J.: Introduction to multi-layer feed-forward neural networks. Chemom. Intell. Lab. Syst. **39**(1), 43–62 (1997)
16. Luong, T., Socher, R., Manning, C.: Better word representations with recursive neural networks for morphology. In: Proceedings of the Seventeenth Conference on Computational Natural Language Learning, pp. 104–113 (2003)
17. Etzioni, O., Cafarella, M., Downey, D., et al.: Unsupervised named-entity extraction from the web: an experimental study. Artif. Intell. **165**(1), 91–134 (2015)
18. Wang, W., Bao, F., Gao, G.: Mongolian named entity recognition with bidirectional recurrent neural networks. In: Proceedings of the 28th IEEE International Conference on Tools with Artificial Intelligence (ICTAI), pp. 495–500 (2016)

Language Resources and Evaluation

CJRC: A Reliable Human-Annotated Benchmark DataSet for Chinese Judicial Reading Comprehension

Xingyi Duan[1(✉)], Baoxin Wang[1], Ziyue Wang[1], Wentao Ma[1], Yiming Cui[1,2], Dayong Wu[1], Shijin Wang[1], Ting Liu[2], Tianxiang Huo[3], Zhen Hu[3], Heng Wang[3], and Zhiyuan Liu[4]

[1] Joint Laboratory of HIT and iFLYTEK (HFL), iFLYTEK Research, Beijing, China
{xyduan,bxwang2,zywang27,wtma,ymcui,dywu2,sjwang3}@iflytek.com
[2] Research Center for Social Computing and Information Retrieval (SCIR),
Harbin Institute of Technology, Harbin, China
{ymcui,tliu}@ir.hit.edu.cn
[3] China Justice Big Data Institute, Beijing, China
{huotianxiang,huzhen,wangheng}@cjbdi.com
[4] Department of Computer Science and Technology, Tsinghua University,
Beijing, China
lzy@tsinghua.edu.cn

Abstract. We present a Chinese judicial reading comprehension (CJRC) dataset which contains approximately 10K documents and almost 50K questions with answers. The documents come from judgment documents and the questions are annotated by law experts. The CJRC dataset can help researchers extract elements by reading comprehension technology. Element extraction is an important task in the legal field. However, it is difficult to predefine the element types completely due to the diversity of document types and causes of action. By contrast, machine reading comprehension technology can quickly extract elements by answering various questions from the long document. We build two strong baseline models based on BERT and BiDAF. The experimental results show that there is enough space for improvement compared to human annotators.

1 Introduction

Law is closely related to people's daily life. Almost every country in the world has laws, and everyone must abide by the law, thereby enjoying rights and fulfilling obligations. Tens of thousands of cases such as traffic accidents, private lending and divorce disputes occurs every day. At the same time, many judgment documents will be formed in the process of handling these cases. The judgment document is usually a summary of the entire case, involving the fact description, the court's opinion, the verdict, etc. The relatively small number of legal staff and the uneven level of judges may lead to wrong judgments. Even the judgments

© Springer Nature Switzerland AG 2019
M. Sun et al. (Eds.): CCL 2019, LNAI 11856, pp. 439–451, 2019.
https://doi.org/10.1007/978-3-030-32381-3_36

Cause of Action	变更抚养关系纠纷
Case Description	经审理查明,原告王x0与被告张1原系夫妻关系,2011年3月16日生育一女王某雯**2016年1月20日**,原告王x0与被告张1协议离婚,约定婚生女王某雯由被告张1抚养,原告王x0每月支付**1000.00元**抚养费直到婚生女王某雯满十八周岁另查明,婚生女王某雯随被告张1现居住在**保定市莲池区永华园小区**,被告张1现在保定吉轩商贸有限公司工作
QA Pairs	Q1: 原告与被告何时离婚? A1: 2016年1月20日 Q2: 王某雯是否是原被告双方亲生女儿? A2: YES Q3: 约定王某雯由谁抚养? A3: 张1 Q4: 原告每个月需要支付多少抚养费? A4: 1000.00元 Q5: 王某雯现在住在哪里? A5: 保定市莲池区永华园小区

Fig. 1. An example from the CJRC dataset. Each case contains cause of action (or called charge for criminal cases), context, and some QA pairs where yes/no and unanswerable question types are included.

in similar cases can be very different sometimes. Moreover, a large number of documents make it challenging to extract information from them. Thus, it will be helpful to introduce artificial intelligence to the legal field for helping judges make better decisions and work more effectively.

Currently, researchers have done amounts of work on the field of Chinese legal instruments, involving a wide variety of research aspects. Law prediction [1,20] and charge prediction [8,13,25] have been widely studied, especially, CAIL2018 (Chinese AI and Law challenge, 2018) [22,26] was held to predict the judgment results of legal cases including relevant law articles, charges and prison terms. Some other researches include text summarization for legal documents [11], legal consultation [15,24] and legal entity identification [23]. There also exists some systems for similar cases search, legal documents correction and so on.

Information retrieval usually only returns a batch of documents in a coarse-grained manner. It still takes a lot of effort for the judges to read and extract information from document. Elements extraction often requires pre-defining element types. Different element types need to be defined for different cases or

Table 1. Comparison of CJRC with existing reading comprehension datasets

	Lang	#Que	Domain	Answer type
CNN/Daily Mail	ENG	1.4M	News	Fill in entity
RACE	ENG	870K	English Exam	Multi. choices
NewsQA	ENG	100K	CNN	Span of words
SQuAD	ENG	100K	Wiki	Span of words, Unanswerable
CoQA	ENG	127K	Children's Sto. etc.	Span of words, yes/no, unanswerable
TriviaQA	ENG	40K	Wiki/Web doc	Span/substring of words
HFL-RC	CHN	100K	Fairy/News	Fill in word
DuReader	CHN	200K	Baidu Search/Baidu Zhidao	Manual summary
CJRC	**CHN**	**50K**	**Law**	**Span of words, yes/no, unanswerable**

crimes. Manual definition and labeling processes are time consuming and labor intensive. These two technologies cannot cater for the fine-grained, unconstrained information extraction requirements. By contrast, reading comprehension technology can naturally extract fine-grained and unconstrained information.

In this paper, we present the first Chinese judicial reading comprehension dataset (CJRC). CJRC consists of about 10K documents which are collected from http://wenshu.court.gov.cn/ published by the Supreme People's Court of China. We mainly extract the fact description from the judgment document and ask law experts to annotate four to five question-answer pairs based on the fact. Eventually, our dataset contain around 50K questions with answers. Since some of the questions cannot be directly answered from the fact description, we have asked law experts to annotate some unanswerable and yes/no questions similar to SQuAD2.0 and CoQA datasets (Fig. 1 shows an example). In view of the fact that the civil and criminal judgment documents greatly differ in the fact description, the corresponding types of questions are not the same. This dataset covers the two types of documents and thereby covers most of the judgment documents, involving various types of charge and cause of action (in the following parts, we will use *casename* to refer to civil cases and criminal charges.).

The main contribution of our work can be concluded as follows:

- CJRC is the first Chinese judicial reading comprehension dataset to fill gaps in the field of legal research.
- Our proposed dataset includes a wide range of areas, specifically 188 causes of action and 138 criminal charges. Moreover, the research results obtained through this dataset can be widely applied, such as information retrieval and factor extraction.
- The performance of some powerful baselines indicates there is enough space for improvement compared to human annotators.

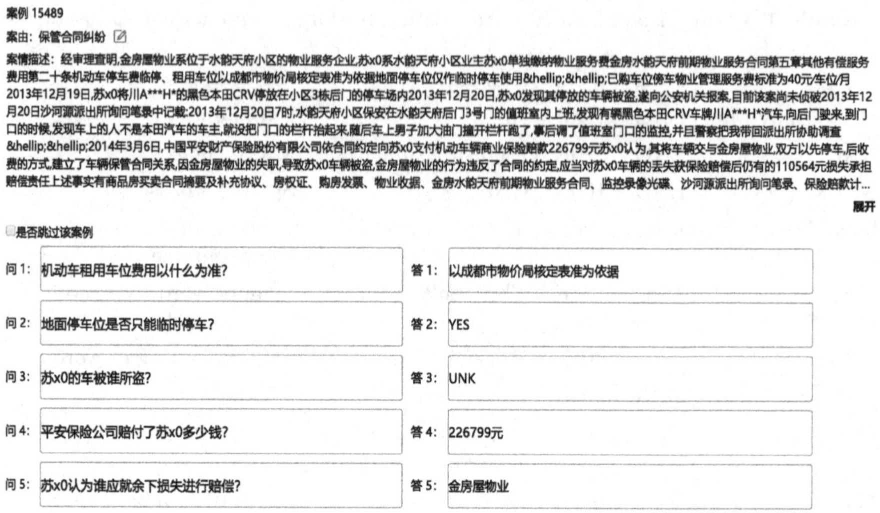

案例 15489

案由: 保管合同纠纷 ✏

案情描述: 经审理查明,金房屋物业系位于水韵天府小区的物业服务企业,苏x0系水韵天府小区业主,苏x0单独缴纳物业服务费金房水韵天府前期物业服务合同第五章其他有偿服务费用第二十条机动车停车费偿停、租用车位位以成都市物价局核定表准为依据地面停车位仅作临时停车使用……已离车位停车物业管理服务费标准为40元/车位/月2013年12月19日,苏x0将川A***H*的黑色本田CRV停放在小区3栋后门的停车位内2013年12月20日,苏x0发现其停放的车辆被盗,遂向公安机关报案,目前该案尚未侦破2013年12月20日沙河源派出所询问笔录中记载:2013年12月20日7时,水韵天府小区保安在水韵天府后门3号门的值班室内上班,发现有辆黑色本田CRV车牌川A***H*汽车,向后门驶来,到门口的时候,发现车上的人不是本田汽车的车主,就没把门口的栏杆抬起来,随后车上男子加大油门撞开栏杆跑了,事后调了值班室门口的监控,并且警察把我带回派出所协助调查……2014年3月6日,中国平安财产保险股份有限公司依合同约定向苏x0支付机动车辆商业保险赔款226799元苏x0认为,其将车辆交与金房屋物业,双方以先停车、后收费的方式,建立了车辆保管合同关系,因金房屋物业的失职,导致苏x0车辆被盗,金房屋物业的行为违反了合同的约定,应当对苏x0车辆的丢失获保险赔偿后仍有的110564元损失承担赔偿责任上述事实有商品房买卖合同摘要及补充协议、房权证、购房发票、物业收据、金房水韵天府前期物业服务合同、监控录像光碟、沙河源派出所询问笔录、保险赔款计...

展开

☐ 是否跳过该案例

问 1:	机动车租用车位费用以什么为准?	答 1:	以成都市物价局核定表准为依据
问 2:	地面停车位是否只能临时停车?	答 2:	YES
问 3:	苏x0的车被谁所盗?	答 3:	UNK
问 4:	平安保险公司赔付了苏x0多少钱?	答 4:	226799元
问 5:	苏x0认为应就余下损失进行赔偿?	答 5:	金房屋物业

Fig. 2. Annotate platform interface

2 Related Work

2.1 Reading Comprehension Datasets

Machine reading comprehension (MRC) has emerged a few datasets for researches. Among these data sets, English reading comprehension datasets occupy a large proportion. Almost each of the mainstream datasets is designed to cater for demands of requiring specific scenes or domains corpus, or to solve one or more certain problems. CNN/Daily mail [7] and NewsQA [21] refer to news field, SQuAD 2.0 [16] focuses on Wikipedia, and RACE [12] concentrates on Chinese middle school students' English reading comprehension examination questions. SQuAD 2.0 [16] mainly introduces the unanswerable questions due to the real situations that we sometimes cannot find a favourable answer according to a given context. CoQA [17] is a large-scale reading comprehension dataset which contains questions that depend on a conversation history. TriviaQA [21] and SQuAD 2.0 [9] pay attention to complex reasoning questions, which means that we need to jointly infer the answers via multiple sentences.

Compared with English datasets, Chinese reading comprehension datasets are quite rare. HFL-RC [3] is the first Chinese Cloze-style reading comprehension dataset, and it is collected from People Daily and Children's Fairy Tale. DuReader [6] is an open-domain Chinese reading comprehension dataset, and it is based on Baidu Search and Baidu Zhidao. Our dataset is the first Chinese judicial reading comprehension dataset, and contains multiple types of questions. Table 1 compares the above datasets with ours, mainly considering the four dimensions: language, scale of questions, domain, and answer type.

2.2   Reading Comprehension Models

Cloze-style and span-extraction are two of the most widely studied tasks of MRC. Cloze-style models are usually designed as classification models to predict which word has the maximum probability. Generally, models need to encode query and document respectively into a sequence of vectors, where each vector denotes a token's representation. The next operations lead to different methods. Stanford Attentive Reader [2] firstly obtains the query vector, and then exploits it to calculate the attention weights on all the contextual embeddings. The final document representation is computed by the weighted contextual embeddings and is used for the final classification. Some other models [5,10,19] are similar with Stanford Attentive Reader.

Span-extraction based reading comprehension models are basically consistent in terms of the goal of calculating the start position and the end position. Some classic models are R-Net [14], BiDAF [18], BERT [4], etc. BERT is a powerful pre-trained model and performs well on many NLP tasks. It is worth noting that almost all the top models on the SQuAD 2.0 leaderboard are integrated with BERT. In this paper, we use BERT and BiDAF as two strong baselines. The gap between human and BERT is 15.2%, indicating that models still have enough room for improvement.

3   CJRC: A New Benchmark Dataset

Our legal documents are all collected from China Judgments Online[1]. We select from a batch of judgment documents, obeying the standard that the length of fact description or plaintiff's claim is not less than 150 words, where both of the two parts are extracted with regular rules. We obtain 5858 criminal documents and 5737 civil documents. We build a data annotation platform (Fig. 2) and ask law experts to annotate QA pairs. In the following subsections, we detail how to confirm the training, development, and test sets by several steps.

In-Domain and Out-of-Domain. Referring to CoQA dataset, we divide the dataset into in-domain and out-of-domain. In-domain means that the data type of test data exists in train sets, and conversely, out-of-domain means the absence. Taking into account that *casename* can be regarded as the natural segmentation attribute, we firstly determine which *casenames* should be included in the training set. Then development set and test set should contain *casenames* in the training set and *casenames* not in the training set. Finally, we obtain totally 8000 cases for training set and 1000 cases respectively for development set and

---

[1] http://wenshu.court.gov.cn/.

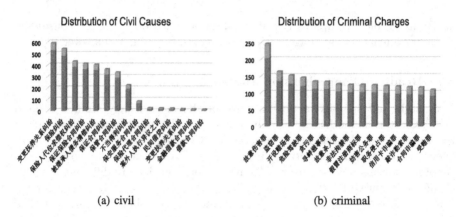

Fig. 3. (a) Distribution of the top 15 civil causes. (b) Distribution of the top 15 criminal charges. Blue area denotes the training set and red area denotes the development set. (Color figure online)

test set. For development and test set, the number of cases is the same whether it is divided by civil and criminal, or by in-domain and out-of-domain. The distribution of *casenames* on the training set is shown in Fig. 3.

Annotate Development and Test Sets. After splitting the dataset, we ask annotators to annotate two extra answers for each question of each example in development and test sets. We obtain three standard answers for each question.

Redefine the Task. Through preliminary experiments, we discovered that the distinction between in-domain and out-of-domain is not obvious. It means that performance of the model trained on training set is almost the same regarding in-domain and out-of-domain, and it is even likely that the latter works better. The possible reasons are as follows:

- *Casenames* inside and outside the domain are similar. In other words, the corresponding cases show some similar case issues. For example, two cases related to the contract, housing sales contract disputes and house lease contract disputes, may involve same issues such as housing agency or housing quality.
- Questions about time, place, etc. are more common. Moreover, due to the existence of the "similar *casenames*" phenomenon, the corresponding questions would also be similar.

However, as we all known, there are remarkable differences between civil and criminal cases. As mentioned in the module **"In-domain and out-of-domain"**, the corpus would be divided by domain or type of cases (civil and criminal). Although we no longer consider the division of in-domain and out-of-domain, it would also make sense to train a model to perform well on both civil and criminal data.

Table 2. Dataset statistics of CJRC

	Civil	Criminal	Total
Train			
Total Cases	4000	4000	8000
Total *Casenames*	126	53	179
Total Questions	19333	20000	40000
Total Unanswerable Questions	617	617	1901
Total Yes/No Questions	3015	2093	5108
Development			
Total Cases	500	500	1000
Total *Casenames*	188	138	326
Total Questions	3000	3000	6000
Total Unanswerable Questions	685	561	1246
Total Yes/No Questions	404	251	655
Test			
Total Cases	500	500	1000
Total *Casenames*	188	138	326
Total Questions	3000	3000	6000
Total Unanswerable Questions	685	577	1262
Total Yes/No Questions	392	245	637

Adjust Data Distribution. Through preliminary experiments, we also discovered that the unanswerable questions are more challenging than the other two types of questions. To increase the difficulty of the dataset, we have increased the number of unanswerable questions in development set and test set. Related experiments will be presented in the experimental section.

Via the processing of the above steps, we get the final data. Statistics of the data are shown in Table 2. The subsequent experiments will be performed on the final data.

4 Experiments

4.1 Evaluation Metric

We use macro-average F1 as our evaluation metric which is consistent with the CoQA competition. For each question, n F1 scores need to be calculated with n standard human answers, and the maximum value is taken as its F1 score. However, in assessing human performance, each standard answer needs to be compared to $n - 1$ other standard answers to calculate the F1 score. In order to compare human indicators more fairly, n standard answers need to be divided into n groups, where each group contains $n - 1$ answers. Finally, the F1 score

Table 3. Experimental results

	Civil	Criminal	Overall
Human	94.9	92.7	93.8
BiDAF	61.1	62.7	61.9
BERT	80.1	77.2	78.6

Table 4. Experimental results of in-domain and out-of-domain on development set and test set

Method	Development			Test		
	Civil	Criminal	Overall	Civil	Criminal	Overall
In-Domain	82.1	78.6	80.3	84.7	80.2	82.5
Out-of-Domain	**82.3**	**83.9**	**83.1**	80.9	**82.9**	81.9

of each question is the average of the n groups' F1. The F1 score of the entire dataset is the average of all questions' F1. The formula is as follow:

$$Lg = len(gold) \tag{1}$$

$$Lp = len(pred) \tag{2}$$

$$Lc = InterSec(gold, pred) \tag{3}$$

$$precision = \frac{Lc}{Lp} \tag{4}$$

$$recall = \frac{Lc}{Lg} \tag{5}$$

$$f1(gold, pred) = \frac{2 * precision * recall}{precision + recall} \tag{6}$$

$$Avef1 = \frac{\sum_{i=0}^{Count_{ref}} (max(f1(gold_{\rightarrow i}, pred)))}{Count_{ref}} \tag{7}$$

$$F1_{macro} = \frac{\sum_{i=1}^{N}(Avef1_i)}{N} \tag{8}$$

Where *gold* denotes standard answers, *pred* denotes answers predicted by models, *len* means to calculate length, *InterSec* means to calculate the number of overlap chars. $Count_{ref}$ represents the total references, $\rightarrow i$ represents that the predicted answer is compared to all standard answers except the current one in a single group described as above.

4.2   Baselines

We implement and evaluate two powerful and typical model architectures: BiDAF proposed by [18] and BERT proposed by [4]. Both of the two models are designed to deal with these three types of questions. These two models learn to predict the probability which is used to judge whether the question is unanswerable. In addition to the way of dealing with unanswerable questions, we concatenate [YES] and [NO] as two tokens with the context for BERT, and concatenate "KYN" as three chars with the context for BiDAF where 'K' denoting "Unknown" means cannot answer the question according to the context. Taking BiDAF for example, during the prediction stage, if start index is equal to 1, then model outputs "YES", and if it is equal to 2, then model outputs "NO".

Some other implementation details: for BERT, we choose the Bert-Base Chinese pre-trained model[2], and then fine-tuning on it with our train data. It is trained on Tesla P30G24, and batch size is set to 8, max sequence length is set to 512, number of epoch is set to 2. For BiDAF, we remove the char embedding, and split string into a sequence of chars, which roles as word in English, like "2019年5月30日". We set embedding size to 300, and other parameters follow the setting in [4].

4.3 Result and Analysis

Experimental results on test set are shown in Table 3. From this table, it is obvious that BERT is 14.5–19% points higher than BiDAF, and Human performance is 14.8–15.5% points higher that BERT. This implies that models could be improved markedly in future research.

Experimental Effect of In-Domain and Out-of-Domain. In this section, we mainly explain why we no loner consider the division of in-domain and out-of-domain described in Sect. 2. We adopts the dataset before adjusting data distribution and select BERT model to verify. Notice that we only train data belong to civil for "Civil", train data belong to criminal for "Criminal", and train all data for "Overall". And type of cases on development set and test set is corresponding to the training corpus. It can be seen from Table 4 that the F1 score of out-of-domain is even higher than that of in-domain, which obviously does not meet the expected result of setting in-domain and out-of-domain.

Comparisons of Different Types of Questions. Table 5 presents fine-grained results of models and humans on the development set and test set, where both of the two sets are not adjusted. We observe that humans maintain high consistency on all types of questions, especially on the "YES" questions. The human agreement on criminal data is lower than that on civil data. This is partly because that we firstly annotate the criminal data, and then have more experience when marking the civil data. It could result in a more consistent granularity of the selected segments on the "Span" questions.

Among the different question types, unanswerable questions are the hardest, and "No" questions are second. We analyze why the performance of unanswer-

---

[2] https://github.com/google-research/bert.

Table 5. Comparisons of different types of questions.

	Bert			BiDAF			Human		
	Civil	Criminal	Overall	Civil	Criminal	Overall	Civil	Criminal	Overall
Development									
Unanswerable	69.5	63.3	68.0	7.6	11.4	8.5	92.0	87.1	90.8
YES	91.7	93.2	92.4	83.5	91.2	86.9	96.9	96.2	96.6
NO	78.0	59.0	73.2	57.9	44.9	54.6	94.2	87.8	92.6
Span	84.8	81.8	83.2	80.1	76.0	77.9	91.6	88.4	89.9
Test									
Unanswerable	67.7	65.6	67.1	10.6	16.0	12.2	91.5	87.7	90.4
YES	91.8	95.6	93.4	77.3	92.8	83.7	97.3	96.5	96.9
NO	72.9	69.7	71.8	47.8	43.3	46.3	96.3	92.5	95.0
Span	84.3	82.4	83.3	79.1	76.2	77.6	93.5	90.9	92.2

Table 6. Comparison data of unanswerable questions and "NO" questions, where unanswerable+ denotes adding extra unanswerable questions on the training set of the civil data.

	Number of Questions (Training set)		Number of Questions (Test set)		Performance (Test set)	
	Civil	Criminal	Civil	Criminal	Civil	Criminal
Unanswerable	617	617	186	77	67.7	65.6
NO	1058	485	134	67	72.9	69.7
Unanswerable+	1284	617	186	77	77.3	67.1
NO	1058	485	134	67	81.6	71.1

able questions is the lowest, and conclude two possible causes: (1) the total number of unanswerable questions on the training set is few; (2) the unanswerable questions are more troublesome than the others.

It is easy to verify the first cause via observing the corpus. To verify the second point, we compare the unanswerable questions and the "NO" questions. Table 6 shows some comparison data of the two types of questions. The first two rows show that unanswerable questions presents a lower performance than the other on the criminal data, even though the former owns more questions. This has basically illustrated that the unanswerable questions are more hard. We have further experimented with increasing the number of unanswerable questions of civil data on the training set. The last two rows in Table 6 demonstrates that increasing unanswerable questions' quantity has an significant impact on performance. However, despite having a larger amount of questions for unanswerable questions, it presents a lower score than "NO" questions.

The above experiments could explain that the unanswerable questions are more challenging than other types of questions. To increase the difficulty of the corpus, we adjusts data distribution through controlling the number of unanswer-

Table 7. Influence of unanswerable questions. Implement BERT and BiDAF on development set and test set. +Train stands for increasing the number of unanswerable questions on the training set. −Dev-Test means no adjusting the number of unanswerable questions on the development set and the test set.

	Bert			BiDAF		
	Civil	Criminal	Overall	Civil	Criminal	Overall
	Development					
Human (before adjust)	92.3	89.0	90.7	–	–	–
Human (after adjust)	93.6	90.8	92.2	–	–	–
CJRC+Train	83.7	77.3	80.5	63.3	62.5	62.9
CJRC−Dev-Test	84.0	81.8	82.9	73.7	75.0	74.3
CJRC+Train−Dev-Test	84.8	81.7	83.3	73.8	74.9	74.4
CJRC	82.0	76.4	79.2	62.8	63.1	63.0
	Test					
Human (before adjust)	93.9	91.3	92.6	–	–	–
Human (after adjust)	94.9	92.7	93.8	–	–	–
CJRC+Train	82.3	77.9	80.1	61.3	61.9	61.6
CJRC−Dev-Test	83.2	82.5	82.8	72.2	74.6	73.4
CJRC+Train−Dev-Test	84.5	82.1	83.3	72.6	74.0	73.3
CJRC	80.1	77.2	78.6	61.1	62.7	61.9

able questions. The following section would show details about the influence of unanswerable questions.

Influence of Unanswerable Questions. In this section, we mainly discuss the impact of the number of unanswerable questions on the difficulty of the entire dataset. **CJRC** represents that we only increase the number of unanswerable answers on the development and the test set without changes on the training set. **CJRC+Train** stands for adjusting all the datasets. **CJRC−Dev-Test** means no adjusting any of the datasets. **CJRC+Train−Dev-Test** means only increasing the number of unanswerable questions of the training set. From Table 7, we can observe the following phenomenon:

- Increasing the number of unanswerable questions in development and test sets can effectively increase the difficulty of the dataset. In terms of BERT, before adjustment, the gap with human indicator is 9.8%, but after adjustment, the gap increases to 15.2%.
- By comparing CJRC+Train and CJRC (or comparing CJRC+Train−Dev-Test and CJRC−Dev-Test), we can conclude that BiDAF cannot handle unanswerable questions effectively.
- Increasing the proportion of unanswerable questions in development and test sets is more effective in increasing the difficulty of the dataset, compared with

reducing the number of unanswerable questions of the training set (get the conclusion by observing CJRC, CJRC+Train and CJRC−Dev-Test).

5   Conclusion

In this paper, we construct a benchmark dataset named CJRC (Chinese Judicial Reading Comprehension). CJRC is the first Chinese judicial reading comprehension, and could fill gaps in the field of legal research. In terms of the types of questions, it involves three types of questions, namely span-extraction, YES/NO and unanswerable questions. In terms of the types of cases, it contains civil data and criminal data, where various of criminal charges and civil causes are included. We hope that researches on the dataset could improve the efficiency of judges' work. Integrating Machine reading comprehension with Information extraction or information retrieval would produce great practical value. We describe in detail the construction process of the dataset, which aims to prove that the dataset is reliable and valuable. Experimental results illustrate that there is still enough space for improvement on this dataset.

Acknowledgements. This work is supported by the National Key R&D Program of China under Grant No. 2018YFC0832103.

References

1. Fawei, B., Pan, J.Z., Kollingbaum, M., Wyner, A.Z.: A methodology for a criminal law and procedure ontology for legal question answering. In: Ichise, R., Lecue, F., Kawamura, T., Zhao, D., Muggleton, S., Kozaki, K. (eds.) JIST 2018. LNCS, vol. 11341, pp. 198–214. Springer, Cham (2018). https://doi.org/10.1007/978-3-030-04284-4_14
2. Chen, D., Bolton, J., Manning, C.D.: A thorough examination of the CNN/daily mail reading comprehension task. CoRR abs/1606.02858 (2016). http://arxiv.org/abs/1606.02858
3. Cui, Y., Liu, T., Chen, Z., Wang, S., Hu, G.: Consensus attention-based neural networks for Chinese reading comprehension, July 2016
4. Devlin, J., Chang, M., Lee, K., Toutanova, K.: BERT: pre-training of deep bidirectional transformers for language understanding. CoRR abs/1810.04805 (2018). http://arxiv.org/abs/1810.04805
5. Dhingra, B., Liu, H., Cohen, W.W., Salakhutdinov, R.: Gated-attention readers for text comprehension. CoRR abs/1606.01549 (2016). http://arxiv.org/abs/1606.01549
6. He, W., et al.: DuReader: a Chinese machine reading comprehension dataset from real-world applications. CoRR abs/1711.05073 (2017). http://arxiv.org/abs/1711.05073
7. Hill, F., Bordes, A., Chopra, S., Weston, J.: The goldilocks principle: reading children's books with explicit memory representations, November 2015
8. Hu, Z., Li, X., Tu, C., Liu, Z., Sun, M.: Few-shot charge prediction with discriminative legal attributes. In: Proceedings of the 27th International Conference on Computational Linguistics, COLING 2018, Santa Fe, New Mexico, USA, 20–26 August 2018, pp. 487–498 (2018). https://aclanthology.info/papers/C18-1041/c18-1041

9. Joshi, M., Choi, E., Weld, D.S., Zettlemoyer, L.: TriviaQA: a large scale distantly supervised challenge dataset for reading comprehension. CoRR abs/1705.03551 (2017). http://arxiv.org/abs/1705.03551

10. Kadlec, R., Schmid, M., Bajgar, O., Kleindienst, J.: Text understanding with the attention sum reader network. CoRR abs/1603.01547 (2016). http://arxiv.org/abs/1603.01547

11. Kanapala, A., Pal, S., Pamula, R.: Text summarization from legal documents: a survey. Artif. Intell. Rev. 1–32 (2017)

12. Lai, G., Xie, Q., Liu, H., Yang, Y., Hovy, E.H.: Race: large-scale reading comprehension dataset from examinations. CoRR abs/1704.04683 (2017). http://arxiv.org/abs/1704.04683

13. Luo, B., Feng, Y., Xu, J., Zhang, X., Zhao, D.: Learning to predict charges for criminal cases with legal basis. In: Proceedings of EMNLP (2017)

14. Natural Language Computing Group, Microsoft Research Asia: R-net: machine reading comprehension with self-matching networks. In: Proceedings of ACL (2017)

15. Quaresma, P., Rodrigues, I.P.: A question answer system for legal information retrieval (2005)

16. Rajpurkar, P., Jia, R., Liang, P.: Know what you don't know: unanswerable questions for squad. CoRR abs/1806.03822 (2018). http://arxiv.org/abs/1806.03822

17. Reddy, S., Chen, D., Manning, C.D.: CoQA: a conversational question answering challenge. CoRR abs/1808.07042 (2018). http://arxiv.org/abs/1808.07042

18. Seo, M.J., Kembhavi, A., Farhadi, A., Hajishirzi, H.: Bidirectional attention flow for machine comprehension. CoRR abs/1611.01603 (2016). http://arxiv.org/abs/1611.01603

19. Sukhbaatar, S., Szlam, A., Weston, J., Fergus, R.: Weakly supervised memory networks. CoRR abs/1503.08895 (2015). http://arxiv.org/abs/1503.08895

20. Tran, A.H.N.: Applying deep neural network to retrieve relevant civil law articles. In: Proceedings of the Student Research Workshop Associated with RANLP 2017, pp. 46–48. INCOMA Ltd., Varna, September 2017. https://doi.org/10.26615/issn.1314-9156.2017_007

21. Trischler, A., et al.: NewsQA: a machine comprehension dataset. CoRR abs/1611.09830 (2016). http://arxiv.org/abs/1611.09830

22. Xiao, C., et al.: CAIL 2018: a large-scale legal dataset for judgment prediction. CoRR abs/1807.02478 (2018). http://arxiv.org/abs/1807.02478

23. Yin, X., Zheng, D., Lu, Z., Liu, R.: Neural entity reasoner for global consistency in NER (2018)

24. Zhang, N., Pu, Y.F., Yang, S.Q., Zhou, J.L., Gao, J.K.: An ontological Chinese legal consultation system. IEEE Access 5, 18250–18261 (2017)

25. Zhong, H., Guo, Z., Tu, C., Xiao, C., Liu, Z., Sun, M.: Legal judgment prediction via topological learning. In: Proceedings of EMNLP (2018)

26. Zhong, H., et al.: Overview of CAIL 2018: legal judgment prediction competition. CoRR abs/1810.05851 (2018). http://arxiv.org/abs/1810.05851

On the Semi-unsupervised Construction of Auto-keyphrases Corpus from Large-Scale Chinese Automobile E-Commerce Reviews

Yang Li[1], Cheng Qian[1], Haoyang Che[2], Rui Wang[3], Zhichun Wang[1], and Jiacai Zhang[1(✉)]

[1] College of Artificial Intelligence, Beijing Normal University, Beijing, China
jiacai.zhang@bnu.edu.cn
[2] Data Intelligence Lab, Auto-Smart Inc., Beijing, China
[3] Princeton International School of Mathematics and Science,
Princeton, NJ 08540, USA

Abstract. The long-standing automobile e-commerce websites in China have accumulated huge amounts of auto reviews, and extracting keyphrases of these reviews can assist researchers and practitioners in obtaining online users' typical opinions and acquiring their underlying motivations. However, there haven't existed any relevant text corpora so far. In this paper, the authors propose a semi-unsupervised scheme to construct a comprehensive auto-keyphrases corpus from online collected reviews in Chinese automobile e-commerce websites by Position Rank, which performs very well in keyphrases extraction from texts in the scenario of scarce labeled data. The iterative annotation process consists of three-round labeling and two-round corrections. During the process of the three-round unsupervised labeling, the computing model will extract seven most important words as the keyphrases of the whole paragraph. Between each labeling phase, there are manual check, correction, re-check and arbitration stages, in which the previous labeling errors are corrected and new vocabulary and rules are summarized up to further improve the unsupervised model. For comparison, the paper runs the experiments using another two unsupervised approaches: TF-IDF and Text Rank, the experimental results also show that Position Rank is a more efficient and effective method for keyphrases extraction. By the time this paper was written, the auto-keyphrases corpus had contained 110,023 entries, and there are still much room for improvement in corpus volume and labeling quality.

Keywords: Auto-keyphrases corpus · Keyphrases corpus · Chinese corpus · E-Commerce website reviews · Position Rank · Semi-unsupervised method

1 Introduction

1.1 Background

According to the survey data of the National Bureau of Statistics, the automobiles sales in China has reached 27.819 million in 2018, and declined 4.1% compared with the previous year [1]. The auto market in China is very huge, but now it has transferred

M. Sun et al. (Eds.): CCL 2019, LNAI 11856, pp. 452–464, 2019.
https://doi.org/10.1007/978-3-030-32381-3_37

from a high-speed growth stage to a low-speed and steady growth stage [2]. Inspired by new technologies such as artificial intelligence, the traditional automobile retail models need innovation and changes. To this end, automobile makers and retailers need to obtain a deep sense of users' requirements quickly and accurately. One effective way to get end users' opinions is to retrieve and distill the user reviews on the auto internet websites. Over the past years, the Chinese automobile e-commerce websites (CAEW) have accumulated huge amounts of user reviews, based on which researchers and practitioners can utilize the up-to-date NLP technologies to gain an exact and instant understanding of auto users' underlying perspectives and the most authentic needs.

Thus, a large-scale and high-quality keyphrases corpus is required to lay the groundwork for further research and experiments. Recent years have seen very few text corpora in the auto field, and even fewer keyphrases corpora extracted from auto reviews. In a short term, to obtain enough labeled data is costly and time-consuming, so this paper devises to establish a semi-unsupervised approach incorporating Position Rank, an unsupervised method and manual efforts to construct the corpus, which widely ingest comments data from the most popular CAEW.

However, for the time being, existing methods relied on supervised methods with labeled data, but few of them adopted the unsupervised method in terms of scenarios with scarce labeled data. What we have constructed in the vertical field can be considered as a beneficial attempt towards the construction of Chinese corpora in an unsupervised method. Prior researchers often applied Kappa or F value to measure the consistency in a supervised construction method, while unsupervised methods were rarely proposed. In this paper, we measure the corpus accuracy with the help of practical manpower.

The labeled review keyphrases will help researchers and practitioners in the auto field access the key concerns of end users and achieve accurate marketing effects in the short term [3].

1.2   Characteristics of CAEW Reviews

Reviews of CAEW have their own characteristics. The Chinese vocabulary lacks morphological changes [4], and the automobile reviews are distinctly different from those of other fields. All of these characteristics indeed play an important role in the process of developing labeling specifications and selecting labeling methods or models. In summarization, there are six main characteristics of CAEW reviews.

Firstly, different online users have posted huge amounts of comments on different car models. Digging out the user concerns hidden behind these reviews can help automotive enterprises achieve accurate marketing effects.

Secondly, online user reviews update very fast. The number of online users in the most popular CAEW is massive, and reviews and discussions are generated frequently. In the process of constructing the review corpus, newly posted reviews need special consideration, to find a good method that can handle these data properly.

Thirdly, comments may include emoticons and internet jargons. Nowadays Internet users like to express their attitudes using emoticons or popular online jargons, which cannot be apt for processing by the old-fashioned model or the existing dictionary generally. This requires brand-new methods or Chinese dictionaries to deal with.

Fourth, the online comments are mostly directed at special car models. Perhaps the comment bodies may not mention the car model explicitly, but an implicit object does exist in default, of which phenomenon needs special focus when generating keyphrases from website reviews.

Fifth, there are some irrelevant comments in the raw review data, such as short comments, date info replies, reviews for location check-in and so on. Instead of providing enough information, such reviews should be deleted from raw prepared data during data preprocessing.

Sixth, labeled data is always lacking. Only a few websites proactively contain keyphrases codified by the website editors, but most majorities of websites only have raw comment contents without keyphrases, which makes it more difficult to use supervised method to construct the review corpus. However, as we all know, data labeling is a time-consuming and labor-intensive activity which may cost a lot of resource investments and require a very slow decision-making cycle.

1.3   Organization

The rest of this paper is organized as follows: Sect. 2 introduces the related work about the construction of CAEW corpus or knowledge base. Section 3 discusses the labeling rules. Section 4 introduces the process of corpus construction and how we use the unsupervised method. Section 5 reveals the labeling results and gives an analysis of the results. Section 6 gives the future work and concludes the paper.

2   Related Work

In the 1990s, the academia began to research on the topics and subjects of the Chinese corpora. After years of development, the Chinese corpora have greatly improved in terms of volume, efficiency, and consistency, which cover a wide range of areas, including education, medicine, culture, engineering and so on. According to the survey results, the relevant Chinese corpora in industries are much less than expected, and reviews corpora of CAEW are currently not available.

This paper proposes a keyphrases corpus drilled from CAEW reviews for the very first time. Prior to this, Zheng et al. [5] proposed a hierarchical diagnostic system of car engines, through the combination of static knowledge base and dynamic knowledge base. Shanghai Translation Network [6] contains bilingual language materials for auto IT industry in Chinese and English, but lacks relevant data for CAEW reviews. Guo et al. [7] put forward to construct the car evaluation knowledge base based on the fuzzy concept, that is, combing with the partial sequence relationship between fuzzy theory and different concepts. They constructed a knowledge base about car evaluation, which is only used to indicate a relatively simple relationship instead of complex relationships between objects. Wang et al. [8] suggested that building a knowledge base of car evaluation based on fuzzy association rules can represent more complex relationships and can be used for knowledge reasoning in related fields. However, the topics and views put forward in the automobile evaluation haven't been deeply excavated. Feng et al. [9] proposed to build an opinion-based ontology base for passenger cars, using

OWL and open source ontology editing tools from Stanford University. They used the following five aspects: the concept, relationship, concept level, non-classification relationship of concepts and axiom, to describe the automotive evaluation ontology and conduct deep mining from the users' emotion and views.

Overall, fewer corpus in the automotive-related sector exists to the public, and hardly any focusing on the CAEW reviews. In terms of the construction methods, researchers generally use labeled data, or combine manual labeling with tool labeling, while the unsupervised labeling method is rarely used at present.

3 Specification for Labeling

3.1 Raw Data Selection

This paper selects the word-of-mouth and forum contents from the most popular inland CAEW as the original data sources, which include BITA (bitauto.com), XCar (www. xcar.com.cn), ATHM (www.autohome.com.cn) and PCauto (www.pcauto.com.cn). The chosen principle is defined primarily as that reviews must come from the most popular CAEW at present with high popularity, justifiable credibility, and huge quantities of user comments. The original corpus contains not only comments, but also information such as the car model, the date of comment generation, and the date of comment publication. The contents of each car models are stored continuously when selected. Following that, it will delete comments with only emoticon or shorter words. The original corpus is stored in a JSON format. In general, the method has totally processed 134,741 word-of-mouth reviews at the outset.

3.2 Granularity of Labeling

The proposed corpus in this paper sets paragraphs as the labeling unit. To extract keyphrases, the method needs an exact understanding of what the statement expresses. It is necessary to grasp the contextual information as well as to understand the fine-grained semantics. Labeling keyphrases for paragraphs can be carried out from two points of views. The first one is a paragraph view, in which each user's entire comment does not split up, just label keyphrases directly for whole paragraph. The second one is a sentence view, in which the paragraph is cut into sentences, and then label keyphrases for each sentence, and finally integrated all phrases as the keyphrases of the whole paragraph.

At this stage, we intended to choose the first approach and regard paragraphs as a unit of research. Another method will be tried as planned in the follow-up work to make a further comparison and scheme upgrade.

3.3 Specification for Labeling

To construct a corpus, it is necessary to confirm the specification for labeling in advance. Specification is helpful to ensure the well-organized development of the labeling work and to ensure the quality of the labeling [10]. It also lays a good foundation for the expanding corpus and constructing comprehensive corpus.

Firstly, create a new property. Comments have been stored according to different attributes for different users, including user name, car model, review, etc. The model creates a new property named "keyphrases" to store keyphrases generated by the unsupervised methods. Secondly, control the number of keyphrases. When labeling according to the whole paragraph, the number of keyphrases is limited to 5–7. Because the reviews in this section is relatively long, we generally extract 7 words or phrases as the initial keyphrases. If we will label keyphrases based on a sentence mode later, each sentence will generate one or two important phrases, and then the model will integrate all words as the keyphrases for the entire paragraph. Thirdly, choose the Part of Speech. We extract notional words as keyphrases as far as possible. Notional words generally refer to nouns, pronouns etc., and these words contain more information. While function words generally refer to adjectives, adverbs, prepositions, conjunctions etc., which contain less information [11]. Notional words help a lot more than function words when mining users' opinions about passenger cars.

3.4   Process for Labeling

For obtaining the keyphrases from CAEW, we will perform the following processes sequentially. First, collect raw data. Choose data from the most popular CAEW, which include forum block name, car model name, reviews, publication date, and access date. Rich information can help analysis the semantics of keyphrases comprehensively, it also plays an important role in mining the typical opinion of users later. Secondly, data pre-processing. Cleaning up the raw data by deleting some bad data, which can ensure the quality of unsupervised labeling. Thirdly, unsupervised labeling and error correction. There are three rounds unsupervised labeling process. Between each two rounds, there is an error correction process to check labeling errors manually. If researchers find a labeling error, they will correct it and summarize the cause of the error, and then add the new principles to original dictionary and rule library. After updating, we will use the new dictionary and rule library to make unsupervised labeling again, until the third round of labeling is completed, as seen in Fig. 1. After comparing the common unsupervised methods: Position Rank, TF-IDF and Text Rank, finally we choose Position Rank as our method for its better results. The other two methods are set as control methods in our research.

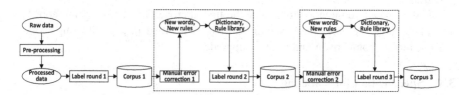

Fig. 1. The process of three-round unsupervised labeling and two-round error correction.

According to the requirements of error corrections, we team up with six other colleagues to complete all of the check, error correction, re-check, and arbitration tasks. Two of them are graduate students majoring in software engineering, responsible for checking and correcting errors. Another three are graduate students we recruited from the Faculty of Arts, responsible for the re-check work. The last one is a data scientist from an auto technology company, responsible for the final arbitration in terms of dispute. Before labeling, team members would conduct labeling training at first. We prepared 100 labeled reviews in CAEW in advance. All six team members labeled these data, and then checked the results and compared their labels with the original labels. When different opinions occurred, they would discuss and confirm which label was the best according to the related reviews, upon confirmation, they would modify the existing specification. The two-round error corrections lie somewhere in between the first round and second round of the unsupervised labeling. In this process, two students were responsible for checking and correcting the errors, respectively. After that, three students would re-check the labeled data. If the two students in charge of error corrections had different opinions about the labeled data, arbitration would be introduced at this time. The arbitration needs that all six colleagues organize a group discussion and judge which results are the best ones finally. At the same time, we would summarize some new vocabularies and extraction rules manually over those wrong items. Then we would add them into the original dictionary and extraction rules in the unsupervised method.

Through an iterative process, the labeling accuracy can be improved continuously. For the subsequent expansion of the reviews corpus, researchers can use these above-mentioned processes to ensure the consistency of the construction methods.

4 Construction of the Corpus

4.1 Data Pre-processing

Before labeling, we started with a cleaning of the 134,741 raw data, with an aim to improve the training efficiency and accuracy. And the following contents need to be deleted. At the beginning, the model should delete the irrelevant information from the comments, including the special symbols, emoticons, deactivated words, pictures, and so on. Secondly, the model should delete the comments that are too short and low-quality. Next, repetitive contents should be deleted too. Then, multiple comments with similar contents from the same user in the same post should be dealt with. Finally, unrelated comments should be deleted completely. After the deletion, 110,023 review items were available to the model.

4.2 Position Rank

Recently, keyphrases methods have already played a very important role in the inductive classification of text information and the theme search [12], this paper also focused on studying the construction method of the keyphrases corpus from the CAEW reviews. The research aims to extract the keyphrases from text, enrich the corpus contents, and facilitate the extensive utilization of the corpus as much as possible. The extraction of keyphrases is the critical step to the construct the keyphrases corpus. As of now, keyphrases extraction methods can be roughly divided into supervised learning methods and unsupervised learning methods [13]. In view of the CAEW reviews, such type of training sets makes supervised learning methods infeasible. Unsupervised learning methods are widely used in the emerging fields of keyphrases generation because they do not definitely require the training sets, and have developed rapidly [14] as a reliable and effective way of learning. This paper adopts a very popular unsupervised learning method based on the graph sorting (see Fig. 2).

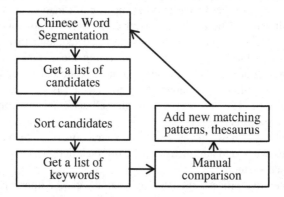

Fig. 2. Unsupervised learning method based on graph sorting, which uses the unlabeled data.

The unsupervised learning method based on graph sorting first needs to confirm the list of candidate words from text [15], out of the complexity of the Chinese language itself, this paper intends to confirm the list of candidate words by means of word labeling. Usually, keyphrases will be nouns or adjectives plus nouns form. However, the grams are more complex in Chinese, and because the comments will include a certain colloquialism, it is necessary to match more word-based labeling methods, such as verbs plus nouns and nouns plus adjectives [16]. When confirming the list of candidates, the model should consider as many lexical dimensions as possible.

Upon getting the list of candidates, the model starts building a graph based on them, where the nodes of the graph are the candidate words, and the model uses a fixed-size sliding window w to slide in the candidate list, if two candidates are in the same sliding window, they will be connected by a line between the nodes in the graph. Between the two candidates, a link exists. The graph can be directed or undirected, in order to facilitate the calculation of the fraction of the serrated nodes, and this paper adopts the undirected graph for computation.

The nodes in the undirected graph are candidates, and in order to find the keyphrases of text information, candidate words need to be evaluated, sorted from highest to lowest in terms of their importance. The evaluation criteria choose to use Position Rank [17], a PageRank [18] algorithm based on location bias. For candidate nodes in an undirected diagram, there are:

$$S(t+1) = \tilde{M} \cdot S(t) \tag{1}$$

S in the upper formula represents the PageRank matrix, $\tilde{M}$ represents the adjacent matrix of the undirected graph, and $S(t+1)$ is the PageRank matrix of $(t+1)$ time, multiplied by the adjacent matrix $\tilde{M}$ and (t) time PageRank matrix.

Where the adjacent matrix $\tilde{M}$ is subject to normalizing before calculation, the values in the matrix $\tilde{m}_{ij}$ are calculated as follows. V represents the undirected graph, $|V|$, which represents the number of nodes, is normalized to $\tilde{m}_{ij}$:

$$\tilde{m}_{ij} = \begin{cases} \tilde{m}_{ij} / \sum_{j=1}^{|V|} m_{ij}, & \text{if } \sum_{j=1}^{|V|} m_{ij} \neq 0 \\ 0, & \text{otherwise} \end{cases} \tag{2}$$

At the same time, in order to ensure that the undirected graph does not fall into the graph loop, the damping factor α is added, and the positional bias $\tilde{p}$ of the word is added, we get (3):

$$S = a \cdot \tilde{M} \cdot S + (1 - a) \cdot \tilde{p} \tag{3}$$

Where the position bias represents the position of candidate words in the text message, because the comments are the views expressed by the user, and according to the characteristics of the opinion selling, people tend to express their views at the beginning of the speech, so the candidate in the text message is more convincing than the candidate words. The specific calculation formula for the position bias $\tilde{p}$ is as follows:

$$\tilde{P} = \left[\frac{p1}{p1 + p2 + \ldots + p|V|}, \frac{p2}{p1 + p2 + \ldots + p|V|}, \ldots, \frac{p|V|}{p1 + p2 + \ldots + p|V|} \right] \tag{4}$$

The initial score of the word, which is represented by p_1, p_2, is inversely proportional to the position of the word in the text, and the frequency of the words is proportional, if the first word appears in the article 5, 6, 7, then the $p_1 = \frac{1}{5} + \frac{1}{6} + \frac{1}{7}$, $p_1 + p_2 + \ldots + p_{|v|}$ Represents the total score of all words, and then divides the total score by p_1 to get the first word's share of all words, and finally gets the position bias $\tilde{p}$.

From these, we can get this:

$$S(v_u) = (1 - a) \cdot \tilde{p} + a \cdot \sum_{v_j \in Adj(v_i)} \frac{w_{ji}}{O(v_j)} S(v_j) \tag{5}$$

The Position Rank score of the v_i node is identified by $S_{(v_i)}$ and α is the damping factor, and $\tilde{p}_i$ is the positional bias of v_i. $S_{(v_j)}$ is the position Rank score of the v_j node, w_{ji} is the weight from the v_j node to the v_i node, $Adj(v_i)$ represents the accompanying matrix of the v_i node, and $O(v_j)$ represents all the out-bound weights of the v_j node.

Through the above Position Rank method, we can get a set of keyphrases for text information. The top 10 words can be included in the keyphrases list. Further, if the two combinations have appeared three or more times in the original text, the synthetic label is added to the keyphrases tag list, the value of Position Rank is the respective position value of the two keyphrases Rank. The resulting list of keyphrases is sorted from highest to lowest by Position Rank values, and the top five or the top seven keyphrases can be marked as keyphrases for comments. Through the processing of a large number of comments, a lot of labels are obtained, so as to build a keyphrases library based on the CAEW comments.

4.3   Process of Constructing Corpus

When the pre-processing is completed, the model needs to select a list of candidates in the comments. The Stanford NLP tool or other Chinese word breakers will be chosen for Chinese word segmentation, and this paper used the jieba word breaker. The matching word-sharing patterns are nouns, but the matching pattern is not fixed and unchangeable, according to the actual use the model can add a variety of matching patterns, such as various pronouns, verbs plus nouns or adjective forms, and so on. After the comments are manually labeled, we can observe the construction of the keyphrases syntax of the manual labels, increase or decrease the corresponding matching patterns, and improve the accuracy of the candidate words.

Once we get the list of candidates, we can implement the Position Rank algorithm. In the course of the experiment, this paper sets the damping factor α mentioned in the previous section to 0.85, and the number of keyphrases extracted is set to 7, wherein the sliding window w is set to 6, the iteration runs five times, and the comments and keyphrases list are combined to get the final keyphrases corpus.

The implementation method is not fixed and unchangeable, the damping factor α and sliding window w size in the algorithm parameters can be changed, the model selects the most appropriate parameter values according to manual evaluation results. At the same time, the model may not necessarily use the Position Rank algorithm or rigidly base on the figure-based sequencing of the unsupervised learning methods. The methods such as the Text Rank [19, 20] algorithm and the topic-based unsupervised learning method [21, 22] can be selected, however, this paper focuses on the idea of how to build a keyphrases corpus from CAEW comments, these alternative methods are not the focal points of this paper. The construction method aims only to gain enlightenment from academia and industry, and there are still many improvements left.

4.4   Quality Assurance

The method strictly controls the labeling quality through multiple rounds of unsupervised labeling and manual corrections. Firstly, it generates primary keyphrases labeled corpus (referred to as the primary corpus) through an unsupervised way. Secondly, it checks and corrects the constituent parts of the primary corpus. We randomly select 1000 comments from the primary corpus, and try to cover a variety of models and avoid the data gathered in a fixed certain mode. Because of the large-scale volume of the corpus, it is difficult to check all the data manually. We randomly retrieve 1000 entries more than once to detect the quality and improve the unsupervised model. Then we summarize the new vocabularies and extraction rules, and add them to the original dictionary and rule library. Thirdly, we use unsupervised method once again with the new dictionary and rule library to get the secondary corpus. Fourth, we randomly extract another 1000 entries, test the quality and summary rules. Finally, we use the unsupervised model once more, and then obtain the final corpus.

It is important to note that the data for the three rounds of unsupervised labeling are the same, and the data for the two random samplings are different. The labeling quality is guaranteed with the "three-round labeling and two-round correction" iteration.

5   Results and Analysis

5.1   Accuracy: Position Rank

We list the accuracy in different rounds and changes between each two adjacent rounds using Position Rank. Besides the accuracy, we record the new words and new rules added to the old model. Table 1 gives the accuracy of Position Rank in different rounds.

Table 1. Accuracy: Position Rank.

Stage	Accuracy of the Position Rank (%)	Lift Rate (%)
1st- unsupervised labeling	33%	–
1st- artificial error correction	Add adjectives and verbs	–
2nd- unsupervised labeling	35%	+2%
2nd- artificial error correction	Add pronouns	–
3rd- unsupervised labeling	40%	+5%

We can see that the first round accuracy is 33%. At the first artificial error correction stage, we summarized adjectives and verbs are important factors for extracting, and added related words and rules into dictionary and rules library. The second round accuracy is 35%, and have an increase 2% from the first round. At the second artificial error correction stage, we add pronouns into rules library. The third round accuracy is 40%, and have an increase 5% from the second round, 7% from the first round. It reveals that iterative annotation is effective for unsupervised method.

5.2   Accuracy: Position Rank, TF - IDF, vs. Text Rank

In this section, we compared the accuracies of the three common unsupervised labeling methods: Position Rank, TF- IDF, and Text Rank. The three-round unsupervised labeling and two-round error corrections stays the same in all the three methods. The experimental results reveal that Position Rank obtain the best score after iterative annotation. Sometimes, the other two methods can get better keyphrases, which deserves in-depth research and discussion. Table 2 gives the accuracy rates of these different methods.

Table 2. Accuracy: Position Rank, TF - IDF, vs. Text Rank.

Stage	Position Rank	TF-IDF	Text Rank
1st-unsupervised labeling	33%	34%	27%
2nd-unsupervised labeling	35%	37%	31%
3rd-unsupervised labeling	40%	35%	33%

6 Future Work and Conclusion

6.1 Future Work

There are several directions for further research and practices. First and foremost, the corpus needs the continuous expansion of the volume and quality. We will prepare several other medium-sized, niche CAEW as the original data sources, and process more forum blocks than the current method does. Secondly, the data preprocessing step needs further optimization. The website reviews may include many misspelled and misused words. If the error correction link is added in the pre-processing stage, we will further improve the labeling accuracy. Thirdly, the corpus asks for an infusion of the network jargons dictionary. The network jargons appeared with the development of the Internet in recent times and were used in high frequency by the net surfers. If the network jargons dictionary can be added into the unsupervised model, it can identify the users' comment semantics more effectively. Finally, it suggests that the corpus should be checked and corrected on the labeled data in a one-by-one basis manually to build a basic high-quality keyphrases corpus, but this may require a huge amount of manpower and time.

6.2 Conclusion

This paper proposes a semi-unsupervised scheme by the aid of Position Rank to construct the desired keyphrases corpus from CAEW reviews and get a relatively satisfactory result. The iterative annotation process consists of three-round unsupervised labeling and two-round manual error correction, which is proven to be an effective strategy. When correcting the error keyphrases, we summary new words and new rules which the original dictionary and rules library do not have. And we also add them into the unsupervised model to improve the accuracy.

At the same time, we tested another two methods: TF-IDF and Text Rank for comparison, and found that Position Rank is better at extracting keyphrases from users' reviews.

Acknowledgments. The work of this paper is funded by the National Key Technologies R&D Program (2017YFB1002502) and the project of Beijing Advanced Education Center for Future Education (BJAICFE2016IR-003). We would also like to thank Mr. Sheng Zhang for his help on correcting the paper and thank all of the teammates for their hard work.

References

1. National Bureau of Statistics Homepage. http://www.stats.gov.cn/tjsj/zxfb/201902/t20190228_1651265.html
2. Shi, J.: What will happen to the low-growth car market? New Energy Vehicle News. Accessed 13 May 2019
3. Thousand City Number zhi Guo Dengli: Opening up a new model for automobiles e-commerce. Internet Econ. **05**, 102–103 (2019)
4. Yu, S., Su, Z., Zhu, X.: The comprehensive knowledge base and its prospect. Chin. J. Inform. **25**(06), 12–20 (2011)
5. Xiaojun, Z., Shuzi, Y., Anfa, Z., et al.: A knowledge-based diagnosis system for automobile engines. In: Proceedings of the 1988 IEEE International Conference on Systems, Man, and Cybernetics. IEEE (1988)
6. Homepage. http://www.e-ging.com/article20150602064627/. Accessed 12 June 2019
7. Guo, X., Wang, S., Li, D.: Building a knowledge base for automotive evaluation based on fuzzy concepts. In: Proceedings of the CCIR 2015 (2015)
8. Wang, S., Guo, X., Zhang, S.: The construction and application of automobile evaluation knowledge based on fuzzy association rules. J. Shanxi Univ. (Nat. Sci.) **39**(03), 423–428 (2016)
9. Feng, S., Wang, S.: The construction of automated ontology knowledge base for perspective mining. Comput. Appl. Softw. **28**(05), 45–47+105 (2011)
10. Yu, S., Zhu, X., Duan, H.: The guideline for segmentation and part-of-speech tagging on very large scale corpus of contemporary Chinese. J. Chin. Inf. Process. **14**(6), 58–64 (2000)
11. Khandelwal, U., He, H., Qi, P., et al.: Sharp nearby, fuzzy far away: how neural language models use context (2018)
12. Frank, E., Paynter, G.W., Witten, I.H., Gutwin, C., Nevill-Manning, C.G.: Domain-specific keyphrase extraction. In: Proceedings of the 16th International Joint Conference on Artificial Intelligence, pp. 668–673 (1999)
13. Hasan, K.S., Ng, V.: Automatic keyphrase extraction: a survey of the state of the art. In: Proceedings of the 27th International Conference on Computational Linguistics, pp. 1262–1273 (2014)
14. Hasan, K.S., Ng, V.: Conundrums in unsupervised keyphrase extraction: making sense of the state-of-the-art. In: Proceedings of the 23rd International Conference on Computational Linguistics, pp. 365–373 (2010)
15. Zhao, J.S., Zhu, Q.M., Zhou, G.D., Zhang, L.: Review of research in automatic keyphrases extraction. Ruan Jian Xue Bao/J. Softw. **28**(9), 2431–2449 (2017). (in Chinese). http://www.jos.org.cn/1000-9825/5301.htm
16. Chang, Y.C., Zhang, Y.X., Wang, H., Wan, H.Y., Xiao, C.J.: Features oriented survey of state-of-the-art keyphrase extraction algorithms. Ruan Jian Xue Bao/J. Softw. **29**(7), 2046–2070 (2018). (in Chinese). http://www.jos.org.cn/1000-9825/5538.htm
17. Florescu, C., Caragea, C.: A position-biased pageRank algorithm for keyphrase extraction. In: Proceedings of the AAAI. AAAI Press, Palo Alto, pp. 4923–4924 (2017)
18. Brin, S., Page, L.: The anatomy of a large-scale hypertextual web search engine. Comput. Netw. **30**(1–7), 107–117 (1998)
19. Mihalcea, R., Tarau, P.: TextRank: bringing order into texts. In: Proceedings Conference on Empirical Methods in Natural Language Processing (2004)
20. Wan, X., Xiao, J.: Single document keyphrase extraction using neighborhood knowledge. In Proceedings of the 23rd AAAI Conference on Artificial Intelligence, pp. 855–860 (2008b)

21. Grineva, M.P., Grinev, M.N., Lizorkin, D.: Extracting key terms from noisy and multitheme documents. In: Proceedings of the 18th International Conference on World Wide Web, WWW 2009, Madrid, Spain, April 20–24, 2009. DBLP (2009)
22. Liu, Z., Li, P., Zheng, Y., et al.: Clustering to find exemplar terms for keyphrase extraction. In: Proceedings of the 2009 Conference on Empirical Methods in Natural Language Processing, EMNLP 2009, 6–7 August 2009, Singapore, A meeting of SIGDAT, a Special Interest Group of the ACL (2009)

Social Computing and Sentiment Analysis

Contextualized Word Representations with Effective Attention for Aspect-Based Sentiment Analysis

Zixuan Cao[1], Yongmei Zhou[1,2(✉)], Aimin Yang[1,3], and Jiahui Fu[3]

[1] School of Information Science and Technology, School of Cyber Security,
Guangdong University of Foreign Studies, Guangzhou, China
niketim@163.com, yongmeizhou@163.com, amyang18@qq.com
[2] Eastern Language Processing Center, Guangdong University of Foreign Studies,
Guangzhou, China
[3] School of Business, Guangdong University of Foreign Studies, Guangzhou, China
chaihui_fu@163.com

Abstract. Aspect-based sentiment analysis (ABSA) aims at identifying sentiment polarities towards aspect in a sentence. Attention mechanism has played an important role in previous state-of-the-art neural models. However, existing attention mechanisms proposed for aspect based sentiment classification mostly focus on identifying the sentiment words, without considering the relevance of such words with respect to the given aspects in the sentence. To solve this problem, we propose a new architecture, self-attention with co-attention (SACA) for aspect-based sentiment analysis. Self-attention is capable of conducting direct connections between arbitrary two words in context from a global perspective, while co-attention can capture the word-level interaction between aspect and context. Moreover, previous works simply averaged aspect vector to learn the attention weights on the context words, which may bring information loss if the aspect has multiple words. To address the problem, we employ the pre-trained contextual word embeddings and character-level word embeddings as word representation. We evaluate the proposed approach on three datasets, experimental results demonstrate that our model outperforms the state-of-the-art on all three datasets.

Keywords: Aspect-based sentiment analysis · Self attention · Co-attention

1 Introduction

Aspect-based sentiment classification is an important task in fine-grained sentiment analysis. The goal of ABSA is to predict the sentiment polarity of the sentence for the given aspect. For example, in the sentence *"The **food** is usually good but it certainly isn't a relaxing **place** to go."*, the user mentions two targets *"**food**"* and *"**place**"*, and expresses positive sentiment over the target *"food"*,

© Springer Nature Switzerland AG 2019
M. Sun et al. (Eds.): CCL 2019, LNAI 11856, pp. 467–478, 2019.
https://doi.org/10.1007/978-3-030-32381-3_38

but negative sentiment over *"place"*. Compared to sentence or document level sentiment analysis, the challenge of aspect level sentiment analysis is to differentiate the emotions of different targets.

Traditional methods [1,2] for ABSA mainly focus on feature engineering (such as bag-of-words and sentiment lexicons) to train a classifier for sentiment classification. However, traditional methods are mostly based on manual work, which requires a lot of time and manpower. Therefore, many neural network-based models have been proposed in recent years. Recurrent Neural Networks (RNNs) with attention mechanism, firstly proposed in machine translation [3], has been successfully used in many NLP tasks. In the task of ABSA, many works [4–7] employ attention mechanism to measure the semantic relatedness between context word and the target. However, many approaches [6,8] simply average aspect vectors to learn the attention weights on the context words. This may work fine for targets that only contain one word but may fail to capture the semantics of more complex expressions. For example, we cannot obtain the representation for *"hot pot"* by averaging the word vectors of *"hot"* and *"pot"*. *"hot"* would be close to words like *"temperature"* and *"pot"* would be close to words like *"cooking tools"*. The averaged word vector could be distant from the actual vector for *"hot pot"*.

To address this problem, we first use the contextualized word embedding named ELMo [9]. The traditional word embeddings, such as Word2vec [14], Glove [15], only have one representation per word, and therefore cannot capture how the meaning of each word can change based on surrounding context. However, ELMo analyses words within the context that they are used and also allowing the model to form representations of out-of-vocabulary (OOV) words. On this basis we further employ a co-attention mechanism to characterize the word-level interactions between aspect and context words.

Though bidirectional recurrent neural networks can model long distance dependencies, it cannot conduct direct connections between arbitrary two words. To better exploit the global dependencies of the sequential input (i.e., context and aspect), we employ the self-attention mechanism, which has been introduced to machine translation by Vaswani [10], and it is very expressive and flexible for modeling long-range and local dependencies.

In this paper, we introduce a novel neural network named **Self-Attention** with **Co-Attention** (SACA) for aspect-term sentiment analysis. It consists of a self-attention blocks to better exploit the global and local dependencies, and an aspect-context co-attention block to attend target and textual information for improving sentence representation. To evaluate the proposed approach, we conduct experiments on three datasets: SemEval 2014 dataset, containing reviews of restaurant domain and laptop domain, the third one is a tweet collection. The experimental results demonstrate the effectiveness of our proposed model.

2   Related Work

Most of the early methods adopted supervised learning methods with extensive manually designed features such as sentiment lexicon, n-grams, and dependency information, then training a sentiment classifier [1,2,11]. Kiritchenko et al. [1] proposed to use SVM based on n-gram features. Vo and Zhang [11] used pooling functions to extract features from sentiment-specific word embeddings and sentiment lexicons. However, these methods are labor-intensive and they failed to model the semantic relatedness between a target and its context information.

With the advances of deep learning methods, various neural models [6,12,13] have been proposed for automatically encoding sentence features as continuous and low-dimensional vectors without feature engineering. Tang et al. [12] proposed the target-dependent LSTM (TD-LSTM) and target connection LSTM (TC-LSTM) to model the interaction between target and the whole sentence. The RNN based models have achieved promising results, but they do not take into account the relatedness between the context words and the given aspect. To solve this problem, attention based neural methods [5,8,13] have been successfully applied to the ABSA problem due to their ability to explicitly capture the importance of context words. Tang et al. [8] computed the sentence representation by stacking multiple layer of attention. Ma et al. [5] proposed a bidirectional attention model IAN to interactively learning attention in the context and target, and generate the representation for target and context separately. More recently, Fan et al. [13] used the fine-grained and coarse-grained attention mechanisms to capture the interaction between aspect and context.

3 Model

3.1 Task Definition

Given a sentence $s = (w_1, w_2, ..., w_n)$ consisting of n words, and an aspect occurring in the sentence $q = (a_1, a_2, ..., a_m)$ consisting of a subsequence of m continuous words from s. Aspect-based sentiment classification aims to determine sentiment polarity of the sentence s towards each aspect q.

We present the overall architecture of the proposed **Self-Attention and Co-Attention (SACA)** in Fig. 1. It consists of the word representation layer, the contextual layer, the attention layer and output layer.

3.2 Word Representation Layer

Our model encodes words of an input sentence or aspect as a combination of three different types of embeddings, which can be listed as follows:

Pre-trained Word Embeddings: We use the pre-trained Glove embedding [15] for initialization and keep fixed during the training process.

Pre-trained Contextual Word Embeddings: We use pre-trained ELMo [9] embeddings. These representations are extracted from the hidden states of a bidirectional language model. ELMo embeddings have been shown to give state-of-the-art results in many NLP tasks.

Character-Level Word Embeddings: We use character-level word embeddings to extract character-level features, which have been shown to be helpful to deal with out-of-vocab (OOV) tokens.

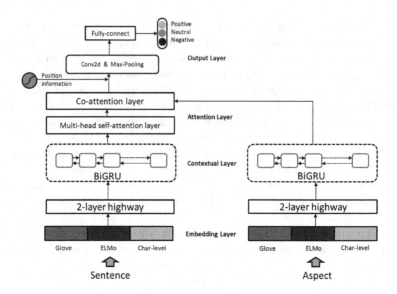

Fig. 1. The architecture of the proposed model.

Formally, given a sentence s, we suppose the Glove, ELMo and character-level word representations of the sentence are $W \in R^{n*d_w}$, $E \in R^{n*d_e}$ and $C \in R^{n*r*d_c}$, respectively. Where n denotes the sentence length, r denotes the word length, d_w, d_e and d_c represent the Glove embedding dimension, ELMo embedding dimension and character-level embedding dimension, respectively. Specifically, we firstly use a convolutional neural network (CNN) to encode the character C, and then concatenate the output of convolutional feature with Glove embedding and ELMo embedding to form the final word representations for the review sentence $e_s \in R^{n*|e|}$ and aspect $e_q \in R^{m*|e|}$, where n, m are the length of the review sentence and aspect, $|e|$ is the final embedding size (including all three components).

$$e = Concat(W, E, Conv2D(C)) \tag{1}$$

After the concatenation of three embedding component, we further feed word embeddings into a two-layer highway network [21] with tanh output activation.

$$\tilde{e} = \tanh(2 \sim highway(e)) \tag{2}$$

3.3 Contextual Layer

We employ a bidirectional recurrent neural network (RNN) as encoder to sequentially process each word in s and q, we chose to use Gated Recurrent Unit (GRU) [16] in our experiment since it performs similarity to LSTM but is computationally cheaper.

$$h_{s_i} = BiGRU(h_{s_{i+1}}, h_{s_{i-1}}, \tilde{e}_{s_i}) \qquad h_{q_j} = BiGRU(h_{q_{j+1}}, h_{q_{j-1}}, \tilde{e}_{q_j}) \tag{3}$$

We employ the BiGRU separately and get the context hidden output $H_s = (h_{s_1}, h_{s_2}, ..., h_{s_n})$ and the aspect hidden output $H_q = (h_{q_1}, h_{q_2}, ..., h_{q_m})$.

3.4   Attention Layer

The attention layer includes three sub-layers: (1) self-attention layer; (2) co-attention layer; (3) position-aware attention layer.

Self-attention Layer. We adopt the multi-head self-attention mechanism [10] to capture long range dependencies and inner structure of the sequential input. We regard Q, K and V as query matrix key matrix and value matrix, respectively. The scaled dot product attention can be described as follows:

$$Att(Q, K, V) = softmax(\frac{QK^T}{\sqrt{d_k}})V \tag{4}$$

Where n is the length of the sequence, and d_k denotes feature dimension.

In order to attend the information from different representation subspaces, multi-head self-attention concatenates the output of all scale l dot product attention models and then project concatenated feature to a fix dimensional feature.

$$head_i = Att(H_s W_i^Q, H_s W_i^K, H_s W_i^V) \tag{5}$$

$$\tilde{H} = (head_1 \oplus ... \oplus head_l)W_o \tag{6}$$

Where $H_s - (h_{s_1}, h_{s_2}, ..., h_{s_n})$ denotes the context output of BiGRU, n is the length of context sequence, W_i and W_o are trainable projection parameters. In order to avoid the loss of information in multi-head self-attention operations, we add a residual connection to $\tilde{H}$ and then apply a layer normalization:

$$O = LayerNorm(\tilde{H} + H_s) \tag{7}$$

Co-attention Layer. Considering the example *"Straight-forward, no surprises, very decent Japanese food"*, the word *"food"* in aspect *"Japanese food"* should have more effect on the context words compared with word *"Japanese"*. Accordingly, the context words should pay more attention on *"food"* instead of *"Japanese"*. To solve this problem, we employ co-attention mechanism that attends to the review sentence and aspect simultaneously.

Formally, we first construct an alignment matrix:
$L_{ij} = W_l([O; H_q; O - H_q; O * H_q])$, where L_{ij} denotes the similarity between i-th context word and j-th aspect word, O is the output of the self-attention layer, H_q is the aspect hidden output, [;] denotes the concatenation operator and * denotes the element-wise multiplication. The alignment matrix is normalized row-wise to produce the context-to-aspect attention weight A_q, and normalized column-wise to produce the aspect-to-context attention A_s across the aspect for each word in the context:

$$A_q = softmax(L) \qquad A_s = softmax(L^T) \tag{8}$$

And then we apply the computed attention map to the A_q to the aspect feature H_q to obtain the attended context feature $C_s = A_q H_q$, we also compute the attended aspect feature $C_q = A_s H_s$. Finally, we define co-attention representation of the aspect and context: $C = [C_s; (A_s^T C_q)]$.

In order to capture all of the information and highlight the significant features, we use a Bi-GRU to get the fusion of temporal information:

$$u_t = BiGRU(u_{t-1}, u_{t+1}, C_i) \tag{9}$$

We define $U = [u_1, ..., u_n]$, which is the aspect-aware sentence representation.

Position-Aware Attention Layer. Following an important observation found in [6,12] that sentiment words towards the aspect is more likely to be expressed near the aspect. For example, "*service*" in "*The price is reasonable although the service is poor.*" may be associated with opinion word "*poor*". Specifically, we calculate the target position weight between the i-th context and the aspect term as follows:

$$p_i = \begin{cases} 1 - \frac{m_0 - i}{n}, & i < m_0 \\ 0, & m_0 \leq i \leq m_0 + m \\ 1 - \frac{i - (m_0 + m)}{n}, & i > m_0 + m \end{cases} \tag{10}$$

Where m_0 is the index of the first aspect word, n and m are the length of sentence and aspect, respectively. We use the position weight to attend the output of the previous layer: $\hat{u}_i = u_i * p_i$. Obviously, the i-th context word closer to a aspect term with a large position weight. To capture the most informative features, we feed the weight $\hat{u}$ to the convolutional layer with max pooling to generate the feature map:

$$c_i = ReLU(w_{conv}^T \hat{u}_{i:i+s-1} + b_{conv}) \tag{11}$$

$$v_{max} = MaxPooling(c_1, c_2, ..., c_n) \tag{12}$$

Where s is the kernel size, w_{conv} and b_{conv} are trainable parameters of convolution layer.

3.5 Output Layer

Finally, we fed v_{max} to a *softmax* layer for determining the aspect sentiment polarity:

$$p = softmax(W_p v_{max} + b_p) \tag{13}$$

Where p is the probability distribution for the polarity of aspect sentiment, W_p and b_p are learnable parameters.

Loss Function. To train our model, we use traditional categorical cross entropy loss with L_2-regularizer as loss function:

$$loss = -\frac{1}{N} \sum_{i=1}^{N} \sum_{j=1}^{C} y_{i,j} \log p_j + \lambda \|\theta\|^2 \tag{14}$$

Where N is the number of samples, C is the number of sentiment categories, λ is the regularization weight, θ is the set of trainable parameters, $y_{i,j}$ is a one hot class vector for the j-th class and p_j is the predicted probability for the j-th class.

4   Experiments

We conduct experiments on three benchmark datasets, as shown in Table 1. The first two are from SemEval 2014 Task 4 [17], containing customer reviews on restaurants and laptop. The third one is a collection of tweets, built by Dong [18]. Each review has one or more aspect with their corresponding polarities. The polarity of an aspect can be *Positive*, *Netural* and *Negative*. Evaluation metrics are Accuracy and Macro-F1, which is widely used in previous works. The main hyper-parameters of our model are listed in Table 2. Adam [19] is adopted as the optimizer with an initial learning rate of 0.0005. In order to prevent overfitting, we apply dropout of 0.3 to all the representation layers. The model is build on Pytorch platform.

Table 1. The statistics of the three datasets

Dataset	#Positive		#Netural		#Negative	
	Train	Test	Train	Test	Train	Test
Restaurant	2164	728	637	196	807	196
Laptop	994	341	464	169	870	128
Twitter	1561	173	3127	346	1560	173

Table 2. Hyper-parameter settings

Symbol	Descriptions	Size
n	Sentence max length	64
m	Aspect max length	5
d_w	Glove word embedding size	300
d_e	ELMo embedding size	1024
d_c	Character embedding size	16
d_h	Bi-GRU hidden size	400
l	Heads of self-attention	8

4.1   Model Comparisons

We compare our model against the following baseline methods:

(1) **Feature + SVM** [1]: The classic SVM model using a series of manual features.
(2) **LSTM** [4]: An LSTM network is built on top of word embeddings, the mean of all the hidden outputs from LSTM is taken as the sentence representation.
(3) **TD-LSTM** [12]: It uses a forward LSTM and a backward LSTM to model context before and after the aspect.
(4) **AE-LSTM and ATAE-LSTM** [4]: AE-LSTM is a simple LSTM model incorporating the target embedding as input. ATAE-LSTM is developed based on AE-LSTM and uses attention mechanism to generate the final representation from hidden states.
(5) **MemNet** [8]: It applies multi-hop attention on the memory stacked by input word embeddings and predicts sentiment based on the top context representation.
(6) **IAN** [5]: It adopts two LSTMs with attention mechanism to generate representations of aspect and context separately by interactive learning.
(7) **IARM** [20]: It adopts a GRU and attention mechanism to generate the aspect-aware sentence representations, and also incorporate the neighboring aspects related information into the sentiment classification of the target aspect using memory networks.
(8) **MGAN** [13]: It adopts coarse-grained and fine-grained attention mechanism for sentence representation.

Table 3. The performance comparisons of different methods on three datasets, where the results of baseline methods are retrieved from the original papers. "-" means this result is not available. The best performances are marked in bold.

Methods	Laptop		Restaurant		Twitter	
	Acc.	Macro-F1	Acc.	Macro-F1	Acc.	Macro-F1
Feature+SVM	0.7049	-	0.8016	-	0.6340	0.6330
LSTM	0.6679	0.6402	0.7523	0.6421	-	-
TD-LSTM	0.6825	0.6596	0.7537	0.6451	0.6662	0.6401
AE-LSTM	0.6890	-	0.7660	-	-	-
ATAE-LSTM	0.6870	-	0.7720	-	-	-
MemNet	0.7033	0.6409	0.7816	0.6583	0.6850	0.6691
IAN	0.7210	-	0.7860	-	-	-
IARM	0.7380	-	0.8000	-	-	-
MGAN	0.7539	0.7247	0.8125	0.7194	0.7254	0.7081
SACA	**0.7633**	**0.7292**	**0.8205**	**0.7310**	**0.7268**	**0.7103**

4.2   Main Results

From Table 3, we can have the following observations: (1) Compared with all other neural baselines, our model achieves significant improvement on both accuracy and macro-F1 scores for three datasets. (2) Feature-based SVM is still a strong baseline, which demonstrates the importance of high quality features for aspect based sentiment analysis. Our approach can achieve competitive results without relying on so many manually-designed features. (3) The attention based models (ATAE-LSTM, MemNet, IAN, IARM, MGAN and SACA) perform better than non-attention based models (LSTM, TD-LSTM and AT-LSTM), one main reason maybe the attention mechanism can make the model notices important parts of a sentence for a given aspect. (4) IARM achieves slightly better results with the previous RNN-based methods, which employ memory network to model the dependency of the target aspect with other aspects in the sentence. (5) MGAN achieves the best performances among the baselines. MGAN does not only use the dot attention mechanism and deep bidirectional LSTM, but also use a fine-grained attention mechanism to capture the word-level interaction between the aspect and context.

4.3   Ablation Study

To investigate the effectiveness of each component of our model, we perform comparison between the full model and its ablations. The result is shown in Table 4.

Table 4. Ablation experiments on three datasets

Models	Laptop		Restaurant		Twitter	
	Acc.	Macro-F1	Acc.	Macro-F1	Acc.	Macro-F1
SACA	**0.7633**	**0.7292**	**0.8205**	**0.7310**	**0.7268**	**0.7103**
- ELMo	0.7555	0.7148	0.8041	0.7124	0.7138	0.6984
- Char.	0.7601	0.7204	0.8189	0.7297	0.7221	0.7017
- Self-Attn.	0.7534	0.7017	0.8091	0.7188	0.7119	0.6914
- Co-Attn.	0.7392	0.6911	0.7828	0.7013	0.7064	0.6768
- Position	0.7609	0.7257	0.8123	0.7299	0.7108	0.6891

Effect of the Embeddings. We perform ablation study on words and character embeddings. As expected, contextualized ELMo embeddings have a noticeable effect on each metric. Removing ELMo leads to ~1 Acc. point drop on each dataset. Furthermore, after removing the character-level word embeddings, the performance degraded, slightly. It shows that the integration of multi-level information (*e.g.*, word and character level) is crucial for good performance.

Why Self-attention? We conduct experiment to study the effect of self-attention mechanism. Firstly, we remove the self-attention and directly use hidden output of BiGRU to conduct experiment. The experimental result is shown in Table 4. After removing the self-attention layer, we find the performance degrade to 75.34 on Laptop dataset, 80.91 on Restaurant dataset and 71.19 on Twitter dataset, which verifies that the self-attention mechanism is effective for ABSA task.

Why Co-attention? We compare the performance of SACA and SACA without co-attention layer in Table 4, we see that co-attention outperforms no co-attention method on three datasets. It shows that the interaction information between sentence and aspect is crucial for good performance.

Effect of Position Information. As for the position information, we removed position-aware attention and convolutional layer, thus retaining only max-pooling. It is worth noting that the performance is not much worse than SACA, especially on Laptop dataset and Restaurant dataset, which shows that the ability of position-aware attention is limited. We argue that it is because the importance of a context word is not only dependent on word order, but also on the information of context and aspect.

4.4   Case Study

We take a sentence in restaurant dataset as example for illustrating the effectiveness of our proposed model. We visualize the weights of self-attention layer and co-attention layer. Figure 2 shows the visualization results of the attention weights from self-attention and co-attention layer. The sentence in Fig. 2 is *"Our waiter was friendly and it is a shame that he didn't have a supportive staff to work with"*. It has two aspects *"waiter"* and *"staff"*, whose sentiment polarities are positive and negative, respectively. For the self-attention, we use the aspect *"waiter"* as query, visualize the scaled product attention weight. From the top bar, we can find that the aspect *"waiter"* has more attention weight

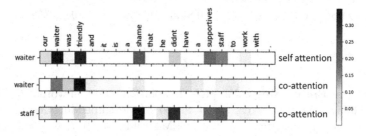

Fig. 2. Attention visualizations of an example sentence. The color depth indicates the importance degree of a word. (Color figure online)

on *"friendly"*, *"supportive"* and *"staff"*. This phenomenon shows that the self-attention can not only learn the important context words for the aspect, but also can extract interactions between words, especially focus on neighboring words.

In addition, we evaluate the effect of co-attention mechanism and visualize the aspect to context attention weights. From the bottom two bars, we can observe that the co-attention can enforce the model to pay more attentions on the important words with respect to the aspect. For example, the words *"shame"*, *"didn't"*, and *"supportive"* which are the most relevant to sentiment polarity of *"staff"* has higher attention weights compared with other words.

5   Conclusion

In this paper, we present a new framework, term **SACA**, for aspect-based sentiment analysis. We re-examine the drawbacks of word representations for ABSA, to solve these issues, contextualized word representation and character-level word embeddings are integrated to word representations. Moreover, three novel attention mechanisms, namely self-attention, co-attention and position-aware attention mechanism have been introduced to our model. The self-attention captures the important information from a global perspective by considering the information of entire sentence. Co-attention mechanism captures the word-level interaction between aspect and context. Experimental results demonstrate the effectiveness of our approach on three datasets. The ablation studies show the efficacy of different modules.

Acknowledgements. This work was supported by the Ministry of Education of Humanities and Social Science project (No. 19YJAZH128), and Guangdong Graduate Education Innovation project (No. 2018JGXM41).

References

1. Kiritchenko, S., Zhu, X., Cherry, C., Mohammad, S., Polosukhin, I.: NRC-Canada-2014: detecting aspects and sentiment in customer reviews. In: Proceedings of the 8th International Workshop on Semantic Evaluation (SemEval 2014) (2014)
2. Wagner, J., et al.: DCU: aspect-based polarity classification for SemEval task 4. In: Proceedings of the 8th International Workshop on Semantic Evaluation (SemEval 2014) (2014)
3. Bahdanau, D., Cho, K., Bengio, Y.: Neural machine translation by jointly learning to align and translate. arXiv preprint. arXiv:1409.0473 (2014)
4. Wang, Y., Huang, M., Zhao, L.: Attention-based LSTM for aspect-level sentiment classification. In: Proceedings of EMNLP (2016)
5. Ma, D., Li, S., Zhang, X., Wang, H.: Interactive attention networks for aspect-level sentiment classification. In: Proceedings of IJCAI (2017)
6. Chen, P., Sun, Z., Bing, L., Yang, W.: Recurrent attention network on memory for aspect sentiment analysis. In: Proceedings of EMNLP (2017)
7. He, R., Lee, W.S., Ng, H.T., Dahlmeier, D.: Effective attention modeling for aspect-level sentiment classification. In: Proceedings of the 27th International Conference on Computational Linguistics (2018)

8. Tang, D., Qin, B., Liu, T.: Aspect level sentiment classification with deep memory network. In: Proceedings of EMNLP (2016b)
9. Peters, M.E., et al.: Deep contextualized word representations. In: Proceedings of NAACL (2018)
10. Vaswani, A., et al.: Attention is all you need. In: Advances in Neural Information Processing Systems (2017)
11. Vo, D.T., Zhang, Y.: Target-dependent twitter sentiment classification with rich automatic features. In: IJCAI (2015)
12. Tang, D., Qin, B., Feng, X., Liu, T.: Effective LSTMs for target-dependent sentiment classification. In: COLING (2016a)
13. Fan, F., Feng, Y., Zhao, D.: Multi-grained attention network for aspect-level sentiment classification. In: EMNLP (2018)
14. Mikolov, T., Sutskever, I., Chen, K., Corrado, G.S., Dean, J.: Distributed representations of words and phrases and their compositionality. In: Advances in neural information processing systems (2013)
15. Pennington, J., Socher, R., Manning, C.: Glove: global vectors for word representation. In: EMNLP (2014)
16. Chung, J., Gulcehre, C., Cho, K., Bengio, Y.: Empirical evaluation of gated recurrent neural networks on sequence modeling. arXiv preprint. arXiv:1412.3555 (2014)
17. Pontiki, M., Galanis, D., Papageorgiou, H., Androutsopoulos, I., Manandhar, S.: SemEval-2014 task 4: aspect based sentiment analysis. In: Proceedings of the 10th International Workshop on Semantic Evaluation (SemEval-2016) (2014)
18. Dong, L., Wei, F., Tan, C., Tang, D., Zhou, M., Xu, K.: Adaptive recursive neural network for target-dependent twitter sentiment classification. In: ACL (2014)
19. Kingma, D.P., Ba, J.: Adam: a method for stochastic optimization. arXiv preprint. arXiv:1412.6980 (2014)
20. Majumder, N., Poria, S., Gelbukh, A., Akhtar, M.S., Cambria, E., Ekbal, A.: IARM: inter-aspect relation modeling with memory networks in aspect-based sentiment analysis. In: EMNLP (2018)
21. Srivastava, R.K., Greff, K., Schmidhuber, J.: Highway network. arXiv preprint. arXiv:1505.00387 (2015)

Multi-label Aspect Classification on Question-Answering Text with Contextualized Attention-Based Neural Network

Hanqian Wu[1,2]([✉]), Shangbin Zhang[1,2], Jingjing Wang[3], Mumu Liu[1,2], and Shoushan Li[3]

[1] School of Computer Science and Engineering, Southeast University, Nanjing, China
hanqian@seu.edu.cn, ternencewind@outlook.com, liudoublemu@163.com
[2] Key Laboratory of Computer Network and Information Integration of Ministry of Education, Southeast University, Nanjing, China
[3] NLP Lab, School of Computer Science and Technology, Soochow University, Suzhou, China
djingwang@gmail.com, lishoushan@suda.edu.cn

Abstract. In the e-commerce websites, such as Taobao and Amazon, interactive question-answering (QA) style reviews usually carry rich aspect information of products. To well automatically analyze the aspect information inside QA style reviews, it's worthwhile to perform aspect classification on them. Unfortunately, until now, there are few papers that focus on performing aspect classification on the QA style reviews. For short, we referred to this novel task as QA aspect classification (QA-AC). In this study, we model this task as a multi-label classification problem where each QA style review is explicitly mapped to multiple aspect categories instead of only one aspect category. To solve this issue, we propose a contextualized attention-based neural network approach to capture both the contextual information and the QA matching information inside QA style reviews for the task of QA-AC. Specifically, we first propose two aggregating strategies to integrate multi-layer contextualized word embeddings of the pre-trained language representation model (i.e., BERT) so as to capture contextual information. Second, we propose a bidirectional attention layer to capture the QA matching information. Experimental results demonstrate the effectiveness of our approach to QA-AC.

Keywords: Aspect classification · Question Answering · Pre-trained language model · Bidirectional attention

1 Introduction

In recent years, a new form of the product review, called Question-Answering (QA) style review, has emerged on some e-commerce platforms, including

© Springer Nature Switzerland AG 2019
M. Sun et al. (Eds.): CCL 2019, LNAI 11856, pp. 479–491, 2019.
https://doi.org/10.1007/978-3-030-32381-3_39

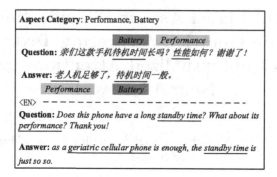

Fig. 1. A translated example of QA style reviews from an e-commerce website.

Taobao, Yelp, and Amazon. As shown in Fig. 1, unlike traditional product reviews, a QA style review consists of a question sentence and an answer sentence. Thus, we also regard QA style reviews as QA pairs. QA aspect classification (QA-AC) is an essential fundamental task in sentiment analysis for e-commerce reviews, which aims to identify the aspect set of the product contained in a QA style review. The customary e-commerce product review is mainly written for subjectively and generally commenting after the transaction completed, which may not answer questions from other consumers. Thus, QA style reviews become widespread and proliferating. Notably, due to the nature of the conversation, questions often focus on some aspects of the product, where answers also aim at. Therefore, QA style reviews are more suitable for aspect-based sentiment analysis tasks, and aspect-based sentiment analysis research on QA style reviews is capturing increasing attention. Aspect classification is an essential sub-task of the aspect-based sentiment analysis task, and it can improve the performance of aspect-based sentiment analysis significantly. However, most of the relevant researches are based on traditional reviews, while few studies focus on the task of QA-AC. In general, the QA-AC exists the following three specific challenges.

First, instead of one individual aspect, the aspect categories of the sample shown in Fig. 1 include *Performance* and *Battery*. It is ubiquitous in QA style reviews that a single review involves more than one aspect. However, classifying a review into multiple aspect labels have more difficulties than classifying a review into a single label. In this study, we regard the QA-AC task as a multi-label classification problem and accommodate our model to this task.

Second, a QA style review is a short text, it's rather difficult to identify the multiple aspect categories inside it due to data sparseness problem. Conventional language representation models perform poorly in obtaining semantic information from short texts. In addition, informal expressions are widespread in QA style reviews, and existing Chinese words segmentation utilities could not correctly recognize some informal expressions. For instance, in Fig. 1, the phrase "老人机 (geriatric cellular phone)" means that this phone does not have many complex functions and excellent performance. After the word segmentation, the phrase will split to "老人 (elder people)" and "机 (machine)".

Obviously, according to this instance, if using word segmentation, it's difficult to capture the performance information of the phone. To address these problems, we adopt the latest outstanding pre-trained language representation model, i.e., BERT, which can effectively alleviate the data sparseness and polysemy problem by modeling the context semantic representations with large-scale external data.

Third, the QA style review has much contextual matching information between its question part and answer part. In QA style reviews, the matching information between question and answer plays a crucial role in QA-AC task. For example, the phrase " 待机时间 (standby time)" in the answer part has a strong correlation with the same phrase in question part than other words. Its correlation provides valuable clues to predict the aspect of this review to *Battery*. Therefore, we construct the bidirectional attention neural network model to detect the important degrees of different characters.

In this paper, with the best of our knowledge, we are the first to define the QA-AC task as a multi-label classification problem, which aims to identify multiple aspect categories inside a given QA style review. Furthermore, we propose a contextualized attention-based Network, i.e., the Bidirectional Attention Neural Network (BANN) based on BERT, for capturing both the contextual information and the QA matching information. In detail, we first adopt BERT as a contextual embedding to alleviate the data sparseness and polysemy problem. Specifically, we employ two strategies for aggregating multi-layer pre-trained semantic representations so as to capture contextual information, including averaging and weighted summation. Then, we encode QA matching information via a bidirectional QA matching attention. Finally, the empirical results demonstrate that our proposed model outperforms several state-of-the-art baselines by larger margins on QA style reviews.

2 Related Work

2.1 Aspect Extraction

Over the last decade, as a fine-grained sentiment analysis task, aspect-level sentiment classification captured enormous attention, especially in the field of reviews texts [1–3]. As a related task to the aspect-level sentiment classification, aspect extraction divide into two sub-tasks, aspect term extraction and aspect category classification, which is also called aspect classification. In the beginning, researchers proposed several rule-based methods and traditional machine learning methods for aspect extraction. For example, Rubtsova et al. [4] used the conditional random field method to extract the aspect term mentioned in the restaurants and automobiles texts. With the prosperity of deep learning, the neural networks based methods are widely employed in natural language processing tasks and remarkably adept in learning complicated feature representations automatically. Liu et al. [5] proposed a Convolutional Neural Network (CNN) model employed two types of pre-trained embeddings for aspect extraction, including general-purpose embedding and domain-specific embedding. However, they focus on general-purpose information and domain information in the embedding layer while ignoring the different importance between two embeddings. He et al. [6] proposed an attention-based model intending to discover coherent aspects and proved the efficiency of attention mechanism.

Different from the above, this paper explores the methods of QA-AC task. However, Wu et al. [7] are the first to conduct the research on QA-AC by impracticably assuming that a QA style review only contains a single aspect. In this paper, we first define the QA-AC task as a multi-label classification problem and design a bidirectional QA attention layer to capture the QA matching information.

2.2 Pre-trained Language Model

Recently, the language representation model has received considerable attention due to improving scores of many natural language processing (NLP) tasks [8, 9]. The pre-trained language model (PLM) is a type of deep language representation model, which is pre-trained on a large unlabelled text corpus. There are two existing strategies in the PLM. Embeddings from Language Models (ELMo) [10] is a feature-based pre-trained language model, and the Generative Pre-trained Transformer (OpenAI GPT) [11] utilizes a fine-tuning approach to apply pre-trained language representations to downstream tasks. Different from Word2vec, PLM can generate deep contextualized word representations. However, these existing PLMs cannot make good use of contextual information because they are unidirectional. Thus, Devlin et al. [12] proposed the Bidirectional Encoder Representations from Transformers (BERT). BERT addresses the mentioned unidirectional constraints and polysemy problem and obtains new state-of-the-art results on eleven NLP tasks. Thus, some researches leverage BERT to improve their models. Adhikari et al. [13] explored BERT on document classification. Their model achieves state-of-the-art across four popular datasets.

In our study, we utilize BERT to capture the contextual information in QA style reviews and leverage dominant context-dependent word representations to expand contextual information and alleviate the data sparseness and polysemy problem. To our knowledge, we are the first to leverage BERT on the QA-AC task.

3 Models

We model the QA-AC task as a multi-label classification task. For introducing our model explicitly, we formulate the task as follows:

Given a QA style review text $C = \{c_1, \cdots, c_n\}$, the target is to predict the aspect set $S = \{s_1, \cdots, s_k\}$ of C, where c_i denotes the i-th character in C which has n characters. As a QA pair, the text $C = (Q, A)$ can split into a question part and an answer part. Besides, there are k aspects that occur in the whole corpus. If the text C has the i-th aspect, its s_i should take the value of 1. As a multi-label problem, the set S may contain more than one aspect, and the value of elements in S which represent other aspects are zeros.

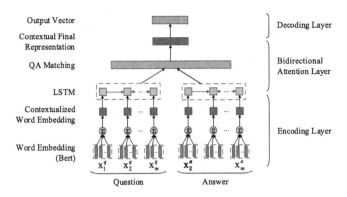

Fig. 2. The overall structure of our model.

3.1 Model Overview

Figure 2 depicts the illustration of our model architecture. In the macroscopic view, our model can decompose into three layers:

Encoding layer encodes the text C into a contextualized sentence vector which consists of word vectors. For obtaining the contextualized information in word embedding, we leverage the BERT contextual embeddings [12] to encode text C. Especially, we apply two strategies, averaging and weighted summation, to aggregate embeddings from twelve BERT transformer layers.

Bidirectional Attention layer captures the contextual matching information between question and answer by a bidirectional QA matching attention. Moreover, we conduct the Long Short-Term Memory (LSTM) to learn the order information hidden in the sentence. Hence, after the whole attention layer, we can get the contextual representation for each text.

Decoding layer compresses the representation vector's dimension and generates the output vector, which is the predicted aspect set of this text. Distinct from the multi-class classification task, we employ the sigmoid function as the activation function.

3.2 Encoding Layer

BERT Word Embedding: The first step of our model is to embed our input QA pair. BERT for Chinese is a character-level model, whose output is a vector with fixed dimensions. Firstly, after padding the question and answer to the same length m, we stitch them into a single sentence $C = \{c_1^q, c_2^q, \cdots, c_m^q, c_1^a, c_2^a, \cdots, c_m^a\}$. Then, we feed the BERT model with vector C directly. BERT generates 12 layers of hidden states for every token in total, especially, which can be all used to present words. After BERT word embedding, we get the vector $X = \{x_1^q, x_2^q, \cdots, x_q^m, x_1^a, x_2^a, \cdots, x_m^a\}$, and x_i represents $\{x_i^1, x_i^2, \cdots, x_i^L\}$, where x_i^j denotes the j-th layer representation of the i-th character.

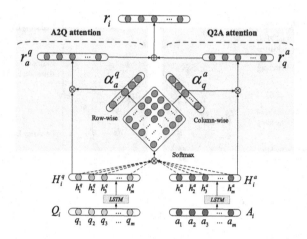

Fig. 3. The structure of bidirectional attention layer.

Contextualized Word Embedding: In the component of layers aggregating, we apply two strategies to aggregate embeddings from these twelve layers. The first strategy is averaging. We compute the mean value of each x_i to encode the i-th character. The other strategy is the weighted summation shown below:

$$q_i = \sum_{j=1}^{L} \omega_j x_{ij}^q, \quad a_i = \sum_{j=1}^{L} \omega_j x_{ij}^a \tag{1}$$

In detail, q_i denotes the word vector of i-th character in question and a_i represents the word vector of i-th character in answer. ω_j is the trainable weight parameter of the j-th layer vector. We compare two strategies in our experiment part, and the results show that the weighted summation outperforms averaging.

3.3 Bidirectional Attention Layer

Undoubtedly, there is sufficient QA information hidden in QA style reviews, which can contribute to the aspect classification task significantly. Figure 3 reveals the detailed structure of the bidirectional attention layer.

LSTM: For efficiently making use of sequential internal correlation, we employ two LSTM networks to capture the contextual order information in question and answer severally. Both question $Q_i = \{q_1, q_2, \cdots, q_m\}$ and answer $A_i = \{a_1, a_2, \cdots, a_m\}$ have m elements. Therefore, we collect entire m hidden states of LSTM, $H_i^q = \{h_1^q, h_2^q, \cdots, h_m^q\}$ and $H_i^a = \{h_1^a, h_2^a, \cdots, h_m^a\}$, where h_j^q denotes the j-th hidden state in question and h_j^a denotes the j-th hidden state in answer.

QA Matching: In this component, we mine the bidirectional QA matching information. Firstly, we calculate the matching matrix by $D = H_i^q (H_i^a)^\mathsf{T}$, which denotes the bidirectional pair-wise matching information. Then, we compute two directional attention matrixes, including Question-to-Answer attention (Q2A attention) and Answer-to-Question attention (A2Q attention).

- **A2Q Attention:** We employ a series of row-wise operations to obtain A2Q attention weight vector α_a^q by Eq. (2).

$$\alpha_{row} = \sum_{i=0}^{m} softmax_{row-wise}(D_i), \quad \alpha_a^q = softmax(\alpha_{row}) \tag{2}$$

where $\alpha_a^q \in \mathbb{R}^m$ is the A2Q attention vector which presents the importance of characters in question Q_i, α_{row} is the row-wise matching vector, and D_i is the i-th row in matrix D. To strengthen the features in each row, we compute the row-wise softmax before the summation operation in matrix D.

- **Q2A Attention:** Different from above, we employ some column-wise operations to obtain Q2A attention weight vector α_q^a by Eq. (3)

$$\alpha_{col} = \sum_{j=0}^{m} softmax_{column-wise}(D^\mathsf{T}{}_j), \quad \alpha_q^a = softmax(\alpha_{col}) \tag{3}$$

where $\alpha_q^a \in \mathbb{R}^m$ is the Q2A attention vector which presents the importance of characters in answer A_i, α_{col} is the column-wise matching vector, and $D^\mathsf{T}{}_j$ is the j-th column in D. Similarly, we compute the column-wise softmax before the column-wise summation operation in matrix D.

Contextual Final Representation: After bidirectional attention, we can get two directional sentence representations, $r_a^q \in \mathbb{R}^d$ and $r_q^a \in \mathbb{R}^d$, where d is the dimension of BERT word vector. We combine A2Q sentence vector r_a^q and Q2A sentence vector r_q^a to present the contextualized final representation as follows:

$$r_a^q = \alpha_a^q H_i^q, \ r_q^a = \alpha_q^a H_i^a, \ r_i = W_r(r_a^q \oplus r_q^a) + B_r \tag{4}$$

where $\oplus$ denotes the concatenate operator, $r_i \in \mathbb{R}^d$ is the contextualized final representation of C_i, $W_r \in \mathbb{R}^{2d \times d}$ is the weighted matrix, and B_r is the bias matrix. After compression, we can obtain a d-dim vector.

3.4 Decoding Layer

After transforming each text C into a contextual representation, we conduct a dense layer to generate the ultimate vector of aspects in C. The whole computational process of decoding layer is shown below:

$$out_i = sigmoid\,(W_l r_i + B_l) \tag{5}$$

where $out_i \in \mathbb{R}^k$ is the predicted vector of text C_i and k is the number of all aspects, W_l is an intermediate weight matrix, B_l is a bias matrix. Notably, different from the multi-class classification problem, we employ sigmoid as the final activation function to evaluate the possibility of each category respectively. In reality, for cooperating with the sigmoid activation function, we set the binary cross-entropy as our objective function, which we will detailedly expatiate later.

Table 1. Statistics of our QA dataset.

Aspect category	Number of all instances	The scale of multi-label instances
Quality	214	20.1%
Battery	303	23.1%
Performance	652	16.9%
Certified product	421	10.7%
IO	969	12.3%
Function	120	7.5%
Computation	126	21.4%
Total	2586	7.9%

3.5 Model Training

Our model can be trained end-to-end by backpropagation. For minimizing the loss of each aspect respectively, the binary cross entropy is selected as our objective function. The Eq. (6) demonstrates how the loss is calculated out.

$$loss = -\frac{1}{n} \sum_{i=0}^{k} (y[i] \times \log(\hat{y}[i]) + (1 - y) \times \log(1 - \hat{y}[i])) \tag{6}$$

where y is the ground truth label set for the given text and $\hat{y}$ is the prediction of our model, k is the number of total aspects, and $y[i]$ denotes that whether the given text contains the $(i + 1)$-th aspect. We average all cross entropy to measure the model loss.

In our experiments, we employ Adam [14] to optimize trainable parameters in our model, which adaptively modify its learning rate during the training process.

4   Experiments

4.1   Experimental Settings

- **Datasets.**   We conduct our experiments on the QA style reviews dataset collected from *"asking all"* in Taobao that Wu et al. [7] provided. The whole corpus has 2580 reviews of *electronic appliances* and contains seven aspects of products. Each review may express more than one aspect, and we aim to identify the whole aspects list of each review. To better illustrate the data distribution, the statistics are reported in Table 1. Notably, the scale of multi-label instances in major aspects exceeds 10%. It means that settling the multi-label problem in QA-AC task is essential and ponderable.
- **Evaluation Metrics.**   In our experiments, three evaluation metrics are employed to measure the performance of each experiment, including hamming loss, accuracy, and F1-measure.

Table 2. Experimental results. The best scores are in bold.

Models	Hamming loss	Accuracy	F1-measure
Binary Relevance [16]	0.064	0.662	0.707
Classifier Chains [16]	0.032	0.836	0.886
LSTM(Word2vec) [3]	0.031	0.843	0.901
BANN(Word2vec)	0.029	0.861	0.904
BANN(BERT)	**0.021**	**0.886**	**0.935**

Hamming loss is the fraction of labels that are incorrectly predicted, and the smaller hamming loss manifests a better performance.

$$hammingloss = 1 - \frac{1}{nk} \sum_{i=1}^{n} \sum_{j=1}^{k} \mathbb{I}(y_i^j = \hat{y}_i^j) \tag{7}$$

where n is the number of all test instances, k is the number of total aspect labels, y_i^j denotes the true value about j-th aspect of the i-th instances, $\hat{y}_i^j$ is the estimated value correspondingly, and $\mathbb{I}(.)$ is an indicator function which equals 1 if the condition in parentheses is true and 0 otherwise.

Accuracy measures how many instances have the right prediction of aspect list. In our experiment, only each element in the ground label set and predicted label set is identical, we regard this instance as the right one.

F1-measure is the harmonic mean between precision and recall. We employ the formula [15] and adapt it to multi-label tasks. The larger accuracy and F1-measure correspond to more outstanding performance.

$$F1 = \frac{2}{n} \sum_{i=1}^{n} \frac{|y_i \cap \hat{y}_i|}{|y_i| |\hat{y}_i|} \tag{8}$$

where y_i presents the ground truth label sets, and $\hat{y}_i$ presents the predicted label sets. For example, a given instance includes two aspects, which means its $|y_i|$ is 2.

- **Hyper-parameters.** In our experiments, we split the corpus into training sets and test sets by the ratio of 4:1, In BERT word embeddings, vector dimension is fixed to 768. The max length of the question and the answer are 30, and the initial learning rate is 0.005. Moreover, all models are trained by mini-batch of 32 instances. Besides, all weight matrix and bias matrix are initialized by sampling from the uniform distribution $U(-0.01, 0.01)$, and the dropout rate is 0.5.

4.2   Experimental Results

To comprehensively verify the effectiveness and advantages of our proposed model, we design a series of models as baselines on the same corpus.

Table 3. Ablation studies results

Models	Hamming loss	Accuracy	F1-measure
BANN(BERT)	0.021	0.886	0.935
- LSTM	0.081	0.559	0.606
- Attention	0.029	0.849	0.911
- Q2A Attention	0.082	0.629	0.675
- A2Q Attention	0.031	0.855	0.911
Using weighted summation instead of averaging	**0.019**	**0.896**	**0.942**

Binary Relevance: This approach transforms the multi-labeled aspect classification into multiple binary classifications without label correlation [16].

Classifier Chains: This approach addresses the multi-labeled aspect classification by a chain of binary classifies, which considers the label correlation [16].

LSTM(Word2vec): This approach leverages Word2vec to transform the inputs, and captures the contextual information via LSTM [3].

BANN(Word2vec): Our model employs a bidirectional attention layer instead of simple LSTM and uses Word2vec embedding.

BANN(BERT): Our final model employs a bidirectional attention layer and uses BERT embedding instead of Word2vec.

We performed experiments of each approach above, and the results are shown in Table 2 with the mean value of 10 times experiments.

From Table 2, we can see that all deep neural network methods outperform traditional machine learning methods including **Binary Relevance** and **Classifier Chains** by a large margin, showing that deep neural network methods can learn more complicated aspect information in our QA-AC task.

In addition, our proposed **BANN(Word2vec)** has a better performance than **LSTM(Word2vec)** with the same embeddings and achieves a reduction of 0.1% Hamming loss and the improvement of 1.8 % (Accuracy) and 0.1 % (F1-measure), which is a popular deep neural network method in a large proportion of NLP tasks. It suggests that our bidirectional QA attention neural network is significantly effective, which highlights the contextual QA matching information.

In the end, we restructure our model by BERT instead of Word2vec and achieve the best scores in all evaluation metrics, markedly with 0.1% reduction in Hamming loss, 4.3% increase in Accuracy and 3.4% in F1-measure, which shows that BERT contextualized representation can bring performance improvement for our aspect classification task and proves the superiority of BERT. Significance test shows that the improvement of our model is significant ($p - value < 0.05$).

4.3   Ablation Studies

We conduct ablation studies on our BANN model and expose the results in Table 3 for evaluating the individual contribution of different components. The

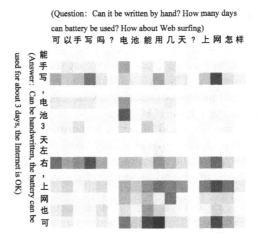

Fig. 4. An attention visualization with the aspects of "Function" and "Battery".

ablation of the LSTM layer results much more Hamming loss and a drop of over 30% on both Accuracy and F1-measure, which shows the consequence of characters' order information. The bidirectional attention accounts for about 3.7% of performance degradation on Accuracy and 2.4% on F1-measure, but the importance of these two directions are not balanced. By comparing the ablations of different directional attention, we can see that the effect of ablating Q2A attention is more severe than ablating A2Q attention. This is because the question contains more aspect-related information while the answer contains more sentiment polarities information in QA style reviews. Finally, we substitute the strategy of weighted summation for averaging in coalescing 12 transformer layers representations from BERT to get the contextualized word embeddings. The result manifests that the weighted summation improves performance by 1% on Accuracy and 0.7% on F1-measure while reduces Hamming loss by 0.2% via generating more rational word representations.

4.4   Visualized Analysis

Figure 4 depicts the A2Q attention weights from our BANN, which are used to measure the importance of characters in the question according to the matching information between question and answer. The sample review contains two aspects: "Function" and "Battery". It is apparent that there are six characters related to aspects in question including " 手 ", " 写 ", " 电 ", " 池 ", " 上 " and " 网 ", which also appears in answer. By carefully inspecting the color depth on these characters, we can see that our BANN emphasizes the importance and relationship between the similar aspect-related characters in question and answer. After bidirectional attention layer, our model can enhance the weights of aspect-related characters in " 手写 ", " 电池 " and " 上网 ", and decrease the value of characters in "3 天 ", " 怎样 " and punctuation characters.

5   Conclusion

In this paper, we propose a contextualized attention-based model for aspect classification on QA style reviews. Firstly we recast the QA-AC task as a multi-label classification problem. Then we capture the contextual information by BERT model, where we employ two strategies to aggregate multiple transformer layers outputs. Further, we use the BANN model to incorporate the QA matching information between questions and answers. Experimental results demonstrate that our approach outperforms several widely-used baselines significantly. For future work, we will explore the effectiveness of joint learning on QA-AC task by jointing aspect classification and aspect-level sentiment classification.

Acknowledgements. This work is supported in part by Industrial Prospective Project of Jiangsu Technology Department under Grant No. BE2017081 and the National Natural Science Foundation of China under Grant No. 61572129.

References

1. Fan, F., Feng, Y., Zhao, D.: Multi-grained attention network for aspect-level sentiment classification. In: Proceedings of the 2018 Conference on Empirical Methods in Natural Language Processing, pp. 3433–3442 (2018)
2. He, R., Lee, W.S., Ng, H.T., Dahlmeier, D.: Exploiting document knowledge for aspect-level sentiment classification. arXiv preprint. arXiv:1806.04346 (2018)
3. Wang, Y., Huang, M., Zhao, L., et al.: Attention-based LSTM for aspect-level sentiment classification. In: Proceedings of the 2016 Conference on Empirical Methods in Natural Language Processing, pp. 606–615 (2016)
4. Rubtsova, Y., Koshelnikov, S.: Aspect extraction from reviews using conditional random fields. In: Klinov, P., Mouromtsev, D. (eds.) KESW 2015. CCIS, vol. 518, pp. 158–167. Springer, Cham (2015). https://doi.org/10.1007/978-3-319-24543-0_12
5. Xu, H., Liu, B., Shu, L., Yu, P.S.: Double embeddings and CNN-based sequence labeling for aspect extraction. arXiv preprint. arXiv:1805.04601 (2018)
6. He, R., Lee, W.S., Ng, H.T., Dahlmeier, D.: An unsupervised neural attention model for aspect extraction. In: Proceedings of the 55th Annual Meeting of the Association for Computational Linguistics (vol. 1: Long Papers), pp. 388–397 (2017)
7. Wu, H., Liu, M., Wang, J., Xie, J., Shen, C.: Question-answering aspect classification with hierarchical attention network. In: Sun, M., Liu, T., Wang, X., Liu, Z., Liu, Y. (eds.) CCL/NLP-NABD -2018. LNCS (LNAI), vol. 11221, pp. 225–237. Springer, Cham (2018). https://doi.org/10.1007/978-3-030-01716-3_19
8. Dai, A.M., Le, Q.V.: Semi-supervised sequence learning. In: Advances in Neural Information Processing Systems, pp. 3079–3087 (2015)
9. Kim, Y., Jernite, Y., Sontag, D., Rush, A.M.: Character-aware neural language models. In: Thirtieth AAAI Conference on Artificial Intelligence (2016)
10. Peters, M.E., et al.: Deep contextualized word representations. arXiv preprint. arXiv:1802.05365 (2018)

11. Radford, A., Narasimhan, K., Salimans, T., Sutskever, I.: Improving language understanding by generative pre-training (2018). https://s3-us-west-2.amazonaws.com/openai-assets/research-covers/languageunsupervised/languageunderstanding paper.pdf
12. Devlin, J., Chang, M.W., Lee, K., Toutanova, K.: BERT: pre-training of deep bidirectional transformers for language understanding. arXiv preprint. arXiv:1810.04805 (2018)
13. Adhikari, A., Ram, A., Tang, R., Lin, J.: DocBERT: BERT for document classification. arXiv preprint. arXiv:1904.08398 (2019)
14. Kingma, D.P., Ba, J.: Adam: a method for stochastic optimization. arXiv preprint. arXiv:1412.6980 (2014)
15. Godbole, S., Sarawagi, S.: Discriminative methods for multi-labeled classification. In: Dai, H., Srikant, R., Zhang, C. (eds.) PAKDD 2004. LNCS (LNAI), vol. 3056, pp. 22–30. Springer, Heidelberg (2004). https://doi.org/10.1007/978-3-540-24775-3_5
16. Szymański, P., Kajdanowicz, T.: A scikit-based Python environment for performing multi-label classification. ArXiv e-prints (February 2017)

NLP Applications

Modeling the Long-Term Post History for Personalized Hashtag Recommendation

Minlong Peng, Yaosong Lin, Lanjun Zeng, Tao Gui, and Qi Zhang[✉]

School of Computer Science, Fudan University, Shanghai, China
{mlpeng16,yslin18,ljzeng18,tgui16,qz}@fudan.edu.cn

Abstract. Hashtag recommendation aims to recommend hashtags when social media users show the intention to insert a hashtag by typing in the hashtag symbol "#" while writing a microblog. Previous methods usually considered the textual information of the post itself or only fixed-length short-term post history. In this paper, we propose to model the long-term post histories of user with a novel neural memory network called the Adaptive neural MEmory Network (AMEN). Compared with existing memory networks, AMEN was specially designed to combine both content and hashtag information from historical posts. In addition, AMEN contains a mechanism to deal with out-of-memory situations. Experimental results on a dataset of Twitter demonstrated that the proposed method significantly outperforms the state-of-the-art methods.

Keywords: Hashtag recommendation · Long-term post history · Neural memory network

1 Introduction

Along with the rapid development of social media, microblogging has experienced tremendous success as a news medium. Every day, hundreds of millions of posts are created. To facilitate the navigation of this huge knowledge base, microblogging services encourage users to insert hashtags, which start with the "#" symbol (e.g., "#CCL2019", "#MentalHealth"), into posts to concisely indicate an object or categorize their posts. However, due to the increasing number of hashtags, it is not easy for authors to find appropriate hashtags matching their intention. Therefore, recommending hashtags to users is imperative.

In recent years, a variety of methods have been proposed to perform hashtag recommendation. These methods can be generally organized into two groups. The first group of methods treat posts of different users without distinction. They mainly focus on modeling the textual content of posts, which is traditionally dominated by topic-model-based methods [4–7,17]. This dominance has been overturned by a resurgence of interest in deep neural network based approaches [2,3,8,14]. The second group of methods additionally model the user's personal information, which is commonly extracted from their post history [10,12,19–22]. Zhang et al. [21] proposed the TPLDA model for this purpose. Their model

© Springer Nature Switzerland AG 2019
M. Sun et al. (Eds.): CCL 2019, LNAI 11856, pp. 495–507, 2019.
https://doi.org/10.1007/978-3-030-32381-3_40

grouped posts by user and introduced an additional parameter for each user into the LDA model. Huang et al. [10] proposed a hierarchical end-to-end memory model (HMemN2N) to encode the post history. For each user, their model constructed the memory using the latest five historical posts and then recursively accessed it with the current post content for the recommendation. These existing methods have already demonstrated the informativeness of the post history. However, all of the above methods can only model short and fixed-length post histories.

In this work, we investigate the effectiveness of utilizing the long-term post history for hashtag recommendation. Modeling long-term post history is structurally inapplicable for the previous methods because of the incrementally increasing size of records. Inspired by the success of neural memory networks in modeling long-term dependency of sequences [1,9,18], we propose a novel neural memory network called the Adaptive neural MEmory Network (AMEN), to tackle the challenge. AMEN was specially designed to treat the content and the hashtag as two different views of posts, rather than treat hashtags as common words. The two different representations are simultaneously combined and encoded into memory. We argue that this design has several benefits. First, it does not require that the representations of words and hashtags in the same vector space. Second, it can reduce the size of the word vocabulary used to represent hashtags. Finally, it highlights the hashtag information, making its modeling more flexible. Despite the significance of post history, considering the immediacy of microblogging, a user may post a tweet about a brand new topic that is almost independent of historical posts. That is what we call the out-of-memory (OOM) problem. To address this situation, we designed a novel gate mechanism. The main contributions of this work are summarized as follows:

- We proposed a novel neural memory network that combines both textual content and hashtags to model users' long-term post history for personalized hashtag recommendation tasks.
- We designed a gate mechanism to deal with OOM situations where the hashtag usage is almost independent of previous posts.
- Comprehensive experiments and quantitative studies demonstrate the effectiveness of the proposed model on a dataset collected from Twitter.

2   Proposed Model

In this work, we formulate the task of recommending hashtags as a matching-based problem. Given a user $\mathbf{u}$ with post history sequence $[(p_1, H_1), (p_2, H_2), \cdots, (p_{t-1}, H_{t-1})]$, for the tth post, our task is to rank the candidate hashtags $H = \{H^1, H^2, \cdots, H^n\}$. To address this problem, we proposed the Adaptive neural MEmory Network (AMEN). The general architecture of the proposed model is depicted in Fig. 1, which is specially designed to highlight the core of this model in modeling the long-term post history.

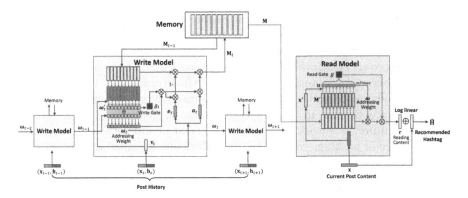

Fig. 1. General architecture of the proposed model.

The proposed model could be separated into three parts. First, we employ a convolution neural network and a recurrent neural network to obtain representation of post content and hashtag. Second, we model users' long-term post history with the core memory network. Lastly, we combine the current post content with memory augmented information to rank candidate hashtags.

2.1 Representation Learning

Content Representation. We utilize a *one-layer convolution neural network* to obtain the post content representation. It is a variant of the traditional convolution network proposed by Kim et al. [11]. Specifically, let $\mathbf{w}_i \in R^{k_w}$ be the k_w-dimensional word vector, corresponding to the ith word of the post. Therefore, A post of length n is represented as:

$$\mathbf{p} = \left[\mathbf{w}_1, \mathbf{w}_2, \cdots, \mathbf{w}_n \right] \tag{1}$$

The one-layer CNN takes the dot product of the filter $\mathbf{m} \in R^{k_w \times h}$ with each h-gram in $\mathbf{p}$ to obtain a sequence $\mathbf{s}$.

$$\mathbf{s}_i = f(\mathbf{m} \cdot \mathbf{p}_{i:i+h-1} + b). \tag{2}$$

Here, $b \in R$ is a bias term, and f is the hyperbolic tangent function. This filter is applied to each possible window of words in the sequence $\{\mathbf{p}_{1:h}, \mathbf{p}_{2:h+1}, \cdots, \mathbf{p}_{n-h+1:n}\}$ to produce a feature map:

$$\mathbf{s} = [\mathbf{s}_1, \mathbf{s}_2, \cdots, \mathbf{s}_{n-h+1}] \tag{3}$$

To deal with various post lengths, max pooling is applied over the feature map. By extending this operation to multiple filters with various window sizes, we obtain multiple features:

$$\mathbf{x} = \left[\max(\mathbf{s}^1) \cdots \max(\mathbf{s}^d) \right] \tag{4}$$

Here d is the filter number and $\mathbf{s}^i$ is the feature map extracted with the ith filter. We use these features as the post content representation, denoted by $\mathbf{x}$.

Hashtag Representation. Most of the previous hashtag recommendation systems treat the hashtags as independent meaningless categories and represent them with orthogonal one-hot vectors or randomly initialized dense vectors. We argue that there are two drawbacks of this practice. First, it neglects the semantic relationship between hashtags. For examples, "Christmas" and "ChristmasEve" are semantically close, so the representation of them should naturally be close in the vector space. Second, in this practice, the size of the hashtag set is unfixed. Different from previous work, we apply a recurrent neural network to the character sequence of hashtags to capture the underlying semantic and obtain their vector representations. The formal definition is specified as follows:

$$\mathbf{h}_t = \tanh(\mathbf{W}_h \cdot \mathbf{h}_{t-1} + \mathbf{W}_c \cdot \mathbf{c}_t + \mathbf{b}_c) \tag{5}$$

where $\mathbf{c}_t \in R^{d_c}$ is the vector representation of the tth character of the hashtag $\mathbf{H}$, $\mathbf{W}_h \in \mathbb{R}^{d_h \times d_h}$, $\mathbf{W}_c \in \mathbb{R}^{d_h \times d_c}$ and $\mathbf{b}_c \in \mathbb{R}^{d_h}$ are trainable parameters. We take the final hidden state $\mathbf{h_n} \in \mathbf{R}^{d_h}$ to represent $\mathbf{H}$. In the following sections, we refer $\mathbf{h}$ to this vector representation if without further explanation.

2.2   Adaptive Neural MEmory Network

In this section, we describe the core parts of the proposed model AMEN. Intuitively, the proposed memory network consists of a memory module, a write module and a read module, which will be introduced in detail.

Memory Module. The memory module consists of several dense vector cells for storing historical post information. Technically, let $\mathbf{M} \in \mathbb{R}^{N_m \times d}$ be the contents of memory matrix for each user, where N_m denotes the number of memory locations, and d is the vector dimension at each location, while $\mathbf{M}_t$ represents the memory state after the write process at time step t.

Note that the memory is empty at the beginning. If no further processing is applied, the memory cells will be read and written equally, acting just like a single cell. To avoid this, in every epoch, we write the first N_m samples of each user into the memory, rather than initialize all memory cells with zero vectors.

Write Module. The write module is designed to encode the post history one-by-one into the memory module. During the writing process, both the post content representation $\mathbf{x}_t$ and the hashtag representation $\mathbf{h}_t$ are written into the memory at time step t. More specifically, we construct the post content $\mathbf{v}_t$ with the combination of $\mathbf{x}_t$ and $\mathbf{h}_t$ as follows:

$$\mathbf{v}_t = \mathrm{MLP}(\mathbf{x}_t \oplus \mathbf{h}_t; \boldsymbol{\theta}_o). \tag{6}$$

This MLP consists of two non-linear layers with sigmoid and hyperbolic tangent activation, respectively, and $\boldsymbol{\theta}_o$ denotes trainable parameters. Then, we generate the unnormalized addressing weight $\mathbf{u}'_t(i)$ by similarity measure between the $\mathbf{v}_t$ with each memory vector $\mathbf{M}_{t-1}(i)$, i denoting the index of memory locations:

$$\mathbf{u}'_t(i) = K[\mathbf{v}_t, \mathbf{M}_{t-1}(i)]. \tag{7}$$

where

$$K[\mathbf{x}, \mathbf{y}] = \frac{\mathbf{x}^T \mathbf{y}}{||\mathbf{x}||^2}.$$

Note that this differs from the conventional NTM model [9] in that we omit the norm of the memory matrix in the denominator, so as to preserve the quantity information of each memory cell. After that, we apply a softmax non-linear operation to obtain the normalized addressing weight $\boldsymbol{\omega}'_t$:

$$\omega'_t(i) = \frac{\exp(\mathbf{u}'_t(i))}{\sum_j^{N_m} \exp(\mathbf{u}'_t(j))}. \tag{8}$$

where N_m is the size of the memory module. By combining $\boldsymbol{\omega}'_t$ with the writing weights of the previous time $\boldsymbol{\omega}_{t-1}$ using a recurrent framework, we obtain the addressing weight $\boldsymbol{\omega}_t$ as follows:

$$\mathbf{u}_t(i) = \mathbf{W}_\omega \left[\boldsymbol{\omega}_{t-1} \oplus \boldsymbol{\omega}'_t \right] + b_\omega$$
$$\omega_t(i) = \frac{\exp(\mathbf{u}_t(i))}{\sum_j^{N_m} \exp(\mathbf{u}_t(j))}. \tag{9}$$

In addition, considering that a user may re-tweet some posts for particular purposes which are not highly related to their interests, we further generate an writing gate g_t to determine whether or not the current example should be written into the memory, based on the unnormalized address weight $\mathbf{u}'_t$:

$$\mathbf{z}_t = \text{sort}(\mathbf{u}'_t)$$
$$g_t = \sigma(\mathbf{s}_\alpha[\mathbf{z}_t \oplus (\mathbf{z}_t - \bar{\mathbf{z}}_t)^2]), \tag{10}$$

where the sort function takes an array as input and returns its ordered counterpart, $\oplus$ represents the concatenating operation, $\bar{\mathbf{z}}_t$ is the mean value of $\mathbf{z}_t$, and $\mathbf{s}_\alpha \in R^{2N}$ is a trainable user-specific gate vector. Here, the sorting operation is critical because of the randomness inherent in the memory reading and writing operations.

Next, we obtain the content to erase from memory $\mathbf{e}_t$ and the content to write into memory $\mathbf{a}_t$, from the combination of $\mathbf{x}_t$ and $\mathbf{h}_t$:

$$\mathbf{e}_t = \sigma(\mathbf{W_e}(\mathbf{x}_t \oplus \mathbf{h}_t) + b_e),$$
$$\mathbf{a}_t = \tanh(\mathbf{W_a}(\mathbf{x}_t \oplus \mathbf{h}_t) + b_a). \tag{11}$$

Finally, we update the memory cells formulated as follows:

$$\mathbf{M}_t(i) = \mathbf{M}_{t-1}(i) \left[\mathbf{1} - g_t \boldsymbol{\omega}_t(i) \mathbf{e}_t \right] + g_t \boldsymbol{\omega}_t(i) \mathbf{a}_t. \tag{12}$$

Read Module. The read model is designed to access the memory with the current post content and output a vector representation of the memory, which is further used to recommend hashtags. However, during the reading process, only the post content representation $\mathbf{x}$ of the post $\mathbf{p}$ can be used for addressing. As we mentioned above, the memory contains both the information of content and hashtag of previous posts. Consequently, the vector space of the memory content $\mathbf{M}(i)$ is not consistent with the space of $\mathbf{x}$. In order to utilize the memory, we first incorporate a transformation process to map the input and the memory cells into a consistent space, formulated as follows:

$$\mathbf{x}' = \mathbf{MLP}(\mathbf{x};\ \boldsymbol{\theta}_p),$$
$$\mathbf{M}'(i) = \mathbf{MLP}(\mathbf{M}(i);\ \boldsymbol{\theta}_q), \tag{13}$$

where $\mathbf{MLP}$ is a two-layer perceptron with *relu* and *tanh* activation functions. We then obtain an unnormalized addressing weight $\mathbf{u}'$:

$$\mathbf{u}(i) = K[\mathbf{x}', \mathbf{M}'(i)], \tag{14}$$

Based on that, we generate the normalized addressing weight $\boldsymbol{\omega}(i)$:

$$\boldsymbol{\omega}(i) = \frac{\exp(\mathbf{u}(i))}{\sum_j^{N_m} \exp(\mathbf{u}(j))}, \tag{15}$$

The addressing weight $\boldsymbol{\omega}(i)$ can now be used to produce an output of the reading module. However, instead of directly using the addressing weight $\boldsymbol{\omega}(i)$ to obtain memory output $\mathbf{r}$, we consider the out-of-memory situations to enhance our read module in the following.

Out-of-Memory Mechanism. Despite the significance of post history, there is a risk of the out-of-memory situations of hashtag recommendation task, where users may write a post about events that just happened. In this case, we hope to ignore the memory and perform the recommendation based merely on the post content itself. With this consideration, we design a mechanism to automatically detect the out-of-memory situations based on the global view of the input and the memory. Mathematically, we generate a gate g to indicate whether it should depend on the memory or not, as follows:

$$\mathbf{z} = \text{sort}(\mathbf{u}(i))$$
$$g = \sigma(\mathbf{s}_\beta[\mathbf{z} \oplus (\mathbf{z} - \bar{\mathbf{z}})^2]), \tag{16}$$

where $\bar{z}$ is the mean value of $\mathbf{z}$, and $\mathbf{s}_\beta \in R^{2N}$ is a trainable user-specific gate vector. Finally, based on the designed gate g and the addressing weight $\boldsymbol{\omega}(i)$, the memory output $\mathbf{r}$ is given by:

$$\mathbf{r} = g \cdot \sum_i^{N_m} \mathbf{M}(i)\boldsymbol{\omega}(i). \tag{17}$$

2.3   Recommendation

Given a post $\mathbf{p}$, the hashtag recommendation of the proposed model is performed based on both the post content $\mathbf{x}$ and the memory augmented output $\mathbf{r}$ using a log-linear model. We redefine the score function of hashtag $\mathbf{h}_i$:

$$\text{score}(\mathbf{p}, \mathbf{h}_i | \mathbf{M}) = \exp(\mathbf{x}^T \mathbf{h}_i + g \cdot \mathbf{r}^T \mathbf{h}_i), \tag{18}$$

where $\mathbf{M}$ denotes the memory content corresponding to $\mathbf{x}$. We apply a softmax operation to obtain the probability of recommending $\mathbf{h}_i$, namely:

$$\mathbf{p}(\mathbf{h}_i | \mathbf{p}, \mathbf{M}) = \frac{\exp\left(\text{score}(\mathbf{p}, \mathbf{h}_i)\right)}{\sum_{j=0}^{N_h} \exp\left(\text{score}(\mathbf{p}, \mathbf{h}_j)\right)}, \tag{19}$$

where N_h is the size of candidate hashtag set. Top-k hashtags with the highest probability are recommended for post $\mathbf{p}$.

2.4   Training and Inference

The training objective function of the model is defined as:

$$J = \frac{1}{\sum_u (N_u - N_m)} \sum_u \sum_{t=N_m}^{N_u} -\log p(\mathbf{H}_t | \mathbf{p}_t, \mathbf{M}_{t-1}; \boldsymbol{\Theta}), \tag{20}$$

where N_u is the number of history posts for user $\mathbf{u}$, N_m is the size of memory locations, and $\boldsymbol{\Theta}$ denotes the parameter set. We sample $\mathbf{N}_m$ posts of each user to initialize the memory, so these posts are excluded from the calculation of training cost.

During inference, for each user, we fix the memory content $\mathbf{M}$ corresponding to the state after writing last training example. For user $\mathbf{u}$ and his/her post $\mathbf{p}$, the probability of hashtag $\mathbf{H}_i$ as the recommendation candidate is defined as:

$$\mathbf{H}_i = p(\mathbf{H}_i | \mathbf{p}, \mathbf{M}_{N_u}; \boldsymbol{\Theta}). \tag{21}$$

Here $\mathbf{M}_{N_u}$ denotes the memory state after writing the last training example.

3   Experiment

3.1   Dataset

Most of previous studies only consider the content of microblogs for the hashtag recommendation, and there is no user and time information within their datasets. To evaluate the proposed model, we constructed a new dataset from Twitter.

We crawled tweets posted within 1/1/2015 to 2/28/2015 and removed users who have less than 30 or more than 300 posts within the duration. The lower bound was used to ensure that our system could learn the user-specific parameters well, and the upper bound was to filter out noisy users like advertisers. From these users, we selected 2,000 users and extracted all the tweets posted by them for training. For development and testing dataset, we crawled tweets posted by these users during 3/1/2015 to 3/10/2015, and applied the same preprocessing. Statistical information about this dataset is listed in Table 1.

Table 1. Statistics of the dataset in experiments.

Item	Train	Develop	Test
#User	2,000	1,000	1,000
#Tweet	127,846	8,086	9,190
#Example	187,247	13,174	13,946
#Hashtag	3,104	1,936	1,952
#Hashtag/User	23.54	6.28	6.29
#Length/Example	13.30	13.26	13.04

3.2   Methods for Comparison

We first compared the proposed model with several baselines and state-of-the-art methods, including methods that ignore the post history and those that only model the short-term post history.

The history-independent models includes the translation model **IBM1**[16], the topical word alignment model **TopicWA**[5], the discriminated lstm model **Tweet2Vec**[3]. And the history-dependent models includes the attention-based LSTM model **LSTM-Attention** [15], the user-specific topic model **TPLDA**[21], the end-to-end memory network **HMemN2N**[10], and the naive neural turing machine model **NTM**[9] which does not separately model the hashtag from the post content and not deal with the out-of-memory problem of this task.

Then to explore the effectiveness of several main components of the proposed model, we implemented the two variants of AMEN. The first variant only writes the post contents into the memory without the hashtag history, called **AMEN w/o Post History**. This was designed to validate the informativeness of historical hashtags. The second one, called **AMEN w/o OOM**, ignores the out-of-memory problem for hashtag recommendation as described above. This is designed to verify the effectiveness of the mechanism for dealing with the out-of-memory (OOM) situations.

3.3   Implementation Details

Parameter. For the proposed model and its variants, the embedding dimensions of the word, character, and memory cell were set to 300, 50, and 200 respectively. We initialized the word embeddings with Word2vec. Hidden size of the recurrent neural network for hashtag encoding was set to 100. Memory size N_m was set to 5. For the post content encoding, we used 200 filters for each h-gram size $\in \{1, 2, 3, 4\}$. Dropout was applied to word embeddings with a probability of 0.5. We used the Adadelta optimizer for optimization.

Evaluation Metric. There are several evaluation metrics for hashtag recommendation, such as $Hits@N$ [13,17], Precision, Recall and F1 [10,17]. Here, we choose the $Hits@N$ because in the setting of our work, there is only one ground

Table 2. Comparison of our method with state-of-the-art methods and two variants.

Models	Hits@1	Hits@5
IBM1 [16]	0.2322	0.3043
TopicWA [5]	0.3023	0.3975
Tweet2Vec [3]	0.3116	0.4021
LSTM-Attention [15]	0.3413	0.4430
TPLDA [21]	0.2737	0.5359
HMemN2N [10]	0.3843	0.5460
NTM [9]	0.4021	0.5936
AMEN (proposed)	**0.4732**	**0.6744**
AMEN w/o Hashtag History	0.4251	0.6237
AMEN w/o OOM	0.4372	0.6414

truth hashtag for every recommendation. A hit occurs when the ground truth hashtag is include among the recommended n hashtags.

$$Hits@(n) = \frac{\text{Number of Hits}}{\text{Recommendation times}}.$$

3.4   Results and Discussion

Table 2 shows that the proposed model outperforms all state-of-the-art methods, which indicates the effectiveness of AMEN. From this table, we can draw the following conclusions.

First, modeling users' historical posts provides valuable information for recommendations. According to Table 2, the methods modeling post history (e.g., TPLDA, HMemN2N, NTM) generally outperform content-only-based methods (e.g., IBM1, TopicWA, Tweet2Vec), especially on Hits@5. This validates the informativeness of the post history for hashtag recommendation.

Next, for methods considering historical posts, leveraging the long-term post history could bring significant improvements. Compared to HMemN2N, AMEN w/o Hashtag History which models the long-term post history instead of the short-term history, improves Hits@1 and Hits@5 by approximately 4% and 8%, respectively. This empirically verifies our assumption that the long-term post history should be informative.

Lastly, compared to its variants, the performance of AMEN can be obviously boosted with the two specially designed components. On the one hand, comparison of AMEN and AMEN w/o Hashtag History shows that modeling users's historical hashtags bring additional improvements of 5% on Hits@1 and Hits@5, revealing the high informativeness of historical hashtags. On the other hand, by incorporating an OOM gating mechanism, AMEN beats AMEN w/o OOM about 4% on Hits@1 and 3% on Hits@5, which justifies the necessity to

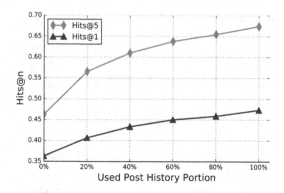

Fig. 2. Performance of the proposed model using different portion of post history.

deal with OOM situations. Combining the two designs together, AMEN yields 47.32% on Hits@1 and 67.44% on Hits@5, showing noticeable improvements.

Impact of Different Parts of Post History. How much influence each part of the post history has on the final recommendation performance? To explore this, we performed a study on different parts of the post history. For each user, we removed different portions of training data and re-generated the memory content M_{N_u} using the trained parameters and the left training data.

Figure 2 shows the results of the proposed model using the latest 20%, 40%, 60%, 80%, and 100% of the post history for each user. Here the latest post history refers to historical posts that are posted at the time closest to the testing dataset. From the figure, we can see that the performance of our proposed model continuously increases as the portion of the post history increases. In addition, the improving speed slowly decreases as more historical posts are considered. For example, with the most recent 20% of the post history, the performance increases approximately 4% on Hits@1 and 10% on Hits@5. While additionally using 20%–40% of the post history, the performance only increases almost 3% on Hits@1 and 4% on Hits@5. We argue that the latest 20% of the post history has a greater influence on the testing data than the 20%–40% ones. Similar observations can be obtained from comparisons between other portions.

Impacts of Filter Number. One hyper-parameter we are interested in is the filter number N_f for each n-gram, so we tried different values ranging from 100 to 250. As shown in Table 3, our proposed model performs quite robustly to the variation of N_f over the given range. Moreover, when the value is smaller than 200, increasing this value can sightly improve the general performance. However, when the value reaches 200, increasing this value can even harm the performance because of more serious over-fitting problem.

Table 3. Performance of our proposed model with different filter numbers.

Filter number	Hits@1	Hits@5
100	0.4454	0.6628
150	0.4663	**0.6768**
200	**0.4732**	0.6744
250	0.4500	0.6637

Impact of Memory Size. We further investigated the impacts of memory size N_m on model performance. Table 4 lists the results for various memory sizes $N_m \in \{2, 3, 5, 7\}$. We can see that the value has great influence on the performance of AMEN. It performed better for $N_m = 3$ and $N_m = 5$. Increasing or decreasing the memory size led to a worse performance. This was not unexpected because if the memory size was too small, it could not effectively remember the post history, while if the memory size was too large, the over-fitting problem would be more serious.

Table 4. Performance of our proposed model with different memory size.

Memory size	Hits@1	Hits@5
2	0.4483	0.6683
3	0.4698	**0.6757**
5	**0.4732**	0.6744
7	0.4555	0.6674

4   Conclusion

In this work, we addressed the problem of recommending hashtags when a user shows the intention to insert a hashtag by typing the hashtag symbol "#". Compared to existing works that did not consider users' post histories or considered only short- and fixed-length post histories, we proposed to leverage the long-term post history for hashtag recommendation. For this purpose, we proposed a novel memory network to model the long-term post history of users, which was specially designed to combine both content and the hashtag information from historical posts. In addition, we designed a mechanism to address the out-of-memory problem. Experimental results on a dataset collected from Twitter showed that the proposed method achieves state-of-the-art performance.

References

1. Bordes, A., Usunier, N., Chopra, S., Weston, J.: Large-scale simple question answering with memory networks. arXiv preprint arXiv:1506.02075 (2015)
2. Dey, K., Shrivastava, R., Kaushik, S., Subramaniam, L.V.: EmTagger: a word embedding based novel method for hashtag recommendation on Twitter. arXiv preprint arXiv:1712.01562 (2017)
3. Dhingra, B., Zhou, Z., Fitzpatrick, D., Muehl, M., Cohen, W.W.: Tweet2Vec: character-based distributed representations for social media. arXiv preprint arXiv:1605.03481 (2016)
4. Ding, Z., Qiu, X., Zhang, Q., Huang, X.: Learning topical translation model for microblog hashtag suggestion. In: Proceedings of the 22nd International Joint Conference on Artificial Intelligence (IJCAI 2013) (2013)
5. Ding, Z., Zhang, Q., Huang, X.: Automatic hashtag recommendation for microblogs using topic-specific translation model. In: Proceedings of COLING 2012: Posters, pp. 265–274 (2012)
6. Godin, F., Slavkovikj, V., De Neve, W., Schrauwen, B., Van de Walle, R.: Using topic models for Twitter hashtag recommendation. In: Proceedings of the 22nd International Conference on World Wide Web, pp. 593–596. ACM (2013)
7. Gong, Y., Zhang, Q., Han, X., Huang, X.: Phrase-based hashtag recommendation for microblog posts. Sci. Chin. Inf. Sci. **60**(1), 012109 (2017)
8. Gong, Y., Zhang, Q.: Hashtag recommendation using attention-based convolutional neural network. In: IJCAI, pp. 2782–2788 (2016)
9. Graves, A., Wayne, G., Danihelka, I.: Neural Turing Machines. Arxiv, pp. 1–26 (2014). https://doi.org/10.3389/neuro.12.006.2007, http://arxiv.org/abs/1410.5401
10. Huang, H., Zhang, Q., Gong, Y., Huang, X.: Hashtag recommendation using end-to-end memory networks with hierarchical attention. In: Proceedings of COLING 2016, the 26th International Conference on Computational Linguistics: Technical Papers, pp. 943–952 (2016)
11. Kim, Y.: Convolutional neural networks for sentence classification. arXiv preprint arXiv:1408.5882 (2014)
12. Kowald, D., Pujari, S.C., Lex, E.: Temporal effects on hashtag reuse in Twitter: a cognitive-inspired hashtag recommendation approach. In: Proceedings of the 26th International Conference on World Wide Web, pp. 1401–1410. International World Wide Web Conferences Steering Committee (2017)
13. Kywe, S., Hoang, T.A., Lim, E.P., Zhu, F.: On recommending hashtags in Twitter networks. In: Social Informatics, pp. 337–350 (2012)
14. Li, Y., Liu, T., Hu, J., Jiang, J.: Topical co-attention networks for hashtag recommendation on microblogs. Neurocomputing **331**, 356–365 (2018)
15. Li, Y., Liu, T., Jiang, J., Zhang, L.: Hashtag recommendation with topical attention-based LSTM. In: COLING (2016)
16. Liu, Z., Chen, X., Sun, M.: A simple word trigger method for social tag suggestion. In: Proceedings of the Conference on Empirical Methods in Natural Language Processing, pp. 1577–1588. Association for Computational Linguistics (2011)
17. She, J., Chen, L.: TOMOHA: topic model-based hashtag recommendation on Twitter. In: Proceedings of the 23rd International Conference on World Wide Web, pp. 371–372. ACM (2014)
18. Sukhbaatar, S., Weston, J., Fergus, R., et al.: End-to-end memory networks. In: Advances in Neural Information Processing Systems, pp. 2440–2448 (2015)

19. Tran, V.C., Hwang, D., Nguyen, N.T.: Hashtag recommendation approach based on content and user characteristics. Cybern. Syst. **49**, 1–16 (2018)
20. Wang, Y., Qu, J., Liu, J., Chen, J., Huang, Y.: What to tag your microblog: hashtag recommendation based on topic analysis and collaborative filtering. In: Chen, L., Jia, Y., Sellis, T., Liu, G. (eds.) APWeb 2014. LNCS, vol. 8709, pp. 610–618. Springer, Cham (2014). https://doi.org/10.1007/978-3-319-11116-2_58
21. Zhang, Q., Gong, Y., Sun, X., Huang, X.: Time-aware personalized hashtag recommendation on social media. In: Proceedings of COLING 2014, the 25th International Conference on Computational Linguistics: Technical Papers, pp. 203–212 (2014)
22. Zhao, F., Zhu, Y., Jin, H., Yang, L.T.: A personalized hashtag recommendation approach using LDA-based topic model in microblog environment. Future Gener. Comput. Syst. **65**, 196–206 (2016)

A Document Driven Dialogue
Generation Model

Ke Li[(✉)], Ziwei Bai, Xiaojie Wang, and Caixia Yuan

Beijing University of Posts and Telecommunications, Beijing, China
{cocolike,bestbzw,xjwang,yuancx}@bupt.edu.cn

Abstract. Most of the current man-machine dialogues are at the two end-points of a spectrum of dialogues, i.e. goal-driven dialogues and non goal-driven chit-chats. Document-driven dialogues provide a bridge between them with the change of documents from structured data to unstructured free texts. This paper proposes a Document Driven Dialogue Generation model (D3G) which generates dialogues centering a given document, as well as answering user's questions. A Doc-Reader mechanism is designed to locate the content related to user's questions in documents. A Multi-Copy mechanism is employed to generate document-related responses. And the dialogue history is used in both mechanisms. Experimental results on the CMU_DOG dataset show that our D3G model can not only generate informative responses that are more relevant to the document, but also answer user's questions better than the baseline models.

Keywords: Document-driven dialogue · Doc-Reader · Multi-copy

1 Introduction

Most of the current man-machine dialogues are at the two end-points of a spectrum of dialogues. The goal-driven dialog [6,9,16] is on one end and chit-chat [13,14] is on the other end. Goal-driven dialogue systems communicate with users based on pre-defined task goals. While chit-chat aims to generate suitable responses without pre-defined task goals.

However, there are various dialogues between the two ends in daily life. For the goal-driven dialogue, it is often difficult to obtain the structured goal for a dialogue. For example, lots of services are illustrated by unstructured or semi-structured documents instead of structured frames, and operators are asked to talk with customers directly according to the information in documents. For the chit-chat, conversations are not completely unconstrained. A typical example is that a reasonable response might be relevant to the news when people chat centering on it. The dialogues centering on free texts are called document-driven.

Some work has been done to make a bridge between the goal-driven dialogue and chit-chat. For example, the goal-driven dialogue drew on the experience of the sequence to sequence framework which is widely used in chit-chat [5,17].

© Springer Nature Switzerland AG 2019
M. Sun et al. (Eds.): CCL 2019, LNAI 11856, pp. 508–520, 2019.
https://doi.org/10.1007/978-3-030-32381-3_41

While the chi-chat attempted to incorporate more structural information [8]. In recent years, people tried to add some external knowledge and information into dialogues, such as the knowledge base [1], the knowledge graph [18], and the image [15]. But there are few studies on the document-driven dialogue. In 2018, Zhou et al. [19] presented a document grounded dataset for conversations and proposed two baseline models on it. They demonstrated the model using the documents outperforms the model without documents. The study on document-driven dialogue bridges the gap between the goal-driven dialogue and chit-chat.

This paper aims to extend the document-driven dialogue, enabling it to answer specific questions as well as chatting with users based on the document. To achieve this goal, we propose a Document Driven Dialog Generation (D3G) model, which has two mechanisms: a Doc-Reader mechanism is designed to locate answers in the document regarding to user's question and a Multi-Copy mechanism is used to help generate document-related responses. Experimental results show that D3G model significantly outperforms baseline methods in several ways.

In general, our contributions are as follows:

- We propose a document-driven dialogue generation model (D3G). Our model can not only chat with users centering on a given document, but also answer user's questions related to the document.
- In the D3G model, we use the Doc-Reader mechanism to help locate the content related to user's questions. We also propose the Multi-Copy mechanism to generate document-related responses. And the dialogue history is used in both mechanisms.
- Experimental results on automatic evaluation show our model can generate much more document-related responses compared to the baseline models.

2   Related Work

There are few studies focused on document-driven dialog generation models. Most of the chat models aim to use the document to increase the diversity of responses, but pay less attention to the relevance between generated responses and documents they are referring, and let alone answer document-related questions.

Zhou et al. [19] is one of the few jobs that focus on the relevance of generated responses to documents. As far as we know, they present the first document grounded dataset for conversations, which called CMU_DOG, along with two baseline models. The conversations in the dataset are about the contents of a specified document. In order to verify the validity of this dataset, Zhou et al. proposes NW score and use BLEU [10] to measure the relevance between documents and the conversations generated by annotators. The two baseline models have similar structures, except for the input of decoder, where SEQS utilizes both current utterance and the document while SEQ only concerns the current utterance. Moreover, Zhou et al. uses human evaluation "Engagement" to measure the response generated by SEQ and SEQS, causing subjective errors.

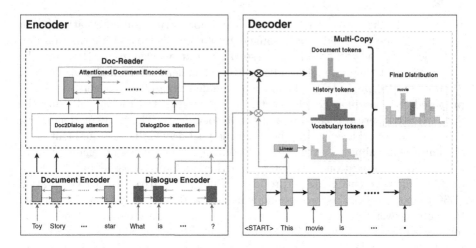

Fig. 1. The Document driven dialogue generation model

Unlike the work of Zhou [19], our proposed D3G model employs the Doc-Reader mechanism and the Multi-Copy mechanism to make full use of the document information. The dialogue history is also used in both mechanisms. Our model can not only generate document-related responses, but also answer user's questions relevant to documents. In addition, we evaluate generated responses automatically.

There are some studies that introduce documents to the chit-chat. Those studies focus on solving safe response problems, such as "emmm..." or "I don't know". In 2017, Ghazvininejad et al. proposes the MTASK-R model to introduce external text information into the chit-chat [2]. Experiments show that the model with external text information can generate more diverse responses. However, such a system can only chat with the user, without the ability to answer user's questions. The evaluation of these tasks also focuses on the fluency and diversity of the responses but pays less attention to the relevance to the document.

Other work such as conversational machine reading comprehension, focuses on whether the machine accurately locates the answers in the document. In 2018, Zhu et al. proposes an attention-based conversational deep neural network SDNet model [20], which helps the system determine the answer by understanding the document and dialogue history. Such research only requires the model to answer user's questions, but cannot chat with the user.

3   Model

Let $\{u_1, u_2, \ldots, u_t\}$ be the dialogue up to t-th utterance and $E^s = \{w_1, w_2, \ldots, w_m\}$ denotes a sequence of m tokens in the document. The response produced by D3G model is defined as $Y = \{y_1, y_2, \ldots, y_T\}$, where T is the number of tokens in the response. As shown in Fig. 1, D3G model mainly consists of two modules: Encoder and Decoder. The Encoder Module first encodes the document and the dialogue, then uses a Doc-Reader mechanism to obtain dialogue-aware representations of document tokens. The Decoder Module proposes a Multi-Copy mechanism to generate document-related responses.

3.1   Encoder Module

Document Encoder: Given the document $E^s = \{w_1, w_2, \ldots, w_m\}$, a bidirectional LSTM is used to get the document contextual representation S,

$$S = BiLSTM(E^s), \tag{1}$$

where $S = [s_1, s_2, \ldots, s_m] \in R^{2d \times m}$, d is the hidden size.

Dialogue Encoder: Given the dialogue history $\{u_1, u_2, \ldots, u_{t-1}\}$ and the current user's utterance u_t, we concatenate the utterances to get $E^h = \{w_1, w_2, \ldots, w_k\}$, where k is the number of tokens in dialogue. Then we use the same bidirectional LSTM to get the dialogue contextual representation H,

$$H = BiLSTM(E^h), \tag{2}$$

where $H = [h_1, h_2, \ldots, h_k] \in R^{2d \times k}$.

Doc-Reader Mechanism: The Doc-Reader mechanism is the core of the Encoder Module. It enables the D3G model with the ability to locate contents which are related to the dialogue. Inspired by BiDAF [12], a classical model for machine reading comprehension, we introduce the Doc-Reader mechanism to indicate the "importance" of tokens in the document. Moreover, we consider the attention from two directions: document to dialogue attention (Doc2Dialog Attention) and dialogue to document attention (Dialog2Doc Attention) to obtain the dialogue-aware representations of document tokens.

We build M, a matrix donates the token-level similarity between the dialogue representation H and document representation S,

$$M = S^{\mathrm{T}} H, \tag{3}$$

where $M \in R^{m \times k}$, and M_{ij} indicates the similarity between i-th document token and j-th dialogue token.

Doc2Dialog Attention: We use the Doc2Dialog Attention to signify which dialogue tokens are more relevant to each document token. First of all, we use a column softmax to get the attention weights α on dialogue tokens,

$$\alpha = softmax_{col}(M), \tag{4}$$

where $\alpha = [\alpha_1, \alpha_2, \ldots, \alpha_m]^{\mathrm{T}} \in R^{m \times k}$, $\alpha_i \in R^k$ indicates the attention weights on dialogue tokens for the i-th document token.

Then, we obtain the attended dialogue context $\widetilde{H}$ for the entire document,

$$\widetilde{H} = H\alpha^{\mathrm{T}}, \tag{5}$$

where $\widetilde{H} = [\widetilde{h}_1, \widetilde{h}_2, \ldots, \widetilde{h}_m] \in R^{2d \times m}$, $\widetilde{h}_i \in R^{2d}$ indicates the attended dialogue context for the i-th document token.

Dialog2Doc Attention: Dialog2Doc Attention signifies which document tokens have more closer similarity to the dialogue context.

First of all, we obtain the attention weights on the document tokens $\beta \in R^m$,

$$\beta = softmax(max_{col}(M)), \tag{6}$$

Then we obtain the attended document vectors $\widetilde{s} \in R^{2d}$.

$$\widetilde{s} = S\beta. \tag{7}$$

To maintain the contextual relevance, we fuse the results of Doc2Dialog Attention and Dialog2Doc Attention, then feed it into a bidirectional LSTM layer. Finally we obtain the dialogue-aware representations of document tokens $D \in R^{2d \times m}$,

$$G = [S \circ \widetilde{H}; S \circ (\widetilde{s} \otimes e_m)] \in R^{4d \times m}, \tag{8}$$

$$D = BiLSTM(G), \tag{9}$$

where the outer product $(\cdot \otimes e_m)$ produces a matrix or row vector by repeating the vector or scalar on the left for m times.

3.2   Decoder Module

In this section, we give a detailed introduction to the proposed Multi-Copy mechanism. Our model is primarily motivated by the Pointer Generator [11], which aims to handle the OOV problems by copying tokens from the dynamic dialogue context and generating tokens from the external vocabulary at the same time. In our task, the response tokens may come from the external vocabulary, the dialogue, and the document. To generate a document related response, we propose the Multi-Copy mechanism which allows generating tokens from vocabulary and copying tokens from both dialogue and document.

Given the dialogue-aware document representation D and dialogue representation H, we use a single layer LSTM as decoder, which receives the word embedding of the previous token, the document attention representation, dialogue attention representation, and the decoder hidden state.

Generate from Vocab: At each time step t, we use a two-layer fully connected network and a softmax function to map the decoder hidden state $d_t \in R^d$ into the probability distribution of each token in the external vocabulary. The probability of token w generated through vocabulary is $p^v(w)$:

$$p^v(w) = \begin{cases} softmax(W_2^v(W_1^v d_t + b_1^v) + b_2^v) \cdot e^w & \text{if } w \in vocabulary, \\ 0 & \text{otherwise.} \end{cases}$$

(10)

where $e^w \in R^v$ is a one-hot vector used to distinguish each token in the vocabulary, v is the external vocabulary size. $W_1^v \in R^{d \times d}$, $b_1^v \in R^d$, $W_2^v \in R^{v \times d}$, and $b_2^v \in R^v$ are parameters to be learned.

Copy from Dialogue: We first obtain the attention weights $\gamma^h \in R^k$ over the contextual dialogue tokens and the dialogue context vector $C^h \in R^{2d}$.

$$F^h = tanh(W^f H + W^b d_t \otimes e_k),$$

(11)

$$\gamma^h = softmax(F^{h^{\mathsf{T}}} W^\gamma + b^f \otimes e_k),$$

(12)

$$C^h = H\gamma^h.$$

(13)

where $W^\gamma \in R^{2d}$, $b^f \in R$, $W^f \in R^{2d \times 2d}$, and $W^b \in R^{2d \times d}$ are parameters to be learned. $(\cdot \otimes e_k)$ follows the same definition as before.

γ_j^h also represents the probability of the j-th dialogue token. In many cases, a token may appear more than one times in dialogue context. $p^h(w)$ is the sum of all probabilities of the token "w".

$$p^h(w) = \begin{cases} \sum_{j:w_j=w} \gamma_j^h & \text{if } w \in dialogue, \\ 0 & \text{otherwise.} \end{cases}$$

(14)

Copy from Document: Similar to Copy from Dialogue context, we obtain the attention weights $\gamma^s \in R^m$, the token distribution in the document p^s and the attention weighted document $C^s \in R^{2d}$.

We feed the decoder state d_t, the attention weighted dialogue context C^h, and the attention weighted document C^s into a nonlinear neural network to obtain $\delta \in R^3$, which indicates the probabilities of choosing tokens from the external vocabulary, dialogue context or document.

$$\delta = softmax(W^d d_t + W^s C^s + W^h C^h + b).$$

(15)

where $W^d \in R^{3 \times d}$. $W^s, W^h \in R^{3 \times 2d}$, and $b \in R^3$.

Then we obtain the final token distribution p,

$$p(w) = \delta_1 p^v(w) + \delta_2 p^h(w) + \delta_3 p^s(w),$$

(16)

where $w \in V \cup S \cup H$. $\delta_1 p^v(w)$, $\delta_2 p^h(w)$, $\delta_3 p^s(w)$ represent probabilities of the token w from vocabulary, dialogue and document respectively. We have a detailed description in Sect. 4.4.

3.3 Training

In the training stage, the loss for timestep t is the negative log likelihood of the target word w_t^* for that timestep:

$$loss_t = -log\ p(w_t^*),\tag{17}$$

and the overall loss for the whole sequence is:

$$loss = \frac{1}{T}\sum_{t=1}^{T} loss_t,\tag{18}$$

4 Experiment

4.1 Experimental Settings

Dataset: We conduct experiments on the Document Grounded Conversations dataset (CMU_DOG). It is built by Carnegie Mellon University in 2018. The conversations in CMU_DOG are about the contents of a specified document(Wikipedia articles about popular movies). The dataset contains 4112 conversations with an average of 21.43 turns per conversation. According to the visibility of the documents by two interlocutors, CMU_DOG is divided into Scenario 1 and Scenario 2[1]. In Scenario 1, only one interlocutor has access to the document. The interlocutor who has no access to the document asks the other one to get information. In Scenario 2, both the interlocutors have access to the same Wiki document. They discuss the content in the document. In this paper, we experiment in Scenario1 and Scenario 2 respectively.

In order to evaluate our model's ability to answer question correctly, we automatically generate 98 questions about movies in Scenario 1. The answers of all questions can be found in the documents.

Baselines: We use the SEQ and SEQS proposed in the Zhou et al. [19] as our baseline models, which are only document-driven dialogue models as far as we know.

Implement Details: In all the models explored in this paper, we keep previous two utterances as dialogue history[2]. We use a two-layer bidirectional LSTM as an encoder. The dropout rate is set to be 0.3. The batch size is 32. The size of hidden units for both LSTMs is 300. The size of the vocabulary is 10000. We cut off the first 100 words in the documents during training. We use the pre-trained 100-dimensional glove embedding[3] and fine-tune it during the training. The models are trained with Adam optimizer [3] with learning rate 0.001 until they converge on the validation set for the valid loss. During the test, we use beam search with size 5.

---

[1] Scenario1 contains 2128 conversations and Scenario 2 contains 1984 conversations.

[2] After many experiments, the results obtained by using the previous two utterances as dialogue history are the best.

[3] https://nlp.stanford.edu/projects/glove/.

Table 1. NW and Doc_BLEU scores in Scenario 1 & 2

Models	Scenario 1		Scenario 2	
	NW	Doc_BLEU	NW	Doc_BLEU
SEQ	0.245	0.153	0.160	0.181
SEQS	0.467	0.218	0.186	0.237
D3G	**0.712**	**0.345**	**0.209**	**0.262**

Table 2. Targe_BLEU and METEOR scores in Scenario 1 & 2

Models	Scenario 1					Scenario 2				
	B-1	B-2	B-3	B-4	METEOR	B-1	B-2	B-3	B-4	METEOR
SEQ	0.086	0.061	0.054	0.052	0.016	0.031	0.021	0.019	0.019	0.016
SEQS	0.097	0.068	0.060	0.057	0.025	0.029	0.020	0.019	0.018	0.019
D3G	**0.142**	**0.084**	**0.069**	**0.065**	**0.044**	**0.059**	**0.042**	**0.037**	**0.036**	**0.024**

Evaluation Metrics: Doc_BLEU[4] and NW [18] are used to measure the relevance between generated responses and documents. Target_BLEU [10] and METEOR [4] scores are employed to measure the similarity between generated responses and standard outputs. We also report the diversity [7] and the average length of the generated responses. We use the accuracy to evaluate the question answering.

4.2   Experimental Results

Table 1 shows the NW and Doc_BLEU scores in Scenario1 and Scenario2. As we can see, our D3G model significantly outperforms the baseline models on both datasets. It demonstrates that our D3G model can generate more document-related responses. The improvements of NW and Doc_BLUE in Scenario1 are more than 24.5% and 12.7% compared to SEQS. Compared to Scenario 1, the conversations in Scenario 2 are freer and much lower related to the document, so the improvements are relatively small in Scenario2.

Table 2 shows the Target_BLEU and the METEOR scores in Scenario1 and Scenario2. Compared to the baseline models, our D3G model slightly improves the BLEU score and METEOR score. In combination with Table 1, it shows that responses generated by our model are not only document-related, but also with high quality.

Table 3 shows the Dist and Avg_len scores in Scenario 1 and Scenario 2. It demonstrates that D3G model can generate longer and more diverse responses than that generated by SEQ and SEQS.

---

[4] We only use the Bleu-1 and ignore the brevity penalty. Moreover, we use nltk to calculate the BLEU and smooth it with SmoothingFunction().method2.

Table 3. Diversity scores and average lengths in Scenario 1 & 2

Models	Scenario 1			Scenario 2		
	Dist-1	Dist-2	Avg_len	Dist-1	Dist-2	Avg_len
SEQ	0.017	0.037	3.85	0.008	0.024	2.83
SEQS	0.024	0.061	4.99	0.016	0.056	2.91
D3G	**0.071**	**0.312**	**7.27**	**0.046**	**0.205**	**4.96**

Table 4. Accuracy rate

Models	Acc (%)
SEQ	8.16
SEQS	11.22
D3G	**16.32**

Table 5. Ablation study in Scenario1

Models	NW	Doc_BLEU	Target_BLEU	METEOR	AVG_LEN	Dist-1	Dist-2	Acc
D3G	0.712	0.345	0.065	0.044	7.27	0.071	0.312	16.32
-Multi-Copy	0.550	0.312	0.059	0.034	6.89	0.061	0.231	13.27
-Doc-Reader	0.651	0.299	0.065	0.039	6.44	0.069	0.305	12.24
-History	0.476	0.293	0.059	0.034	5.65	0.064	0.253	13.27

Table 6. Ablation study in Scenario2

Models	NW	Doc_BLEU	Target_BLEU	METEOR	AVG_LEN	Dist-1	Dist-2
D3G	0.209	0.262	0.036	0.024	4.96	0.046	0.205
-Multi-Copy	0.189	0.255	0.031	0.021	4.49	0.043	0.196
-Doc-Reader	0.174	0.261	0.029	0.022	4.91	0.044	0.193
-History	0.181	0.234	0.024	0.021	4.55	0.037	0.148

Table 4 shows the accuracy rates on our 98 automatically generated questions. It also validates the superiority of our D3G model. There is an improvement of 5% in the accuracy rate. Our D3G model can generate better responses relative to the baseline models.

In addition, our model performs better in Scenario 1 than in Scenario 2. One possible reason is that both the interlocutors in Scenario 2 can see the content of the document, which has high degree of freedom and relatively complex sentences.

4.3   Ablation Study

To validate the effectiveness of our D3G model, we conduct three ablation experiments in Scenario1 and 2 separately since two datasets have different characteristics.

Table 7. Examples of question answering

ID	Document	Inputs	Responses
1	**Name** Toy Story **Year** 1995 **Genre** Comedy, Animation, Adventure **Director** John Lasseter **Introduction** Toy Story is a 1995 American... comedy adventure film ... as they evolve from rivals ... Catmull	What is the movie name?	**SEQ:** the wolf of wall street **SEQS:** the movie is the shape of water **Ours:** toy story
2	**Name** The Wolf of Wall Street **Year** 2013 **Genre** black comedy **Director** Martin Scorsese **Introduction** The Wolf of Wall Street is a 2013 American black comedy[3] ... Leonardo DiCaprio (who was also a producer) stars as Belfort... also star	What is the movie name?	**SEQ:** have you seen it? **SEQS:** the movie is based on a true story **Ours:** the wolf of wall street
3	**Name** Zootopia **Year** 2016 **Genre** Comedy, Adventure, Animation **Director** Byron Howard, Rich Moore **Introduction** In a world populated ... her childhood dream of becoming ... Nick Wilde and Finnick	Awesome! I bet my kids would like it, is it old?	**SEQ:** nod **SEQS:** i do n't know **Ours:** it came out in 2016

- The full D3G model with all of the components.
- D3G model without Multi-Copy mechanism.[5]
- D3G model without Doc-Reader mechanism.
- D3G model without dialog history.

The results are shown in Tables 5 and 6 respectively. Table 6 removes the accuracy since the document is accessible for both interlocutors in Scenario 2, there is no need to ask questions.

We can draw that the methods proposed by D3G are indispensable for all evaluation metrics on both Scenarios. Firstly, Multi-Copy mechanism is very important to D3G model. All scores decrease when Multi-Copy is moved from D3G model. Especially, the NW score decreases significantly from 0.712 to 0.550 in Scenario 1. It shows that Multi-Copy not only has a great impact on generating document-related responses, but also has some influences on diversity and fluency of responses.

Secondly, the Doc-Reader mechanism leads to a significant improvement on accuracy as shown in Table 5. The accuracy of questions answering drops significantly from 16.32 to 12.24. This shows that the Doc-Reader mechanism can effectually help locate the content related to user's questions.

Finally, as shown in Tables 5 and 6, removing the dialogue history causes drop on all scores. Because the dialogue history is used in both of the above mechanisms, removing history will have an impact on them.

4.4   Case Study

In this part, we show typical examples of generated responses of answering question (as shown in 7) and chatting with users (as shown in Table 8).

---

[5] We follow SEQ, only copy tokens from dialogue.

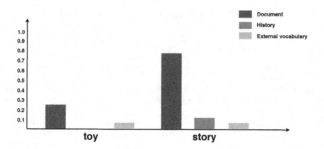

Fig. 2. Probabilities of the token in different sources

Table 8. Examples of chatting with users

ID	Document	Inputs	Responses
1	**Name** The Avengers **Year** 2012 **Genre** Drama, Superhero **Director** Joss Whedon **Introduction** Marvel's The Avengers is a 2012 American ... Robert Downey Jr., Chris Evans, ... form a team that must stop Thor's brother Loki from subjugating Earth	No way. Robert Downey Jr makes the character so funny, but so cool	**SEQ:** good **SEQS:** it no on then not together **Ours:** it do a very cool film really best
2	**Name** Imitation Game **Year** 2014 **Genre** Drama, biopic **Director** Morten Tyldum **Introduction** The Imitation Game is a ... Benedict Cumberbatch decrypted German intelligence codes for the British government during the Second World War ... also star	Actually, what I enjoyed most was Benedict Cumberbatch's performance He was excellent!	**SEQ:** en **SEQS:** yeah, it is good **Ours:** He made a great supporting role

Table 7 shows the responses to user's questions. For example, in Item1, our proposed D3G model can generate correct movie name "Toy Story". The SEQ model can just generate the high frequency movie name in the train set. The SEQS model that uses the document also doesn't generate the correct answer.

We visually analyze the reasons why the D3G model accurately answers user's questions. As shown in the Fig. 2, D3G allows copying tokens from multi-sources (dialogue, document, external vocabulary). The token "Toy" has a low probability in the external vocabulary, but highest in the document. Combining the above probabilities, the final output token is "Toy".

Table 8 shows the responses to chat with users. For example, in Item 2, the SEQ model can only generate short and general responses, such as "en". Our D3G model can generate "He made a great supporting role", which is meaningful, diverse and natural.

5 Conclusion

In this paper, we propose a Document Driven Dialogue Generation model, called D3G. Experimental results on CMU_DOG dataset show that D3G model outperforms state-of-art approaches. Ablation study shows that Doc-Reader mechanism effectively helps to locate the user's question related to document contents. Multi-Copy mechanism plays an important role to generate document-related responses. Finally, we visually analyze the reasons why our model works.

References

1. Cui, W., Xiao, Y., Wang, H., Song, Y., Wei, W.: KBQA: learning question answering over QA corpora and knowledge bases. Proc. VLDB Endow. **10**(5), 565–576 (2017)
2. Ghazvininejad, M., et al.: A knowledge-grounded neural conversation model. In: Thirty-Second AAAI Conference on Artificial Intelligence (2018)
3. Kingma, D.P., Ba, J.: Adam: a method for stochastic optimization. Computer Science (2014)
4. Lavie, A., Agarwal, A.: METEOR: an automatic metric for MT evaluation with high levels of correlation with human judgments. In: Proceedings of the Second Workshop on Statistical Machine Translation, pp. 228–231. Association for Computational Linguistics (2007)
5. Lei, W., Jin, X., Kan, M.Y., Ren, Z., He, X., Yin, D.: Sequicity: simplifying task-oriented dialogue systems with single sequence-to-sequence architectures. In: Proceedings of the 56th Annual Meeting of the Association for Computational Linguistics (vol. 1: Long Papers), pp. 1437–1447 (2018)
6. Levin, E., Narayanan, S.S., Pieraccini, R., Zeljkovic, I.: Method of using a natural language interface to retrieve information from one or more data resources. AT & T (1999)
7. Li, J., Galley, M., Brockett, C., Gao, J., Dolan, B.: A diversity-promoting objective function for neural conversation models. arXiv preprint. arXiv:1510.03055 (2015)
8. Lian, R., Xie, M., Wang, F., Peng, J., Wu, H.: Learning to select knowledge for response generation in dialog systems (2019)
9. McTear, M.F.: Modelling spoken dialogues with state transition diagrams: experiences with the CSLU toolkit. In: Fifth International Conference on Spoken Language Processing (1998)
10. Papineni, K., Roukos, S., Ward, T., Zhu, W.J.: BLEU: a method for automatic evaluation of machine translation. In: Proceedings Meeting of the Association for Computational Linguistics (2002)
11. See, A., Liu, P.J., Manning, C.D.: Get to the point: summarization with pointer-generator networks (2017)
12. Seo, M., Kembhavi, A., Farhadi, A., Hajishirzi, H.: Bidirectional attention flow for machine comprehension (2016)
13. Serban, I.V., Sordoni, A., Bengio, Y., Courville, A., Pineau, J.: Building end-to-end dialogue systems using generative hierarchical neural network models. In: Thirtieth AAAI Conference on Artificial Intelligence (2016)
14. Vinyals, O., Le, Q.: A neural conversational model. Computer Science (2015)
15. Yang, Z., He, X., Gao, J., Li, D., Smola, A.: Stacked attention networks for image question answering (2016)
16. Young, S., Gasic, M., Thomson, B., Williams, J.D.: Pomdp-based statistical spoken dialog systems: a review. Proc. IEEE **101**(5), 1160–1179 (2013)
17. Zhao, T., Lu, A., Lee, K., Eskenazi, M.: Generative encoder-decoder models for task-oriented spoken dialog systems with chatting capability (2017)

18. Zhou, H., Young, T., Huang, M., Zhao, H., Xu, J., Zhu, X.: Commonsense knowledge aware conversation generation with graph attention. In: IJCAI, pp. 4623–4629 (2018)
19. Zhou, K., Prabhumoye, S., Black, A.W.: A dataset for document grounded conversations (2018)
20. Zhu, C., Zeng, M., Huang, X.: SDNet: contextualized attention-based deep network for conversational question answering (2018)

Capsule Networks for Chinese Opinion Questions Machine Reading Comprehension

Longxiang Ding[1], Zhoujun Li[1(✉)], Boyang Wang[1], and Yueying He[2]

[1] State Key Laboratory of Software Development Environment,
Beihang University, Beijing, China
{antdlx,lizj,wangboyang}@buaa.edu.cn
[2] National Computer Network Emergency Response Technical
Team/Coordination Center of China, Beijing, China
hyy@cert.edu.cn

Abstract. In recent years, machine reading comprehension is becoming a more and more popular research topic. Promising results were obtained when the machine reading comprehension task had only two inputs, context and query. In this paper, we propose a capsule networks based model for Chinese opinion machine reading comprehension task which has three inputs: context, query and alternatives. First, we use a bi-directional LSTM to encode the three inputs. Second, model the complex interactions between context and query with a multiway attention layer. In addition to the attention mechanism used in BiDAF, the other two attention functions are designed to match the relationship between inputs. Finally, we present a capsule networks layer to route the right alternative. Specifically, we use two strategies to improve the dynamic routing process to filter noisy capsules, which may contain useless information such as stop words. Our single model achieves competitive results compared to the baseline methods on a Chinese dataset and obtains a significant improvement of 2.45% accuracy.

Keywords: Capsule networks · Machine reading comprehension · Multiway attention

1 Introduction

The tasks of machine reading comprehension (MRC) and automated question answering (QA) have gained great popularity. In this paper, we mainly focus on opinion questions MRC task which needs to use the information of multiple sentences in the whole article for comprehensive analysis to get the correct answer. The differences between them can be listed as follows. First, the standard MRC is a QA task, while the opinion questions MRC is essentially an answer selection task. Second, standard MRC generally only has two inputs, context and query. Whereas opinion questions MRC has alternative inputs additionally. To better explain the difference, we present two examples in Table 1. There are two kinds of opinion questions, first kind of these are the alternatives which are appeared in context like example-1, the other alternatives are "YES" and "NO" which need models do some reasoning like example-2.

© Springer Nature Switzerland AG 2019
M. Sun et al. (Eds.): CCL 2019, LNAI 11856, pp. 521–532, 2019.
https://doi.org/10.1007/978-3-030-32381-3_42

Over the past few years, lots of end-to-end models show promising results on a variety of tasks. One of the latest advancements is BERT [1]. However, Chinese pre-trained BERT only supports 512 characters as inputs. For many Chinese MRC tasks, the input length limitation is so short and the cost is too huge to re-train it. It is thus necessary to design a light-weight and powerful model. In addition, other successful models generally have three key components. First, using an RNN or CNN based embedding layer to process sequential inputs. Second, an attention layer to model the interactions between inputs. Third, a neural based encoder to match specific tasks, such as BiDAF [2] v r-net [3] and Qanet [4]. These standard models basically support only two inputs but opinion questions MRC task has three inputs: context, query and alternatives. Naturally, attention mechanism plays an important role in models, it is meaningful to design a new attention mechanism which could extract interactions as many as possible. Meanwhile, with the growth of complex information extracted, the noise is growing and the encoder of primitive works can't separate them clearly.

Table 1. Examples of MRC and opinion questions MRC

Task	Data
MRC	Context: Super Bowl 50 was an American football game to determine the champion of ...... Query: Which NFL team represented the AFC at Super Bowl 50? Outputs: start span, end span
Opinion questions MRC-example1	Context: 科学研究表明,在各种水果中...... Query: 吃什么水果降血脂? Alternatives: 苹果, 葡萄, 无法确定 Outputs: 苹果
Opinion questions MRC-example2	Context: 有研究表明,长期熬夜会导致...... Query: 长期熬夜对身体是否有害? Alternatives:是, 否, 无法确定 Outputs: 是

In this paper, aiming to match the opinion questions MRC task, we use three Bi-LSTMs [5] to encode the inputs. Subsequently, we design a multiway attention layer including the attention mechanism used in BiDAF, bilinear attention function used in [6] and an element-wise dot product attention function used in [7]. Then we design a capsule networks structure introduced in [8] to filter noise. The main thought behind the design of these networks is the following: convolution network captures the feature of the interactions between query and context, while the dynamic routing can aggregate noise together. Inspired by [9], we add two strategies to eliminate the interference of noisy capsules. Eventually, we encode each alternative to a capsule and pass them through a gate. By multiplying these capsules with attention mechanism outputs, our model will choose the right answer.

We conduct experiments on the latest Chinese opinion questions MRC dataset[1] with three competitive models (BiDAF, r-net, Qanet) as baselines. Our single model obtains an accuracy of 74.03% compared to the best baseline result of 71.58% (Qanet) on evaluation set. In summary, the contribution of this paper are as follows:

- We propose a strategy with capsule networks to filter noise in our model. When you concatenate lots of vectors which are come out from attentions or other algorithms, you may get more noise as well. It is meaningful to alleviate the disturbance of noise.
- We design a model based on capsule networks to deal with Chinese Opinions MRC task and get a good result.

2 Related Work

The MRC and QA have gained a significant improvement, an important contributor to the improvement is the high quality datasets, such as SQuAD [10], CNN/Daily News [11], Wikireading [12], Dureader [13] and the AI-Challenger2018 dataset [15] we used. A great number of end-to-end reading comprehension works achieve promising results, including BiDAF, r-net, Qanet, DCN [14], Document Reader [16] and Interactive AoA Reader [17].

Most of the successful models generally use attention mechanism. Basically, the interactions between two sentences can be modeled in sentence level and word level. Sentence level framework decides the relationship between sentences solely based on sentence vectors [20–22]. However, this kind of strategy ignores the lower interactions. Therefore, the word level framework proposes matching two sentence at word level. [23] used this strategy to improve LSTM-based network and achieved a better result. Our model uses two widely-used attention mechanisms [6, 7] as part of attention layer. Specific to MRC model, attention mechanism can be separated in two distinct ways. The first way is dynamic attention [22]. Given the context, query vectors and previous attention, the attention weights will be updated dynamically. Hermann made some experiments on CNN/Daily News and the result showed that dynamic attention actually made sense. The other way is computing attention weights only one time [23], when given a similarity matrix between query and context [17]. Then we can use the attention in subsequent modeling layers. BiDAF uses a memory-less dynamic attention mechanism which means it is not directly depending on the attention at previous time steps and we use this as part of our attention layer.

For opinion questions MRC, after modeling the interactions between query and context, we can use the attention information and alternative vectors to select the right answer. This challenge can be considered as a text classification task. Primitive algorithms used typical features such as n-grams, POS tags as inputs. Specifically, Support vector machine(SVM) [24], random forest [25] and xgboost [26] are common used on

---

[1] This dataset was published by AI-Challenger2018 and available at: challenger.ai/competition/oqmrc2018.

text classification. In recent years, neural networks, especially LSTMs [5] and CNNs [27], have largely improved the performance of text classification task. Kim migrated CNN from computer vision domain to natural language processing domain on sentence classification [27]. [28] proposed fastText model which could be trained on a huge dataset in a few minutes. However, CNN and RNN have lots of weakness. [29] firstly introduced the concept of capsule to cope with the representational limitations of CNNs and RNNs. Then [8] designed a capsule network on MNIST task. They replaced the CNN scalar outputs with capsule vectors and max-pooling with dynamic routing algorithm. At the same time, [8] designed two methods to filter noisy capsules. Recently, [9] investigated capsule networks with dynamic routing for text classification and the result showed its potential on classification task.

3   Our Model

In this section, we first describe the opinion questions MRC task and then introduce our model in detail.

3.1   Task Definition

The opinion questions MRC task which is investigated in this paper, is defined as follows. Given a context passage with n words $C = \{c_1, c_2,...,c_n\}$, a query with m words $Q = \{q_1, q_2, ..., q_m\}$ and a group of k alternatives $A = \{A_1, A_2,...,A_k\}$, A_i is an answer segment with L words $A_i = \{a_1, a_2, ..., a_L\}$, output a label y representing the best alternative to answer the query from the context. Specifically, $y \in \{1,2,...,k\}$.

3.2   Model Overview

Our model can be separated to five components in high level structure (Fig. 1): an embedding layer, a contextual embedding layer, an attention layer, a model encoder layer and an output layer, it's a common strategy for most MRC models. However, the major differences between ours and other models are as follows: we use multiway attention mechanism and aggerate attentions information for subsequent layers. As a result, our networks model the interactions between query and context better. Otherwise, we use capsule networks in the model encoder layer, this makes our model filters lots of noise and gets a 2.45% accuracy gain in our experiments. Compared to the other MRC tasks, we have three inputs, so we cope with the information from model encoder layer and alternative capsules in output layer.

Input Embedding Layer. Nowadays, obtaining the embedding of each word by concatenating its word embedding and character embedding is a popular strategy, we directly use this approach to get embeddings. For both word and character embedding, we produce the *dim* = 300 dimension Word2vec [30] word vectors by training data. All the out-of-vocabulary words are mapped to <unk> token which is random initialized and trainable. Specifically, every word can be viewed as the concatenation of each characters. Most of Chinese words have two to four characters, so we truncate or pad each word to the length of four. Following [27], we use a convolutional neural networks and a max-pooling to model the character level embeddings of words, then we can get an embedding X_c. We finally pass the concatenation of X_c and word embedding to a Highway Network [31] like what BiDAF did. Now, we have matrix $C \in \mathbb{R}^{d \times N}$ for the context, where d is the hidden size of highway network, matrix $Q \in \mathbb{R}^{d \times M}$ for the query and K matrices for each alternative $A_k \in \mathbb{R}^{d \times L}$.

Contextual Embedding Layer. We simply apply a bi-directional LSTM on the top of embeddings provided by previous layers. Here, we attain matrix $H \in \mathbb{R}^{2d \times N}$ from C, matrix $U \in \mathbb{R}^{2d \times M}$ from Q and K matrices $P_k \in \mathbb{R}^{2d \times L}$ for each alternative. By concatenating both directions of LSTM outputs, the first dimension of matrices is $2d$ now.

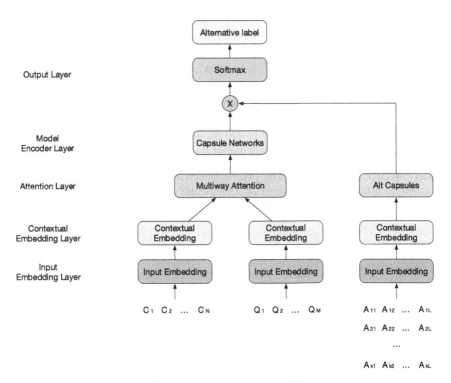

Fig. 1. Overview of our model

Attention Layer. This component is standard in almost every previous MRC models. BiDAF has made a significant improvement on attention mechanism, we directly use its structure as part of this layer. BiDAF attention uses H and U as inputs and outputs $B \in \mathbb{R}^{8d \times N}$ which represents the complex interactions between context and query. To extract more interactions, we also use another two attention mechanisms which will be introduced below. Therefore, we obtain a vector $\hat{G} \in \mathbb{R}^{12d \times N}$ by concatenating these vectors together. Finally, the layer uses a Bi-LSTM to fuse multiway attention interactions $'G$ and then outputs $G \in \mathbb{R}^{2d \times N}$.

We use h_t represents the t-th vector in H, u_i represents the i-th vector in U and W represents the parameters. The major functions of bilinear and dot attention are listed as follows.

Bilinear Attention:

$$s_i^t = Relu\left(u_i^T W_b h_t\right)$$

$$a_j^t = Softmax\left(s_j^t\right)$$

$$b\_att_t = a^t u$$

Dot Attention:

$$s_i^t = Relu(W_d(u_i \odot h_t))$$

$$a_j^t = Softmax\left(s_j^t\right)$$

$$d\_att_t = a^t u$$

where s_i^t represents the similarity of u_i and h_t, then we attain a_j^t by adopting softmax used to scale s_i^t. Finally, our functions output the attention results b_att_t and d_att_t.

Model Encoder Layer. Inspired by [8], we design a capsule networks layer to encode the attention information. It consists of 3 elements: convolutional layer, primary capsule layer and dynamic routing. The structure can be illustrated in Fig. 2.

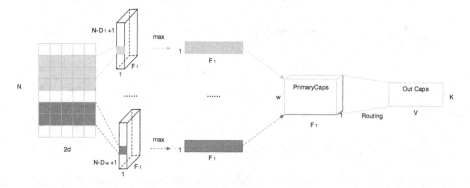

Fig. 2. Capsule networks structure

Convolutional Layer. This layer is just a standard convolutional layer aimed to extract n-gram features from G. Suppose $W_b \in R^{D \times 2d}$ denotes one filter where the n-gram size is D, we can produce a feature vector $m_i \in R^{(N-D+1) \times 1}$ with stride of 1 for each filter.

When we have F_1 filters and w kinds of filter sizes, the outputs are $\boldsymbol{Cov} \in \mathbb{R}^{(N-D+1) \times 1 \times F_1 \times w}$.

Primary Capsule Layer. This is a layer to produce our capsules called "primary capsule" from convolution outputs. Capsule vectors contain more information than scalars, such as position and semantic information of texts. We take the maximum of first dimension to represent outputs of one filter. For F_1 filters of each filter size, we can adopt cov $\in \mathbb{R}^{1 \times F_1}$. Therefore, this layer outputs primary capsules $\boldsymbol{Pri} \in \mathbb{R}^{1 \times w \times F_1}$ by concatenating w kinds of filter size cov s.

Dynamic Routing. As argued in [8], dynamic routing is a more effective method than primitive routing strategies. It allows the networks to automatically learn relationships between the child capsule i to parent capsule j. First, generate prediction vectors (votes) by computing:

$$\hat{\mu}_{j|i} = W_{ij}^y \mu_i + \hat{b}_{j|i}$$

where $\hat{b}_{j|i}$ is bias item, and W_{ij}^y is the weights between two capsule layers. Then algorithm in an iterative manner would route each vote to an appropriate parent in the subsequent layer, meanwhile, noisy capsules can also be routed together. Initially, the votes routed to each parent and the networks can increase or decrease the weights by dynamic routing. Inspired by [9], we use two approaches to eliminate the interference of noisy capsules. First is the orphan category. To separate the noise such as stop words, we add an orphan category when computing softmax in dynamic routing and this would help our networks model the relationships between two capsule layers more efficiently. Second is using leaky-softmax which was implemented by Sara [32] to take the place of standard softmax. This function adds extra dimension to routing logits. When active capsule is not a good fit for any of the capsules in layer above, they will be routed to the extra dimension. The algorithm can be depicted in Fig. 3.

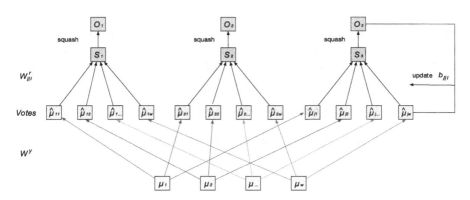

Fig. 3. The dynamic routing algorithm

Procedure 1: Dynamic Routing Algorithm

1: **procedure** Routing ($\hat{\mu}_{j|i}$, r, l)

2:   for all capsule i in layer l and capsule j in layer ($l+1$), initialize $b_{j|i} = 0$

3:   **for** r iterations **do**:

4:     for all capsule i in layer l: $W_{j|i}^r = |O_j| \cdot leaky\_softmax(b_{j|i})$

5:     for all capsule j in layer $l+1$: $O_j = squash(\sum W_{j|i}^r \hat{\mu}_{j|i})$

6:     for all capsule i in layer l and capsule j in layer $l+1$: $b_{j|i} = b_{j|i} + \hat{\mu}_{j|i} O_j$

7: **return** O_j

Given each vote $\hat{\mu}_{j|i}$ and its existence probability $|O_j|$, the connection weights can be updated by:

$$W_{j|i}^r = |O_j| \cdot leaky\_softmax(b_{j|i})$$

Then the output capsules are computed by:

$$O_j = squash\left(\sum W_{j|i}^r \hat{\mu}_{j|i}\right)$$

We view $|O_j|$ as the existence probability of j-th label. In detail, squash is a non-linear function to ensure that short vectors get shrunk to almost zero length and long vectors get shrunk to a length slightly below 1 [8]. Next, we can produce the coefficients used to update W^r by:

$$b_{j|i} = b_{j|i} + \hat{\mu}_{j|i} O_j$$

Finally, a new round of iteration will be started by the steps mentioned above. And the dynamic routing algorithm is summarized in Procedure 1.

Output Layer. This layer is application-specific. Formally, in opinion questions MRC task, we concatenate alternative capsules together to get **alt** capsules and initialize a **aux** capsule which has the same shape with alternative capsules. Then, a gate is designed to compute the **alt** and **aux** by:

$$gated = gate \cdot alt + (1 - gate) \cdot aux$$

Next, we use **gated** and **O** to obtain the similarity matrix:

$$Sim = gate \cdot O$$

To get the probability distribution of the alternative label, we add a softmax layer on the top of **Sim**. Finally, the objective function is:

$$L(W) = -\frac{1}{N}\sum_{N}^{i} log(softmax(\textbf{Sim}_i))$$

4   Experiments

4.1   Dataset and Implementation Details

In order to evaluate our model, we conduct a series of experiments on AI-Challenger2018 MRC dataset. It contains 250 K training data and 30 K evaluation data. Each data has a context, a query and three alternatives.

In the experiments, we produce Word2vec vectors in 300 dimension by training data to initialize embedding vectors. The batch size is 64 and we use Adam optimization. Meanwhile, we adopt a cosine decay restart algorithm [33] to transform the learning rate. The context sequence length is limited to 300 words, the query sequence length is limited to 50 words and each alternative sequence length is limited to 2 words.

4.2   Baselines

We choose three strong MRC models as baselines: QANet [4], BiDAF [2] and r-net [3]. For each baseline, we use their own embedding layer to encode the alternatives, then we add a softmax behind the product of alternative embeddings and attentions. All the experiments use the same word2vec vectors and learning rate transform function.

QANet. We find an open source code[2] from NLPLearn's github. The batch size is 64, learning rate is 0.001, dropout is 0.1 and epochs is 10.

BiDAF. DuReader had released an BiDAF baseline[3] on github and we just add alternatives embedding behind that. The batch size is 64, learning rate is 0.001, dropout is 0.05 and epochs is 8.

R-net. We still used NLPLearn's code[4] from github. The batch size is 64, learning rate is 0.001, dropout is 0.2 and epochs is 10.

In this paper, we prefer to focus on the performance of existing data and be limited by the inputs length, we do not choose pretrained model BERT [1] as a baseline.

4.3   Results and Discussion

Evaluation. In all of our experiments, the evaluation metric is accuracy. We summarize the results in Table 2. The *r-net*, *BiDAF* and *Qanet* represent the results of three

---

[2] This github url is: https://github.com/NLPLearn/QANet.

[3] This release can be found at: https://github.com/baidu/DuReader/tree/master/tensorflow.

[4] This github url is: https://github.com/NLPLearn/R-net.

baselines respectively. Moreover, we replace the attention layer with our multiway attention mechanism on *BiDAF-relu*. The *Capsule-O* model only consists the standard structure without any dynamic routing optimization methods and multiway attention mechanism. Based on *Capsule-O*, *Capsule-relu* adds bilinear and dot attentions. In addition, *Capsule-leaky* uses leaky-softmax to take the place of the standard one in dynamic routing on the base of *Capsule-relu*. Compared to *Capsule-leaky*, *Capsule-orphan* only optimize dynamic routing by add an orphan category. Finally, we use *Capsule-final* to test all the methods together.

Table 2. Results of our models and baselines

Model	Accuracy
r-net	70.22%
BiDAF	71.15%
BiDAF-relu	70.94%
Qanet	71.58%
Capsule-O	73.10%
Capsule-relu	73.47%
Capsule-leaky	73.67%
Capsule-orphan	73.92%
Capsule-final	**74.03%**

Discussion. From the table, we have shown capsule networks without any optimization achieves an impressive 1.52% gain than the best baseline, which verifies the effectiveness of it. Applying multiway attention mechanism and capsule networks to study opinion questions MRC is reliable. We find that multiway attention mechanism extracts more useful information but also noise than single attention mechanism. Furthermore, our results demonstrate that capsule networks filter noise better than traditional methods.

Comparing with *BiDAF* and *BiDAF-relu*, there is a slight decrease in accuracy after we used the multiway attention mechanism. This result means that multiway attention mechanism extracts more noise than single attention mechanism and the primitive model encoder can't filter noise well. Thus, the accuracy is reduced. Furthermore, results of *Capsule-relu* prove that capsule networks are better at noise filtering. We can also make a conclusion that multiway attention mechanism can give more useful interactions than noise comparing the results of *Capsule-relu* with *Capsule-O*. Moreover, both leaky-softmax and orphan category which are used in dynamic routing to eliminate the interference of noisy capsules get significant improvement. This result tells us that routing child capsules to appropriate parent layer and filtering the noisy capsules are both important to opinion questions MRC task.

5   Conclusion and Future Work

In this paper, we propose a new end-to-end model for opinion reading comprehension task. Specifically, we explore a multiway attention mechanism to extract more useful interactions between context and query, two methods for dynamic routing process to eliminate the interference of noisy capsules. In addition, we investigate a strategy to cope with more than two inputs MRC tasks.

One direction of future work is using more powerful embedding vectors to enhance word representation such as ELMO [34]. Otherwise, RNN-based networks are often slow, maybe we can adopt other structure to replace it in contextual embedding layer such as what Qanet did.

Acknowledgments. This work is supported in part by the National Natural Science Foundation of China (Grand Nos. U1636211, 61672081, 61370126), and the National Key R&D Program of China (No. 2016QY04W0802).

We would like to thank lixinsu, sarasra, freefuiiismyname and andyweizhao. Their open source projects on github reduce our work on coding, thus we can take more time to focus on studying.

References

1. Devlin, J., et al.: Bert: pre-training of deep bidirectional transformers for language understanding. arXiv preprint arXiv:1810.04805 (2018)
2. Seo, M., et al.: Bidirectional attention flow for machine comprehension. arXiv preprint arXiv:1611.01603 (2016)
3. Wang, W., et al.: R-NET: machine reading comprehension with self-matching networks. Natural Language Computer Group, Microsoft Reserach. Asia, Beijing, China, Technical Report 5 (2017)
4. Yu, A.W., et al.: Qanet: combining local convolution with global self-attention for reading comprehension. arXiv preprint arXiv:1804.09541 (2018)
5. Hochreiter, S., Schmidhuber, J.: Long short-term memory. Neural Comput. 9(8), 1735–1780 (1997)
6. Chen, D., Jason, B., Manning, C.D.: A thorough examination of the cnn/daily mail reading comprehension task. arXiv preprint arXiv:1606.02858 (2016)
7. Wang, S., Jiang, J.: A compare-aggregate model for matching text sequences. arXiv preprint arXiv:1611.01747 (2016)
8. Sabour, S., Frosst, N., Hinton, G.E.: Dynamic routing between capsules. In: Advances in Neural Information Processing Systems (2017)
9. Zhao, W., et al.: Investigating capsule networks with dynamic routing for text classification. arXiv preprint arXiv:1804.00538 (2018)
10. Rajpurkar, P., et al.: Squad: 100,000+ questions for machine comprehension of text. arXiv preprint arXiv:1606.05250 (2016)
11. Hermann, K.M., et al.: Teaching machines to read and comprehend. In: Advances in Neural Information Processing Systems (2015)
12. Hewlett, D., et al.: Wikireading: a novel large-scale language understanding task over wikipedia. arXiv preprint arXiv:1608.03542 (2016)

13. He, W., et al.: Dureader: a chinese machine reading comprehension dataset from real-world applications. arXiv preprint arXiv:1711.05073 (2017)
14. Xiong, C., Zhong, V., Socher, R.: Dynamic coattention networks for question answering. arXiv preprint arXiv:1611.01604 (2016)
15. AI-Challenger2018 Homepage. https://challenger.ai/competition/oqmrc2018. Accessed 17 May 2019
16. Chen, D., et al.: Reading wikipedia to answer open-domain questions. arXiv preprint arXiv: 1704.00051 (2017)
17. Cui, Y., et al.: Attention-over-attention neural networks for reading comprehension. arXiv preprint arXiv:1607.04423 (2016)
18. Bowman, S.R., et al.: A large annotated corpus for learning natural language inference. arXiv preprint arXiv:1508.05326 (2015)
19. Yang, Y., Yih, W.T., Meek, C.: Wikiqa: a challenge dataset for open-domain question answering. In: Proceedings of the 2015 Conference on Empirical Methods in Natural Language Processing (2015)
20. Tan, M., et al.: Improved representation learning for question answer matching. In: Proceedings of the 54th Annual Meeting of the Association for Computational Linguistics, (Long Papers), vol. 1 (2016)
21. Rocktäschel, T., et al.: Reasoning about entailment with neural attention. arXiv preprint arXiv:1509.06664 (2015)
22. Bahdanau, D., Cho, K., Bengio, Y.: Neural machine translation by jointly learning to align and translate. arXiv preprint arXiv:1409.0473 (2014)
23. Kadlec, R., et al.: Text understanding with the attention sum reader network. arXiv preprint arXiv:1603.01547 (2016)
24. Joachims, T.: Text categorization with support vector machines: learning with many relevant features. In: Nédellec, C., Rouveirol, C. (eds.) ECML 1998. LNCS, vol. 1398, pp. 137–142. Springer, Heidelberg (1998). https://doi.org/10.1007/BFb0026683
25. Liaw, A., Wiener, M.: Classification and regression by randomForest. R News 2(3), 18–22 (2002)
26. Chen, T., Carlos G.: Xgboost: a scalable tree boosting system. In: Proceedings of the 22nd Acm Sigkdd International Conference on Knowledge Discovery and Data Mining. ACM (2016)
27. Kim, Y.: Convolutional neural networks for sentence classification. arXiv preprint arXiv: 1408.5882 (2014)
28. Joulin, A., et al.: Bag of tricks for efficient text classification. arXiv preprint arXiv:1607. 01759 (2016)
29. Hinton, Geoffrey E., Krizhevsky, A., Wang, Sida D.: Transforming auto-encoders. In: Honkela, T., Duch, W., Girolami, M., Kaski, S. (eds.) ICANN 2011. LNCS, vol. 6791, pp. 44–51. Springer, Heidelberg (2011). https://doi.org/10.1007/978-3-642-21735-7_6
30. Mikolov, T., et al.: Efficient estimation of word representations in vector space. arXiv preprint arXiv:1301.3781 (2013)
31. Srivastava, R.K., Greff, K., Schmidhuber, J.: Highway networks. arXiv preprint arXiv:1505. 00387 (2015)
32. Sara Github. https://github.com/Sarasra/models/tree/master/research/capsules. Accessed 20 May 2019
33. Loshchilov, I., Hutter, F.: Sgdr: Stochastic gradient descent with warm restarts. arXiv preprint arXiv:1608.03983 (2016)
34. Peters, M.E., et al.: Deep contextualized word representations. arXiv preprint arXiv:1802. 05365 (2018)

Table-to-Text Generation via Row-Aware Hierarchical Encoder

Heng Gong, Xiaocheng Feng, Bing Qin[(✉)], and Ting Liu

Research Center for Social Computing and Information Retrieval,
Harbin Institute of Technology, Harbin, China
{hgong,xcfeng,qinb,tliu}@ir.hit.edu.cn

Abstract. In this paper, we present a neural model to map structured table into document-scale descriptive texts. Most existing neural network based approaches encode a table record-by-record and generate long summaries by attentional encoder-decoder model, which leads to two problems. (1) portions of the generated texts are incoherent due to the mismatch between the row and corresponding records. (2) a lot of irrelevant information is described in the generated texts due to the incorrect selection of the redundant records. Our approach addresses both problems by modeling the row representation as an intermediate structure of the table. In the encoding phase, we first learn record-level representation via transformer encoder. Afterwards, we obtain each row's representation according to their corresponding records' representation and model row-level dependency via another transformer encoder. In the decoding phase, we first attend to row-level representation to find important rows. Then, we attend to specific records to generate texts. Experiments were conducted on ROTOWIRE, a dataset which aims at producing a document-scale NBA game summary given structured table of game statistics. Our approach improves a strong baseline's BLEU score from 14.19 to 15.65 (+10.29%). Furthermore, three extractive evaluation metrics and human evaluation also show that our model has the ability to select salient records and the generated game summary is more accurate.

Keywords: Table-to-Text generation · Seq2Seq · Hierarchical encoder

1 Introduction

We focus on table-to-text generation that maps structured statistical data to document-level natural language texts [8,19,22]. In this task, STAT[1] is multi-row multi-column table that consists of multiple records, which would be transformed to long descriptive texts in encoding-decoding process. Datasets are the main driver of progress for statistical approaches in table-to-text [10]. In recent years, Wiseman et al. [22] release ROTOWIRE, a more complex table-to-text generation dataset, which aims at producing a document-scale game news based

---

[1] We abbreviate the statistics table as STAT.

© Springer Nature Switzerland AG 2019
M. Sun et al. (Eds.): CCL 2019, LNAI 11856, pp. 533–544, 2019.
https://doi.org/10.1007/978-3-030-32381-3_43

Team	PTS	PTS_QTR1	PTS_QTR2	WIN	LOSS	...
Raptors	104	28	28	21	8	...
Jazz	98	29	22	18	13	...

Player	PTS	AST	REB	STL	...
Raptors					
DeMar DeRozan	24	1	6	4	...
Kyle Lowry	36	5	4	2	...
Jonas Valanciunas	14	0	7	1	...
Terrence Ross	10	1	1	0	...
Cory Joseph	4	3	1	1	...
Fred VanVleet	N/A	N/A	N/A	N/A	...
...	...	...	...	...	...
Jazz					
Rudy Gobert	13	1	14	1	...
Gordon Hayward	23	5	1	1	...
Trey Lyles	19	1	7	0	...
Shelvin Mack	17	5	4	1	...
Rodney Hood	2	1	0	0	...
Joe Johnson	7	0	0	0	...
Raul Neto	0	0	1	0	...
George Hill	N/A	N/A	N/A	N/A	...
...	...	...	...	...	...

STAT

Baseline results

The Toronto Raptors defeated the Utah Jazz , 104 - 98 ... The Raptors were without **DeMar DeRozan** (ankle) , who missed the previous two games with an ankle injury . **Kyle Lowry** led the way for the Raptors with a game - high 36 points , which he supplemented with five assists , four rebounds and two steals . **DeMar DeRozan** followed with 24 points , six rebounds , an assist and four steals . **Jonas Valanciunas** posted a 14 - point , 14 - rebound double - double that also included a block , a steal and a block . Cory Joseph led the bench with 20 points , three rebounds , three assists and a steal . Fred VanVleet paced the reserves with 17 points , six rebounds , an assist and a steal . The Raptors were led by a pair of 24 - point efforts from Rodney Hood and Joe Johnson , with the former adding seven rebounds , an assist and a steal , and the latter supplying seven rebounds , an assist and a steal . George Hill was right behind him with 36 points , five assists , four rebounds and a steal . **Rudy Gobert** was next with a 13 - point , 14 - rebound double - double that also included an assist , a steal and a block . George Hill was next with 17 points , five assists , four rebounds , two steals and a block . Rudy Gobert posted a 13 - point , 14 - rebound double - double that also included an assist , a steal and a block . Raul Neto was productive in a reserve role as well with 17 points , five rebounds , an assist and a steal ...

Fig. 1. Generated example on ROTOWIRE by using a strong baseline (conditional copy [22]). Red words mean they are in line with statistics in the table. Blue words mean they contradict table's information or they are not the salient players to be mentioned. Underscored player names indicate that the model correctly select those salient players in line with reference. (Color figure online)

on plenty of NBA game records. Figure 1 shows an example of parts of a game's statistics and its corresponding computer generated summary. In STAT, each row is a player or team's name and corresponding records, each column is the player or team's attributes, such as points, steals, rebounds, etc. This dataset has an order of magnitude longer target text length with significantly more records than conventional datasets, such as WIKIBIO [8] or WEATHERGOV [11].

In table-to-text task, a lot of neural models [2,13,15,20] are developed, which encode a table record-by-record and generate a long descriptive summary by a decoder with attention mechanism [1,14]. However, we claim that these two general designs are problematic toward complex statistics. (1) portions of generated texts are incorrect due to the mismatch between the row and the corresponding records. As shown in Fig. 1, player *Fred VanVleet* and *George Hill* didn't play in this game, but baseline model wrongly believe they did and use others' scores to describe them. Also, player *DeMar DeRozan* did play in this game, but baseline model thinks he was absent due to injury at the beginning of the texts. Baseline model also use others' information to describe *Jonas Valanciunas, Cory Joseph,* etc. (2) a large portion of records in a table are redundant and attention mechanism is insufficient to properly select salient information. Based on extractive evaluation results on reference there are 628 records per game in ROTOWIRE corpus, but only an average of 24.14 records are mentioned in the summary. As shown in Figs. 1 and 3, base model fails to include players with impressive performance such as *Gordon Hayward* and *Trey Lyles* in the generated texts while include players like *Cory Joseph, Fred VanVleet, Rodney Hood, Joe Johnson, George Hill* and *Raul Neto* who don't appear in the reference.

To address the aforementioned problems, we present a Seq2Seq model with hierarchical encoder that considers the rows of the table as an intermediate representation structure. The approach is partly inspired by the success of structure-aware neural network approaches in summarization [4]. Since records in the table are not sequential, we use transformer encoder [21] to learn record-level representation of the table. Then, we obtain each row's representation considering the corresponding records' representation, modeling dependencies between rows via row-level transformer encoder. The advantage is that our encoding strategy can model the dependencies between players. We conduct experiments on ROTOWIRE. Results show that our approach outperforms existing systems, improving the strong base model's BLEU to 15.65 (+1.46), RG Precision to 93.26 (+18.46) , CS F1% to 38.46 (+5.97) and CO to 18.24 (+2.82).

2 Background

2.1 Task Definition

We model the document-scale data-to-text generation task in an end-to-end fashion. Statistics STAT consists of multiple records $\{r_{1,1}, ..., r_{i,j}..., r_{R,C}\}$ where R is the number of rows and C is the number of columns. Each record is in form of tuple $(row_{i,j}, column_{i,j}, cell_{i,j}, feat_{i,j})$ where $row_{i,j}$ refers to the row's name like player's name or team's name, $column_{i,j}$ refers to the tuple's type like points, $cell_{i,j}$ represents the value of this tuple and $feat_{i,j}$ indicates whether the player or team plays in their home court or not. Given STAT about one game, the model is expected to generate a document-scale summary of this game SUM $= \{w_1, ..., w_j, ..., w_N\}$ with w_j being the jth word in the summary. N is the number of words in the summary.

2.2 Base Model

We use Seq2Seq model with attention mechanism [14] and conditional copy mechanism [6] as the base model because it achieves best performance across BLEU and extractive evaluation metrics according to Wiseman et al. [22]. First, given STAT about one game, the encoder utilizes 1-layer MLP to map the representation of each record $r_{i,j} = (emb(row_{i,j}) \circ emb(column_{i,j}) \circ emb(cell_{i,j}) \circ emb(feat_{i,j}))$ into a dense representation $h_{i,j}$ separately. Then, given the representation of STAT $\tilde{H} = \{h_{1,1}, ..., h_{i,j}, ..., h_{R,C}\}$, a LSTM decoder decomposes the probability of choosing each word at time step t into two parts as shown in Eq. 1. The binary variable z_t indicates whether the word is copied from table or generated from vocabulary. $z_t = 1$ means the word is copied and $z_t = 0$ indicates the word is generated. In this paper, we use a 1-layer MLP with sigmoid function as its activation function to compute the probability $P(z_t = 1|y_{<t}, S)$. $P_{gen}(y_t|y_{<t}, \tilde{H})$ is computed by the LSTM decoder considering previous generated word and attentional representation of $\tilde{H}$. We reuse the attention weight $\alpha_{t,i',j'}$ as the probability of copying each word in the table. The model is trained to minimize the negative log-likelihood of words in reference in an end-to-end fashion.

$$P(y_t, z_t|y_{<t}, \tilde{H}) =$$
$$\begin{cases} P(z_t = 1|y_{<t}, \tilde{H}) \sum_{y_t \leftarrow r_{i',j'}} \alpha_{t,i',j'} & z_t = 1 \\ P(z_t = 0|y_{<t}, \tilde{H}) P_{gen}(y_t|y_{<t}, \tilde{H}) & z_t = 0 \end{cases} \tag{1}$$

3   Approach

In this section, we present a row-aware Seq2Seq model with hierarchical encoder. Given a STAT(table) as input, the model output corresponding descriptive texts. We divide the model into following parts. Figure 2 presents a detailed illustration of our model.

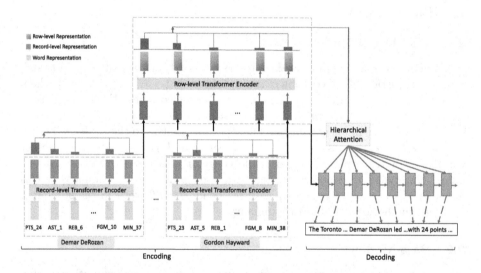

Fig. 2. The architecture of our proposed model. The left part is the hierarchical transformer encoder that encode table at both record-level and row-level. Right part is the LSTM decoder with hierarchical attention that generate words by first attending to rows then attending to each records.

3.1   Hierarchical Encoder

We explore two hierarchical encoder structures to encode game statistics on both record-level and row-level in this paper.

Firstly, we explore hierarchical bi-LSTM encoder structure, which can take other rows or records from the same row into consideration when modeling each row or record. On record level, we use a bi-LSTM to model records of the same row. For example, for row i, we take $r_{i,j}$, which is described in Sect. 2.2, as input for each time step, then obtain records' representation of row i $\{\tilde{h}_{i,1},..., \tilde{h}_{i,j},..., \tilde{h}_{i,C}\}$ by bi-LSTM.

On row level, we concatenate $\overrightarrow{\tilde{h}_{i,C}}$ (C is the number of columns) and $\overleftarrow{\tilde{h}_{i,1}}$ to represent row_i's overall performance h_i. $\overrightarrow{\tilde{h}_{i,T}}$ is the last time step for forward encoder for row i. $\overleftarrow{\tilde{h}_{i,1}}$ is also the last time step for backward encoder for row i. Then we use another bi-LSTM to obtain row-level representation $\tilde{h}_i$.

Secondly, despite the wide use of sequential model to learn representation of table, since the data in the table are not sequential, using sequential model is inadequate. Therefore, we propose to use a hierarchical transformer encoder structure which can model the records representation in the context of other records in a non-sequential way. On record level, we use a transformer encoder [21] to learn the record's representation considering other records in the same row. First, we use 1-layer MLP to obtain record representation $h_{i,j}$ similar to the base model. Then, we use transformer encoder to take records in the same row into consideration. Transformer encoder contains multiple layers. As described in Eqs. 2 and 3, each layer consists of two components: multi-head self attention and feed-forward network. K represents the layer number of transformer encoder. We use $\tilde{h}_{i,j} = h_{i,j}^K$ from last layer as the record-level representation.

$$t_{i,j}^k = LayerNorm(h_{i,j}^k + MultiHeadSelfAttention(h_{i,j}^k)) \tag{2}$$

$$h_{i,j}^{k+1} = LayerNorm(t_{i,j}^k + FeedForwardNetwork(t_{i,j}^k)) \tag{3}$$

Then, we use mean pooling over $\{\tilde{h}_{i,1}, ..., \tilde{h}_{i,j}, ..., \tilde{h}_{i,T}\}$ to obtain i^{th} row representation h_i. T is the number of records in the ith row. On row level, we use another transformer encoder to model dependency between rows $\{h_1, ..., h_N\}$. N is the number of rows. Then, we can obtain row-level representation $\tilde{h}_i$, which is similar to the record-level transformer encoder.

3.2   Decoder with Hierarchical Attention

Based on our observation, sentences in game summary intend to focus on few player or team. Also, with hierarchical encoder proposed above, we now have both row-level and record-level representation about the table. Therefore, we use hierarchical attention for decoder to first decide which row will be focused on, then attend to relevant records for generating words.

As shown in Eq. 4, the decoder takes previous LSTM hidden state, previous generated word and context vector to update the LSTM hidden state. Then, the decoder will first attend to row-level representation in order to identify important row to be mentioned according to Eq. 5. Afterwards, the decoder will attend to record-level representation to find relevant records as shown in Eq. 6. Please note that the record-level attention weight is normalized among other records in the same row. Then, the decoder will use row-level attention weight as a guidance to re-weight record-level attention weight and obtain context vector c_t as shown in Eq. 7. Please note that the re-weighted attention weight sum to 1 across all records in the table.

$$s_t = LSTM(\tilde{s}_{t-1}, c_{t-1}, y_{t-1}) \tag{4}$$

$$\alpha_{t,i} = \frac{exp(score(s_t, \tilde{h}_i))}{\sum_{i'} exp(score(s_t, \tilde{h}_{i'}))} \tag{5}$$

$$\beta_{t,i,j} = \frac{exp(score(s_t, \tilde{h}_{i,j}))}{\sum_{j'} exp(score(s_t, \tilde{h}_{i,j'}))} \tag{6}$$

$$c_t = \sum_i \sum_j \alpha_{t,i} \beta_{t,i,j} \tilde{h}_{i,j} \tag{7}$$

With the context vector c_t that contains relevant records' information, the decoder obtains an attentional hidden state $\tilde{s}_t$ by applying 1-layer MLP on the concatenation of context vector c_t and hidden state s_t as described in Eq. 8. W_c is a trainable parameter and [;] denotes vector concatenation. Then we use Eq. 9 to replace the base model's $p_{gen}(y_t|y_{<t}, \tilde{H})$ described in Sect. 2.2. Also, we use $\gamma_{t,i,j} = \alpha_{t,i} \beta_{t,i,j}$ as the probability of copying each word in the table.

$$\tilde{s}_t = tanh(W_c[c_t; s_t]) \tag{8}$$

$$p_{gen}(y_t|y_{<t}, \tilde{H}) = softmax(W_s \tilde{s}_t) \tag{9}$$

3.3   Training

During training, the model is optimized by minimizing the negative log-likelihood given the gold descriptive texts $y^\star$ as described in Eq. 10.

$$L = -\sum_{t=1}^{T} log P(y_t^\star|y_{<t}, \tilde{H}) \tag{10}$$

4   Experiments

4.1   Dataset and Evaluation Metrics

We conducted experiments on ROTOWIRE dataset [22]. For each example, it provides 628 records relating to teams and players' statistics and its corresponding long game summary. The average length of game summary is 337.1 and the number of record types is 39. In the end, we use training, validation, test set with 3398, 727, 728 summaries respectively for comparing models' performance.

We follow Wiseman et al. [22] and use BLEU [17] and three extractive evaluation metrics [22] RG, CS and CO for automatic evaluation. The main idea is using an IE (Information Extraction) model to extract record tuples from reference and generated texts. Then, CS (Content Selection) measures model's ability on content selection by comparing records in generated texts with records in reference, CO (Content Ordering) measures ability on content placement by calculating normalized Damerau-Levenshtein Distance [3] between records from generated texts and from reference and RG (Relation Generation) compares summary's records with the table and measures both ability. Furthermore, we also conducted human evaluation study to compare model's ability on content selection and producing coherent texts.

4.2 Implementation Details

Following conditional copy model proposed by Wiseman et al. [22], we use 1-layer encoder to map statistics into R^{600} and use 2-layer LSTM with hidden size of 600 for decoder of the base model. Also, we use input-feeding for attention, following Luong et al. [14]. We set bi-LSTM layer as 1 and hidden size as 600 for both bi-LSTM in hierarchical bi-LSTM encoder. For hierarchical transformer encoder, we set the number of head as 8 and layer as 5 for both transformer encoder. For training, we use Adagrad optimizer [5] with learning rate of 0.15, truncated BPTT (block length 100) and a batch size of 5. During inference, we use beam search to generate texts given tables' information. We set beam size to 5.

4.3 Results

Automatic Evaluation. Automatic evaluation results are shown in Table 1. We compare our models with reference. Also we adopt a template system constructed in the same way as the one in [22]. It consists three parts: one introductory sentence generally describe the match which includes the score of both

Table 1. Automatic evaluation results using the updated IE model by [18]. (Our model refers to the hierarchical transformer encoder).

Model	RG		CS			CO	BLEU
	P%	#	P%	R%	F1%	DLD%	
Reference	94.89	24.14	100.00	100.00	100.00	100.00	100.00
Template	**99.94**	**54.21**	27.02	**58.22**	36.91	15.07	8.58
Conditional copy [22]	74.80	23.72	29.49	36.18	32.49	15.42	14.19
OpAtt [16]	-	-	-	-	-	-	14.74
Hierarchical bi-LSTM encoder	89.15	33.81	31.44	45.84	37.30	**18.24**	15.36
Our model	93.26	30.38	**34.67**	43.17	**38.46**	**18.24**	**15.65**

teams and who wins the match, six sentences describe six players statistics in the table who get most scores in the match, ranked by their score from high to low and a conclusion sentence. We defer readers to [22] for more details.

Since conditional copy (CC) model achieves best performance according to Wiseman et al. [22], we use it as base model. We also compare our models with OpAtt [16].

We compare two hierarchical encoder structure mentioned in approaches, and find that our model (hierarchical transformer encoder) achieves better performance in terms of CS P%, CS F1%. This shows that it has better ability to select important information than hierarchical bi-LSTM encoder. In addition, compared to hierarchical bi-LSTM Encoder, our model achieves higher RG P% and BLEU, which indicates our model can generate more accurate information. Compared with base model and other models on this task, our model also achieves significant improvement on RG, which indicates that our model can generate more high fidelity texts with respect to information in the table. One possible reason is that using hierarchical transformer encoder can capture dependency on both row-level and record-level which help the decoder find relevant records in the table more accurately.

Human Evaluation. In this section, we present human evaluation studies, assessing models' ability on choosing salient information and generating coherent texts. We randomly sampled 20 games from test test, shuffled models' produced texts and gave raters $\langle reference, generation \rangle$ pair for review. Raters were asked to identify how many players mentioned in reference was also mentioned in model's output and how many were missed. Also raters were asked to give 1–3 score on model's generated texts. 1 stands for extremely incoherent texts with extensive repetition. 2 stands for moderately coherent texts with little repetition. 3 stands for coherent texts with almost no repetition. Results are presented in Table 2. The results align with the content selection (CS) metrics for automatic evaluation. The higher # include and the lower # miss compared to base model (CC) indicate our model performs better on selecting salient information. Also, our model can produce more coherent texts than base model.

Case Study. Figure 3 shows reference and texts generated by the base model and our model. Both models can produce fluent and understandable texts. We can see that base model includes many incorrect claims about player or team's statistics as shown in Fig. 1.

It generates many wrong information for *Cory Joseph, Rodney Hood, Joe Johnson* and *Raul Neto*. It also mistakenly use others' information to describe some of the absent players such as *Fred VanVleet* and *George Hill*. In comparison, most of the statistics information in our model's generated texts are correct. Furthermore, base model mentioned six players who don't appear in reference which indicates they are not salient information.

Team	PTS	PTS_QTR1	PTS_QTR2	WIN	LOSS	...
Raptors	104	28	28	21	8	...
Jazz	98	29	22	18	13	...

Player	PTS	AST	REB	STL	...
Raptors					
DeMar DeRozan	24	1	6	4	...
Kyle Lowry	36	5	4	2	...
Jonas Valanciunas	14	0	7	1	...
Terrence Ross	10	1	1	0	...
Cory Joseph	4	3	1	1	...
Fred VanVleet	N/A	N/A	N/A	N/A	...
...	...	...	...	...	...
Jazz					
Rudy Gobert	13	1	14	1	...
Gordon Hayward	23	5	1	1	...
Trey Lyles	19	1	7	0	...
Shelvin Mack	17	5	4	1	...
Rodney Hood	2	1	0	0	...
Joe Johnson	7	0	0	0	...
Raul Neto	0	0	1	0	...
George Hill	N/A	N/A	N/A	N/A	...
...	...	...	...	...	...

STAT

On Friday , the Raptors saw a combined 60 points from their All-Star guards . Point guard **Kyle Lowry** had one of his best nights of the season , leading the team with 36 points on ridiculous 15 - of - 20 shooting . **DeMar DeRozan** , meanwhile , scored 24 and recorded four steals on the defensive end . Despite playing just 16 minutes and facing stiff competition in Utah center **Rudy Gobert** , Toronto big man **Jonas Valanciunas** scored 14 points on perfect 5 - of - 5 shooting and had seven rebounds . For Utah , forward **Gordon Hayward** led the way with 23 points . **Hayward** shot 50 percent from the field and had five assists . **Gobert** scored 13 and led the team with 14 rebounds while shooting a solid 5 - of - 7 . Sophomore **Trey Lyles** impressed off the bench , scoring 19 points including four threes .

Reference

The Toronto Raptors defeated the Utah Jazz , 104 - 98 , at Vivint Smart Home Arena on Monday . The Raptors (21 - 8) checked in to Wednesday 's contest looking to snap a five - game losing streak , as they 'd lost four of their last five contests . **DeMar DeRozan** led the way for the Raptors (21 - 8) with 24 points , six rebounds , an assist and four steals . **Kyle Lowry** followed with 36 points , five assists , four rebounds and two steals . **Trey Lyles** was next with a bench - leading 19 points , which he supplemented with seven rebounds and an assist . **Rudy Gobert** turned in a 13 - point , 14 - rebound double - double that also included an assist and a steal . Shelvin Mack led the bench with 17 points , five assists , four rebounds and a steal . Shelvin Mack led the bench with 17 points , five assists , four rebounds and a steal . Terrence Ross paced the bench with 10 points , a rebound and an assist . The Jazz remain in second place in the Western Conference 's Northwest Division . They head to Utah to take on the Jazz on Monday .

Our Model

Fig. 3. The generated game summary for *Raptors v.s. Jazz* and its corresponding game statistics. Red words mean it is in line with statistics in the table. Blue words mean it contradicts table's information or it is not the salient player to be mentioned. Underscored player name indicates that the model correctly select this salient player in line with reference. (Color figure online)

Table 2. Human evaluation results.

Model	# include	# miss	coherency
Reference	5.63	0.13	2.95
CC	2.73	3.03	2.73
Our model	3.78	1.75	2.85

Instead, our model only includes two players who are not in the reference while still include four important players in the game. This indicates that our model has better ability on content selection than base model.

5   Related Work

Given preprocessed data, the traditional data-to-text systems usually treat the generating task as a pipeline. The first step is to perform document planning, then the traditional models perform microplanning and generating actual sequence of texts via realisation [19]. Recently, neural data-to-text systems perform this task in a end-to-end fashion. Mei et al. [15] propose a pre-selector to generate weather forecast, which strengthens model's content selection ability and obtains considerable improvement over previous models. Some studies focus on transforming Wikipedia infobox into introductory sentences. Sha et al. [20] propose a hybrid attention mechanism to enhance the model's ability on choosing the order of content when generating texts. Liu et al. [13] propose a field-gating encoder focusing on modeling table structure. Meanwhile, Bao et al. [2] develop a table-aware Seq2Seq model. In recent years, a document-scale data-to-text dataset has been introduced, which contains significantly more records and longer target texts by one order of magnitude than previously mentioned datasets. It provides two strong baselines with different copy mechanism to improve model's ability to generate texts with correct record information. There are some studies on this task. Nie et al. [16] introduce pre-executed operation such as minus and argmax in order to help model generate higher fidelity texts. Li et al. [9] decompose the generating process into generating templates at first and then fill in the slots. Meanwhile, Puduppully et al. [18] decompose the process into two stages: selecting a sequence of important records from the table and generating texts according to those important records. However, they treat the table as a set or sequence of records without exploiting the table's multi-row structure. In addition, different from models that decompose data-to-text generation into multiple stages, our model generate texts in one stage.

Some studies encode long input texts in a hierarchical fashion. Cohan et al. [4] propose to use a hierarchical encoder to obtain input's representation on word-level and section-level for single, long-form documents. Also, they propose a discourse-aware decoder that attends to both discourse section and words in the document. Ling et al. [12] also propose to hierarchically encode document with hierarchical attention. Recently, Jain et al. [7] propose a hierarchical encoder and mixed hierarchical attention on data-to-text task.

6   Conclusion

In this work, we develop a hierarchical encoder model that automatically maps the structured table to natural language texts. In detail, we consider the row of the table as an intermediate representation structure to improve the table encoding component. Also these row-level and record-level representations can be used to better focus on important records when generating texts. Experiments were conducted on the ROTOWIRE dataset. Both automatic and human evaluation results show that our neural hierarchical encoder architecture achieves significant improvement over the base model.

Acknowledgements. We would like to thank the anonymous reviewers for their helpful comments. This work was supported by the National Key R&D Program of China via grant 2018YFB1005103 and National Natural Science Foundation of China (NSFC) via grants 61632011 and 61772156.

References

1. Bahdanau, D., Cho, K., Bengio, Y.: Neural machine translation by jointly learning to align and translate. In: International Conference on Learning Representations (2015)

2. Bao, J., et al.: Table-to-text: describing table region with natural language. In: The Thirty-Second AAAI Conference on Artificial Intelligence, pp. 5020–5027. Association for the Advancement of Artificial Intelligence (2018)
3. Brill, E., Moore, R.C.: An improved error model for noisy channel spelling correction. In: Proceedings of the 38th Annual Meeting of the Association for Computational Linguistics, pp. 286–293. Association for Computational Linguistics (2000)
4. Cohan, A., et al.: A discourse-aware attention model for abstractive summarization of long documents. In: Proceedings of the 2018 Conference of the North American Chapter of the Association for Computational Linguistics: Human Language Technologies, pp. 615–621. Association for Computational Linguistics (2018)
5. Duchi, J.C., Hazan, E., Singer, Y.: Adaptive subgradient methods for online learning and stochastic optimization. J. Mach. Learn. Res. **12**, 2121–2159 (2010)
6. Gulcehre, C., Ahn, S., Nallapati, R., Zhou, B., Bengio, Y.: Pointing the unknown words. In: Proceedings of the 54th Annual Meeting of the Association for Computational Linguistics, pp. 140–149. Association for Computational Linguistics (2016)
7. Jain, P., Laha, A., Sankaranarayanan, K., Nema, P., Khapra, M.M., Shetty, S.: A mixed hierarchical attention based encoder-decoder approach for standard table summarization. In: Proceedings of the 2018 Conference of the North American Chapter of the Association for Computational Linguistics: Human Language Technologies, pp. 622–627. Association for Computational Linguistics (2018)
8. Lebret, R., Grangier, D., Auli, M.: Neural text generation from structured data with application to the biography domain. In: Proceedings of the 2016 Conference on Empirical Methods in Natural Language Processing, pp. 1203–1213. Association for Computational Linguistics (2016)
9. Li, L., Wan, X.: Point precisely: towards ensuring the precision of data in generated texts using delayed copy mechanism. In: Proceedings of the 27th International Conference on Computational Linguistics, pp. 1044–1055. Association for Computational Linguistics (2018)
10. Liang, P.: Learning executable semantic parsers for natural language understanding. Commun. ACM **59**(9), 68–76 (2016)
11. Liang, P., Jordan, M., Klein, D.: Learning semantic correspondences with less supervision. In: Proceedings of the Joint Conference of the 47th Annual Meeting of the ACL and the 4th International Joint Conference on Natural Language Processing of the AFNLP, pp. 91–99. Association for Computational Linguistics (2009)
12. Ling, J., Rush, A.: Coarse-to-fine attention models for document summarization. In: Proceedings of the Workshop on New Frontiers in Summarization, pp. 33–42. Association for Computational Linguistics (2017)
13. Liu, T., Wang, K., Sha, L., Chang, B., Sui, Z.: Table-to-text generation by structure-aware seq2seq learning. In: The Thirty-Second AAAI Conference on Artificial Intelligence, pp. 4881–4888. Association for the Advancement of Artificial Intelligence (2018)
14. Luong, T., Pham, H., Manning, C.D.: Effective approaches to attention-based neural machine translation. In: Proceedings of the 2015 Conference on Empirical Methods in Natural Language Processing, pp. 1412–1421. Association for Computational Linguistics (2015)
15. Mei, H., Bansal, M., Walter, M.R.: What to talk about and how? selective generation using LSTMs with coarse-to-fine alignment. In: Proceedings of the 2016 Conference of the North American Chapter of the Association for Computational Linguistics: Human Language Technologies, pp. 720–730. Association for Computational Linguistics (2016)

16. Nie, F., Wang, J., Yao, J.G., Pan, R., Lin, C.Y.: Operation-guided neural networks for high fidelity data-to-text generation. In: Proceedings of the 2018 Conference on Empirical Methods in Natural Language Processing, pp. 3879–3889. Association for Computational Linguistics (2018)
17. Papineni, K., Roukos, S., Ward, T., Zhu, W.J.: Bleu: a method for automatic evaluation of machine translation. In: Proceedings of 40th Annual Meeting of the Association for Computational Linguistics, pp. 311–318. Association for Computational Linguistics (2002)
18. Puduppully, R., Dong, L., Lapata, M.: Data-to-text generation with content selection and planning. In: Proceedings of the AAAI Conference on Artificial Intelligence, pp. 6908–6915. Association for the Advancement of Artificial Intelligence (2019)
19. Reiter, E.: An architecture for data-to-text systems. In: Proceedings of the Eleventh European Workshop on Natural Language Generation, pp. 97–104. Association for Computational Linguistics (2007)
20. Sha, L., et al.: Order-planning neural text generation from structured data. In: The Thirty-Second AAAI Conference on Artificial Intelligence, pp. 5414–5421. Association for the Advancement of Artificial Intelligence (2018)
21. Vaswani, A., et al.: Attention is all you need. In: Advances in Neural Information Processing Systems, pp. 5998–6008. Curran Associates Inc., Long Beach (2017)
22. Wiseman, S., Shieber, S., Rush, A.: Challenges in data-to-document generation. In: Proceedings of the 2017 Conference on Empirical Methods in Natural Language Processing, pp. 2253–2263. Association for Computational Linguistics (2017)

Dropped Pronoun Recovery in Chinese Conversations with Knowledge-Enriched Neural Network

Jianzhuo Tong, Jingxuan Yang, Si Li$^{(\boxtimes)}$, and Sheng Gao

Beijing University of Posts and Telecommunications, Beijing, China
{tongjianzhuo,yjx,lisi,gaosheng}@bupt.edu.cn

Abstract. Dropped pronoun recovery, which aims to detect the type of pronoun dropped before each token, plays a vital role in many applications such as Machine Translation and Information Extraction. Recently, deep neural networks have been applied to this task. Though promising improvements have been observed, these methods recover dropped pronouns from the limited context in a small-size window and lack common sense to connect the referred entity to a proper pronoun. In this paper, we propose a knowledge-enriched neural attention framework for Chinese dropped pronoun (DP) recovery. A structured attention mechanism is used to capture the semantics of DP referents from the wider context. External knowledge, which consists of a knowledge base and a hierarchical pronoun-category assumption, is also incorporated in our model to provide pronoun classification information of referred entity and contextual dependency degree. Results on three different conversational genres show that our approach achieves a convincing improvement over the current state of the art.

Keywords: Chinese dropped pronoun · Semantic modeling · External knowledge

1 Introduction

Chinese is a typical pro-drop language which allows a pronoun to be omitted according to some pragmatic constraints. The pro-drop sentence is compact yet comprehensible since the identity of dropped pronoun can be easily understood from the context. Recovering dropped pronoun is crucial in many applications such as Machine Translation [1–3], since these dropped pronouns influence not only sentence structure but also the semantics of output sentence when Chinese is the source language. Existing research [3] shows that around 26.55% of pronouns are dropped from the Chinese side, and over 92.65% of them are dropped in informal genres. Therefore, recovering dropped pronouns in conversations is significant.

As one line of work that is closely related to our task, zero pronoun (ZP) resolution aims to resolve an anaphoric pronoun to its antecedent, assuming that

© Springer Nature Switzerland AG 2019
M. Sun et al. (Eds.): CCL 2019, LNAI 11856, pp. 545–557, 2019.
https://doi.org/10.1007/978-3-030-32381-3_44

> A : 你在那边过的怎么样啊？(你)生活还都顺利吧？
> How are you doing in there? Are (you) OK?
>
> B : (我)过得挺好的。
> (I) am fine.
>
> A : (天气)怎么样啊？
> What's about (the weather)?
>
> B : 它变化莫测。
> It changes a lot.
>
> B : (它)一会刮风下雨。(它)一会晴朗无云。
> (It) will be windy and rainy. Or (it) will be clear and cloudless for a while.

Fig. 1. An example conversation. The dropped pronoun refers to "the weather" and is translated into "it" in English. It needs to be recovered from large scale context.

the position of the zero pronoun has already been determined. But in DP recovery, we focus on identifying the position of dropped pronoun and recovering the dropped pronoun to one type of the predefined pronouns but not the antecedent. According to Yang et al. [4], the type of dropped pronoun can either be a concrete possible pronouns in Chinese, or an abstract one that does not correspond to any actual pronouns.

Dropped pronoun recovery is traditionally formulated as a sequence labeling problem, and each token in the sentence has a tag indicating which pronoun is dropped before it. Yang et al. [4] leverage a set of elaborately designed features to detect the dropped pronoun. Giannella et al. [5] utilize a linear-chain conditional random field (CRF) classifier to predict DP. These featured-based methods need labor-intensive feature engineering. Recently, neural network methods have been attempted on this task and have achieved some advanced results. Zhang et al. [6] use a multi-layer perceptron model to recover dropped pronouns from tokens in a fixed-length window. However, the semantics of DP referent is usually expressed in wider context. As shown in Fig. 1, the recovered third person singular pronoun 它 ("it") refers to 天气 ("weather"), which appears in the previous utterance outside the token window. Therefore, long-distance contextual information needs to be captured. Yang et al. [7] employ an attention based neural network to model the referent of dropped pronoun using contextual information. The model lacks some critical common sense, which incurs failure to connect DP with its referents. For example, the referent 天气 ("weather") should be recovered as 它 ("it") is a kind of essential knowledge that should be incorporated to help the model recover DP.

In this paper, we propose an knowledge-enriched neural dropped pronoun recovery framework (KNDP), as illustrated in Fig. 3. It recovers the dropped pronoun by modeling the semantics of DP referent with attention mechanism. External knowledge is incorporated to improve the referent representation. Specifically, we construct a knowledge base from a public corpus to provide connections between DP and its potential referent. We also propose a hierarchical DP category assumption and leverage it as a supervised signal to predict DP. We demonstrate the effectiveness of our proposed model on three Chinese conversational data sets and experimental results show that our approach outperforms current state-of-the-art methods. We also perform ablative studies to explore the contribution of external knowledge in our model.

Our contributions are summarized as follows:

- We build a knowledge base which provide necessary common sense for dropped pronoun recovery, and present a hierarchical pronoun-category assumption according to the contextual dependency of each pronoun.
- We propose a novel knowledge-enriched neural network to recover dropped pronouns by incorporating these two kinds of external knowledge. Results show that our model outperforms the state-of-the-art methods.
- We perform ablative experiments to demonstrate the effectiveness of external knowledge, and visualize attention distribution to make a proper interpretation of referent modeling process.

2 Related Work

Dropped pronoun recovery originates from Empty Category (EC) detection and resolution [8–10], which aim to recover dropped elements in syntactic treebanks. DP recovery task was firstly introduced in [4], which leveraged a set of hand-crafted features to detect DPs before each word from the full list of pronouns. Giannella et al. [5] utilized a linear-CRF classifier to determine DP. Zhang et al. [6] used a multi-layer perceptron model by concatenating embedding of word tokens in a small fixed-length window to recover dropped pronoun. Yang et al. [7] employed a neural network to model the referent of dropped pronoun using contextual semantics. There has also been some work of DP recovery in Chinese text to help Machine Translation [1–3]. As another related line of research, zero pronoun (ZP) resolution aims to resolve pronouns to their antecedents. Converse and Palmer [11] used the Hobbs algorithm to resolve ZP. Zhao et al. [12] and Kong et al. [13] presented systems using decision trees and context-sensitive convolution tree kernels for ZP resolution. There were also some works [14–16] resolve zero pronouns with neural networks.

Attention mechanisms [17–19] have been used in many natural language processing tasks. Traditional models focused on applying attention mechanism on top of convolutional neural networks or recurrent neural networks [20,21]. Shen et al. [22] proposed a self-attention network which is CNN/RNN free. Yang et al. [20] proposed a word and sentence-level hierarchical attention network for document classification. Xing et al. [23] also presented a hierarchical recurrent attention network (HRAN) to model the hierarchy of conversation context. Weston et al. [24] firstly proposed the concept memory networks reasoning with inference components. Sukhbaatar et al. [25] trained the network end-to-end. Miller et al. [26] employed a Key-Value memory network on reading comprehension systems.

3 External Knowledge

3.1 External Knowledge Base

Traditional attention mechanism fails to build necessary connections between DP and its referents since the patterns are sparse. For example, it cannot learn

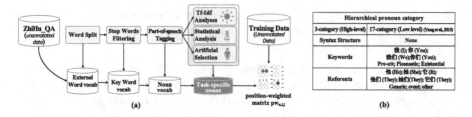

Fig. 2. External knowledge includes (a) *External knowledge base* and (b) *Hierarchical pronoun-category assumption* .

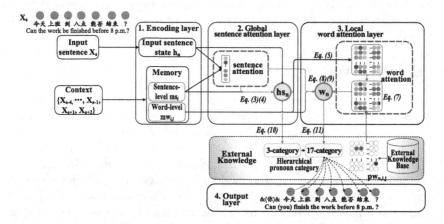

Fig. 3. Knowledge-enriched neural dropped pronoun recovery framework.

that the referent 工厂 ("factory") should be recovered as 它 ("it") since 工厂 ("factory") are rarely appeared in training samples. To solve this problem, we use Chinese conversational corpus Zhihu_QA [27], which contains 32,137 questions and 3,239,114 answers to construct an unsupervised knowledge base. Concretely, the construction process of knowledge base (KB) is shown in Fig. 2(a), which consists of part-of-speech tagging, statistical analysis, and artificial selection. The target output of knowledge base is task-specific nouns such as 工厂 ("factory"). With the guidance of unannotated training data, the model constructs a position-weighted matrix $pw_{n,i,j}$. The top-ranked nouns represent local semantics of referent with a high probability and should be allocated with an additional weight. This idea is inspired by that nouns semantically similar to specific pronouns are more likely to be referred to.

3.2   Hierarchical Pronoun-Category Assumption

We propose a hierarchical pronoun-category assumption (HCA) which divides the full list of 17 fine grained pronouns into 3 coarse grained categories. The 3 categories are concluded based on the contextual dependency extent of each specific pronoun. Concretely, category "none", which means no pronouns are

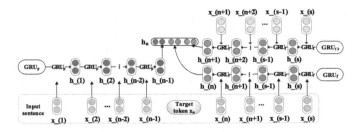

Fig. 4. PCG-BiGRU encoder.

dropped, are mostly in accordance to the *syntax structure*. For example, tokens after a verb or words before a punctuation. Categories like "previous utterance", "pleonastic" and "existential" are often dropped before *key words* like 好的 ("OK"), 就是 ("must be"), 存在 ("exist"). Categories like 他 ("he"), 她 ("she"), 它 ("it") depend on semantics of *referents* more than previous two categories. The degree of contextual dependency is in turn decrease on these 3 types: syntax structure, keywords, and referents. The details of hierarchical pronoun-category assumption are shown in Fig. 2.

4   Approach

We formulate DP recovery as a sequential tagging problem same as [4]. The input is raw sentence $X = (x_1, x_2, ..., x_s)$ and its context $\mathbf{C} = (X_1, ..., X_m)$. The task aims to model the conditional distribution $P(Y|X, \mathbf{C})$ and predict labels $Y = (y_1, y_2, ..., y_s)$ indicating the type of dropped pronoun before each word. We incorporate external knowledge base in local word attention and utilize hierarchical DP-category assumption in the the output layer. Our KNDP model consists of three components.

4.1   Encoding Layer

Given the input sentence $X = \{x_1, x_2, ..., x_s\}$ and its context $C = \{X_1, X_2, ..., X_m\}$, encoding layer transforms them into distributed representations.

Sentence X, composed of s tokens, are converted into a sequence of d-dimensional word embeddings, and then encoded by a bidirectional recurrent neural networks (RNN) [28], yieldig a set of representations $\boldsymbol{h}_n$. There are two different encoders:

(i) **PC-BiGRU** [15]: PC means pronoun-centered. Token $\boldsymbol{x}_n$ is represented as $\boldsymbol{h}_n = [\overrightarrow{\boldsymbol{h}}_{n-1}, \overleftarrow{\boldsymbol{h}}_{n+1}]$ by concatenating results of a preceding GRU_p which encodes the words before DP and a following GRU_f which encodes words after DP.

(ii) **PCG-BiGRU**: PCG means pronoun-centered with 1-gram information. The PCG-BiGRU contains three independent GRU networks as shown in Fig. 4. $\boldsymbol{x}_n$ is finally represented as $\boldsymbol{h}_n = [\overrightarrow{\boldsymbol{h}}_{n-1}, \overleftrightarrow{\boldsymbol{h}}_n, \overleftarrow{\boldsymbol{h}}_{n+1}]$.

Context C, consists of five preceding and two following utterances[1] of the input sentence X, is also encoded by the same BiGRU as sentence X, yielding two-granularity memory modules: (1) sentence-level: concatenate the final states of the forward and backward GRU. Each sentence i is represented as $\boldsymbol{ms}_i = [\overleftarrow{\boldsymbol{ms}}_i, \overrightarrow{\boldsymbol{ms}}_i]$. (2) word-level: concatenate the forward and backward hidden states at each token j, which is represented as $\boldsymbol{mw}_{i,j} = [\overleftarrow{\boldsymbol{mw}}_{i,j}, \overrightarrow{\boldsymbol{mw}}_{i,j}]$.

4.2 Global Sentence Attention Layer

This layer identifies which utterances introduce DP referents and aggregate global contextual information of DP. Interactions between the DP representations $\boldsymbol{h}_n$ and its sentence memory states $\boldsymbol{ms}_i$ are explored through vanilla attention mechanism. Global background information $\boldsymbol{s}_n$ is further obtained by a weighted sum.

$$as_{n,i} = softmax(\boldsymbol{h_n}^T \cdot \boldsymbol{ms}_i) \tag{1}$$

$$\boldsymbol{s}_n = \sum_{i=1}^{m} as_{n,i} \cdot \boldsymbol{ms}_i \tag{2}$$

$$\boldsymbol{hs}_n = W^{2d \times 4d}[\boldsymbol{h}_n, \boldsymbol{s}_n] + b^{2d} \tag{3}$$

4.3 Local Word Attention Layer

This layer aims to identify words describe DP referents and express local-referent semantics. External **knowledge base (KB)** is added to provide referent related common sense in an unsupervised way. We cast global-context representation $\boldsymbol{hs}_n$ into the same space as word-level memory $\boldsymbol{mw}_{i,j}$, and explore relevance between them. After passing it through the Softmax function, we get the word-level attention distribution $aw_{n,i,j}$. Then, we utilize a position-weighted matrix $pw_{n,i,j}$ obtained from knowledge base to modify the traditional word-level distribution by multiply a certain coefficient β, yielding $kw_{n,i,j}$. Finally, We utilize $kw_{n,i,j}$ to compute a weighted sum of word memory $\boldsymbol{mw}_{i,j}$, yielding the referent representation $\boldsymbol{tw}_{n,i}$. Then the local representation is limited by the global sentence attention as_i to further filters out irrelevant referents, yielding the local referent representation $\boldsymbol{w}_n$.

$$aw_{n,i,j} = softmax(W^{1 \times 2d}(\boldsymbol{hs}_n \odot \boldsymbol{mw}_{i,j}) + b^1) \tag{4}$$

$$kw_{n,i,j} = aw_{n,i,j} + \beta \cdot pw_{n,i,j} \tag{5}$$

$$\boldsymbol{tw}_{n,i} = \sum_{j=1}^{k} kw_{n,i,j} \cdot \boldsymbol{mw}_{i,j} \tag{6}$$

$$\boldsymbol{w}_n = \sum_{i=1}^{m} as_{n,i} \cdot \boldsymbol{tw}_{n,i} \tag{7}$$

---

[1] We utilize this setting since previous study [7] shows that over 97% of the dropped pronouns can be inferred from 7 neighborhood utterances.

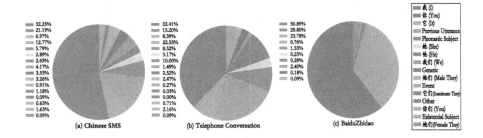

Fig. 5. Distribution of dropped pronouns on different conversational genres.

4.4 Output Layer

This layer predicts the final recovery result for each word using global context information hs_n, DP representation h_n and local referent representation w_n. We first recover each DP as one of three coarse grained types using global information hs_n. This determine according to our proposed **hierarchical pronoun-category assumption (HCA)**. We then further classify it into final 17 fine grained types utilizing the concatenation of DP state h_n and referent representation w_n.

$$\alpha_{n1} = tanh(W_1 \cdot hs_n + b) \tag{8}$$

$$\alpha_{n2} = tanh(W_2 \cdot [h_n; w_n] + b) \tag{9}$$

$$P(y_n|x_n,C) = softmax[(W_3 \cdot \alpha_{n1} + b) \cdot W_4 \cdot softmax(W_5 \cdot \alpha_{n2} + b)] \tag{10}$$

4.5 Training Objective

The training objective is defined by minimizing cross-entropy between the predicted label distributions and the annotated labels as follow:

$$loss = -\sum_{l \in N}\sum_{n=1}^{s} \delta(y_n|x_n,C)\log(P(y_n|x_n,C)) \tag{11}$$

where N represents training samples, s represents the number of tokens in each utterance; $\delta(y_n|x_n,C)$ represents the annotated label of each x_n in the input sentence X.

5 Experimental Setups

5.1 Datasets

We demonstrate the effectiveness of our method on three different conversational genres. *Chinese text message (SMS)* contains 40,279 sentences of 684 message files. Same as previous studies on Chinese DP recovery [4,7], we split the data set as training, development and test set, which consists of 487, 98 and

Table 1. Results of the three baseline systems, two variants and two ablation models compared with our proposed KNDP model on three different conversational genres. For NRM* [6], we implement the proposed model as introduced in the paper.

Model	Chinese SMS			TC of OntoNotes			BaiduZhidao		
	P(%)	R(%)	F	P(%)	R(%)	F	P(%)	R(%)	F
MEPR [4]	37.27	45.57	38.76	-	-	-	-	-	-
NRM* [6]	37.11	44.07	39.03	23.12	26.09	22.80	26.87	46.55	34.54
NDPR [7]	49.39	44.89	46.39	39.23	43.09	39.77	41.04	49.44	42.29
KNDP-rand	46.47	43.23	43.58	28.98	41.50	33.38	35.44	43.82	37.79
KNDP-PCG	46.34	46.21	46.27	36.69	40.12	38.33	38.42	45.04	41.68
KNDP(wo/HCA)	49.29	45.89	47.09	37.51	**44.09**	40.56	40.05	45.55	42.37
KNDP(wo/KB)	46.28	44.92	45.59	38.23	40.47	39.25	39.34	47.26	41.63
KNDP	**49.69**	**46.89**	**48.17**	**39.63**	43.45	**40.86**	**41.67**	**49.51**	**42.94**

99 conversations respectively. For *OntoNotes Release 5.0*, we use the Chinese telephone conversation (TC) portion of this data set, which contains 9,607 sentences. *BaiduZhidao* contains 11,160 sentences with only 10 types of concrete dropped pronoun categories. The pie charts in Fig. 5 show the distribution of different types of pronouns in each genre.

5.2   Evaluation Metrics

Following previous studies [4,7], we use the same evaluation metrics as : precision (P), recall (R) and F-score (F).

5.3   Implementation Details

Our model is implemented with Tensorflow. The vocabulary is generated from the corpus which contains 17,199 words. The number of hidden units in the GRU cell is 150. The batch size is 8 and the dropout [29] is 0.2. We use the Adam optimizer [30] with a learning rate of 0.0003. The coefficient β acting on of position-weighted matrix is 1.2. We train the model of 16 epochs and do test according to the highest F-score on the development set.

5.4   Baselines and Experiment Settings

We consider the following three baselines, two parallel experiments and two ablation experiments for performance comparisons with our KNDP. (1) *MEPR*: This model [4] uses a set of hand-crafted features in Chinese SMS data. (2) *NRM*: This model [6] employs a multi-layer perceptron to do identification and resolution of DP. (3) *NDPR*: This model [7] models the referent and determine the type of dropped pronouns. (4) *KNDP*: Our proposed model uses a knowledge-enriched neural network with PC-BiGRU encoder to recover DP. The word embeddings are initialized with pre-trained 300-D Zhihu QA vectors [27] and fine-tuned during the training process. (5) *KNDP-rand*: Same as KNDP except that the word embeddings are randomly initialized. (6) *KNDP-PCG*: The encoder of utterance X is replaced with PCG-BiGRU. (7) *KNDP (wo/HCA)*: Same as KNDP without using the hierarchical pronoun-category assumption (HCA). (8) *KNDP (wo/KB)*: Same as KNDP but the local-referent semantic layer only utilize original word-attention without the external knowledge base (KB).

6   Results and Discussion

6.1   Main Results

Table 1 shows the results of baseline methods and our proposed KNDP model on three data sets. We compute the weighted sum of F-score on 16 types of pronouns except "None" category as the final F-score. Our proposed model KNDP outperforms all baseline models such as MEPR and NRM. Compared with the best baseline model NDPR, our model also gains improvements by 1.78%, 1.09% and 0.65% on the Chinese SMS, the OntoNotes and BaiduZhidao data sets, in terms of F-score. Results of KNDP-rand and KNDP-PCG in Table 1 demonstrate the effectiveness of different mechanisms applied in our model. First, the pre-trained word embedding makes a positive impact on our framework. Moreover, the PCG-BiGRU encoder seems perform not as well as KNDP as shown by the slightly lower F-score of the KNDP-PCG model. However, it alleviates the Local Pronoun Repetition Problem which is illustrated in Fig. 7.

Table 2. F-score of two ablation models compared with KNDP on the test set of Chinese SMS.

Tag	KNDP (wo/KB)	KNDP (wo/HCA)	KNDP
我 (I)	50.90	51.21	**52.13**
你 (You)	44.15	44.21	**44.51**
他 (He)	31.20	31.62	**35.42**
她 (She)	31.72	33.82	**34.14**
它 (It)	26.67	26.81	**34.15**
我们 (We)	33.57	35.31	**38.16**
你们 (plural You)	0.00	2.81	**6.17**
他们 (male They)	23.28	19.34	**23.87**
她们 (female They)	33.33	33.33	**33.33**
它们 (inanimate They)	19.15	18.78	**21.89**
previous utterance	84.08	86.24	**87.65**
existential	25.29	29.25	**33.54**
pleonastic	26.71	28.52	**29.44**
generic	13.08	14.21	**14.73**
event	7.68	8.34	**11.24**

6.2 Ablation Study

In this section, we analyze the impact of two external knowledge strategies. Table 2 shows the results of our KNDP model and its two variants KNDP (wo/KB) and KNDP (wo/HCA). We can see that the KNDP makes a 3.30% improvement than KNDP(wo/KB) model, in terms of f-score, which demonstrate the effectiveness of knowledge base Concrete pronouns which have explicit referents such as 它 ("it"), 他 ("he"), 她 ("she"), 我们 ("we") and 你们 ("you")get more benefits from the knowledge base since the referent recognition process needs more common sense. Moreover, the abstract pronouns which generally do not have specific referents still get little improvement from knowledge base. Secondly, Table 2 also indicates that hierarchical pronoun-category assumption is also significant since KNDP is better than KNDP (wo/HCA). The high-level category provide contextual dependency information, which is added as a supervised signal to make a more accurate final prediction. Categories like 它 ("it"), 他们 ("male they") and existential improve most from this knowledge strategy.

6.3 Attention Visualization

Figure 6 shows the attention distribution of a snippet conversation where the dropped pronoun is correctly recovered as 他 ("he") in the pro-drop sentence. The model firstly give higher attention weights to three more relevant sentences with darker color, as shown in the blue pattern. The model then gives higher word level attention weights to the specific words that describe the referent of the dropped pronoun such as 他 ("he") and 弟弟 ("brother"). The attention heat map suggests that our model attends to the relevant utterances and words as we have expected. Furthermore, the word-level weight of 弟弟 ("brother") is amplified by our unsupervised external knowledge base, since it exists in our *position weight matrix*. Intuitively, sentence-level attention plays a role of filter which selects the correlative utterances from context. Then, the word-level attention figures out the words describe referents of corresponding dropped pronouns. Words in knowledge base that mention the referent of the pronoun receive more attention.

6.4 Case Study

In this section, we analyze two concrete cases to demonstrate the effectiveness of our external knowledge base and how the PCG-BiGRU resolve the local pronoun repetition problem. As shown in Fig. 7, in the first example, we compare the result of our proposed KNDP with baseline model NDPR. Traditional NDPR model allocates a small word-level attention weight on 工厂 ("factory") but a bigger one on 我 ("I"). As a result, the model recovers the DP token as 我 ("I") through a matching strategy, since the pronoun 我 ("I") appears in the context closed to DP. Our KNDP, however, amplifies the attention weight on 工厂 ("factory") with the help of external knowledge base and strengthens the connections between 工厂 ("factory") and 它 ("it"), which helps the referent

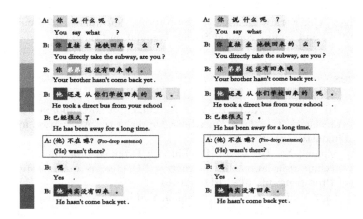

Fig. 6. Visualization of attention distribution between traditional word attention and knowledge-enriched word attention. The Blue pattern shows the distribution of sentence-level attention and the Pink pattern shows the distribution of word-level attention. Darker color indicates higher attention weight. (Color figure online)

External Knowledge Base	Context:	B: 我去过一家工厂.	I have been to a factory.
		B: 不过名字我忘记了。	But I forgot the name.
		A: 那个工厂在哪儿。	Where is the factory?
	NDPR:	B: 我记得(我)是在上海。	I remembered that (I) locates in Shanghai.
	KHDP:	B: 我记得(它)是在上海。	I remembered that (it) locates in Shanghai.
PCG-BiGRU	KHDP:	(我)开始(我)忙了。	(I) begin (I) busy.
	KHDP-PCG:	(我)开始忙了。	(I) begin busy.

Fig. 7. Two typical cases which respectively show the effectiveness of our external knowledge base and PCG-BiGRU encoder.

modeling process to get to the right prediction. The second example shows that the PCG-BiGRU with extra 1-gram information can solve the pronoun repetition problems.

7   Conclusions

We propose a knowledge-enriched neural network architecture to recover dropped pronouns in Chinese conversational data. It utilizes a structured attention mechanism to capture the global-context and local-referent semantics of the dropped pronoun. External knowledge including a knowledge base and a hierarchical pronoun-category assumption, provides referent related common-sense and contextual dependency degree to help the final recovery of DP. Results show that our model makes a significant improvement over baseline models on all three data sets. We further investigate the effectiveness of our external knowledge by performing ablative experiments and demonstrate the interpretability of our model using attention visualizations.

Acknowledge. This work was supported by National Natural Science Foundation of China (61702047) and Beijing Natural Science Foundation (4174098).

References

1. Wang, L., Tu, Z., Zhang, X., Li, H., Way, A., Liu, Q.: A novel approach to dropped pronoun translation. In: North American Chapter of the Association for Computational Linguistics, pp. 983–993 (2016)
2. Wang, L., Zhang, X., Tu, Z., Li, H., Liu, Q.: Dropped pronoun generation for dialogue machine translation. In: International Conference on Acoustics, Speech and Signal Processing, pp. 6110–6114 (2016)
3. Wang, L., Tu, Z., Shi, S., Zhang, T., Graham, Y., Liu, Q.: Translating pro-drop languages with reconstruction models. In: AAAI Conference on Artificial Intelligence (2018)
4. Yang, Y., Liu, Y., Xue, N.: Recovering dropped pronouns from Chinese text messages. In: Meeting of the Association for Computational Linguistics, pp. 309–313 (2015)
5. Giannella, C., Winder, R.K., Petersen, S.: Dropped personal pronoun recovery in Chinese SMS. Nat. Lang. Eng. **23**(6), 905–927 (2017)
6. Zhang, W., Liu, T., Yin, Q., Zhang, Y.: Neural recovery machine for chinese dropped pronoun. arXiv: Computation and Language (2016)
7. Yang, J., Tong, J., Li, S., Gao, S., Guo, J., Xue, N.: Recovering dropped pronouns in Chinese conversations via modeling their referents. In: North American Chapter of the Association for Computational Linguistics, pp. 892–901 (2019)
8. Chung, T., Gildea, D.: Effects of empty categories on machine translation. In: Conference on Empirical Methods in Natural Language Processing, pp. 636–645 (2010)
9. Cai, S., Chiang, D., Goldberg, Y.: Language-independent parsing with empty elements. In: Proceedings of the 49th Annual Meeting of the Association for Computational Linguistics: Human Language Technologies, pp. 212–216 (2011)
10. Xue, N., Yang, Y.: Dependency-based empty category detection via phrase structure trees. In: North American Chapter of the Association for Computational Linguistics, pp. 1051–1060 (2013)
11. Converse, S.P., Palmer, M.S.: Pronominal anaphora resolution in Chinese. Citeseer (2006)
12. Zhao, S., Ng, H.T.: Identification and resolution of chinese zero pronouns: a machine learning approach. In: Conference on Empirical Methods in Natural Language Processing, pp. 541–550 (2007)
13. Kong, F., Zhou, G.: A tree kernel-based unified framework for Chinese zero anaphora resolution. In: Conference on Empirical Methods in Natural Language Processing, pp. 882–891 (2010)
14. Ng, V.: Semantic class induction and coreference resolution. In: Meeting of the Association for Computational Linguistics, pp. 536–543 (2007)
15. Yin, Q., Zhang, Y., Zhang, W., Liu, T.: Chinese zero pronoun resolution with deep memory network. In: Conference on Empirical Methods in Natural Language Processing, pp. 1309–1318 (2017)
16. Yin, Q., Zhang, Y., Zhang, W., Liu, T., Wang, W.Y.: Deep reinforcement learning for Chinese zero pronoun resolution. In: Meeting of the Association for Computational Linguistics, pp. 569–578 (2018)

17. Bahdanau, D., Cho, K., Bengio, Y.: Neural machine translation by jointly learning to align and translate. Comput. Sci. **1409**, 09 (2014)
18. Hermann, K., et al.: Teaching machines to read and comprehend. In: Neural Information Processing Systems, pp. 1693–1701 (2015)
19. Vinyals, O., Fortunato, M., Jaitly, N.: Pointer networks. Comput. Sci. **28** (2015)
20. Yang, Z., Yang, D., Dyer, C., He, X., Smola, A., Hovy, E.: Hierarchical attention networks for document classification. In: North American Chapter of the Association for Computational Linguistics, pp. 1480–1489 (2016)
21. Er, M.J., Zhang, Y., Wang, N., Pratama, M.: Attention pooling-based convolutional neural network for sentence modelling. Inf. Sci. **373**, 388–403 (2016)
22. Shen, T., Zhou, T., Long, G., Jiang, J., Pan, S., Zhang, C.: Disan: directional self-attention network for rnn/cnn-free language understanding. In: National Conference on Artificial Intelligence, pp. 5446–5455 (2018)
23. Xing, C., Wu, Y., Wu, W., Huang, Y., Zhou, M.: Hierarchical recurrent attention network for response generation. In: National Conference on Artificial Intelligence (2018)
24. Weston, J., Chopra, S., Bordes, A.: Memory networks. In: International Conference on Learning Representations (2015)
25. Sukhbaatar, S., Weston, J., Fergus, R., et al.: End-to-end memory networks. In: Neural Information Processing Systems, pp. 2440–2448 (2015)
26. Miller, A.H., Fisch, A., Dodge, J., Karimi, A. H., Bordes, A., Weston, J.: Key-value memory networks for directly reading documents. In: Conference on Empirical Methods in Natural Language Processing, pp. 1400–1409 (2016)
27. Li, S., Zhao, Z., Hu, R., Li, W., Liu, T., Du, X.: Analogical reasoning on Chinese morphological and semantic relations. In: Meeting of the Association for Computational Linguistics, pP. 138–143 (2018)
28. Elman, J.L.: Distributed representations, simple recurrent networks and grammatical structure. Mach. Learn. **7**(2–3), 195–225 (1991). https://doi.org/10.1007/BF00114844
29. Srivastava, N., Hinton, G.E., Krizhevsky, A., Sutskever, I., Salakhutdinov, R.: Dropout: a simple way to prevent neural networks from overfitting. J. Mach. Learn. Res. **15**(1), 1929–1958 (2014)
30. Kingma, D.P., Ba, J.: Adam: a method for stochastic optimization. In: International Conference on Learning Representations (2015)

Automatic Judgment Prediction via Legal Reading Comprehension

Shangbang Long[1] ⓘ, Cunchao Tu[2] ⓘ, Zhiyuan Liu[2(✉)] ⓘ, and Maosong Sun[2]

[1] Department of Machine Learning, Carnegie Mellon University, Pittsburgh, USA
longshangbang@cmu.edu
[2] Department of CST, Tsinghua University, Beijing, China
tucunchao@gmail.com, {liuzy,sms}@tsinghua.edu.cn

Abstract. Automatic judgment prediction aims to predict the judicial results based on case materials. It has been studied for several decades mainly by lawyers and judges, considered as a novel and prospective application of artificial intelligence techniques in the legal field. Most existing methods follow the text classification framework, which fails to model the complex interactions among complementary case materials. To address this issue, we formalize the task as **L**egal **R**eading **C**omprehension according to the legal scenario. Following the working protocol of human judges, LRC predicts the final judgment results based on three types of information, including *fact description*, plaintiffs' *pleas*, and *law articles*. Moreover, we propose a novel LRC model, **AutoJudge**, which captures the complex semantic interactions among facts, pleas, and laws. In experiments, we construct a real-world civil case dataset for LRC. Experimental results on this dataset demonstrate that our model achieves significant improvement over state-of-the-art models. We have published all source codes of this work on https://github.com/thunlp/AutoJudge.

1 Introduction

Automatic judgment prediction is to train a machine judge to determine whether a certain *plea* in a given civil case would be supported or rejected. In countries with *civil law system*, e.g. mainland China, such process should be done with reference to related *law articles* and the *fact description*, as is performed by a human judge. The intuition comes from the fact that under *civil law system*, law articles act as principles for juridical judgments. Such techniques would have a wide range of promising applications. On the one hand, legal consulting systems could provide better access to high-quality legal resources in a low-cost way to legal outsiders, who suffer from the complicated terminologies. On the other hand, machine judge assistants for professionals would help improve the efficiency of the judicial system. Besides, automated judgment system can help in improving juridical equality and transparency. From another perspective, there are currently 7 times much more civil cases than criminal cases in mainland

© Springer Nature Switzerland AG 2019
M. Sun et al. (Eds.): CCL 2019, LNAI 11856, pp. 558–572, 2019.
https://doi.org/10.1007/978-3-030-32381-3_45

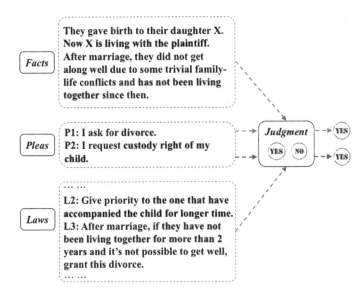

Fig. 1. An example of LRC.

China, with annual rates of increase of 10.8% and 1.6% respectively, making judgment prediction in civil cases a promising application [40].

Previous works [1,12,23,33] formalize judgment prediction as the text classification task, regarding either charge names or binary judgments, i.e., support or reject, as the target classes. These works focus on the situation where only one result is expected, e.g., the US Supreme Court's decisions [12], and the charge name prediction for criminal cases [23]. Despite these recent efforts and their progress, automatic judgment prediction in *civil law system* is still confronted with two main challenges:

One-to-Many Relation between Case and Plea. Every single civil case may contain multiple pleas and the result of each plea is co-determined by related law articles and specific aspects of the involved case. For example, in divorce proceedings, judgment of *alienation of mutual affection* is the key factor for *granting divorce* but *custody of children* depends on which side can provide better an environment for children's growth as well as parents' financial condition. Here, different pleas are independent.

Heterogeneity of Input Triple. Inputs to a judgment prediction system consist of three heterogeneous yet complementary parts, i.e., *fact description*, *plaintiff's plea*, and *related law articles*. Concatenating them together and treating them simply as a sequence of words as in previous works [1,12] would cause a great loss of information. This is the same in *question-answering* where the dual inputs, i.e., *query* and *passage*, should be modeled separately.

Despite the introduction of the neural networks that can learn better semantic representations of input text, it remains unsolved to incorporate proper mechanisms to integrate the complementary triple of *pleas, fact descriptions,* and *law articles* together.

Inspired by recent advances in question answering (QA) based reading comprehension (RC) [7,28,29,37] , we propose the **Legal Reading Comprehension (LRC)** framework for automatic judgment prediction. LRC incorporates the reading mechanism for better modeling of the complementary inputs abovementioned, as is done by human judges when referring to legal materials in search of supporting *law articles.* Reading mechanism, by simulating how human connects and integrates multiple text, has proven an effective module in RC tasks. We argue that applying the reading mechanism in a proper way among the triplets can obtain a better understanding and more informative representation of the original text, and further improve performance . To instantiate the framework, we propose an end-to-end neural network model named **AutoJudge**.

For experiments, we train and evaluate our models in the civil law system of mainland China. We collect and construct a large-scale real-world data set of 100, 000 case documents that the Supreme People's Court of People's Republic of China has made publicly available . *Fact description, pleas,* and *results* can be extracted easily from these case documents with regular expressions, since the original documents have special typographical characteristics indicating the discourse structure. We also take into account law articles and their corresponding juridical interpretations. We also implement and evaluate previous methods on our dataset, which prove to be strong baselines.

Our experiment results show significant improvements over previous methods. Further experiments demonstrate that our model also achieves considerable improvement over other off-the-shelf state-of-the-art models under classification and question answering framework respectively. Ablation tests carried out by taking off some components of our model further prove its robustness and effectiveness.

To sum up, our contributions are as follows:

(1) We introduce reading mechanism and re-formalize judgment prediction as Legal Reading Comprehension to better model the complementary inputs.
(2) We construct a real-world dataset for experiments, and plan to publish it for further research.
(3) Besides baselines from previous works, we also carry out comprehensive experiments comparing different existing deep neural network methods on our dataset. Supported by these experiments, improvements achieved by LRC prove to be robust.

2   Related Work

2.1   Judgment Prediction

Automatic judgment prediction has been studied for decades. At the very first stage of judgment prediction studies, researchers focus on mathematical and statistical analysis of existing cases, without any conclusions or methodologies on how to predict them [13,16,17,27,31,35].

Recent attempts consider judgment prediction under the text classification framework. Most of these works extract efficient features from text (e.g., N-grams) [1,18,21,22,33,39] or case profiles (e.g., dates, terms, locations and types) [12]. All these methods require a large amount of human effort to design features or annotate cases. Besides, they also suffer from generalization issue when applied to other scenarios.

Motivated by the successful application of deep neural networks, Luo et al. [23] introduce an attention-based neural model to predict charges of criminal cases, and verify the effectiveness of taking law articles into consideration. Nevertheless, they still fall into the text classification framework and lack the ability to handle multiple inputs with more complicated structures.

2.2 Text Classification

As the basis of previous judgment prediction works, typical text classification task takes a single text content as input and predicts the category it belongs to. Recent works usually employ neural networks to model the internal structure of a single input [2,14,34,38].

There also exists another thread of text classification called entailment prediction. Methods proposed in [10,26] are intended for complementary inputs, but the mechanisms can be considered as a simplified version of reading comprehension.

2.3 Reading Comprehension

Reading comprehension is a relevant task to model heterogeneous and complementary inputs, where an *answer* is predicted given two channels of inputs, i.e. a textual *passage* and a *query*. Considerable progress has been made [7,8,37]. These models employ various attention mechanism to model the interaction between *passage* and *query*. Inspired by the advantage of reading comprehension models on modeling multiple inputs, we apply this idea into the legal area and propose legal reading comprehension for judgment prediction.

3 Legal Reading Comprehension

3.1 Conventional Reading Comprehension

Conventional reading comprehension [9,11,28,29] usually considers reading comprehension as predicting the *answer* given a *passage* and a *query*, where the *answer* could be a single word, a text span of the original *passage*, chosen from answer candidates, or generated by human annotators.

Generally, an instance in RC is represented as a triple $\langle p, q, a \rangle$, where p, q and a correspond to *passage*, *query* and *answer* respectively. Given a triple $\langle p, q, a \rangle$,

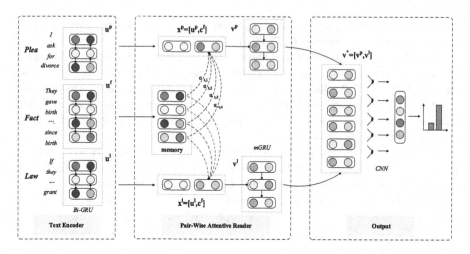

Fig. 2. An overview of AutoJudge.

RC takes the pair $\langle p, q \rangle$ as the input and employs attention-based neural models to construct an efficient representation. Afterwards, the representation is fed into the output layer to select or generate an *answer*.

3.2   Legal Reading Comprehension

Existing works usually formalize judgment prediction as a text classification task and focus on extracting well-designed features of specific cases. Such simplification ignores that the judgment of a case is determined by its fact description and multiple pleas. Moreover, the final judgment should act up to the legal provisions, especially in civil law systems. Therefore, how to integrate the information (i.e., *fact descriptions, pleas,* and *law articles*) in a reasonable way is critical for judgment prediction.

Inspired by the successful application of RC, we propose a framework of **L**egal **R**eading **C**omprehension (**LRC**) for judgment prediction in the legal area. As illustrated in Fig. 1, for each *plea* in a given case, the prediction of judgment *result* is made based the *fact description* and the potentially relevant *law articles*.

In a nutshell, LRC can be formalized as the following quadruplet task:

$$\langle f,\ p,\ l,\ r \rangle, \tag{1}$$

where f is the *fact description, p* is the *plea, l* is the *law articles* and r is the *result*. Given $\langle f, p, l \rangle$, LRC aims to predict the judgment result as

$$r = \underset{r \in \{support,\ reject\}}{\arg\max} P(r|f,p,l). \tag{2}$$

The probability is calculated with respect to the interaction among the triple $\langle f,\ p,\ l \rangle$, which will draw on the experience of the interaction between $\langle passage, question \rangle$ pairs in RC.

To summarize, LRC is innovative in the following aspects:

(1) While previous works fit the problem into text classification framework, LRC re-formalizes the way to approach such problems. This new framework provides the ability to deal with the heterogeneity of the complementary inputs.

(2) Rather than employing conventional RC models to handle pair-wise text information in the legal area, LRC takes the critical law articles into consideration and models the facts, pleas, and law articles jointly for judgment prediction, which is more suitable to simulate the human mode of dealing with cases.

4   Methods

We propose a novel judgment prediction model **AutoJudge** to instantiate the LRC framework. As shown in Fig. 2, AutoJudge consists of three flexible modules, including a text encoder, a pair-wise attentive reader, and an output module.

In the following parts, we give a detailed introduction to these three modules.

4.1   Text Encoder

As illustrated in Fig. 2, **Text Encoder** aims to encode the word sequences of inputs into continuous representation sequences.

Formally, consider a fact description $f = \{w_t^f\}_{t=1}^m$, a plea $p = \{w_t^p\}_{t=1}^n$, and the relevant law articles $l = \{w_t^l\}_{t=1}^k$, where w_t denotes the t-th word in the sequence and m, n, k are the lengths of word sequences f, p, l respectively.

First, we convert the words to their respective word embeddings to obtain $\mathbf{f} = \{\mathbf{w}_t^f\}_{t=1}^m$, $\mathbf{p} = \{\mathbf{w}_t^p\}_{t=1}^n$ and $\mathbf{l} = \{\mathbf{w}_t^l\}_{t=1}^k$, where $\mathbf{w} \in R^d$. Afterwards, we employ bi-directional GRU [3,5,6] to produce the encoded representation $\mathbf{u}$ of all words as follows:

$$
\begin{aligned}
\mathbf{u}_t^f &= \text{BiGRU}_F(\mathbf{u}_{t-1}^f, \mathbf{w}_t^f), \\
\mathbf{u}_t^p &= \text{BiGRU}_P(\mathbf{u}_{t-1}^p, \mathbf{w}_t^p), \\
\mathbf{u}_t^l &= \text{BiGRU}_L(\mathbf{u}_{t-1}^l, \mathbf{w}_t^l).
\end{aligned}
\tag{3}
$$

Note that, we adopt different bi-directional GRUs to encode fact descriptions, pleas, and law articles respectively (denoted as BiGRU_F, BiGRU_P, and BiGRU_L). With these text encoders, f, p, and l are converting into $\mathbf{u}^f = \{\mathbf{u}_t^f\}_{t=1}^m$, $\mathbf{u}^p = \{\mathbf{u}_t^p\}_{t=1}^n$, and $\mathbf{u}^l = \{\mathbf{u}_t^l\}_{t=1}^k$.

4.2   Pair-Wise Attentive Reader

How to model the interactions among the input text is the most important problem in reading comprehension. In AutoJudge, we employ a pair-wise attentive reader to process $\langle \mathbf{u}^f, \mathbf{u}^p \rangle$ and $\langle \mathbf{u}^f, \mathbf{u}^l \rangle$ respectively. More specifically, we propose to use pair-wise mutual attention mechanism to capture the complex semantic interaction between text pairs, as well as increasing the interpretability of AutoJudge.

Pair-Wise Mutual Attention. For each input pair $\langle \mathbf{u}^f, \mathbf{u}^p \rangle$ or $\langle \mathbf{u}^f, \mathbf{u}^l \rangle$, we employ pair-wise mutual attention to select relevant information from fact descriptions $\mathbf{u}^f$ and produce more informative representation sequences.

As a variant of the original attention mechanism [3], we design the pair-wise mutual attention unit as a GRU with internal memories denoted as **mGRU**.

Taking the representation sequence pair $\langle \mathbf{u}^f, \mathbf{u}^p \rangle$ for instance, mGRU stores the fact sequence $\mathbf{u}^f$ into its memories. For each timestamp $t \in [1, n]$, it selects relevant fact information $\mathbf{c}_t^f$ from the memories as follows,

$$\mathbf{c}_t^f = \sum_{i=1}^{m} \alpha_{t,i} \mathbf{u}_i^f. \tag{4}$$

Here, the weight $\alpha_{t,i}$ is the softmax value as

$$\alpha_{t,i} = \frac{\exp(a_{t,i})}{\sum_{j=1}^{m} \exp(a_{t,j})}. \tag{5}$$

Note that, $a_{t,j}$ represents the relevance between $\mathbf{u}_t^p$ and $\mathbf{u}_j^f$. It is calculated as follows,

$$a_{t,j} = \mathbf{V}^T \tanh(\mathbf{W}^f \mathbf{u}_j^f + \mathbf{W}^p \mathbf{u}_t^p + \mathbf{U}^p \mathbf{v}_{t-1}^p). \tag{6}$$

Here, $\mathbf{v}_{t-1}^p$ is the last hidden state in the GRU, which will be introduced in the following part. $\mathbf{V}$ is a weight vector, and $\mathbf{W}^f$, $\mathbf{W}^p$, $\mathbf{U}^p$ are attention metrics of our proposed pair-wise attention mechanism.

Reading Mechanism. With the relevant fact information $\mathbf{c}_t^f$ and $\mathbf{u}_t^p$, we get the t-th input of mGRU as

$$\mathbf{x}_t^p = \mathbf{u}_t^p \oplus \mathbf{c}_t^f, \tag{7}$$

where $\oplus$ indicates the concatenation operation.

Then, we feed $\mathbf{x}_t^p$ into GRU to get more informative representation sequence $\mathbf{v}^p = \{\mathbf{v}_t^p\}_{t=1}^n$ as follows,

$$\mathbf{v}_t^p = \text{GRU}(\mathbf{v}_{t-1}^p, \mathbf{x}_t^p). \tag{8}$$

For the input pair $\langle \mathbf{u}^f, \mathbf{u}^l \rangle$, we can get $\mathbf{v}^l = \{\mathbf{v}_t^l\}_{t=1}^k$ in the same way. Therefore, we omit the implementation details Here.

Similar structures with attention mechanism are also applied in [3,30,36, 37] to obtain mutually aware representations in reading comprehension models, which significantly improve the performance of this task.

4.3   Output Layer

Using text encoder and pair-wise attentive reader, the initial input triple $\langle f, p, l \rangle$ has been converted into two sequences, i.e., $\mathbf{v}^p = \{\mathbf{v}_t^p\}_{t=1}^n$ and $\mathbf{v}^l = \{\mathbf{v}_t^l\}_{t=1}^k$, where $\mathbf{v}_t^l$ is defined similarly to $\mathbf{v}_t^p$. These sequences reserve complex semantic information about the pleas and law articles, and filter out irrelevant information in fact descriptions.

With these two sequences, we concatenate $\mathbf{v}^p$ and $\mathbf{v}^l$ along the *sequence length* dimension to generate the sequence $\mathbf{v}^* = \{\mathbf{v}_t\}_{t=1}^{n+k}$. Since we have employed several GRU layers to encode the sequential inputs, another recurrent layer may be redundant. Therefore, we utilize a 1-layer CNN [14] to capture the local structure and generate the representation vector for the final prediction.

Assuming $y \in [0,1]$ is the predicted probability that the plea in the case sample would be supported and $r \in \{0,1\}$ is the gold standard, AutoJudge aims to minimize the cross-entropy as follows,

$$\mathcal{L} = -\frac{1}{N}\sum_{i=1}^{N}[r_i ln y_i + (1 - r_i)ln(1 - y_i)], \tag{9}$$

where N is the number of training data. As all the calculation in our model is differentiable, we employ Adam [15] for optimization.

5 Experiments

To evaluate the proposed LRC framework and the AutoJudge model, we carry out a series of experiments on the divorce proceedings, a typical yet complex field of civil cases. Divorce proceedings often come with several kinds of *pleas*, e.g. *seeking divorce*, *custody of children*, *compensation*, and *maintenance*, which focuses on different aspects and thus makes it a challenge for judgment prediction.

5.1 Dataset Construction for Evaluation

Data Collection. Since none of the datasets from previous works have been published, we decide to build a new one. We randomly collect $100,000$ cases from China Judgments Online[1], among which $80,000$ cases are for training, $10,000$ each for validation and testing. Among the original cases, 51% are granted *divorce* and others not. There are $185,723$ valid *pleas* in total, with 52% supported and 48% rejected. Note that, if the *divorce plea* in a case is not granted, the other pleas of this case will not be considered by the judge. Case materials are all natural language sentences, with averagely 100.08 tokens per *fact description* and 12.88 per *plea*. There are 62 relevant *law articles* in total, each with 26.19 tokens averagely. Note that the case documents include special typographical signals, making it easy to extract labeled data with regular expression.

---

[1] http://wenshu.court.gov.cn.

Data Pre-Processing. We apply some rules with legal prior to preprocess the dataset according to previous works [4,19,20], which have proved effective in our experiments.

Name Replacement[2]: All names in case documents are replaced with marks indicating their roles, instead of simply anonymizing them, e.g. <*Plantiff*>, <*Defendant*>, <*Daughter_x*> and so on. Since *"all are equal before the law"*[3], names should make no more difference than what role they take.

Law Article Filtration: Since most accessible divorce proceeding documents do not contain ground-truth *fine-grained* articles[4], we use an unsupervised method instead. First, we extract all the articles from the law text with regular expression. Afterwards, we select the most relevant 10 articles according to the fact descriptions as follows. We obtain sentence representation with CBOW [24,25] weighted by inverse document frequency, and calculate cosine distance between cases and law articles. Word embeddings are pre-trained with Chinese Wikipedia pages[5]. As the final step, we extract top 5 relevant articles for each sample respectively from the main marriage law articles and their interpretations, which are equally important. We manually check the extracted articles for 100 cases to ensure that the extraction quality is fairly good and acceptable.

The filtration process is automatic and fully unsupervised since the original documents have no ground-truth labels for fine-grained law articles, and coarse-grained law-articles only provide limited information. We also experiment with the ground-truth articles, but only a small fraction of them has fine-grained ones, and they are usually not available in real-world scenarios.

5.2   Implementation Details

We employ Jieba[6] for Chinese word segmentation and keep the top $20,000$ frequent words. The word embedding size is set to 128 and the other low-frequency words are replaced with the mark <*UNK*>. The hidden size of GRU is set to 128 for each direction in Bi-GRU. In the pair-wise attentive reader, the hidden state is set to 256 for mGRu. In the CNN layer, filter windows are set to 1, 3, 4, and 5 with each filter containing 200 feature maps. We add a dropout layer [32] after the CNN layer with a dropout rate of 0.5. We use Adam [15] for training and set learning rate to 10^{-4}, β_1 to 0.9 , β_2 to 0.999, ϵ to 10^{-8}. We employ *precision, recall, F1* and *accuracy* for evaluation metrics. We repeat all the experiments for 10 times, and report the average results.

---

[2] We use regular expressions to extract *names* and *roles* from the formatted case header.

[3] Constitution of the People's Republic of China.

[4] *Fine-grained* articles are in the *Juridical Interpretations*, giving detailed explanation, while *the Marriage Law* only covers some basic principles.

[5] https://dumps.wikimedia.org/zhwiki/.

[6] https://github.com/fxsjy/jieba.

Table 1. Experimental results(%). Precision/Recall/F1 are reported for positive samples and calculated as the mean score over 10-time experiments. Acc is defined as the proportion of test samples classified correctly, equal to micro-precision. MaxFreq refers to always predicting the most frequent label, i.e. *support* in our dataset. * indicates methods proposed in previous works.

Models	Precision	Recall	F1	Accuracy
MaxFreq	52.2	100	68.6	52.2
SVM*	57.8	53.5	55.6	55.5
CNN	76.1	81.9	79.0	77.6
CNN+law	74.4	79.4	77.0	76.0
GRU	79.2	72.9	76.1	76.6
GRU+law	78.2	68.2	72.8	74.4
GRU+Attention*	79.1	80.7	80.0	79.1
AoA	79.3	78.9	79.2	78.3
AoA+law	79.0	79.2	79.1	78.3
r-net	79.5	78.7	79.2	78.4
r-net+law	79.3	78.8	79.0	78.3
AutoJudge	**80.4**	**86.6**	**83.4**	**82.2**

5.3   Baselines

For comparison, we adopt and re-implement three kinds of baselines as follows:

Lexical Features SVM. We implement an SVM with lexical features in accordance with previous works [1,18,21,22,33] and select the best feature set on the development set.

Neural Text Classification Models. We implement and fine-tune a series of neural text classifiers, including attention-based method [23] and other methods we deem important. CNN [14] and GRU [5,38] take as input the concatenation of *fact description* and *plea*. Similarly, *CNN/GRU+law* refers to using the concatenation of *fact description, plea* and *law articles* as inputs.

RC Models. We implement and train some off-the-shelf RC models, including r-net [37] and AoA [7], which are the leading models on SQuAD leaderboard. In our initial experiments, these models take *fact description* as *passage* and *plea* as *query*. Further, *Law articles* are added to the *fact description* as a part of the reading materials, which is a simple way to consider them as well.

From Table 1, we have the following observations:

(1) AutoJudge consistently and significantly outperforms all the baselines, including RC models and other neural text classification models, which shows the effectiveness and robustness of our model.

(2) RC models achieve better performance than most text classification models (excluding *GRU+Attention*), which indicates that reading mechanism is a better way to integrate information from heterogeneous yet complementary inputs. On the contrary, simply adding *law articles* as a part of the reading materials makes no difference in performance. Note that, *GRU+Attention* employ similar attention mechanism as RC does and takes additional law articles into consideration, thus achieves comparable performance with RC models.

(3) Comparing with conventional RC models, AutoJudge achieves significant improvement with the consideration of additional law articles. It reflects the difference between LRC and conventional RC models. We re-formalize LRC in legal area to incorporate law articles via the reading mechanism, which can enhance judgment prediction. Moreover, CNN/GRU+law decrease the performance by simply concatenating original text with law articles, while *GRU+Attention/AutoJudge* increase the performance by integrating law articles with attention mechanism. It shows the importance and rationality of using attention mechanism to capture the interaction between multiple inputs.

The experiments support our hypothesis as proposed in the Introduction part that in civil cases, it's important to model the interactions among case materials. Reading mechanism can well perform the matching among them.

5.4   Ablation Test

AutoJudge is characterized by the incorporation of pair-wise attentive reader, law articles, and a CNN output layer, as well as some pre-processing with legal prior. We design ablation tests respectively to evaluate the effectiveness of these modules. When taken off the attention mechanism, AutoJudge degrades into a GRU on which a CNN is stacked. When taken off law articles, the CNN output layer only takes $\{v_t^P\}_{t=1}^{L_P}$ as input. Besides, our model is tested respectively without name-replacement or unsupervised selection of law articles (i.e. passing the whole law text). As mentioned above, we system use law articles extracted with unsupervised method, so we also experiment with ground-truth law articles.

Results are shown in Table 2. We can infer that:

(1) The performance drops significantly after removing the attention layer or excluding the law articles, which is consistent with the comparison between AutoJudge and baselines. The result verifies that both the reading mechanism and incorporation of law articles are important and effective.

(2) After replacing CNN with an LSTM layer, performance drops as much as 4.4% in accuracy and 5.7% in F1 score. The reason may be the redundancy of RNNs. AutoJudge has employed several GRU layers to encode text sequences. Another RNN layer may be useless to capture sequential dependencies, while CNN can catch the local structure in convolution windows.

Table 2. Experimental results of ablation tests (%).

Models	F1	Accuracy
AutoJudge	83.4	82.2
w/o reading mechanism	78.9($\downarrow$ 4.5)	78.2($\downarrow$ 4.0)
w/o law articles	79.6($\downarrow$ 3.8)	78.4($\downarrow$ 3.8)
CNN$\rightarrow$LSTM	77.6($\downarrow$ 5.8)	77.7($\downarrow$ 4.5)
w/o Pre-Processing	81.1($\downarrow$ 2.3)	80.3($\downarrow$ 1.9)
w/o law article selection	80.6($\downarrow$ 2.8)	80.5($\downarrow$ 1.7)
with GT law articles	**85.1**($\uparrow$ 1.7)	**84.1**($\uparrow$ 1.9)

(3) Motivated by existing rule-based works, we conduct data pre-processing on cases, including name replacement and law article filtration. If we remove the pre-processing operations, the performance drops considerably. It demonstrates that applying the prior knowledge in legal filed would benefit the understanding of legal cases.

Performance over Law Articles. It's intuitive that the quality of the retrieved *law articles* would affect the final performance. As is shown in Table 2, feeding the whole law text without filtration results in worse performance. However, when we train and evaluate our model with ground truth articles, the performance is boosted by nearly 2% in both F1 and Acc. The performance improvement is quite limited compared to that in previous work [23] for the following reasons: (1) As mentioned above, most case documents only contain coarse-grained articles, and only a small number of them contain fine-grained ones, which has limited information in themselves. (2) Unlike in criminal cases where the application of an article indicates the corresponding crime, *law articles* in civil cases work as reference, and can be applied in both the cases of *supports* and *rejects*. As law articles cut both ways for the judgment result, this is one of the characteristics that distinguishes civil cases from criminal ones. We also need to remember that, the performance of 84.1% in accuracy or 85.1% in F1 score is unattainable in real-world setting for automatic prediction where ground-truth articles are not available.

Reading Weighs More Than Correct Law Articles. In the area of civil cases, the understanding of the case materials and how they interact is a critical factor. The inclusion of law articles is not enough. As is shown in Table 2, compared to feeding the model with an un-selected set of law articles, taking away the reading mechanism results in greater performance drop[7]. Therefore, the ability to read, understand and select relevant information from the complex multi-sourced case materials is necessary. It's even more important in real world since we don't have access to ground-truth law articles to make predictions.

---

[7] 3.9% vs. 1.7% in Acc, and 4.4% vs. 2.8% in F1.

5.5   Case Study

Visualization of Positive Samples. We visualize the heat maps of attention results[8]. As shown in Fig. 3, deeper background color represents larger attention score.

The attention score is calculated with Eq. (5). We take the average of the resulting $n \times m$ attention matrix over the time dimension to obtain attention values for each word.

The visualization demonstrates that the attention mechanism can capture relevant patterns and semantics in accordance with different *pleas* in different *cases*.

Failure Analysis. As for the failed samples, the most common reason comes from the anonymity issue, which is also shown in Fig. 3. As mentioned above, we conduct name replacement. However, some critical elements are also anonymized by the government, due to the privacy issue. These elements are sometimes important to judgment prediction. For example, determination of the key factor *long-time separation* is relevant to the explicit dates, which are anonymized.

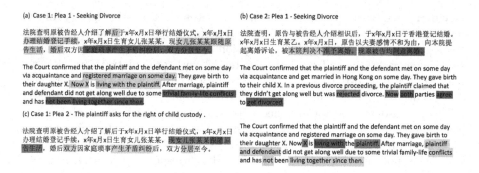

Fig. 3. Visualization of attention mechanism.

6   Conclusion

In this paper, we explore the task of predicting judgments of civil cases. Comparing with conventional text classification framework, we propose **L**egal **R**eading **C**omprehension framework to handle multiple and complex textual inputs. Moreover, we present a novel neural model, **AutoJudge**, to incorporate law articles for judgment prediction. In experiments, we compare our model on divorce proceedings with various state-of-the-art baselines of various frameworks. Experimental results show that our model achieves considerable improvement than all the baselines. Besides, visualization results also demonstrate the effectiveness and interpretability of our proposed model.

---

[8] Examples given here are all drawn from the test set whose predictions match the real judgment.

In the future, we can explore the following directions: (1) Limited by the datasets, we can only verify our proposed model on divorce proceedings. A more general and larger dataset will benefit the research on judgment prediction. (2) Judicial decisions in some civil cases are not always binary, but more diverse and flexible ones, e.g. compensation amount. Thus, it is critical for judgment prediction to manage various judgment forms.

Acknowledgments. This work is supported by the National Key Research and Development Program of China (No. 2018YFC0831900).

References

1. Aletras, N., Tsarapatsanis, D., Preotiuc-Pietro, D., Lampos, V.: Predicting judicial decisions of the European court of human rights: a natural language processing perspective. PeerJ Comput. Sci. **2**, e93 (2016)
2. Baharudin, B., Lee, L.H., Khan, K.: A review of machine learning algorithms for text-documents classification. JAIT **1**(1), 4–20 (2010)
3. Bahdanau, D., Cho, K., Bengio, Y.: Neural machine translation by jointly learning to align and translate. In: Proceedings of ICLR (2015)
4. Bian, G.W., Shun-yuan, T.: Integrating query translation and text classification in a cross-language patent access system. In: Proceedings of NTCIR-7 Workshop Meeting, pp. 252–261 (2005)
5. Cho, K., et al.: Learning phrase representations using rnn encoder-decoder for statistical machine translation. Computer Science (2014)
6. Chung, J., Gulcehre, C., Cho, K., Bengio, Y.: Empirical evaluation of gated recurrent neural networks on sequence modeling. In: Proceedings of NIPS (2014)
7. Cui, Y., Chen, Z., Wei, S., Wang, S., Liu, T., Hu, G.: Attention-over-attention neural networks for reading comprehension. In: Proceedings of ACL (2017)
8. Dhingra, B., Liu, H., Yang, Z., Cohen, W.W., Salakhutdinov, R.: Gated-attention readers for text comprehension. In: Proceedings of ACL (2017)
9. He, W., et al.: Dureader: a chinese machine reading comprehension dataset from real-world applications. arXiv preprint arXiv:1711.05073 (2017)
10. Hu, B., Lu, Z., Li, H., Chen, Q.: Convolutional neural network architectures for matching natural language sentences. In: Proceedings of NIPS, pp. 2042–2050 (2014)
11. Joshi, M., Choi, E., Weld, D.S., Zettlemoyer, L.: Triviaqa: a large scale distantly supervised challenge dataset for reading comprehension. In: Proceedings of ACL (2017)
12. Katz, D.M., Bommarito II, M.J., Blackman, J.: A general approach for predicting the behavior of the supreme court of the united states. PloS One **12**(4), e0174698 (2017)
13. Keown, R.: Mathematical models for legal prediction. Computer/LJ **2**, 829 (1980)
14. Kim, Y.: Convolutional neural networks for sentence classification. In: Proceedings of EMNLP (2014)
15. Kingma, D., Ba, J.: Adam: a method for stochastic optimization. In: Proceedings of ICLR (2015)
16. Kort, F.: Predicting supreme court decisions mathematically: a quantitative analysis of the "right to counsel" cases. Am. Polit. Sci. Rev. **51**(1), 1–12 (1957)
17. Lauderdale, B.E., Clark, T.S.: The supreme court's many median justices. Am. Polit. Sci. Rev. **106**(4), 847–866 (2012)

18. Lin, W.C., Kuo, T.T., Chang, T.J., Yen, C.A., Chen, C.J., Lin, S.D.: Exploiting machine learning models for Chinese legal documents labeling, case classification, and sentencing prediction. In: Processdings of ROCLING, p. 140 (2012)
19. Liu, C.L., Chang, C.T., Ho, J.H.: Case instance generation and refinement for case-based criminal summary judgments in Chinese. JISE (2004)
20. Liu, C.L., Ho, J.H., Ho, J.H.: Classification and clustering for case-based criminal summary judgments. In: Proceedings of the International Conference on Artificial Intelligence and Law, pp. 252–261 (2003)
21. Liu, C.L., Hsieh, C.D.: Exploring phrase-based classification of judicial documents for criminal charges in chinese. In: Proceedings of ISMIS, pp. 681–690 (2006)
22. Liu, Y.H., Chen, Y.L.: A two-phase sentiment analysis approach for judgement prediction. J. Inf. Sci. **44**(5), 594–607 (2017)
23. Luo, B., Feng, Y., Xu, J., Zhang, X., Zhao, D.: Learning to predict charges for criminal cases with legal basis. In: Proceedings of EMNLP (2017)
24. Mikolov, T., Chen, K., Corrado, G., Dean, J.: Efficient estimation of word representations in vector space. arXiv preprint arXiv:1301.3781 (2013)
25. Mikolov, T., Sutskever, I., Chen, K., Corrado, G.S., Dean, J.: Distributed representations of words and phrases and their compositionality. In: Proceedings of NIPS, pp. 3111–3119 (2013)
26. Mitra, B., Diaz, F., Craswell, N.: Learning to match using local and distributed representations of text for web search. In: Proceedings of WWW, pp. 1291–1299 (2017)
27. Nagel, S.S.: Applying correlation analysis to case prediction. Texas Law Rev. **42**, 1006 (1963)
28. Nguyen, T., et al.: Ms marco: a human generated machine reading comprehension dataset. arXiv preprint arXiv:1611.09268 (2016)
29. Rajpurkar, P., Zhang, J., Lopyrev, K., Liang, P.: Squad: 100,000+ questions for machine comprehension of text. In: Proceedings of EMNLP (2016)
30. Rocktaschel, T., Grefenstette, E., Hermann, K.M., Kocisky, T., Blunsom, P.: Reasoning about entailment with neural attention. In: Proceedings of ICLR (2016)
31. Segal, J.A.: Predicting supreme court cases probabilistically: The search and seizure cases, 1962–1981. Am. Polit. Sci. Rev. **78**(4), 891–900 (1984)
32. Srivastava, N., Hinton, G.E., Krizhevsky, A., Sutskever, I., Salakhutdinov, R.: Dropout: a simple way to prevent neural networks from overfitting. JMLR **15**(1), 1929–1958 (2014)
33. Sulea, O.M., Zampieri, M., Vela, M., Genabith, J.V.: Exploring the use of text classification in the legal domain. In: Proceedings of ASAIL workshop (2017)
34. Tang, D., Qin, B., Liu, T.: Document modeling with gated recurrent neural network for sentiment classification. In: Proceedings of EMNLP, pp. 1422–1432 (2015)
35. Ulmer, S.S.: Quantitative analysis of judicial processes: some practical and theoretical applications. Law Contemp. Probs. **28**, 164 (1963)
36. Wang, S., Jiang, J.: Learning natural language inference with LSTM. In: Proceedings of NAACL (2016)
37. Wang, W., Yang, N., Wei, F., Chang, B., Zhou, M.: Gated self-matching networks for reading comprehension and question answering. Proc. ACL **1**, 189–198 (2017)
38. Yang, Z., Yang, D., Dyer, C., He, X., Smola, A.J., Hovy, E.H.: Hierarchical attention networks for document classification. In: Proceedings of NAACL, pp. 1480–1489 (2016)
39. Zhong, H., Guo, Z., Tu, C., Xiao, C., Liu, Z., Sun, M.: Legal judgment prediction via topological learning. In: Proceedings of EMNLP (2018)
40. Zhuge, P.: Chinese Law Yearbook. The Chinese Law Yearbook Press, Shanghai (2016)

Legal Cause Prediction with Inner Descriptions and Outer Hierarchies

Zhiyuan Liu[1]([⊠])🆔, Cunchao Tu[2]🆔, Zhiyuan Liu[2]🆔, and Maosong Sun[2]

[1] School of Science, Xi'an Jiaotong University, Xi'an, China
acharkq@gmail.com
[2] Department of CST, Tsinghua University, Beijing, China
tucunchao@gmail.com, {liuzy,sms}@tsinghua.edu.cn

Abstract. Legal Cause Prediction (LCP) aims to determine the charges in criminal cases or types of disputes in civil cases according to the fact descriptions. The research to date takes LCP as a text classification task and fails to consider the outer hierarchical dependencies and inner text information of causes. However, this information is critical for understanding causes and is expected to benefit LCP. To address this issue, we propose the **H**ierarchical **L**egal **C**ause **P**rediction (HLCP) model to incorporate this crucial information within the seq2seq framework. Specifically, we employ an attention-based seq2seq model to predict the cause path and utilize the inner text information to filter out noisy information in fact descriptions. We conduct experiments on 4 real-world criminal and civil datasets. Experimental results show that our model achieves significant and consistent improvements over all baselines.

1 Introduction

With the release of more than 60 million legal documents from China Judgment Online[1], the analysis, and research on these well-structured and informative legal documents have attracted a wide range of researchers from legal and NLP fields. Among existing works, cause prediction is a representative and fundamental task which aims to predict charges in criminal cases or types of disputes in civil cases. It can provide an effective reference for judges and benefit a series of real-world applications, such as automatic sentencing system and intelligent judgment assistant.

At the early stage, researchers utilize shallow textual features (e.g., characters, words, and phrases) [20,21] or well-designed features (locations, terms, and dates) [12] to predict charges in criminal cases.

With the successful application of deep learning methods in NLP area [14, 32,36], researchers propose to employ deep learning techniques to predict causes according to the fact descriptions in legal cases. For example, Luo et al. [23] employ attention mechanism to predict charges with the consideration of relevant law articles. Ye et al. [37] utilize seq2seq model for court view generation with

---

[1] http://wenshu.court.gov.cn/.

© Springer Nature Switzerland AG 2019
M. Sun et al. (Eds.): CCL 2019, LNAI 11856, pp. 573–586, 2019.
https://doi.org/10.1007/978-3-030-32381-3_46

additional charge information. Hu et al. [10] introduce several discriminative attributes as internal mappings to predict few-shot and confusing charges.

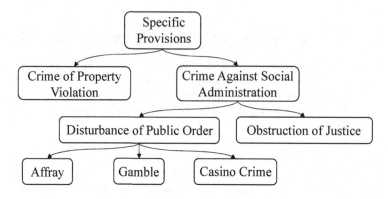

Fig. 1. Hierarchical structure of causes.

It is worth pointing out that all the existing works take cause prediction as a typical text classification task which is usually confronted with two major challenges. First, existing works ignore the implicit relations among causes. As shown in Fig. 1, there exists a hierarchical dependency among both criminal and civil causes. The hierarchy provides effective information for all causes, especially for the confusing and few-shot ones. For each leaf cause node, a unique path connects it with the root node. Besides, each cause is treated as a plain symbol by existing classification models and thus the inner information hidden in their names is missed. However, the causes in the legal area are usually well-defined and each of their names can be treated as an accurate and refined description. According to our statistics, the average lengths of criminal and civil cause names are 4.65 and 4.24 words. This critical text information is expected to be used to filter out the noisy information and retain the relevant parts in fact descriptions.

To address these issues, we propose **H**ierarchical **L**egal **C**ause **P**rediction (HLCP) to incorporate the outer hierarchical relations and inner text information of causes. Specifically, we first transform each cause into a path from the root node to leaf cause node. Then, we propose an inner text attention-based seq2seq model to predict cause paths according to the fact descriptions. Compared with traditional hierarchical classification models, HLCP trains only one *global* classifier hierarchically and is able to share knowledge among all nodes in the hierarchy. Experimental results on several large-scale real-world datasets demonstrate that our model outperforms all existing charge prediction models [10,23] and hierarchical classification models [4,34,38].

We publish our code at https://github.com/acharkq/HLCP for further explorations.

2   Related Work

2.1   Cause Prediction

Automatic legal judgment has become a research topic for decades. Kort [17] pioneers the quantitative study on judicial decision prediction by analyzing factual elements numerically. Segal [25] constructs a probabilistic model with variables to evaluate the fairness of the court decisions. These early works [13,18,24,33] usually focus on statistical and mathematical analysis on specific legal scenarios.

Recent works formalize legal judgment as a text classification problem. They focus on scenarios where the judgment result is selected from a fixed label set, e.g. the charges prediction for criminal cases. Machine learning methods are widely utilized as it shows great effectiveness in many areas. These works usually focus on extracting efficient features from facts. Shallow textual features and manually designed patterns are drawn from fact descriptions and annotations (e.g., type, location, term) of cases [1,19,21,22,28]. Due to the human efforts required for pattern design and annotation, these methods suffer from the issue of scalability.

With the rapid development of deep learning techniques, various neural models [2,32,36] show up with a promising performance in NLP tasks. In the legal area, Luo et al. [23] incorporate law article knowledge for cause prediction with an attention-based neural model. Inspired by this work, we utilize the names of causes as the attention queries to filter out noisy information in fact description. Moreover, we introduce the direct hierarchical dependencies among causes into LCP, which could alleviate the data imbalance issue and help to distinguish confusing causes under different parent nodes. Ye et al. [37] employ a seq2seq model in charge prediction task. Different from our work, their seq2seq model is employed for the generation of court views. Hu et al. [10] achieve decent performance promotion on few-shot charge prediction with the annotations of 10 discriminative legal attributes of criminal charges. However, their annotations only cover 32% criminal charges in China. Our method utilizes the off the shelf dependencies among all causes.

2.2   Hierarchical Classification

Hierarchical classification methods have been developed in various application domains. Silla et al. [26] summarizes and classifies these methods into, *flat*, *local*, and *global* ones. *Flat* classifiers ignore the hierarchy structure and treat the problem as a simple multi-class classification. *Local* methods can be classifies into: *local per node*, training binary classifier on each class [8], *local per parent*, training one multi-class classifier for all child nodes of a parent node [27], and *local per level*, training one multi-class classifier at each level instead of each node [7]. These methods enforce the inference to be consistent with the hierarchy, whereas our model is trained under the constraint of the hierarchy. The last method, *global* [16], trains a multi-class classifier which is responsible for the classification of all causes in the hierarchy and employs a loss function that reflects the similarity among labels. In a similar way, we minimize the loss for parent nodes and leaf nodes at the same time and expect this could help to capture the semantic dependencies among causes.

Neural models are also employed for hierarchical classification in NLP tasks. Cerri et al. [5] employs perceptron as classifier for each parent node of the *local per parent node* method. Karn et al. [11] propose to use RNN based encoder-decoder model [6,30] for entity mention classification task. This model follows the hierarchical path and utilizes attention to filter out noise information level by level. Inspired by the advantage of the seq2seq model for hierarchical dependency modeling, we propose the Hierarchical Legal Cause Prediction model for legal cause prediction.

3   Hierarchical Legal Cause Prediction

In this section, we first give a definition of the LCP task. Next, we introduce the HLCP model.

3.1   Problem Formulation

As shown in Table 1, a fact description refers to the plain description part of a legal document, which is nearly independent of the court's opinion. We regard it as a word sequence $\mathbf{x} = \{x_1, \ldots, x_m, \ldots, x_M\}$, where each word $x_m \in V$. The LCP task aims to predict its corresponding legal cause y, which locates at the bottom of the cause hierarchy.

Table 1. Example of the fact description.

In the early morning of March 28, 2014, the defendant Jia came into Kaiwen restaurant in Shijiazhuang City after drinking. He beat one waiter for no reason. Then, the other waiters of the restaurant started to fight with him...

As shown in Fig. 1, by tracing along the tree-structured hierarchy of causes, we transform the cause label y into a path from root node to cause node, i.e., a label sequence $\mathbf{y} = \{y_1, \ldots, y_i, \ldots, y_I\}$, where $y_i \in Y$. Y denotes the set of all causes in the hierarchy. Note that, each cause y_i in this hierarchy owns its name $s_y = \{x_1, \ldots, x_l, \ldots, x_{L_y}\}$, which can be regarded as a short description of this cause.

With the above denotations, HLCP defines the prediction probability of $\mathbf{y}$ as follows:

$$p(\mathbf{y}|\mathbf{x}) = \prod_{i=1}^{I} p(y_i|y_{1:i-1}, \mathbf{x}), \tag{1}$$

As shown in Fig. 2, HLCP consists of two components, i.e., *fact encoder* and *cause predictor*, which will be introduced in the following parts.

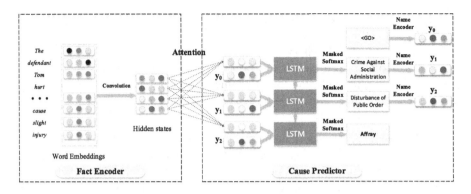

Fig. 2. The framework of HLCP.

3.2   Fact Encoder

As shown in Fig. 2, *fact encoder* transforms the word sequence of the fact description into vector representation as the input of *cause predictor*.

Word Representation. For a given fact description $\mathbf{x} = \{x_1, \ldots, x_m, \ldots, x_M\}$, we first convert each word x_m to a k-dimensional embedding $\mathbf{x_m}$ by looking up a table $W \in \mathbb{R}^{|V| \times k}$, where $|V|$ denotes the size of the vocabulary. Thus, the input is represented by an embedding sequence:

$$\hat{\mathbf{x}} = \{\mathbf{x}_1, \ldots, \mathbf{x}_m, \ldots, \mathbf{x}_M\}. \tag{2}$$

Convolution. Convolutional Neural Network is used after word representation. For the embedding sequence, let $\mathbf{u}_m \in \mathbb{R}^{w \times k}$ denotes the concatenation of w word embeddings $\mathbf{x}_{m:m+w-1}$, where w is the filter width. We then generate $\mathbf{v}_m \in \mathbb{R}^{k_f}$ by

$$\mathbf{v_m} = \mathrm{Relu}(\mathbf{W} \cdot \mathbf{u}_m + \mathbf{b}) \tag{3}$$

where $\mathbf{W} \in \mathbb{R}^{k_f \times (w \times k)}$ and $\mathbf{b} \in \mathbb{R}^{k_f}$. k_f denotes the filter size of CNN. After the convolution step, we obtain $\hat{\mathbf{v}} = \{\mathbf{v}_1, \ldots, \mathbf{v}_m, \ldots, \mathbf{v}_M\}$. This hidden state sequence is used as the attention values of **cause attention**.

Max-Pooling. We max pooled over $\hat{\mathbf{v}}$ alone the sequence length axis to obtain the initial state $\mathbf{h}_0$ for *cause predictor* according to the equation:

$$h_{0,j} = \max(\mathbf{v}_{1,j}, \ldots, \mathbf{v}_{M,j}), \forall j \in [1, k_f]. \tag{4}$$

3.3   Cause Predictor

Owing to the successful usage of LSTM [9] on sequence generation, it is adopted for cause sequence prediction in HLCP.

Specifically, the max-pooling output of *fact encoder* is used as the initial state for the LSTM cell in *cause predictor*.

The input of LSTM cell at the i-th step consists of two parts, i.e., **cause representation** and **cause-aware fact representation**. Here, cause representation is the representation for the name of the cause, which is predicted at the previous step. It is calculated by **Name Vectorizer** as follows.

Name Vectorizer. As mentioned above, we treat the name of each cause as its short description, which reflects how it differs from other causes. Thus, for the i-th cause y_i, we define **Name Vectorizer** to get its representation y_i according to its name s_{y_i}, i.e., $y_i = \text{vec}(s_{y_i})$. In practice, for each name s_{y_i}, we employ an LSTM to encode the corresponding word sequence and take the final hidden state as the cause representation.

Cause Attention. Fact description usually contains a large amount of irrelevant and noisy information which could mislead the predictor. The obtained cause representation provides a suitable way to address this issue. Therefore, for each step i, we utilize the cause representation y_{i-1} at step $i-1$ to select the most relevant information from the fact vectors $\hat{v}$. In a nutshell, we employ the following cause attention mechanism [3] to get **cause-aware fact representation**.

As illustrated in Fig. 2, we employ Bahdanau Attention to calculate the weight vector α_i of hidden states $\hat{v}$ as follows:

$$e(\mathbf{y}_{i-1}, \mathbf{v}_j) = \mathbf{U}\tanh(\mathbf{W_0}\mathbf{y}_{i-1} + \mathbf{W_1}\mathbf{v}_j + \mathbf{b}),$$
$$\alpha_{i,j} = \frac{\exp(e(\mathbf{y}_{i-1}, \mathbf{v}_j))}{\sum_{t\in[1,M]}\exp(e(\mathbf{y}_{i-1}, \mathbf{v}_t))}. \tag{5}$$

Here, $\mathbf{U}$, $\mathbf{W_o}$, $\mathbf{W_1}$, and $\mathbf{b}$ are trained parameters. With the obtained weight vector α_i, we can calculate cause-aware fact representation $\mathbf{c}_i$ at step i as follows:

$$\mathbf{c}_i = \sum_{j=1}^{M} \alpha_{ij}\mathbf{v}_j. \tag{6}$$

With the cause representation $\mathbf{y}_{i-1}$ and the cause-aware fact representation $\mathbf{c}_i$ at step i, we concatenate and feed them into LSTM cell to get current hidden state $\mathbf{h}_i$, which will be used for predicting the cause in the current level later.

Masked Classification. To ensure that the predicted cause sequence is consistent with the tree structure, we employ masked classification to restrict the prediction scope. That means the causes in the current prediction scope must be the child nodes of the previously predicted cause. For the output $\mathbf{h}_i$ of predictor cell at each step, we employ a weight shared fully connected layer to project it into the legal cause space. Then, the logit of each cause, that is not a child node of the parent cause y_{i-1}, is masked to negative infinity. Thus, the probability of these causes will be 0 after the softmax operation. The above operations can be formalized as:

$$p(y_i|y_{1:i-1}, \mathbf{x}) = \text{softmax}(\text{mask}(\mathbf{W}_p\mathbf{h}_i + \mathbf{b}_p, y_{i-1})),$$

$$\text{mask}(\mathbf{x}_j, y) = \begin{cases} \mathbf{x}_j & \pi(Y_j) = y \\ -\infty & \pi(Y_j) \neq y. \end{cases} \tag{7}$$

where $\mathbf{W}_p$ and $\mathbf{b}_p$ are parameters of the fully connected layer. $\pi(y)$ is the parent cause of y.

3.4   Optimization

For each instance, the training objective of HLCP is to minimize the cross-entropy between predicted cause sequences and the ground-truth as follows:

$$\mathcal{L} = -\frac{1}{I}\sum_{i=1}^{I}\hat{y}_i \log(p(y_i|y_{1:i-1}, \mathbf{x})), \tag{8}$$

where $\hat{y}_i$ is an indicator vector, i.e., $\hat{y}_{ij} = 1$ if $y_i = j$ and $\hat{y}_{ij} = 0$ otherwise.

4   Experiments

To demonstrate the effectiveness of our proposed HLCP model, we conduct experiments on several criminal and civil datasets.

4.1   Datasets

Criminal Case. We conduct experiments on the following previous released criminal datasets: CAIL, PKU, and FSC. CAIL (Chinese AI and Law) [35] is a large-scale criminal dataset released for legal competition, we employ the released small version. Zhong et al. [39] collect and construct PKU with criminal cases from Peking University Law Online[2]. FSC (Few-Shot Charge) is a criminal dataset collected by Hu et al. [10] from China Judgment Online[3] for the purpose of few-shot learning. As the cases which contain multiple defendants and multiple causes will increase the difficulty of matching facts to different defendants or causes in LCP task, we filter out these cases in all criminal datasets so that we can focus on the exploration of the validity of introducing hierarchy and text information among causes into LCP task. We set the minimum cause frequency to 30 for CAIL and PKU and 10 for FSC.

---

[2] http://www.pkulaw.com/.
[3] http://wenshu.court.gov.cn/.

Civil Case. In civil cases, legal cause refers to the type of disputes between parties. Courts will apply proper civil causes to cases based on the fact description and the claim of the plaintiff. Since there are no publicly available civil datasets in previous works for cause prediction, we collect 599, 400 cases from China Judgment Online[5]. We firstly extracted fact descriptions with regular templates from documents automatically. Further, we checked the extraction results of randomly sampled cases and there were few extraction errors. We thereafter filtered out civil cases with multiple causes for the same reason as criminal cases. However, the fact descriptions of civil cases always focus on the interactions among multiple people, instead of recording the action of the defendant in criminal cases. Therefore, we keep civil cases with multiple defendants. The minimum cause frequency is set to 10. The detailed statistics of all datasets are shown in Table 2.

Table 2. The statistics of different datasets.

Datasets	CAIL	PKU	FSC	CIVIL
#Cases	147992	170143	383697	599400
Ave case len	225	344	297	390
#Leaf causes	171	80	149	328
#Parent causes	26	22	23	95
Min freq	30	30	10	10

4.2   Baselines

For comparison, we employ 3 typical text classification models: TFIDF+SVM (SVM) [31], TextCNN (CNN) [14], and LSTM [9]; 2 hierarchical classification models: CSSA [4] and Top-Down SVM (TDS) [34]; 2 charge prediction models: Fact-Law Attention (FACT) [23] and Attribute Charge (ATCH) [10] as baselines. We only employ Attribute Charge as baseline on FSC, as it does not cover the cause annotation on other datasets.

4.3   Experimental Settings

As corresponding word embedding is released together with FSC, we conduct experiments followed the experimental settings of Attribute Charge [10] rigorously on FSC. Thus, we can compare with the published performance of Attribute Charge directly.

For other datasets, we employ THULAC [29] for word segmentation. The maximum document length is set to 500 words. Pre-trained 256 dimensional word embedding is employed for word representation. We set the hidden size to 256 for all LSTM cells and set filter widths in CNN to {2,3,4,5} with each filter size to 64. For our model, we set the filter width to 5, filter size to 256, beam width to 5.

For training, we employ Adam [15] as the optimizer and the learning rate is set to 0.001. The batch size is set to 128 and drop out rate is set to 0.5.

Widely used *accuracy* (AC), *macro-precision* (MP), *macro-recall* (MR) and *macro-F_1* (F_1) as evaluation metrics.

4.4   Results and Analysis

As shown in Table 3, we compare HLCP with baselines on 4 LCP datasets. We can observe that:

Table 3. Experimental results (- indicates the model can not converge in 150 epochs; * indicates the model is not applicable on the dataset).

Data	CAIL				PKU				CIVIL				FSC			
Metric	AC	MP	MR	F_1	AC	MP	MR	F_1	AC	MP	MR	F_1	AC	MP	MR	F_1
SVM	79.8	76.0	69.7	70.4	95.6	**83.3**	72.5	76.4	85.7	65.6	47.8	52.9	94.4	65.5	54.9	57.7
TDS	77.8	73.8	67.0	68.0	95.1	82.3	69.2	73.1	84.5	63.4	45.9	50.1	93.8	65.4	54.5	57.8
CSSA	63.5	41.2	34.7	38.9	78.5	65.7	62.3	59.4	57.5	42.9	39.1	38.3	83.9	58.8	56.9	54.5
CNN	68.7	62.1	55.5	56.3	96.2	81.4	77.0	76.8	85.4	**67.8**	46.1	52.3	94.5	66.8	58.7	60.7
LSTM	82.5	**79.2**	73.3	73.5	**96.5**	82.4	77.5	78.1	86.3	62.6	49.4	52.9	95.0	68.0	66.7	65.5
FACT	75.9	63.8	61.9	60.4	94.7	76.3	71.4	72.2	-	-	-	-	95.7	73.3	67.1	68.6
ATCH	*	*	*	*	*	*	*	*	*	*	*	*	95.8	75.8	**73.7**	73.1
HLCP	**83.1**	77.4	**74.7**	**74.6**	**96.5**	82.8	**77.8**	**79.3**	**87.3**	66.6	**56.2**	**59.3**	**95.9**	**78.8**	73.5	**74.7**

(1) HLCP significantly outperforms the baselines on most metrics and datasets, which proves the robustness and practicability of our model;

(2) Specifically, HLCP improves F_1 by 6% in the CIVIL dataset, which contains the most parent causes. Moreover, HLCP outperforms Attribute Charge which requires human designed charge attributes. The experiments demonstrate the effectiveness of utilizing pre-existing legal cause hierarchy and text knowledge in LCP task;

(3) Conventional hierarchical classification methods perform poorly, while HLCP performs much better with a *global* classifier and the consideration of inner text information;

(4) The improvements on "AC" are limited compared with F_1. The reason is that the distribution of causes is extremely unbalanced. According to our statistics, the most frequent 10 causes cover $17.3\%, 86.9\%, 78.1\%$, and 71.6% cases in CAIL, PKU, FSD, and CIVIL datasets, respectively. While the least common 10 causes only cover $0.22\%, 0.20\%, 0.027\%$, and 0.023% cases. From another perspective, it verifies that HLCP outperforms other methods mainly on the few-shot causes.

By tracing along hierarchy branches, we can transform the output of all models into a cause sequence. Thus, we are enabled to compare the performance of HLCP and baselines at different hierarchy level. As shown in Table 4, HLCP defeats baselines in most situations. This affirms our idea of training one *global*

Table 4. Results of causes on different levels.

CAIL	First Level				Second Level				Third level			
Metric	AC	MP	MR	F_1	AC	MP	MR	F_1	AC	MP	MR	F_1
CNN	83.4	78.4	74.9	76.1	75.1	62.9	56.1	57.0	73.3	65.4	60.0	60.6
LSTM	91.4	88.2	**87.6**	87.8	86.1	**78.1**	72.5	72.8	84.1	79.9	**78.9**	77.5
SVM	89.6	83.0	78.9	79.9	83.7	73.2	68.3	68.7	82.3	81.0	74.4	75.3
TDS	88.1	84.9	77.5	78.9	81.9	72.9	65.7	66.9	80.7	77.4	71.7	72.6
HLCP	**91.8**	**89.8**	86.4	**87.9**	**86.6**	76.7	**74.2**	**74.3**	**85.1**	**81.2**	78.1	**77.8**

Table 5. Ablation test.

Datasets	CAIL		CIVIL	
Metric	AC	F_1	AC	F_1
w/o name vectorizer	82.7	74.8	87.2	58.5
w/o cause attention	82.6	72.9	**87.5**	59.2
w/o mask	82.1	69.8	86.6	57.0
HLCP	**82.9**	**75.0**	87.3	**59.3**

classifier for causes in the whole hierarchy benefits the performance on the causes of all levels.

Ablation Test: To verify the importance of different modules in HLCP, we conduct ablation test as shown in Table 5. Note that, when the name vectorizer is removed, we use randomly initialized vectors as the cause representations. We can observe that all the components in HLCP, including the name vectorizer, cause attention and masked classification, benefit the performance of HLCP. The joint utilization of all modules guarantees the effectiveness of our model.

4.5   Case Study

To give an intuitive illustration of how the name vectorizer and cause attention mechanism works, we select a representative case and visualize the attention results in Fig. 3. The fact description records how the defendant conflicted with the public servant. The cause path of this case is "crime against social administration→disturbance of public order→interference with public servant". As shown in Fig. 3, while the cause becomes more and more specific, the attention results turn to be more and more focused. It is consistent with our assumption that irrelevant information, e.g. the word "injured", in fact descriptions is filtered out according to the cause names level by level.

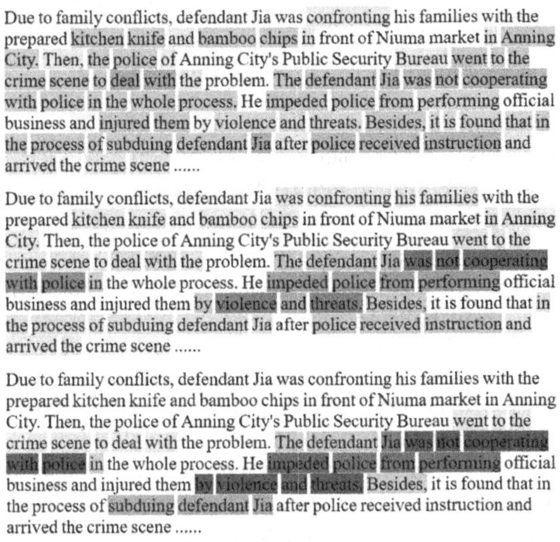

Due to family conflicts, defendant Jia was confronting his families with the prepared kitchen knife and bamboo chips in front of Niuma market in Anning City. Then, the police of Anning City's Public Security Bureau went to the crime scene to deal with the problem. The defendant Jia was not cooperating with police in the whole process. He impeded police from performing official business and injured them by violence and threats. Besides, it is found that in the process of subduing defendant Jia after police received instruction and arrived the crime scene ......

Due to family conflicts, defendant Jia was confronting his families with the prepared kitchen knife and bamboo chips in front of Niuma market in Anning City. Then, the police of Anning City's Public Security Bureau went to the crime scene to deal with the problem. The defendant Jia was not cooperating with police in the whole process. He impeded police from performing official business and injured them by violence and threats. Besides, it is found that in the process of subduing defendant Jia after police received instruction and arrived the crime scene ......

Due to family conflicts, defendant Jia was confronting his families with the prepared kitchen knife and bamboo chips in front of Niuma market in Anning City. Then, the police of Anning City's Public Security Bureau went to the crime scene to deal with the problem. The defendant Jia was not cooperating with police in the whole process. He impeded police from performing official business and injured them by violence and threats. Besides, it is found that in the process of subduing defendant Jia after police received instruction and arrived the crime scene ......

Fig. 3. Attention results of various causes.

Table 6 aims to explain why doing prediction along the hierarchy would be better compared with doing it flat. The target and confusing causes of the listed three cause pairs all belong to different parent causes. However, the causes of each pair still share similarities on fact descriptions. For example, "Gather to disturb social order" belongs to "Disturbance of public order", and "Destruction of production" belongs to "Encroachment of property". But the defendant who breaks the social order usually impacts the production of adjacent businesses. Thus, the causes of this pair are hard to be distinguished from each other. The key difference between them actually refers to the difference between their parent causes. What's more, the classification between parent causes is usually easier. Thus, by doing the prediction hierarchically, we could distinguish confusing causes which belong to different branches at their parent level and avoid the situation in which to make choice between target cause and the confusing one which belongs to another parent cause. As shown in Table 6, HLCP achieves significant performance improvements compared with a flat classifier.

Table 6. Example for confusing causes. (Recall)

Target cause	Confusing cause	HLCP	LSTM
Gather to disturb social order	Destruction of production	71.4%	46.4%
Loan fraud	Financial documents fraud	66.7%	53.3%
Kidnap	Racketeering	73.6%	71.7%

4.6   Error Analysis

We summarized the prediction error to 3 reasons:

Data Imbalance. This would be the primary reason for failed predictions. The hierarchy we introduced could alleviate this phenomenon by dividing causes into smaller groups so that the sample amount for each cause is competing in a smaller region. However, causes like "the crime of privately carving up state-owned property" and "the dispute of duplicate contracts" which only appear around ten times, are still hard to be learned.

Fuzzy Boundary. Some causes, like theft and the crime of forcible seizure, are hard to be distinguished in their nature. The main difference between them is that theft is conducted secretly. However, the secrecy of a crime is sometimes hard to be judged in practical application. There are a few such confusing cause pairs, e.g., (embezzlement, duty encroachment), and (embezzle public money, corruption). As these confusing pairs belong to the same parents, HLCP is unable to distinguish them by predicting cause paths.

Incomplete Information. We follow existing LCP works and use fact description as the input. However, when multiple causes are applicable, the court would apply a cause with the consideration of the plaintiff's claim, which is missed in fact description and cannot always be inferred implicitly. We leave the civil cause prediction task which inputs both the fact description and the plaintiff's claim as a future work.

5   Conclusion and Future Work

In this work, we propose the HLCP model for LCP task. HLCP builds a novel variation of seq2seq model to capture the dependencies among legal causes. A cause vectorizer is employed to encode legal cause names for noise elimination. Experimental results on four large-scale datasets show that HLCP outperforms conventional text classification models and charge prediction models consistently, which demonstrate the effectiveness and robustness of our model.

In the future, we will explore legal intelligence in the following directions: (1) Legal cause prediction with multiple causes and defendants; (2) Incorporating a plaintiff's claim into civil cause prediction task; (3) Incorporate the logic rules defined by law articles into legal judgment prediction task.

Acknowledgments. This work is supported by the National Key Research and Development Program of China (No. 2018YFC0831900).

References

1. Aletras, N., Tsarapatsanis, D., Preotiuc-Pietro, D., Lampos, V.: Predicting judicial decisions of the european court of human rights: a natural language processing perspective. PeerJ Comput. Sci. **2**, e93 (2016)
2. Baharudin, B., Lee, L.H., Khan, K.: A review of machine learning algorithms for text-documents classification. JAIT **1**(1), 4–20 (2010)
3. Bahdanau, D., Cho, K., Bengio, Y.: Neural machine translation by jointly learning to align and translate. In: Proceedings of ICLR (2015)
4. Bi, W., Kwok, J.T.: Multi-label classification on tree-and dag-structured hierarchies. In: Proceedings of ICML, pp. 17–24 (2011)
5. Cerri, R., Barros, R.C., De Carvalho, A.C.: Hierarchical multi-label classification using local neural networks. Comput. Syst. Sci. **80**(1), 39–56 (2014)
6. Cho, K., et al.: Learning phrase representations using rnn encoder-decoder for statistical machine translation. Computer Science (2014)
7. Clare, A., King, R.D.: Predicting gene function in saccharomyces cerevisiae. Bioinformatics **19**(suppl-2), ii42–ii49 (2003)
8. Fagni, T., Sebastiani, F.: On the selection of negative examples for hierarchical text categorization. In: Proceedings of LTC, pp. 24–28 (2007)
9. Hochreiter, S., Schmidhuber, J.: Long short-term memory. Neural Comput. **9**(8), 1735–1780 (1997)
10. Hu, Z., Li, X., Tu, C., Liu, Z., Sun, M.: Few-shot charge prediction with discriminative legal attributes. In: Proceedings of COLING (2018)
11. Karn, S., Waltinger, U., Schütze, H.: End-to-end trainable attentive decoder for hierarchical entity classification. In: Proceedings of EACL, vol. 2, pp. 752–758 (2017)
12. Katz, D.M., Bommarito II, M.J., Blackman, J.: A general approach for predicting the behavior of the supreme court of the united states. PloS One **12**(4), e0174698 (2017)
13. Keown, R.: Mathematical models for legal prediction. Computer/LJ **2**, 829 (1980)
14. Kim, Y.: Convolutional neural networks for sentence classification. In: Proceedings of EMNLP (2014)
15. Kingma, D., Ba, J.: Adam: a method for stochastic optimization. In: Proceedings of ICLR (2015)
16. Kiritchenko, S., Matwin, S., Nock, R., Famili, A.F.: Learning and evaluation in the presence of class hierarchies: application to text categorization. In: Proceedings of CSCSI, pp. 395–406 (2006)
17. Kort, F.: Predicting supreme court decisions mathematically: a quantitative analysis of the "right to counsel" cases. Am. Polit. Sci. Rev. **51**(1), 1–12 (1957)
18. Lauderdale, B.E., Clark, T.S.: The supreme court's many median justices. Am. Polit. Sci. Rev. **106**(4), 847–866 (2012)
19. Lin, W.C., Kuo, T.T., Chang, T.J., Yen, C.A., Chen, C.J., Lin, S.d.: Exploiting machine learning models for Chinese legal documents labeling, case classification, and sentencing prediction. In: Processdings of ROCLING, p. 140 (2012)
20. Liu, C.L., Chang, C.T., Ho, J.H.: Case instance generation and refinement for case-based criminal summary judgments in Chinese. JISE (2004)
21. Liu, C.L., Hsieh, C.D.: Exploring phrase-based classification of judicial documents for criminal charges in chinese. In: Proceedings of ISMIS, pp. 681–690 (2006)
22. Liu, Y.H., Chen, Y.L.: A two-phase sentiment analysis approach for judgement prediction. J. Inf. Sci. **44**(5), 5494–607 (2017)

23. Luo, B., Feng, Y., Xu, J., Zhang, X., Zhao, D.: Learning to predict charges for criminal cases with legal basis. In: Proceedings of EMNLP (2017)
24. Nagel, S.S.: Applying correlation analysis to case prediction. Tex. L. Rev. **42**, 1006 (1963)
25. Segal, J.A.: Predicting supreme court cases probabilistically: the search and seizure cases, 1962–1981. Am. Polit. Sci. Rev. **78**(4), 891–900 (1984)
26. Silla, C.N., Freitas, A.A.: A survey of hierarchical classification across different application domains. Data Min. Knowl. Discov. **22**(1–2), 31–72 (2011)
27. Silla, Jr., C.N., Freitas, A.A., et al.: Novel top-down approaches for hierarchical classification and their application to automatic music genre classification. In: SMC, pp. 3499–3504 (2009)
28. Sulea, O.M., Zampieri, M., Vela, M., Genabith, J.V.: Exploring the use of text classification in the legal domain. In: Proceedings of ASAIL workshop (2017)
29. Sun, M., Chen, X., Zhang, K., Guo, Z., Liu, Z.: Thulac: an efficient lexical analyzer for Chinese (2016)
30. Sutskever, I., Vinyals, O., Le, Q.V.: Sequence to sequence learning with neural networks. In: Advances in Neural Information Processing Systems, pp. 3104–3112 (2014)
31. Suykens, J.A., Vandewalle, J.: Least squares support vector machine classifiers. Neural Process. Lett. **9**(3), 293–300 (1999)
32. Tang, D., Qin, B., Liu, T.: Document modeling with gated recurrent neural network for sentiment classification. In: Proceedings of EMNLP, pp. 1422–1432 (2015)
33. Ulmer, S.S.: Quantitative analysis of judicial processes: some practical and theoretical applications. Law Contemp. Probs. **28**, 164 (1963)
34. Vateekul, P., Kubat, M., Sarinnapakorn, K.: Top-down optimized svms for hierarchical multi-label classification: a case study in gene function prediction. Intelligent Data Analysis (2013)
35. Xiao, C., et al.: Cail 2018: a large-scale legal dataset for judgment prediction. arXiv preprint arXiv:1807.02478 (2018)
36. Yang, Z., et al.: Hierarchical attention networks for document classification. In: Proceedings of NAACL, pp. 1480–1489 (2016)
37. Ye, H., Jiang, X., Luo, Z., Chao, W.: Interpretable charge predictions for criminal cases: learning to generate court views from fact descriptions. In: Proceedings of NAACL (2018)
38. Zeng, X., Yang, C., Tu, C., Liu, Z., Sun, M.: Chinese liwc lexicon expansion via hierarchical classification of word embeddings with sememe attention. In: Proceedings of AAAI (2018)
39. Zhong, H., Guo, Z., Tu, C., Xiao, C., Liu, Z., Sun, M.: Legal judgment prediction via topological learning. In: Proceedings of EMNLP (2018)

Natural Language Inference Based on the LIC Architecture with DCAE Feature

Jie Hu, Tanfeng Sun$^{(\boxtimes)}$, Xinghao Jiang, Lihong Yao, and Ke Xu

Shanghai Jiaotong University, Shanghai, China
{hujie_92781,tfsun,xhjiang,
yaolh,xuke900708}@sjtu.edu.cn

Abstract. Natural Language Inference (NLI), which is also known as Recognizing Textual Entailment (RTE), aims to identify the logical relationship between a premise and a hypothesis. In this paper, a DCAE (Directly-Conditional-Attention-Encoding) feature based on Bi-LSTM and a new architecture named LIC (LSTM-Interaction-CNN) is proposed to deal with the NLI task. In the proposed algorithm, Bi-LSTM layers are used to modeling sentences to obtain a DCAE feature, then the DCAE feature is reconstructed into images through an interaction layer to enrich the relevant information and make it possible to be dealt with convolutional layers, finally the CNN layers are applied to extract high-level relevant features and relation patterns and the discriminant result obtained through a MLP (Multi-Layer Perceptron). Advantages of LSTM layers in sequence information processing and CNN layers in feature extraction are fully combined in this proposed algorithm. Experiments show this model achieving state-of-the-art results on the SNLI and Multi-NLI datasets.

Keywords: Natural Language Inference · Recognizing Textual Entailment · Attention · Bi-LSTM · Reconstructed Interaction · CNN

1 Introduction

NLI, also known as RTE (Recognizing Textual Entailment), is a fundamental and challenging task in the field of NLP (Natural Language Processing). Textual Entailment is defined as the directed inference relationship between a pair of texts [1], if people believe that the semantic meaning of H (Hypothesis) can be inferred from the semantic meaning of P (Premise) according to their common sense, it is said that there is an entailment relation between P and H. The goal of RTE/NLI is to identify logical relationship (entailment, neutral, or contradiction) between a pair of P-H.

With the development of deep learning methods and large annotated datasets like SNLI [2] and Multi-NLI [3] was published, many researchers applied deep learning models to solve the NLI tasks and achieve lots of successful results [4–6].

The previous deep learning methods can be divide into two categories: (1) methods based on RNN models [5, 7–9], this kind of methods use RNN layers such as LSTM or GRU to encode premise and hypothesis separately, then simply concatenate the outputs of RNN layers as the feature of relation between the text pair, attention mechanism can also be used in the encoding phase in these methods; (2) methods based on interactive

© Springer Nature Switzerland AG 2019
M. Sun et al. (Eds.): CCL 2019, LNAI 11856, pp. 587–599, 2019.
https://doi.org/10.1007/978-3-030-32381-3_47

images or spaces [6, 10], this kind of methods convert the expressions of two sentences to an interactive image or space, then use image recognition ways to solve textual relation identification problem.

Researchers using methods in the first category believe that the structure of RNN is better suited for processing temporal information such as texts, while too little interactive information is considered in their framework. Indeed, attention mechanism can obtain some interactive features to help optimize those RNN models, but the effect is limited. And researchers using methods in the second category preserve most of the interactive information between texts, while some temporal features are alleviated in the process.

In this paper, a new DCAE feature based on Bi-LSTM and a new architecture named LIC (LSTM-Interaction-CNN) is proposed to solve the NLI task. The DCAE feature fuses three mainstream encoding methods include directly Bi-LSTM encoding, conditional encoding and encoding with attention. In the LIC architecture, Bi-LSTM layers are used to modeling sentences since their design characteristics are very suitable for the modeling of sequential data, then the encoding vectors are reconstructed into images through an interaction layer to enrich the relevant information between inputs and make it possible to be dealt with convolutional layers, finally the CNN layers are applied to extract high-level relevant features and relation patterns. Advantages of LSTM layers in sequence information processing and CNN layers in feature extraction are fully combined in this proposed LIC architecture. And based on the proposed DCAE feature and the LIC architecture, a complete algorithm for the NLI task was built with a MLP to obtain the relationship label.

The main contributions of this paper include: (1) propose a new DCAE (Directly-Conditional-Attention-Encoding) feature based on multi-features obtained by different Bi-LSTM encoding to incorporate these complementary encoding methods (2) propose a LIC (LSTM-Interaction-CNN) architecture fully combined the advantages of LSTM layers in sequence information processing and CNN layers in feature extraction to deal with the NLI task.

2   Proposed Algorithm

2.1   Directly-Conditional-Attention-Encoding Feature

LSTM (Long-Short-Term Memory), which is known as a kind of RNN (Recurrent Neural Network), is very suitable for the modeling of sequential data, such as text data due to its design characteristics. Bi-LSTM is a variant of LSTM, which is composed of forward directional LSTM and backward directional LSTM and is often used in NLP tasks to model context information. While in the actual sentence modeling phase, there are some different choices such as directly Bi-LSTM encoding [2], conditional encoding [11] and encoding with attention features [12] and so on. All of these methods were demonstrated to be effective in some specific NLP tasks. In this paper a DCAE feature based on all of these Bi-LSTM encoding methods is proposed, experiments in Sect. 3 show the effectiveness of the DCAE feature. And then, how to obtain the DCAE feature will be elaborated.

Directly Bi-LSTM Encoding. Bi-LSTM layers are used to directly encoding the texts at first, and the process of a traditional Bi-LSTM encoding an input is illustrated as Fig. 1.

In NLI task, an input includes a pair of premise and hypothesis, then they are encoded to:

$$(\bar{p}, h_p) = BiLSTM(p, 0) \tag{1}$$

$$(\bar{h}, h_h) = BiLSTM(h, 0) \tag{2}$$

Corresponding to Fig. 1, the compositions of $(\bar{p} \in \mathbb{R}^{n \times 2d}, h_p \in \mathbb{R}^{2d})$ are:

$$(\bar{p}, h_p) = ([hf_1 + hb_1, hf_2 + hb_2, hf_3 + hb_3, \ldots, hf_n + hb_n], (hf_n, hb_n)) \tag{3}$$

The compositions of $(\bar{h}, h_h)$ are similar to $(\bar{p}, h_p)$. And it's worthy to mention that the Bi-LSTM encoding premise and the Bi-LSTM encoding hypothesis shared weight matrices while training.

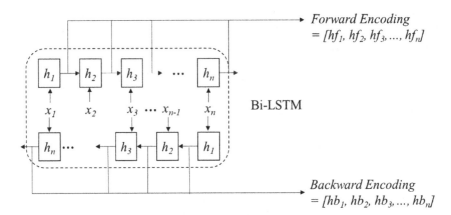

Fig. 1. Process of a Bi-LSTM encoding an input

Conditional Encoding. Conditional encoding is led to obtaining some relation features between premise and hypothesis then. It was inspired by [11]. In this section, the final hidden state from the Bi-LSTM which encoded the premise is used to initialize the Bi-LSTM which will encode the hypothesis, and vice versa. Then some relevant information about the premise and hypothesis was led to the encoding results.

To conditional encode premise and hypothesis, the outputs from directly Bi-LSTM encoding will be used. Conditional encoding results are obtained as following:

$$(\tilde{p}, -) = BiLSTM(p, h_h) \tag{4}$$

$$(\tilde{h}, -) = BiLSTM(h, h_p) \tag{5}$$

where $-$ means the value generated here is not in consideration, $\tilde{p} \in \mathbb{R}^{n \times 2d}, \tilde{h} \in \mathbb{R}^{m \times 2d}$.

Attention Encoding. A soft alignment attention mechanism was introduced into this section to obtain more relevant information between the original texts. Here an unconventional attention weight is computed to express the similarity of hidden states of the premise and the hypothesis [5]:

$$e_{ij} = \tilde{p}_i \tilde{h}_j^T, \quad i \in [1, n], j \in [1, m] \tag{6}$$

where n, m respectively are the length of premise and hypothesis, and $\tilde{p}_i \in \mathbb{R}^{2d}, \tilde{h}_j \in \mathbb{R}^{2d}$ are the components of $\tilde{p}$, $\tilde{h}$, which are the results from conditional encoding. The unconventional attention weight is more simple than conventional attention weight but still effective due to its "soft alignment" by using corresponding word vectors from premise and hypothesis to do the dot product operation. And the final attention encoding results $\hat{p}_i \in \mathbb{R}^{n \times 2d}, \hat{h}_i \in \mathbb{R}^{m \times 2d}$ are obtained as following:

$$\hat{p}_l = \sum_{j=1}^{l_p} \frac{\exp(e_{ij})}{\sum_{k=1}^{l_h} \exp(e_{ik})} \tilde{h}_j, i \in [1, n] \tag{7}$$

$$\hat{h}_l = \sum_{i=1}^{l_h} \frac{\exp(e_{ij})}{\sum_{k=1}^{l_p} \exp(e_{ik})} \tilde{p}_i, j \in [1, m] \tag{8}$$

The DCAE Feature. After the above three steps of encoding, the directly Bi-LSTM encoding vectors $(\bar{p}, \bar{h})$, the conditional encoding vectors $(\tilde{p}, \tilde{h})$ and the attention encoding vectors $(\hat{p}, \hat{h})$ were obtained. Experiments on individual feature will show in Sect. 3. Then how to fuse these separate features into an effective composite feature should be considered.

Common methods of fusion between two vectors include element-wise subtraction, element-wise production and so on. Element-wise subtraction is usually applied to extrude the different part between the two features, while element-wise production is applied to reinforce the relevant parts of the two features.

In the situation mentioned in this paper, the directly Bi-LSTM encoding vectors just contain the information of a single sentence, while conditional encoding and the attention encoding are both based on a pair of texts and contain various degrees of relevant information. Then element-wise subtraction can be applied to extract the difference between the directly Bi-LSTM vectors and the conditional vectors, and the difference between the directly Bi-LSTM vectors and the attention vectors to maximum the retention of relevant information and remove the redundancy. And the element-wise production can be applied to reinforce the relevant part of the conditional vectors and the attention vectors to maintain and strengthen the relevant information.

And then a DCAE feature based on these three encoding features is proposed:

$$\vec{p_l} = [\bar{p_l}, \tilde{p_l} - \bar{p_l}, \hat{p_l} - \bar{p_l}\tilde{p_l} \odot \hat{p_l}] \tag{9}$$

$$\vec{h_j} = [\bar{h_j}, \tilde{h_j} - \bar{h_j}, \hat{h_j} - \bar{h_j}, \tilde{h_j} \odot \hat{h_j}] \tag{10}$$

where $-$ represents element-wise subtraction, and $\odot$ means element-wise multiplication. And the whole process of how to obtain the DCAE feature proposed in this paper is illustrated in Fig. 2.

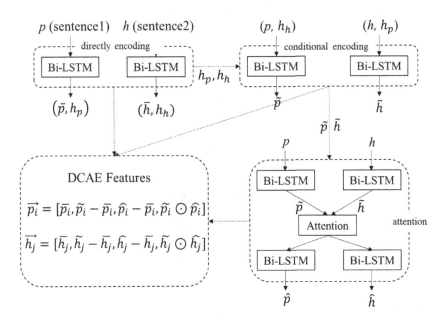

Fig. 2. The process of obtaining the DCAE features

2.2 LSTM-Interaction-CNN Architecture

The proposed LIC architecture contains three main parts: the LSTM encoding part, the reconstructed interaction part, and the CNN feature extraction part, aiming to establish an effective model for classification or discrimination tasks of a pair of texts or other coupled temporal input. The superiority of the LIC is that advantages of LSTM layers in sequence information processing and CNN layers in feature extraction are fully combined in this architecture. In a LIC, LSTM layers are used to modeling sentences to feature vectors, then the feature vectors are reconstructed into images through an interaction layer to enrich the relevant information and make it possible to be dealt with convolutional layers, finally the CNN layers are applied to extract high-level features for subsequent classification or discrimination tasks. And then how to apply the LIC architecture to NLI task will be elaborated.

Bi-LSTM Encoding Layer. In this phase, a pair of premise and the corresponding hypothesis will be encoding to be feature vectors. And in this paper, the P-H pair are encoded to two 4-dimentional feature vectors (the DCAE feature proposed in this paper) shown in Sect. 2.1.

Reconstructed Interaction Layer. In this phase, DCAE features of the premise and the corresponding hypothesis will be converted into images to enrich the interactive information and make it possible to extract high level features by convolutional layers. It was mentioned in Sect. 2.1, a DCAE feature contains four elements, for each element of DCAE features for a pair of P-H, an interactive image will generate.

To illustrate this process more specifically, directly encoding vectors are used to be an example, and the process of interaction shows in Fig. 3.

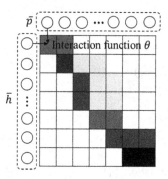

Fig. 3. Directly encoding results of P-H generate an interactive image

M is used to represent the interactive image, then:

$$M = f(\theta; \bar{p}, \bar{h}) \tag{11}$$

$$m_{ij} = \theta(\bar{p}_i, \bar{h}_j) \tag{12}$$

where θ represents the interaction function, and in this paper, dot product was chosen to be θ.

After the reconstructed interaction layer, four images were generated corresponding to a pair of premise and hypothesis due to the DCAE feature contains four elements.

Convolutional Layer. Convolutional layers are implicated to extract high-level relevant information features and relation patterns, and in this paper, a two-layers CNN model is used to extract hierarchical features, since the first layer was supposed to obtain phase-level interaction patterns, and the second layer was supposed to obtain sentence-level interaction patterns.

The framework of how this two-layer CNN model works showed in the Fig. 4. It's worth mentioning that activation function ReLU was introduced to activate feature maps generated in the process. Then the entire feature extraction process can be expressed as:

$$M_i^1 = \sum_{k=0}^{c-1} weigt\left(M_i^1, k\right) \otimes M_k^0 + bias\left(M_i^1\right) \qquad (13)$$

$$N^1 = maxpooling\left[ReLU\left(M^1\right)\right] \qquad (14)$$

$$M_i^2 = \sum_{k=0}^{c-1} weigt\left(M_i^2, k\right) \otimes N_k^1 + bias\left(M_i^2\right) \qquad (15)$$

$$N^2 = maxpooling\left[ReLU\left(M^2\right)\right] \qquad (16)$$

where M^0 is the multi-channel interactive image obtained in last section, c (which should be 4 since 4 interactive images were generated for each P-H) is the number of channels, M^* represent the outputs of convolutional layers, N^* represent the outputs of max-pooling layers, and weight, bias are weight matrices to be learned.

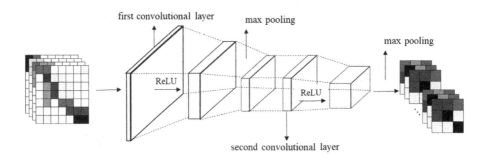

Fig. 4. The two-layer CNN model extract hierarchical features

3   Framework of the Proposed Algorithm

In this section, the complete framework to solve the NLI task based on the proposed algorithms is presented and can be illustrated in Fig. 5. In addition to the LIC architecture proposed in Sect. 2.2, MLP (Multi-Layer Perception) is also introduced into the framework to obtain the final discriminant result. Then there are four main steps: DCAE Feature Obtainment, Reconstructed Interaction, CNN Feature Extraction and the MLP Prediction.

Step 1. DCAE Feature Obtainment. In this step, a pair of premise and the corresponding hypothesis are encoded to two 4-dimentional DCAE feature vectors shown in Sect. 2.1.

Step 2. Reconstructed Interaction. In this step, DCAE features of the premise and the corresponding hypothesis are converted into images to enrich the interactive information and make it possible to extract high-level features by convolutional layers. And four images were generated corresponding to a pair of premise and hypothesis since the DCAE feature contains four elements.

Step 3. CNN Feature Extraction. In this step, a two-layers CNN model is used to extract hierarchical relevant information features and matching patterns, since the first layer was supposed to obtain phase-level matching patterns, and the second layer was supposed to obtain sentence-level matching patterns.

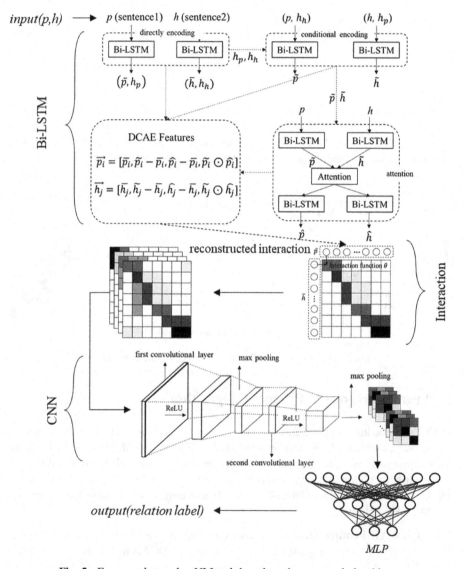

Fig. 5. Framework to solve NLI task based on the proposed algorithm

Step 4. MLP Prediction. Several interactive feature maps contain phase-level and sentence-level patterns were obtained after the process of convolution layers, and in this step, the final discriminant result based on those feature maps will be obtained through an MLP (Multi-Layer Perception):

$$Result = MLP(N^2) \tag{17}$$

4 Experiments and Analysis

4.1 Setup

Data. In this paper, two datasets focused on NLI task were evaluated. They are SNLI (Stanford Natural Language Inference Corpus) dataset [2] and the Multi-NLI (Multi-Genre Natural Language Inference Corpus) dataset [3].

The SNLI dataset contains 570K sentence pairs with human annotated as entailment, neutral, contradiction or -, where - indicates a lack of consensus from human annotators. The premises were collected from the Flickr30k corpus [13], and the hypotheses were then manually adapted corresponding to each relation type.

The Multi-NLI, which considered to be an upgrade version for the SNLI, has 433K sentence pairs with similar collection process as SNLI. The difference is the premises were collected from maximally broad range of genre of American English included written non-fiction genres, spoken genres, less formal written genres and a specialized one for 9/11. And even more subtly, half of these selected genres appear in training set while another half are not, creating in domain (matched) and cross-domain (mismatched) test sets to validate the generalization capabilities of models.

Training. The model proposed in this paper is implemented with Pytorch framework. Pre-trained 300-D Glove 840B vectors [14] are used to initialize the word embedding vectors from the original texts and the vectors are updated during training. Two layers of Bi-LSTM are applied to directly encoding and conditional encoding while two attention units (three layers Bi-LSTM totally) are applied to encode sentences with attention. Then two layers of convolutional layers are used to obtain phase-level and sentence-level patterns, while kernel size is set to be (5, 5) and (3, 3). And finally, two fully connected layers are used to generate the discriminant result. The initial learning rate is set to be 0.0004 during the training, and dropout [15] with a rate of 0.4 for regularization is applied to all feedforward connections to avoid overfitting.

4.2 Compare with Other Works

Table 1 shows the accuracy of the models on test sets of SNLI and Multi-NLI. Result in the first row is a baseline classifier presented by Bowman et al. [2], who is the founder of the SNLI dataset and utilizes handcrafted features. Deep learning methods are used in all of the rest models listed in the Table 1, then their effectiveness can be seen from the accuracy gap between the traditional methods and deep learning methods.

And as is shown in the Table 1, the proposed model achieves competitive results between the deep learning models. The first three models are based on Bi-LSTM. Gated-Att BiLSTM [16] uses Bi-LSTM with gated attention feature to encode the sentence pairs, then obtained the result by a MLP. DR-BiLSTM [5] uses multi-features include dependent encoding and encoding with attention which are the same as conditional encoding and encoding with attention in this paper, then uses another Bi-LSTM for inference and named this mechanism as Deep Reading. And HIM (Hybrid Inference Model) [9] model consists ESIM (Enhanced Sequential Inference Model) which also uses multi-level features based on Bi-LSTM to encode sentences and a tree-LSTM model to collect local inference information and obtain the result.

The biggest difference between this paper and those methods that separate the two processes of feature extraction and relationship inference like the DR-BiLSTM and the HIM is that CNN is used to extract high-level relation patterns in the proposed model since CNN structure has advantages in image feature extraction, while Bi-LSTM is used in the inference phase in DR-BiLSTM model to maintain the continuity of feature vectors and a tree-LSTM structure is used in the HIM model to collect local inference information. Experiments show that the proposed model is 0.5% points more accurate than the DR-BiLSTM and 0.4% points more accurate than the HIM when test on the SNLI dataset.

DIIN (Densely Interactive Inference Network) [6] model is based on interaction space, which uses pre-trained embedding vectors and other statistical characteristic like POS (one-hot part-of speech) and EM (binary exact match) features to generate an interaction space, then use a CNN-based model to extract relation features to obtain the final result. Experiments show that the proposed model is 1.0% point more accurate than DIIN when test on the SNLI dataset, and 0.2% points on Matched Multi-NLI, 0.7% points on Mismatched Multi-NLI.

And the DMAN (Discourse Marker Augmented Network) [17] mainly uses reinforcement learning mechanism (distinguished from the proposed algorithms but still belongs to deep learning methods) to deal with the NLI task, while the proposed algorithm is still 0.% points more accurate than DMAN on average.

Table 1. Accuracy of the models on the test sets of SNLI and Multi-NLI

Model	SNLI	Multi-NLI	
		Matched	Mismatched
Handcrafted features [2]	78.2%	–	–
Gated-Att BiLSTM [16]	85.5%	76.8%	75.8%
DR-BiLSTM (single) [5]	88.5%	–	–
HIM [9]	88.6%	–	–
DIIN (single) [6]	88.0%	78.8%	77.8%
DMAN (single) [17]	88.8%	78.9%	78.2%
DCAE feature + LIC architecture	**89.0%**	**79.0%**	**78.5%**

In a nutshell, the proposed model achieves better results than both models based on Bi-LSTM and models based on interactive image or space and achieves competitive results when compares to other deep learning methods.

4.3   Performance Experiments

Table 2 shows the accuracy of the different features or parts of the proposed algorithm.

Attention Bi-LSTM is supposed to be the best single feature for sentence encoding in NLI task based on the experiments, and the DCAE feature proposed in this paper is 1.2% points more accurate than the best single feature. The addition of DCAE features improves the accuracy by 14.4% points on the SNLI dataset since the model use Pre-trained vectors (Pre-trained 300-D Glove 840B vectors [14], are considered to contain no relation features) and the LIC architecture achieves an accuracy of 74.6% while the model use DCAE feature and LIC architecture achieves an accuracy of 89.0%. On the Multi-NLI dataset, the addition of DCAE feature even improves the accuracy by nearly 17% points. By the same token, it can be seen from the Table 2 that the addition of LIC architecture improves the accuracy by 2.7% points on SNLI dataset and nearly 2.5% points on Multi-NLI dataset.

Table 2. Accuracy of different encoding methods and architecture

Model	SNLI	Multi-NLI	
		Matched	Mismatched
Directly Bi-LSTM + SVM	80.6%	69.7%	68.6%
Conditional Bi-LSTM + SVM	83.2%	72.4%	71.8%
Attention Bi-LSTM + SVM	85.1%	74.6%	74.3%
DCAE feature + SVM	86.3%	76.6%	76.0%
Pre-trained vectors + LIC architecture	74.6%	62.4%	61.5%
DCAE feature + LIC architecture	**89.0%**	**79.0%**	**78.5%**

It can be learned from these data that the DCAE feature does incorporate three complementary encoding methods, and the LIC architecture can also improve the effectiveness of the model.

5   Conclusion

A new DCAE feature based on Bi-LSTM and a new architecture named LIC (LSTM-Interaction-CNN) is proposed to deal with the NLI task in this paper. The DCAE feature fuses three mainstream encoding methods include directly Bi-LSTM encoding, conditional encoding and encoding with attention. And in the algorithm based on the LIC architecture, Bi-LSTM layers are used to modeling sentences to obtain the DCAE feature, then the DCAE features are reconstructed into images through an interaction layer to enrich the relevant information and make it possible to be dealt with convolutional layers, finally the CNN layers are applied to extract high-level relevant features and relation patterns and the discriminant result obtained through a MLP. The advantages of Bi-LSTM layers in sequence information processing and the advantages of CNN layers in feature extraction are fully combined in the proposed algorithm. And experiments show the proposed algorithm achieves better results than models based on Bi-LSTM or interactive images and even other deep learning methods.

Acknowledgements. This work is funded by National Key Research and Development Projects of China (2018YFC0830703). It is also supported by National Natural Science Foundation of China (Grant No. 61572320 & 61572321).

References

1. Dagan, I., Glickman, O.: Probabilistic textual entailment: generic applied modeling of language variability. Learn. Methods Text Underst. Min. **2004**, 26–29 (2004)
2. Bowman, S.R., Angeli, G., Potts, C., Manning, C.D.: A large annotated corpus for learning natural language inference. In: Proceedings of the 2015 Conference on Empirical Methods in Natural Language Processing, pp. 632–642 (2015)
3. Williams, A., Nangia, N., Bowman, S.: A broad-coverage challenge corpus for sentence understanding through inference. In: Proceedings of the 2018 Conference of the North American Chapter of the Association for Computational Linguistics: Human Language Technologies, Volume 1 (Long Papers), pp. 1112–1122 (2018)
4. Tay, Y., Luu, A.T., Hui, S.C.: Compare, compress and propagate: enhancing neural architectures with alignment factorization for natural language inference. In: Proceedings of the 2018 Conference on Empirical Methods in Natural Language Processing, pp. 1565–1575 (2018)
5. Ghaeini, R., et al.: DR-BiLSTM: dependent reading bidirectional LSTM for natural language inference. In: Proceedings of the 2018 Conference of the North American Chapter of the Association for Computational Linguistics: Human Language Technologies, Volume 1 (Long Papers), pp. 1460–1469 (2018)
6. Gong, Y., Luo, H., Zhang, J.: Natural language inference over interaction space. arXiv preprint arXiv:1709.04348 (2017)
7. Liu, Y., Sun, C., Lin, L., Wang, X.: Learning natural language inference using bidirectional lstm model and inner-attention. arXiv preprint arXiv:1605.09090 (2016)
8. Wang, S., Jiang, J.: Learning natural language inference with LSTM. In: Proceedings of NAACL-HLT, pp. 1442–1451 (2016)
9. Chen, Q., Zhu, X., Ling, Z.H., Wei, S., Jiang, H., Inkpen, D.: Enhanced LSTM for natural language inference. In: Proceedings of the 55th Annual Meeting of the Association for Computational Linguistics (Volume 1: Long Papers), pp. 1657– 1668 (2017)
10. Pang, L., Lan, Y., Guo, J., Xu, J., Wan, S., Cheng, X.: Text matching as image recognition. In: Proceedings of the Thirtieth AAAI Conference on Artificial Intelligence, pp. 2793–2799. AAAI Press (2016)
11. Rocktäschel, T., Grefenstette, E., Hermann, K.M., Kočiský, T., Blunsom, P.: Reasoning about entailment with neural attention. arXiv preprint arXiv:1509.06664 (2015)
12. Cheng, J., Dong, L., Lapata, M.: Long short-term memory-networks for machine reading. In: Proceedings of the 2016 Conference on Empirical Methods in Natural Language Processing, pp. 551–561 (2016)

13. Plummer, B.A., Wang, L., Cervantes, C.M., Caicedo, J.C., Hockenmaier, J., Lazebnik, S.: Flickr30k entities: collecting region-to-phrase correspondences for richer image-to-sentence models. In: Proceedings of the IEEE International Conference on Computer Vision, pp. 2641–2649 (2015)

14. Pennington, J., Socher, R., Manning, C.: GloVe: global vectors for word representation. In: Proceedings of the 2014 Conference on Empirical Methods in Natural Language Processing (EMNLP), pp. 1532–1543 (2014)

15. Srivastava, N., Hinton, G., Krizhevsky, A., Sutskever, I., Salakhutdinov, R.: Dropout: a simple way to prevent neural networks from overfitting. J. Mach. Learn. Res. **15**(1), 1929–1958 (2014)

16. Chen, Q., Zhu, X., Ling, Z.H., Wei, S., Jiang, H., Inkpen, D.: Recurrent neural network-based sentence encoder with gated attention for natural language inference. In: Proceedings of the 2nd Workshop on Evaluating Vector Space Representations for NLP, pp. 36–40 (2017)

17. Pan, B., Yang, Y., Zhao, Z., Zhuang, Y., Cai, D., He, X.: Discourse marker augmented network with reinforcement learning for natural language inference. In: Proceedings of the 56th Annual Meeting of the Association for Computational Linguistics (Volume 1: Long Papers), pp. 989–999 (2018)

Neural CTR Prediction for Native Ad

Mingxiao An[1(✉)], Fangzhao Wu[2], Heyuan Wang[3], Tao Di[4],
Jianqiang Huang[3], and Xing Xie[2]

[1] University of Science and Technology of China, Hefei 230026, China
anmx@mail.ustc.edu.cn
[2] Microsoft Research Asia, Beijing 100080, China
wufangzhao@gmail.com, xingx@microsoft.com
[3] Peking University, Beijing 100871, China
{hy.wang,1701210864}@pku.edu.cn
[4] Microsoft, Redmond 98052, USA
taodi@microsoft.com

Abstract. Native ad is an important kind of online advertising which has similar form with the other content in the same platform. Compared with search ad, predicting the click-through rate (CTR) of native ad is more challenging, since there is no explicit user intent. Learning accurate representations of users and ads that can capture user interests and ad characteristics is critical to this task. Existing methods usually rely on single kind of user behavior for user modeling and ignore the textual information in ads and user behaviors. In this paper, we propose a neural approach for native ad CTR prediction which can incorporate different kinds of user behaviors to model user interests, and can fully exploit the textual information in ads and user behaviors to learn accurate ad and user representations. The core of our approach is an ad encoder and a user encoder. In the ad encoder we learn representations of ads from their titles and descriptions. In the user encoder we propose a mult-view framework to learn representations of users from both their search queries and their browsed webpages by regarding different kinds of behaviors as different views of users. In each view we learn user representations using a hierarchical model and use attention to select important words, search queries and webpages. Experiments on a real-world dataset validate that our approach can effectively improve the performance of native ad CTR prediction.

Keywords: Native ad · User modeling · CTR prediction

1 Introduction

Native ad is an important type of online advertising, which is embedded by simulating the form and function of other native content in the display platform [14]. An example of native ad on MSN homepage[1] is illustrated in Fig. 1. The content

---

[1] https://www.msn.com/en-us.

This work was done when the first author was an intern in Microsoft Research Asia.

M. Sun et al. (Eds.): CCL 2019, LNAI 11856, pp. 600–612, 2019.
https://doi.org/10.1007/978-3-030-32381-3_48

in the lower right corner is in fact a native ad of Tripinsider.com, but it has almost the same form with the news articles which makes it easier to attract users' attentions. Native ads are getting more and more popular in online platforms and have become an important source of revenue for many online websites. Thus, accuretaly predicting the click-through rate (CTR) of native ads is critical for online platforms.

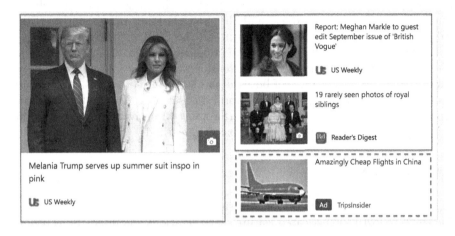

Fig. 1. An illustrative example of native ad on MSN, which is outlined using the red line. (Color figure online)

Different from search ads which place online advertisements according to users' queries posted to search engines, predicting the CTR of native ads is more challenging since there is no explicit user intent such as search queries. Thus, it is very important to accurately model users' interests from their previous behaviors for native ad CTR prediction. To the best of our knowledge, there is no published study on native ad CTR prediction. However, there are several related works on display ad CTR prediction. For example, Chen et al. [3] proposed to learn the representations of display ads from their images using CNN networks, and represent users using handcrafted features such as gender. Then they predicted the CTR of display ads by combining the user features and ad representations using two-layer fully-connected network. However, this method cannot capture personalized information of each user. Zhou et al. [25] proposed a deep interest network (DIN) for display ads CTR prediction. They used handcrafted features such as goods ID, shop ID and so on to represent ads, and learned the representations of users from their goods visting histories. However, this method only utilizes one kind of user behavior to infer users' interests, which may be insufficient. In addition, all these existing methods ignore the textual information in ads (such as ad title and description) and user behaviors, which are very useful for learning user and ad representations.

In this paper, we propose an effective neural approach for native ad CTR prediction. Our approach contains two core components, i.e., an ad encoder and a user encoder. In the ad encoder, we learn ad representations from both the titles and descriptions, since they are complementary and can provide more comprehensive information. In the user encoder, we learn representations of users by incorporating both their historical search queries and browsed webpages, which are regared as different kinds of views reflecting users' potential interests and intents. Besides, the large quantity of textual information on the websites usually contain irrelavant noises, and the multiple kinds of ad characteristics and user behaviors have different impacts on the CTR prediction. To address this problem, we use a hierarchical attention mechanism, which includes the word-level attention, the record-level attention and the view-level attention, to quantify different importances of both ad and user representations. We conduct comprehensive experiments on a real-world dataset. The results show that our approach can effectively improve the performance of native ad CTR prediction and can consistently outperform many baseline methods.

2   Related Work

CTR prediction of online display advertising has previously been widely studied. The key is to learn sophisticated feature representations and interactions behind user history behavior. Existing methods use either shallow or deep-order interactions, or require expertise feature engineering. In [20] and [2], the logistic regression (LR), which is a well-known machine learning method for its easy implementation and high efficiency, is applied to predict CTR based on many handcrafted features. In [6], the boosted decision trees are used to build a prediction model. Then in [9], a model combining decision trees with logistic regression is proposed, which outperforms either of the above two models. Another representative method is the Factorization Machine (FM) [19], which employs embedding layer on sparse inputs, then estimates parameters and captures the interactive relationships with specially designed transformation functions for target fitting. In [17], a Hierarchical Importance-aware Factorization Machine is designed, which provides a generic latent factor framework that incorporates importance weights and hierarchical learning. LS-PLM [7], namely the Large Scale Piece-wise Linear Model, formulates CTR prediction as a non-convex and non-smooth optimization problem. A novel algorithm based on directional derivatives and quasi-Newton method is then used to solve the problem. DSSM [10] uses click-through data to learn deep structured semantics of word hashing through character trigrams and multiple dense layers. The YouTube Recommendation CTR model [5] extends the idea of factorization machine by replacing the transformation function with feed forward networks, in order to improve the representation capability of the model. The Wide & Deep [4] is a well-known model for recommendation systems in industry and academia. The wide linear network can effectively memorize sparse feature interactions using cross-product feature transformations, while the deep neural network can generalize to previously unseen feature interactions through low dimensional embeddings. In DeepFM [8], the power of factorization machines is introduced as the "wide" module in Wide & Deep [4] architecture, with no need of feature engineering besides raw features. Besides the ID-based features, a Convnet is proposed in [3] for extracting high-order features of ad images to enhance CTR prediction. Inspired by the successful application of attention mechanism in Neural Machine Translation (NMT) [1], a Deep Intent model [24] is proposed, which first encodes the text sequence with RNN [23], then takes the weighted sum of all the annotations as a global hidden vector. The DIN model [25] deployed in Alibaba's online display advertising system is designed with a local activation unit to adaptively learn the representation vector of user interests that varies according to different ads. Parsana et al. [18] proposed an event embedding scheme to map the events from users' previous browsing activities to a latent space, and then applies a recurrent neural network to learn the user representations.

Different from these existing methods, our approach has the ability to learn more informative representations of both users and ads from different views, and selectively integrate salient information for better interactions as well as reducing nosies. Our model avoids the huge cost of creating millions of handcrafted features, and can be easily extented to incorporate more behavior views.

3   Our Approach

In this section, we present our attentive multi-view neural approach. The architecture is illustrated in Fig. 2, which includes three sub-modules. The first one is the *text encoder*, which aims to encode the semantic information of the user behavior or ad text into a dense vector. The second one is a *user encoder*, which is used to learn user representations based on the multiple representation series of his/her previous behaviors. The third one is the *ad encoder*, which formulates ad representations by combining different advertising information views. In all the three modules, we apply personalized attention networks at each level to selectively envolve useful information for CTR prediction. We will introduce the details of our approach in the following sections.

3.1   Text Encoder

The text encoder is used to distill textual information in both user and advertising representations, which has the same architecture but different parameters for each view. There are three layers in the text encoder, as shown in the left yellow box of Fig. 2. The first layer is word embedding, which transforms the textual word sequence into a sequence of dense semantic vectors. Given a word sequence $[w_1, w_2, \ldots, w_K]$ with length K, it is converted into $[\mathbf{e}_1, \mathbf{e}_2, \ldots, \mathbf{e}_K]$ through a word embedding matrix. The second layer is the convolutional neural network (CNN) [11,13], which interacts local context to extract n-gram semantic representations. For each word w_i, its contextual representation $\mathbf{c}_i$ is formulated as:

$$\mathbf{c}_i = \mathrm{ReLU}(\mathbf{W}_f \times \mathbf{e}_{[i-L:i+L]} + \mathbf{b}_f), \tag{1}$$

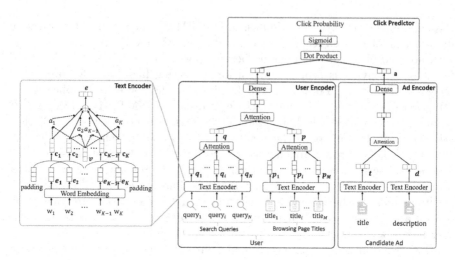

Fig. 2. The framework of our approach for native ad CTR prediction.

where $\mathbf{e}_{[i-L:i+L]}$ is the concatenation of word embedding vectors from position $i - L$ to $i + L$. $\mathbf{W}_f$ and $\mathbf{b}_f$ are parameters of convolutional filters. For the third layer, a word-level attention mechanism [1] is applied to select more important words from a global perspective. We argue that different words in a text have various influence in the understanding of user intents or ad characteristics. For instance, in the search query *"place to buy latest Nike shoes"*, the words *"place"* - *"buy"* - *"shoes"* are important indicators reflecting the user intent, while *"latest"* and *"Nike"* are additional restrictions that the user promised. However, the word *"to"* has little information and can be discarded.

The attention weights for each contextual embedding $\mathbf{c}_i$ are calculated and aggregated as follows:

$$a_i^w = \tanh(\mathbf{v}_w \times \mathbf{c}_i + b_w), \tag{2}$$

$$\alpha_i^w = \frac{\exp(a_i^w)}{\sum_{j=1}^{K} \exp(a_j^w)}, \tag{3}$$

$$\mathbf{e} = \sum_{i=1}^{K} \alpha_i^w \mathbf{c}_i, \tag{4}$$

where $\mathbf{v}_w$ and b_w are trainable parameters, and $\mathbf{e}$ is the final fused vector for the text sequence.

3.2   User Encoder

The user encoder module is shown in the blue box of Fig. 2. There are three compoments, i.e., the text encoder, the behavior-level attention, and the view-level attention.

Given a sequence of user previous search queries $[q_1, q_2, \ldots, q_N]$ with length N, it is first converted into $[\mathbf{q}_1, \mathbf{q}_2, \ldots, \mathbf{q}_N]$ through the text encoder introduced above. Then, we apply a behavior-level attention to select important queries which are more informative to model the user intents:

$$a_i^b = \tanh(\mathbf{v}_b \times \mathbf{q}_i + b_b), \tag{5}$$

$$\alpha_i^b = \frac{\exp(a_i^b)}{\sum_{j=1}^{N} \exp(a_j^b)}, \tag{6}$$

$$\mathbf{q} = \sum_{i=1}^{N} a_i^b \mathbf{q}_i, \tag{7}$$

where $\mathbf{v}_b$ and b_b are attention parameters for user search queries. $\mathbf{q}$ is the final representation of user interest in search query behaviors. Simultaneously, to model the user's browsed behavior, we apply an attention layer with the identical structure and different parameters to calculate the fused vector $\mathbf{p}$ for the sequence of browsed page titles $[t_1, t_2, \ldots, t_M]$ with length M.

Finally, in order to aggregate the multi-view representations of user behaviors and interests, we leverage a view-level attention scheme to selectively combine the extracted search query and browsed page information, formulated as:

$$a_q = \tanh(\mathbf{v}_v \times \mathbf{q} + b_v), \tag{8}$$

$$a_p = \tanh(\mathbf{v}_v \times \mathbf{p} + b_v), \tag{9}$$

$$\alpha_q = \frac{\exp(a_q)}{\exp(a_q) + \exp(a_p)}, \tag{10}$$

$$\alpha_p = \frac{\exp(a_p)}{\exp(a_q) + \exp(a_p)}, \tag{11}$$

$$\mathbf{u} = \mathbf{W_u}(\alpha_q \mathbf{q} + \alpha_p \mathbf{p}) + \mathbf{b_u}, \tag{12}$$

where $\mathbf{v}_v$ and b_v are view-level attention parameters, $\mathbf{W}_u$ and $\mathbf{b}_u$ are projection parameters, and $\mathbf{u}$ is the final aggregated user representations. Our user encoder can be easily extended to incorporate more user behavior views.

3.3   Ad Encoder

Our ad encoder module is illustrated in the green box of Fig. 2. There are two parts in the module. The first part is the text encoder introduced in Sect. 3.1. We apply the text encoders with different parameters to both advertising title t and description d, in order to get the semantic dense representations $\mathbf{t}$ and $\mathbf{d}$ respectively. The second part is the view-level attention. Similar to Eqs. 8–11, we calculate the attention weights α_t and α_d separately for ad title and description view. Then the final ad representation is obtained as follows:

$$\mathbf{a} = \mathbf{W_a}(\alpha_t \mathbf{t} + \alpha_d \mathbf{d}) + \mathbf{b_a}, \tag{13}$$

where $\mathbf{W}_a$ and $\mathbf{b}_a$ are ad projection parameters.

3.4 CTR Prediction

In this section, we will first introduce our click predictor, then explain how we optimize the model.

The click predictor, as shown in the red box in Fig. 2, is used to predict the probability of a user clicking a candidate advertise based on the user representation and ad representation. The prediction score is calculated by the inner product of user representation $\mathbf{u}$ and ad representation $\mathbf{a}$, and we use sigmoid function to get the click-through probability p for each user-ad pair $<u-a>$ as follows:

$$p(u,a) = \text{sigmoid}(\mathbf{u}^\top \mathbf{a}) \tag{14}$$

For model training, we use the negative log-likelihood function as the objective function, defined as:

$$L = -\frac{1}{|S|} \sum_{(u,a,y) \in S} (y \log p(u,a) + (1-y) \log(1 - p(u,a))) \tag{15}$$

where S is the training set with size $|S|$, and $y \in \{0,1\}$ is the binary label indicating whether the user clicked the advertisement.

4 Experiments

4.1 Dataset

Our experiments were conducted on a real-world native ad dataset, which was collected from the logs of native ads displayed on MSN homepage from January 1st, 2019 to January 31st, 2019. Each native ad impression log consists of several native ads that are shown to a user in the same page and the labels indicating whether they are clicked or not. For the user in each native ad impression log, we collected her search queries and browsed webpages in the last 30 days from Bing.com. The detailed statistics of the dataset are summarized in Table 1. We use the data in the last week as testing set, and the rest data as training set. In addition, we randomly sampled 10% of the training data for validation.

Table 1. Statistics of our dataset.

# training impressions	300,000	avg. # words per ad title	11.95
# testing impressions	100,000	avg. # words per ad description	15.80
# users	374,584	avg. # words per search query	3.82
# ads	4,159	avg. # words per page title	10.23
# ad clicking events	364,281	avg. # user's browsed pages	27.23
# ad non-clicking events	568,716	avg. # user's search queries	125.08

4.2   Experiment Settings

In our experiments, we use Skip-gram [15,16] model to pre-train the word embeddings for search queries, browsing page titles and advertising texts respectively. The dimension of word embedding is set to 200. The number of CNN filters is set to 400, and the window size is 3. Adam [12] with learning rate 0.001 is used as the optimization algorithm. The batch size is 200. To mitigate overfitting, we apply dropout [22] with 0.2 to each layer. The hyperparameters are all selected according to the performance of validation set. The evaluation metrics in our experiments include area under the ROC curve (AUC) and Average Precision (AP) over all impressions. We independently repeat each experiment for 10 times and report the average performance.

4.3   Performance Evaluation

We evaluate the performance of our approach by comparing it with several baseline methods. The methods to be compared include:

- LR [2,20]: the logistic regression, which is widely used due to its easy implementation and high efficiency.
- LibFM [19]: a state-of-the-art feature-based matrix factorization method, which takes the combination of categories of features as input to predict in CTR scenarios.
- DeepFM [8]: a general deep model that integrates a component of factorization machines and a component of neural networks.
- DSSM [10]: a deep structured semantic model using word hashing via character trigrams and multiple dense layers.

For all compared algorithms, we extract the TF-IDF features [21] for each view of textual information, and use the concatenation of user features and ad features as input. To evaluate the robustness of our model in terms of different size of training data, we also perform comparative experiments using only 20% and 50% of the dataset for training. The experimental results of different methods are summarized in Table 2.

Table 2. The results of different methods on native ad CTR prediction under different ratios of training data, i.e., 20%, 50% and 100%.

Dataset	20%		50%		100%	
	AUC	AP	AUC	AP	AUC	AP
LR	0.7098	0.7215	0.7125	0.7343	0.7211	0.7404
LibFM	0.7131	0.7244	0.7184	0.7362	0.7271	0.7491
DeepFM	0.7166	0.7382	0.7179	0.7436	0.7206	0.7497
DSSM	0.7147	0.7322	0.7191	0.7463	0.7265	0.7519
Ours	**0.7295**	**0.7534**	**0.7303**	**0.7551**	**0.7346**	**0.7573**

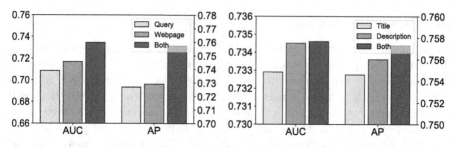

(a) Effectiveness of different user behaviors.   (b) Effectiveness of different ad information.

Fig. 3. Effectiveness of the multi-view learning framework in our approach.

We can find that our approach always outperforms all the other baselines for different sizes of training data. Even with only a small amount of historical impressions, our model can also perform superior predictive ability. Since the text features utilized in baseline methods are based on TF-IDF, they can not distinguish the different meanings and importances of words in different kinds of textual contexts and behavior views. Instead, our approach employs a multi-view learning framework with a hierarchical attention mechanism to identify useful words and behavior signals. Therefore more accurate and comprehensive user and ad representations can be built for the CTR prediction.

4.4   Model Effectiveness

In this section, we conduct several experiments to validate how each component of our neural approach contributes to the performance.

In order to explore the effectiveness of different kinds of user behaviors and ad characteristics, we perform four ablation tests. Specifically, we first compare using only search queries or browsed webpages for modeling user behaviors. Figure 3a shows the results. We can find that the performance will significantly decrease when any behavior view is discarded. In addition, using webpages performs better than using search queries. We think there might be two reasons. On the one hand, the page title is more detailed and contains more implicit semantic information. On the other hand, the behavior of browsing a webpage can reflect relatively more stable and long-term interests of the user, while a user is likely to conduct a search query simply due to a instant news event or a short-term demand. In general, combining these two kinds of information can enhance a better understanding of the user's profile. For constructing the ad representations, we examine using only the ad title or description seperately. Results are shown in Fig. 3b. Removing any of the ad view will have a negative impact on the performance. Using descriptions to form ad embeddings is better than using titles, which might be that titles usually reflect the central information while contents can provide more background and details. Note that although the content may incorporate more irrelevant noises, our model can effectively address the problem by applying the hierarchical attention mechanism.

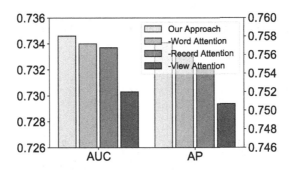

Fig. 4. Effectiveness of different kinds of attentions.

We furthur explore the effectiveness of the attention mechanism at different levels (word-, record- and view- level) through testing multiple variants. Specifically, we replace the attention layer in word- and record- level with average pooling layer. For view-level attention, we replace it with a concatenate layer. As shown in Fig. 4, employing hierarchical attention mechanism at all levels achieves the best performance compared with other variants. Removing the view-level attention will lead to the worst results, indicating the importance of identifying different reference values of multiple kinds of history information. We believe that the multi-level attention mechanism is helpful not only to the advertising CTR prediction scenario, but also to many other applications, such as news recommendation, review-based product recommendation systems, etc.

4.5 Attention Visualization

In this section, we provide visualization results of our attention mechanism, in order to better understand how these attention networks in our framework select important information for native ad CTR prediction.

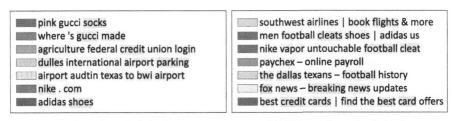

(a) Search query behavior.   (b) Webpage browsing behavior.

Fig. 5. Attention visualization for an example user. Deeper colors stand for higher attention weights. In this example, the weight of query view for the user is 0.1238, and the weight of webpage browsing view is 0.8762.

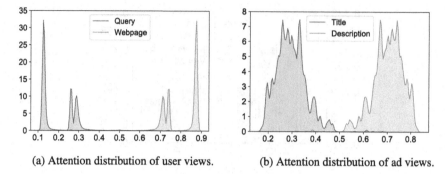

(a) Attention distribution of user views.   (b) Attention distribution of ad views.

Fig. 6. Distribution of attention weights for different user views and different ad views.

Figure 5 shows the visualization results of attention weights at each level from a randomly selected user. As demonstrated, important words and behavior records are effectively recognized and assigned with higher weights. On the word level, for example, in the search query "nike.com", the word "nike" is significantly highlighted, indicating the potential user intent on sports wear. In the webpage title "best credit cards—find the best card offers", the model paid more attention on words "credit" and "cards", which might lead to credit card ad impression. The visualization results also validate the effectiveness of record attention. For instance, although search query "pink gucci socks" and "where's gucci made" both mention a brand name "gucci", the former query is more informative because it implies more concrete user intent of purchasing pink socks. Thus, the former query gets a higher weight from our model. Both the two webpage titles "the dallas texans - football history" and "men football cleats shoes—adidas us" descibe the user as a football fan, while the latter is more important for CTR prediction as it indicates the user's willing of purchasing cleats shoes for football. In addition, the browsed webpages for this user are more informative than search queries according to the weights of view attention. This is reasonable because the user's browsed webpages contain more details, e.g., interests on football cleats shoes. Hoowever, the queries lack of the details about the "shoes".

In order to learn which kinds of information views are more informative, we further present the distribution of attention weights for both users and ads. Specifically, we conduct gaussian kernel density estimation with band width 0.05 to plot the weights distribution for both figures in Fig. 6. According to Fig. 6a, the webpage browsing behavior is usually more important, as page titles usually contain rich detailed information. Note that there are still a few users who have higher weights on search queries, thus the informativeness of each view varies over target users. That's why we need to learn the view attention for different users. Meanwhile, ad descriptions usually get higher weights as shown in Fig. 6b. This is also reasonable since the descriptions contain rich detailed information of the ad, and our framework is capable to capture the important details from raw text. This is consistent with the fact that descriptions contain more detailed textual information of the ads. Our approach can effectively capture important features and reduce noises.

5   Conclusion

In this paper, we propose a neural approach for native ad CTR prediction. Our approach incorporates different kinds of user behaviors and ad characteristics, and can fully exploit the textual information in ads and user behaviors to contrust better representations. The user encoder in the framework can learn user profiles from their search queries and browsed webpages, while the ad encoder can build ad embeddings from both titles and descriptions. In addition, a hierarchical multi-layer attention mechanism is adopted to identify and integrate important words, behavior records and information views. Experimental results demonstrate the significant superiority of our approach compared with many baseline methods for improving the performance of native ad CTR prediction.

References

1. Bahdanau, D., Cho, K., Bengio, Y.: Neural machine translation by jointly learning to align and translate. In: ICLR (2015)
2. Chakrabarti, D., Agarwal, D., Josifovski, V.: Contextual advertising by combining relevance with click feedback. In: WWW, pp. 417–426 (2008)
3. Chen, J., Sun, B., Li, H., Lu, H., Hua, X.S.: Deep CTR prediction in display advertising. In: MM, pp. 811–820 (2016)
4. Cheng, H.T., et al.: Wide and deep learning for recommender systems. In: Proceedings of the 1st Workshop on Deep Learning for Recommender Systems, pp. 7–10 (2016)
5. Covington, P., Adams, J., Sargin, E.: Deep neural networks for YouTube recommendations. In: RecSys, pp. 191–198 (2016)
6. Dave, K.S., Varma, V.: Learning the click-through rate for rare/new ads from similar ads. In: SIGIR, pp. 897–898 (2010)
7. Gai, K., Zhu, X., Li, H., Liu, K., Wang, Z.: Learning piece-wise linear models from large scale data for ad click prediction. arXiv preprint arXiv:1704.05194 (2017)
8. Guo, H., Tang, R., Ye, Y., Li, Z., He, X.: DeepFM: a factorization-machine based neural network for CTR prediction. arXiv preprint arXiv:1703.04247 (2017)
9. He, X., et al.: Practical lessons from predicting clicks on ads at Facebook. In: Proceedings of the Eighth International Workshop on Data Mining for Online Advertising, pp. 1–9 (2014)
10. Huang, P.S., He, X., Gao, J., Deng, L., Acero, A., Heck, L.: Learning deep structured semantic models for web search using clickthrough data. In: CIKM, pp. 2333–2338 (2013)
11. Kim, Y.: Convolutional neural networks for sentence classification. arXiv preprint arXiv:1408.5882 (2014)
12. Kingma, D.P., Ba, J.: Adam: a method for stochastic optimization. arXiv preprint arXiv:1412.6980 (2014)
13. LeCun, Y., Bengio, Y., Hinton, G.: Deep learning. Nature **521**(7553), 436 (2015)

14. Matteo, S., Zotto, C.D.: Native advertising, or how to stretch editorial to sponsored content within a transmedia branding era. In: Siegert, G., Förster, K., Chan-Olmsted, S.M., Ots, M. (eds.) Handbook of Media Branding, pp. 169–185. Springer, Cham (2015). https://doi.org/10.1007/978-3-319-18236-0_12

15. Mikolov, T., Chen, K., Corrado, G., Dean, J.: Efficient estimation of word representations in vector space. arXiv preprint arXiv:1301.3781 (2013)

16. Mikolov, T., Sutskever, I., Chen, K., Corrado, G.S., Dean, J.: Distributed representations of words and phrases and their compositionality. In: NIPS, pp. 3111–3119 (2013)

17. Oentaryo, R.J., Lim, E.P., Low, J.W., Lo, D., Finegold, M.: Predicting response in mobile advertising with hierarchical importance-aware factorization machine. In: WSDM, pp. 123–132 (2014)

18. Parsana, M., Poola, K., Wang, Y., Wang, Z.: Improving native ads CTR prediction by large scale event embedding and recurrent networks. arXiv preprint arXiv:1804.09133 (2018)

19. Rendle, S.: Factorization machines with libFM. ACM Trans. Intell. Syst. Technol. **3**(3), 57 (2012)

20. Richardson, M., Dominowska, E., Ragno, R.: Predicting clicks: estimating the click-through rate for new ads. In: WWW, pp. 521–530 (2007)

21. Salton, G., McGill, M.J.: Introduction to modern information retrieval (1986)

22. Srivastava, N., Hinton, G., Krizhevsky, A., Sutskever, I., Salakhutdinov, R.: Dropout: a simple way to prevent neural networks from overfitting. J. Mach. Learn. Res. **15**(1), 1929–1958 (2014)

23. Williams, R.J., Zipser, D.: A learning algorithm for continually running fully recurrent neural networks. Neural Comput. **1**(2), 270–280 (1989)

24. Zhai, S., Chang, K., Zhang, R., Zhang, Z.M.: Deepintent: learning attentions for online advertising with recurrent neural networks. In: KDD, pp. 1295–1304. ACM (2016)

25. Zhou, G., et al.: Deep interest network for click-through rate prediction. In: KDD, pp. 1059–1068 (2018)

Depression Detection on Social Media
with Reinforcement Learning

Tao Gui, Qi Zhang$^{(\boxtimes)}$, Liang Zhu, Xu Zhou, Minlong Peng,
and Xuanjing Huang

Shanghai Key Laboratory of Intelligent Information Processing, Fudan University,
School of Computer Science, Fudan University, 825 Zhangheng Road, Shanghai,
China
{tgui16,qz,liangzhu17,xuzhou16,mlpeng16,xjhuang}@fudan.edu.cn

Abstract. Depression detection is a significant issue for human well-being. Conventional diagnosis of depression requires a face-to-face conversation with a doctor, which limits the likelihood of the identification of potential patients. We instead explore the potential of using only the textual information to detect depression based on the content users posted on social media sites. Since users may post a variety of different kinds of content, only a small number of posts are relevant to the signs and symptoms of depression. We propose the use of reinforcement learning method to automatically select the indicator posts from the historical posts of users. Our experimental results demonstrate that the proposed method outperforms both feature-based and neural network-based methods (over 14.6% error reduction). In addition, a series of experiments demonstrate that our model can deal with the noise of data effectively and can generalize to more complex situations.

Keywords: Depression · Social media · Reinforcement learning

1 Introduction

Depression is a worldwide prevailing mental disease and a major contributor to the overall global disease burden. A recent fact sheet provided by World Health Organization shows that more than 300 million people of all ages suffer from depression globally[1]. The conventional clinical diagnosis of depression requires a face-to-face conversation between a doctor and patient, which is not available to many potential patients, especially in the early stages. On the other hand, social media is continuously growing and is set to be the communication medium of choice for most people. According to a report published by *The Next Web*, there are over 3 billion social media users around the world[2]. Users post large quantities of content about their daily lives and feelings. Hence, in recent years,

---

[1] http://www.who.int/mediacentre/factsheets/fs369/en/.

[2] https://thenextweb.com/contributors/2017/08/07/number-social-media-users-passes-3-billion-no-signs-slowing/.

M. Sun et al. (Eds.): CCL 2019, LNAI 11856, pp. 613–624, 2019.
https://doi.org/10.1007/978-3-030-32381-3_49

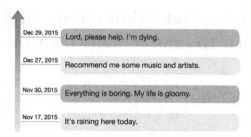

Fig. 1. An example of a user's historical posts. Depression indicator posts are usually sparse on social media. Only the tweets with red highlights may be regarded as the indicators of depression. (Color figure online)

the task of detecting depression via harvesting social media data has received considerable attention [3,12,15].

Previous researchers studied the task of detecting depression via social media using various features, including language, emotion, style and user engagement [3,10]. [12] proposed the use of well-defined discriminative depression-oriented feature groups and a multimodal depressive dictionary learning method to detect depressed users. These methods have proved that the diagnosis of depression through the content published by users on social media is reasonable and feasible. However, most of the existing methods used hand-crafted feature groups to perform the task. In addition to the content users posted on social media, features extracted from user behaviors were also taken into consideration, e.g., hospital attendance [15]. In some cases these user behaviors were hard to be captured [12], which limited the usability of these methods.

In this work, we propose a method to achieve the task using only the historical posts of users. Because the content posted by users on social media are diverse and multi-faceted, depression indicator posts are usually sparse on social media. Figure 1 illustrates an example. From this example, we can observe that there is only two tweets related to the indicators of depression. The other ones are related to the music and weather. If the entire posting history of a user is used as inputs, these content may negatively impact the depression detection. Hence, the model should extract the indicator posts separately from the posting history of a user. However, since it is difficult and a time-consuming task to label each post, most of the benchmark datasets contain only labels at the user level.

To overcome this issue, in this work, we propose a reinforcement learning-based method to achieve the task. Even though we only have the label at the user level, we can evaluate the utility of the selected posts based on the classification accuracy. Our key insight is that the post selection policies can be learned from the utility of the selected posts. Intuitively, a good policy selects posts in a way that allows a classifier trained on these posts to achieve high classification accuracy. Although selecting posts is a non-differentiable action, it can be naturally achieved in a reinforcement learning setting, where actions correspond to the selection of posts and the reward is the effect on the downstream classifier accuracy. Inspired by the work [1], the proposed method consists of two components: a policy gradient agent, which selects depression indicator posts from the entire posting history of users, and a depression classifier trained using the selected posts. Experimental results show that the proposed method can achieve a much better performance than existing state-of-the-art methods.

2 Related Work

During the last decade, social media have become extremely popular, on which billions of users write about their thoughts and lives on the go. Therefore, researchers began analyzing the online behaviors of users to identify depression. [8] explored the potential benefits of using online social network data for clinical studies on depression. They utilized the real-time moods of users captured on the Twitter social network and explored the use of language in describing depressive moods. In their later work, [9] found that depressed individuals tended to perceive Twitter as a tool for social awareness and emotional interaction. Recently, [18] attempted to explain how web users discuss depression-related issues from the perspective of the social networks and linguistic patterns revealed by the members' conversations. In this work, we studied the problem from a text classifier perspective. Inspired by these works, we also proposed the use of only the textual posts of a user to detect depression.

There is a growing body of research focusing on the use of machine learning to analyze and detect depression via social media. [3] used crowdsourcing to collect gold standard labels and applied an SVM to predict depression of an individual. [10] studied the use of supervised topic models in the analysis of linguistic signal for detecting depression, and provided promising results using several models. Most recently, [12] released a well-labeled depression and non-depression dataset on Twitter, and proposed a multimodal depressive dictionary learning model to detect depressed users on Twitter. [19] proposed a model based on a convolutional network to effectively identify depressed users based on textual information. In contrast to these research, we applied reinforcement learning to select indicator posts, and obtained better results. In addition, the proposed method demonstrated a strong and stable performance in realistic scenarios.

3 Approach

In this work, we propose to study the task of detecting depression based on the content of users posted on the social media. We denote the historical posts of the i-th user as $P_{hist}^i = \{p_1^i, p_2^i, ..., p_T^i\}$, where p_t^i is the text of t-th post. Each user has one label y_i corresponding to whether the user is depressed or not. Based on the description given in the previous section, we know that only a small number of posts may related to signs and symptoms of depression. Hence, we try to select a subset P_{indi}^i, which contains depression indicator posts, from the entire posts of users. The depression classifier is trained based on P_{indi}^i.

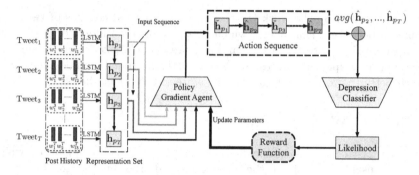

Fig. 2. Architecture of reinforcement learning-based depression detection network. w refers to the word embedding, and $\oplus$ refers to average operation. In the process of training, the policy gradient agent selects post representations in sequence, and then the selected posts (with red highlights) are used to train a better depression classifier. The policy gradient agent computes the rewards based on the likelihood of ground truth to update its parameters. (Color figure online)

The architecture of the proposed method is shown in Fig. 2. It consists of two components: (1) a policy gradient agent [17] that selects depression indicator posts from P_{hist}, and (2) a depression classifier trained using the indicator posts for classification and returning the rewards to the agent. Our goal for training is to optimize the parameters of the depression classifier, which are denoted as θ_d, together with the agent parameters θ_a. The two components should interact with each other to update the parameters during the training process.

3.1   Policy Gradient Agent

We wish to select a subset of depression indicator posts. However, it becomes a key challenge when encountering the diverse content of posts. In addition, the nature of discrete selection decisions makes the loss no longer differentiable. To overcome this problem, we propose the use of reinforcement learning for the task.

The i-th user's historical posts P_{hist}^i correspond to the sequential inputs of one episode. At each step, the agent chooses an action a_t (selecting the current post or not) after observing the state s_t, which is represented by the current post, selected posts and irrelevant posts. When all of the selections are made, the depression classifier will give a delayed reward to update the parameters of agent θ_a.

Next, we will introduce several key points of the agent, including the state representation s_t, the action a_t, and the reward function.

State Representation. We suppose that each post is made up of a sequence of words $p_t = [w_1, w_2, ..., w_l]$, where l is the max length of the post. We use long short-term memory (LSTM) [4] to model each post text. Then, the last hidden state $\mathbf{h}_t$ of LSTM turns into the post representation that will be transferred to the agent, i.e., $\mathbf{h}_t = LSTM(p_t)$. At the step t, the model has obtained t posts as inputs, which are denoted by $P_{1:t}$. Given $P_{1:t}$, the policy gradient agent could make the following observations: the current post representation $\mathbf{h}_t$, the indicator post set $\mathbf{H}_{indi} = [\hat{\mathbf{h}}_1, \hat{\mathbf{h}}_2, ...]$, and the irrelevant post set $\mathbf{H}_{irre} = [\check{\mathbf{h}}_1, \check{\mathbf{h}}_2, ...]$. The notations $\hat{\mathbf{h}}$ and $\check{\mathbf{h}}$ will be defined in the **action** part. Note that at the initial time, $\mathbf{H}_{indi}$ and $\mathbf{H}_{irre}$ are empty set. We use zero vectors to initialize these two sets. We thereby formulate the agent's state s_t as follows:

$$s_t = [\mathbf{h}_t \otimes avg(\mathbf{H}_{indi}) \otimes avg(\mathbf{H}_{irre})], \tag{1}$$

where avg refers to the average pooling operation, and $\otimes$ is the concatenation operation.

Action. The agent takes an action a_t at step t using policy $a_t \sim \pi(s_t, a_t; \theta_a)$, which is attained by sampling from the multinomial distribution. We define action $a_t \in \{1, 0\}$ to indicate whether the agent will select the current post p_t. Therefore, we could adopt a logistic function to sample the actions from the policy function as follows:

$$\begin{aligned} \pi(a_t|s_t; \theta_a) &= Pr(a_t|s_t) \\ &= a_t * \sigma(MLP(s_t)) + (1 - a_t) * (1 - \sigma(MLP(s_t))), \end{aligned} \tag{2}$$

where MLP represents the multilayer perceptron used to map the state s_t to a scalar, and $\sigma(.)$ is the sigmoid function. If the agent takes an action to select the post ($a_t = 1$), then the hidden state $\mathbf{h}_t$ will be rewritten as $\hat{\mathbf{h}}$ and be appended in $\mathbf{H}_{indi}$. Otherwise, it will be rewritten as $\check{\mathbf{h}}$ and be appended in $\mathbf{H}_{irre}$.

Reward Function. After executing a series of actions, the agent will construct a depression indicator post representation set $\mathbf{H}_{indi}$. The set $\mathbf{H}_{indi}$ is used for classification and will be described in Sect. 3.2. Note that we set the reward to be the likelihood of the ground truth after finishing all the selections of the i-th user. In addition, to encourage the model to delete more posts, we include a regularization to limit the number of selected posts as follows:

$$r_i = Pr(y_i|\mathbf{H}_{indi}; \theta_d) - \lambda T'/T, \tag{3}$$

where T' refers to the number of selected posts and λ refers to a hyperparameter to balance the reward. By setting the reward to be the likelihood of the ground truth, we capture the intuition that optimal selections will promote the probability of the ground truth. Therefore, by interacting with the classifier through the rewards, the agent is incentivized to select the optimal posts from P_{hist} for training a good classifier.

3.2 Depression Classifier

Depression classification is a universal binary classification problem. As previously mentioned, at the end of each episode, the post representation subset $\mathbf{H}_{indi}$ is further used to predict the depression label.

We merged $\mathbf{H}_{indi}$ to create a representation of the user's activity across all of the depression related posts. Various merging methods can be applied, such as summation and the attention mechanism, and so on. In this work, we adopted the average operation. This representation is then processed by two fully connected layers (i.e., multilayer perceptron) with the dropout [14] operation. The output at the last layer will be followed by a sigmoid non-linear layer that predicts the probability distribution over two classes.

$$o_t = MLP(avg(\mathbf{H}_{indi}))$$
$$Pr(\hat{y}_i|\mathbf{H}_{indi}; \theta_d) = \hat{y}_i\sigma(o_t) + (1 - \hat{y}_i)(1 - \sigma(o_t)), \qquad (4)$$

where $\hat{y}_i$ represents the prediction probabilities, and o_t is the output unit of the fully connected layers.

3.3 Optimization

We train the agent using a standard reinforcement learning algorithm called REINFORCE [17]. The objective of training the agent is maximizing the expected reward under the distribution of the selection policy:

$$J_1(\theta_a) = \mathbb{E}_{\pi(a_{1:T})}[r], \qquad (5)$$

where $\pi(a_{1:T}) = \prod_{t=1}^{T} Pr(a_t|s_t; \theta_a)$.

However, the gradient is intractable to obtain because of the discrete actions and high dimensional interaction sequences. Following the REINFORCE algorithm, an approximated gradient can be computed as follows:

$$\nabla_{\theta_a} J_1(\theta_a) = \sum_{t=1}^{T} \mathbb{E}_{\pi(a_{1:T})}[\nabla_{\theta_a} \log(Pr(a_t|s_t; \theta_a)) * r]$$
$$\approx \frac{1}{N} \sum_{n=1}^{N} \sum_{t=1}^{T} [\nabla_{\theta_a} \log(Pr(a_t|s_t; \theta_a)) * r^n], \qquad (6)$$

where N denotes the quantity of sampling on one user. In our experiment, $N = 1$ is enough to obtain great performance. By applying the above algorithm, the loss $J_1(\theta_a)$ can be computed by standard backpropagation.

Optimizing the classifier is straightforward, and can be treated as a classification problem. Because the cross entropy loss $J_2(\theta_d)$ is differentiable, we can apply backpropagation to minimize it as follows:

$$J_2(\theta_d) = -[y_i \log \hat{y}_i + (1 - y_i) \log(1 - \hat{y}_i)], \qquad (7)$$

where $\hat{y}_i$ is the output of the classifier. Then, we can get the final objective by minimizing the following function:

$$J(\theta_a, \theta_d) = \frac{1}{M} [\sum_{m=1}^{M} (-J_1(\theta_a) + J_2(\theta_d)], \qquad (8)$$

where M denotes the quantity of the minibatch, and the objective function is fully differentiable.

Table 1. Statistical details of the datasets used in our experiments, where **# Users** and **# Tweets** represent the number of users and tweets, respectively.

Dataset		# Users	# Tweets
D_1	Depressed	1,402	292,564
	Non-depressed	5,160	3,953,183
D_2	Candidate	36,993	35,076,667

4   Experimental Setup

In this section, we first describe the datasets used for experiments. Then, we detail describe several baseline methods and the hyperparameters of our model.

4.1   Datasets

We used the depression datasets introduced by [12]. They constructed a well-labeled depression dataset on Twitter. They also constructed an unlabeled depression-candidate dataset. The statistics of these datasets are summarized in Table 1.

Depression Dataset D_1. The depression dataset D_1 was constructed based on the tweets between 2009 and 2016. This dataset contained 1,402 depressed users and 5,160 non-depressed users with 4,245,727 tweets within one month. According to [2], users were labeled as depressed if their anchor tweets satisfied the strict pattern *"(I'm/ I was/ I am/ I've been) diagnosed with depression"*. The non-depressed users were labeled if they had never posted any tweet containing the character string "depress".

Depression-Candidate Dataset D_2. The unlabeled depression-candidate dataset D_2 was constructed based on the tweets on December 2016. The users in D_2 were obtained if their anchor tweets loosely contained the character string "depress". By this method, D_2 would contain more depressed users than randomly sampling. Finally, D_2 contained 36,993 depression-candidate users and over 35 million tweets within one month, which will be used for indicator posts discovery.

4.2   Comparison Methods

We applied several classic and state-of-the-art methods for comparison. In addition, we used a series of deep learning methods as baselines for comparison.

Feature-Based Methods. The feature-based methods used various features and a lot of external resources, such as social network features, user profile features, visual features, emotional features, topic-level features, and domain-specific features as shown in [12]. The methods of Naive Bayes (NB), multiple social networking learning (MSNL) [13], Wasserstein dictionary learning

(WDL) [11], and multimodal depressive dictionary learning (MDL) [12] are used as baseline models.

Neural Network Methods. We also made a comparison to a series of neural network methods. These methods just used the context information to identify depression, i.e., the users' posts were the only resource for all the methods.

- **Convolutional neural networks (CNN):** CNN has been widely applied to text classification [5]. We used CNN to model each post of users to obtain the post representations, which would be merged to identify depression [19].
- **Long short-term memory (LSTM):** Similar to CNN, we applied LSTM to obtain the representations of posts, which were then used for classification.
- **SDP-attention and MPSDP-attention:** We introduced two self-attention mechanisms on post level as our baselines. One was defined as $Attention(Q, K, V) = softmax(d_k^{-\frac{1}{2}}QK^T)V$ called Scaled Dot-Product Attention (SDP-attention) [16]. The other one could be achieved by average over all the attention vectors, and then normalizing the resulting weight vector to sum up to 1, i.e., $Attention(Q, K, V) = softmax(avg(d_k^{-\frac{1}{2}}QK^T))V$ [7], denoted by Mean Pooling Scaled Dot-Product Attention (MPSDP-attention).
- **Random sampling:** We also randomly sampled half of posts from each user to train CNN and LSTM model.

4.3   Initialization and Hyperparameter

More difficult than [12], we did not apply emoji processing, stemming, irregular words processing and pretraining word2vec. The word embeddings and other parameters for all the deep learning models were initialized by randomly sampling from a standard normal distribution and a uniform distribution in [−0.05, 0.05], respectively. We set the dimensionality of the word embedding to 128. In addition, we use one layer of the LSTM to model the post text, and set the hidden neurons of LSTM to 200. The policy agent used a two fully connected layers with 100 and 20 units for each layer.

Our model could be trained end-to-end with backpropagation, and gradient-based optimization was performed using the Adam update rule [6], with a learning rate of 0.0001.

5   Results and Analysis

In this section, we detail the performance of the proposed and baseline models, and present the results of various experiments to demonstrate the effectiveness of the proposed model from different aspects.

5.1   Method Comparison

For a fair comparison, we constructed the training and test set in the same way as reported in [12]. With 1,402 depressed users in total, we randomly selected

Table 2. Comparison of performance in terms of four selected measures. CNN/LSTM+RL refers to the proposed model.

Methods	Accuracy	Precision	Recall	F1
NB	0.724	0.727	0.728	0.728
MSNL [13]	0.818	0.818	0.818	0.818
WDL [11]	0.768	0.769	0.768	0.768
MDL [12]	0.848	0.848	0.850	0.849
CNN [19]	0.843	0.843	0.843	0.844
CNN + Random sampling	0.789	0.789	0.788	0.785
CNN + SDP-attention [16]	0.836	0.836	0.836	0.837
CNN + MPSDP-attention [7]	0.849	0.850	0.849	0.849
CNN + RL	**0.871**	0.871	**0.871**	**0.871**
LSTM	0.828	0.830	0.828	0.828
LSTM + Random sampling	0.760	0.760	0.757	0.756
LSTM + SDP-attention [16]	0.847	0.848	0.847	0.847
LSTM + MPSDP-attention [7]	0.850	0.850	0.850	0.850
LSTM + RL	0.870	**0.872**	0.870	**0.871**

1,402 non-depressed users on D_1 to make the scale of depressed users 50%, but in a more difficult manner by removing all the anchor tweets [2]. After obtaining the dataset, we trained and tested these methods using 5-fold cross validation.

We compared the depression detection performance of the proposed model with the baselines in terms of the four selected measures, i.e., Accuracy, Macro-averaged Precision, Macro-averaged Recall, and Macro-averaged F1-Measure. The comparison results are summarized in Table 2.

In the table, the first four lines list the results of the classic methods reported in [12], which use various features for training. MDL achieved the previous state-of-the-art performance with 0.849 in F1-Measure, indicating that combining the multimodal strategy and dictionary learning strategy is effective in depression detection.

The remaining part of the table lists the results of the neural network methods, which only use the users' posts for training. From the results, we can see that just using the posts, the CNN and LSTM model can achieve the accuracies of greater than 82.8%, which indicates the post text contains valuable information and is reasonable to use for depression detection. To give more attention to depression indicator posts, we evaluated two different self-attention models on the dataset. The results of the attention models showed that post level attention is effective in depression detection in most cases. The performance of the MPSDP-attention model is better than that of the SDP-attention model. We can see that the MPSDP-attention model may be more suitable for a task of this nature. Because there are an average of 396.6 tweets per user in the dataset, attention mechanism may be hard to effectively model the users to obtain an

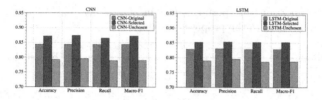

Fig. 3. Comparison between the models trained on original posts, selected posts, and unchosen posts.

obvious improvement. The LSTM + random sampling model has the lowest performance, which shows that the important effect of the post selection strategy. The poor selection strategy may be harmful the model. If we used the RL model to select posts, the CNN/LSTM + RL methods achieve the best performance, with a value of more than 87% for the F1-measure compared with both the CNN-based model and LSTM-based model, indicating that the RL post selection strategy was the most effective in depression detection. Next, we will show why our RL selection strategy was more effective than other methods.

5.2   Utility of Selected Posts

In order to verify the effectiveness of the method, we compared the baseline models trained on the original dataset, selected dataset, and unchosen dataset. We first trained the policy gradient agent to provide depression indicator posts and unchosen posts from the original dataset. Then, these indicator posts and unchosen posts made up the selected dataset and unchosen dataset, respectively. We compared the baseline models with three settings. One setting was training the model on the original dataset, which was denoted as model-original. The other settings denoted as model-selected and model-unchosen were training on the selected dataset and unchosen dataset, respectively.

The comparisons are shown in Fig. 3. From the results, we can observe that both of the models could benefit from the selected posts. The baseline models trained on selected dataset can achieve almost 2.4% better than those on original dataset. The error reduction rate was more than 9%. Inevitably, the unchosen dataset achieves poor performance. The results also indicate that the agent can select depression indicator posts that are more beneficial for depression classification.

5.3   Robustness Analysis in Realistic Scenarios

For a fair comparison, in the Table 2, we constructed the training/test set the same as [12]'s setting, where they made a balanced data set, and made sure 50% of the data contained depressed users. This does not seem a realistic scenario, as the real world data set may only contain a small number of depressed users.

With 1,402 depressed users in total, we fixed the capacity of our dataset to 1,500 and varied the scale of depressed users from 10% to 90% with increment of 10%. Figure 4 shows the trend of detection performance with different proportions of depressed users. It can be found that our method achieved a stable and outstanding performance even though there is only a very low proportion of users with depression. However, when the depression users' scale does not laid at 50%, we retrieved a seriously decent performance of MDL under imbalanced scales. Therefore, our method is more instructive in detecting the depression than MDL in the realistic scenario.

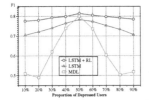   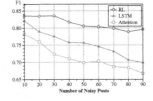

Fig. 4. Comparison between the models trained on the datasets with different scales of depressed users. The total number of users is 1,500.

Fig. 5. Effect of different number of noisy data. The average post number of one user is 396.6.

5.4   Analysis of Noisy Data

Because social media are full of noisy data, we also evaluated all the models in a situation where different number of noisy posts were inserted in the dataset. We randomly selected 252,360 posts from the depression-candidate dataset D_2, and added from 10 to 90 posts to each user. We wanted to verify if the RL model can select indicator posts from noisy data.

As shown in Fig. 5, the performance of the models decreased to various degrees. However, as the number of posts increased, the advantage of the proposed model became increasingly obvious, and the RL model remarkably outperformed the other models. The other models suffered more form noisy posts. Espectively, at the 90 point, our model outperforms attention-based model over 13% in F1 score. The results indicated that our proposed model could obtain better performance when encountering the noisy data.

6   Conclusion

In this study, we investigated the problem of detecting depression based on the content users posted on social media, and verified the feasibility of using only the contextual information to detect depression. To overcome the problem of discrete selection, we proposed a reinforcement learning-based method to select indicator posts and remove other posts. Experimental results demonstrated that the proposed method could achieve better performance than previous methods. Through several experiments, we found that other detection models could benefit from the newly selected dataset. The further experiments demonstrated that our model could obtain a strong and stable performance in realistic scenarios.

References

1. Reinforcement learning for relation classification from noisy data. In: AAAI (2018)
2. Coppersmith, G., Dredze, M., Harman, C.: Quantifying mental health signals in Twitter. In: Proceedings of the Workshop on Computational Linguistics and Clinical Psychology: From Linguistic Signal to Clinical Reality, pp. 51–60 (2014)
3. De Choudhury, M., Counts, S., Horvitz, E.: Social media as a measurement tool of depression in populations. In: Proceedings of the 5th Annual ACM Web Science Conference, pp. 47–56. ACM (2013)
4. Hochreiter, S., Schmidhuber, J.: Long short-term memory. Neural Comput. **9**(8), 1735–1780 (1997)
5. Kim, Y.: Convolutional neural networks for sentence classification. arXiv preprint arXiv:1408.5882 (2014)
6. Kingma, D., Ba, J.: Adam: a method for stochastic optimization. arXiv preprint arXiv:1412.6980 (2014)
7. Liu, Y., Sun, C., Lin, L., Wang, X.: Learning natural language inference using bidirectional LSTM model and inner-attention. arXiv preprint arXiv:1605.09090 (2016)
8. Park, M., Cha, C., Cha, M.: Depressive moods of users portrayed in Twitter. In: Proceedings of the ACM SIGKDD Workshop on Healthcare Informatics (HI-KDD), vol. 2012, pp. 1–8. ACM, New York (2012)
9. Park, M., McDonald, D.W., Cha, M.: Perception differences between the depressed and non-depressed users in twitter. In: ICWSM, vol. 9, pp. 217–226 (2013)
10. Resnik, P., Armstrong, W., Claudino, L., Nguyen, T., Nguyen, V.A., Boyd-Graber, J.: Beyond LDA: exploring supervised topic modeling for depression-related language in Twitter. In: Proceedings of the 2nd Workshop on Computational Linguistics and Clinical Psychology: From Linguistic Signal to Clinical Reality, pp. 99–107 (2015)
11. Rolet, A., Cuturi, M., Peyré, G.: Fast dictionary learning with a smoothed wasserstein loss. In: Artificial Intelligence and Statistics, pp. 630–638 (2016)
12. Shen, G., et al.: Depression detection via harvesting social media: a multimodal dictionary learning solution. In: IJCAI, pp. 3838–3844 (2017)
13. Song, X., Nie, L., Zhang, L., Akbari, M., Chua, T.S.: Multiple social network learning and its application in volunteerism tendency prediction. In: SIGIR, pp. 213–222. ACM (2015)
14. Srivastava, N., Hinton, G.E., Krizhevsky, A., Sutskever, I., Salakhutdinov, R.: Dropout: a simple way to prevent neural networks from overfitting. J. Mach. Learn. Res. **15**(1), 1929–1958 (2014)
15. Suhara, Y., Xu, Y., Pentland, A.: DeepMood: forecasting depressed mood based on self-reported histories via recurrent neural networks. In: WWW, pp. 715–724. International World Wide Web Conferences Steering Committee (2017)
16. Vaswani, A., et al.: Attention is all you need. arXiv preprint arXiv:1706.03762 (2017)
17. Williams, R.J.: Simple statistical gradient-following algorithms for connectionist reinforcement learning. Mach. Learn. **8**(3–4), 229–256 (1992)
18. Xu, R., Zhang, Q.: Understanding online health groups for depression: social network and linguistic perspectives. J. Med. Internet Res. **18**(3), e63 (2016)
19. Yates, A., Cohan, A., Goharian, N.: Depression and self-harm risk assessment in online forums. arXiv preprint arXiv:1709.01848 (2017)

How Important Is POS to Dependency Parsing? Joint POS Tagging and Dependency Parsing Neural Networks

Hsuehkuan Lu[1,2,3], Lei Hou[1,2,3(✉)], and Juanzi Li[1,2,3]

[1] DCST, Tsinghua University, Beijing 100084, China
s810142000@gmail.com, {houlei,lijuanzi}@tsinghua.edu.cn
[2] KIRC, Institute for Artificial Intelligence, Tsinghua University, Beijing, China
[3] Beijing National Research Center for Information Science and Technology, Beijing, China

Abstract. It is widely accepted that part-of-speech (POS) tagging and dependency parsing are highly related. Most state-of-the-art dependency parsing methods still rely on the results of POS tagging, though the tagger is not perfect yet. Inevitably, dependency parsing model will encounter performance degradation due to the error propagation problems. And it still remains uncertain about how important POS tagging is to dependency parsing. In this work, we propose a method to jointly learn POS tagging and dependency parsing so as to alleviate the error propagation problems. Our proposed method is based on transition system, which is capable to produce dependency tree efficiently and accurately. The results reported in the experiments support our idea that POS tagging is a crucial syntactic component for dependency parsing.

Keywords: Dependency parsing · Part-of-speech tagging · Joint learning

1 Introduction

Both POS tagging and dependency parsing are fundamental tasks of natural language processing (NLP), which are beneficial to various downstream applications such as relation extraction [5,8], text summarization [14,23], machine translation [7], and named entity recognition [11,13]. These two tasks are highly related, and many dependency parsing methods use POS tags as essential features. However, using automatically predicted POS tags triggers the error propagation problem, which degrades the performance of parser. In order to relieve the error propagation problem, we propose a method to jointly learn dependency parsing and POS tagging.

Either traditional or neural network methods use the information of POS tags, which are generated automatically by taggers. Currently, POS taggers are not perfect yet, and the taggers may be irrelevant to the sentences to be parsed.

© Springer Nature Switzerland AG 2019
M. Sun et al. (Eds.): CCL 2019, LNAI 11856, pp. 625–637, 2019.
https://doi.org/10.1007/978-3-030-32381-3_50

Some works try to avoid using POS tags for dependency parsing, instead they explore the usage of lexical information [2,10]. However, the parsers achieving the highest performance still rely on the usage of POS tags [1,9,16]. Therefore, joint modeling POS tagging and dependency parsing is then proposed to improve both tagging and parsing results [4,12,17,19,25–27]. The work of Li [17] reports that dependency accuracy drops around 6% on Chinese applying automatically labelled POS tags instead of correct POS tags.

Typically, dependency parsing models can be classified into graph-based [18] and transition-based [20] methods. Graph-based method is able to traverse all the potential solutions, while transition-based method restricts the search of solutions. Overall, graph-based method is able to achieve higher performance than transition-based method in most cases, but with extra cost of computation. Traditional graph-based or transition-based models define a set of features such as lexical features, POS tagging features, dependency label features, and word correlation features [3,18,22,28]. Nevertheless, defining feature template is time-consuming and requires linguistic expertise's efforts. In the work reported by Bohet et al. [3], their parser spends 99% of its time processing feature extraction, though they use standard efficient ways. Recent state-of-the-art models adopt neural architectures to replace feature-engineering [1,6,9,10,16,24,29].

In this work, we propose a neural networks architecture to jointly learn POS tagging and dependency parsing based on transition-based algorithm. We first treat POS tagging as downstream task, and stack dependency parsing on the results of POS tagging. Our method extends BIST transition-based dependency parser [16] with extra POS tagging component. Our proposed method is relatively lightweight which uses shallow bidirectional-LSTMs (BiLSTMs) and applies greedy strategy to generate dependency parsing. Despite the simplicity of our neural architecture, our method is capable to capture both lexical and syntactic information. In the experimental results, our method can achieve comparable results (-1.4% in LAS) with automatically generated POS tags on Universal Dependency (UD) 1.2, and outperforms baseline on UD 2.0 by 3.3% (LAS). Moreover, our model obtains the state-of-the-art scores on UD 1.2 with ground-truth POS tags by leading 2% (LAS), and nearly close to the overall state-of-the-art results on UD 2.0 (-0.5% in LAS). At last, the ablation tests reveal the importance of each component (n-gram, character, and POS) to dependency parsing, and validate our idea of joint learning.

2   Method

In this section, we present our joint model of POS tagging and transition-based dependency parsing. Given a sentence $T = \langle w_1, w_2, \ldots, w_K \rangle$ of length K, where w is word token, our goal is to predict a sequence of POS tags $P = \langle p_1, p_2, \ldots, p_K \rangle$ and dependency labels $L = \langle l_1, l_2, \ldots, l_K \rangle$, as shown in Fig. 1.

Our proposed neural architecture is shown in Fig. 2, which is composed of three stages to generate the word representations used in dependency parsing, i.e., **word representations**, **POS tagging** and **joint representations**.

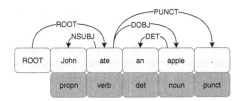

Fig. 1. The example sentence "John ate an apple." including Part-of-speech tags and dependency relations.

2.1   Joint Model of POS Tagging and Dependency Parsing

The joint model starts with a BiLSTM layer to learn vectors representing word tokens for POS tagging purpose. In the stage of tagging, we adopt multi-layer perceptron (MLP) as classifier to predict POS tags. Then we integrate POS information to lexical information. Based on the idea of multi-task learning, we hypothesize that sharing the bottom neural layer is beneficial for model to learn more general vector representations across multiple tasks, which are POS tagging and dependency parsing in our case.

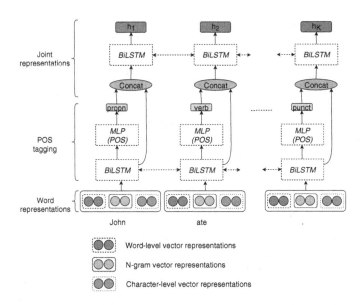

Fig. 2. The overall architecture of our joint model to generate word representations for dependency parsing.

Word Representation. The first stage generates word features, including word token, n-gram, and characters. As shown in Fig. 2, we follow Kiperwasser [16] to learn $v_{w_i}(i = 1, 2, \ldots)$ using BiLSTM, where v is the vector representations of word token w_i. Additionally, we attempt to extend lexical information by adding n-gram features. In the inputs we use the concatenations of word embeddings, n-gram representations, and character-level representations. For instance, given a word "John", we extract three types of features, including word features [John], n-gram features [⟨Jo, Joh, ohn, hn⟩], and character features [J, o, h, n]. In this example, we use 3-gram as illustration, and we add a start token ⟨and an end token⟩ in n-gram features as to better capture the positional information. In the experiments, all the parameters are initialized from uniform distribution randomly, and we do not use pre-trained embeddings.

We apply respective BiLSTMs to learn the representations of n-gram and character, and we simply adopt the concatenation of last hidden states in forward and backward LSTM as representations. Afterwards, the designed word representations are applied in POS tagging task to capture the syntactic information during training.

POS Tagging. We consider POS tagging task as downstream task, and stack dependency parsing upon the results of tagging. As shown in Fig. 2, we treat POS tagging as a sequence labeling task, and each process of tagging is an N-way classification. In order to simplify the task, we use MLP as classifier. BiLSTM cell is capable to model neighboring information, and suppose to obtain reliable results.

Given a sentence $T = \langle w_1, w_2, \ldots, w_K \rangle$, POS tagging task is to label a sequence of POS tags $P = \langle p_1, p_2, \ldots, p_K \rangle$. The prediction is formally defined in Eq. 1

$$\hat{p}_i = \mathrm{MLP}_{pos}(v_{w_i}), \hat{p}_i \in \mathbb{R}^{N_{pos}} \tag{1}$$

where $\hat{p}_i$ is the predicted POS tag of word w_i, and MLP_{pos} is the MLP classifier for POS tagging. Here $\hat{p}_i$ is represented by probability distribution of all possible categories of POS tags. The last layer of MLP output uses softmax function to produce the probability distribution. Based on the Eq. 1, we define the loss of POS tagging by cross-entropy loss in Eq. 2

$$\mathcal{L}_{pos} = \text{cross-entropy}(p_i, \hat{p}) = -\sum p_i \log \hat{p} \tag{2}$$

where p_i is the golden POS tag.

Joint Representation. Based on the POS tagging results, we design a method to jointly represent lexical and syntactic information. As stated by Zhang [27], stack-propagation is beneficial to strongly-connected tasks, even without directly using the results of downstream task. In addition to share the bottom neural layer for multiple tasks, we further integrate the results of downstream task (POS tagging) to upstream task (dependency parsing). Similar to POS tagging, we stack another BiLSTM layer for dependency parsing, where the inputs are concatenations of outputs of BiLSTM for POS tagging and predicted POS tags embeddings. The pipeline of joint representations is illustrated in Fig. 2.

Given a sentence $T = \langle w_1, w_2, \ldots, w_K \rangle$, and corresponding predicted POS tags $\hat{P} = \langle \hat{p}_1, \hat{p}_2, \ldots, \hat{p}_K \rangle$. The formal definition of joint representations is in Eq. 3

$$h_i = \text{BiLSTM}_{dep}(v_{w_i}, v_{\hat{p}_i}), i \in \{1, 2, \ldots, K\} \tag{3}$$

where h_i is the concatenation of BiLSTM_{dep} outputs of word w_i, BiLSTM_{dep} is BiLSTM for dependency parsing, $v_{\hat{p}_i}$ is embedding of predicted tag $\hat{p}_i$.

2.2   Transition-Based Dependency Parsing

Transition-based dependency parsing aims to predict a transition sequence from an initial configuration to certain terminal configuration, and generates a target dependency parsing tree. During the process of transition, the action is determined by the current configuration. In this paper, we implement our parsing process in a greedy fashion, which is much more efficient and achieves remarkable accuracy as well.

We apply **arc-standard** transition system [21] as dependency parsing method. Arc-standard is a simple and efficient method to generate dependency parsing tree, though it is incapable to handle non-projective trees which are not discussed in this work. There are three main components in arc-standard system to represent transition configuration. The formal definition of *configuration* is $C = (S, B, A)$, where

1. S: a *stack* holding words to parse dependency labels.
2. B: a *buffer* holding words to be processed.
3. A: a set of *dependency arcs* to record the parsed dependency labels.

The initial configuration for a sentence $T = \langle w_1, w_2, ..., w_K \rangle$ is $S = [\text{ROOT}], B = [w_1, w_2, ..., w_K], A = \varnothing$. The terminal configuration is that stack contains only ROOT node and buffer is empty, then transition system collects all the dependency labels stored in A to generate parsing tree. Let $s_i(i = 1, 2, ...)$, $b_i(i = 1, 2, ...)$ denote the ith element in S and B respectively, the arc-standard system defines three types of transitions:

1. *Left-arc(l)*: add an arc $s_1 \rightarrow s_2$ with dependency label l, and pops s_2 out from S. Precondition: $|S| > 2$.
2. *Right-arc(l)*: add an arc $s_2 \rightarrow s_1$ with dependency label l, and pops s_1 out from S. Precondition: $|S| \geq 2$.
3. *Shift*: pops b_1 out from B and add to the top of S. Precondition: $|B| \geq 1$.

The whole parsing process costs computations in linear time complexity. Each word requires 2 transitions *(Shift, Left- or Right-arc)*, while the last word with ROOT label requires only *Shift* transition. In Table 1, a complete process of transition sequence is illustrated.

As to the representations of configuration C, we uses the joint representations of word tokens defined in Eq. 3.

Transition Actions. The parsing process can be deemed as a sequence of transitions, and determined by current configuration C. In this paper, we define a

Table 1. An example of the process of transition-based dependency parsing. Use the illustrated sentence "John ate an apple." from Fig. 1.

Transition	Stack	Buffer	A
	[ROOT]	[John, ate, an, apple, .]	∅
Shift	[ROOT, John]	[ate, an, apple, .]	
Shift	[ROOT, John, ate]	[an, apple, .]	
Left-arc	[ROOT, ate]	[an, apple, .]	A ∪ (ate, NSUBJ, John)
Shift	[ROOT, ate, an]	[apple, .]	
Shift	[ROOT, ate, an, apple]	[.]	
Left-arc	[ROOT, ate, apple]	[.]	A ∪ (apple, DET, an)
Right-arc	[ROOT, ate]	[.]	A ∪ (ate, DOBJ, apple)
Shift	[ROOT, ate, .]	[]	
Right-arc	[ROOT, ate]	[]	A ∪ (ate, PUNCT, .)
Right-arc	[ROOT]	[]	A ∪ (ROOT, ROOT, ate)

simple feature template to represent C. There are 4 features in the template, including 3 from S and 1 from B, and the formal definition is in Eq. 4

$$v_{arc} = [v_{s_1}; v_{s_2}; v_{s_3}; v_{b_1}]$$
$$\hat{a} = \text{MLP}_{arc}(v_{arc}), \hat{a} \in \mathbb{R}^{N_{arc}} \tag{4}$$
$$\mathcal{L}_{arc} = \text{cross-entropy}(a_i, \hat{a}) = -\sum a_i \log \hat{a}$$

where v_{arc} stands for representations of transition action, $\hat{a}$ is the probability distribution of categories of transition action, MLP_{arc} is MLP classifier for transition action, and $\mathcal{L}_{arc}$ is cross-entropy loss of transition action.

Dependency Labels. As to the prediction of dependency labels, we hypothesize that label is strongly-connected to the pair of words containing dependency relation. Therefore, we define a feature template so as to better explore the correlation of words. There are 4 features in the template, including first and second words from S, absolute difference of words, and element-wise production of words. The formal definition is in Eq. 5

$$v_{dep} = [v_{s_1}; v_{s_2}; |v_{s_1} - v_{s_2}|; v_{s_1} \odot v_{s_2}]$$
$$\hat{l} = \text{MLP}_{dep}(v_{dep}), \hat{l} \in \mathbb{R}^{N_{dep}} \tag{5}$$
$$\mathcal{L}_{dep} = \text{cross-entropy}(l, \hat{l}) = -\sum l \log \hat{l}$$

which is similar to Eq. 4, the only difference comes from the definition of feature v_{dep}.

2.3 Training

In the training stage, we apply Adam optimizer [15] as our gradient stochastic optimizer. Based on the loss occurred in the classifications of POS tagging from

Eq. 2, transition action from Eq. 4, and dependency label from Eq. 5, the final loss is defined in Eq. 6

$$\mathcal{L}_{total} = \mathcal{L}_{pos} + \mathcal{L}_{arc} + \mathcal{L}_{dep} + \frac{\lambda}{2}||\theta||^2 \tag{6}$$

where $\frac{\lambda}{2}||\theta||^2$ is the L2-regularization term for all the parameters in the model. In order to accelerate training, we do not tune our model on development dataset, instead we simply use the model training 30 epochs as final model. The size of mini-batch is 150 and learning rate is 0.001.

All embeddings (words, POS tags, n-gram tokens, characters) are initialized from uniform distribution randomly with 50 dimensions. The lengths of character, n-gram are both restricted to 20, and the maximal length of sentence is 100. We use 3-gram as n-gram feature. Dropout is 0.33 for all layers. Each LSTM cell contains 256 units. Each MLP classifier is composed of 2 layers, first layer is fully-connected layer (512 units) with ReLU activation function, and second layer is corresponding categories for each task with softmax function.

3   Experiments

In this section, we first introduce the experiment settings, including datasets, baseline methods and comparison metrics, then report the overall comparison results, and finally analyze the feature contributions via ablation studies.

3.1   Experiment Settings

Datasets. We evaluate our approach on two datasets: Universal Dependencies[1] 1.2, and 2.0, which both contain abundant multi-lingual dependency trees.

Baselines. For UD 1.2 dataset, we report the results from Yang [25], and use seven languages including German (de), English (en), Spanish (es), French (fr), Italian (it), Portuguese (pt) and Swedish (sv). For UD 2.0 dataset, we report the results of 37 datasets from big treebank, and compare with the results reported in CoNLL 2017[2], including baseline method UDPipe 1.1.

Metrics. For POS tagging, we use accuracy based on words as evaluation metric. As to dependency parsing, two evaluation metrics are adopted, which are unlabeled attachment score (UAS) and labeled attachment score (LAS). UAS only counts the correctness of head words, while LAS considers both head words and dependency labels.

---

[1] http://universaldependencies.org.
[2] http://universaldependencies.org/conll17/results.html.

3.2　Results

For UD 1.2 dataset, we follow the experimental settings in [27], and report results in Table 2. Compared with baseline methods, our proposed method with automatically generated POS tags performs slightly worse (-1.4%) than the state-of-the-art method. Nevertheless, with the aid of golden POS tags, it achieves a significant improvement by 3.6%, and obtains the state-of-the-art LAS score with over 2% leading. Besides, our proposed method is able to achieve the highest performance in all languages.

Table 2. Dependency parsing results on UD 1.2 dataset with LAS metric. All values are reported in %, and the table includes other 3 results from Yang [25].

Method	de	en	es	fr	it	pt	sv	AVG
Ballesteros et al. [2]	73.00	77.90	77.80	78.00	84.20	80.40	74.50	77.97
Zhang and Weiss [27]	74.20	80.70	80.70	80.00	85.80	80.40	77.50	79.90
Yang et al. [25]	77.10	82.50	82.50	81.20	87.00	83.10	80.40	81.97
Our method (*auto-POS*)	74.60	81.71	81.98	79.82	86.37	81.34	77.77	80.51
Our method (*gold-POS*)	**79.21**	**85.69**	**85.76**	**82.17**	**88.77**	**84.25**	**83.35**	**84.17**

The results for UD 2.0 dataset are reported in Table 3, from which we have the following observations:

- Our method with automatically generated POS tags is better than baseline UDPipe 1.1, with about 3.3% leading in average.
- Because the results from CoNLL 2017 vary from languages (i.e., the leading teams are not always the same), we do not compute the overall best performance. Through comparison on each language, our proposed method can achieve comparable, even better performance. The results show the good generalization capacity of the proposed method.
- The method with golden POS taggers outperforms those of auto-predicted ones by an average increase of 5.7% on all languages, which validates our hypothesis that POS information is beneficial for dependency parsing, and the quality of POS taggers largely influences the performance.

3.3　Ablation Test

In order to analyze the feature contributions to the joint model, we conduct ablation tests on UD 1.2 dataset. Taking the results in Table 2 as baseline, we remove features respectively, including n-gram, character and POS tags features. The results of ablation tests are shown in Table 4.

In this paper, our goal is to analyze the importance of POS tagging to dependency parsing, and it can be observed from the table that after removing the POS tags, the overall performance decreases 3% in UAS, and 3.6% in LAS. The results suggest that lexical features are important to dependency parsing.

Table 3. Dependency parsing results on UD 2.0 dataset with LAS metric. All values are reported in %, and in each language we report the top 3 results from CoNLL 2017 competition. The first result is marked bold and underlined, while the second result is marked bold.

Method	ar	bg	ca	cs	cu	da	de	el	en	es	et	eu	fa
1st	72.90	89.81	90.70	90.17	76.84	82.97	80.71	87.38	82.23	87.29	71.65	81.44	86.31
2nd	71.96	88.39	88.27	86.52	72.35	81.55	77.17	86.90	79.94	85.22	69.71	79.61	84.90
3rd	70.70	87.65	88.09	86.50	71.84	79.52	75.47	84.96	79.64	85.16	67.60	77.97	83.34
UDPipe1.1	65.30	83.64	85.39	82.87	62.76	73.38	69.11	79.26	75.84	81.47	58.79	69.15	79.24
auto-POS	75.18	85.38	85.16	85.84	72.27	72.96	73.26	77.40	80.88	84.01	54.37	66.76	78.83
gold-POS	77.71	88.64	89.89	87.35	77.05	79.85	79.24	82.81	85.57	86.52	69.36	74.94	84.68

Method	fi	fr	gl	got	he	hi	hr	id	it	ja	ko	lv	nl
1st	85.64	85.51	83.23	71.36	68.16	91.59	85.25	79.19	90.68	91.13	82.49	74.01	80.48
2nd	82.38	84.36	83.22	68.34	63.94	90.41	83.15	78.55	89.08	80.85	81.10	71.35	75.50
3rd	81.21	83.82	81.60	66.82	63.72	90.40	82.51	77.70	87.85	80.01	79.51	68.03	75.07
UDPipe1.1	73.75	80.75	77.31	59.81	57.23	86.77	77.18	74.61	85.28	72.21	59.09	59.95	68.90
auto-POS	78.64	82.61	78.81	66.58	78.82	87.36	76.94	72.68	86.76	92.67	68.17	58.06	69.54
gold-POS	83.17	86.50	82.24	72.68	82.11	90.98	80.03	79.73	90.11	95.98	77.46	67.96	77.31

Method	pl	pt	ro	ru	sk	sl	sv	tr	ur	vi	zh	AVG
1st	90.32	87.65	85.92	83.65	86.04	91.51	85.87	62.79	82.28	47.51	68.56	–
2nd	87.15	85.11	84.40	83.50	81.75	88.24	84.98	62.66	81.06	42.52	65.88	–
3rd	86.75	85.01	83.50	81.49	80.53	87.08	82.28	62.39	80.93	42.13	65.15	–
UDPipe1.1	78.78	82.11	79.88	74.03	72.75	81.15	76.73	53.19	76.69	37.47	57.40	72.14
auto-POS	80.82	81.90	79.76	74.00	75.27	81.78	77.25	54.73	76.35	49.46	71.23	75.47
gold-POS	86.32	86.64	82.27	78.50	81.17	88.29	83.85	57.52	84.36	63.02	80.19	81.14

Table 4. Ablation tests on UD 1.2 dataset. All values are reported in %, "all lexical" stands for removing both character and n-gram features.

Method	de			en			es			fr		
	UAS	LAS	POS	UAS	LAS	POS	UAS	LAS	POS	UAS	LAS	POS
gold-POS	83.78	79.21	–	88.26	85.69	–	88.64	85.76	–	85.77	82.17	–
auto-POS	80.92	74.60	92.08	85.68	81.71	93.38	85.82	81.98	95.90	84.31	79.82	95.07
- POS	77.17	70.45	–	83.63	78.91	–	84.24	74.18	–	82.43	77.04	–
- character	80.02	73.60	91.58	84.69	80.56	92.98	84.79	80.97	95.51	84.16	79.24	94.90
- n-gram	79.39	73.29	91.46	85.08	80.88	93.23	85.11	80.89	95.17	84.24	79.38	94.67
- all lexical	74.28	66.88	86.98	82.40	77.30	89.56	82.17	77.21	92.66	81.75	75.86	92.72
- all	70.91	62.77	–	80.43	74.86	–	74.18	74.18	–	79.79	73.74	–

Method	it			pt			sv			AVG		
	UAS	LAS	POS	UAS	LAS	POS	UAS	LAS	POS	UAS	LAS	POS
gold-POS	90.91	88.77	–	86.72	84.25	–	86.91	83.35	–	87.28	84.17	–
auto-POS	89.37	86.37	96.36	85.43	81.34	95.74	82.18	77.77	95.01	84.82	80.51	94.79
- POS	86.54	83.20	–	82.19	77.55	–	77.38	71.85	–	81.94	76.17	–
- character	88.36	85.21	96.43	85.09	80.72	95.62	81.60	76.74	94.80	84.10	79.58	94.55
- n-gram	88.29	85.18	96.24	85.21	80.79	95.36	80.66	75.82	94.24	84.00	79.46	94.34
- all lexical	85.49	81.20	94.10	80.02	73.70	90.32	73.46	65.95	86.50	79.94	74.01	90.41
- all	82.28	77.84	–	76.56	69.76	–	69.69	60.86	–	76.26	70.57	–

Specifically, n-gram and character features performs quite similarly, and the overall performance decreases 5% in UAS, 5.5% in LAS, and 4.4% in POS accuracy after removing both features. Although n-gram and character features are close, using them jointly can slightly improve the performance for 1% in both UAS and LAS, and 0.4% in POS accuracy.

The influence of POS tagging varies across different languages. As the languages with weakest POS taggers, de (92.08%) and en (93.38%) are affected largely due to the inadequacy of POS tagger for about 3% drop in UAS and LAS. Furthermore, the importance of POS information can be observed from the table, model with perfect POS tagger (100%) is able to reach 87.28% in UAS and 84.17% in LAS, while without tagger model obtains 81.94% in UAS and 76.17% in LAS. The performance of model declines 5.2% in UAS and 7% in LAS with only the replacement of POS information. The final results validate our hypothesis that POS tagging is vital to dependency parsing.

4   Related Work

In this section, we review the literatures related to our work, i.e., joint learning of POS tagging and dependency parsing and transition-based dependency parsing

Most traditional joint models are implemented with feature templates, and are difficult to analyze the contribution of individual feature [4,12,17]. There is another line of works to avoid using POS tagging in dependency parsing [2, 10], e.g., Ballesteros et al. introduce character features to better capture lexical information [2], but it is not contradictive to explore lexical and POS information simultaneously, and in our work we revealed the contributions of these features with extensive experiments.

Our work is inspired by neural networks methods of dependency parsing [1, 6,9,10,16,24,29]. The work from Li et al. [17] is the pioneer to jointly learn POS tagging and dependency parsing, but in this work about 6% performance decrease is reported in Chinese. Most recently, Zhang et al. [27], Yang et al. [25], and Nguyen et al. [19] propose joint models to improve both POS tagging and dependency parsing. However, in their research, the impact of POS information to dependency parsing is not explained sufficiently.

5   Conclusion

In this work, we propose a method to jointly learn POS tagging and dependency parsing so as to alleviate the impact of error propagation problem. The experimental results indicate that POS information is significantly influential to dependency parsing, with about 5.2% and 7% differences in UAS and LAS respectively across languages. We also explore the impact of lexical features, and achieve a slight improvement by introducing n-gram feature. Our proposed method is capable to be applied in multiple languages, and is lightweight with shallow neural networks but still obtains high accuracy. Our code are publicly available for further research[3].

---

[3]   https://github.com/hsuehkuan-lu/JointParser.

Acknowledgement. The work is supported by NSFC key projects (U1736204, 61533018, 61661146007), Ministry of Education and China Mobile Joint Fund (MCM20170301), a research fund supported by Alibaba Group, and THUNUS NExT Co-Lab.

References

1. Andor, D., et al.: Globally normalized transition-based neural networks. In: Proceedings of the 54th Annual Meeting of the Association for Computational Linguistics (Volume 1: Long Papers), pp. 2442–2452. Association for Computational Linguistics (2016)
2. Ballesteros, M., Dyer, C., Smith, N.A.: Improved transition-based parsing by modeling characters instead of words with LSTMs. In: Proceedings of the 2015 Conference on Empirical Methods in Natural Language Processing, pp. 349–359. Association for Computational Linguistics (2015)
3. Bohnet, B.: Top accuracy and fast dependency parsing is not a contradiction. In: Proceedings of the 23rd International Conference on Computational Linguistics (Coling 2010), pp. 89–97. Coling 2010 Organizing Committee (2010)
4. Bohnet, B., Nivre, J.: A transition-based system for joint part-of-speech tagging and labeled non-projective dependency parsing. In: Proceedings of the 2012 Joint Conference on Empirical Methods in Natural Language Processing and Computational Natural Language Learning, pp. 1455–1465. Association for Computational Linguistics (2012)
5. Bunescu, R.C., Mooney, R.J.: A shortest path dependency kernel for relation extraction. In: Proceedings of the Conference on Human Language Technology and Empirical Methods in Natural Language Processing, pp. 724–731. Association for Computational Linguistics (2005)
6. Chen, D., Manning, C.D.: A fast and accurate dependency parser using neural networks. In: Proceedings of the 2014 Conference on Empirical Methods in Natural Language Processing, pp. 740–750 (2014)
7. Cho, K., et al.: Learning phrase representations using RNN encoder-decoder for statistical machine translation. In: Proceedings of the 2014 Conference on Empirical Methods in Natural Language Processing, pp. 1724–1734. Association for Computational Linguistics (2014)
8. Culotta, A., Sorensen, J.: Dependency tree kernels for relation extraction. In: Proceedings of the 42nd Annual Meeting on Association for Computational Linguistics. Association for Computational Linguistics (2004)
9. Dozat, T., Manning, C.D.: Deep biaffine attention for neural dependency parsing. In: 5th International Conference on Learning Representations (2017)
10. Dyer, C., Ballesteros, M., Ling, W., Matthews, A., Smith, N.A.: Transition-based dependency parsing with stack long short-term memory. In: Proceedings of the 53rd Annual Meeting of the Association for Computational Linguistics and the 7th International Joint Conference on Natural Language Processing (Volume 1: Long Papers), pp. 334–343. Association for Computational Linguistics (2015)
11. Finkel, J.R., Manning, C.D.: Joint parsing and named entity recognition. In: Proceedings of Human Language Technologies: The 2009 Annual Conference of the North American Chapter of the Association for Computational Linguistics, pp. 326–334. Association for Computational Linguistics (2009)

12. Hatori, J., Matsuzaki, T., Miyao, Y., Tsujii, J.: Incremental joint POS tagging and dependency parsing in Chinese. In: Proceedings of 5th International Joint Conference on Natural Language Processing, pp. 1216–1224. Asian Federation of Natural Language Processing (2011)

13. Jie, Z., Muis, A.O., Lu, W.: Efficient dependency-guided named entity recognition. In: Proceedings of the Thirty-First AAAI Conference on Artificial Intelligence, pp. 3457–3465. AAAI Press (2017)

14. Kikuchi, Y., Hirao, T., Takamura, H., Okumura, M., Nagata, M.: Single document summarization based on nested tree structure. In: Proceedings of the 52nd Annual Meeting of the Association for Computational Linguistics (Volume 2: Short Papers), pp. 315–320. Association for Computational Linguistics (2014)

15. Kingma, D.P., Ba, J.: Adam: a method for stochastic optimization. arXiv:1412.6980 (2014)

16. Kiperwasser, E., Goldberg, Y.: Simple and accurate dependency parsing using bidirectional LSTM feature representations. Trans. Assoc. Comput. Linguist. **4**, 313–327 (2016)

17. Li, Z., Zhang, M., Che, W., Liu, T., Chen, W., Li, H.: Joint models for Chinese pos tagging and dependency parsing. In: Proceedings of the Conference on Empirical Methods in Natural Language Processing, pp. 1180–1191. Association for Computational Linguistics (2011)

18. McDonald, R., Crammer, K., Pereira, F.: Online large-margin training of dependency parsers. In: Proceedings of the 43rd Annual Meeting of the Association for Computational Linguistics (ACL 2005), pp. 91–98. Association for Computational Linguistics (2005)

19. Nguyen, D.Q., Verspoor, K.: An improved neural network model for joint POS tagging and dependency parsing. In: Proceedings of the CoNLL 2018 Shared Task: Multilingual Parsing from Raw Text to Universal Dependencies, pp. 81–91. Association for Computational Linguistics (2018)

20. Nivre, J.: An efficient algorithm for projective dependency parsing. In: Proceedings of the Eighth International Workshop on Parsing Technologies, pp. 149–160 (2003)

21. Nivre, J.: Incrementality in deterministic dependency parsing. In: Proceedings of the Workshop on Incremental Parsing: Bringing Engineering and Cognition Together, pp. 50–57. Association for Computational Linguistics (2004)

22. Nivre, J., et al.: MaltParser: a language-independent system for data-driven dependency parsing. Nat. Lang. Eng. **13**(2), 95–135 (2007)

23. Rush, A.M., Chopra, S., Weston, J.: A neural attention model for abstractive sentence summarization. In: Proceedings of the 2015 Conference on Empirical Methods in Natural Language Processing, pp. 379–389. Association for Computational Linguistics (2015)

24. Weiss, D., Alberti, C., Collins, M., Petrov, S.: Structured training for neural network transition-based parsing. In: Proceedings of the 53rd Annual Meeting of the Association for Computational Linguistics and the 7th International Joint Conference on Natural Language Processing (Volume 1: Long Papers), pp. 323–333. Association for Computational Linguistics (2015)

25. Yang, L., Zhang, M., Liu, Y., Sun, M., Yu, N., Fu, G.: Joint pos tagging and dependence parsing with transition-based neural networks. IEEE/ACM Trans. Audio Speech Lang. Process. **26**(8), 1352–1358 (2018)

26. Zhang, Y., Li, C., Barzilay, R., Darwish, K.: Randomized greedy inference for joint segmentation, POS tagging and dependency parsing. In: Proceedings of the 2015 Conference of the North American Chapter of the Association for Computational

Linguistics: Human Language Technologies, pp. 42–52. Association for Computational Linguistics (2015)

27. Zhang, Y., Weiss, D.: Stack-propagation: improved representation learning for syntax. In: Proceedings of the 54th Annual Meeting of the Association for Computational Linguistics (Volume 1: Long Papers), pp. 1557–1566. Association for Computational Linguistics (2016)

28. Zhang, Y., Nivre, J.: Transition-based dependency parsing with rich non-local features. In: Proceedings of the 49th Annual Meeting of the Association for Computational Linguistics: Human Language Technologies: Short Papers - Volume 2, pp. 188–193. Association for Computational Linguistics (2011)

29. Zhou, H., Zhang, Y., Huang, S., Chen, J.: A neural probabilistic structured-prediction model for transition-based dependency parsing. In: Proceedings of the 53rd Annual Meeting of the Association for Computational Linguistics and the 7th International Joint Conference on Natural Language Processing (Volume 1: Long Papers), pp. 1213–1222. Association for Computational Linguistics (2015)

Graph Neural Net-Based User Simulator

Xinrui Nie, Zehao Lin, Xinjing Huang, and Yin Zhang$^{(\boxtimes)}$

Zhejiang University, Hangzhou 310007, China
{3150104023, georgelin, huangxinjing, zhangyin98}@zju.edu.cn,
http://www.dcd.zju.edu.cn

Abstract. User Simulators are major tools that enable offline training of task-oriented dialogue systems. To efficiently utilize semantic dialog data and generate natural language utterances, user simulators based on neural network architectures are proposed. However, existing neural user simulators still rely on hand-crafted rules, which is difficult to ensure the effectiveness of feature extraction. This paper proposes the Graph Neural Net-based User Simulator (GUS), which constructs semantic graphs from the corpus and uses them to build Graph Convolutional Network (GCN) to extract feature vectors. We tested our model on examined public dataset and also made conversation with real human directly to verify the effectiveness. Experimental results show GUS significantly outperforms several state-of-the-art user simulators.

Keywords: User simulator · Graph neural networks · Dialogue systems

1 Introduction

The Dialogue Manager of Spoken Dialogue System (SDS) is generally trained by reinforcement learning [2]. Through the control of the reward function, the dialog manager learns optimal policy by maximizing the cumulative rewards. The process of selecting the next action is called the policy. To train a well-functioning policy, a large amount of dialog data is needed. But considering the low effectiveness and high cost, it is unrealistic to obtain large-scale dataset from interactions with real users. Either we can't directly learn policy from the corpus since it is static and the accessed dialog state space is very limited. Therefore, user simulator came into being. User simulator is trained from the corpus to learn how real users will respond in a given context. It transforms a static corpus into a dynamic tool that can generate any number of dialogues and explore the unrecorded state space.

In recent years, user simulators have evolved from hand-crafted rules-based models to deep learning models. Works [1,13,14] try to introduce neural network and expect a better performance with less human efforts than rule-based models. However, these user simulators still use hand-crafted rules and thus suffers from migrating to other domains. To relieve this problem, we adopt Graph Neural

© Springer Nature Switzerland AG 2019
M. Sun et al. (Eds.): CCL 2019, LNAI 11856, pp. 638–650, 2019.
https://doi.org/10.1007/978-3-030-32381-3_51

Network (GNN) to automatically learn to represent not only utterance sequence but also semantic structure as dialogue state.

In this paper, we propose the Graph Neural Net-based User Simulator (GUS), which takes the semantic data as inputs and generates natural language. Its core consists of a graph generator, a GCN-based feature extractor, and a sequence-to-sequence model. The graph generator constructs the semantic graphs from the semantic data in the corpus by our designed mechanism. The GCN feature extractor extracts feature from the dialogue history by applying convolution on the semantic graphs. The sequence-to-sequence model contains an RNN encoder, which encodes the feature vector sequence, and an RNN decoder, which outputs natural language. In addition, GUS can generate its own goals and change goals during the dialogue. This helps to train a more complex dialog manager policy.

GUS is trained on DSTC2 (Dialog State Tracking Challenge 2) dataset, which is built on the restaurant recommendation domain. Its ontology defines: (1) informable slots, which are user-selectable restaurant attributes (such as food, area, price range) (2) requestable slots, which are attributes that users can request (such as address, phone number, zip code) (3) all restaurant names. Each attribute of the restaurant appears as a slot-value pair (e.g. $\texttt{area} = \text{north}$).

We use BLEU to measure the performance of GUS and visually verify its validity by establishing a real dialogue history and observing the model outputs. We compare the performance of GUS with NUS, Sequence-to-Sequence User Simulator (S2SUS) and Graph Convolutional Encoded User Simulator (GCEUS). In the two evaluation tasks, GUS is the best in all models.

2   Related Work

User simulator can be classified according to the abstraction level of modeling dialogues [4], which is roughly divided into semantic level and word level. Up to now, most user simulators work on the semantic level, which needs to handle user dialog acts and corresponding slot-value pairs. The first user simulator [5] proposed a simple bi-gram model to predict the user action a_u based on the system action a_s as $p = P(a_u|a_s)$. The advantage lies in pure probability modeling and complete domain independence. However, it can't produce consistent user behavior, since it only focuses on the last action of the system and the full dialogue history is not modeled. This leads to user utterances often being meaningless in a longer context. Scheffler [6] tried to solve the problem of inconsistency by introducing a fixed goal structure. But it requires a large amount of domain-specific knowledge. Pietquin [7] combined the characteristics of the former two models. The core idea is to condition the probability to explicit user goals: $P(provideA_x|requestA_x, goal)$ (A_x represents an attribute). This allows it to explicitly model the dependencies between user acts and goals. Georgila [8] used the Markov model to describe the user's current state with a large feature vector. However, the model is not conditioned on user goals, which makes it impossible to know whether the goal has been completed, and therefore cannot be used to train the dialogue manager's policy. Chandramohan [9] regarded the user simulation as an inverse reinforcement learning problem, and modeled users as a decision-making agent. But the model does not incorporate user goals either. The most prominent user simulator is ABUS [10] (Agenda-based User Simulator), which models state transition and user action generation through a stack structure called agenda. Based on hand-crafted rules, ABUS explicitly encodes user goals and dialogue history, and has a corresponding update strategy when the system is unable to complete the goal.

3   Proposed Framework

The above user simulators are all built on the semantic level. Many problems arise from this approach. First, research [11] shows that the Spoken Language Understanding and Belief Tracking module in SDS should be jointly trained as a single entity, which is impossible when user simulator does not output natural language. Second, all user sentences must be correctly semantically labeled, which is costly. Third, user simulator is sometimes incremented with an error model which reproduces the speech recognizer errors. SDS performs better when the error model is more consistent with the characteristics of the speech recognizer [12]. Since speech recognition errors are word-level, the error model based on semantic modeling may be worse than the error model based on natural language modeling.

The study of word-level user simulators has begun to reheat in recent years. Crook [13] used the sequence-to-sequence model to train word-level user simulators, but did not incorporate user goals, so it could not be used for policy optimization. Li [14] added a natural language generation module to ABUS to convert it into a word-level user simulator. Kreyssig [1] proposed the NUS architecture, which takes the semantic input of SDS and generates feature vectors through hand-crafted rules. Then natural language is generated through the sequence-to-sequence model.

The overall architecture of GUS is shown in Fig. 1. At the beginning of the dialogue, an initial random goal G_0 is generated by the goal generator. In dialogue turn T, the semantic input of the system statement, the current user

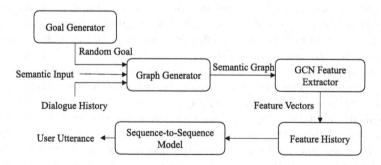

Fig. 1. General architecture of the graph neural net-based user simulator (GUS)

goal G_T, and the dialogue history are fed into the graph generator to generate a series of semantic graphs: $g_{1:T} = g_1...g_T$, which are then fed into the GCN feature extractor and used to generate a sequence of feature vectors $\mathbf{v}_{1:T} = \mathbf{v}_1...\mathbf{v}_T$. Note these vectors can be viewed as a summary of each turn, thus dialogue history can be modeled by applying sequence-to-sequence model on the vector sequence to generate a user utterance. $\mathbf{u}_T = \omega_0...\omega_{n_T}$ with length n_T. Values corresponding to all slots that appear in the user utterances are replaced with tokens, a process called delexicalisation. When the system statement indicates that no venue matches the current user goal, the goal generator randomly modifies the goal.

3.1   Goal Generator

At the beginning of the dialogue, the goal generator generates an initial random goal $G_0 = C_0$, which represents a set of constraints imposed by the user on the final venue, such as (food = indian, pricerange = cheap). The probability of occurrence of each informable slot in C_0 is defined by the ontology. In DSTC2, C_0 can contain up to three constraints: food, area and pricerange, with a sampled probability of 0.66, 0.62, and 0.58, respectively. Therefore, when generating the initial random goal, each constraint is subjected to the independent probability to determine whether it appears.

When training GUS, we did not use the goal generator, but the goal-label in DSTC2. DSTC2 provides turn-specific goal labels for each turn of the dialogue, which contain all constraints imposed by the user by the current turn.

When the system outputs the canthelp dialogue act, it means that there is no venue that matches the current user goal. The goal generator will regenerate a random goal G_T to replace the previous one.

3.2   Graph Generator

The graph generator constructs semantic graphs based on semantic inputs. For certain dialogue turn, the structure of the semantic graph is shown in Fig. 2. The whole graph is a tree structure with three layers. The motivation is to preserve all necessary value-independent information in a turn of dialogue.

The first layer contains only one node turn_index, whose feature (i.e., node value) is the index of the dialog turn. This node is the central node of the whole graph, which is used to store the aggregation of features of all other nodes in the current turn. Its node index is fixed to 0.

The second layer contains four nodes, which are dialog act node, requested slot node, inconsistent slot node, and user goal node respectively (Fig. 2 nodes 1–4). The features of these four nodes are fixed for any turn of dialogue, namely the strings 'dialog_act', 'requested_slots', 'inconsistency_slots' and 'goal_labels'. They are designed mainly to store the information of respective child nodes.

The child nodes of the dialog act node store information related to system acts, which are divided into two classes. The first class stores all dialog acts. Each

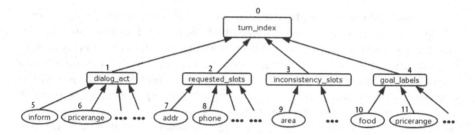

Fig. 2. Semantic graphs generated using certain turn of dialogue data

act is a node by itself (Fig. 2 node 5). The second class stores slot information related to system acts (Fig. 2 node 6). When system act is `request` or `select`, or is `inform` or `expl-conf` with a correct pair of informable slot-value, a node is created for each corresponding slot. The motivation is that the difference in user responses required for different slots is often large for the same system action. For example, the structure of the user utterance that responds to `request(food)` and `request(area)` can be quite different.

The child nodes of the requested slot node store all request information that has not been fulfilled (Fig. 2 nodes 7, 8). In DSTC2, user requests are stored in `requested-slots`, from which we can directly construct nodes. When the system act is `inform` or `offer`, the slot corresponding to the system act will be marked as fulfilled, and then be removed.

The child nodes of the inconsistent slot node store the inconsistency between the user goal and the system response (Fig. 2 node 9). When system act is `inform` or `expl-conf` or `impl-conf`, the system will propose one or more informable slots accordingly. Some of these slots may not appear in the user goal, which are added to the branch. The storage of inconsistent information is necessary, which helps GUS to correct the system.

The child nodes of the user goal node store all constraints contained in the current user goal (Fig. 2 nodes 10, 11). From the turn-specific goal label in DSTC2, all constraint information can be directly obtained.

3.3   GCN Feature Extractor

To encode node features in the semantic graph, GUS adopts the GCN architecture. The purpose of GCN is to calculate the representation of each node in the graph structure [18]. Formally, given a directed graph $\mathcal{G} = (\mathcal{V}, \mathcal{E})$, where $\mathcal{V}$ is the set of all vertices, $\mathcal{E}$ is the set of all edges. For each node $v_i \in \mathcal{V}$ in the graph, use $h_i^{(l)}$ as the representation of v_i at the lth layer of GCN, then $h_i^{(l+1)}$ is [17]:

$$h_i^{(l+1)} = \sigma\left(\sum_{j \in \mathcal{N}_i} \frac{1}{c_{ij}} h_j^{(l)} W^{(l)} + b^{(l)}\right) \tag{1}$$

where $\mathcal{N}_i$ is the set of adjacent nodes of v_i. c_{ij} is the normalized constant corresponding to the edge (v_i, v_j). $W^{(l)}$ is the weight matrix of layer l. $b^{(l)}$ is the bias of layer l. σ is a nonlinear differentiable activation function (ReLU). As pointed out by Marcheggiani [19], a self-loop should be added for each node in the graph to ensure that $h_i^{(l)}$ will affect the generation of $h_i^{(l+1)}$. $h_i^{(0)}$ is the initial node feature of v_i in $\mathcal{G}$.

Considering that the semantic graph is a three-layer tree structure, and the information of all nodes in the third layer needs to be aggregated into the central node in the first layer (i.e., node 0), GCN is designed as two layers. The initial representation of each node is the word embedding of node features. After two-layer convolution, node 0 stores the information of the entire graph. So we use the last representation of node 0 as the feature vector of the graph. That is, for semantic graph g_T, its corresponding feature vector $\mathbf{v}_T$ is:

$$\mathbf{v}_T = h_{0_T}^{(2)} \tag{2}$$

where 0_T is node 0 of graph g_T. Each turn of dialogue generates its own feature vector, which is combined in a turn order as the feature history $\mathbf{v}_{1:T} = \mathbf{v}_1...\mathbf{v}_T$. It will then be inputted into the sequence-to-sequence model.

3.4 Sequence-To-Sequence Model

The sequence-to-sequence model contains an RNN encoder and an RNN decoder, as shown in Fig. 3. The definition of RNN is as follows:

$$(\mathbf{h}_t, \mathbf{s}_t) = RNN(\mathbf{x}_t, \mathbf{s}_{t-1}) \tag{3}$$

At time step t, RNN takes the input $\mathbf{x}_t$ and the last hidden state $\mathbf{s}_{t-1}$, outputs $\mathbf{h}_t$ and the new hidden state $\mathbf{s}_t$. RNN contains many variants. In GUS, both the encoder and decoder use the single-layer and single-direction LSTM [20].

The encoder takes one feature vector as input at each time step t, i.e. $\mathbf{x}_t^E = \mathbf{v}_t$. If the current turn is T, then the final output of the encoder is $\mathbf{h}_T^E$, which is also called the context vector (Fig. 3 **c** in the dotted box).

The sequence-to-sequence model must define a probability distribution over different sequences on **c** so that when GUS is sampled on this distribution, different sets of utterances can be generated from the same context. The conditional probability distribution of a length L sentence is defined as:

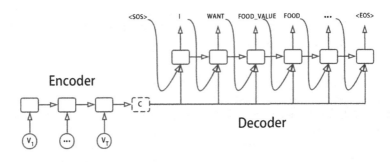

Fig. 3. The sequence-to-sequence model in GUS

$$P(\mathbf{u}|\mathbf{c}) = P(\omega_0|\mathbf{c}) \prod_{t=1}^{L} P(\omega_t|\omega_{t-1}...\omega_0, \mathbf{c}) \tag{4}$$

The decoder is used to model $P(\omega_t|\omega_{t-1}...\omega_0, \mathbf{c})$. Its input at a time is the concatenation of the word embedding $\mathbf{w}_{t-1}$ of the previous step output word ω_{t-1} and the context vector $\mathbf{c}$, that is, $\mathbf{x}_t^D = [\mathbf{w}_{t-1}\mathbf{c}]$. For $P(\omega_0|\mathbf{c})$, the sentence start tag <SOS> is used as ω_{-1}. When the decoder outputs the sentence end tag <EOS>, the decoding process is finished. The output $\mathbf{h}_t^D$ forms $P(\omega_t|\omega_{t-1}...\omega_0, \mathbf{c})$ through a projection layer defined as follows:

$$P(\omega_t|\omega_{t-1}...\omega_0, \mathbf{c}) = Softmax(W_\omega \mathbf{h}_t^D + \mathbf{b}_\omega) \tag{5}$$

The word ω_t is obtained by taking the word with the highest probability from the distribution. During the training of GUS, ω_t is not used as the input of the next time step, but directly using the ground-truth words in the dataset. This technique is called teacher-forcing. In addition, the user utterances in the dataset are delexicalized. It can be seen in Fig. 3 that the value in the output utterance of the decoder has been replaced with the corresponding slot.

The training goal of the sequence-to-sequence model is to maximize the output sequence probability conditioned on the input sequence, so the log probability corresponding to formula (4) can be used as the loss function:

$$\mathcal{L} = -\sum_{n=1}^{N} log P(\omega_0|\mathbf{c}) \sum_{t=1}^{L_n} P(\omega_t|\omega_{t-1}...\omega_0, \mathbf{c}) \tag{6}$$

4   Experimental Setup and Results

4.1   Experimental Setup

Dataset. GUS is trained and tested on the DSTC2 dataset, which is collected by Amazon Mechanical Turk on the restaurants recommendation domain. The dialogue state in DSTC2 is very variable, and users often change their goals during the dialogue. DSTC2 is manually recorded, so it contains many spelling and recording errors. We corrected some common spelling errors to ensure correct delexicalization. The training set contains 2,118 dialogues, and the test set contains 1,117 dialogues. After delexicalization, the maximum sequence length is 22. The maximum dialogue length is 30 turns.

Evaluation Method. The evaluation of the user simulator is an ongoing research area, and various technical methods can be found in related work. We use BLEU [22] (Bilingual Evaluation Understudy) and real dialogue output to measure the performance.

BLEU is a method for automatic evaluation of machine translation. BLEU has been widely employed in evaluating sequence generation including machine translation, text summarization and dialogue systems. BLEU calculates the n-gram precision which is the fraction of n-grams in the candidate text present in any of the reference texts.

In addition to BLEU, we also verify the validity of the model by establishing a real dialogue history and observing the model output utterance. This is a relatively straightforward way to evaluate the model.

Baseline and Setup. All models use the full teacher-forcing mode and maximum decoding steps is 15. The experimental setup is in Table 1.

- GCEUS: This model directly uses GCN as the encoder and discards the original LSTM structure. The encoder combines the semantic graphs of all turns into a large graph, and uses the representation of the central node after two-layer convolution as the final output. The decoder is the same as GUS.
- S2SUS: The semantic input data is sequenced and directly inputted into the sequence-to-sequence model, which omits the step of extracting the feature vector.
- NUS: This model is proposed by Kreyssig [1]. It outperforms the classic ABUS and is the reference architecture of GUS. It extracts 35-dimensional feature vectors from the dialog data based on hand-crafted rules.

Table 1. Experimental setup of all models

Model name	Encoder input dimension	Decoder output dimension	Hidden size	Batch size	Learning rate
GCEUS	256	256	512	64	0.0004
S2SUS	256	256	512	32	0.0003
NUS	35	128	256	16	0.001
GUS	256	256	512	64	0.0007

4.2   Results

As Table 2 shows, the BLUE of GUS is 0.211, which is the highest among all models. It reflects that GUS can generate user utterances that best fit the context and contain the necessary information. Table 3 shows the user utterance

Table 2. The highest BLEU of all models on test set after 20 epochs training

Model name	BLEU
GUS	.211
NUS	.169
S2SUS	.136
GCEUS	.099

Table 3. Outputs of all models on the same real dialogue history

Dialogue history	System: How may I help you?
	User: I'm looking for food_value food in the area_value part of town
	System: There are 12 restaurants in the west of town. What type of food do you want?
	User: food_value food
	System: There are 4 restaurants serving Indian in the west of town. What price range would you like?
Model name	Outputs
GUS	pricerange_value
NUS	what is the address
S2SUS	food_value
GCEUS	address

that all models output on the same dialogue history. The history was not taken from DSTC2, but was artificially constructed. Since all models can only accept input in semantic form rather than natural language, the dialogue history was converted into json data of a specific format, which contained the semantics of each turn of dialogue. The data was then inputted into each model. In the final turn of the dialogue, the system asked the user what price range he liked. GUS correctly answered the price range. NUS requested the address of the restaurant, which was obviously wrong. S2SUS incorrectly answered what food to eat. GCEUS was the same as NUS which also asked for the restaurant address. Only GUS answered with the correct user utterance, which verifies its validity again.In both evaluation tasks, GUS works best for all models. In order to verify that it is able to finish a complete and successful dialogue, we asked a human as the system to have a real dialogue with GUS. The dialogue is shown in Table 4. In the 4th turn, the system said there was no venue that met the requirements, and GUS successfully changed the user goal in the 5th turn and began to re-inform the system of its constraints on the new venue. When the system proposed a venue in the 7th turn, GUS asked for the address, phone number and zip code in order. When the system did not correctly inform the phone number in the 9th turn, GUS re-inquired in the 10th turn. Finally, in the 12th turn, GUS determined that the task had been completed, and output the thank-you statement. The dialogue ended. It can be seen that GUS can output context-compliant user utterances in practical applications.

Table 4. Dialogue between GUS as the user and human as the system

Turn	GUS	Human
		how may I help you?
0	i'm looking for food_value food	what kind of food would you like?
1	food_value	you're looking for a restaurant serving chinese food, right?
2	yes	what part of town do you have in mind?
3	area_value	there are 3 restaurants in west of the town that serve chinese food. what price range do you want?
4	pricerange_value	I'm sorry but there is no expensive restaurant
5	food_value food	what area do you prefer?
6	area_value	and what price range would you like?
7	do nt care	golden house is a great restaurant that serves indian food
8	address	golden house is a nice place. The address is on Ohoho Street
9	phone number	golden house is a nice place. The address is on Ohoho Street
10	phone number	the phone number of golden house is 213124512
11	post code	the post code of golden house is A.7 B. C1
12	thank you	goodbye

4.3   Analysis

First, the BLEU of all models is relatively low. The reason is presumed as follows: All models are trained on DSTC2, which is a very noisy dataset containing many spelling errors. Although the commonly occurred errors are corrected, there are also quite a few accidental errors. Besides, DSTC2 often misses or mislabels semantic data when annotating. These errors cannot be solved except for manual correction. In addition, each dialogue in DSTC2 has a corresponding customer feedback, many of which are marked as "disagree". But these noise dialogues are not removed. Furthermore, the amount of data is relatively small, which may also result in a decrease in the performance of the model on the test set. And as shown in Tables 2 and 3, the user utterances output by the model are mostly very short. Although the necessary information is provided, this also leads to a lower BLEU.

Compared with NUS, GUS has a large improvement (from 0.169 to 0.211). This shows that the effect of feature extraction using GCN is better than that of hand-crafted rules. The semantic graph design of GUS refers to the design of the feature vector of NUS, both of which divides the semantics into four parts. In NUS, a simple binary vector is used to represent the semantics of each part, and the four parts are simply concatenated. In GUS, the four-part semantics are explicitly distinguished by the nodes in the middle layer. Through training, the corresponding weights are learned for the nodes on each branch, which is equivalent to a more elaborate combination of semantics and leads to a better effect.

S2SUS's BLEU is lower than GUS and NUS, which means that it is necessary to extract feature vectors from semantic data. There are some information that does not affect the result (such as user acts) in the input. Performing feature extraction operation on the semantic data can help the sequence-to-sequence model learn the input features better, thereby improving its effect.

The BLEU of GCEUS is the lowest among all models, which is even lower than S2SUS (0.136 to 0.099). It indicates that the GCN encoder is less effective than the RNN encoder in tasks that require modeling dialogue history. Because the nature of dialogue is a sequence of data, the content of the last turn is significantly more important than the previous history, which coincides with the fact that the later input in the RNN encoder has a greater impact on the results. In the GCN encoder, the order of different turns is not explicitly distinguished, and relies on the network to learn by itself. So the final effect is very poor.

5   Conclusion

We propose the Graph Neural Net-based User Simulator (GUS) architecture, which takes the semantic response of the system as input and outputs user utterances in natural language form. Compared with Neural User Simulator (NUS), GUS uses GCN to extract feature vectors, which makes the extraction process more detailed and efficient, and helps the sequence-to-sequence model better learn the dialogue features. This work is the first time to introduce the GNN architecture to the field of user simulator. It not only explores the feasibility of GNN applications on semantic data, but also achieves good results. In the BLEU evaluation task, GUS has a great improvement than NUS, and is much higher than S2SUS and GCEUS. In the evaluation of establishing real dialogue history and observing output, GUS can correctly reply to system statements and complete a successful dialogue.

Acknowledgement. We thank the anonymous reviewers for their insightful comments on this paper. This work was supported by the NSFC (No. 61402403), Alibaba-Zhejiang University Joint Institute of Frontier Technologies, Chinese Knowledge Center for Engineering Sciences and Technology, and the Fundamental Research Funds for the Central Universities.

References

1. Kreyssig, F., Casanueva, I., Budzianowski, P., Gasic, M.: Neural user simulation for corpus-based policy optimisation for spoken dialogue systems. In: Proceedings of the 19th Annual SIGdial Meeting on Discourse and Dialogue (2018)
2. Gasic, M., Young, S.: Gaussian processes for POMDP-based dialogue manager optimization. IEEE/ACM Trans. Audio, Speech, Lang. Process. **22**(1), 28–40 (2014)

3. Zhou, J., et al.: Graph Neural Networks: A Review of Methods and Applications. arXiv preprint arXiv:1812.08434v3 (2019)
4. Schatzmann, J., Georgila, K., Young, S.: Quantitative evaluation of user simulation techniques for spoken dialogue systems. In: SIGdial6, pp. 45–54 (2005)
5. Eckert, W., Levin, E., Pieraccini, R.: User modeling for spoken dialogue system evaluation. In: 1997 IEEE Workshop on Automatic Speech Recoginition and Understanding, pp. 80–87 (1997)
6. Scheffler, K., Young, S.: Probabilistic simulation of human-machine dialogues. In: Speech, and Signal Processing, Acoustics (2000)
7. Pietquin, O., Dutoit, T.: A probabilistic framework for dialog simulation and optimal strategy learning. IEEE Trans. Audio Speech Lang. Process. **14**(2), 589–599 (2006)
8. Georgila, K., Henderson, J., Lemon, O.: Learning user simulations for information state update dialogue systems. In: Ninth European Conference on Speech Communication and Technology (2005)
9. Chandramohan, S., Geist, M., Lefevre, F., Pietquin, O.: User simulation in dialogue systems using inverse reinforcement learning. In: Proceedings of the Twelfth Annual Conference of the International Speech Communication Association (2011)
10. Schatzmann, J., Thomson, B., Weilhammer, K., Ye, H., Young, S.: Agenda-based user simulation for bootstrapping a pomdp dialogue system. In: Human Language Technologies 2007: The Conference of the North American Chapter of the Association for Computational Linguistics, Companion Volume, Short Papers, pp. 149–152 (2007)
11. Mrksic, N., Seaghdha, D.O., Wen, T.-H., Thomson, B., Young, S.: Neural belief tracker: data-driven dialogue state tracking. In: Proceedings of the 55th Annual Meeting of the Association for Computational Linguistics (Volume 1: Long Papers), vol. 1, pp. 1777–1788 (2017)
12. Williams, J.D.: Evaluating user simulations with the Cramer-von Mises divergence. Speech Commun. **50**(10), 829–846 (2008)
13. Crook, P., Marin, A.: Sequence to sequence modeling for user simulation in dialog systems. In: Proceedings of the 18th Annual Conference of the International Speech Communication Association (2017)
14. Li, X., Chen, Y.-N.., Li, L., Gao, J., Celikyilmaz, A.: End-to-end task-completion neural dialogue systems. In: Proceedings of the Eighth International Joint Conference on Natural Language Processing (Volume 1: Long Papers) (2017)
15. Serras, M., Torres, M.I., Pozo, A.: Regularized neural user model for goal oriented spoken dialogue systems. In: International Workshop on Spoken Dialogue Systems (2017)
16. Liu, B., Lane, I.: Iterative policy learning in end-to-end trainable task-oriented neural dialog models. In: 2017 IEEE Automatic Speech Recognition and Understanding Workshop, ASRU, pp. 482–489 (2017)
17. Kipf, T.N., Welling, M.: Semi-supervised classification with graph convolutional networks. In: Proceedings of the International Conference on Learning Representations, ICLR (2016)
18. Marcheggiani, D., Perez-Beltranchini, L.: Deep graph convolutional encoders for structured data to text generation. In: The 11th International Conference on Natural Language Generation, INLG (2018)
19. Marcheggiani, D., Titov, I.: Encoding sentences with graph convolutional networks for semantic role labeling. In: Proceedings of the 2017 Conference on Empirical Methods in Natural Language Processing, EMNLP, pp. 1506–1515 (2017)

20. Hochreiter, S., Schmidhuber, J.: Long short-term memory. Neural Comput. **9**(8), 1735–1780 (1997)
21. Kingma, D.P., Ba, J.: Adam: a method for stochastic optimization. In: Proceedings of the ICLR (2015)
22. Papineni, K., Roukos, S., Ward, T., Zhu, W.: BLEU: a method for automatic evaluation of machine translation. In: Proceedings of the 40th Annual Meeting of the Association for Computational Linguistics, pp. 311–318 (2002)

Pinyin as a Feature of Neural Machine Translation for Chinese Speech Recognition Error Correction

Dagao Duan[✉], Shaohu Liang, Zhongming Han, and Weijie Yang

Beijing Technology and Business University, Haidian, Beijing, China
duandg@th.btbu.edu.cn

Abstract. Text correction after automatic speech recognition (ASR) is an important method to improve the speech recognition system. We regard the speech error correction as a translation task—from the language of bad Chinese to the language of good Chinese. We propose a speech recognition error correction algorithm based on neural machine translation (NMT) model. The algorithm is characterized by Chinese Pinyin coding, using a multilayer convolutional encoder-decoder with attention neural network. In the WeChat speech transcription data set we collected, our model substantially outperforms all prior neural approaches on this data set as well as the strong statistical machine translation-based systems. Our analysis shows the superiority of convolutional neural networks in capturing the local context via attention and thereby improving the coverage in speech transcription errors. By boosting multiple modes, using data augmentation and 3-gram language model tricks, our novel algorithm makes the error rate on the test set decreased by 26.2% on average. Our results show that using a multilayer convolutional encoder-decoder with Pinyin feature is able to achieve state-of-the-art performance in text correction after speech recognition.

Keywords: Automatic speech recognition · Neural machine translation · Attention mechanism · Pinyin encoding · Chinese error correct

1 Introduction

In the speech recognition system, the text after the speech recognition is difficult to be understood and there are many errors in the text due to the noisy background environment of the speaker or the phenomenon of swallowing, dragging, accent and dialect of the speaker. Generally speaking, there are four types of text errors, including redundant words (R), missing words (M), bad word selection (S) and mixing errors (Mi). Table 1 lists examples of each type of error. Because of the speaker's dragging, the speech recognition system separated the "pian" vowels into "ping" and "an" syllable, resulting in redundancy errors. Because of the speaker's accent or dialect, the speech recognition system recognizes the "na" vowel "a" as "ei", resulting in the incorrect selection of the Chinese character "那" with "内". Because the speaker speaks fast, or some zero initial Chinese characters, it is easy to join the first Chinese character and the second Chinese syllable together, and the speech recognition system identifies

© Springer Nature Switzerland AG 2019
M. Sun et al. (Eds.): CCL 2019, LNAI 11856, pp. 651–663, 2019.
https://doi.org/10.1007/978-3-030-32381-3_52

two syllables as one syllable, which leads to deletion errors. For example, "xi'an" is identified as "xian". The last type of error is a combination of the first three types, which will result in mixing errors due to the speaker's non-standard tone and speaking speed. For example, in the last column of Table 1, " xué" can be translated as " xuě", and "yǐ yǐ" can be translated as " yǐ", resulting in mixed errors in selection and deletion. Due to the complexity of speech recognition errors in Chinese language, the common methods based on rule matching and simple statistics are difficult to effectively solve practical problems.

Table 1. Error types

Error type	Bad sentence	Good sentence
Redundant error	有没有平安一点的衣服啊 Do you have anything safe clothes yǒu méi yǒu píng ān yì diǎn de yī fú a	有没有便宜点的衣服啊 Do you have any cheaper clothes yǒu méi yǒu pián yí diǎn de yī fú a
Replacement error	内个人是我妈妈 Inner one women is my mother nèi gè rén shì wǒ mā mā	那个人是我妈妈 That woman is my mother nà gè rén shì wǒ mā mā
Missing error	暑假我去先玩了 I went to play first in the summer vacation shǔ jià wǒ qù xiān wán le	暑假我去西安玩了 I went to xi'an during the summer vacation shǔ jià wǒ qù xī ān wán le
Mixing error	雪不可以 Snow can not xué bù kě yǐ	学不可以已 You cannot learn by halves xué bù kě yǐ yǐ

2   Related Works

In recent years, with the development of automatic speech recognition, speech recognition has been significantly improved. However, due to noise and speaker accent, the results of speech recognition are still unsatisfactory. Therefore, text correction after speech recognition is an auxiliary task to improve speech recognition. Text error correction is an important task in natural language processing. After Ng et al. organized the CoNLL-2013 open task [1], many methods based on statistics and neural network were proposed, which greatly improved the research on English error correction. Brockett [2] put forward the concept of "translation" in the first time, by putting the wrong sentences "translate" the correct sentence to achieve the purpose of syntax error correction. The use of channel noise statistical machine translation model (SMT) came to correct errors in English. In addition, SMT can be combined with other methods, such as integrating rule-based methods to build complex and efficient systems [3]. Recently, the SMT-based method proposed by Chollampatt [4] has achieved fairly good error correction results. With the development of deep learning, neural machine translation (NMT) as a new paradigm has been put forward and applied to machine translation. Compared with SMT, NMT has greatly improved the translation quality. On the basis of previous work, Yuan [5] applied NMT to English error correction tasks, specifically, they used the classical translation model: a bidirectional RNN encoder and decoder model containing attention mechanism. In order to solve the problem of Out

Of Vocabulary (OOV), Ji [6] applied a hybrid NMT model that combines word-level and character-level information. Chollampatt [7] firstly proposed the encoder-decoder model of convolution based on attention mechanism, which can extract local context information better. Later, Junczys [8] demonstrated the similarities between grammatical error correct (G(EC) and low-resource neural machine translation, and transformed some low-resource neural machine translation methods into grammatical error correction methods, achieving the best results in GEC tasks. Recently, Grundkiewicz [9] proposed a hybrid system by combining SMT and NMT and achieved a level close to human. Chinese scholars focus on Chinese grammatical error diagnosis (CGED), Yu [10] et al. organized a CGED task, that goal was to develop a computer aided tool for CGED. The diagnosis types included redundant words, missing words, word order error and word choice error. Zheng [11], Xie [12] and Tan [13] regarded CGED as a sequence labeling problem. Their solution was to classify words in combination with CRF and LSTM. The CGED task prompted researchers to focus on the detection of grammatical errors in computational linguistics. The task still solely concentrates on the detection of the grammatical errors rather than the automatic generation of corrections. This limits its application. Hu [14] et al. proposed a method of error correction for search engine query by replacing the parts of the original query that need to be modified, then sorting all the candidate error correction queries according to various judgments to find the most suitable correction. Since most of the search engine queries are phrase errors and keyboard input errors, the speech errors in the sentences cannot be well corrected. Wu [15] used phrase-based statistical machine translation model to convert ASR Pinyin outputs to translated Pinyin. Then they used the beam search algorithm based on dynamic tree to convert translated Pinyin to error-corrected Chinese characters. Since one syllable can correspond to multiple Chinese characters, in the process of converting pinyin into Chinese characters, errors may occur.

This paper presents a neural network error correction model based on machine translation. We use word (character) vector as the features of the machine translation model, the Chinese Pinyin coding as an additional feature to join the network model. Since there is a one-to-one correspondence between word (character) vector and word (character). The Chinese Pinyin contains speech information, which better solves the problem of changing from pinyin to Chinese character and correcting the text that was wrong due to speech.

In sum, this paper makes three significant contributions:

1. We subtly employ a convolutional encoder-decoder model to achieve state-of-art performance for Chinese speech recognition error correction. To our knowledge, ours is the first work to use full convolutional neural networks for end-to-end Chinese speech recognition error correction.
2. We add Pinyin as an additional feature to the convolutional encoder-decoder model. Besides, we design 8 kinds of Pinyin coding schemas and detailly report corresponding performance. Our work may guide further development in Pinyin coding.
3. In order to get better results, we propose a data augmentation method. Specifically, we train a reverse generative model, that input are good sentences while labels are bad sentences. The reverse generative model can learn how to transform good

sentences to bad sentences. And we verify 3-gram language model which can further reduce the error rate.

3 Error Correction Translation Model

3.1 Deep CNN Error Correction Model

The neural network error correction model in the continuous vector space after speech recognition sentences $\mathbf{x} = (x_1, x_2, \ldots, x_m)$ map to the target sentence after error correction $\mathbf{y} = (y_1, y_2, \ldots, y_n)$. Short sentences are filled with a special marker <PAD> to keep the same length, and the end of each sentence is marked with <EOS>. Considering that most speech errors are caused by local words in a sentence, we use convolutional neural network (CNN), which can better extract the local context information of a sentence compared with RNN. Given the window size, the convolutional network glided on the input sequence to calculate the local characteristics of the window size. The wider context information and information between distant of Chinese characters can also be captured by a multi-level convolutional hierarchy. In addition, when predicting target words, we also use an attention mechanism, which assigns weights to source words according to the correlation between source words and target words.

The deep convolutional neural error correction model uses the encoder-decoder model. The encoder network is used to encode error-prone source statements in a vector space, and the decoder network generates a corrected output statement by using the source code. Our model is based on an encoder and decoder architecture with multilayer convolution and attention mechanism. Network structure is described in detail below.

Given an input source statement $\mathbf{S}$ is composed of m segmentation $s_1, s_2, \ldots, s_i \ldots, s_m$ $s_i \in V_s$, where V_s is the source vocabulary, the last segmentation marker s_m is a special statement end tag, source statement $\mathbf{S}$ in the continuous space of the vector $s_1, s_2, \ldots, s_m$, $s_i \in \mathbb{R}^d$, $s_i = \mathbf{w}(s_i) + P(i)$, where $\mathbf{w}(s_i)$ is the embedded segmentation, $P(i)$ is the position of s_i in the source statement. These two kinds of embedding are obtained by training the embedding matrix together with other parameters of the network. The encoder and decoder are respectively composed of L layers. The network main frame is shown in Fig. 1. Source segmentation encode $s_1, s_2, \ldots, s_m$, by affine transformation for the first layer coding input $h_1^1, h_2^1, \ldots, h_m^1 \, h_i^1 \in \mathbb{R}^h$ h is all encoder decoder input/output vector dimensions. The affine transformation is obtained through formula (1).

$$h_i^1 = W s_1 + b \tag{1}$$

Where, the weight $W \in \mathbb{R}^{h \times d}$, bias $b \in \mathbb{R}^h$.

In the first coding layer, there are a number of 2 h convolution kernels with $3 \times h$ dimensions in windows size 3 to map the input vector to the eigenvector $f_i^2 \in \mathbb{R}^{2h}$. Adding <PAD> at the beginning and end of the source sentence (as shown in Fig. 1) is to preserve the output vector of the same length as the source sentence before the convolution operation.

$$f_i^2 = conv(h_{i-1}^1, h_i^1, h_{i+1}^1) \tag{2}$$

Where $conv(\cdot)$ represents the convolution operation. Then pass the gate linear unit (GLU):

$$\text{GLU}(f_i^2) = f_{i,1:h}^2 \circ \sigma\left(f_{i,h+1:2h}^2\right) \tag{3}$$

Where $\text{GLU}(f_i^2) \in \mathbb{R}^h$, $\circ$ Means multiply by elements, $\sigma(\cdot)$ Represents the sigmoid activation function, $f_{i,m:u}^2$ Represents f_i^2 from m to u elements. Finally, the input vector of the encoder layer is added as a residual join. The output vector of the l encoder layer is given by formula (4):

$$h_i^l = GLU(f_i^l) + h_i^{l-1} \quad i = 1, 2, \ldots, m \tag{4}$$

The last layer encoder for each of the output vector $h_i^L \in \mathbb{R}^h$ is obtained by linear mapping eventually encoder output vector $e_i \in \mathbb{R}^d$, as shown in (5):

$$e_i = W_e h_i^L + b_e \quad i = 1, 2, \ldots, m \tag{5}$$

Consider the decoding process of the decoder, Given $n - 1$ generated target segment, to decode the target segment t_n at time step n. First, <PAD> is added at the beginning of the target output, followed by <s>, They are embedded as $t_{-2}, t_{-1}, t_0, t_1, \ldots, t_{n-1}$ vector, the method of embedding is the same as that of source statement segmentation embedding. Each embedded $t_j \in \mathbb{R}^d$ is linearly mapped to $g_j^1 \in \mathbb{R}^h$ as input to the first decoding layer. Each decoding layer, like the encoding layer, contains a nonlinear gate unit GLU and a convolution operation with a window size 3, as shown in Eq. (6):

$$y_j^l = \text{GLU}\left(conv\left(g_{j-3}^{l-1}, g_{j-2}^{l-1}, g_{j-1}^{l-1}\right)\right) j = 1, 2, \ldots, n \tag{6}$$

Where g_j^{l-1} represents the output vector of the previous decoding layer. y_j^l represents the decoding state of the l layer decoder at the j time step, and the number and size of convolution kernels are the same as that of the encoder. Each layer decoder has an attention module that computes the attention at the time step n predicted by the l layer decoder. The decoder state $y_n^l \in \mathbb{R}^h$ is linearly mapped to a d-dimensional vector, as shown in Eq. (7):

$$z_n^l = W_z y_n^l + b_z + t_{n-1} \tag{7}$$

Where the weight is $\boldsymbol{W_z} \in \mathbb{R}^{d \times h}$ and the bias is $\boldsymbol{b_z} \in \mathbb{R}^d$, $\boldsymbol{t_{n-1}}$ Represents the target segmentation embedding of the previous time step, Attention weight $\alpha_{n,i}^l$ is obtained by dot product operation of output vectors $\boldsymbol{z_n^l}$ and $\boldsymbol{e_1}, \ldots, \boldsymbol{e_m}$ of the encoder output and then normalized through softmax, as shown in Eq. (8):

$$\alpha_{n,i}^l = \frac{\exp\left(\boldsymbol{e_i^T z_n^l}\right)}{\sum_{k=1}^m \exp\left(\boldsymbol{e_k^T z_n^l}\right)} i = 1, 2, \ldots, m \tag{8}$$

The context vector $\boldsymbol{x_n^l}$ weighted the encoder output vector and source shard embedding by attention weight to obtain formula (9). The additional source shard embedding can better retain the source shard information without being diluted after multi-layer coding.

$$\boldsymbol{x_n^l} = \sum_{i=1}^m \alpha_{n,i}^l (\boldsymbol{e_i} + \boldsymbol{s_i}) \tag{9}$$

The context vector $\boldsymbol{x_n^l}$ is linearly mapped to $\boldsymbol{c_n^l} \in \mathbb{R}^h$, and the output $\boldsymbol{g_n^l}$ of the l-layer decoder is obtained by adding $\boldsymbol{c_n^l}, \boldsymbol{y_n^l}$ and the previous layer's output vector $\boldsymbol{g_n^{l-1}}$, as shown in Eq. (10):

$$\boldsymbol{g_n^l} = \boldsymbol{y_n^l} + \boldsymbol{c_n^l} + \boldsymbol{g_n^{l-1}} \tag{10}$$

The final decoding layer output vector $\boldsymbol{g_n^L}$ is linearly mapped to $\boldsymbol{d_n} \in \mathbb{R}^d$. The Dropout [16] is used in each layer. The probability that the decoder output vector is finally mapped into the vector with the number of shards to get the target shard through softmax function, as shown in formula (11) and (12):

$$\boldsymbol{o_n} = \boldsymbol{W_o d_n} + \boldsymbol{b_o} \tag{11}$$

Where $\boldsymbol{W_o} \in \mathbb{R}^{|V_t| \times d}$, $\boldsymbol{b_o} \in \mathbb{R}^{|V_t|}$.

$$p(t_n = w_i | t_1, \ldots, t_{n-1}, S) = \frac{\exp\left(o_{n,i}\right)}{\sum_{k=1}^{|V_t|} \exp\left(o_{n,k}\right)} \tag{12}$$

w_i means the i-th token in the target token V_t.

We use the negative logarithmic likelihood loss function to train the model:

$$\mathcal{L} = -\frac{1}{N} \sum_{i=1}^N \frac{1}{T_i} \sum_{j=1}^{T_i} \log\left(p\left(t_{i,j} | t_{i,1}, \ldots, t_{i,j-1}, S\right)\right) \tag{13}$$

Where N is the number of training instances in batch processing, T_i is the number of segmentations in the i-th sentence, and $t_{i,j}$ is the j-th target segmentation in the target sentence of the i-th training instance.

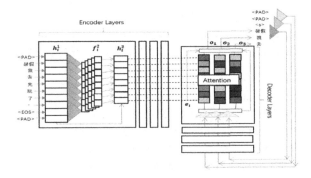

Fig. 1. Deep CNN architecture of four-layer encoder-decoder network

3.2 Pinyin Encoding

In Chinese speech recognition system, Chinese Pinyin as the main feature for speech to text conversion, Speech recognition error is also caused by Chinese pinyin error. There is evidence that the use of pinyin as an additional feature in Chinese-English translation of neural networks has an effect on translation performance [17]. So we put the Chinese Pinyin as feature of phonetic error correction. To study the impact of different pinyin coding schemes on error correction models, we designed different segmentation granularity: character segmentation, word segmentation. Different Pinyin coding schemes are designed according to different segmentation granularity. It be named as CPinyinCh, CPinyinChT, CInitials, CInitialsT, WPinyinCh WPinyinChT, WInitials, WInitialsT. these design in detail are:

CPinyinCh: Based on Chinese characters, each character of the pinyin string is encoded

CPinyinChT: Based on Chinese characters, each character of a pinyin string with tones is encoded.

CInitials: Based on Chinese characters, the phonetic alphabet is divided into initials and finals for encoding respectively.

CInitialsT: Based on Chinese characters, the pinyin with tone is divided into initial vowel and final vowel to be coded separately.

WPinyinCh: Token the sentence (using jieba) and encode each character of the word's pinyin string.

WPinyinChT: Token the sentence, and encode each character of the word's pinyin string with tone.

WInitials: Token the sentence, the phonetic strings of words are coded separately.

WInitialsT: Token the sentence, the phonetic strings of words with tones are coded separately.

The following are specific examples to illustrate different coding schemes (Table 2):

Table 2. Pinyin coding

Source sentence：内个人是我妈妈。   Token：内 个人 是 我 妈妈 。
CPinyinCh：n e i g e r e n s h i w o m a m a 。
CPinyinChT：n e i 4 g e 4 r e n 2 s h i 4 w o 3 m a 1 m a 1 。
CInitials：n e i g e r en sh i w o m a m a 。
CInitialsT：n e i 4 g e 4 r en 4 sh i 4 w o 3 m a 1 m a 1 。
WPinyinCh：n e i ｜ g e r e n｜s h i ｜ w o ｜ m a m a 。
WPinyinChT：n e i 4 ｜ g e 4 r e n 2 ｜ s h i 4 ｜ w o 3 ｜ m a 1 m a 1 。
WInitials：n ei ｜ g e r en ｜ sh i ｜ w o ｜ m a ｜ m a 。
WInitialsT：n ei 4 ｜ g e 4 r en 4 ｜ sh i 4 ｜ w o 3 ｜ m a 1 ｜ m a 1 。

We spliced the embedded $\mathbf{w}(\mathbf{s}_i)$ of $\mathbf{s}_i$ segmentation mentioned in Sect. 3.1 with the pinyin coding. The pinyin vector obtained by different pinyin coding schemes is denoted as $\mathbf{Pin}(\mathbf{s}_i)$, the neural network input is $\mathbf{s}_i = [\mathbf{w}(\mathbf{s}_i); \mathbf{Pin}(\mathbf{s}_i)] + P(i)$, where $P(i)$ is location coding.

4 Experiments

4.1 Datasets

We used two datasets, one is the WeChat speech transcription data set and the other is the public Chinese grammar correction (CGEC) data set. WeChat speech transcriptional data set collects WeChat speech and transcribes it into text through an open source speech recognition system ASRT[1]. The transcribed text is corrected by manual combination with recording. Chinese grammar correction data set is the data set given by NLPCC2018 Shared Task2 [18], which is the first parallel expected data set for Chinese grammar correction. Since Chinese grammatical errors contain spelling errors, the proposed model can be tested.

Among them, the WeChat transcribed data set contains 455,821 sentence pairs, which are randomly scrambled, and the data set is divided into three parts. The Validation set contained 60,772 sentence pairs, the training set contained 364,663 sentence pairs, and the test set contained 30,386 sentence pairs.

4.2 Evaluation

In the speech transcription data set of WeChat, because the errors mainly occurred in the local words of the sentence, we used the editing distance to measure. In other words, the sentences predicted by the model are converted into the same as tags after at least n times of deletion, addition and substitution of characters. The edit distance is n.

---

[1] https://github.com/nl8590687/ASRT_SpeechRecognition.

For the Chinese grammar error correction data set, the source sentence may have multiple [2]correct error correction results. Therefore, edit distance is not suitable for syntax correction. We use the MaxMatch(M2) score as the evaluation index [19]. M2 is widely used in the evaluation of grammar correction. The main idea is to calculate the matching degree of the model output and tags at the phrase level. A sentence is first segmented before evaluation is carried out on a set of sentences. The metrics measured at the testing stage are: Precision, Recall and $F_{0.5}$.

4.3   Experimental Configuration and Training Details

In the CNN error correction model, we extended the fairseq (See footnote 2) model based on pytorch. The specific parameters are as follows: the encoder and decoder use a stack of 5 layers of convolutional networks with windows size of 3. Each layer of the decoder has an attention mechanism. The number of layers is set according to the performance of the network in the validation set. The output of encoder and decoder of each layer is 512 dimensions, dropout is 0.2. We train the model on 4 NVIDIA 1080Ti GPUs. The training epoch time is about 1.5 h, and the beam width is set to 12 in the decoding stage.

4.4   Experimental Results and Analysis

Baseline refers to the error rate of speech recognition system, namely the ratio of the edit distance and the length of the corrected text after speech recognition and manual error correction, representing the effect of speech recognition system. "None" refers to no pinyin coding, only word (character) vector was used (Table 3).

Table 3. WeChat data set result

Model	Segmentation granularity	Number	Pinyin coding	Typo rate
Baseline	–	1	–	20.41%
Phrase translation model	phrase	2	–	18.82%
Deep CNN model	character	3	None	17.32%
		4	CPinyinCh	16.51%
		5	CPinyinChT	16.02%
		6	CInitials	17.16%
		7	CInitialsT	16.47%
Deep CNN model	word	8	None	16.85%
		9	WPinyinCh	16.47%
		10	WPinyinChT	16.27%
		11	WInitials	16.23%
		12	WInitialsT	16.18%

---

[2] https://github.com/pytorch/fairseq.

SMT NMT. First, we compared the previously proposed error correction model based on phrase translation. Our best experimental results have increased by 14.8% (line 2 and 5). According to our analysis, (1) the modeling ability of phrase translation error correction model based on statistics in translation tasks is not as good as that based on neural network, which also verifies that NMT proposed by Mahata [20] is superior to SMT in large data sets. In the industry, Google, baidu, youdao and others have migrated their translation models to NMT. (2) the phrase translation model first converts Chinese characters into pinyin for error correction, and then converts the corrected pinyin into Chinese characters. It is easy to make mistakes in translating pinyin into Chinese characters. We use word (character) vector and pinyin coding feature, which not only solve the problem of errors in speech recognition due to pinyin errors, but also directly "translate" Chinese characters to reduce the error of pinyin into Chinese characters.

Pinyin. In order to verify the validity of the proposed pinyin as an additional code, we designed a comparison experiment using only word (character) vectors and additional pinyin features. It is worth noting that the difference in the experimental results is due to the dimension of the input features. Differently caused (we tend to think that the larger the feature vector dimension is, the better the effect), we train the different dimensional word vectors (350-D and 300-D) in the same corpus, we splicing the 50-D Pinyin vector into 300-D. The dimensions of the different word vectors are guaranteed to be the same. The experimental results show that the Pinyin coding has a significant improvement effect under different granularity divisions, which fully proves our idea: in the speech recognition text error correction task, the Pinyin feature is important.

Tone. In deep CNN model, among different encoding schemes, the encoding scheme with tone is always significantly better than the encoding scheme without tone (compared with the relative improvement of 2.9% in line 5 and 4). It is not difficult to understand that tone plays a very important role in putonghua. The commonly used Chinese characters are about 3000 words, and there are about 400 syllables without tone. On average, one syllable corresponds to about 7 Chinese characters, which is very easy to be confused.

Word Character. The best experimental result of segmentation is that in the case of character segmentation, the word segmentation effect of the word splitter is not good when there are spelling errors in the source sentences, which increases the difficulty of error correction. The errors in the source sentences are mainly errors of local words, and word segmentation will lead to "distorted" results. Since most Chinese characters are misspelled, it seems more reasonable to use Chinese characters for segmentation.

In the Chinese grammar correction data set, the results in the test set are shown in Table 4.

Table 4. Chinese grammar correction dataset results

Model	Number	P	R	F0.5
CNN	1	21.28	11.36	18.11
CNN + CPinyinChT	2	21.61	11.49	18.37
CNN + WInitialsT	3	21.84	11.69	18.61

In the data set of Chinese grammar correction, our purpose is not to get the SOTA model, but to verify that adding pinyin features is also helpful for grammar correction, that is, pinyin features have certain generalization ability. CNN refers to the deep CNN model with attention mechanism that only use word vectors as features. In the second line, word vectors and pinyin alphabet coding schemes are used as features. In the third line, word vectors and vowel codes are used as features. Comparing lines 1 and 2, lines 1 and 3, we can see that the scheme with the addition of Pinyin coding improves the effect to varying degrees. We see that the finer-grained word segmentation scheme is worse than the word segmentation scheme. We analyze it is believed that Chinese grammar correction is more focused on a higher level of semantic level, while speech text correction is more focused on lower level word levels, words are the smallest morpheme units, and word segmentation reduces the length of sentences. Chinese grammar correction also contains spelling errors but mainly grammatical errors. Therefore, in Chinese grammar correction, models that achieve better results [21, 22] use word vectors instead of character vectors.

4.5   Boosting

Data Augmentation. Theoretically, on large data sets, the neural network error correction model will have better results, and expanding the WeChat transcribed data set is expensive. We propose a data enhancement method, since we regard the error correction task as a "translation" task. We can train a reverse model, that is, the model input is the corrected sentence. The tag is a sentence with errors, and the error sentence generated by the reverse model is added to the original data set, and the original data set is expanded to 655821 pairs.

Language Model (LM). In order to further improve the effect of our proposed error correction model, we improve the error correction capability by adding a 3-gram-based language model error corrector. The language model is widely used in Chinese spell checking. We use the kenlm[3] language model training tool to train the 3-gram speech model on the Baidu Encyclopedia dataset. First, the jieba tokenizer is used to segment the sentence. The sentence contains spelling errors. As a result, there is a case where a segmentation error occurs, so errors are detected from both the word granularity and the character granularity, and the two granularity-like erroneous results are integrated, and the error position candidate set is obtained, and then all the suspected error positions are traversed, and the sound shape dictionary is replaced. The wrong position local word, the sentence confusion score is calculated by training the good speech model, and the candidate set results are sorted to obtain the optimal word.

The experimental effects on the WeChat speech transcription data set by adding various schemes are shown in Table 5.

---

[3] https://github.com/kpu/kenlm.

Table 5. Boosting result

Model	Typo rate
CNN CPinyinChT	16.02%
CNN CPinyinChT + Data Augmentation	15.21%
CNN CPinyinChT + Data Augmentation + Language model	15.07%

Experimental analysis showed that after adding data augmentation and 3-gram language model, the model was significantly improved. Finally, our experimental result showed that the error rate was 15.07%, which was 26.2% lower than Baseline.

5 Summary

In this paper, a text error correction model based on pinyin is proposed. Compared with the Chinese speech recognition error correction algorithm based on phrase translation model proposed previously, our model is improved by 20%. On WeChat transcribed dataset, our model makes the typo rate reduced 26%, we verify the effectiveness of the pinyin as coding, at the same time, we confirmed the tones in the importance of the speech recognition, confirmed in speech text correction, fine-grained character segmentation is better than word segmentation. We propose a reverse-transformed data augmentation method that incorporates a language model and eventually achieves a good level. Future work integrates word (character) vector and pinyin coding to further enhance the model's effects.

References

1. Ng, H.T., et al.: The CoNLL-2013 shared task on grammatical error correction. In: Proceedings of the Eighteenth Conference on Computational Natural Language Learning: Shared Task, pp. 15–23(2013)
2. Brockett, C., Dolan, W.B., Gamon, M.: Correcting ESL errors using phrasal SMT techniques. In: Proceedings of the 21st International Conference on Computational Linguistics and the 44th annual meeting of the Association for Computational Linguistics. Association for Computational Linguistics, pp. 249–256 (2006)
3. Susanto, R.H., Phandi, P., Ng, H.T.: System combination for grammatical error correction. In: Proceedings of the 2014 Conference on Empirical Methods in Natural Language Processing (EMNLP), pp. 951–962 (2014)
4. Chollampatt, S., Ng, H.T.: Connecting the dots: towards human-level grammatical error correction. In: Proceedings of the 12th Workshop on Innovative Use of NLP for Building Educational Applications, pp. 327–333 (2017)
5. Yuan, Z., Briscoe, T.: Grammatical error correction using neural machine translation. In: Proceedings of the 2016 Conference of the North American Chapter of the Association for Computational Linguistics: Human Language Technologies (2016)
6. Ji, J., Wang, Q., Toutanova, K., et al.: A nested attention neural hybrid model for grammatical error correction. arXiv preprint arXiv:1707.02026 (2017)

7. Chollampatt, S., Ng, H.T.: A multilayer convolutional encoder-decoder neural network for grammatical error correction. arXiv preprint arXiv:1801.08831 (2018)

8. Junczys-Dowmunt, M., Grundkiewicz, R., Guha, S., et al.: Approaching neural grammatical error correction as a low-resource machine translation task. arXiv preprint arXiv:1804.05940 (2018)

9. Grundkiewicz, R., Junczys-Dowmunt, M.: Near human-level performance in grammatical error correction with hybrid machine translation. arXiv preprint arXiv:1804.05945 (2018)

10. Yu, L.C., Lee, L.H., Chang, L.P.: Overview of grammatical error diagnosis for learning Chinese as a foreign language. In: Proceedings of the 1st Workshop on Natural Language Processing Techniques for Educational Applications (NLP-TEA 2014), pp. 42–47 (2014)

11. Zheng, B., Che, W., Guo, J., et al.: Chinese grammatical error diagnosis with long short-term memory networks. In: Proceedings of the 3rd Workshop on Natural Language Processing Techniques for Educational Applications (NLPTEA2016), pp. 49–56 (2016)

12. Xie, P.: Alibaba at IJCNLP-2017 task 1: embedding grammatical features into LSTMs for chinese grammatical error diagnosis task. In: Proceedings of the IJCNLP 2017, Shared Tasks, pp. 41–46 (2017)

13. Tan, Y., Yang, Y., Yang, L., et.al.: Grammatical Error correction using LSTM and N-gram. J. Chin. Inform. Process., 32(6), 19–27 (2018)

14. Hu, Y., Liu, Y., Yang, H., et al.: An online system for chinese query correction in search engine. J. Chin. Inform. Process. 30(1), 71–79 (2016)

15. Wu, L.: Chinese speech recognition results correction based on phrase-based statistical machine translation model. In: The 14th National Conference on Human-Computer Voice Communication, p. 6 (2017)

16. Srivastava, N., Hinton, G., Krizhevsky, A., et al.: Dropout: a simple way to prevent neural networks from overfitting. J. Mach. Learn. Res. 15(1), 1929–1958 (2014)

17. Du, J., Way, A.: Pinyin as subword unit for chinese-sourced neural machine translation. In: AICS, pp. 89–101 (2017)

18. Zhao, Y., Jiang, N., Sun, W., Wan, X.: Overview of the NLPCC 2018 shared task: grammatical error correction. In: Zhang, M., Ng, V., Zhao, D., Li, S., Zan, H. (eds.) NLPCC 2018. LNCS (LNAI), vol. 11109, pp. 439–445. Springer, Cham (2018). https://doi.org/10.1007/978-3-319-99501-4_41

19. Dahlmeier, D., Ng, H.T.: Better evaluation for grammatical error correction. In: Proceedings of the 2012 Conference of the North American Chapter of the Association for Computational Linguistics: Human Language Technologies. Association for Computational Linguistics, pp. 568–572 (2012)

20. Mahata, S.K., Mandal, S., Das, D., et al.: SMT vs NMT: a comparison over Hindi & Bengali simple sentences. arXiv preprint arXiv:1812.04898 (2018)

21. Ren, H., Yang, L., Xun, E.: A sequence to sequence learning for Chinese grammatical error correction. In: Zhang, M., Ng, V., Zhao, D., Li, S., Zan, H. (eds.) NLPCC 2018. LNCS (LNAI), vol. 11109, pp. 401–410. Springer, Cham (2018). https://doi.org/10.1007/978-3-319-99501-4_36

22. Zhou, J., Li, C., Liu, H., Bao, Z., Xu, G., Li, L.: Chinese grammatical error correction using statistical and neural models. In: Zhang, M., Ng, V., Zhao, D., Li, S., Zan, H. (eds.) NLPCC 2018. LNCS (LNAI), vol. 11109, pp. 117–128. Springer, Cham (2018). https://doi.org/10.1007/978-3-319-99501-4_10

Neural Gender Prediction from News Browsing Data

Chuhan Wu[1(✉)], Fangzhao Wu[2], Tao Qi[1], Yongfeng Huang[1], and Xing Xie[2]

[1] Department of Electronic Engineering, Tsinghua University, Beijing 100084, China
{wu-ch19,qit16,yfhuang}@mails.tsinghua.edu.cn
[2] Microsoft Research Asia, Beijing 100080, China
{fangzwu,xing.xie}@microsoft.com

Abstract. Online news platforms have attracted massive users to read digital news online. The demographic information of these users such as gender is critical for these platforms to provide personalized services such as news recommendation and targeted advertising. However, the gender information of many users in online news platforms is not available. Fortunately, male and female users usually have different pattern in reading online news. Thus, the news browsing data of users can provide useful clues for inferring their genders. In this paper, we propose a neural gender prediction approach based on the news browsing data of users. Usually a news article has different kinds of information such as title, body and categories. However, the characteristics of these components are very different, and they should be processed differently. Thus, we propose to learn unified user representations for gender prediction by incorporating different components of browsed news as different views of users. In each view, we use a hierarchical framework to first learn news representations and then learn user representations from news representations. In addition, since different words in news titles and bodies usually have different informativeness for learning news representations, we use attention mechanisms to select important words. Besides, since different news articles may also have different informativeness for gender prediction, we use news-level attentions to attend to important news articles for learning informative user representations. Extensive experiments on a real-world dataset validate the effectiveness of our approach.

Keywords: Gender prediction · News browsing · Multi-view learning · Attention mechanism

1 Introduction

Online news platforms such as Google News have attracted massive users for online news reading [4]. The demographics of users such as gender are very important for these platforms to provide personalized services to their users [3]. For example, with the gender information of users, online news platforms can make more personalized new recommendations to them [23], e.g., recommending

© Springer Nature Switzerland AG 2019
M. Sun et al. (Eds.): CCL 2019, LNAI 11856, pp. 664–676, 2019.
https://doi.org/10.1007/978-3-030-32381-3_53

NBA news to male users and fashion news to female users. In addition, advertisers can display their ads to users on news platforms more effectively [22]. For example, advertisers can display ads of dress and makeups to female users, and ads of shaver and tie to male users. Without the gender information of users, advertisers may display the ads of dress to both female and male users, which may be less effective [29]. Thus, the gender information of users is critical for news platforms to provide personalized services [15].

However, the gender information of many users in online news platforms is not available, making it difficult to provide personalized services for them [24]. Luckily, many users click and browse news articles displayed on these platforms, and there are usually some differences in news reading patterns between male and female users. For example, as illustrated in Fig. 1, a female user may browse a news article about dressing fashion styles, while a male user may browse a news article about NBA. Thus, the news browsing data of users can provide useful clues for inferring their genders. In this paper, we explore the problem of gender prediction of users on news platforms based on their browsed news.

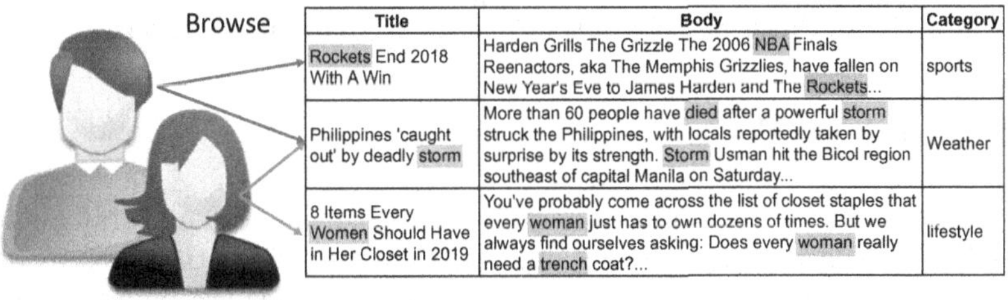

Title	Body	Category
Rockets End 2018 With A Win	Harden Grills The Grizzle The 2006 NBA Finals Reenactors, aka The Memphis Grizzlies, have fallen on New Year's Eve to James Harden and The Rockets...	sports
Philippines 'caught out' by deadly storm	More than 60 people have died after a powerful storm struck the Philippines, with locals reportedly taken by surprise by its strength. Storm Usman hit the Bicol region southeast of capital Manila on Saturday...	Weather
8 Items Every Women Should Have in Her Closet in 2019	You've probably come across the list of closet staples that every woman just has to own dozens of times. But we always find ourselves asking: Does every woman really need a trench coat?...	lifestyle

Fig. 1. Several example news articles browsed by a female user and a male user. Important words are highlighted using color bars. (Color figure online)

Our work is inspired by several observations. First, a news article usually contains different kinds of information such as title, body and topic categories, which are all useful for inferring the gender of users. For example, as shown in Fig. 1, the title of the third news indicates that this news is about fashions, and the body of this news contains the details. Moreover, the topic categories of news articles are also informative for gender prediction. For instance, if a user frequently browses news about lifestyles and fashions but rarely browses sports news, then we can infer this user is probably a female user. Thus, incorporating different fields of news has the potential to learn more accurate user representations for gender prediction. Second, the characteristics of different news fields are very different. For example, news titles are usually short and concise sentences, while news bodies are usually long documents with detailed information, and topic categories are usually tags. Thus, they should be handled differently. Third, different words in the same news usually have different informativeness in learning news representations. For example, in Fig. 1 the word "Woman" is more informative than the word "2019" for learning informative news representations.

In addition, different news articles browsed by the same user may also have different informativeness in learning user representations for gender prediction. For example, the third news in Fig. 1 is more informative than the second news in inferring the gender of this user.

In this paper, we propose a neural gender prediction approach to infer the genders of online news platform users from their news browsing data. In our approach, we propose to learn unified user representations via a multi-view learning framework from the titles, bodies and topic categories of their browsed news articles as different views of these users. In each view, we use a hierarchical model to first learn hidden news representations, and then learn user representations from representations of browsed news. In addition, since different words in the same news title or body usually have different informativeness in representing this news, we apply attention mechanism to select important words. Besides, since different news articles may also have different informativeness in learning user representation for gender prediction, we apply news-level attention networks in each view to select important news articles. Extensive experiments are conducted on a real-world dataset, and the results show that our approach can achieve satisfactory performance in predicting the genders of online news platform users and consistently outperform many baseline methods.

2   Related Work

User demographic prediction is an important task in both natural language processing and data mining fields, and has attracted wide attentions in recent years [12]. Existing user demographic prediction methods are mainly based on online behaviors and user generated content such as blogs, forum posts and social media messages [6,8,11,13,14,16–19,25,31]. For example, Rosenthal and McKeown [25] proposed to use logistic regression to infer user ages from blogs. They used many kinds of features including blog content features, stylistic features, behavior features and user interest features. Nguyen et al. [17] proposed to use linear regression with Lasso regularization to predict user ages. They used several different kinds of texts, such as blogs, forum posts and transcribed telephone speeches to build user representations. Peersman et al. [19] applied SVM to predict genders of Twitter users based on word n-gram and character n-gram features extracted from their tweets. However, these methods cannot effectively utilize contextual information. In addition, these methods cannot distinguish informative contexts and records from uninformative ones.

In recent years, several deep learning based methods have been proposed for user demographic prediction [1,2,5,26–29,31]. For example, Zhang et al. [31] used LSTMs to predict the genders and ages of social media users based on their microblogging messages, retweeted messages, comments from others and the comments to others. Wang et al. [28] proposed a CNN based method to jointly predict the ages and genders of social media users based on their messages in a collaborative manner. Wu et al. [29] proposed a hierarchical user representation model to predict the ages and genders of users in commercial search engines based on their search queries. Different from these methods which rely on blogs, social media data or search queries, our approach predicts the genders of users on news platforms based on their news browsing data. In addition, existing methods for demographic prediction usually aggregate different kinds of user information together and ignore the differences of their characteristics, which may be suboptimal for user demographic prediction.

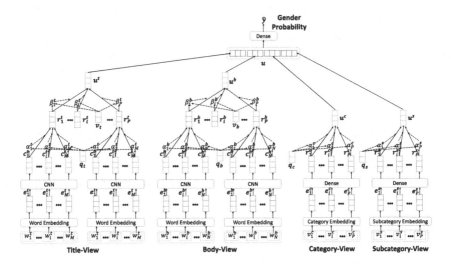

Fig. 2. The framework of our approach.

There are only a few approaches to predict user demographics based on news browsing data [8,21]. For example, Phuong et al. [21] proposed to use SVM to predict genders of users based on their news website browsing histories. They used various hand-crafted features such as news categories, topic features extracted by LDA, access time, and sequential features extracted by webpage k-grams. However, these methods cannot effectively utilize the contexts and orders of words in news, which are very important for gender prediction. Different from all the aforementioned methods, our approach predicts the genders of users on news platforms based on their news browsing use a multi-view learning framework to learn unified user representations by regarding different kinds of news information as different views of users. Experiments on benchmark dataset show our approach can outperform these baseline methods.

3   Our Approach

In this section, we introduce our approach in detail. There are two major modules in our model. The first one is *user representation*. Which learns representations of users from their browsed news. The second one is *gender classification*, which classifies the genders of users according to their representations. In the *user representation* module, we use a multi-view learning framework to learn unified user representations by incorporating news titles, bodies and topic categories as different views of users. The framework of our approach is shown in Fig. 2.

3.1   User Representations from Title View

The first view in the *news representation* module is *title-based user representation*. It is used to learn representations of users from the titles of their browsed news. In this module, we first use a *word encoder* to learn title representations from words, and then use a *title encoder* to learn title-based user representations from news titles. As shown in Fig. 2, there are three layers in the *word encoder*.

The first one is word embedding. It is used to convert a news title with M words from a word sequence $[w_1^t, w_2^t, ..., w_M^t]$ into a vector sequence $[\mathbf{e}_1^t, \mathbf{e}_2^t, ..., \mathbf{e}_M^t]$.

The second layer is a convolutional neural network (CNN) [9]. Local contexts within a news title are very important for understanding this news. For example, in the title of the first news in Fig. 1, the local contexts of "deadly" such as "storm" is very useful for learning accurate representations of this title. Thus, we employ a CNN network at word-level to learn contextual word representations by capturing local contexts. It takes the aforementioned embedding sequence as input, and outputs a contextual word representation sequence $[\mathbf{c}_1^t, \mathbf{c}_2^t, ..., \mathbf{c}_M^t]$.

The third layer is a word-level attention network. Usually, different words in the same news title may have different informativeness in representing this news. For example, in the title of the third news in Fig. 1 the word "woman" is much more informative than "Items" for learning gender discriminative news representations. Thus, we propose to apply attention mechanism at word-level to select important words for learning informative news representations. The attention weight α_i^t of the i-th word in the same news title is formulated as:

$$a_i^t = \mathbf{q}_t \times \tanh(\mathbf{U}_t \times \mathbf{c}_i^t + \mathbf{u}_t), \quad \alpha_i^t = \frac{\exp(a_i^t)}{\sum_{j=1}^{M} \exp(a_j^t)}, \tag{1}$$

where $\mathbf{V}_t$ and $\mathbf{v}_t$ are linear transformation parameters, $\mathbf{q}_t$ is the attention query vector. The final title representation is the summation of the word contextual representations weighted by their attention weights, i.e., $\mathbf{r}^t = \sum_{j=1}^{M} \alpha_j^t \mathbf{c}_j^t$.

In the *title encoder* module, we learn title-based user representations based on the representations of browsed news titles. The titles of different news usually have different informativeness for learning user representations for gender prediction. For example, the title of the first and third news are very informative for learning gender discriminative user representations, while the second news is uninformative since it is widely browsed by both male and female users. Thus, we propose to use a news-level attention network to select important news for learning informative user representations. Denote the attention weight on the title of the i-th news browsed by a user as β_i^t, which is calculated as follows:

$$b_i^t = \mathbf{v}_t \times \tanh(\mathbf{W}_t \times \mathbf{r}_i^t + \mathbf{w}_t), \quad \beta_i^t = \frac{\exp(b_i^t)}{\sum_{j=1}^{P} \exp(b_j^t)}, \tag{2}$$

where P is the number of browsed news, $\mathbf{W}_t$, $\mathbf{w}_t$ and $\mathbf{v}_t$ are parameters. The final representation of a user in this view is the summation of the title representations weighted by their attention weights, i.e., $\mathbf{u}^t = \sum_{i=1}^{P} \beta_i^t \mathbf{r}_i^t$.

3.2   User Representations from Body View

The second view is *body-based user representation*, which is used to learn user representations from the bodies of browsed news.

Similar with the title view, there are also two major components in this view, i.e., a *word encoder* to learn body representations from words, and a *body encoder* to learn user representations from news bodies. The *word encoder* also has three layers, i.e., a shared word embedding layer to convert the word sequence (denoted as $[w_1^b, w_2^b, ..., w_N^b]$, where N is the number of words) of a news body into a vector sequence, a word-level CNN network to learn contextual word representations (denoted as $[\mathbf{c}_1^t, \mathbf{c}_2^t, ..., \mathbf{c}_M^t]$) by capturing local contexts in news bodies, and a word-level attention network to select important words for learning informative representations of news bodies. Denote the bodies of the news browsed by users as $[b_1, b_2, ..., b_P]$. We apply the *word encoder* in this view to each body to obtain the hidden representation sequence $[\mathbf{r}_1^b, \mathbf{r}_2^b, ..., \mathbf{r}_P^b]$ of news bodies.

The *body encoder* is used to learn representations of users from the body representations of their browsed news. Different news bodies may also have different informativeness for learning user representations for gender prediction. For example, the body of the first news in Fig. 1 is more informative than the second one in inferring the gender of a user. Thus, we apply a news-level attention network to select important news based on their body representations. Denote the attention weight of the i-th body as β_i^b, which is computed as:

$$b_i^b = \mathbf{v}_b \times \tanh(\mathbf{W}_b \times \mathbf{r}_i^b + \mathbf{w}_b), \quad \beta_i^b = \frac{\exp(b_i^b)}{\sum_{j=1}^{P} \exp(b_j^b)}, \tag{3}$$

where $\mathbf{v}_b$, $\mathbf{W}_b$ and $\mathbf{w}_b$ are parameters. The final news body based representation of a user is the summation of the contextual body representations weighted by their attention weights, i.e., $\mathbf{u}^b = \sum_{j=1}^{P} \beta_j^b \mathbf{r}_j^b$.

3.3   User Representations from Category/Subcategory View

The third view is *category-based user representations*, which is used to learn user representations from the topic categories of browsed news. On many online news platforms such as MSN News, a news article is classified into a general topic category (e.g., "sports") and a finer-grained subcategory (e.g., "basketball_nba") to target user interests more effectively. Usually, news categories are important clues for gender prediction. For example, news articles about fashions are more likely to be clicked by female users than male users. Thus, we propose to incorporate the categories and subcategories of news to learn gender discriminative user representations. There are three layers in this module.

The first one is category/subcategory embedding. Denote the input category ID sequence as $[v_1^c, v_2^c, ..., v_P^c]$ and the subcategory ID sequence as $[v_1^s, v_2^s, ..., v_P^s]$. We use a category/subcategory embedding layer to convert the sequences of discrete category and subcategory IDs into sequences of low-dimensional vectors, which are respectively denoted as $[e_1^c, e_2^c, ..., e_P^c]$ and $[e_1^s, e_2^s, ..., e_P^s]$.

The second one is a dense layer. It is used to learn hidden representations of categories and subcategories as follows:

$$\mathbf{r}_i^c = \text{ReLU}(\mathbf{W}_c \times \mathbf{e}_i^c + \mathbf{w}_c), \quad \mathbf{r}_i^s = \text{ReLU}(\mathbf{W}_s \times \mathbf{e}_i^s + \mathbf{w}_s), \tag{4}$$

where $\mathbf{W}_c$, $\mathbf{w}_c$, $\mathbf{W}_s$ and $\mathbf{w}_s$ are linear transformation parameters.

The third one is a news-level attention network. Different categories and subcategories usually have different importance for gender prediction. For example, the news in the "sports" category is more informative than the news in the "weather" category, since the latter one is usually browsed by both male and female users. Thus, we apply a news-level attention network to select important news according to their topic categories. Denote the attention weight of the i-th topic category as α_i^c, which is calculated as:

$$a_i^c = \mathbf{q}_c \times \tanh(\mathbf{V}_c \times \mathbf{r}_i^c + \mathbf{v}_c), \quad \alpha_i^c = \frac{\exp(a_i^c)}{\sum_{j=1}^P \exp(a_j^c)}, \tag{5}$$

where $\mathbf{q}_c$, $\mathbf{V}_c$ and $\mathbf{v}_c$ are parameters. The category based user representations are the summation of the hidden category representations weighted by their attention weights, i.e., $\mathbf{u}^c = \sum_{j=1}^P \alpha_j^b \mathbf{r}_j^c$. The subcategory based user representations $\mathbf{u}^s$ can be computed in a similar way. The final unified user representations is the concatenation of the user representations learned from different views, i.e., $\mathbf{u} = [\mathbf{r}^c; \mathbf{r}^{sc}; \mathbf{r}^t; \mathbf{r}^b]$.

3.4   Gender Classification

The *gender classifier* is used to predict the probability $\hat{\mathbf{y}}$ of a user in different gender categories from his/her representations, which is computed by: $\hat{\mathbf{y}} = softmax(\mathbf{W}_y \times \mathbf{u} + \mathbf{w}_y)$, where $\mathbf{W}_y$ and $\mathbf{w}_y$ are parameters. The loss function we use is cross entropy, which is formulated as follows:

$$\mathcal{L} = -\frac{1}{N_g} \sum_{i=1}^{N_g} \sum_{k=1}^{K_g} y_{i,k} \log(\hat{y}_{i,k}), \tag{6}$$

where N_g is the number of user with gender labels, K_g is the number of gender categories, $y_{i,k}$ and $\hat{y}_{i,k}$ respectively denote the ground-truth and predicted probability of the i-th user in the k-th gender category.

4   Experiments

4.1   Datasets and Experimental Settings

Since there is no off-the-shelf dataset for news based gender prediction study, we built one by crawling 10,000 users from MSN News[1]. Among them, the gender information of 4,228 users is available which were used for our experiments. We also collected their news browsing histories in a month, i.e., from Dec. 13, 2018 to Jan. 12, 2019. The statistics of our dataset are shown in Table 1. We randomly sampled 80% of users for training, 10% for validation, and 10% for test.

Table 1. Statistics of our dataset.

# male users	2,484	avg. # words per news title	11.29
# female users	1,744	avg. # words per news body	730.72
# categories	15	# subcategories	284

In our experiments, we used the 300-dimensional pre-trained Glove embedding [20] to initialize the word embeddings. The CNN networks had 400 filters, and their window size was 3. The dimensions of attention query vectors were 200. The optimizer we used was Adam [10]. We added 20% dropout to each layer. The batch size was set to 50. These hyperparameters were tuned on the validation set. We independently repeated each experiment 10 times and reported the average results in terms of accuracy and macro F1-score.

4.2   Performance Evaluation

We evaluate the performance of our approach by comparing it with several baseline methods, including: (1) *LinReg* [17], linear regression with Lasso regularization. (2) *SVM* [19], support vector machine; (3) *LR* [25], logistic regression; (4) *CNN* [9], convolutional neural network. (5) *LSTM* [7], long short-term memory network. (6) *CNN-Att*, CNN with attention mechanism. (7) *LSTM-Att*, LSTM with attention mechanism. (8) *HAN* [30], a hierarchical attention network for document classification. (9) *HURA* [29], a hierarchical attention network for document classification. (10) *Ours*, our neural gender prediction approach with news browsing data. In traditional methods (1-3), we used features including news category/subcategory IDs and TF-IDF features extracted from news titles and bodies as the input. In baseline methods based on neural networks (4-9), we used the combination of news titles, bodies, categories and subcategories by aggregating them into a long document. The results of these methods are reported in Table 2. According to the results, we have several observations.

First, the methods based on neural networks outperform traditional methods such as *SVM*, *LR* and *Linreg*. This is probably because neural network based

---

[1] https://www.msn.com/en-us/news.

methods can learn more accurate user representations for gender prediction. In addition, traditional methods usually rely on bag-of-words features, while the contexts and orders of words cannot be fully captured. Second, the methods using attention mechanism usually outperform the methods without mechanism. This is probably because different words and news usually have different informativeness for learning user representations, and selecting important contexts in news can benefit user representation learning. Third, the methods based on hierarchical models (*HAN*, *HURA* and *Ours*) outperform methods based on flatten models (e.g., *CNN-Att* and *LSTM-Att*). This may be because learning user representations in a hierarchical manner can utilize the document structures of news, which can benefit news and user representation learning. Fourth, our approach can outperform other baselines such as *HAN* and *HURA*. This is probably because different kinds of news information have different characteristics, and aggregating different news fields is usually not optimal. In our approach we use a multi-view learning framework to incorporate titles, bodies and categories as different views of news, which can learn better user representations.

Table 2. The performance of different methods under different ratios of training data.

Method	25%		50%		100%	
	Accuracy	Fscore	Accuracy	Fscore	Accuracy	Fscore
Linreg [17]	63.89 ± 1.14	62.54 ± 1.19	65.16 ± 1.21	63.86 ± 1.22	66.22 ± 1.22	65.37 ± 1.24
SVM [19]	66.10 ± 0.78	64.98 ± 0.79	66.85 ± 0.82	66.23 ± 0.84	67.26 ± 0.88	66.89 ± 0.90
LR [25]	65.26 ± 0.61	63.89 ± 0.63	66.43 ± 0.66	65.67 ± 0.67	67.33 ± 0.71	66.84 ± 0.72
CNN [9]	66.22 ± 0.87	65.33 ± 0.89	67.35 ± 0.77	66.78 ± 0.77	68.12 ± 0.66	67.68 ± 0.68
LSTM [7]	66.11 ± 0.74	65.05 ± 0.75	67.23 ± 0.61	66.71 ± 0.62	67.99 ± 0.54	67.60 ± 0.55
CNN-Att	66.84 ± 0.68	65.76 ± 0.70	68.01 ± 0.62	67.26 ± 0.62	68.79 ± 0.53	68.32 ± 0.55
LSTM-Att	66.57 ± 0.73	65.44 ± 0.77	67.67 ± 0.64	67.01 ± 0.65	68.45 ± 0.56	68.03 ± 0.57
HAN [30]	67.22 ± 0.58	65.89 ± 0.60	68.54 ± 0.47	68.03 ± 0.49	69.88 ± 0.42	69.67 ± 0.44
HURA [29]	67.95 ± 0.64	66.66 ± 0.66	69.77 ± 0.52	68.89 ± 0.55	70.45 ± 0.41	70.01 ± 0.43
Ours*	$\mathbf{70.03 \pm 0.53}$	$\mathbf{68.89 \pm 0.55}$	$\mathbf{71.28 \pm 0.41}$	$\mathbf{70.78 \pm 0.42}$	$\mathbf{72.25 \pm 0.37}$	$\mathbf{71.89 \pm 0.39}$

*Our approach v.s. baselines significantly different at $p < 0.01$

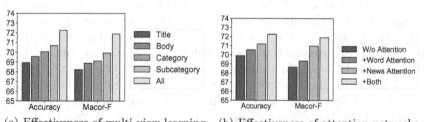

(a) Effectiveness of multi-view learning.   (b) Effectiveness of attention networks.

Fig. 3. Effectiveness of the multi-view learning framework and attention networks.

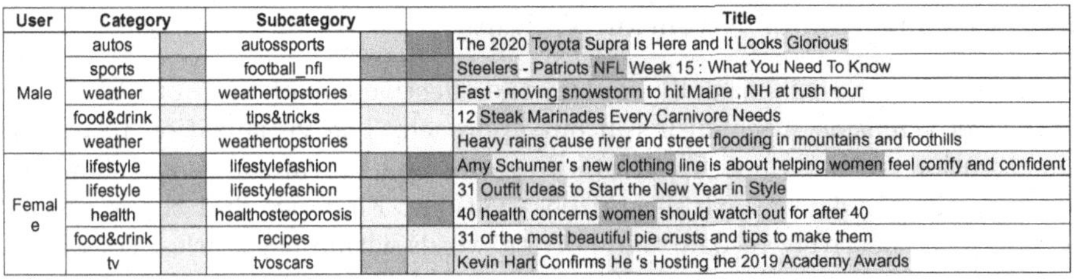

User	Category	Subcategory	Title
Male	autos	autossports	The 2020 Toyota Supra Is Here and It Looks Glorious
	sports	football_nfl	Steelers - Patriots NFL Week 15 : What You Need To Know
	weather	weathertopstories	Fast - moving snowstorm to hit Maine , NH at rush hour
	food&drink	tips&tricks	12 Steak Marinades Every Carnivore Needs
	weather	weathertopstories	Heavy rains cause river and street flooding in mountains and foothills
Female	lifestyle	lifestylefashion	Amy Schumer 's new clothing line is about helping women feel comfy and confident
	lifestyle	lifestylefashion	31 Outfit Ideas to Start the New Year in Style
	health	healthosteoporosis	40 health concerns women should watch out for after 40
	food&drink	recipes	31 of the most beautiful pie crusts and tips to make them
	tv	tvoscars	Kevin Hart Confirms He 's Hosting the 2019 Academy Awards

Fig. 4. Visualization of the word- and news-level attention weights on the titles, categories and subcategories of several news browsed by a randomly selected male and female users. Darker colors represent higher attention weights. (Color figure online)

4.3 Model Effectiveness

In this section, we will explore the effectiveness of different components of our approach. First, we want to verify the effectiveness of multi-view learning framework in our approach. We compare the performance of our approach and its variants with different combinations of news views, and the results are shown in Fig. 3(a). According to Fig. 3(a), we have several observations. First, we find the category view or subcategory view are very important for our approach. This is intuitive because the topics of news browsed by male and female users usually have huge differences. Thus, incorporating the information of news topic categories can learn more gender discriminative user representations. In addition, we find the model with the subcategory view slightly outperform that with the category view. This may be because subcategories usually contain finer-grained topic information than categories. Second, the model with the body view can achieve better performance than that with the title view only. This is probably because news bodies usually contain the details of news, which can provide richer information than news titles for learning informative news representations. Third, combining all four views can further improve the performance of our approach. These results validate the effectiveness of our multi-view learning framework.

Then, we conducted experiments to validate the influence of the word-level and news-level attention mechanism on our approach. The performance of our approach and its variant with different combinations of attention networks is shown in Fig. 3(b). From Fig. 3(b), we have several observations. First, the news-level attention network has an important impact on our approach. This is probably because different news usually have different informativeness for learning user representations, and selecting important news can benefit user representation learning. Second, the word-level attention networks are also important for our approach. This is probably because different words usually have different informativeness for learning informative news and user representations for gender prediction, and our approach can attend to important words in news titles and bodies. Third, combining both kinds of attention networks can further improve the performance of our approach. These results validate the effectiveness of attention mechanism in our approach.

4.4 Visualization of Attention Weights

In this section, we will visually explore the effectiveness of the word-level and news-level attention mechanisms in our approach. The visualization results of the attention weights on news titles, categories and subcategories are shown in Fig. 4. From Fig. 4, we have several observations. First, we find the attention network can effectively recognize important words within news titles. For example, the words "Toyota" and "NFL" are assigned high attention weights, since these words are very informative for learning gender discriminative news and user representations, while the words such as "2020" and "Week" are assigned low attention weights since they are less informative. These results show that our approach can learn informative news representations by selecting important words. Second, we find our approach can effectively recognize important news according to their categories and subcategories. For example, the news in the "autos" and "lifestyles" categories are gained more attentions than those in the "weather" and "food&drink" categories. These results show that our approach can learn informative user representations by modeling the importance of different news topics for gender prediction. Third, we find approach can help learn more informative user representations by selecting news according to their title representations. For example, the news of "40 health concerns..." is very informative for inferring the gender of users. However, the topic categories of this news are not very gender discriminative, and the category/subcategory views fail to recognize this news. Fortunately, our approach can still highlight this news by utilizing the content of news title, which is useful for learning more informative user representations. These results validate the effectiveness of the word- and news-level attention networks in our approach.

5   Conclusion

In this paper, we propose a neural gender prediction approach to infer the genders of online news platform users from their news browsing data. In our approach, we use a multi-view learning framework to incorporate different kinds of new information such as title, body, category and subcategory as different views of users to learn accurate user representations for gender prediction. In addition, since different words and news usually have different informativeness for gender prediction, we apply attention mechanism at both word and news levels to select important words and news articles for user representation learning. Extensive experiments on a real-world dataset validate the effectiveness of our approach.

Acknowledgments. The authors would like to thank Microsoft News for providing technical support and data in the experiments, and Jiun-Hung Chen (Microsoft News) and Ying Qiao (Microsoft News) for their support and discussions. This work was supported by the National Key Research and Development Program of China under Grant number 2018YFC1604002, the National Natural Science Foundation of China under Grant numbers U1836204, U1705261, U1636113, U1536201, and U1536207, and the Tsinghua University Initiative Scientific Research Program.

References

1. Buraya, K., Farseev, A., Filchenkov, A.: Multi-view personality profiling based on longitudinal data. In: Bellot, P., et al. (eds.) CLEF 2018. LNCS, vol. 11018, pp. 15–27. Springer, Cham (2018). https://doi.org/10.1007/978-3-319-98932-7_2
2. Ciccone, G., Sultan, A., Laporte, L., Egyed-Zsigmond, E., Alhamzeh, A., Granitzer, M.: Stacked gender prediction from tweet texts and images notebook for pan at CLEF 2018. In: CLEF, 11 p. (2018)
3. Culotta, A., Kumar, N.R., Cutler, J.: Predicting the demographics of Twitter users from website traffic data. In: AAAI, pp. 72–78 (2015)
4. Das, A.S., Datar, M., Garg, A., Rajaram, S.: Google news personalization: scalable online collaborative filtering. In: WWW, pp. 271–280. ACM (2007)
5. Farnadi, G., Tang, J., De Cock, M., Moens, M.F.: User profiling through deep multimodal fusion. In: WSDM, pp. 171–179 (2018)
6. Filippova, K.: User demographics and language in an implicit social network. In: EMNLP, pp. 1478–1488 (2012)
7. Hochreiter, S., Schmidhuber, J.: Long short-term memory. Neural Comput. $9(8)$, 1735–1780 (1997)
8. Hu, J., Zeng, H.J., Li, H., Niu, C., Chen, Z.: Demographic prediction based on user's browsing behavior. In: WWW, pp. 151–160 (2007)
9. Kim, Y.: Convolutional neural networks for sentence classification. In: EMNLP, pp. 1746–1751 (2014)
10. Kingma, D.P., Ba, J.: Adam: a method for stochastic optimization. arXiv preprint arXiv:1412.6980 (2014)
11. Li, W., Dickinson, M.: Gender prediction for Chinese social media data. In: RANLP, pp. 438–445 (2017)
12. Mac Kim, S., Xu, Q., Qu, L., Wan, S., Paris, C.: Demographic inference on Twitter using recursive neural networks. In: ACL, vol. 2, pp. 471–477 (2017)
13. Malmi, E., Weber, I.: You are what apps you use: demographic prediction based on user's apps. In: ICWSM, pp. 635–638 (2016)
14. Mislove, A., Lehmann, S., Ahn, Y.Y., Onnela, J.P., Rosenquist, J.N.: Understanding the demographics of Twitter users. In: 2011 5th ICWSM, vol. 25 (2011)
15. Mukherjee, S., Bala, P.K.: Gender classification of microblog text based on authorial style. IseB $15(1)$, 117–138 (2017)
16. Nguyen, D., Gravel, R., Trieschnigg, D., Meder, T.: "How old do you think i am?" a study of language and age in Twitter. In: ICWSM, pp. 439–448 (2013)
17. Nguyen, D., Smith, N.A., Rosé, C.P.: Author age prediction from text using linear regression. In: Proceedings of the 5th ACL-HLT Workshop on Language Technology for Cultural Heritage, Social Sciences, and Humanities, pp. 115–123 (2011)
18. Nguyen, D., Trieschnigg, D., Doğruöz, A.S., Gravel, R., Theune, M., Meder, T., De Jong, F.: Why gender and age prediction from tweets is hard: lessons from a crowdsourcing experiment. In: COLING, pp. 1950–1961 (2014)
19. Peersman, C., Daelemans, W., Van Vaerenbergh, L.: Predicting age and gender in online social networks. In: SMUC, pp. 37–44 (2011)
20. Pennington, J., Socher, R., Manning, C.: GloVe: global vectors for word representation. In: EMNLP, pp. 1532–1543 (2014)
21. Phuong, T.M., et al.: Gender prediction using browsing history. In: Huynh, V., Denoeux, T., Tran, D., Le, A., Pham, S. (eds.) Knowledge and Systems Engineering, vol. 244, pp. 271–283. Springer, Cham (2014). https://doi.org/10.1007/978-3-319-02741-8_24

22. Rangel, F., Rosso, P., Montes-y Gómez, M., Potthast, M., Stein, B.: Overview of the 6th author profiling task at pan 2018: multimodal gender identification in Twitter. Working Notes Papers of the CLEF (2018)
23. Rangel Pardo, F.M., Celli, F., Rosso, P., Potthast, M., Stein, B., Daelemans, W.: Overview of the 3rd author profiling task at pan 2015. In: CLEF, pp. 1–8 (2015)
24. Reddy, T.R., Vardhan, B.V., Reddy, P.V.: N-gram approach for gender prediction. In: IACC, pp. 860–865. IEEE (2017)
25. Rosenthal, S., McKeown, K.: Age prediction in blogs: a study of style, content, and online behavior in pre-and post-social media generations. In: ACL, pp. 763–772 (2011)
26. Sezerer, E., Polatbilek, O., Sevgili, Ö., Tekir, S.: Gender prediction from tweets with convolutional neural networks: notebook for pan at CLEF 2018. In: CLEF (2018)
27. Wang, J., Li, S., Zhou, G.: Joint learning on relevant user attributes in micro-blog. In: IJCAI, pp. 4130–4136 (2017)
28. Wang, L., Li, Q., Chen, X., Li, S.: Multi-task learning for gender and age prediction on chinese microblog. In: Lin, C.-Y., Xue, N., Zhao, D., Huang, X., Feng, Y. (eds.) ICCPOL/NLPCC -2016. LNCS (LNAI), vol. 10102, pp. 189–200. Springer, Cham (2016). https://doi.org/10.1007/978-3-319-50496-4_16
29. Wu, C., Wu, F., Liu, J., He, S., Huang, Y., Xie, X.: Neural demographic prediction using search query. In: WSDM, pp. 654–662. ACM (2019)
30. Yang, Z., Yang, D., Dyer, C., He, X., Smola, A., Hovy, E.: Hierarchical attention networks for document classification. In: NAACL, pp. 1480–1489 (2016)
31. Zhang, D., Li, S., Wang, H., Zhou, G.: User classification with multiple textual perspectives. In: COLING, pp. 2112–2121 (2016)

A Top-Down Model for Character-Level Chinese Dependency Parsing

Yuanmeng Chen, Hang Liu, Yujie Zhang[✉], Jinan Xu,
and Yufeng Chen

School of Computer and Information Technology,
Beijing Jiaotong University, Beijing, China
yjzhang@bjtu.edu.cn

Abstract. This paper proposes a novel transition-based algorithm for character-level Chinese dependency parsing that straightforwardly models the dependency tree in a top-down manner. Based on the stack-pointer parser, we joint Chinese word segmentation, part-of-speech tagging, and dependency parsing in a new way. We recursively build the character-based dependency tree from root to leaf in a depth-first fashion, by searching for candidate dependents through the sentence and predicting relation type at each step. We introduce intra-word dependencies into the relation types for word segmentation, and the inter-word dependencies with POS tags for part-of-speech tagging. Since the top-down model provides a global view of an input sentences, the information of the whole sentence and all previously generated arcs are available for action decisions, and all characters of the sentence are considered as candidate dependencies. Experimental results on the Penn Chinese Treebank (CTB) show that the proposed model outperformed existing neural joint parsers by 0.81% on dependency parsing, and achieved the F1-scores of 95.97%, 91.72%, 80.25% for Chinese word segmentation, part-of-speech tagging, and dependency parsing.

Keywords: Dependency parsing · Chinese · Joint model · Pointer networks · Stack

1 Introduction

Dependency parsing is a fundamental technique for natural language processing, whose accuracy obviously affects downstream tasks such as machine translation, question answering, text generation, and so on [1–3]. Word segmentation is needed before Chinese dependency parsing due to lack of obvious delimiters between words. For this problem, the joint models for word segmentation, part-of-speech tagging, and dependency parsing have been proposed by researches [4–7]. Compared with the pipeline models, the joint models integrate the three tasks into one framework to solve the error propagation problem between three tasks, and to be able to use multiple levels (character, word, and phrase level) information for each task. The character-level dependency parsers generally use the classical transition-based framework and improve performance significantly.

© Springer Nature Switzerland AG 2019
M. Sun et al. (Eds.): CCL 2019, LNAI 11856, pp. 677–688, 2019.
https://doi.org/10.1007/978-3-030-32381-3_54

However, the weakness of the classical transition-based framework is the lack of a global view of the input sentence. In greedy or beam search, only local information is available for action decisions, and two top nodes of the stack and the first token in the buffer are considered as analysis objects at each step. This restriction usually results in error propagation, particularly in long dependencies that require a larger number of transitions to be built [8].

In order to address this problem, we propose a novel top-down joint model based on stack-pointer networks [9]. Stack-pointer networks realize parsing in a top-down manner, which uses pointer networks [10] to directly find dependencies in sentence for a given word, and predicts the relation types with a multiclass classifier. On the basis of stack-pointer networks, we construct a character-level Chinese dependency framework by designing the intra-word dependencies and predicting of dependency types expanded with POS tags, to integrate Chinese word segmentation, part-of-speech tagging and dependency parsing. With the global view of input sentence, this architecture can capture information from the whole sentence, and make it possible to directly generate arcs between any characters.

We evaluate our model on the Penn Chinese Treebank (CTB version 5.0) and compare with other joint architectures. Comparison results show that our model surpasses the best results of the neural joint models on dependency parsing task. Our contributions are summarized as follows: (1) we propose a top-down algorithm for character-level Chinese dependency parsing, (2) we integrate word segmentation and part-of-speech tagging tasks into a top-down parsing framework, (3) our model surpasses the best neural joint model for dependency parsing.

2 Word Segmentation and POS Tagging as Top-Down Parsing

2.1 Stack-Pointer Networks

Stack-Pointer Networks (StackPTR) [9] is a dependency framework which has a global view of the input sentence during parsing and a lower computation complexity compared with graph-based parsers. The framework builds dependency arcs in a tree structure without the restriction of left-to-right in classical transition-based parsers, and therefore can consider all words of the sentence as candidates.

For an input sentence $\mathbf{x} = \{w_1, \ldots, w_n\}$, the outputs of StackPTR is $\mathbf{y} = \{p_1, \ldots, p_i, \ldots, p_k\}$, where $p_i = \$, w_{i,1}, w_{i,2}, \ldots, w_{i,li}$ represents the path from the dummy root "\$" to the leaf $w_{i,li}$, and the two adjacent words $w_{i,j}$, $w_{i,j+1}$ on the path is a head-child pair, representing a dependency arc with its type. Figure 1(a) shows the construction of the dependency tree for the sentence "叛逆是青少年普遍的特质 (Rebellion is a common characteristic of teenagers.)".

StackPTR combines pointer networks with an internal stack. The stack tracks the status of the top-down search and the pointer networks select one child for the word at the top of the stack at each step. Given a stack-top word w_t, the pointer network returns the position p in the input sentence, and then the following actions are conducted according to position p [11]:

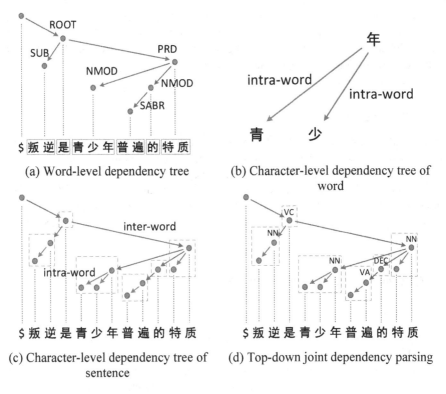

(a) Word-level dependency tree

(b) Character-level dependency tree of word

(c) Character-level dependency tree of sentence

(d) Top-down joint dependency parsing

Fig. 1. Illustration of top-down character-level Chinese dependency parsing

- Shift-Attach-p: if $p \neq t$, then the pointed word w_p is considered as a child of w_t, then the parser moves w_p into the stack and build an arc $w_t \rightarrow w_p$, and the relation type will be predicted;
- Reduce: if $p = t$, then w_t is considered to have found all its children, so the parser pops w_t out from the stack.

2.2   Word Segmentation

Since traditional word segmentation makes decisions from left to right as sequence labeling [12, 13], it is difficult to directly integrate it into the top-down parsing framework. In order to address this problem, we define an intra-word dependency structure for a Chinese word that consists of a few characters. We take the last character of the word as head (because we know that the segmentation of Chinese words usually performs slightly better in the reverse direction), and each other character as its child. As shown in Fig. 1(b), for the word "青少年(teenagers)", two intra-word arcs "青←年" and "少←年" are constructed. In this way, we transform a word to a character-level dependency tree and incorporate such dependency parsing into the top-down framework to implement word segmentation.

As shown in Fig. 1(c), we utilize the prediction mechanism of relation types in StackPTR to predict "intra-word" and "inter-word". The "intra-word" is used to indicate character-level arc for word segmentation and the "inter-word" is used to indicate word-level arc for dependency parsing. A head and all its children with "intra-word" relation type make up a word.

2.3 POS Tagging

In the prediction mechanism of relation types, we further introduce POS tagging task. We tag each word with POS (as shown in Fig. 1(d)) by adding POS tag to its head character. When the relation type of a dependency arc is predicted as "inter-word", the child of the arc, which is a head of some word, will be assigned with a POS tag. For simplicity, we predict the relation types and the POS tags simultaneously, so the prediction mechanism can be implemented for both relation types and POS tags by using one classifier.

3 Chinese Joint Dependency Parsing

3.1 Overview

We constructed a top-down joint dependency parsing model based on stack-pointer networks, by integrating word segmentation and POS tagging described in Sect. 2. As shown in Fig. 2, our model consists of three parts: encoder, decoder, and decision layer. The encoder calculates the representation for each character of the input sentence. The decoder regards the top character of the stack S as the head and get its representation at each step. And the decision layer contains a Biaffine attention and a Biaffine classifier. We calculate the correlation scores of the head and all candidates by the attention mechanism and select one with the highest score as child. Then Shift-Attach-p or Reduce is executed according to the position of the selected child with the highest score in the sentence. And the classifier predicts the relation type according to the feature vectors of the head and the child.

We use tag "IN" and "OUT" to indicate "intra-word" and "inter-word" relation types, respectively, and add the POS tag to the "OUT" tag. Specially, we use a meaningless tag "PAD" for Reduce action because there is no arc formed.

3.2 Encoder

The encoder of our model is a deep bidirectional LSTM (Bi-LSTM) [14]. For a given sequence of characters x $= \{c_1, \ldots, c_n\}$, we lookup embedding vector e_i from the pre-trained embedding matrix for each character c_i, and then feed them into the deep BiLSTM to obtain representations containing context information.

The Bi-LSTM learns the representation $\{s_1, \ldots, s_n\}$ of every character from two directions respectively as follows:

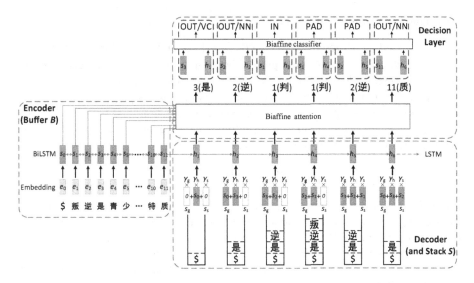

Fig. 2. Joint Chinese dependency parsing model based on StackPTR

$$\overrightarrow{s_i} = \overrightarrow{\mathrm{LSTM}}(\overrightarrow{s_{i-1}}, e_i) \tag{1}$$

$$\overleftarrow{s_i} = \overleftarrow{\mathrm{LSTM}}(\overleftarrow{s_{i+1}}, e_i) \tag{2}$$

$$s_i = \overrightarrow{s_i} \oplus \overleftarrow{s_i} \tag{3}$$

where $\overrightarrow{s_i}$ and $\overleftarrow{s_i}$ represent the hidden states of the forward and the backward LSTM, respectively. $\oplus$ represents the vector concatenating operation. We use the hidden state of the last layer as a representation of each character.

3.3 Decoder

The decoder consists of a uni-directional LSTM. At each time, the vector s_h of the stack-top character looked as a head is fed into the LSTM, and then the hidden state h_t is obtained. Considering that LSTM's hidden state can be passed over time, previously generated information can be implicitly used in every step of decisions.

Higher-Order Information We also use the higher-order information to add to the head as the input of LSTM like StackPTR [9]: grandfather and sibling. Furthermore, we introduce three weights γ_h, γ_g and γ_s for the head, grandfather and sibling, respectively, to control the utilization of the three kinds of information.

3.4   Biaffine Attention and Biaffine Classifier

At each step, all the represents $s_j(s_j \in \{s1, \ldots, sn\})$ of children and the h_t of the head are used in attention mechanism to predict the position p for the action decisions, and the classifier use the h_t and s_p to predict the relation type.

There are different types for attention functions such as dot-product, concatenation (or nonlinear transformation), general (bilinear transformation) and bi-affine [15, 16]. We adopt the biaffine attention mechanism following Ma et al. [9]. The calculation for bi-affine attention score function is as follows:

$$\alpha_i^t = h_t^{\mathrm{T}} \mathbf{W} s_i + \mathbf{U}^{\mathrm{T}} h_t + \mathbf{V}^{\mathrm{T}} s_i + b \tag{4}$$

where $\mathbf{W}, \mathbf{U}, \mathbf{V}$ and b are parameters, denoting the weight matrix of the bilinear term, the two weight vectors of the linear terms, and the bias, respectively; h and s represent vector representations of head and child, respectively. It can be seen that the bi-affine function combines the bilinear and the linear transformation, and therefore the advantages of both are brought about. The first three terms correspond to the relationship between h_t and s_i, the information of h_t, and the information of s_i, Therefore, the prediction can be made from a more comprehensive perspective [17].

The Eq. (4) is also used in the Biaffine classifier to accept the two inputs of the head and its child. Since the different sets of parameters are set up for every type, the scores of every type are available.

3.5   Training

Children Order. When a head character has multiple children, it is possible that there is more than one valid selection at each step. In order to define a deterministic decoding process to make sure there is only one ground-truth choice at each step, a predefined order for selecting children for this head needs to be introduced. The predefined order of children can have different alternatives, such as *left-to-right* or *inside-out*. In this paper, we adopt them both and the *inside-out* performs better.

LOSS. Given the input sentence $\mathbf{x} = \{c_1, \ldots, c_n\}$, the top-down parsing model is to search for a dependency tree $\mathbf{y} = \{p_1, \ldots, p_k\}$ with a highest probability. We train the model by minimizing the cross-entropy loss as follows:

$$
\begin{aligned}
L(\theta) &= - \sum_{i=1}^{k} \log P_\theta(p_i | p_{<i}, \mathbf{x}) \\
&= - \sum_{i=1}^{k} \sum_{j=1}^{l_i} \log P_\theta(c_{i,j} | c_{i,<j}, p_{<i}, \mathbf{x})
\end{aligned}
\tag{5}
$$

where θ represents all parameters of the model. $p_{<i}$ represents paths that have been built completely, k is the total number of paths, l_i represents the total number of characters in p_i, $c_{i,\,j}$ denotes the j-th character in p_i, $c_{i,<j}$ denotes all the proceeding characters that has been built in p_i. We define $P_\theta(c_{i,j} | c_{i,<j}, p_{<i}, \mathbf{x}) = \alpha^A + \alpha^C$, where α^A

and α^C are scores of the attention and the classifier, respectively. The prediction of dependency arc and type is taken as joint learning at each time.

3.6 Testing

In order to guarantee a valid dependency tree at test time, each character should have only one head except the dummy root "$". We introduce a list of "available" characters in the decoder. At each decoding step, the model selects a child for the current head, and removes the child from the list of available characters to make sure that it cannot be selected as a child of other head. For implementation of the Reduce action, the head character is temporarily recovered considering it has been removed before.

4 Experiments

4.1 Setup

We evaluate our model on the Penn Chinese Treebank (CTB version 5.0). At first, we process the data to obtain intra-word dependency structure. For one word, each character except the last, is annotated as the child of the last character with relation type tag "IN". The tag of the last character of the word is annotated as "OUT" suffixed with the POS tag of the word. The statistical information of the processed data is shown in Table 1.

Table 1. The statistics of the processed dataset

Dataset	Sentence	Word (inter-word arc)	Intra-word arc
Training	16k	494k	311k
Development	352	6.8k	4.7k
Test	348	8.0k	5.7k

We use standard metrics of word-level F1 score to evaluate word segmentation, POS tagging and dependency parsing. F1 score is calculated according to the precision P and the recall R as $F = 2PR/(P+R)$ [18]. Dependency parsing task is evaluated with the unlabeled attachment scores excluding punctuations. The output of POS tags and dependency arcs cannot be correct unless the corresponding words are segmented correctly.

4.2 Hyper-parameter

We use word2vec to pre-train the character vectors on the Gigaword, and the dimension of character vectors is 500 dimensions. The dimension of LSTM hidden layer is 512. The layers' number in encoder is 3, and in decoder is 1. We set batch size is 32. Adam is used in the optimization algorithm, with the initial learning rate is 0.001, and dropout value is 0.33. The learning rate is annealed by multiplying a fixed decay rate

0.75 when parsing performance stops increasing on development sets. To reduce the effects of "gradient exploding", we use gradient clipping of 5.0 [19].

4.3 Results

Higher-order Information. In order to choose the best model, we first compare the accuracies with and without grandfather and sibling in decoder. We further investigate the way of weight setting, one case with the frozen weights $\gamma_h = \gamma_g = \gamma_s = 1$, the other case with the unfrozen weights by adjusted with model training. The results are shown in the Table 2, where the "base" means a model without grandfather and sibling. From the table, we found that the best results on three tasks are those obtained by the model "base" using an unfrozen weight for $\gamma_h = 0.08$. The low value of the weight γ_h implies that previously generated information is more important in prediction. And the accuracies of the model using higher-order information are not improved like Ma et al. [9]. This best model will be used for subsequent comparisons with other works.

Table 2. Accuracy with/without higher-order information

Model	Weight	SEG	POS	DEP
Base	Freeze	95.73	91.40	79.81
+grand	$(\gamma_h = \gamma_g = \gamma_s = 1)$	95.66	91.44	79.33
+sibling		95.64	91.06	78.97
+grand+sibling		95.63	91.21	79.22
Base	$\gamma_h = 0.08$	**95.97**	**91.72**	**80.25**
+grand	$\gamma_h = 0.11;\ \gamma_g = 0.01$	95.88	91.56	79.94
+sibling	$\gamma_h = 0.08;\ \gamma_s = 0.04$	95.67	91.39	79.78
+grand+sibling	$\gamma_h = 0.10;\ \gamma_g = 0.01;\ \gamma_s = 0.05$	95.66	91.44	79.66

Main Comparison. We conduct comparison of our model with other joint parsing models and show the results in Table 3. The comparison models include Hatori12 [4] and Zhang14 (character-level dependency parser based on feature engineering) [5], Kurita17 (joint dependency model combined with feature engineering for feature extraction and neural network for decision), Kurita17 (4-gram), Kurita17 (8-gram) [6] and Li19 (pure neural joint model) [7].

Table 3. Comparison results with the existing joint dependency parsing models

Model	SEG	POS	DEP
Hatori12	97.75	94.33	81.56
Zhang14	97.67	94.28	**81.65**
Kurita17	**98.24**	**94.49**	80.15
Kurita17 (4-gram)	**97.72**	93.12	79.03
Kurita17 (8-gram)	97.70	**93.37**	79.38
Li18	96.64	92.88	79.44
Ours (base)	95.92	91.65	**80.25**

The table is divided into two parts. The models in the top part are based on feature engineering and those in the bottom part are based entirely on neural network. From the table, we see that our model outperformed the existing neural joint model in dependency parsing tasks by 0.81%, and is closer to the feature engineering methods using a large number of artificially designed feature templates. At the same time, we also found that the accuracies of word segmentation and POS tagging are lower than other neural models. The observed results show that the proposed top-down character-level dependency architecture can significantly improve the accuracy of dependency parsing despite the lower accuracies on word segmentation and POS tagging.

Length. We compare our model with the reproduced Kurita17 (8-gram) model according to dependency length and sentence length respectively. The results are shown in Fig. 3. From the figure, we can see that our model has better accuracy on long-distance dependency and long sentences, which shows that our model effectively extract global information and identify and construct long-distance dependency. At the same time, we find that our dependency accuracy on short sentences is lower. We investigate the results and analyze the reason is the low accuracy of word segmentation.

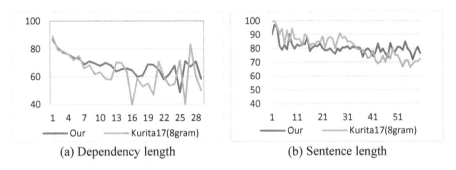

(a) Dependency length   (b) Sentence length

Fig. 3. The influence of length on dependency parsing

Case Analysis. In this part, we try to analyze the difference between the top-down model(our model) and the classical transition-based model(Kurita17) through an instance. As shown in Fig. 4, we take the sentence "中国缺粮大省贵州农村初步实现粮食自给 (In Guizhou, China's major grain shortage province, rural areas have initially achieved self-sufficiency in grain.)" as an example. The following are the results of word segmentation, POS tagging and dependency parsing respectively got from the gold standard, Kurita17 and our model.

Firstly, these orange lines for Kurita17 represent the prediction errors. The errors are due to that Kurita17 analyzes "中国缺粮大省贵州农村 (Guizhou rural areas, China's major grain shortage province)" as a sub-sentence, and "初步实现粮食自给 (have initially achieved self-sufficiency in grain)" as the other sub-sentence. Specifically, Kurita17 predicts the relationship between "缺 (lack)" and "农村 (rural areas)" as a right arc "缺→农村", but "缺粮 (grain shortage)" should be an attributive modifier of "省 (province)" (as seen in Gold). In the prediction between "缺 (lack)" and "实现

(achieve)", because the first verb is usually the root of a sentence when there are multiple verbs in a sentence in CTB corpus.

Secondly, our model does not have the above errors due to the top-down model framework that enables it to correctly identify the root. However, since verbs rarely modify nouns, our model predicts "缺粮 (grain shortage)" as a word with POS tag "NN". This may be the reason why our model performs poorly in word segmentation task.

Finally, the blue arcs are the dependency arcs that Kurita17 and our model both predicted wrong. This mistake is mainly caused by the omission of the sentence, because the complete expression of the original sentence should be "在中国缺粮大省 贵州, 农村初步实现粮食自给", while the continuous nouns "贵州 (Guizhou)" and "农村 (rural areas)" in the sentence are usually generated a left arc "贵州←农村", so neither model can correctly predict.

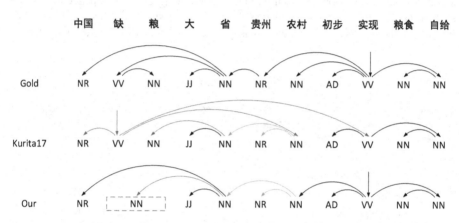

Fig. 4. Forecast Results of Different models for "中国缺粮大省贵州农村初步实现粮食自给"

5   Related Work

Hatori et al. [4] proposed a character-level dependency architecture for the first time, which combines word segmentation, part-of-speech tagging and dependency parsing. They combined the key feature templates on the basis of the previous feature engineering research on the three tasks, and realized the synchronous processing of the three tasks. Zhang et al. [5] annotated the internal structure of words for the Penn Chinese Treebank (CTB5), and regarded the word segmentation task as dependency parsing within characters to jointly process with dependency parsing. And they achieved the best accuracy in dependency parsing task at present.

Kurita et al. [6] applied neural network to the character-level dependency parsing for the first time. They used feature templates, pre-trained character and word vectors to extract distributed features, which were input into a multi-layer perceptron (MLP) to make decisions on parsing actions. They achieved the best results at present in word segmentation and part-of-speech tagging tasks. At the same time, they also used

BiLSTM to encode the N-gram information to realize automatic feature extraction, with accuracies close to the best results of feature engineering on three tasks. Li et al. [7] used the tagged part of speech and intra-word dependency of Chinese characters to process character-level dependency parsing with Stack-LSTM [20]. Their model outperformed the models of Kurita17(8-gram) on dependency parsing task.

The above described models are based on classical transition-based framework, which has a restriction of left-to-right. Ma et al. [9] proposed a novel transition-based model performing parsing in an incremental, top-down, depth-first fashion. The model recursively searches for child nodes from the root downward, making full use of global information, and breaking through the limitation of the classical transition-based models that can only consider the top two words of the stack and the first word of the buffer for decision.

Inspired by the work of Ma et al. [9], we propose a top-down character-level Chinese dependency parsing architecture to joint three tasks together, which overcomes the drawbacks of the traditional transition-based dependency parsers, and solves the problem of global information utilization and error propagation in long-distance dependence.

6 Conclusion

This paper proposes a top-down character-level Chinese dependency framework. The word segmentation task is realized by constructing the intra-word relation type, and the POS tagging task is realized by introducing POS tags into the inter-word relation type. In this way, a top-down joint model of Chinese word segmentation, POS tagging and dependency parsing is realized. Since our model has a global view for the input sentence, the information of the whole sentence and all previously generated dependency arcs are available for action decisions, and all characters of the sentence are considered as candidate children, without the left-to-right restriction. Our evaluation results on the CTB5 dataset outperform the existing neural network models in the dependency parsing. The model is also evaluated on higher-order information, sentence and dependency length, and children order. In the future, we will improve the joint model under the top-down dependency parsing framework, and make our model more robust for word segmentation and part-of-speech tagging, so as to further improve the accuracy of dependency parsing.

Acknowledgments. The authors are supported by the National Nature Science Foundation of China (Nos. 61876198, 61370130 and 61473294), the Fundamental Research Funds for the Central Universities (2015JBM033), and the International Science and Technology Cooperation Program of China (No. K11F100010).

References

1. Ma, X., Liu, Z., Hovy, E.: Unsupervised ranking model for entity coreference resolution. In: Proceedings of NAACL-2016, San Diego, California, USA (2016)
2. Bastings, J., Titov, I., Aziz, W., Marcheggiani, D., Simaan, K.: Graph convolutional encoders for syntax-aware neural machine translation. In: Proceedings of EMNLP-2017. Copenhagen, Denmark, pp. 1957–1967 (2017)
3. Peng, N., Poon, H., Quirk, C., Toutanova, K., Yih, W.: Cross-sentence n-ary relation extraction with graph lstms. Trans. Assoc. Comput. Linguist. **5**, 101–115 (2017)
4. Hatori, J., Matsuzaki, T., Miyao, Y., Tsujii, J.: Incremental joint approach to word segmentation, POS tagging, and dependency parsing in Chinese. In: Proceedings of ACL, pp. 1045–1053. Association for Computational, Jeju (2012)
5. Zhang, M., Zhang, Y., Che, W., Liu, T.: Character-level Chinese dependency parsing. In: Proceedings of ACL, pp. 1326–1336. Association for Computational, Baltimore (2012)
6. Kurita, S., Kawahara, D., Kurohashi, S.: Neural joint model for transition-based Chinese syntactic analysis. In: Proceedings of ACL, pp. 1204–1214. Association for Computational, Vancouver (2018)
7. Li, H., Zhang, Z., Ju, Y., Zhao, H.: Neural character-level dependency parsing for Chinese. In: Proceedings of 32nd AAAI Conference on Artificial Intelligence (2018)
8. Mcdonald, R., Nivre, J.: Analyzing and integrating dependency parsers. Comput. Linguist. **37**(1), 197–230 (2011)
9. Ma, X., Hu, Z., Liu, J., Peng, N., Neubig, G., Hovy, E.: Stack-pointer networks for dependency parsing. In: Proceedings of ACL, pp. 1403–1414. Association for Computational, Melbourne (2018)
10. Vinyals, O., Fortunato, M., Jaitly, N.: Pointer networks. In: Advances in Neural Information Processing Systems, pp. 2692–2700 (2015)
11. Fernández-González, D., Gómez-Rodríguez, C.: Left-to-right dependency parsing with pointer networks. arXiv preprint arXiv:1903.08445 (2019)
12. Zheng, X., Chen, H., Xu, T.: Deep learning for Chinese word segmentation and POS tagging. In: Proceedings of the 2013 Conference on Empirical Methods in Natural Language Processing, pp. 647–657 (2014)
13. Shao, Y., Hardmeier, C., Tiedemann, J., Nivre, J.: Character-based joint segmentation and POS tagging for Chinese using bidirectional RNN-CRF. arXiv preprint arXiv:1704.01314 (2017)
14. Hochreiter, S., Schmidhuber, J.: Long short-term memory. Neural Comput. **9**(8), 1735–1780 (1997)
15. Vaswani, A., et al.: Attention is all you need. In: Advances in Neural Information Processing Systems, pp. 5998–6008 (2017)
16. Luong, M.T., Pham, H., Manning, C.D.: Effective approaches to attention-based neural machine translation. arXiv preprint arXiv:1508.04025 (2015)
17. Dozat, T., Manning, C.D.: Deep biaffine attention for neural dependency parsing. arXiv preprint arXiv:1611.01734 (2016)
18. Jiang, W., Huang, L., Liu, Q., Lü, Y.: A cascaded linear model for joint Chinese word segmentation and part-of-speech tagging. In: Proceedings of ACL. Association for Computational Linguistics, Columbus (2008)
19. Pascanu, R., Mikolov, T., Bengio, Y.: On the difficulty of training recurrent neural networks. In: International Conference on Machine Learning, pp. 1310–1318 (2013)
20. Dyer, C., Ballesteros, M., Ling, W., Matthews, A., Smith, N.A.: Transition-based dependency parsing with stack long short-term memory. arXiv preprint arXiv:1505.08075 (2015)

A Corpus-Free State2Seq User Simulator for Task-Oriented Dialogue

Yutai Hou[1], Meng Fang[2], Wanxiang Che[1(✉)], and Ting Liu[1]

[1] Research Center for Social Computing and Information Retrieval,
Harbin Institute of Technology, Harbin, China
{ythou,car,tliu}@ir.hit.edu.cn
[2] Tencent Robotics X, Shenzhen, China
mfang@tencent.com

Abstract. Recent reinforcement learning algorithms for task-oriented dialogue system absorbs a lot of interest. However, an unavoidable obstacle for training such algorithms is that annotated dialogue corpora are often unavailable. One of the popular approaches addressing this is to train a dialogue agent with a user simulator. Traditional user simulators are built upon a set of dialogue rules and therefore lack response diversity. This severely limits the simulated cases for agent training. Later data-driven user models work better in diversity but suffer from data scarcity problem. To remedy this, we design a new corpus-free framework that taking advantage of their benefits. The framework builds a user simulator by first generating diverse dialogue data from templates and then build a new State2Seq user simulator on the data. To enhance the performance, we propose the State2Seq user simulator model to efficiently leverage dialogue state and history. Experiment results on an open dataset show that our user simulator helps agents achieve an improvement of 6.36% on success rate. State2Seq model outperforms the seq2seq baseline for 1.9 F-score.

1 Introduction

Task-oriented dialogue systems assist users to achieve specific goals such as finding restaurants or booking flights [25]. To learn such a system is very challenging. Recently, reinforcement learning (RL) methods have been introduced due to their advantages in sequential decision [7,18,24,25]. An RL based dialogue agent can learn from dialogue data or reward signals by interacting with real users. Unfortunately, interacting with real users is costly and there is often no enough data or even no data for new domains. To overcome these obstacles, building user simulators is studied for training RL dialogue algorithms [14,22].

User simulators can be divided into two categories: traditional and data-driven user simulator. Traditional user simulators are agenda-based or rule-based [13,20]. A rule-based user simulator can be built without data, but needs domain-specific knowledge and hard to generalize to new contexts. Besides, the rule-based model lacks response diversity, which largely limits the effectiveness of RL

© Springer Nature Switzerland AG 2019
M. Sun et al. (Eds.): CCL 2019, LNAI 11856, pp. 689–702, 2019.
https://doi.org/10.1007/978-3-030-32381-3_55

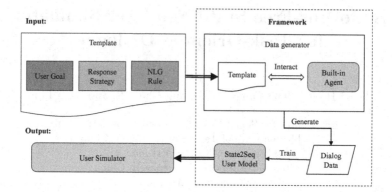

Fig. 1. The proposed corpus-free framework for building user simulators.

training. Latter data-driven user simulators ease the problem of diversity and depend less on expert knowledge. They imitate user behaviors from datasets with statistical models, such as Bayesian models [8,17], hidden Markov models [4] and seq2seq models [1]. Statistical user simulators are inherently diverse and often require a large amount of expert-labeled data for training. However, they can not cope with limited data situations.

In this paper, to overcome the data scarcity problem and build a user simulator with sufficient diversity, we propose a new corpus-free framework for building user simulators. It combines the ideas of rule-based and data-driven user simulators. As shown in Fig. 1, the framework generates dialogue data from templates and train data-driven user model on it. The template consists of user goals, response strategy and natural language generation rules. An example of the template is shown in Table 1. In addition to templates, data generation uses an RL-based built-in agent to improve data diversity and explore more dialogue cases. The statistical nature of data-driven user simulator provides more diversity than rule-based ones. Diverse responses allow covering more situation for policy training.

To enhance the user simulator's quality, we propose a novel attention based State2Seq user simulator to leverage the dialogue state and history better. The model first learns representations for dialogue context items. Dialogue context contains structured data of dialogue states, user goals and agent response. Then for each dialogue turn, the model predicts user actions sequentially with the attention on context. Attention helps to pick action more accurately. For example, suppose agent response is (request:movie, inform:date=today), and user goal is (request=[ticket], constraint=[date:tomorrow, movie:Deadpool]). The user model can easily output actions (inform:movie, deny:date) by attending to the agent request and the states of constraint inconsistency respectively.

Experiments are conducted on the movie booking dataset [13] and an in-house restaurant domain dataset. We evaluate both the user simulator model itself and the policy trained by it. On movie booking dataset, our policy achieves

Table 1. Template example for movie booking domain. G is user goal, V is response strategy and N is NLG rules. `req` is request.

Template name	Template content
G	$g_0 = \begin{bmatrix} C = [\texttt{movie} = \text{ Godfather}, \texttt{time} = 5 \text{ pm }], \\ R = [\texttt{ticket}, \texttt{theater}] \end{bmatrix}, g_1, g_2, \dots$
V	$r_0 = \begin{bmatrix} \textbf{if } A_{t-1} = \{\texttt{req:time}\} \text{ and time in } g.C \\ \textbf{then } A_t = \texttt{inform:time} = g.C.\texttt{time} \end{bmatrix}, r_1, r_2, \dots$
N	$l_0 = \begin{bmatrix} \textbf{if } A = \{\texttt{inform:time} = \text{g.C.time}\} \\ \textbf{then } L = \text{ "I want to see it at 5 pm"} \end{bmatrix}, l_1, l_2, \dots$

an improvement of 6.36 points on the success rate over the strongest baseline. And proposed State2Seq model outperforms the seq2seq baseline for 1.9 F-score on response accuracy.

This paper has 3 main contributions: 1. To solve data scarcity, we design a new corpus-free framework for building a user simulator with response diversity. 2. We introduce the attention mechanism to task-oriented user simulator and propose a State2Seq model to get better user behavior modeling. 3. Experiments show that response diversity and attention on dialogue context improve user model and agent policy.

Our code is available at: https://github.com/AtmaHou/UserSimulator

2 Proposed Framework

We focus on developing a user simulator, which is diverse to cover enough dialogue situations. To solve the data scarcity problem, our framework builds a user simulator with only templates and no dialogue data. There are two main components in the framework: a template based data generator and a neural user simulator. The framework (1) first generate data from the templates using a built-in agent, (2) then train a data-driven user simulator on it.

2.1 Template Definition

Template T is the input to our framework and is used to generate data. We define a template as: $T = (G, V, N)$, which includes user goals G, response strategies V and natural language generation rules N. An example of movie booking domain template is illustrated in Table 1.

G is a set of predefined user goals and defined as: $G = \{g_i\}_i^\alpha$, where α is the number of goals. Each user goal g is defined as $g = (C, R)$ which includes a set of user constraints C and a set of user requests R [20].

V is a set of rules for response strategies, which is relatively easy to obtain [13]. It is defined as: $V = \{r_i\}_i^\beta = V_a \cup V_u$, where V_a and V_u are response rules for user and agent respectively. For each $r \in V$, we define it as a function that maps dialogue context to response:

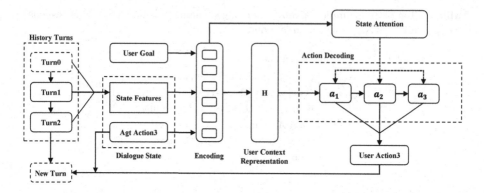

Fig. 2. State2Seq user simulator.

$$\begin{cases} r : f(A_1, A_2, ..., A_{t-1}) \rightarrow A_t, \quad r \in V_a \\ r : f(A_1, A_2, ..., A_{t-1}, G) \rightarrow A_t, \, r \in V_u \end{cases}$$

where A_i is the response for the i_{th} turn. Specifically, as user and agent may take multiple actions in one turn, we define response A as a set of single actions: $A = \{a_i\}_0^k$.

$N = \{l_i\}_i^\gamma$ is a set of rules for natural language generation(NLG): Each rule l maps actions and user goal to natural language L and is defined as $l : f(a_1, a_2, a_3, ..., a_n, g) \rightarrow L$.

2.2   Data Generation

The data generator generates conversation log with templates T as input, as shown in Fig. 1. There are two steps for the generation: i. Generate basic data with templates only. ii. Generate diverse data with a built-in RL agent and templates.

Generate Basic Data. To start from no corpus, we first collect some rule-based conversation as basic data, which is a common warm-up option [13]. When generating one dialogue, we first pick a user goal g from G as background and construct a random starting utterance. Then, for each turn, we search the suitable rule $r \in V$ to generate the response actions for user and agent. NLG rules N is then used to render actions to utterance.

Generate Diverse Data. On the basis of basic data, we generate diverse data. A built-in RL based agent $\mathcal{M}^*$ is used here. $\mathcal{M}^*$ is warmed up with basic data initially and further trained by interacting with V_u. When generating each dialogue, all process are same to basic data generation, except for agent response actions are given by $\mathcal{M}^*$'s policy.

We enhance the data diversity by the following operations: i. Leverage the RL exploration mechanism during data generation. ii. Generate data with the built-in agents of different training stages. So both weak and strong policy are used, which allows collecting both clumsy and fluent dialogue.

Table 2. Feature definition

Feature	Description
Constraint status	Status about whether user constraint slots have been informed by user
Request status	Status about whether user request slots have been satisfied by agent
Slot consistency	Status of whether the slot values provided by agent are consistent to user constraints
Dialog status	Dialogue status of success, failed and no outcome yet

2.3 User Model Training

After collecting the dialogue data, we train the user simulator on it in supervised style. Given a dialogue context, the user model is trained to predict a set of user actions as response. Only user policy is learned from data here.

3   State2Seq User Simulator

We aim to propose a user simulator that makes better use of dialogue context.

User simulator mimics human responses to the dialogue system output. A user model predicts user response with dialogue context. Given user goal g and dialogue history $< A_{t-1}, A_{t-2}, ..., A_1 >$, it predict t_{th} turn response as :

$$A_t = \arg \max_A \ p(A \mid g, A_{t-1}, A_{t-2}, ..., A_1)$$

where $A = \{a_i\}_0^k$. Following [1], we formulate the action selection as a sequence generation problem:

$$p(A \mid g, A_{t-1}, A_{t-2}, ..., A_1) = \prod_i^n p(a_i \mid a_1, ..., a_{i-1}, g, A_{t-1}, A_{t-2}, ..., A_1)$$

However, dialogue history can be very long. So it is very hard for the user model to leverage history information directly. To remedy this, we extract the key information as dialogue state S from the dialogue history. We then define the dialogue context as a combination of dialogue state and user goals.

For better usage of dialogue context, we learn vector representations for each context items. And attention mechanism is proposed to leverage context more clearly.

Figure 2 shows the structure of the State2Seq user simulator. The State2Seq model maintains a dialogue state S for tracking the dialogue history. It uses an encoding module to provide vector representations for dialogue context items. For each turn, the model integrates those representations into a context representation **H**. Then the model decodes **H** into a sequence of actions as output. Attention helps the model use dialogue context during decoding.

3.1   Dialogue Context Representation

The main idea of dialogue context representation is to refine history information. Forgetting useless information is proven to be important for data-driven model.

The dialogue context consists of dialogue state and user goal. Dialogue state S includes last turn agent response and state features. Following [1], we extract the state features explained in Table 2. Constraint Status, Request Status, and Dialog Status are used to help the user simulator track the progress of the current conversation, and Slot Consistency allows the user simulator to correct the agent on wrong information. Each feature is recorded as a status vector $\{status\}_0^m$, where m is the number of slot types and $status$ could be 1, 0, -1 for active, irrelevant and negative.

Attention mechanism relies on a good representation of items. Representation is also important for sequence decoding, as its need a good initial state $\mathbf{H}$. So we propose an encoding module to learn vector representations for context items. To help to learn of representation, we share representations for the common slots in context items. Some negative status is rare in the corpus, which makes it hard to learn good representations. To remedy this, we only learn vector representations $\mathbf{e}$ for positive slot status, and then use the corresponding inverse vector $-\mathbf{e}$ to represent negative status.

H is obtained by dimension reduction of context representation, which can further forget irrelevant information. Formally, given an agent response A_{t-1}, the user goal g and current dialogue history $< A_{t-1}, A_{t-2}, ..., A_1 >$, State2Seq updates dialogue state S and represent context with vectors. $\mathbf{H}$ is then obtained as:

$$\mathbf{H} = \mathbf{W}_c \cdot ([\mathbf{E}(A_{t-1}); \mathbf{E}(g); \mathbf{E}(S)]) + \mathbf{b}_c$$

where $[;]$ denotes vectors concatenation and $\mathbf{E}(\mathbf{x})$ means fetching vector representations for items in x.

3.2   Action Generation with Attention

We formulate actions selection as sequence generation process with attention. Sequence generation provides diversity in selecting actions. Attention mechanism helps to use dialogue context information directly. Specifically, we generate responses by attending on items in user goal g, dialogue state S and the last agent response A_{t-1}.

During actions decoding, for each time step t, LSTM provides hidden representation $\mathbf{h_k} = \mathrm{LSTM}(\mathbf{h_{k-1}}, \mathbf{c_{k-1}}, \mathbf{x_k})$, where $\mathbf{h_k}$ denotes the LSTM hidden state at time step k, the $\mathbf{c_{k-1}}$ is cell state and $\mathbf{x}$ is input.

The attention weight of i_{th} item in dialogue context is calculated as: $\mathbf{att_i} = \frac{\exp(\mathbf{e_i} \cdot \mathbf{h_k})}{\sum_j \exp(\mathbf{e_j} \cdot \mathbf{h_k})}$ where $\mathbf{e_i}$ is semantic representation vector of i_{th} item. Attention is used to calculate the decoding output $\mathbf{h'_t} = \tanh(\mathbf{w_a} \cdot (\mathbf{att} \cdot \mathbf{Emb}) + \mathbf{b_a} \cdot \mathbf{h_k})$, where $\mathbf{Emb}$ is the representations of all items in context.

Then we model the distribution the time step k's action a_k:

$$P(a \mid g, S, A_{t-1}, a_{k-1}, ..., a_1) = \mathrm{Softmax}(\mathbf{W_p}\mathbf{h'_t} + \mathbf{b_p})$$

And the user action a_k is predicted as:

$$a_k = \arg\max_a \ p(a \mid g, S, A_{t-1}, a_{k-1}, ..., a_1)$$

4 Experiment

4.1 Dataset

We used two datasets in our experiments: movie ticket booking data and restaurant reservation data. The movie ticket booking data is an open task-completion dialogue dataset proposed by [13]. For each dialogue, the system gathers information about the customer's desires and books the movie tickets. The success or failure of dialogue is assessed based on (1) whether a movie is booked, and (2) whether the movie satisfies the user's constraints. The data includes 11 dialogue acts, 29 slots, 277 user goal templates. Rule templates for response strategy and NLG are also included in [13]'s work.

To test the method's generation ability, we also build a dataset for restaurant reservation domain. In each dialogue, the user reserves a table under his/her requirements. The data includes 11 dialogue acts, 24 slots and 184 user goal templates. We design rule templates for response strategy and NLG based on the ones in [13]'s work.

4.2 Evaluation

We evaluate a user simulator by: i. evaluation of the user simulator itself. ii. evaluation of the policy trained with it.

Evaluation of Agent Policy. The main value of simulator is to train agent policy. We use both human and automatic evaluation for policy here. A DQN model is used as the agent to learn the policy.

We adopt cross-model evaluation proposed by [19]. $\mathcal{N}$ user models are first used to train $\mathcal{N}$ policies. Each policy is then tested against $\mathcal{N}$ different user simulators. Finally, we calculate the average of $\mathcal{N} \times \mathcal{N}$ scores. A policy trained by a good simulator can still perform well on poor simulators [11,19]. A higher average score indicates a better simulator ability for training agent. For metric, we use success rate, average reward and average turn number, which have been widely accepted as a standard metric of multi-turn agent [2,6,13].

Evaluation of User Simulator. We evaluate the user simulator itself from two aspects.

Firstly, following [1], we evaluate the accuracy of predicting actions. F-score is used as the metric.

Secondly, a user simulator's generalization ability is also important. It has more tolerance for exploration of training agent. We compare different user simulators' such ability by making conversation against a same rule agent. User model with better ability should achieve more success rate.

4.3 Model Details

For the State2Seq model, we set embedding size as 300. We use a 2 layer LSTMs for decoding with hidden state size of 256. During the training, we set the batch size as 32, a dropout as 0.8 and teacher forcing rate of 0.5. The learning rate is set as 0.001 and we set a learning rate decay of 0.9.

For the RL agent model used in data generation and evaluation, we use the DQN model. We set experience pool size as 1000, hidden layer size as 80. Experience replay redesigned for dialogue setting is applied. We use ϵ-greedy exploration of 0.01. The learning batch size is 16. We use 100 warm-up epochs and 500 training epochs. Model simulates 100 dialogues for each epoch.

We also simply extend our model by replacing the sequence decoder with a multi-label classifier. We name it as State2MLC. State2MLC takes dialogue state, user goal, agent response as input and predict multiple actions. State2MLC is trained with Multi-Label Soft Margin Loss.

4.4 Baselines

We compared with the following baselines:

- **Seq2Seq** is user simulator proposed by [1]. It extracts history turns' features as input sequence and decodes action sequence.
- **Seq2Seq-att** is based on Seq2Seq model and adds attention mechanism over input sequence.
- **Agenda** based user simulator is proposed by [20]. It is corpus free and generates user response by maintaining a user agenda with rule. We use the agenda user simulator provided by [13].

4.5 Performance of Agent Policy

We compare the policy trained by different user simulators with cross-model evaluation.

Table 3. Evaluation of agent policy trained by different user model. Results above dash-line are from our model, which achieve best performance in most task.

Model	Movie booking			Restaurant reservation		
	Avg. Succ.	Avg. Rwd.	Avg. Turns.	Avg. Succ.	Avg. Rwd.	Avg. Turns.
State2MLC	0.487	8.55	**21.82**	0.305	−17.22	29.69
State2Seq	**0.551**	**14.17**	25.85	**0.524**	**11.77**	24.21
Seq2Seq	0.412	−3.49	27.77	0.501	6.23	29.83
Seq2Seq-Att	0.430	−2.59	30.39	0.514	9.67	26.20
Agenda	0.438	−2.88	32.88	0.508	10.88	**22.17**

Table 4. Human evaluation of trained agent

Model	Avg. Succ.	Avg. Rwd.	Avg. Turns.
State2Seq (Ours)	**0.778**	**53.88**	**11.88**
Agenda	0.571	22.29	14.57

Results on Movie Booking Data. Table 3 shows the evaluation results on movie booking domain. The results show that the policy obtained by our model outperforms baselines.

On average success rate, policy trained by our State2Seq model outperforms the Agenda model for 6.36 points. As the Agenda [13] model is rule-based, the main difference between State2Seq and it is that State2Seq has more diverse responses. This demonstrates that user simulator diversity improves policy ability for finishing task and generalization. Policy trained by our model outperforms the Agenda model on success rate and average reward.

The results show that the policy trained by other statistical methods underperform the Agenda based model. Because those user simulators lack for response accuracy, which will mislead and confuse the policy training. Comparing Seq2Seq-att to Seq2Seq, the results show the effectiveness of attention mechanism. And State2Seq's improvements over the Seq2Seq-att show that the refined context representation of the State2Seq model does help the response generation.

Results on Restaurant Data. Table 3 shows the cross evaluation results for restaurant reservation domain. The results show our method could work consistently well in different domains. Most models score higher on data in the restaurant domain, because the field is relatively simpler and has fewer slots. The State2MLC model does not perform well. This is due to the fact that State2MLC model has a much simpler structure, so it is likely to overfit to generated data in a simple domain and limits the generalization ability of the policy.

Human Evaluation. We perform a human evaluation on the movie domain, and each agent is tested by chatting with 2 domain experts for 50 dialogues. Table 4 shows that the agents trained in our model can be better adapted to the real situation.

Analysis on Policy Training Process. Table 5 shows the evaluation of training process. We perform testing at each training epoch, and report the averaged score. The policy is evaluated against the environment for training. Policy trained by our model outperforms the ones from statistical user simulator, which reflects that our user simulator improves the training performance of agent policy. It is because our user simulator has better generalization to respond to agent's unreasonable actions in the early training stage. On the other hand, diversity helps RL algorithm training. Agenda achieve good scores as rule environment is relatively easy for overfitting.

Table 5. Analysis of agent policy performance during training on movie domain. At beginning of each policy training epoch, we test the policy's performance. The results reflect the models' overfitting to training set.

Model	Succ	Avg. Rwd	Avg. Turns
State2MLC (Ours)	0.628	26.19	16.36
State2Seq (Ours)	0.800	48.39	17.23
Seq2Seq	0.462	2.98	28.92
Seq2Seq-Att	0.480	2.11	32.98
Agenda	**0.814**	**50.36**	**15.66**

Table 6. Evaluation of user simulator model.

Model	Action accuracy		Generalization ability test			
	Movie	Restaurant	Movie		Restaurant	
	F1	F1	Avg. Succ.	Avg. Rwd.	Avg. Succ.	Avg. Rwd.
State2MLC (Ours)	0.704	**0.695**	**0.442**	**6.06**	0.436	5.34
State2Seq (Ours)	**0.711**	0.683	0.400	5.51	**0.484**	**11.09**
Seq2Seq	0.692	0.662	0.063	−31.87	0.126	−31.87
Seq2Seq-Att	0.705	0.677	0.000	−46.99	0.000	−46.99
Agenda	N/A	N/A	0.392	0.04	0.410	2.20

4.6   Performance of User Simulator

User model's performance is mainly reflected by the ability of predicting user responses.

Action Accuracy. Table 6 shows the models' accuracy of predicting user actions. Our model achieves the best performance, outperforming the seq2seq model [1] for 1.9 points on F-score. The results also show a correlation between the evaluation of action prediction and the agent policy's performance, which demonstrates that user simulator quality affects agents performance.

The improvements mainly come from two aspects: Firstly, the attention mechanism provides specific context information. Secondly, refined context representation filters the useless information. The Seq2Seq-att model outperforms the Seq2Seq model by adding the attention mechanism. This reflects the effectiveness of the attention mechanism. By comparing State2Seq to Seq2Seq-att, improvement shows that due to forgetting mechanism, refined context representation is more effective than sequence encoder.

Analysis of Generalization Ability. The generalization ability is also important for a user simulator in agent training. We compare different user simulators' generalization by making them chat with a same rule-based agent. Table 6 shows the results and State2Seq and MLC2Seq are optimal in all user simulators. Our best model outperforms the Agenda for 5.0 and 7.4 points success rate on the 2 domains. As the user simulators are trained to mimic user rules in the template, the improvement reflects that our framework can generate new diverse dialogues data to avoid user simulator overfitting to user response strategy in the template. Other user simulators perform worse on this test, we address this to the fact that these methods are less accurate in generating user actions and rule-based agent has a low allowance for response error.

Table 7. Case study of the difference between our user simulator and agenda user simulator. Here, user is requesting for ticket and theater, and user's constraints are {movie name: deadpool, city: Seattle , num: 2, date: tomorrow }

Agenda user simulator	
...	
usr: Which theater is available?	act:req, req slots: {theater}
agt: Which theater would you like?	act:req, req slots: {theater}
usr: Which theater is available?	act:req, req slots: {theater}
agt: Which theater would you like?	act:req, req slots: {theater}
(loop...)	
Our user simulator	
...	
usr: Which theater is available?	act:req, req slots:{theater}
agt: Which theater would you like?	act:req, req slots:{theater}
usr: I want to watch at Seattle	act:inform, inform slots:{city: Seattle}
agt: Seattle is available	act:inform, inform: {city: Seattle}
usr: Which theater is available?	act:req, req slots: {theater}
agt: The Pearl Theater is available	act:inform, inform slots: {theater: The Pearl Theater}
...	

4.7 Case Study and Visualization

To find out the difference between rule-based model and the proposed model. We perform case study on dialogues between the user simulator and an agent. The comparison is shown in Table 7. When agents and users are unable to satisfy each other, the fixed rules of Agenda user simulator may be trapped in the loop shown in the example. But the response from our user simulator is diverse and uncertain. It can try other actions to jump out of the endless loop. The difference shows that our proposed framework for building user simulators has successfully improved the response diversity.

To demonstrate the effectiveness of the attention mechanism over context, we provide a visualization example shown in Fig. 3. The figure shows that the attention mechanism successfully learns the correlation between the generated actions and the context. Specifically, when generating the dialogue action of `inform` and the inform slots, the proposed model pays a higher attention to context items of `sys_request_slots` and `user_goal`.

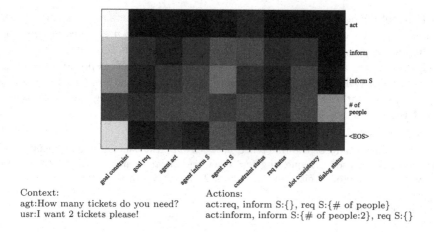

Context:
agt:How many tickets do you need?
usr:I want 2 tickets please!

Actions:
act:req, inform S:{}, req S:{# of people}
act:inform, inform S:{# of people:2}, req S:{}

Fig. 3. Visualization of attention. The vertical axis is the generated user response and the horizontal axis is the dialogue context. The lighter color means higher attention. S denotes slots. (Color figure online)

5   Related Work

There is little work solving the data insufficient problem of user simulator. [11] proposed a user simulator that generates user utterance directly, which could ease the effort of user semantic annotation. However, their work is corpus-based and still needs a large corpus to train on it. Other data scarcity problems for task-oriented dialogue system are also investigated. [26] leverage the idea of zero shot learning [16]. They solve the problem of dialogue generation for a new domain by mapping actions to latent space. [12] and [9] provided data augmentation methods for language understanding. [23] proposed to build new domain agent efficiently by machine self-play and crowdsourcing.

The first user simulator was proposed by [5]. There are two kinds of user simulators in terms of working levels. User simulators of semantic level interact to agent with dialogue acts and corresponding slot-value pairs. [1,3,6,21]. User simulators of utterance level communicates to agent with utterance directly [10, 11,15]. Our user simulator can work on both of the two levels.

6   Conclusion

In this paper, we study the problem of building user simulators for task-oriented dialogue from templates with no corpus. We solve the data scarcity and increase simulator response diversity by proposing a corpus-free framework. In our framework, we generate diverse data with only templates and trains a data-driven user simulator on it. To predict user response more accurately, we proposed a novel State2Seq user model. It predicts user response with attention on refined dialogue context representations. Experiment results show that with more response diversity, our user simulator improves the agent policy by 6.36% success rate. Attention and refined context representation help the State2Seq model outperform Seq2Seq baseline for 1.9 F-score.

Acknowledgments. We are grateful for helpful comments and suggestions from the anonymous reviewers. This work was supported by the National Natural Science Foundation of China (NSFC) via grant 61632011, 61772153 and 61772156.

References

1. Asri, L.E., He, J., Suleman, K.: A sequence-to-sequence model for user simulation in spoken dialogue systems. In: Proceedings of Interspeech, pp. 1151–1155 (2016). https://doi.org/10.21437/Interspeech.2016-1175
2. Casanueva, I., et al.: A benchmarking environment for reinforcement learning based task oriented dialogue management. CoRR abs/1711.11023 (2017). http://arxiv.org/abs/1711.11023
3. Chandramohan, S., Geist, M., Lefevre, F., Pietquin, O.: User simulation in dialogue systems using inverse reinforcement learning. In: Proceedings of Interspeech (2011)
4. Cuayáhuitl, H., Renals, S., Lemon, O., Shimodaira, H.: Human-computer dialogue simulation using hidden markov models. In: Proceedings of ASRU, pp. 290–295. IEEE (2005)
5. Eckert, W., Levin, E., Pieraccini, R.: User modeling for spoken dialogue system evaluation. In: Proceedings of ASRU, vol. 97 (1997)
6. Gao, J., Wong, K., Peng, B., Liu, J., Li, X.: Deep dyna-Q: integrating planning for task-completion dialogue policy learning. In: Proceedings of ACL, pp. 2182–2192 (2018). https://aclanthology.info/papers/P18-1203/p18-1203
7. Gašić, M., Young, S.: Gaussian processes for POMDP-based dialogue manager optimization. IEEE/ACM Trans. Audio, Speech Lang. Process. **22**(1), 28–40 (2014)
8. Georgila, K., Henderson, J., Lemon, O.: Learning user simulations for information state update dialogue systems. In: Proceedings of Eurospeech (2005)
9. Hou, Y., Liu, Y., Che, W., Liu, T.: Sequence-to-sequence data augmentation for dialogue language understanding. In: Proceedings of COLING (2018)
10. Jung, S., Lee, C., Kim, K., Jeong, M., Lee, G.G.: Data-driven user simulation for automated evaluation of spoken dialog systems. Comput. Speech Lang. **23**(4), 479–509 (2009)
11. Kreyssig, F., Casanueva, I., Budzianowski, P., Gasic, M.: Neural user simulation for corpus-based policy optimisation of spoken dialogue systems. In: Proceedings of SIGdial, pp. 60–69 (2018). https://aclanthology.info/papers/W18-5007/w18-5007
12. Kurata, G., Xiang, B., Zhou, B.: Labeled data generation with encoder-decoder LSTM for semantic slot filling. In: INTERSPEECH 2016, pp. 725–729 (2016). https://doi.org/10.21437/Interspeech.2016-727
13. Li, X., Chen, Y.N., Li, L., Gao, J.: End-to-end task-completion neural dialogue systems. In: Proceedings of IJCNLP (2017). http://arxiv.org/abs/1703.01008
14. Li, X., Lipton, Z.C., Dhingra, B., Li, L., Gao, J., Chen, Y.N.: A user simulator for task-completion dialogues. arXiv preprint arXiv:1612.05688 (2016)
15. Liu, B., Lane, I.: Iterative policy learning in end-to-end trainable task-oriented neural dialog models. In: ASRU Workshop, pp. 482–489. IEEE (2017)

16. Palatucci, M., Pomerleau, D., Hinton, G.E., Mitchell, T.M.: Zero-shot learning with semantic output codes. In: Bengio, Y., Schuurmans, D., Lafferty, J.D., Williams, C.K.I., Culotta, A. (eds.) Proceedings of NIPS, pp. 1410–1418. Curran Associates, Inc. (2009). http://papers.nips.cc/paper/3650-zero-shot-learning-with-semantic-output-codes.pdf

17. Pietquin, O., Dutoit, T.: A probabilistic framework for dialog simulation and optimal strategy learning. IEEE Trans. Audio Speech Lang. Process. **14**(2), 589–599 (2006)

18. Roy, N., Pineau, J., Thrun, S.: Spoken dialogue management using probabilistic reasoning. In: Proceedings of ACL (2000). http://www.aclweb.org/anthology/P00-1013

19. Schatzmann, J., Stuttle, M.N., Karl, W., Young, S.: Effects of the user model on simulation-based learning of dialogue strategies. In: Proceedings of ASRU (2005)

20. Schatzmann, J., Thomson, B., Weilhammer, K., Ye, H., Young, S.: Agenda-based user simulation for bootstrapping a POMDP dialogue system. In: Proceedings of NAACL, pp. 149–152. Association for Computational Linguistics (2007)

21. Schatzmann, J., Thomson, B., Young, S.: Statistical user simulation with a hidden agenda. Proc. SIGDial **273282**(9) (2007)

22. Schatzmann, J., Weilhammer, K., Stuttle, M., Young, S.: A survey of statistical user simulation techniques for reinforcement-learning of dialogue management strategies. knowl. Eng. Rev. **21**(2), 97–126 (2006)

23. Shah, P., et al.: Building a conversational agent overnight with dialogue self-play. arXiv preprint arXiv:1801.04871 (2018)

24. Williams, J.D., Young, S.: Partially observable Markov decision processes for spoken dialog systems. Comput. Speech Lang. **21**(2), 393–422 (2007)

25. Young, S., Gašić, M., Thomson, B., Williams, J.D.: POMDP-based statistical spoken dialog systems: a review. Proc. IEEE **101**(5), 1160–1179 (2013)

26. Zhao, T., Eskénazi, M.: Zero-shot dialog generation with cross-domain latent actions. CoRR abs/1805.04803 (2018). http://arxiv.org/abs/1805.04803

Utterance Alignment in Custom Service by Integer Programming

Guirong Bai[1,2], Shizhu He[1], Kang Liu[1,2], and Jun Zhao[1,2(✉)]

[1] National Laboratory of Pattern Recognition Institute of Automation,
Chinese Academy of Sciences, Beijing 100190, China
{guirong.bai,shizhu.he,kliu,jzhao}@nlpr.ia.ac.cn
[2] University of Chinese Academy of Sciences, Beijing 100049, China

Abstract. In customer service (CS), customers pose questions that will be answered by customer service staff, and the communication in CS is a typical multi-round conversation. However, there are no explicit correspondences among conversational utterances, and obtaining the explicit alignments of those utterances not only contributes to dialogue analysis but also provides valuable data for learning intelligent dialogue systems. In this paper, we first present a study on utterance alignment (UA) in CS. We divide the alignment of utterances into four types: *None, One-to-One, One-to-Many* and *Jump*. The direct design models such as rule-based and matching-based methods are often only good at solving part of types, and the major reason is that they ignore the interactions of different utterances. Therefore, to model the mutual influence of different utterances as well as their alignments, we propose a joint model which models the UA as a task of joint disambiguation and resolved by integer programming. We conduct experiments on a dataset of an in-house online CS. And the results indicate that it performs better than baseline models, especially for *One-to-Many* and *Jump* alignments.

Keywords: Utterance alignment · Integer programming · Customer service

1 Introduction

Customer service (**CS**), which provides an irreplaceable platform for sellers to answer the questions and solve the problems of customers in the shopping, plays important roles in e-commerce websites. Recently, with the development of the artificial intelligence techniques, automatic CS conversation analysis of CS has attracted more and more attention, including intention analysis [3], emotion identification [6], suggestion mining [12], etc.

In general, in CS, **customers** pose questions that will be answered by customer service staff (**servers**), and the communication between a customer and a server is a typical multi-round conversation. In this paper, we focus on the utterance alignment (**UA**), which is to align the utterances between different

© Springer Nature Switzerland AG 2019
M. Sun et al. (Eds.): CCL 2019, LNAI 11856, pp. 703–714, 2019.
https://doi.org/10.1007/978-3-030-32381-3_56

sides (customer and server) in a dialogue. This task is very important and useful for practical online services but was seldom addressed before in our knowledge. As shown in Fig. 1, based on the alignments, the questions posed by a customer are connected with the corresponding responses from the server. It could benefit the server to perform quality control and check whether the customer is satisfied with the service. Moreover, most current intelligent dialogue models such as deep learning methods [5,9] need sufficient aligned question-answer (response) pairs to train the model. Obviously, such question-answer pairs could be automatically acquired through UA.

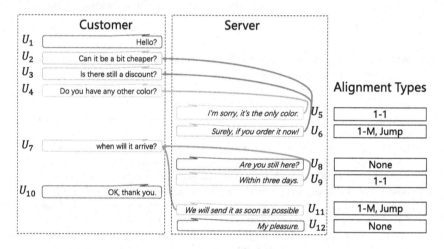

Fig. 1. Utterance alignments of a sample dialogue in CS. The left and right utterances are raised by customer and server respectively.

Although UA is a valuable task, it is full of challenges. In the conversation of CS, a response is may not adjacent the corresponding customer's question (e.g., as shown in Fig. 1, U_8 does not align with U_7, and U_{11} does not align with U_{10}). And the orders of the utterances are possibly out of turns. Moreover, sometimes the customer even consults one consultation with multiple similar questions. According to our observation from the server side, there are usually four types of alignments among utterances, including *None*, *One-to-One* (*1-1*), *One-to-Many* (*1-M*) and *Jump*.

Intuitively, a response usually follows a corresponding question, and it is easy to think of employing the position information to align utterances as other alignment models such as IBM models [2]. However, in the types of *1-M* and *Jump*, there are freedom and disorder in a dialogue, which make the position information is too weak to align the correct utterances. It is naturally to further consider the content of utterances, where two utterances in an aligned pair must have a semantic connection or relatedness [8,18]. Nevertheless, solely considering the semantic connection and position information of a pair may still cause a local optimization. The information of different utterances and their alignments in a dialogue should be considered globally. In fact, the alignments of different utterances are correlated and interactional with each other. For example, if question Q_1 (e.g., U_2 in Fig. 1) and Q_2 (U_3) are very *similar* and the question Q_2 (U_3) aligns with answer A_1 (U_6), the question Q_1 (U_2) should also *align* with answer A_1 (U_6). But if question Q_1 (U_3) and Q_2 (U_4) are very *dissimilar* and the question Q_2 (U_4) aligns with answer A_1 (U_5), the question Q_1 (U_3) should *align* with the answer A_1 (U_6) rather than A_1 (U_5), even though A_1 (U_5) is more closer to Q_1 (U_3) than A_1 (U_6).

To this end, we propose to find all alignments of a conversation in a joint model. We model the utterance alignment as the progress of joint disambiguation (whether U_5 aligns with U_4 or not), consider the correlatives of different utterances as well as their relationship by integer constraints (if U_2 and U_3 are similar, and U_3 aligns with U_6, U_2 should align with U_6), and resolve them by integer programming (**IP**). Moreover, two neural models are proposed to capture the semantic representation of the utterance content, which are also able to incorporate the position information. Based on the learned semantic representation, the content-based alignments are calculated. Finally, we fuse the above information of all pairs in a dialogue into an integer programming algorithm. In this way, all possible alignments could affect each other and the final results are optimized globally.

We create a dataset to verify the feasibility of the proposed model, and the experimental results demonstrate the effectiveness of the proposed model. Compared with the best matching model, the F1 is increased by 7% in totally. In special, it obtains 3.6%, 3.7% improvements on the more challenging alignments of *1-M* and *Jump*, respectively.

In brief, the main contributions are as follows:

- We propose a new task, named utterance alignment (UA), which contributes to dialogue analysis and provides valuable data for learning intelligent dialogue systems.
- We propose a joint model for UA by integer programming (IP), which considers the correlatives of different utterances as well as their relationship by integer constraints and make effects on experiments.
- We collect dialogues from a real CS and construct a dataset for UA with human-annotation.

2 Problem Definition

2.1 Utterance Alignment

In most cases, the customer poses the questions that will be answered by the server. In this work, we mainly focus on helpful and crucial question-answer pairs. Therefore, we only need to consider which customer utterances (question, Q for short) are aligned for each server utterance (answer, A for short). We formulate the task as a joint disambiguation task based on classification models: whether A_i aligns with Q_j or not.

2.2 Alignment Types

There are a number of consultations in a CS conversation as shown in Fig. 1. Considering the dialogue process, we can align utterances by their posing orders (e.g., U_5 aligns with U_4, U_8 aligns with U_7). However, in the actual scenario, this simply alignment strategy is not enough. A customer may pose a number of questions at a time (e.g., U_2, U_3, U_4), and the server may answer them in different orders. In addition, each server may need to talk to multiple customers at the same time, and each customer's question may not be answered immediately, it will aggravate the above situation. On the other hand, people may express the same intention in a number of short and simple utterances in the oral communicating environment (e.g., U_2 and U_3 express the similar meaning), therefore, some answers should align with more than one questions (e.g., U_6 should align with both U_2 and U_3). Moreover, there may be some chat messages (e.g., U_8 and U_{12}) interspersed in the consulting process, and they should not be aligned with any utterance.

Therefore, considering the different alignments which may be suitable for different models, We divide the alignments into four types: *None*, *One-to-One* (1-1), *One-to-Many* (1-M) and *Jump*. The *None* type means that there is no alignment for a given utterance of the server (e.g., U_8 and U_{12} in Fig. 1). The *1-1* means a response aligns with only one question (e.g., U_5, U_9 and U_{11}), which is the simplest and most intuitive alignment type. The *1-M* means a response aligns with more than one questions (e.g., U_6 should align with two utterances: U_2 and U_3). And the *Jump* means a response replies to a question which is posed serval turns ago, and their alignments cross some other questions (e.g., U_{11} should align with U_7, which crosses the closest question: U_{10}). The *1-M* and *Jump* alignments violate the regular order in a dialogue and provide main difficulties for UA, which are the main focus of this paper.

2.3 Data

We create a Chinese dataset from an online CS. We first sample 10,000 conversations from a human-to-human customer service system, which owns about 6-20 utterances for each conversational episode. we invited five annotators for the explicit alignments. For example, U_8 should be independently annotated whether or not to align with one or more of U_1, U_2, U_3, U_4 and U_7. If the server utterance is a meaningless utterance or cannot answer any customer question, it will be annotated a *None* label. For example, in Fig. 1, U_{12} is a meaningless utterance as a chat message, and none of the customer utterances can semantically match U_8. So both of them are annotated *None* label. The coincidence rate of the five annotators is about 85% with another annotator reviewing.

In the end, we obtain 5,741 labeled conversations with average 6.0 turns from the server and 4.5 turns from the customer. Every turn has average 22.7 and 6.2 words on each side respectively. Moreover, for a given utterance of the server, there are average 2.8 customer utterances as alignment candidates. And the alignments of *None*, *1-1*, *1-M* and *Jump* account for 57%, 31%, 12%, and 9% respectively.

3   Utterance Alignment Models

Let $CS = [(U_1, t_1), (U_2, t_2), ..., (U_n, t_n)]$ denotes the conversation of CS, $U_i = [w_1, ..., w_{L_{U_i}}]$ indicates the word sequence of a utterance and t_i indicates the role of speaker (Customer (C) and Server (S)). For each server utterance $(U_i, t_i), t_i = S$, we should find all customer raised utterances before i ($\{j | j < i, t_j = C\}$) which could be answered by U_i.

3.1   Matching-Based Alignment

We first learn different Q-A matching models that utilize deep neural networks scoring the matching degree between a customer utterance and a server utterance. The matching score $s_{q,a}$ of a customer's question q and a server's answer a is calculated as: $s_{q,a} = \mathbf{q}^T \cdot \mathbf{M} \cdot \mathbf{a}$, where $\mathbf{q}$ and $\mathbf{a}$ are the semantic representation of them, and the matrix $\mathbf{M}$ is the parameter of the matching model. Then, we can obtain the alignments of utterance pairs when their matching scores large than a threshold. We utilize the following margin-based ranking loss to train the matching models: $L = max(0, s_{q,a'} + \gamma - s_{q,a})$ (or $L = max(0, s_{q',a} + \gamma - s_{q,a})$) , where q' and a' indicate the random selected nonaligned utterances. In specific, we obtain the representation of utterance by Convolutional Neural Networks (CNN) [1] and Long Short-Term Memory (LSTM) [7].

Position Embedding: We consider position information among utterances and design the following three features to encode the position information: (1) Index: indicates the absolute position of a server utterance in a dialogue; (2) All-Distance (A-$Dist$): records the number of utterances in the dialogue between two alignment utterances; (3) Customer-Distance (C-$Dist$): records the number of customer utterances in the dialogue between two alignment utterances. For example, as shown in Fig. 1, the index, A-Dist, and C-Dist are 6, 3, 2 while judging the alignment between U_6 and U_2 (the candidate alignment utterances of U_6 are from U_1 to U_4) respectively. Each position feature is represented by a fixed-dimension vector and concatenated with sentence representation.

3.2   Utterance Alignment with Joint Disambiguation

The above-mentioned methods independently judge the alignment of each utterance pair, which cause the alignments are local optimum results. In fact, the alignments of different utterance pairs in a dialogue have coherence and interaction. As shown in Fig. 2, similar questions are usually aligned with the same answer and vice versa. Therefore, we propose a joint disambiguation model with some global constraints by integer programming (IP).

We define three types of 0–1 variables ($\in \{0, 1\}$): (1) A_{ij} indicates whether the i-th customer utterance align with j-th server utterance (final results). (2) MQ_{ij} indicates whether the i-th customer utterance is semantically similar with j-th customer utterance. (3) MA_{ij} indicates whether the i-th server utterance is semantically similar with j-th server utterance.

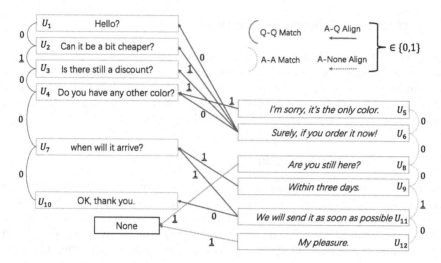

Fig. 2. A dialogue example of utterance alignments with joint disambiguation. The relations of Q-Q, A-A and Q-A/A-Q with 0/1 variables denote matching or not.

Basic Models: We train three scoring models for the above three variables. The first one is adopt the matching-based methods. We define $s_{q,q}$ and $s_{a,a}$ to indicate the similarity of two questions and two answers. And they are modeled by the following function: $s_{q,q}(i,j) = \sigma(\mathbf{W}_{qq}[\mathbf{q}_i, \mathbf{q}_j])$ and $s_{a,a}(i,j) = \sigma(\mathbf{W}_{aa}[\mathbf{a}_i, \mathbf{a}_j])$, where σ indicates the *sigmoid* function, $\mathbf{W}_{qq}$ and $\mathbf{W}_{aa}$ are the model parameters trained by minimizing the cross entropy. In addition, we define $n(a) = \sigma(\mathbf{a}\mathbf{W}_n)$ to indicate the probability of the server's answer a not aligning with any utterance by minimizing the cross entropy.

Objective of the Joint Disambiguation Model: The objective contains four parts as follows:

(1) The question-answer alignment scores: the probability questions align with answers: $T_1 = \sum_{j=1}^{J} \sum_{k=1}^{K} A_{ij} \cdot (s_{q,a}(i,j) - \beta_1)$, where J and K indicate the utterance numbers of customer and server, respectively.
(2) The question similarity scores: the probability questions are aligned with a same server's answer: $T_2 = \sum_{(q_j,q_{j'}) \in L_q} MQ_{jj'} \cdot (s_{q,q}(j,j') - \beta_2)$, where L_q contains all candidate question pairs from the customer utterances.
(3) The answer similarity scores: the probability answers are to be connected to a same customer's question: $T_3 = \sum_{(a_k,a_{k'}) \in L_a} MA_{kk'} \cdot (s_{a,a}(k,k') - \beta_3)$, L_a contains all candidate answer pairs from the server utterances.
(4) The *None* alignment probabilities: the probability they have no alignment with any question. $T_4 = \sum_{j=1}^{J} \sum_{k=1}^{K} A_{jk} \cdot (n(k) - \beta_4)$.

Then, the final objective function is as follows:

$$maximize \quad T = \alpha_1 T_1 + \alpha_2 T_2 + \alpha_3 T_3 + \alpha_4 T_4 \tag{1}$$

where α_1, α_2, α_3, α_4, β_1, β_2, β_3, β_4 are the hyper-parameters.

Global Constraints: To model the interaction among different decisions, we additionally set a series of global constraints on the three types of binary variables:

(C1). A response could only reply to the former questions. The constraint could be formulated as:

$$A_{jk} = 0, \ \forall index(j) >= index(k) \tag{2}$$

where $Index(U)$ denotes the index of the utterance U.

(C2). If two customer utterances are similar, they must be aligned to at least one same server utterance. Conversely, they could not be aligned to a same server utterance if they are dissimilar.

$$MQ_{jj'} \cdot QQ_{jj'} + (1 - MQ_{jj'}) \cdot (1 - QQ_{jj'}) \geq 1 \tag{3}$$

where $QQ_{jj'} = \sum_{k=1}^{K} A_{jk} \cdot A_{j'k}$. In fact, it is an XNOR gate between $MQ_{jj'}$ and $QQ_{jj'}$. Therefore, C2 is a nonlinear operation, which is different from integer linear programming (ILP) models in other NLP tasks.

(C3). Similar with **(C2)** about another side.

$$MA_{kk'} \cdot AA_{kk'} + (1 - MA_{kk'}) \cdot (1 - AA_{kk'}) \geq 1 \tag{4}$$

where $AA_{kk'} = \sum_{j=1}^{J} A_{jk} \cdot A_{j'k}$.

4 Experiment

In this section, we present our experiment settings and results, which devote to answering the following questions: (1) *Is the joint disambiguation model able to obtain a better performance of utterance alignments compared with rule-based and matching-based methods?* (2) *Is the proposed model able to resolve the types of 1-M and Jump alignments?*

4.1 Configurations

The dataset is randomly split into training (4741 dialogues, about 80%), validation (500 dialogues, about 10%) and testing set (500 dialogues, about 10%). The utterances are segmented into word sequences with Jieba[1] tool after some basic preprocessing such as convert all URLs to a special label. Hyper-parameters γ, α_1, α_2, α_3, α_4, β_1, β_2, β_3, β_4 are set to 0.5, 1.0, 1.0, 0.05, 1.0, 0.3, 0.5, 0.5, 0.5 respectively. The word embedding size and the hidden size are 128 for all deep learning models. Each position embedding size is 4. For a fair comparison, CNN based models set filter sizes as [3, 4, 5] and employ 42 filters. We used the Adam with learning rate 0.001.

---

[1] https://github.com/fxsjy/jieba.

4.2   Baselines

The multi-round conversation in CS has some specific characteristics: (1) there are only two participants (customer and server); (2) the customer mainly poses the questions; and (3) the server mainly answer customer's questions. Therefore, we can simply obtain the utterance alignments using some heuristic rules based on the posing order of utterances. We utilize the following manual rules to align utterances which only consider the position information.

Rule-1: For a given server utterance, we choose the closest customer utterance as its alignment.

Rule-2: Rule-1 lacks the ability to handle *None* alignment type, which occupies a large proportion of alignments. Thus, another rule is proposed: if the closest customer utterance has been aligned to other utterances, we directly give the current server utterance the *None* label.

Table 1. Sample utterance alignments obtained by heuristic rules.

Rules	Utterance ID	Golden alignments	Predicted alignments
Rule-1	U_5	$[U_4]$	$[U_4]$
Rule-1	U_{12}	[None]	$[U_{10}]$
Rule-2	U_{12}	[None]	[None]
Rule-2	U_6	$[U_2,U_3]$	[None]

Two rules are adopted and their sampling results for the dialogue in Fig. 1 are given in Table 1.

Matching-Based Model: We utilize semantic composition models such as CNN and RNN (LSTM) for learning the representations of utterances, which also incorporate the position embeddings of the multi-turn dialogue into the representations.

4.3   Evaluation Metrics

Based on the human-labeled alignments for each server utterance, we could calculate the precision (P), recall (R) and F1 for utterance alignments. Considering that there are multiple alignments, we utilize the micro averaging to obtain the overall metrics for equally treating all utterance pairs.

4.4   Results and Discussion

The overall experimental results are shown in Table 2. The last two rows are the results of our proposed joint disambiguation models with integer programming (IP).

From the overall results, we can observe that: (1) The rule-based methods are not very bad, the overall F1 even exceeds the results of Match-CNN. (2) The

Table 2. The precision (P), recall (R) and F1 (%) for overall and different alignment types on test data.

	Overall			None			One-to-One			One-to-Many			Jump		
	P	R	F1	P	R	F1	P	R	F1	P	R	F1	P	R	F1
Rule-1	45.2	42.1	43.6	14.7	14.7	14.7	**93.1**	**93.1**	**93.1**	**96.1**	42.1	58.6	0	0	0
Rule-2	58.0	54.0	55.9	**55.0**	**55.0**	**55.0**	61.9	61.9	61.9	66.9	29.3	40.8	0	0	0
Match(CNN)	52.4	59.7	55.8	35.5	39.9	37.6	69.8	92.2	**79.5**	92.8	70.2	79.9	**31.1**	53.9	39.4
Match(LSTM)	55.7	60.9	58.2	42.4	46.1	44.2	68.9	87.0	76.9	92.0	64.8	76.0	26.3	43.1	32.7
IP(CNN)	**58.9**	**66.6**	**62.5**	48.5	53.1	50.7	67.1	87.7	76.0	88.6	**76.8**	**82.3**	32.4	65.4	**43.3**
IP(LSTM)	**61.5**	**69.3**	**65.2**	**55.8**	**60.3**	**58.0**	62.8	82.3	71.3	89.8	**78.1**	**83.5**	29.8	**66.0**	41.1

matching-based methods have a better recall, that is, they have an advantage in obtaining more valid alignments. (3) From the above four rows, we believe that it is very hard to obtain a satisfactory result merely relying on the position information or the utterance texts. (4) The proposed methods obviously exceed other methods, which demonstrates that the alignments of different utterances are correlated and interactional with each other. (5) In most cases, LSTM has a better composition semantics on spoken utterances.

From the result of different alignment types, we can observe that: (1) For the None type, the Rule-1 is very bad because it always obtain an alignment for all utterances in any case. The Rule-2 is the most competitive model which even better than all matching models except the proposed method. It indicates that the extra information such as utterance texts will help to work on *None* type. (2) For the 1-1 type, the Rule-1 is outstanding. It is because that an answer usually follows its corresponding question in a dialogue. Our proposed joint model still outperforms better. The extra classification information such as restrictions on other utterances can help to judge whether to choose an alignment or not. (3) For the 1-M type, the Rule-1 has the best precision but a worse recall. The proposed joint model performs best for recall and F1. It demonstrates that joint models are able to capture more potential alignments by utilizing global restrictions. (4) For the Jump type, the rule-based methods are broken because their assumption always chooses a nearest one. By modeling the utterance contents, the matching-based methods are able to deal with this alignment type in some extent. The proposed models outperform other models, which indicate that joint models are able to consider all relations among different utterance pairs.

In total, the proposed joint model obtain the best performance for overall results , especially for $1 - M$ and *Jump* alignments which are very hard for rule-based and matching-based methods.

4.5 Detailed Analysis

In this section, we analyze the effects of some core components in joint models.

At first, we validate the importance of position embeddings in matching-based methods. The experimental results in Table 3 compare the models with and without (w/o) it. Because of they can absorb the advantages of manual rules based on position information, the results with them perform better on most types (*Overall, One-to-One, One-to-Many* and *Jump*), except *None*.

Table 3. The effects of position embeddings in matching-based methods.

	Overall			None			One-to-One			One-to-Many			Jump		
	P	R	F1	P	R	F1	P	R	F1	P	R	F1	P	R	F1
CNN o pos	45.1	**62.3**	52.3	44.1	58.0	50.1	42.5	71.6	53.3	63.2	58.9	61.0	14.6	**61.4**	23.5
LSTM o pos	45.8	61.4	52.4	**45.6**	**58.3**	**51.2**	42.3	68.9	52.4	63.2	56.7	59.7	13.9	56.9	22.4
CNN w pos	52.4	59.7	55.8	35.5	39.9	37.6	**69.8**	**92.2**	**79.5**	**92.8**	70.2	**79.9**	**31.1**	53.9	**39.4**
LSTM w pos	**55.7**	60.9	**58.2**	42.4	46.1	44.2	68.9	87.0	76.9	92.0	64.8	76.0	26.3	43.1	32.7

Table 4. The effects of modeling coherence among utterances in UA. C and L denote CNN and LSTM respectively.

	Overall			None			One-to-One			One-to-Many			Jump		
	P	R	F1	P	R	F1	P	R	F1	P	R	F1	P	R	F1
IP(C) o None	52.9	60.2	56.3	34.6	39.0	36.7	**73.3**	**92.9**	**81.9**	**91.0**	76.8	83.3	**38.8**	63.7	**48.2**
IP(L) o None	55.0	62.1	58.3	39.6	44.4	41.9	70.9	89.1	79.0	90.7	76.8	83.2	35.9	62.7	45.7
IP(C)+Pipe	62.7	65.5	64.0	70.9	73.5	72.2	46.5	55.3	50.5	74.4	53.7	62.4	16.9	45.1	24.6
IP(L)+Pipe	**62.8**	65.9	64.3	**70.9**	**73.6**	**72.2**	46.7	56.0	50.9	74.6	54.9	63.3	17.7	48.4	25.9
IP(C) w None	58.9	66.6	62.5	48.5	53.1	50.7	67.1	87.7	76.0	88.6	76.8	82.3	32.4	65.4	43.3
IP(L) w None	61.5	**69.3**	**65.2**	55.8	60.3	58.0	62.8	82.3	71.3	89.8	**78.1**	**83.5**	29.8	**66.0**	41.1

With *None* type occupying the largest proportion 57%, we then compare different methods for it in IP. We compare the model with and without (w/o) considering such part in IP (contains T_4 or not). In addition, we design a pipeline model (*+Pipe*), which first judges whether the utterance should align with *None* based on a threshold, and next utilize the IP models. Table 4 shows the experimental results. It demonstrates the IP models effectively deal with *None* alignment and overcome the problem of *1-M* and *Jump* alignments in other models.

5   Related Work

There are many tasks on dialogue analysis such as dialogue analysis state tracking [20], dialogue act classification [15], the speaker and addressee recognition [13], response generation [16] and other tasks. However, there is little research work paying attention to utterance alignments.

Utterance alignment relates to other alignment tasks in NLP, such as word alignment in machine translation. However, it is to align words rather than utterances and has fixed word size with infinite space of generated utterances. As a result, related approaches such as the HMM model [19] could not be directly applied to our task. Some previous approaches transform words into continuous space to achieve it [17,23], but the utterances in dialogues still have different distributions from words. Moreover, the utterance alignment in CS deal has a larger linguistic unit and focus more on conversation analysis rather than sentence analysis.

And IP the proposed models employ to combine local features and global restrictions has received wide attention in other NLP tasks, such as semantic role labeling [14], syntactic and semantic dependency parsing [4], named entity disambiguation [10], sentiment analysis [11], summarization [21] and question answering [22,24], etc. However, most of the aforementioned approaches apply linear constraints in joint disambiguation models. By contrast, there are nonlinear constraints in our model.

6 Conclusion

In this paper, we define a new task in real CS, utterance alignment, which devotes to aligning the utterances between customer and server in a dialogue. Where utterance alignments are divided into four types: *None*, *One-to-One*, *One-to-Many* and *Jump*. To model the mutual influence of different utterances as well as their alignments for *One-to-Many* and *Jump* alignments, we propose a joint model for UA, which models the task as a joint disambiguation problem with integer programming resolving and obtain better results.

Acknowledgement. This work is supported by the National Natural Science Foundation of China (No.61533018), the Natural Key R&D Program of China (No.2017YFB1002101), the National Natural Science Foundation of China (No.61702512, No.61806201) and the independent research project of National Laboratory of Pattern Recognition. This work was also supported by CCF-DiDi BigData Joint Lab and CCF-Tencent Open Research Fund.

References

1. Blunsom, P., Grefenstette, E., Kalchbrenner, N.: A convolutional neural network for modelling sentences. In: Proceedings of the 52nd Annual Meeting of the Association for Computational Linguistics (2014)
2. Brown, P.E., Pietra, S.A.D., Pietra, V.J.D., Mercer, R.L.: The mathematics of statistical machine translation: parameter estimation. Comput. Linguist. **19**(2), 263–311 (1993). http://www.aclweb.org/anthology/J93-2003
3. Carlos, C.S., Yalamanchi, M.: Intention analysis for sales, marketing and customer service. In: Proceedings of COLING 2012: Demonstration Papers, pp. 33–40 (2012)
4. Che, W., et al.: A cascaded syntactic and semantic dependency parsing system. In: CoNLL 2008: Proceedings of the Twelfth Conference on Computational Natural Language Learning, pp. 238–242. Coling 2008 Organizing Committee (2008). http://www.aclweb.org/anthology/W08-2134
5. Cui, L., Huang, S., Wei, F., Tan, C., Duan, C., Zhou, M.: SuperAgent: a customer service chatbot for e-commerce websites. In: Proceedings of ACL 2017, System Demonstrations, pp. 97–102 (2017)
6. Herzig, J., et al.: Classifying emotions in customer support dialogues in social media. In: Meeting of the Special Interest Group on Discourse and Dialogue (2016)
7. Hochreiter, S., Schmidhuber, J.: Long short-term memory. Neural Comput. **9**(8), 1735 (1997)
8. Hu, B., Lu, Z., Li, H., Chen, Q.: Convolutional neural network architectures for matching natural language sentences. In: Advances in Neural Information Processing Systems, pp. 2042–2050 (2014)

9. Ji, Z., Lu, Z., Li, H.: An information retrieval approach to short text conversation. Comput. Sci. (2014)
10. Kulkarni, S., Singh, A., Ramakrishnan, G., Chakrabarti, S.: Collective annotation of wikipedia entities in web text. In: Proceedings of the 15th ACM SIGKDD International Conference on Knowledge Discovery and Data Mining, KDD 2009, pp. 457–466. ACM, New York (2009). https://doi.org/10.1145/1557019.1557073
11. Lu, Y., Castellanos, M., Dayal, U., Zhai, C.: Automatic construction of a context-aware sentiment lexicon: an optimization approach. In: Proceedings of the 20th International Conference on World Wide Web, pp. 347–356. ACM (2011)
12. Negi, S., Buitelaar, P.: Towards the extraction of customer-to-customer suggestions from reviews. In: Proceedings of the 2015 Conference on Empirical Methods in Natural Language Processing, pp. 2159–2167. Association for Computational Linguistics (2015). https://doi.org/10.18653/v1/D15-1258
13. Ouchi, H., Tsuboi, Y.: Addressee and response selection for multi-party conversation. In: EMNLP, pp. 2133–2143 (2016)
14. Punyakanok, V., Roth, D., Yih, W.T., Zimak, D.: Semantic role labeling via integer linear programming inference. In: COLING 2004: Proceedings of the 20th International Conference on Computational Linguistics (2004). http://www.aclweb.org/anthology/C04-1197
15. Reithinger, N., Klesen, M.: Dialogue act classification using language models. In: EuroSpeech (1997)
16. Sutskever, I., Vinyals, O., Le, Q.V.: Sequence to sequence learning with neural networks. In: Advances in Neural Information Processing Systems, pp. 3104–3112 (2014)
17. Tamura, A., Watanabe, T., Sumita, E.: Recurrent neural networks for word alignment model. ACL 1(52), 1470–80 (2014)
18. Tan, M., dos Santos, C., Xiang, B., Zhou, B.: Improved representation learning for question answer matching. In: Proceedings of the 54th Annual Meeting of the Association for Computational Linguistics (Volume 1: Long Papers), vol. 1, pp. 464–473 (2016)
19. Vogel, S., Ney, H., Tillmann, C.: Hmm-based word alignment in statistical translation. In: Proceedings of the 16th Conference on Computational Linguistics-Volume 2, pp. 836–841. Association for Computational Linguistics (1996)
20. Williams, J., Raux, A., Ramachandran, D., Black, A.: The dialog state tracking challenge. In: Proceedings of the SIGDIAL 2013 Conference, pp. 404–413 (2013)
21. Woodsend, K., Lapata, M.: Automatic generation of story highlights. In: Proceedings of the 48th Annual Meeting of the Association for Computational Linguistics, pp. 565–574. Association for Computational Linguistics (2010). http://www.aclweb.org/anthology/P10-1058
22. Yahya, M., Berberich, K., Elbassuoni, S., Ramanath, M., Tresp, V., Weikum, G.: Natural language questions for the web of data. In: Joint Conference on Empirical Methods in Natural Language Processing and Computational Natural Language Learning, pp. 379–390 (2012)
23. Yang, N., Liu, S., Li, M., Zhou, M., Yu, N.: Word alignment modeling with context dependent deep neural network. In: Proceedings of the 51st Annual Meeting of the Association for Computational Linguistics (Volume 1: Long Papers), pp. 166–175. Association for Computational Linguistics (2013). http://www.aclweb.org/anthology/P13-1017
24. Zhang, Y., He, S., Liu, K., Zhao, J.: A joint model for question answering over multiple knowledge bases. In: Thirtieth AAAI Conference on Artificial Intelligence, pp. 3094–3100 (2016)

Author Index

Printed in the United States
By Bookmasters